ELECTRICAL ENGINEERING

ELECTRICAL ENGINEERING

(As per Prescribed Syllabus of Anna University, Tamil Nadu and Various Other Universities)

IN

S.I. UNITS

For

1st Year B.E. Students

(Common to Computer Science and Engineering and Information Technology Disciplines)

By

R.K. RAJPUT

M.E. (Hons.), (Gold Medallist) ; Grad. (Elect. Engg. & Mech. Engg.)
M.I.E. (India) ; M.S.E.S.I. ; M.I.S.T.E. ; C.E. (India)
PATIALA, PUNJAB

LAXMI PUBLICATIONS (P) LTD

BANGALORE • CHENNAI • COCHIN • GUWAHATI • HYDERABAD
JALANDHAR • KOLKATA • LUCKNOW • MUMBAI • RANCHI
NEW DELHI

Published by :
LAXMI PUBLICATIONS (P) LTD
113, Golden House, Daryaganj,
New Delhi-110002
Phone : 011-43 53 25 00
Fax : 011-43 53 25 28

www.laxmipublications.com
info@laxmipublications.com

Price : **Rs. 295.00** *Only*

First Edition : 2007
Reprint : 2008, 2009

OFFICES

© **Bangalore**	080-26 61 15 61	© **Chennai**	044-24 34 47 26
© **Cochin**	0484-239 70 04	© **Guwahati**	0361-254 36 69, 251 38 81
© **Hyderabad**	040-24 65 23 33	© **Jalandhar**	0181-222 12 72
© **Kolkata**	033-22 27 43 84	© **Lucknow**	0522-220 95 78
© **Mumbai**	022-24 91 54 15, 24 92 78 69	© **Ranchi**	0651-221 47 64

EEE-0637-295-ELECTRICAL ENGINEERING (TN)

C. 1414 /010/04

Typeset at : Goswami Printers, Delhi.

Printed at : Ajit Printers, Delhi.

SYLLABUS

ANNA UNIVERSITY, TAMIL NADU

Electrical Engineering—EE1X11

(Common to B.E. Computer Science & Engineering & B. Tech. Information Technology)

UNIT-I. D.C. AND A.C. CIRCUITS **15**

Basics of electricity—Electric energy and power—Circuit elements and sources—Kirchhoff's laws—Series and parallel combination of resistances—Mesh analysis—Nodal analysis—Network theorems.

Sinusoidal excitation—RMS, average and peak values—Phasor representation—RC, RL and RLC circuits—Complex power—Resonance—Three phase circuits—Line and phase values.

UNIT-II. MAGNETIC CIRCUITS **15**

Magnetic effects of electric current—Magnetic circuits—Magnetization characteristics of materials—Electromagnetic induction and force—Self and mutual inductance—AC operation of magnetic circuits and energy losses.

UNIT-III. D.C. MACHINES AND TRANSFORMERS **20**

D.C. machines—Constructional features—EMF and torque—Circuit model—Characteristics of D.C. motors—Speed control.

Transformers—Constructional features—Transformer operation—Circuit model of transformer—Voltage regulation—Efficiency—Introduction to 3 phase transformers.

UNIT-IV. A.C. MACHINES **20**

Synchronous machines—Circuit model—Armature leakage reactance—Synchronous reactance—Voltage regulation—Synchronizing to mains—Operating characteristics.

Induction machines—Construction—Circuit model—Power across airgap, torque and power output—Torque-slip characteristic—Starting arrangements—Speed control of induction motor.

Single phase induction motors—A.C. series motor.

UNIT-V. CONTROL SYSTEMS **20**

Introduction to Control Systems—Closed loop control—Examples—Mathematical models of simple physical systems—Transfer function—Control components—D.C. and A.C. servo motors—Potentiometers—Encoders—Stepper motors—Time response of first and second order systems—Desirable pole locations of transfer functions and system stability.

Simple problems in all the units.

Total = 90 Hrs.

SYLLABUS

ANNA UNIVERSITY, TAMIL NADU

Electrical Engineering—EE1X11

(Common to B.E. Computer Science & Engineering & B.Tech. Information Technology)

UNIT I: D.C. AND A.C. CIRCUITS 15

Basics of electricity—Electric energy and power—Circuit elements and sources—Kirchhoff's laws—Series and parallel combination of resistances—Mesh analysis—Nodal analysis—Network theorems.

Sinusoidal excitation—RMS, average and peak values—Phasor representation—RC, RL and RLC circuits—Complex power—Resonance—Three phase circuits—Line and phase values.

UNIT II: MAGNETIC CIRCUITS 15

Magnetic effects of electric current—Magnetic circuits—Magnetisation characteristics of materials—Electromagnetic induction and force—Self and mutual inductance—AC operation of magnetic circuits and energy losses.

UNIT III: D.C. MACHINES AND TRANSFORMERS 20

D.C. machines—Constructional features—EMF and torque—Circuit model—Characteristics of D.C. motors—Speed control.

Transformers—Constructional features—Transformer operation—Circuit model of transformer—Voltage regulation—Efficiency—Introduction to 3 phase transformers.

UNIT IV: A.C. MACHINES 20

Synchronous machines—Circuit model—Armature leakage reactance—Synchronous reactance—Voltage regulation—Synchronising to mains—Capability characteristics.

Induction machines—Construction—Circuit model—Power across airgap, torque and power output—Torque-slip characteristics—Starting arrangements—Speed control of induction motors.

Single phase induction motors—Universal motor.

UNIT V: CONTROL SYSTEMS 20

Introduction to Control systems—Closed loop control—Examples—Mathematical models of simple physical systems—Transfer function—Control components—D.C. and A.C. servo motors—Potentiometers—Encoders—Stepper motors—Time response of first and second order systems—Desirable pole locations of transfer functions and system stability.

Simple problems in all the units.

Total = 90 Hrs

CONTENTS

Chapters *Pages*

Chapters *Pages*

UNIT-II : MAGNETIC CIRCUITS

UNIT-III : D.C. MACHINES AND TRANSFORMERS

Chapters *Pages*

Chapters | Pages

UNIT-IV : A.C. MACHINES

Chapters		Pages

Chapters *Pages*

UNIT-V : CONTROL SYSTEMS

Chapters *Pages*

PREFACE TO THE FIRST EDITION

This book has been written specifically to meet exhaustively the requirements of the **latest syllabus** of the subject **"Electrical Engineering"** for B.E., B. Tech. Examinations (*Common to Computer Science and Engineering and Information Technology Disciplines*) of **Anna University, Tamil Nadu.**

The book consists of five units consisting of chapters namely *D.C. circuits ; Single-phase A.C. circuits ; Three-phase A.C. Circuits ; Magnetic circuits ; D.C. machines ; Transformers ; Synchronous generator ; Poly-phase induction motors ; Single-phase induction motors.*

The book includes comprehensive treatment of subject matter under wide range of topics mentioned in the syllabus, with a large number of solved examples to support the text. Besides this, *Highlights*, *Objective Type Questions*, *Theoretical Questions* and *Unsolved Examples* have been added at the end of each chapter to make the book a comprehensive unit in all respects.

Any suggestions for the improvement of this book will be thankfully acknowledged and incorporated in the next edition.

—AUTHOR

Introduction to SI Units and Conversion Factors

A. INTRODUCTION TO SI UNITS

SI, the international system of units are divided into three classes :

1. Base units

2. Derived units

3. Supplementary units.

From the scientific point of view division of SI units into these classes is to a certain extent arbitrary, because it is not essential to the physics of the subject. Nevertheless the General Conference, considering the advantages of a single, practical, world-wide system for international relations, for teaching and for scientific work, decided to base the international system on a choice of six well-defined units given in Table 1 below :

Table 1. SI Base Units

Quantity	*Name*	*Symbol*
length	metre	m
mass	kilogram	kg
time	second	s
electric current	ampere	A
thermodynamic temperature	kelvin	K
luminous intensity	candela	cd
amount of substance	mole	mol

The second class of SI units contains derived units, *i.e.,* units which can be formed by combining base units according to the algebraic relations linking the corresponding quantities. Several of these algebraic expressions in terms of base units can be replaced by special names and symbols can themselves be used to form other derived units.

Derived units may, therefore, be classified under three headings. Some of them are given in Table 2, 3 and 4.

Table 2. Examples of SI Derived Units Expressed in terms of Base Units

Quantity	*SI Unit*	
	Name	*Symbol*
area	square metre	m^2
volume	cubic metre	m^3
speed, velocity	metre per second	m/s
acceleration	metre per second squared	m/s^2
wave number	1 per metre	m^{-1}
density, mass density	kilogram per cubic metre	kg/m^3
concentration (of amount of substance)	mole per cubic metre	mol/m^3
activity (radioactive)	1 per second	s^{-1}
specific volume	cubic metre per kilogram	m^3/kg
luminance	candela per square metre	cd/m^2

Table 3. SI Derived Units with Special Names

Quantity	*SI Units*			
	Name	*Symbol*	*Expression in terms of other units*	*Expression in terms of SI base units*
frequency	hertz	Hz	—	s^{-1}
force	newton	N	—	$m.kg.s^{-2}$
pressure	pascal	Pa	N/m^2	$m^{-1}.kg.s^{-2}$
energy, work, quantity of heat power	joule	J	N.m	$m^2.kg.s^{-2}$
radiant flux quantity of electricity	watt	W	J/S	$m^2.kg.s^{-3}$
electric charge	coloumb	C	A.s	s.A
electric tension, electric potential	volt	V	W/A	$m^2.kg.s^{-3}.A^{-1}$
capacitance	farad	F	C/V	$m^{-2}.kg^{-1}.s^4$
electric resistance	ohm	Ω	V/A	$m^2.kg.s^{-3}.A^{-2}$
conductance	siemens	S	A/V	$m^{-2}.kg^{-1}.s^3.A^2$
magnetic flux	weber	Wb	V.S.	$m^2.kg.s^{-2}.A^{-1}$
magnetic flux density	tesla	T	Wb/m^2	$kg.s^{-2}.A^{-1}$
inductance	henry	H	Wb/A	$m^2.kg.s^{-2}.A^{-2}$
luminous flux	lumen	lm	—	cd.sr
illuminance	lux	lx	—	$m^{-2}.cd.sr$

Table 4. Examples of SI Derived Units Expressed by means of Special Names

Quantity	*SI Units*		
	Name	*Symbol*	*Expression in terms of SI base units*
dynamic viscosity	pascal second	Pa-s	$m^{-1}.kg.s^{-1}$
moment of force	metre newton	N.m	$m^2.kg.s^{-2}$
surface tension	newton per metre	N/m	$kg.s^{-2}$
heat flux density, irradiance	watt per square metre	W/m^2	$kg.s^{-2}$
heat capacity, entropy	joule per kelvin	J/K	$m^2.kg.s^{-2}.K^{-1}$
specific heat capacity, specific entropy	joule per kilogram kelvin	J/(kg.K)	$m^2.s^{-2}.K^{-1}$
specific energy	joule per kilogram	J/kg	$m^2.s^{-2}$
thermal conductivity	watt per metre kelvin	W/(m.K)	$m.kg.s^{-3}.K^{-1}$
energy density	joule per cubic metre	J/m^3	$m^{-1}.kg.s^{-2}$
electric field strength	volt per metre	V/m	$m.kg.s^{-3}.A^{-1}$
electric charge density	coloumb per cubic metre	C/m^3	$m^{-3}.s.A$
electric flux density	coloumb per square metre	C/m^2	$m^{-2}.s.A$
permitivity	farad per metre	F/m	$m^{-3}.kg^{-1}.s^4.A^4$
current density	ampere per square metre	A/m^2	—
magnetic field strength	ampere per metre	A/m	—
permeability	henry per metre	H/m	$m.kg.s^{-2}.A^{-2}$
molar energy	joule per mole	J/mol	$m^2.kg.s^{-2}mol^{-1}$
molar heat capacity	joule per mole kelvin	J/(mol.K)	$m^2.kg.s^{-2}.K^{-1}.mol^{-1}$

The SI units assigned to third class called "Supplementary units" may be regarded either as base units or as derived units. Refer Table 5 and Table 6.

Table 5. SI Supplementary Units

Quantity	*SI Units*	
	Name	*Symbol*
plane angle	radian	rad
solid angle	steradian	sr

Table 6. Examples of SI Derived Units Formed by Using Supplementary Units

Quantity	*SI Units*	
	Name	*Symbol*
angular velocity	radian per second	rad/s
angular acceleration	radian per second squared	rad/s^2
radiant intensity	watt per steradian	W/sr
radiance	watt per square metre steradian	$W\text{-}m^{-2}.sr^{-1}$

Table 7. SI Prefixes

Factor	*Prefix*	*Symbol*	*Factor*	*Prefix*	*Symbol*
10^{12}	tera	T	10^{-1}	deci	d
10^{9}	giga	G	10^{-2}	centi	c
10^{6}	mega	M	10^{-3}	milli	m
10^{3}	kilo	k	10^{-6}	micro	μ
10^{2}	hecto	h	10^{-9}	nano	n
10^{1}	deca	da	10^{-12}	pico	p
			10^{-15}	fasnto	f
			10^{-18}	atto	a

B. DEFINITIONS

The SI *seven* base units and two supplementary units are defined below :

(*i*) **Metre.** The metre is the length equal to 1,650, 76373 wavelengths in vacuum of the radiation corresponding to the transition between the levels $2p_{10}$ and $5d_5$ of the krypton - 86 atom.

(*ii*) **Kilogram.** One kilogram is equal to the mass of the international prototype of the kilogram.

(*iii*) **Second.** The second is the duration of 9,192,631,770 periods of the radiation corresponding to the transition between the two hyperfine levels of the ground state of the cesium-133 atom.

(*iv*) **Ampere.** One ampere is that constant current which, if maintained in two straight parallel conductors of infinite length, of negligible circular cross-section and placed one metre apart in vacuum, would produce between these conductors a force equal to 2×10^{-7} newton per metre of length.

(*v*) **Kelvin.** The kelvin is the fraction $\frac{1}{273.16}$ of thermodynamic temperature of the triple point of water.

(*vi*) **Candela.** The candela is the luminous intensity, in the perpendicular direction, of a surface of a $\frac{1}{600,000}$ square metre of a black body at a temperature of freezing platinum under a pressure of 101, 325 newton per square metre.

(*vii*) **Mole.** The mole is the amount of substance of a system which contains as many elementary entities as there are atoms in 0.012 kg, carbon 12. When the mole is used, the elementary entities must be specified and may be atoms, molecules, ions, electrons, other particles or specified groups of such particles.

(*viii*) **Radian.** The radian is the plane angle between two radii of a circle that cut off on the circle an arc equal in length to the radius.

(*ix*) **Steradian.** The steradian is the solid angle which having its vertex in the centre of a sphere, cuts off an area of the surface of the sphere equal to that of a square with sides of length equal to the radius of the sphere.

Some other Definitions

Newton. The newton (N) is a derived unit of force and is defined as the unit of force which when acting on a mass of 1 kilogram gives it an acceleration of one metre per second. Since acceleration due to gravity equals 9.81 m/s^2, one kilogram force equals 9.81 newtons.

Joule. The joule (J) is a derived unit of energy, work or quantity of heat and is defined as the work done when a force of one newton acts so as to cause a displacement of one metre. Energy is defined as the capacity to do work. A unit of energy in nuclear physics is the electron volt (eV) which is defined as the energy gained by an electron in rising through a potential difference of one volt.

$$1 \text{ eV} = 1.6021 \times 10^{-19} \text{ J}.$$

Watt. The watt (W) is a unit of power (*i.e.*, rate of doing work)

$$\text{Power in watts} = \frac{\text{work (or energy) in joules}}{\text{time in seconds}}$$

Thus 1 watt equals 1 Joule/sec.

1 kilo watt-hour (kWh) = 1000 watt-hours = 36000000 joules.

Coulomb. The coulomb (C) is the derived unit of charge. It is defined as *the quantity of electricity passing a given point in a circuit when a current of 1 A is maintained for 1 second.*

$$Q = l.t$$

where Q = charge in coulombs,
l = current in amperes, and
t = time in seconds.

l coulomb represents 6.24×10^{18} electrons.

Ohm. The ohm (Ω) is the unit of electric resistance and is defined as *the resistance in which a constant current of 1 A generates heat at the rate of 1 watt.*

Siemen. The siemen is a unit of electric conductance (*i.e.*, reciprocal of resistance). If a circuit has a resistance of 5 ohms, its conductance is 0.2 siemens. A more commonly used name for siemen is *mho* (Ω).

Volt. The volt is a unit of potential difference and electromotive force. It is defined as *the difference of potential across a resistance of 1 ohm carrying a current of 1 ampere.*

Hertz. The hertz (Hz) is a unit of frequency. 1 Hz = 1 cycle per second.

Horse-power. It is a practical unit of mechanical output. BHP (British horse power or brake horse power) equals 746 watts. The metric horse power equals 735.5 watts. To avoid confusion between BHP and metric horse power, the mechanical output of machines in SI units, is expressed in watts or kilowatts.

C. SALIENT FEATURES OF SI UNITS

The salient features of SI units are as follows :

1. It is a coherent system of units, *i.e.*, product or quotient of any two base quantities results in a unit resultant quantity. For example, unit length divided by unit time gives unit velocity.
2. It is a rationalised system of units, applicable to both, magnetism and electricity.
3. It is a non-gravitational system of units. It clearly distinguishes between the units of mass and weight (force) which are kilogram and newton respectively.
4. All the units of the system can be derived from the base and supplementary units.
5. The decimal relationship between units of same quantity makes possible to express any small or large quantity as a power of 10.
6. For any quantity there is one and only one SI unit. For example, joule is the unit of energy of all forms such as mechanical, heat, chemical, electrical and nuclear. However, kWh will also continue to be used as unit of electrical energy.

Advantages of SI Units :

1. Units for many different quantities are related through a series of simple and basic relationship.
2. Being an absolute system, it avoids the use of factor 'g' *i.e.*, acceleration due to gravity in several expressions in physics and engineering which had been a nuisance in all numericals in physics and engineering.
3. Being a rationalised system, it ensures all the advantages of rationalised MKSA system in the fields of electricity, magnetism, electrical engineering and electronics.
4. Joule is the only sole unit of energy of all forms and watt is the sole unit of power hence a lot of labour is saved in calculations.
5. It is a coherent system of units and involves only decimal co-efficients. Hence it is very convenient and quick system for calculations.
6. In electricity, all the practical units like volt, ohm, ampere, henry, farad, coulomb, joule and watt accepted in industry and laboratories all over the world for well over a century have become absolute in their own right in the SI system, without the need for any more practical units.

Disadvantages :

1. The non-SI time units 'minute' and 'hour' will still continue to be used until the clocks and watches are all changed to kilo seconds and mega seconds etc.
2. The base unit kilogram (kg) includes a prefix, which creates an ambiguity in the use of multipliers with gram.
3. SI units for energy, power and pressure (*i.e.*, joule, watt and pascal) are too small to be expressed in science and technology, and, therefore, in such cases the use of larger units, such as MJ, kW, kPa, will have to be made.
4. There are difficulties with regard to developing new SI units for apparent and reactive energy while joule is the accepted unit for active energy in SI systems.

D. CONVERSION FACTORS

1. Force :

$$1 \text{ newton} = \text{kg-m/sec}^2 = 0.012 \text{ kgf}$$

$$1 \text{ kgf} = 9.81 \text{ N}$$

2. Pressure :

$$1 \text{ bar} = 750.06 \text{ mm Hg} = 0.9869 \text{ atm} = 10^5 \text{ N/m}^2$$

$$= 10^3 \text{ kg/m-sec}^2$$

$$1 \text{ N/m}^2 = 1 \text{ pascal} = 10^{-5} \text{ bar} = 10^{-2} \text{ kg/m-sec}^2$$

$$1 \text{ atm} = 760 \text{ mm Hg} = 1.03 \text{ kgf/cm}^2 = 1.01325 \text{ bar}$$

$$= 1.01325 \times 10^5 \text{ N/m}^2$$

3. Work, Energy or Heat :

$$1 \text{ joule} = 1 \text{ newton metre} = 1 \text{ watt-sec}$$

$$= 2.7778 \times 10^{-7} \text{ kW-h} = 0.239 \text{ cal}$$

$$= 0.239 \times 10^{-3} \text{ kcal}$$

$$1 \text{ cal} = 4.184 \text{ joule} = 1.1622 \times 10^{-6} \text{ kWh}$$

$$1 \text{ kcal} = 4.184 \times 10^3 \text{ joule} = 427 \text{ kgf m}$$

$$= 1.1622 \times 10^{-3} \text{ kWh}$$

$$1 \text{ kWh} = 8.6042 \times 10^5 \text{ cal} = 860 \text{ kcal}$$

$$= 3.6 \times 10^6 \text{ joule}$$

$$1 \text{ kgf-m} = \left(\frac{1}{427}\right) \text{kcal} = 9.81 \text{ joules}$$

4. Power :

1 watt = 1 joule/sec = 0.860 kcal/h
1 h.p. = 75 m kgf/sec = 0.1757 kcal/sec = 735.3 watt
1 kW = 1000 watts = 860 kcal/h

5. Specific heat :

1 kcal/kg-°K = 0.4184 joules/kg-K

6. Thermal conductivity :

1 watt/m-K = 0.8598 kcal/h-m-°C
1 kcal/h-m-°C = 1.163 watt/m-K = 1.163 joules/s-m-K.

7. Heat transfer co-efficient :

1 watt/m^2-K = 0.8598 kcal/m^2-h-°C
1 kcal/m^2-h-°C = 1.163 watt/m^2-K.

The following conversion factors may be used to convert the quantities in non-SI units into SI units.

To convert	*To*	*Multiply by*
angstroms	m	10^{-10}
atmospheres	kg/m^2	10332
bars	kg/m^2	1.02×10^4
Btu	joules	1054.8
Btu	kwh	2.928×10^{-4}
circular mils	m^2	5.067×10^{-10}
cubic feet	m^2	0.02831
dynes	newtons	10^{-5}
ergs	joules	10^{-7}
ergs	kWh	0.2778×10^{-13}
feet	m	0.3048
foot-pounds	joules	1.356
foot-pounds	kg-m	0.1383
gauss	tesla	10^{-4}
grams (force)	newton	9.807×10^{-3}
horse power (metric)	watts	735.5
lines/sq. inch	tesla	1.55×10^{-5}
Maxwell	webers	10^{-8}
mho	siemens	1
micron	metre	10^{-6}
miles	km	1.609
miles	cm	2.54×10^{-3}
poundals	newton	0.1383
pounds	kilogram	0.494
pounds (force)	newtons	0.448
pounds/sq. ft.	N/m^2	47.878
pounds/sq. inch	N/m^2	6894.43

E. IMPORTANT ENGINEERING CONSTANTS AND EXPRESSIONS

Engineering Constants and Expressions	*M.K.S. System*	*SI Units*
1. Value of g_0	9.81 kg-m/kgf-sec²	1 kg-m/N-sec²
2. Universal gas constant	848 kgf-m/kg mole-°K	848 × 9.81 = 8314 J/kg-mole-K (∵ 1 kgf-m = 9.81 joules)
3. Gas constant (R)	29.27 kgf-m/kg-°K	$\frac{8314}{29}$ = 287 joules/kg-K
4. Specific heats (for air)	for air c_v = 0.17 kcal/kg-°K c_p = 0.24 kcal/kg-°K	for air c_v = 0.17 × 4.184 = 0.71128 kJ/kg-K c_p = 0.24 × 4.184 = 1 kJ/kg-K
5. Flow through nozzle-Exit velocity (C_2)	91.5 $\sqrt{U}$ where U is in kcal	44.7 $\sqrt{U}$ where U is in kJ
6. Refrigeration 1 ton	= 50 kcal/min	= 210 kJ/min
7. **Heat transfer** The Stefan Boltzman Law is given by :	$Q = \sigma T^4$ kcal/m²-h when $\sigma = 4.9 \times 10^{-8}$ kcal/h-m²-°K⁴	$Q = \sigma T^4$ watts/m²-h when $\sigma = 5.67 \times 10^{-8}$ W/m²K⁴

F. DIMENSIONS OF QUANTITIES

Different units can be represented dimensionally in terms of units of length L, mass M, time T and current I. The dimentions can be derived as under :

1. Velocity = length/time $= L/T = LT^{-1}$
2. Acceleration = velocity/time $= LT^{-1}/T = LT^{-2}$
3. Force = mass × acceleration $= MLT^{-2}$
4. Charge (coulomb) = current × time $= IT$
5. work or energy = force × distance $= ML^2T^{-2}$
6. EMF or potential = work/charge $= ML^2T^{-2}/IT = ML^2I^{-1}T^{-3}$
7. Power = work/time $= ML^2T^{-2}/T = ML^2T^{-3}$
8. Current density = current/area $= I/L^2 = IL^{-2}$
9. Resistance = emf/current $= ML^2I^{-1}T^{-3}/I = ML^2I^{-2}T^{-3}$
10. Electric flux density = electric flux or charge/area $= IT/L^2 = ITL^2$
11. MMF = current × number of turns $= I$
12. Conductance = I/resistance $= I/ML^2I^{-2}T^{-3} = I^2T^3M^{-1}L^{-2}$
13. Electric field intensity = volt/metre $= ML^2I^{-1}T^{-3}/L = MLI^{-1}T^{-3}$
14. Resistivity $= \dfrac{\text{resistant} \times \text{area}}{\text{length}} = (ML^2I^{-2}T^{-3})(L^2)/L = ML^3I^{-2}T^{-3}$

15. Magnetic field intensity (H) = MMF/length
$= I/L = IL^{-1}$

16. Magnetic flux = emf × time = $(ML^2I^{-1}T^{-3})(T) = ML^2I^{-1}T^{-2}$

17. Magnetic flux intensity = magnetic flux/area
$= (ML^2I^{-1}T^{-2})/L^2 = MI^{-1}T^{-2}$

18. Impedence = emf/current = $ML^2I^{-1}T^{-3}$

19. Admittance = I/impedence = $I^2T^3M^{-1}L^{-2}$

20. Inductance = magnetic flux/current
$= ML^2T^{-2}I^{-1}/I = ML^2T^{-2}I^{-2}$

21. Capacitance = electric charge/potential
$= IT/ML^2T^{-3}l^{-1} = M^{-1}L^{-2}T^4l^2$

UNIT-I : D.C. AND A.C. CIRCUITS

Chapters :

1. D.C. Circuits
2. Single-phase A.C. Circuits
3. Three-phase Circuits

1

D.C. Circuits

1. Basic concepts and Ohm's law—Electricity—Electron theory—Electric current—Electromotive force and potential—Resistance—Laws of resistance—Volume resistivity—Conductance—Conductivity—Temperature coefficient of resistance —Variation of resistivity with temperature —Ohm's law—An electric circuit—Resistances in series—Voltage divider rule—Resistances in parallel—Current divider rule—Superconductivity—Resistors—Insulation resistance of a cable. 2. Definitions of important terms relating network. 3. Limitations of Ohm's law. 4. Kirchhoff's laws. 5. Applications of Kirchhoff's laws—Branch-current method—Maxwell's loop (or mesh) current method—Nodal voltage method. 6. Source conversion. 7. Solving equations by determinants—Cramer's rule. 8. Principle of a generator. 9. Elementary generator. 10. Advantages and applications of electrical energy. 11. Types of supply. 12. Work, power and energy—Work—Power—Energy—Efficiency—Heating effect of electric current—Highlights—Objective Type Questions—Theoretical Questions—Unsolved Examples.

1. BASIC CONCEPTS AND OHM'S LAW

1.1. Electricity

It is not easy to define electricity.

- Electricity may be *defined* as a *form of energy*. It involves making and using energy.
- It may also be *defined* as a *way in which materials behave.*

Sometimes people use the term 'electricity' as the name for a material that *flows through a solid wire,* motion of this strange material is called *electric current.*

1.2. Electron Theory

- An *element* is defined as a substance which cannot be decomposed into other substances. The smallest particle of an element which takes part in chemical reaction is known as an *atom.*
- All matter is composed of atoms which are *infinitesimally small.*
- The atom, itself, is composed of *electrons, protons* and *neutrons.* The number and arrangement of these particles determines the type of atom : oxygen, carbon, copper, lead or any other element.
- Weight, colour, density, and all other characteristics of an element are *determined by the structure of the atom.* Electrons from lead would be the same as electrons from any other element.
- The *'electron'* is a *very light particle* that *spins around the centre* of the atom. Electrons move in an orbit. The number of electrons orbiting around the centre or nucleus of the atom varies from element to element. The electron has a *negative* (–) electric charge.
- The *'proton'* is a *very large* and *heavy particle* in relationship to the electron. One or more protons will form the centre or nucleus of the atom. The proton has a *positive* (+) electrical charge.

- The *'neutron'* consists of an electron and proton bound tightly together. Neutrons are located near the centre of the atom. The neutron is electrically neutral ; it has *no electrical charge.*

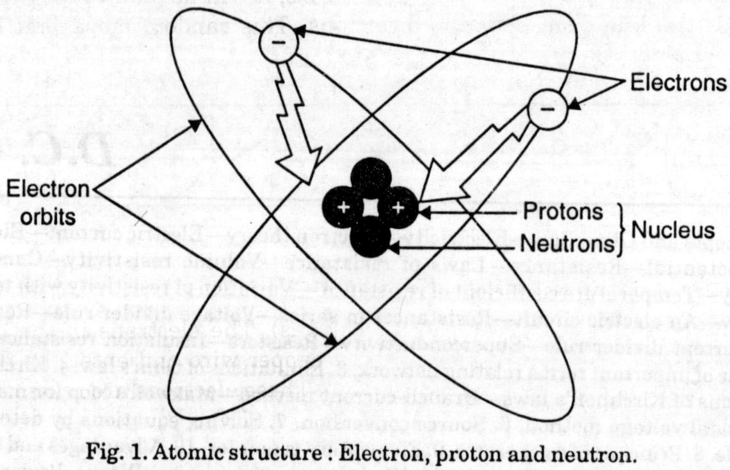

Fig. 1. Atomic structure : Electron, proton and neutron.

- Normally, the atoms are electrically neutral, that, the number of electrons and protons are the same, cancelling out each other's electrical force. Atoms "stay together" because *unlike electrical charges attract each other.* The electrical force of the protons holds the electrons in their orbits. *Like electrical charges 'repel each other* so negatively charged electrons will not collide with each other.

1.3. Electric Current

- In order to have electric current, *electrons must move from atom to atom* (Fig. 2). There are quite a few substances in which it is relatively easy for an electron to jump out of its orbit and begin to orbit in an adjoining or nearby atom. Substances which permit this movement of electrons are called *conductors of electricity* (*e.g.* copper, aluminium, silver, etc.)

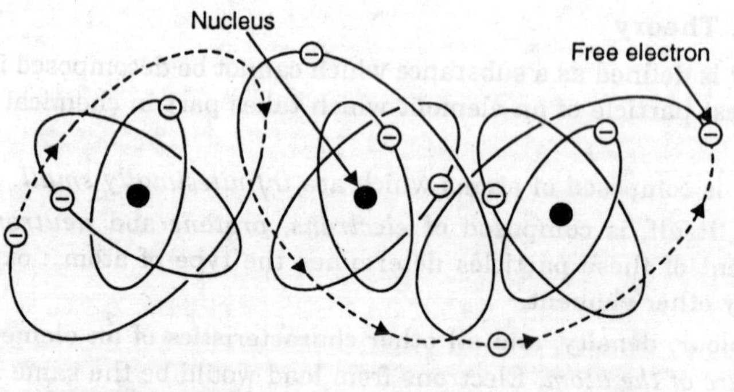

Fig. 2. Current : Flow of electrons within a conductor.

- The controlled *movement of electrons,* (or drift) through a substance is called *current.* Current occurs only when a difference of potential (e.m.f. or voltage) is present. For example, we can get a difference of potential by connecting a battery to the ends of a length of copper wire. The pressure from the battery will then move the electrons. (Fig. 3).

Refer Fig. 3 (a). The electrons of copper are free to drift in random fashion through the copper. If an imaginary line is set up in the copper wire, it will be found that the same number of electrons will cross the line from both the directions. This random movement however will not result in electric current.

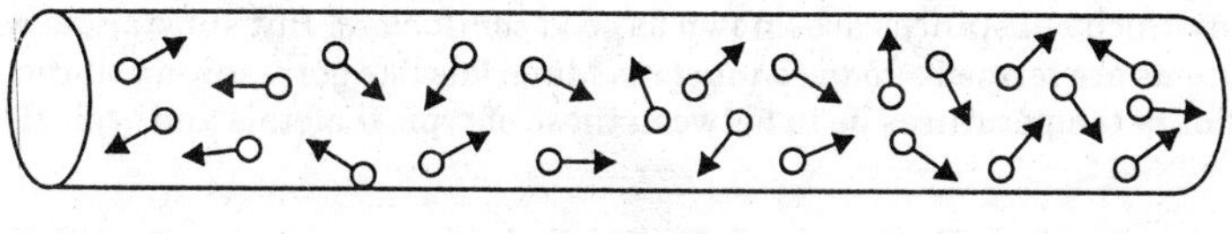

Fig. 3. (*a*) With no voltage applied : Electrons drift aimlessly.

Refer Fig. 3 (b). One end of the wire attracts electrons because it is connected to the battery terminal which has a positive charge (lack of electrons). The electrons in the copper wire drift towards this positive charge. As electrons leave the copper wire and enter the positive terminal, more electrons enter the other end of the copper wire. These electrons are taken from the negative terminal of the battery.

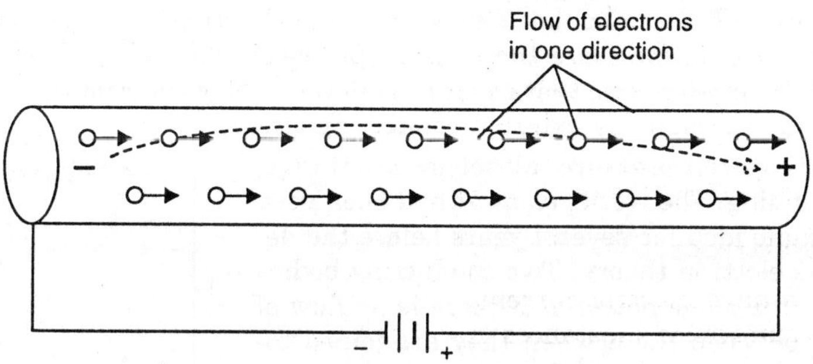

Fig. 3. (*b*) When voltage is applied to the wire :
A controlled drift of electrons takes place.

Current is the rate at which electrons move. One ampere (unit of current) represent 6.28 $\times$ 10^{18} electrons passing a point each second (1 coloumb past a point in 1 second).

A moving stream of positive charges also constitutes an electric current, and in the case of flow through *ionized gases* and *electrolytes* the current consists partly of positively charged particles moving in one direction and partly of negatively charged particles moving in the opposite direction. (Fig. 4). In all other cases the electric current consists of *solely of moving electrons.*

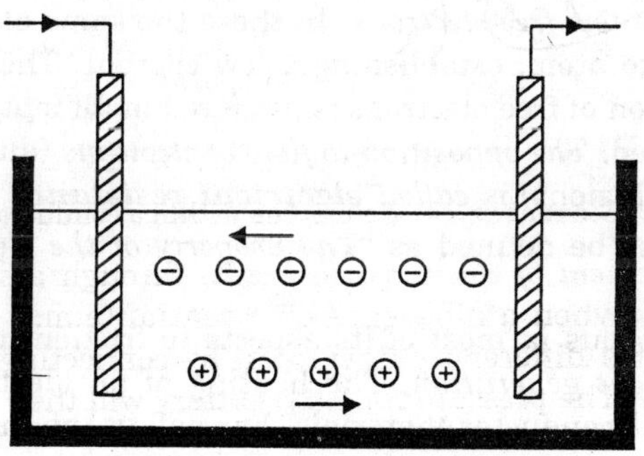

Fig. 4

One ampere is that *constant current, which* if *maintained in two straight parallel conductors of infinite length, of negligible circular cross-section and placed one metre apart in vacuum, would produce between these conductors a force equal to* 2×10^{-7} *newton per metre of length.*

Note. Those substances whose atoms have their outermost orbit incomplete act as good conductors of electricity *i.e.*, they permit an easy detachment of their outermost electrons and offer very little hinderance to their flow through atoms. Such substances are known as *good conductors.* But substances whose electrons are rigidly attached to their atoms are termed as *bad conductors.* Materials like germanium, silicon, and silicon carbide whose resistances at ordinary temperatures lie in between those of typical metals and typical insulators are called *semi-conductors.*

1.4. Electromotive Force and Potential

Electromotive force (e.m.f.) *is the force that causes a current* of *electricity to flow.*

- If we have two compressed-air storage tanks, one of which is at a higher pressure than the other, and we connect the two tanks together by means of a pipe, air will flow along this pipe from the higher pressure tank to the low-pressure tank. The force that causes this current of air to flow is the *pressure difference* between the two tanks.
- Similarly, in the zinc-copper voltaic cell, the zinc and copper electrodes both contain vast quantities of electrons, but owing to chemical action the zinc electrode is charged to a greater *electron pressure* than the copper electrode. The *electron-pressure difference* is the *electromotive force* that causes the flow of electrons from the zinc electrode to copper electrode through the external connecting circuit.

The term electron pressure, although descriptive, has no social standing. The word *potential* had been used to express the same idea for several years before the development of the electron theory. Two conducting bodies are said to be at the *same potential* if there is no flow of electric current between them when they are joined together by a conducting wire.

The ***potential difference (p.d.)*** *V, between two points in a circuit is the electrical pressure or voltage required to drive the current between them* (Fig. 5).

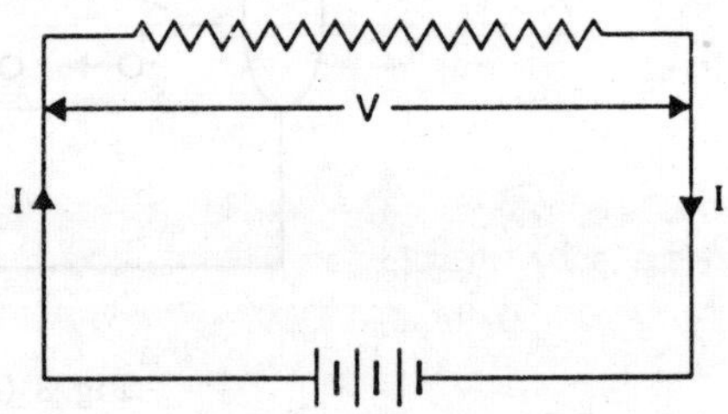

Fig. 5. Current and p.d. in a simple circuit.

The volt is unit of potential difference and electromotive force. *It is defined as the difference of potential across a resistance of 1 ohm carrying a current of 1 ampere.*

1.5. Resistance

Some materials have an *abundance of free electrons,* which require a low pressure to move them from atom to atom, and establish a high current. Such materials are known as *good conductors.* Other materials have *few free electrons.* In these the same electric pressure can move only a few electrons from atom to atom, establishing a low current. These are considered *poor conductors.* The progressive motion of free electrons is hindered in all materials, because they collide with atoms of the substance used. *The opposition to flow of electrons* (due to bonds between protons and electrons, as well as to collisions) is *called* ***electrical resistance*** ***(R).***

Resistance may also be defined as *"The property of the electric circuit which opposes the flow of current".*

Resistance is analogous in most of its aspects to friction in mechanics or hydraulics and, like friction, results in *heat generation.* The heating of an electric iron or stove is due to the resistance of the heat-unit conductor materials. For safety the material and cross-section of the conductors must be such that the temperature is kept well below a value which would result in damage to the conductor or its protective coating of insulating material.

The practical unit of electric resistance is *ohm* (Ω). It (ohm) is defined *as the resistance in which a constant current of 1 ampere generates heat at the rate of 1 watt. One volt* applied across *1 ohm* will produce *1 ampere.*

$$1 \text{ Mega-ohm } (\text{M }\Omega) = 10^6\ \Omega$$
$$1 \text{ kilo-ohm } (\text{k }\Omega) = 10^3\ \Omega$$
$$1 \text{ milli-ohm } (\text{m }\Omega) = 10^{-3}\ \Omega$$
$$1 \text{ micro-ohm } (\mu\ \Omega) = 10^{-6}\ \Omega.$$

1.6. Laws of Resistance

The resistance of a conductor, such as a wire, of uniform cross-section depends on the following factors :

(*i*) ***Length (l)......***Varies *directly* as its length, l.

(*ii*) ***Cross-section (A)......***Varies *inversely* as the cross-section, A of the conductor.

(*iii*) **Nature** *of the material.*

(*iv*) ***Temperature*** *of the conductor.*

Neglecting the last factor (*iv*) for the time being, we can say that

$$R \propto \frac{l}{A} \quad \text{or} \quad R = \rho\,\frac{l}{A} \qquad ...(1)$$

where, l = Length of the conductor,

A = Area of cross-section of the conductor, and

ρ = A constant depending on the nature of the material of the conductor and is known as its *specific resistance* or *resistivity.*

If in eqn. (1), l = 1 metre, A = 1 metre, then $R = \rho$.

Hence *specific resistance* or *resistivity* of a material may be defined as "the *resistance between the opposite faces of a metre cube of that material* " (See Fig. 6).

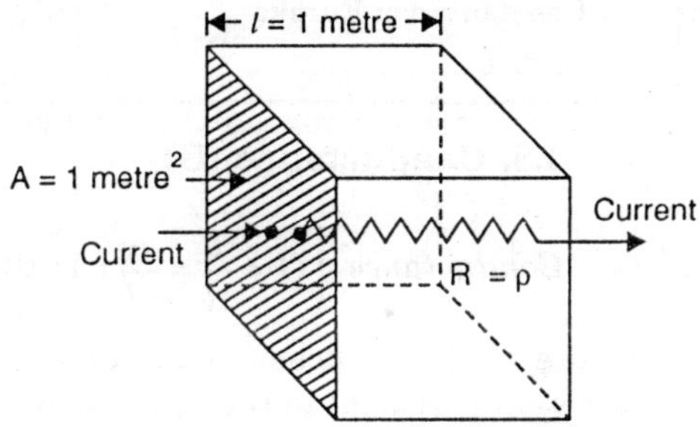

Fig. 6

Unit of resistivity. From eqn. (1), we have

$$\rho = \frac{RA}{l}$$

In S.I. system of units

$$\rho = \frac{R \text{ ohm} \times A \text{ m}^2}{l \text{ m}} = \frac{RA}{l} \text{ ohm-m}$$

Hence the unit of resistivity is *ohm-metre* (Ω-*m*).

1.7. Volume Resistivity

We know that $R = \dfrac{\rho l}{A}$

$$= \frac{\rho l \times A}{A \times A} = \frac{\rho V}{A^2} \qquad ...(i)$$

$\left[\because \quad V = lA \right.$, where V = volume, l = length, and A = uniform cross-sectional area of a conductor.$\left.\right]$

Also,
$$R = \frac{\rho \times l \times l}{A \times l} = \frac{\rho l^2}{V} \quad ...(ii)$$

i.e.,
$$R = \frac{\rho V}{A^2} = \frac{\rho l^2}{V} \quad \text{[from (i) and (ii)]} \quad ...(2)$$

The above eqn. signifies that for a *given volume* :

1. Resistance of a conductor varies *inversely as square of its cross-section.*
2. Resistance varies as *square of its length.*

The values of resistivity and temperature co-efficients for some materials are given in Table 1.

Table 1. Resistivities and Temperature Co-efficients

Material	Resistivity in Ω-m at 20° C	Temperature co-efficient at 20° C
Copper	1.59×10^{-8}	0.00428
Aluminium	2.8×10^{-8}	0.0020
Silver	1.52×10^{-8}	0.00377
Platinum	11×10^{-8}	0.00340
Iron	$(9 \text{ to } 15) \times 10^{-8}$	0.0070
Mercury	19.9×10^{-8}	0.00089
German silver (4 Cu ; 2Ni ; 1 Zn)	20.2×10^{-8}	0.00027
Constantan or Eureka	49×10^{-8}	– 0.00004 to + 0.00001
Carbon	7000×10^{-8}	– 0.0005

1.8. Conductance (G)

Conductance $\left(G = \frac{1}{R} = \frac{A}{\rho l}\right)$ is the *reciprocal of resistance* and therefore be utilised as a measure of the inducement to current flow offered by the circuit. The conductance of circuit is directly proportional to the cross-section of the conductor and inversely proportional to its length.

The unit of conductance is reciprocal ohm, or *mho* (℧).

1.9. Conductivity (σ)

The reciprocal of specific resistance $\left(\sigma = \frac{1}{\rho}\right)$ of a material is called its *conductivity*. The *unit of conductivity* $\left(\sigma = G\frac{l}{A}\right)$ is *mho/metre* (℧/m).

In practice conductivities are used chiefly to compare one conductor material with another. It is therefore convenient to choose one material as the standard and to compare others with it. Accordingly, *pure annealed copper has been chosen as the standard and its conductivity is taken as 100 percent.*

Example 1. *Find the resistance of copper wire at 20°C whose cross-sectional area is 0.02 cm² and length 400 metres. Take resistivity of copper at 20°C as 1.7 × 10^{-8} Ω-m. What is the conductivity of wire ?*

Solution. Resistivity of copper at 20° C, $\rho_{20} = 1.7 \times 10^{-8}$ Ω-m

Cross-sectional area, $A = 0.02\ cm^2 = 0.02 \times 10^{-4}\ m^2$

Length of wire, $l = 400$ m

Resistance of copper wire at 20° C, R_{20} :

Using the relation, $$R = \frac{\rho l}{a}$$

or $$R_{20} = \frac{\rho_{20} l}{a} = \frac{1.7 \times 10^{-8} \times 400}{0.02 \times 10^{-4}} = 3.4\ \Omega. \text{ (Ans.)}$$

Conductivity of wire, σ :

Conductivity (σ) $$= \frac{1}{\rho} = \frac{1}{1.7 \times 10^{-8}}$$

$$= 58.8 \times 10^6 \text{ mho/metre. (Ans.)}$$

Example 2. *Find the specific resistance of a conductor with the following particulars :*

Length .... 250 metres

Volume .... 0.04 m³

Resistance .... 0.04 Ohm.

Solution. Length, $l = 250$ m, Volume, $V = 0.04\ m^3$, Resistance, $R = 0.04$ ohm

We know that $$R = \frac{\rho l^2}{V} \quad \text{...[Eqn. (2)]}$$

Substituting the values of R, l and V in the above expression, we get

$$0.04 = \frac{\rho \times (250)^2}{0.04} \quad \text{or} \quad \rho = \frac{0.04 \times 0.04}{(250)^2}$$

$$= \frac{0.0016}{(250)^2} = 2.56 \times 10^{-8}\ \Omega\ .\ \text{m.}$$

Hence specific resistance of the conductor = 2.56×10^{-8} Ω-m (Ans.)

Example 3. *Find the relative diameters of two wires having resistance 16 Ω and 25 Ω respectively. Assume that they are of equal length and are made up of same material.*

Solution. Resistance of wire 1, $R_1 = 16\ \Omega$

Resistance of wire 2, $R_2 = 25\ \Omega$

$\mathbf{\frac{d_1}{d_2}}$:

Now, $$R = \frac{\rho l}{A} = \frac{\rho l}{\pi/4\, d^2} \quad \text{(where } d = \text{diameter of a wire)}$$

As the wires are of equal length and are made up of same material, l and ρ are constant.

$$\therefore \quad R \propto \frac{1}{d^2} \quad \textit{i.e.,} \quad R_1 \propto \frac{1}{d_1^2} \quad \text{...}(i)$$

and $$R_2 \propto \frac{1}{d_2^2} \quad \text{...}(ii)$$

From (*i*) and (*ii*), we have

$$\frac{R_1}{R_2} = \frac{d_2^2}{d_1^2} \quad \text{or} \quad \left(\frac{d_1}{d_2}\right)^2 = \frac{R_2}{R_1}$$

or $$\frac{d_1}{d_2} = \sqrt{\frac{R_2}{R_1}} = \sqrt{\frac{25}{16}} = 1.25$$

Hence $$\frac{\mathbf{d_1}}{\mathbf{d_2}} = \mathbf{1.25.} \quad \textbf{(Ans.)}$$

1.10. Temperature Coefficient of Resistance

Consider a metallic conductor having a resistance of R_0 at 0°C. Let it be heated to 1°C and R_t be its corresponding resistance at this temperature. Then it is found that within ordinary temperature ranges ; $R_t - R_0$ depends :

(*i*) directly on its initial resistance,

(*ii*) directly on the temperature rise, and

(*iii*) on the nature cf material of conductor.

In other words, $R_t - R_0 \propto R_0 . t$ or $R_t - R_0 = \alpha . R_0 . t$

where α = a constant, known as *"temperature co-efficient of resistance"* of the conductor.

or $$\alpha = \frac{R_t - R_0}{R_0 . t} \qquad ...(3)$$

In case $R_0 = 1\ \Omega$, $t = 1°C$

then $$\alpha = R_t - R_0$$

Hence temperature co-efficient of resistance at 0°C may be defined as :

"The change in resistance per ohm for change in temperature of 1°C from 0°C."

Over *large temperature ranges* the simple formula

$$R_t - R_0 = \alpha R_0 t \quad \text{or} \quad R_t = R_0 (1 + \alpha t) \qquad ...[3(a)]$$

does not completely fit, but a formula of the type

$$R_t = R_0 (1 + \alpha t + \beta t^2) \qquad ...(4)$$

(where β is a smaller co-efficient)

applies.

- At *very low temperatures*, a few degrees above absolute zero, many metals become *superconductors* ; that is, the resistance becomes so close to zero that currents once started circulate for a very long time without generating appreciable heat or needing an external e.m.f. to maintain them.
- In case of many metals it is not too inaccurate for simple rough calculations to suppose that the resistance is directly proportional to absolute temperature. Among pure metals iron and mercury cannot be included in this rough generalisation, which is also, of course, quite inapplicable to alloys.

Effect of temperature on resistance :

The following points are worth noting :

1. The resistance of *metal conductors increases* (α being positive) with rise of temperature ; the rate of increase is very considerable for most pure metals, being as much as about $\frac{1}{150}$ of the total resistance for each centigrade rise in the case of iron ; the effect is smaller in case of alloys, and very small indeed for materials such as manganin and constantan which are therefore very suitable for making *standard resistances.*

2. The resistance of *semiconductors* such as carbon, and all *electrolytes* 'decreases' as the temperature rises (α being negative).

Value of α at different temperatures :

Let R_1, R_2, and R_3 be resistance of a conductor at temperatures t_1, t_2 and t_3 °C respectively and α_1 and α_2 be the temperature co-efficients of resistance at temperatures t_1 and t_2 °C respectively.

Then, we know that,
$$R_2 = R_1 [1 + \alpha_1 (t_2 - t_1)] \quad ...(i)$$
$$R_3 = R_1 [1 + \alpha_1 (t_3 - t_1)] \quad ...(ii)$$
$$R_3 = R_2 [1 + \alpha_2(t_3 - t_2)] \quad ...(iii)$$

Equation (*iii*) can be rearranged as follows :
$$\frac{R_3}{R_2} = 1 + \alpha_2 (t_3 - t_2) \quad ...(iv)$$

Also by dividing (*ii*) by (*i*), we get
$$\frac{R_3}{R_2} = \frac{1+\alpha_1 (t_3 - t_1)}{1+\alpha_1 (t_2 - t_1)} \quad ...(v)$$

Comparing (*iv*) and (*v*), we have
$$1 + \alpha_2 (t_3 - t_2) = \frac{1+\alpha_1 (t_3 - t_1)}{1+\alpha_1 (t_2 - t_1)}$$

or
$$\alpha_2(t_3 - t_2) = \frac{1+\alpha_1 (t_3 - t_1)}{1+\alpha_1 (t_2 - t_1)} - 1$$
$$= \frac{[1+\alpha_1 (t_3 - t_1)] - [1+\alpha_1 (t_2 - t_1)]}{1+\alpha_1 (t_2 - t_1)} = \frac{\alpha_1 (t_3 - t_2)}{1+\alpha_1 (t_2 - t_1)}$$

or
$$\alpha_2 = \frac{\alpha_1}{1+\alpha_1 (t_2 - t_1)} \quad ...(5)$$

or
$$\alpha_2 = \frac{1}{\frac{1}{\alpha_1} + (t_2 - t_1)} \quad ...[5(a)]$$

Example 4. *A semi-circular ring of copper has an inner radius of 12 cm, radial thickness 6 cm and axial thickness 8 cm. Find the resistance of the ring at 60° C between its two end faces.*

Assume : Specific resistance of copper at 20° C = 1.724 × 10⁻⁸ ohm-metre. Temperature co-efficient of resistance of copper at 0°C = 0.0043/°C.

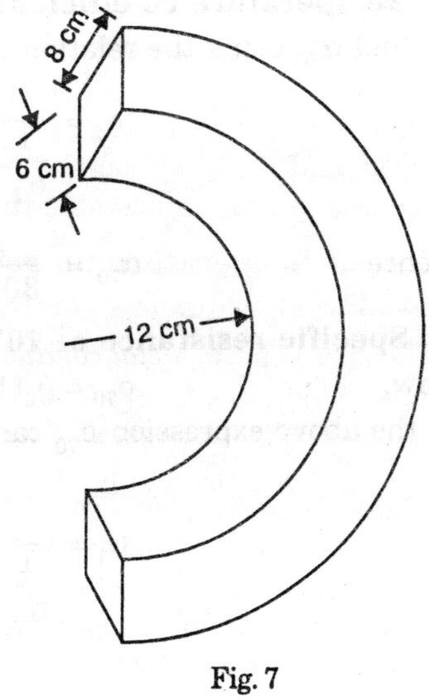

Fig. 7

Solution. Refer Fig. 7.

Mean radius of the ring
$$= \frac{18 + 12}{2} = 15 \text{ cm.}$$

Mean length between the faces 15 π cm
$$= 0.47 \text{ m.}$$

Cross-section of the ring
$$= 8 \times 6 = 48 \text{ cm}^2 = 48 \times 10^{-4} \text{ m}^2$$

Now $\alpha_0 = 0.0043/°\text{ C}$;
$$\alpha_{20} = \frac{\alpha_0}{1+\alpha_0(20-0)} = \frac{0.0043}{1+0.0043 \times 20}$$
$$= 0.00396$$

$$\rho_{60} = \rho_{20}\,[1 + \alpha_{20}(60 - 20)]$$
$$= 1.724 \times 10^{-8}\,[1 + 0.00396 \times 40] = 1.997 \times 10^{-8}\ \Omega\text{-m.}$$

Now, $R_{60} = \dfrac{\rho_{60} \times l}{A} = \dfrac{1.997 \times 10^{-8} \times 0.47}{48 \times 10^{-4}}$ = $\mathbf{1.955 \times 10^{-6}\ \Omega}$. **(Ans.)**

Example 5. *The resistance of a field coil measures 55 ohm at 25° C and 65 ohm at 75° C. Find the temperature co-efficient of conductor at 0° C.*

Solution. $R_1 = 55\ \Omega$, $t_1 = 25°C$, $R_2 = 65\ \Omega$, $t_2 = 75°C$

Temperature co-efficient of conductor at 0° C, α_0 :

Now, $R_1 = R_0\,[1 + \alpha_0 \cdot t_1]$

i.e., $55 = R_0\,[1 + \alpha_0 \times 25]$ ...(*i*)

and $R_2 = R_0\,(1 + \alpha_0 \cdot t_2)$

i.e., $65 = R_0\,(1 + \alpha_0 \times 75)$ ...(*ii*)

Dividing (*ii*) by (*i*), we get

$$\frac{65}{55} = \frac{1 + 75\alpha_0}{1 + 25\alpha_0} \quad \text{or} \quad \frac{13}{11} = \frac{1 + 75\alpha_0}{1 + 25\alpha_0}$$

or $13(1 + 25\,\alpha_0) = 11(1 + 75\,\alpha_o)$ or $13 + 325\,\alpha_0 = 11 + 825\,\alpha_o$

or $500\,\alpha_0 = 2$ or $\alpha_0 = \dfrac{2}{500} = \dfrac{1}{250} = 0.004/°C$

Hence α_0 = 0.004 per °C rise of temperature. (Ans.)

Example 6. *A specimen copper wire has a specific resistance of 1.7 × 10⁻⁸ ohm-metre at 0°C and has temperature co-efficient of $\dfrac{1}{254.5}$ at 20°C, find :*

(i) The temperature co-efficient at 70°C.

(ii) The specific resistance at 70°C.

Solution. Specific resistance at 0°C, $\rho_0 = 1.7 \times 10^{-8}\ \Omega\text{-m}$

Temperature co-efficient at 20°C, $\alpha_{20} = \dfrac{1}{254.5}$

(*i*) **Temperature co-efficient at 70°C, α_{70} :**

To find α_{70} using the relation :

$$\alpha_2 = \frac{1}{\dfrac{1}{\alpha_1} + (t_2 - t_1)} \quad \text{or} \quad \alpha_{70} = \frac{1}{\dfrac{1}{\alpha_{20}} + (70 - 20)} = \frac{1}{254.5 + 50} = \frac{1}{304.5}$$

Hence $\alpha_{70} = \dfrac{1}{304.5}$. **(Ans.)**

(*ii*) **Specific resistance at 70° C, ρ_{70} :**

Now, $\rho_{70} = \rho_0(1 + \alpha_0 \times 70)$ ...(*i*)

In the above expression ρ_{70} can be calculated only if α_0 is known, α_0 will be calculated as follows :

$$\alpha_1 = \frac{1}{\dfrac{1}{\alpha_0} + (t_1 - 0)}$$

$$\frac{1}{254.5} = \frac{1}{\frac{1}{\alpha_0} + 20} \quad \text{or} \quad 254.5 = \frac{1}{\alpha_0} + 20 \quad \text{or} \quad \alpha_0 = \frac{1}{234.5}$$

By putting the value of α_0 in expression (*i*), we get

$$\rho_{70} = 1.7 \times 10^{-8} \left(1 + \frac{1}{234.5} \times 70\right) = 2.21 \times 10^{-8} \ \Omega\text{-m.}$$

Hence $\rho_{70} = 2.21 \times 10^{-8}$ **Ω-m. (Ans.)**

Example 7. *The specific resistance of platinum at 0°C is 10.3 × 10^{-8} Ω-m. How long must a wire of No. 32 S.W.G. (diameter = 0.0274 cm) platinum be to have a resistance of 4 Ω at 0°C.*

What will be resistance of a wire at 100°C, if the temperature co-efficient of platinum is 0.0038/°C?

Solution. Specific resistance of platinum at 0°C, $\rho_0 = 10.3 \times 10^{-8}$ Ω-m.

Diameter of the platinum wire, d = 0.0274 cm = 0.000274 m.

∴ Area of the platinum wire, $A = \pi/4 \times 0.000274^2 = 5.896 \times 10^{-8}$ m²

Resistance of the wire at 0°C, $R_0 = 4$ Ω

Now, using the relation :

$$R = \frac{\rho l}{A}$$

$$\therefore \quad l = \frac{RA}{\rho} = \frac{4 \times 5.896 \times 10^{-8}}{10.3 \times 10^{-8}} = 2.289 \text{ m}$$

Hence length of the wire = **2.289 m. (Ans.)**

Resistance of the wire at 100°C, R_{100} :

$$R_{100} = R_0 (1 + \alpha_0 \times 100)$$
$$= 4 (1 + 0.0038 \times 100) = 5.52 \ \Omega$$

Hence $\mathbf{R_{100} = 5.52 \ \Omega}$ **. (Ans.)**

Example 8. *A potential difference of 200 V is applied to a copper field coil at a temperature of 15° C and the current is 10A. What will be the mean temperature of the coil when current has fallen to 5 A, the applied voltage being the same as before ?*

$$\alpha = \frac{1}{234.5} \text{ per °C at 0°C.}$$

Solution. Potential difference, $V = 200$ V

$$R_{15} = \frac{200\,(\text{V})}{10\,(\text{I})} = 20 \ \Omega$$

$$R_{t_2} = \frac{200}{5} = 40 \ \Omega$$

where t_2 is the temperature when current falls to 5 A

$$\alpha_0 = \frac{1}{234.5} \quad \text{...(Given)}$$

Temperature co-efficient of resistance at 15° C,

$$\alpha_{15} = \frac{1}{\frac{1}{\alpha_0} + (15 - 0)} = \frac{1}{234.5 + 15} = \frac{1}{249.5} /°\text{C}$$

Also, $R_{t_2} = R_{15}\,[1 + \alpha_{15}\,(t_2 - 15\,)]$

$$40 = 20\left[1 + \frac{1}{249.5}(t_2 - 15)\right]$$

$$2 = 1 + \frac{1}{249.5}(t_2 - 15)$$

$$1 = \frac{t_2 - 15}{249.5}$$

i.e., $t_2 = 264.5°\text{ C.}$

Hence mean temperature of the coil when current has fallen to 5 A

= 264.5° C. (Ans.)

☞**Example 9.** *The moving coil of a permanent magnet instrument has a temperature co-efficient of resistance 0.004 / ° C. A swamping resistance of temperature coefficient of resistance 0.0002 / ° C is to be added such that the overall temperature coefficient of resistance is 0.0006 / °C. Assuming that all measurements are given for 0°C, find the value of resistance if the coil resistance is 2 ohms.*

Solution. Let R_c = Resistance of the coil,

R_s = Swamping resistance,

R_T = Total resistance,

α_c = Temperature co-efficient of resistance for coil,

α_s = Temperature co-efficient of resistance for swamping resistance, and

α_T = Overall temperature co-efficient of resistance.

Here, α_c (at 0° C) = 0.004/°C

α_s (at 0° C) = 0.0002/° C

α_T (at 0° C) = 0.0006/° C

Total resistance at 0°C $= R_c + R_s$

Total resistance at any temperature t° C,

$$R_T = R_c\,(1 + \alpha_c\,.\,t) + R_s\,(1 + \alpha_s\,.\,t) \quad ...(i)$$

Also, $$R_T = (R_c + R_s)\,(1 + \alpha_T\,.\,t) \quad ...(ii)$$

Comparing eqns. (*i*) and (*ii*), we get

$$R_c(1 + \alpha_c\,.\,t) + R_s\,(1 + \alpha_s\,.\,t) = (R_c + R_s)\,(1 + \alpha_T\,.\,t)$$

$$R_c\,(1 + 0.004\,t) + R_s\,(1 + 0.0002\,t) = (R_c + R_s)\,(1 + 0.0006t)$$

$$R_c + 0.\,004 + R_c\,.\,t + R_s + 0.0002\,R_s\,.\,t = R_c + 0.0006\,R_c.\,t + R_s + 0.0006\,R_s.t$$

$$0.0034\,R_c = 0.0004\,R_s$$

$$R_s = \frac{0.0034\,R_c}{0.0004} = 8.5\,R_c.$$

i.e., Swamping resistance = 8.5 × coil resistance

= 8.5 × 2 = 17 Ω

Hence swamping resistance = **17 Ω . (Ans.)**

☞ **Example 10.** *Calculate the resistance at the working temperature of 75° C of a 4-pole, lap connected armature winding from the following data :*

Number of slots *100*

Conductors per slot *12*

Mean length of turn .... *3 m*

Cross-section of each conductor .... *1.5 cm × 0.2 cm*

Specific resistance of copper of 20° C .... *1.72 × 10^{-8} ohm metre*

Temperature co-efficient of resistivity at 0°C .... *0.00427/° C.*

Solution. Total number of conductors, $Z = 100 \times 12 = 1200$

Length of each conductor, $l = \frac{3}{2} = 1.5$ m

Cross-sectional area of each conductor = 1.5 cm × 0.2 cm = 0.3 cm² = 0.3×10^{-4} m²

Specific resistance of copper at 20° C, $\rho_{20} = 1.72 \times 10^{-8}$ ohm-metre.

Temperature co-efficient of resistivity at 0°C, $\alpha_0 = 0.00427$/°C

If '*a*' is the number of parallel paths,

Resistance of each parallel path, $R_0 = \frac{\rho_0 l Z}{A \times a}$

As there are such '*a*' paths in parallel, hence equivalent resistance,

$$R_0 = \frac{\rho_0 l . Z}{A \times a \times a} = \frac{\rho_0 l . Z}{A \times a^2} \quad ...(i)$$

Also,

$$\rho_t = \rho_0 (1 + \alpha_0 . t)$$

$$1.72 \times 10^{-8} = \rho_0 (1 + 0.00427 \times 20) = 1.0854 \rho_0$$

or

$$\rho_0 = \frac{1.72 \times 10^{-8}}{1.0854} = 1.585 \times 10^{-8} \ \Omega\text{-m}$$

Now by substituting the different values in eqn. (*i*), we get,

$$R_0 = \frac{1.585 \times 10^{-8} \times 1.5 \times 1200}{0.3 \times 10^{-4} \times 4 \times 4} = 0.0594 \ \Omega$$

Resistance of winding at 75° C,

$$R_{75} = R_0 (1 + \alpha_0 \times 75)$$
$$= 0.0594 (1 + 0.00427 \times 75) = 0.0784 \ \Omega$$

Hence resistance of winding at 75° C = **0. 0784 Ω. (Ans.)**

Example 11. *The base of an incandescent lamp is marked 120 V, 50 W. Measurement of resistance of the lamp on a Wheatstone bridge indicates 20 ohms at an ambient temperature of 20° C. What is the normal operating temperature of incandescent of the filament, if the resistance temperature co-efficient of tungsten is 5 × 10^{-3}/° C at 20° C ?*

Solution. Marked voltage of incandescent lamp, $V = 120$

Marked wattage of incandescent lamp, $W = 50$

$$\therefore \quad \text{Current, } I = \frac{W}{V} = \frac{50}{120} = \frac{5}{12} \text{ A}$$

and, Resistance of lamp at operating temperature (t),

$$R_t = \frac{V}{I} = \frac{120}{5/12} = 288\ \Omega$$

$$R_{20} = 20\ \Omega \quad \text{(given)}$$

$$\alpha_{20} = 5 \times 10^{-3}/^\circ \text{C (given)}$$

Also,

$$R_t = R_{20}\ [1 + \alpha_{20}\ (t - 20)]$$

$$288 = 20\ [1 + 5 \times 10^{-3}\ (t - 20)]$$

$$14.4 = 1 + 5 \times 10^{-3}\ (t - 20)$$

$$13.4 = 5 \times 10^{-3}\ (t - 20)$$

i.e.,

$$t = \frac{13.4}{5 \times 10^{-3}} + 20 = 2680 + 20 = 2700^\circ \text{ C}$$

Hence normal operating temperature = **2700° C. (Ans.)**

Example 12. *A coil has a resistance of 21.6 ohms when its mean temperature is 20°C and of 24 ohms when its mean temperature is 50° C. Find its mean temperature rise, when its resistance is 25.2 ohms and the surrounding temperature is 16° C.*

Solution. *Given :* $R_1 = 21.6\ \Omega$, $t_1 = 20$°C ; $R_2 = 24\ \Omega$; $t_2 = 50$°C ; $R_3 = 25.2\ \Omega$, $t_{surr.} = 16$°C.

Mean temperature rise :

Now,

$$R_2 = R_1\ [1 + \alpha_1\ (t_2 - t_1)]$$

or

$$24 = 21.6\ [\ 1 + \alpha_1(50 - 20)\ ]$$

$\therefore$

$$\alpha_1 = \frac{1}{270} \text{ per}^\circ \text{ C} \quad \text{(The temperature coefficient of resistance at 20°C)}$$

Again,

$$R_3 = R_1\ [1 + \alpha_1\ (t_3 - t_1)]$$

$$25.2 = 21.6\left[1 + \frac{1}{270}\ (t_3 - 20)\right]$$

$\therefore$

$$t_3 = 65^\circ \text{ C}$$

$\therefore$ Temperature rise $= t_3 - t_{surr.} = 65 - 16 =$ **49° C. (Ans.)**

Example 13. *A potential difference of 250 V is applied to a copper field coil at a temperature of 15° C, and the current is 5A. What will be the mean temperature of the coil when the current has fallen to 4A, the applied voltage being the same as before ? Assume* $\alpha_0 = \frac{1}{234.5}$ *per°C.*

Solution. *Given* : $V = 250$ V ; $t_1 = 15$°C, $I_1 = 5$ A ; $I_2 = 4$ A ;

$$\alpha_0 = \frac{1}{234.5} \text{ per}^\circ \text{ C}$$

At temperature 15° C,

$$R_1 = \frac{V}{I_1} = \frac{250}{5} = 50\ \Omega$$

Let, t_2 be the temperature when the current is 4 A.

$\therefore$

$$R_2 = \frac{V}{I_2} = \frac{250}{4} = 62.5\ \Omega$$

We know that,

$$R_2 = R_1\ [1 + \alpha_1(t_2 - t_1)]$$

$$62.5 = 50\ [1 + \alpha_{15}\ (t_2 - 15)] \qquad ...(i)$$

Also,
$$\alpha_1 = \frac{\alpha_0}{1+\alpha_0 \cdot t_1}$$

$$\therefore \qquad \alpha_{15} = \frac{\frac{1}{234.5}}{1+\frac{1}{234.5}\times 15} = \frac{1}{249.5} \text{ per °C}$$

Inserting the value of α_{15} in eqn. (1), we get

$$62.5 = 50\left[1+\frac{1}{249.5}(t_2-15)\right]$$

$$\therefore \qquad \mathbf{t_2 = 77.37 \text{ °C.}} \quad \textbf{(Ans.)}$$

1.11. Variation of Resistivity with Temperature

Like resistance, the specific resistance or resistivity of a metallic conductor increases with rise in temperature and *vice versa*. Within normal ranges of temperatures ρ_t and ρ_0 follow the relation given below :

$$\rho_t = \rho_0 (1 + \alpha_0 \cdot t) \qquad \text{...(6)}$$

where ρ_t and ρ_0 are the resistivities at $t°$ and 0° C respectively.

1.12. Ohm's Law

Ohm's law can be stated as follows :

"The ratio of potential difference between any two points of a conductor to the current flowing through it, is constant provided the physical conditions (i.e., temperature etc.) do not change.

In other words, Ohm's law states that in a closed circuit (Fig. 8), when a voltage V is applied across a conductor, then current I flowing through it is directly proportional to the applied voltage,

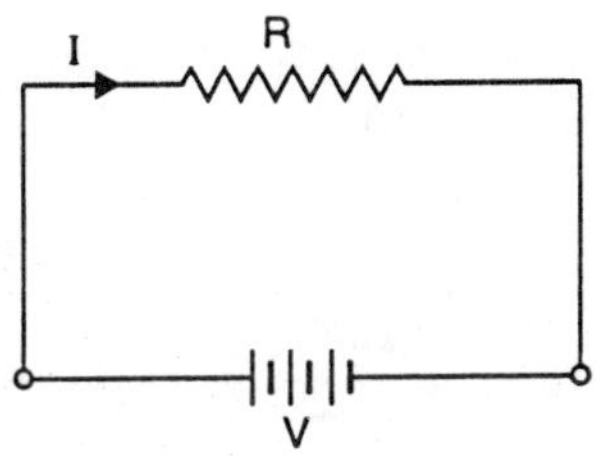

Fig. 8

i.e.,
$$I \propto V \quad \text{or} \quad \frac{V}{I} = \text{constant}$$

$$\therefore \qquad \frac{V}{I} = R \text{ ohms} \qquad \text{...[7]} \qquad \text{or} \qquad V = IR \text{ volts} \qquad \text{...[7 (a)]}$$

or
$$I = \frac{V}{R} \text{ amp.} \qquad \text{...[7 (b)]}$$

where, V = Voltage between two points,
I = Current flowing in the circuit, and
R = The resistance of the conductor.

The units of voltage, current and resistance are volt (V), ampere (A) and ohm (Ω) respectively.

- The *linear relationship ($I \propto V$) does not apply to all non-metallic conductors.* For example, for silicon carbide, the relationship is given by :

$$V = k\,I^x$$

where, k and x are *constants* and x is *less than unity.*

1.12.1. Conditions for applying Ohm's law. Ohm's law is applicable under the following *conditions* :

1. Ohm's law can be applied either to the entire circuit or to a part of the circuit.

2. When Ohm's law is applied to a part of a circuit, part resistance and the potential difference across that part resistance should be used.

3. Ohm's law can be applied to both D.C. and A.C. circuits. However, in case of A.C. circuits, impedance Z, is used in place of resistance.

Then, $$I = \frac{V}{Z} = \frac{\text{Applied voltage}}{\text{Impedance in the circuit}}.$$

- ***Electrical power.*** It *is the rate of doing work.* In other words, the *amount of work done in one second is called "power".*

or $$P = \frac{W}{t} \quad ...(8)$$

where P = The power in watts,

W = The work done in joules, and

t = Time in second.

Power is equal to the product of voltage (V) and current (I) in a particular circuit

i.e., $$^*P = V \times I \quad ...(9)$$

The following relations hold good :

(*i*) $$^*P = VI = I^2 R = \frac{V^2}{R}$$

where, P = power in watts, V = voltage in volts, I = current in amperes, and R = resistance in ohms.

(*ii*) $$I = \frac{P}{V} = \sqrt{\frac{P}{R}}$$

(*iii*) $$R = \frac{P}{I^2} = \frac{V^2}{P}$$

(*iv*) $$V = \frac{P}{I} = \sqrt{PR}$$

Power is expressed in terms of kW (kilowatt = 1000 W) or MW (megawatt = 1000 kW or 10^6 W).

- ***Electrical energy.*** It *is the total amount of work done in an electric circuit.* In other words, it is measured by the product of *power and time.*

i.e., $$W = P \times t \quad ...[\text{From eqn. (8)}]$$

or $$W = VI\,t = VQ \text{ joules}$$

where Q = The quantity of electricity passing through the circuit in coulombs.

The unit of electrical energy is *joules or watt-sec.*

It is expressed in ***kWh*** (kilowatt hours)

1 ***kWh*** (commercial unit) = 1 kW × 1 hour = 1000 watt-hours

= 1000 × 60 × 60 watt-sec

= 3.6×10^6 watt-sec. or joules.

1.12.2. Linear and non-linear resistors

A ***linear resistor*** is *one which obeys ohm's law.* A circuit which contains only linear components is called a *linear circuit.*

Such elements in which V/I (volt-ampere) plots are not straight lines but curves are called ***non-linear resistors*** or ***non-linear elements.***

Examples of non-linear elements : Filaments of incandescent lamps, diodes, thermistors and varistors.

"Varistor (Non-linear resistor)" :

- It is a voltage-dependent metal-oxide material whose resistance decreases sharply with increasing voltage.
- The *zinc oxide-based varistors* are primarily used for protecting solid-state power supplies from low and medium size voltage in the supply line.
- *Silicon carbide varistors* provide protection against high-voltage surges caused by lightning and by discharge of electromagnetic energy stored in the magnetic fields of large coils.

1.12.3. Limitations of Ohm's law. Ohm's law *does not apply under the following conditions* :

1. Electrolytes where enormous gases are produced on either electrode.
2. Non-linear resistors like vacuum radio valves, semiconductors, gas filled tubes etc.
3. Arc lamps.
4. Metals which get heated up due to flow of current.
5. Appliances like metal rectifiers, crystal detectors, etc. in which operation depends on the direction of current.

1.13. An Electric Circuit

An *"electric circuit" is a conducting path through which either an electric current flows, or is intended to flow.* It can be divided into *four* categories :

(*i*) Closed circuit (*ii*) Open circuit

(*iii*) Short circuit (*iv*) Earth or leakage circuit.

(*i*) **Closed circuit.** It is the *complete path* for flow of electric current through the load.

Example. The glowing of a bulb.

(*ii*) **Open circuit.** In case any one of the supply wires is disconnected or the fuse burns out, then the current will *not flow* through the bulb, which is an *example of open circuit.*

(*iii*) **Short circuit.** If the supply mains are *connected directly* by *a piece of wire without any load,* then the value of current will be much greater than that in closed circuit. Hence, the fuse gets blown off and this circuit is known as *short circuit.*

(*iv*) **Earth or leakage circuit.** If any wire of supply mains *touches the 'body of an appliance',* then it is known as earth or leakage circuit.

Depending upon the type of current flowing in it, an electric circuit may be further classified as :

(*i*) D.C. circuit (*ii*) A.C. circuit.

1.14. Resistances in Series

Fig. 9 shows three resistance connected in series. Obviously *current flowing through each resistance will be same* but *voltage drop across* each of them will *vary* as per value of individual resistance.

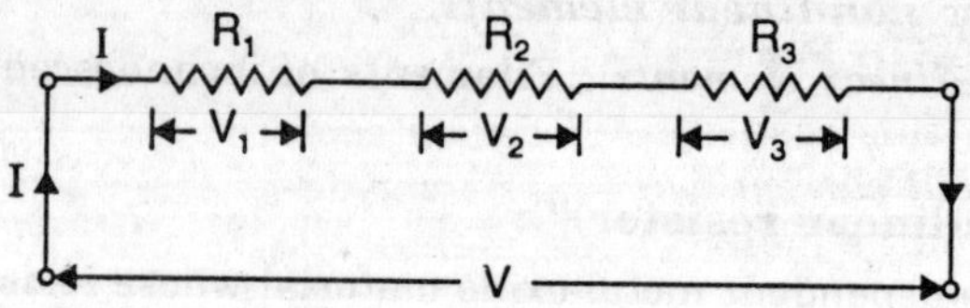

Fig. 9. Resistances in series.

Also the sum of all the voltage drops ($V_1 + V_2 + V_3$) is *equal to* the applied voltage (V).

i.e., $V = V_1 + V_2 + V_3$

$IR = IR_1 + IR_2 + IR_3$ (Using Ohm's law : $V = IR$)

i.e., $R = R_1 + R_2 + R_3$ ...(10)

where R is the *equivalent resistance* of series combination.

Also $\frac{1}{G} = \frac{1}{G_1} + \frac{1}{G_2} + \frac{1}{G_3}$ ...[10(*a*)]

As seen from above, the *main characteristics of a series circuit* are :

1. Same current flows through all parts of the circuit.
2. Different resistors have their individual voltage drops.
3. Voltage drops are additive.
4. Applied voltage equals the sum of different voltage drops.
5. Resistances are additive.
6. Powers are additive.

1.14.1. Voltage divider rule. Since in a series circuit, same the same current flows in each of the given resistors, voltage drop varies directly with its resistance. Fig. 10 shows a 24 V battery connected across a series combination of three resistors.

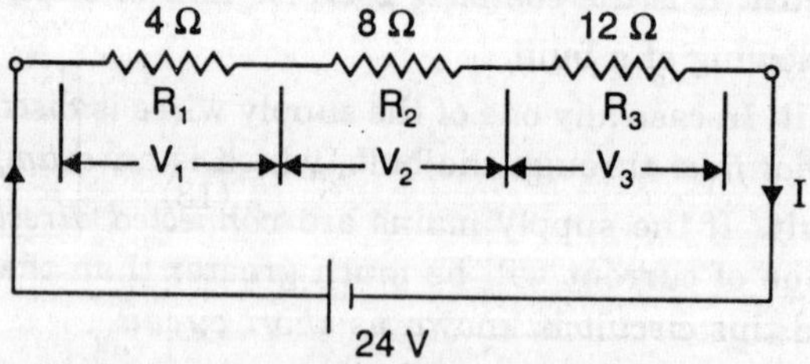

Fig. 10

Total resistance, $R = R_1 + R_2 + R_3 = 4 + 8 + 12 = 24\ \Omega$

According to **voltage divider rule,** various voltage drops are :

$$V_1 = V \times \frac{R_1}{R} = 24 \times \frac{4}{24} = 4\text{ V}$$

$$V_2 = V \times \frac{R_2}{R} = 24 \times \frac{8}{24} = 8 \text{ V}$$

$$V_3 = V \times \frac{R_3}{R} = 24 \times \frac{12}{24} = 12 \text{ V}$$

1.15. Resistances in Parallel

Refer Fig. 11. In this case *voltage across each resistance will be same but current will be different* depending upon the value of the individual resistance.

i.e., $I = I_1 + I_2 + I_3$

$$\frac{V}{R} = \frac{V}{R_1} + \frac{V}{R_2} + \frac{V}{R_3}$$

$$\frac{1}{R} = \frac{1}{R_1} + \frac{1}{R_2} + \frac{1}{R_3} \quad ...(11)$$

where R is the *equivalent resistance* of the parallel combination.

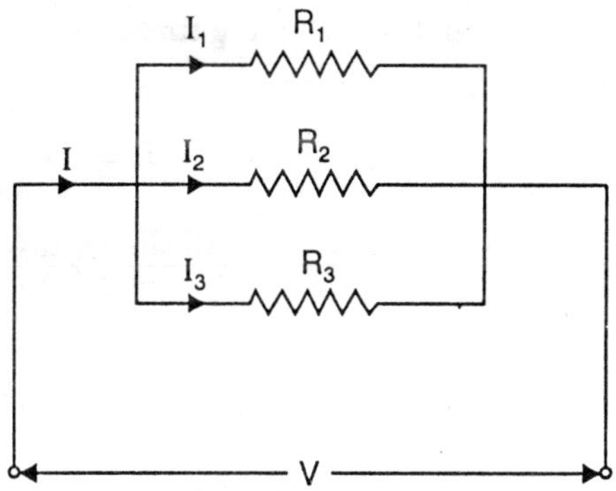

Fig. 11. Resistances in parallel.

$$R = \frac{R_1R_2R_3}{R_1R_2 + R_2R_3 + R_3R_1} \quad ...[11(a)]$$

$$G = G_1 + G_2 + G_3 \quad ...(12)$$

The *main characteristics of a parallel circuit* are :

1. Same voltage acts across all parts of the circuit.
2. Different resistors have their individual current.
3. Branch currents are additive.
4. Conductances are additive.
5. Powers are additive.

1.15.1. Division of current in parallel circuits–Current divider rule. Fig. 12 shows two resistances R_1 and R_2 connected in parallel across a voltage V. According to Ohm's law, the current in each branch is given as :

$$I_1 = \frac{V}{R_1} \; ; \quad I_2 = \frac{V}{R_2}$$

$$\therefore \quad \frac{I_1}{I_2} = \frac{R_2}{R_1} \quad ...(13)$$

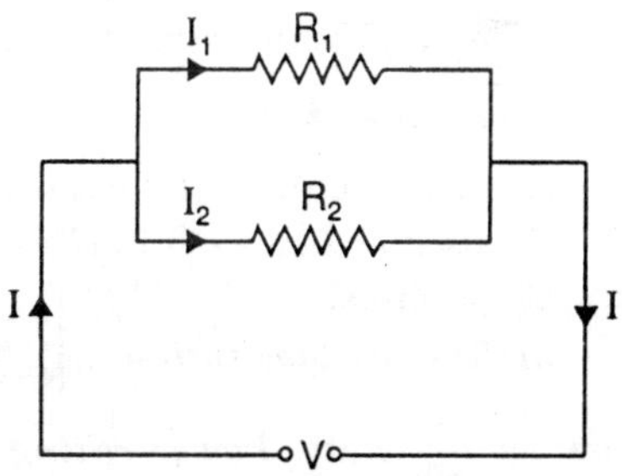

Fig. 12

Hence, the division of current in the branches of a parallel circuit is *inversely proportional to their resistances.*

This Current Divider Rule has direct application in solving electric circuits by Norton's theorem. We may also express the branch currents in terms of the total circuit current as follows :

Now, $I_1 + I_2 = I$

$\therefore \quad I_2 = I - I_1$

$$\therefore \quad \frac{I_1}{I_2} = \frac{I_1}{I - I_1} = \frac{R_2}{R_1}$$

or $$I_1 R_1 = R_2 (I - I_1) = IR_2 - I_1 R_2$$

or $$I_1(R_1 + R_2) = IR_2$$

or $$I_1 = I \times \frac{R_2}{R_1 + R_2} \quad ...(14)$$

Similarly, $$I_2 = \frac{R_1}{R_1 + R_2} \quad ...(15)$$

Refer Fig. 11 where three resistors R_1, R_2 and R_3 are connected in parallel across a voltage V, we have :

$$I_1 = I \times \left(\frac{R_2R_3}{R_1R_2 + R_2R_3 + R_3R_1} \right) \quad ...(16)$$

$$I_2 = I \times \left(\frac{R_3R_1}{R_1R_2 + R_2R_3 + R_3R_1} \right) \quad ...(17)$$

$$I_3 = I \times \left(\frac{R_1R_2}{R_1R_2 + R_2R_3 + R_3R_1} \right) \quad ...(18)$$

1.16. Superconductivity

Equation $R_t = R[1 + \alpha(t - 20)]$ holds good for temperature below 20°C. But at *very low temperature, some metals acquire zero electrical resistance and zero magnetic induction ; the property known as* ***superconductivity.***

Superconducting elements. Zinc, cadmium, mercury, lead.

Typical superconducting compounds and alloys : PbAu, $PbTl_2$, SnSb, CuS, NbN, NbB, ZrC.

The superconductivity will *disappear* if

(*i*) *The temperature of the material is raised above its critical temperature,*

Or

(*ii*) *a sufficiently strong magnetic field or current density is employed.*

1.17. Resistors

A resistor entails the following two main characteristics :

(*i*) Its *resistance (R)* in ohms. ...... The resistors are available from a fraction of an ohm to many mega ohms.

(*ii*) The *wattage rating*. ...... The power rating may be as high as several hundred watts or as low as $\frac{1}{10}$ watt. *Power rating indicates the maximum wattage the resistor can dissipate without excessive heat* (Too much heat can make the resistor burn open).

Classification of resistors :

The resistors are classified as follows :

1. *Fixed resistors.* The fixed resistor is the simplest type of resistor. Fixed means that the unit is so constructed that its resistance value is *constant* and *unchangeable.* These are made of a *carbon composition* and have a cover of black or brown hard plastics.

2. *Tapped resistors.* A tapped resistor is a resistor which has a tap, or connection, somewhere along the resistance material. These resistors are usually wire wound type. If they have more than one tap, they will have a separate terminal for each.

3. *Variable resistors.* A variable resistor has a movable contact that is used to adjust or select the resistance value between two or more terminals. A variable resistor is commonly called a *control.*

4. *Special resistors.* The most common type of special resistors is the fusible type. A *fusible resistor* have a definite resistance value and it protects the circuit much like a fuse. Another special resistor is the *temperature compensating unit.* Such resistors are used to provide special control of circuit that must be extremely stable in their operation.

Schematic symbols for various resistors are shown in Fig. 13.

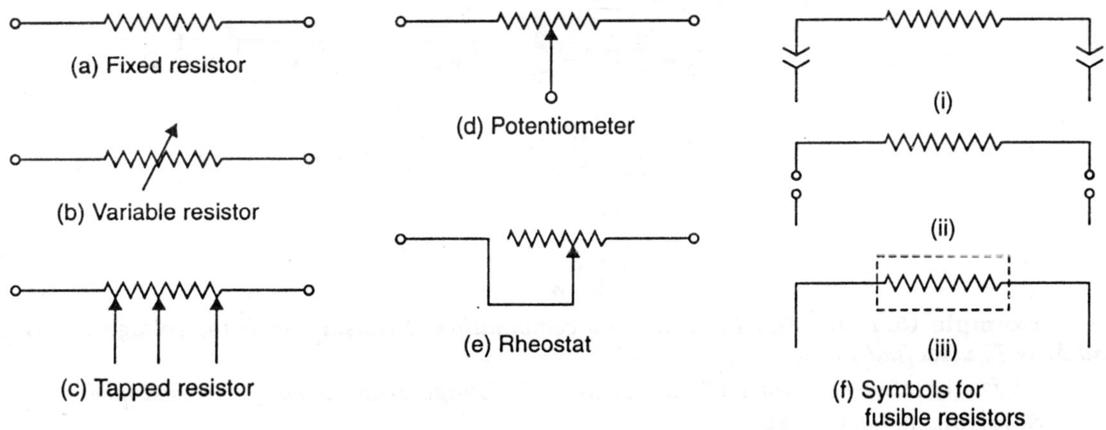

Fig. 13. Schematic symbols for various resistors.

The following types of resistors are used in electrical circuits :

(*i*) Carbon resistors.

(*ii*) Wire-wound resistors on ceramic or plastic forms (as in case of rheostats etc.)

(*iii*) Deposited carbon resistors on ceramic base.

(*iv*) Deposited metal resistors on ceramic base.

(*v*) Printed, painted or etched circuit resistors.

Example 14. *Three resistors are connected in series across a 12 V battery. The first resistance has a value of 2 Ω, second has a voltage drop of 4 V and third has a power dissipation of 12 W. Calculate the value of the circuit current.*

Solution. *Given* : $V = 12$ V ; $R_1 = 1\ \Omega$, $V_{R_2} = 4$ V ; $W_{R_3} = 12$ W

Let R_2, R_3 = Unknown resistances.

Circuit current, I :

$$I^2 R_3 = 12 \quad ...(i)$$

$$IR_2 = 4 \quad \text{or} \quad I = \frac{4}{R_2} \quad ...(ii)$$

and
$$I^2 R_2^2 = 16 \quad ...(iii)$$

∴
$$R_3 = \frac{3}{4} R_2^2 \quad \text{[From } (i) \text{ and } (ii)\text{]}$$

Also, $$I = \frac{4}{R_2} \qquad ...(iv)$$

Now, $$I(R_1 + R_2 + R_3) = 12 \qquad ...(v)$$

Inserting the values of I and R_3, we get

$$\frac{4}{R_2}\left(1 + R_2 + \frac{3}{4}R_2^2\right) = 12$$

or $$4 + 4R_2 + 3R_2^2 = 12R_2 \quad \text{or} \quad 3R_2^2 - 8R_2 + 4 = 0$$

or $$R_2 = \frac{8 \pm \sqrt{64 - 48}}{6} = \frac{8 \pm 4}{6} \text{ or } 2\ \Omega, \frac{2}{3}\ \Omega$$

$$\therefore \quad R_3 = \frac{3}{4}R_2^2 = \frac{3}{4} \times 2^2 = 3\ \Omega \text{ or } \frac{3}{4} \times \left(\frac{2}{3}\right)^2 = \frac{1}{3}\ \Omega$$

$$\therefore \quad I = \frac{12}{R_1 + R_2 + R_3} = \frac{12}{1 + 2 + 3} = 2\text{ A. } \textbf{(Ans.)}$$

or $$I = \frac{12}{1 + \frac{2}{3} + \frac{1}{3}} = 6\text{ A. } \textbf{(Ans.)}$$

Example 15. *In the Fig. 14 is shown a combination of resistances. If the voltage across L and M is 75 volts find :*

(i) The effective resistance of the circuit. *(ii) Voltage drop across each resistance.*

Solution. Refer Fig. 14.

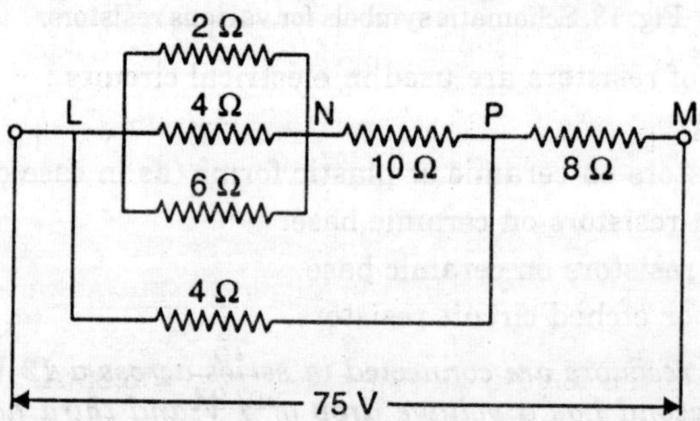

Fig. 14

(*i*) **Effective resistance of the circuit :**

Resistance between L and N,

$$\frac{1}{R_{LN}} = \frac{1}{2} + \frac{1}{4} + \frac{1}{6} = \frac{6 + 3 + 2}{12} = \frac{11}{12}\ \Omega$$

$$\therefore \quad R_{LN} = \frac{12}{11}\ \Omega \qquad ...(\text{Fig. 14})$$

Resistance of branch LNP,

$$R_{LNP} = \frac{12}{11} + 10 = \frac{122}{11} = 11.09\ \Omega \qquad ...(\text{Fig. 15})$$

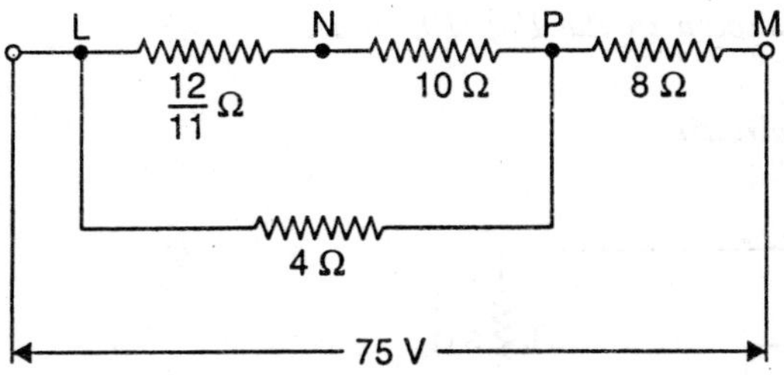

Fig. 15

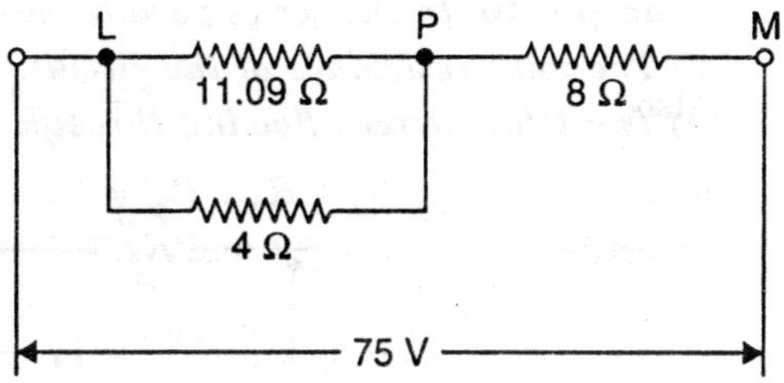

Fig. 16

Resistance between L and P

$$\frac{1}{R_{LP}} = \frac{1}{11.09} + \frac{1}{4}$$ [As there are two parallel paths between points L and P of resistances 11.09 Ω and 4 Ω]

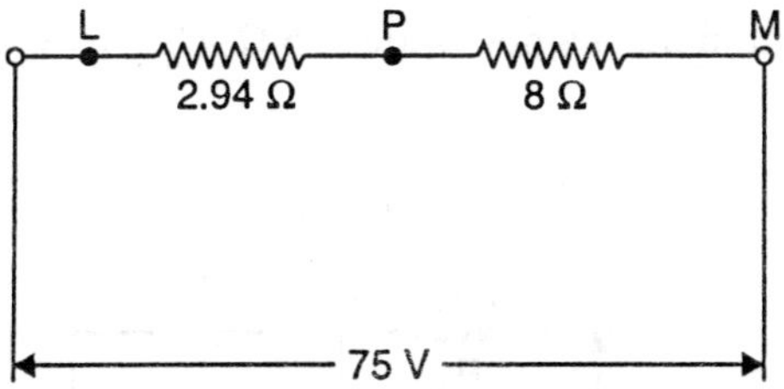

Fig. 17

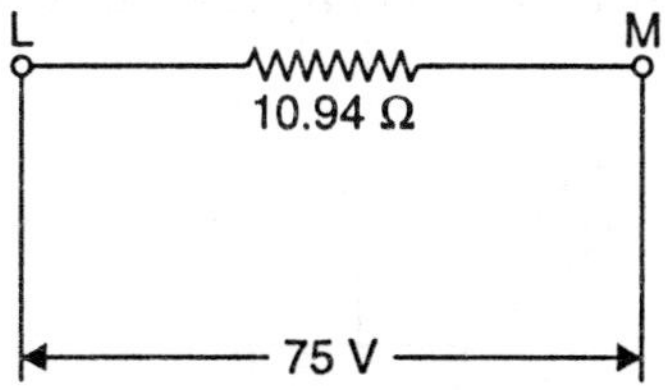

Fig. 18

$$= \frac{4 + 11.09}{44.36}$$

or $$R_{LP} = \frac{44.36}{15.09} = 2.94\ \Omega$$ ...(Fig. 16)

Effective resistance of the circuit,

$$R_{LM} = R_{LP} + R_{PM}$$
$$= 2.94 + 8 = 10.94\ \Omega$$

Hence effective resistance of the circuit = **10.94 Ω. (Ans.)**

Total current through the circuit $= \dfrac{V}{R_{LM}} = \dfrac{75}{10.94} = 6.85$ A.

Current through 4 Ω resistor $= 6.85 \times \dfrac{11.09}{11.09 + 4} = 5.03$ A

Current in branch LNP $= 6.85 \times \dfrac{4}{11.09 + 4} = 1.82$ A

∴ *Voltage drop across 2, 4 and 6 Ω resistors*

$= 1.82 \times R_{LN} = 1.82 \times \dfrac{12}{11} =$ **1.98 V. (Ans.)**

Voltage drop across 10 Ω resistor = 1.82 × 10 = **18.2 V. (Ans.)**

Voltage drop across 4 Ω resistor = 5.03 × 4 = **20.12 V. (Ans.)**

Voltage drop across 8 Ω resistor = 6.85 × 8 = **54.8 V. (Ans.)**

Example 16. *In the series-parallel circuit shown in the Fig. 19, find :*

(*i*) *The total resistance of the circuit.*

(*ii*) *The total current flowing through the circuit.*

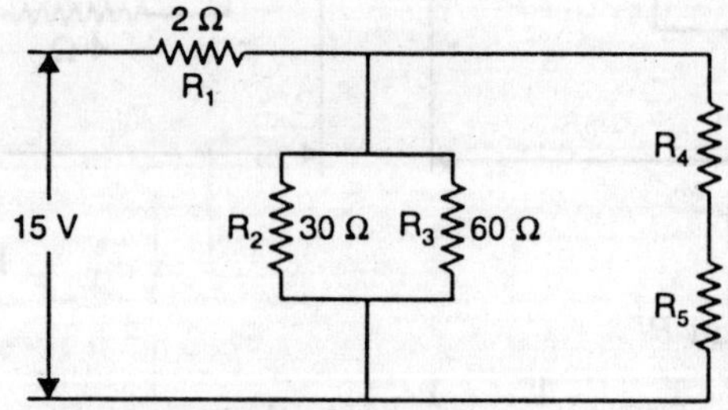

Fig. 19

Solution. (*i*) **Total resistance R_T :**

Resistors R_4 and R_5 are connected in series and have a net value of 5 Ω ($R_4 + R_5 = 3 + 2 = 5$) ...(Fig. 19)

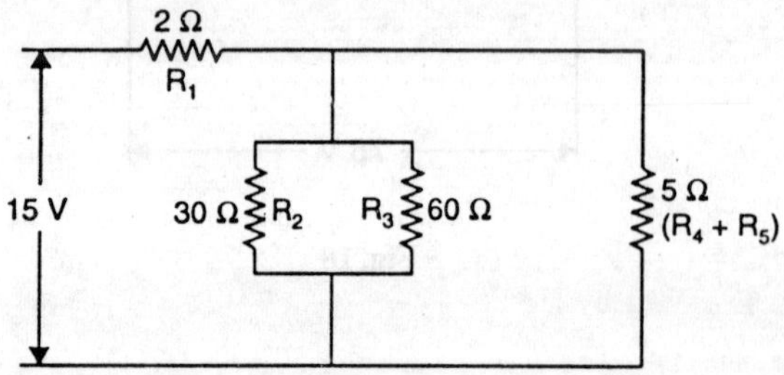

Fig. 20

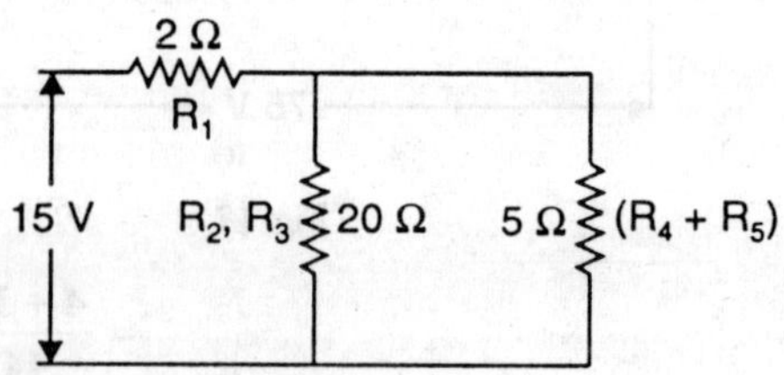

Fig. 21

The equivalent resistance of resistor R_2 and R_3 (connected in parallel)

$$= \frac{R_2 \times R_3}{R_2 + R_3} = \frac{30 \times 60}{30 + 60} = 20\ \Omega \qquad \text{...(Fig. 20)}$$

To simplify the circuit even more, we can combine the 20 Ω equivalent of R_2 and R_3, and 5 Ω equivalent of R_4 and R_5 (Fig. 21) :

$$R_{eq} = \frac{20 \times 5}{20 + 5} = \frac{100}{25} = 4\ \Omega$$

We now find the simple two-resistor series circuit shown in Fig. 22. This circuit includes a single resistance (4 Ω) which is equivalent of the series and parallel arrangement shown in Fig. 21 and R_2, R_3, R_4 and R_5. The total resistance of the circuit, R_T can now be found by addition :

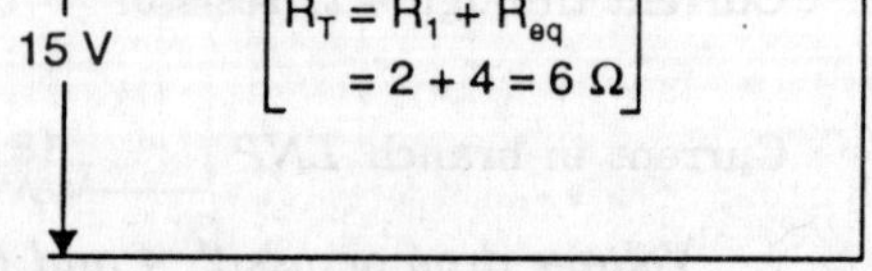

Fig. 22

$$R_T = R_1 + R_{eq} = 2 + 4 = \mathbf{6\ \Omega}. \quad \textbf{(Ans.)}$$

(*ii*) **Total current, I_T :**

$$I_T = \frac{V}{R_T} = \frac{15}{6} = 2.5\ \text{A}$$

Hence total current = **2.5 A. (Ans.)**

Example 17. *A resistance of 20 ohms is connected in series with a combination of two resistances arranged in parallel each of value 40 ohms. Determine the resistance R_3 which should be shunted across the parallel combination so that the total current drawn by the circuit is 1.5 A with applied voltage of 40 V.*

Solution. Refer Fig. 23.

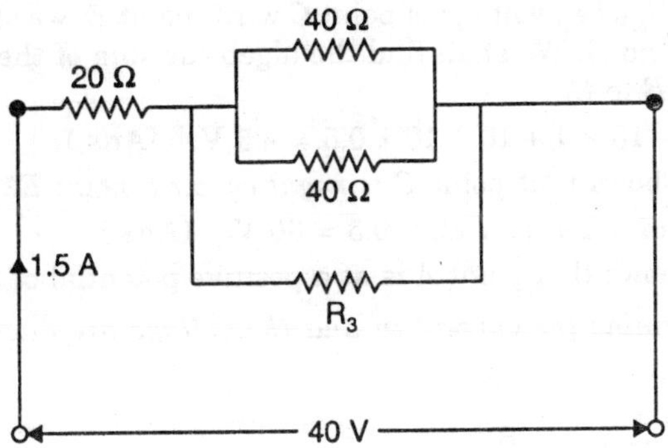

Fig. 23

Total resistance of the circuit, $R_T = \frac{V}{I} = \frac{40}{1.5} = 26.66$ ohms.

Series resistance $R = 20$ ohms

$\therefore$ Resistance of the parallel combination of resistances

$$R_T - R = 26.66 - 20 = 6.66 \text{ ohms}$$

Also resistance of parallel combination is given by :

$$\frac{1}{40} + \frac{1}{40} + \frac{1}{R_3} = \frac{1}{6.66} = 0.15$$

$\therefore$
$$\frac{1}{R_3} = 0.15 - \frac{1}{40} - \frac{1}{40} = 0.1$$

i.e., $\mathbf{R_3 = 10}$ **ohms. (Ans.)**

Example 18. *For the circuit shown in Fig. 24, find V_{CE} and V_{AG}.*

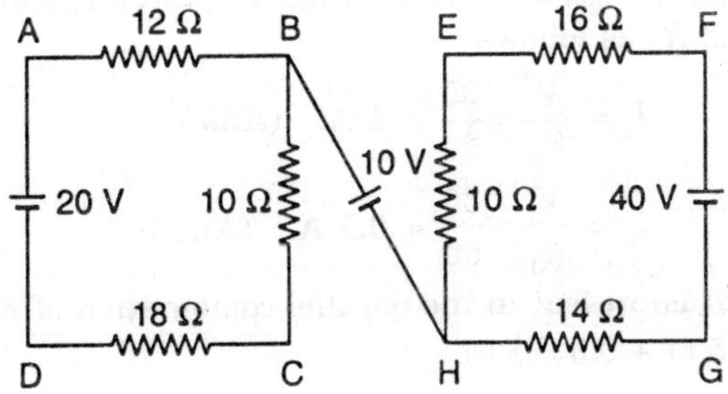

Fig. 24

(Pune University)

Solution. Refer Fig. 24.

- Let us consider the two battery circuits separately.

Current in 20 V battery circuit $ABCD = \dfrac{20}{(12 + 10 + 18)} = 0.5$ A

Current in 40 V battery circuit $EFGH = \dfrac{40}{10 + 16 + 14} = 1$ A

Voltage drops over different resistors can be found by using Ohm's law.

- In order to find V_{CE} *i.e.*, voltage of point C w.r.t. point E, we shall start from E and go to C via points H and B. We shall find the algebraic sum of the voltage drops met on the way from point E to C.

$$V_{CE} = -10 \times 1 + 10 - 10 \times 0.5 = \mathbf{-5\ V.} \quad \textbf{(Ans.)}$$

The negative sign shows that point C *is negative w.r.t. point E.*

$$V_{AG} = 14 \times 1 + 10 + 12 \times 0.5 = \mathbf{30\ V.} \quad \textbf{(Ans.)}$$

The positive sign shows that point A is at a positive potential of 30 V w.r.t. point G.

Example 19. *Determine the current in and the voltage drop across each one of the resistors in Fig. 25.*

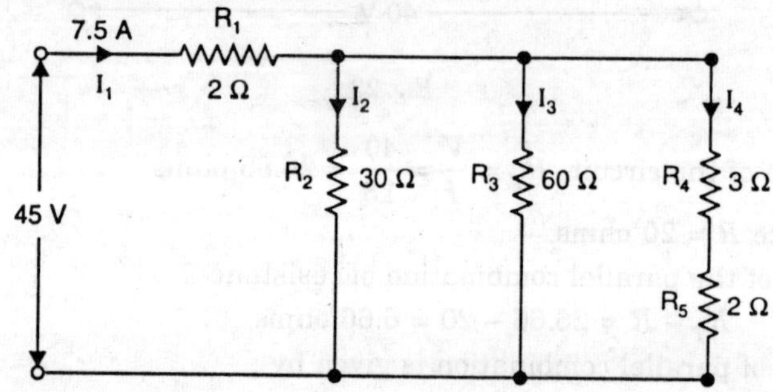

Fig. 25

Solution. The current from the source is *7.5 A* (I_1). This is the amount of current present in resistor R_1. The voltage drop across resistor R_1 is **15 V** (7.5 × 2). **(Ans.)**

- Now, subtract this voltage (15 V) from the supply voltage (45 V) to determine the voltage across the three parallel branches on the right side of resistor R_1. This is of course 30 V (45 – 15).
- Knowing the voltage across the parallel branches, we can determine the currents present in resistors R_2 and R_3 as follows :

$$I_2 = \frac{V}{R_2} = \frac{30}{30} = \mathbf{1\ A.} \quad \textbf{(Ans.)}$$

$$I_3 = \frac{V}{R_3} = \frac{30}{60} = \mathbf{0.5\ A.} \quad \textbf{(Ans.)}$$

The total current that is present in the parallel combination of R_2 and R_3 is
1.5 (1 + 0.5 = 1.5).

- The current present in the series combination $R_4 + R_5$ is

$$I_4 = \frac{V}{R_4 + R_5} = \frac{30}{3 + 2} = \mathbf{6\ A.} \quad \textbf{(Ans.)}$$

Note that the sum of the three currents (three parallel branches) is equal the supply current of 7.5 A (1 + 0.5 + 6 = 7.5).

- When the series current in the $R_4 + R_5$ combination is known, it is possible to determine the voltage drop across each resistor as follows :

$$V_{R_4} = I_4 \times R_4 = 6 \times 3 = \mathbf{18\ V.} \quad \textbf{(Ans.)}$$

$$V_{R_5} = I_5 \times R_5 = 6 \times 2 = \mathbf{12\ V.} \quad \textbf{(Ans.)}$$

Check the sum of the two voltage drops to be sure they equal 30 V (18 + 12 = 30).

1.18. Insulation Resistance of a Cable

Fig. 26 shows a single-core cable of inner and outer radii r_1 and r_2 respectively and length l. Consider an annular ring of radius 'x' and radial thickness 'dx'. If the volume resistivity of the material is ρ, then resistance of this narrow ring (dR) will be :

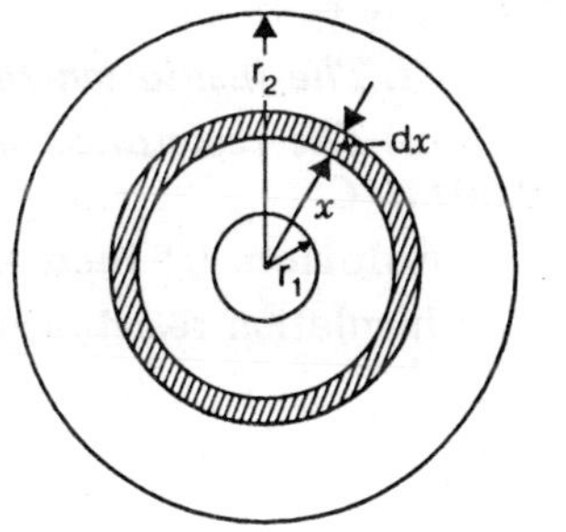

Fig. 26. Single-core cable.

$$dR = \frac{\rho dx}{2\pi x l}$$

and the insulation resistance of the whole cable will be :

$$\int dR = \int_{r_1}^{r_2} \frac{\rho dx}{2\pi x l}$$

or
$$R = \frac{\rho}{2\pi l} \Big| \log_e x \Big|_{r_1}^{r_2}$$

or
$$R = \frac{\rho}{2\pi l} \cdot \log_e \frac{r_2}{r_1}\ \Omega \qquad ...(19)$$

$$= \frac{2.3\rho}{2\pi l} \cdot \log_{10} \frac{r_2}{r_1} \qquad ...[19\ (a)]$$

In the above equation it may be noted that insulation resistance varies *inversely as the length* (whereas the *conductor resistance varies directly as the length*).

Example 20. *A single cable has a conductor (copper) diameter of 1 cm and an overall diameter of 2 cm. If the resistivity of the insulating material is 3 × 10⁶ M Ω-m, calculate the insulation resistance of 3 km length of the cable.*

What will be the conductor resistance of this cable length at a temperature of 20°C if the resistivity of copper is 1.73 × 10⁻⁸ Ω-m at 20°C ?

Solution. Inner radius of the cable, $r_1 = \frac{1}{2} = 0.5$ cm = 0.005 m.

Outer radius of the cable, $r_2 = 2/2 = 1$ cm = 0.01 m.

Length of the cable, $l = 3$ km = 3000 m.

Resistivity of insulating material, $\rho_i = 3 \times 10^6$ MΩ-m.

Resistivity of copper, $\rho_c = 1.73 \times 10^{-8}$ Ω-m (at 20°C)

Insulation resistance, R_i :

$$\mathbf{R_i} = \frac{\rho_i}{2\pi l} \log_e \frac{r_2}{r_1}$$

$$= \frac{3 \times 10^6}{2\pi \times 3000} \log_e \left(\frac{0.01}{0.005}\right) = \mathbf{110.3\ M\ \Omega.} \quad \textbf{(Ans.)}$$

Conductor resistance at 20°C, R :

$$R = \frac{\rho l}{a} = \frac{1.73 \times 10^{-8} \times 3000}{\pi/4 \times (0.01)^2} = \mathbf{0.66\ \Omega.} \quad \textbf{(Ans.)}$$

☞**Example 21.** *At 20°C, the insulation resistance of a cable, 600 metres in length is 480 mega-ohms and copper conductor resistance of the cable is 0.7 ohm. If the insulating material has its resistance halved by a temperature rise of 10°C, find for 1600 metres length of the cable at 36°C.*

(i) The insulation resistance.

(ii) The resistance of copper conductor. The temperature co-efficient of copper at 20°C = 0.0043/°C.

Solution. (*i*) **Insulation resistance at 36°C :**

Insulation resistance of 1600 metres long cable at 20°C,

$$R_1 = \frac{480 \times 600}{1600} = 180\ \text{M}\Omega$$

(∵ Insulation resistance is *inversely* proportional to *length*)

Now, $t_1 = 20°\text{C}$ and $t_2 = 36°\text{C}$

The insulation resistance R_2 at temperature t_2°C can be found out by using the expression :

$$\log_e R_2 = \log_e R_1 - k(t_2 - t_1) \qquad \text{...(i)}$$

where $k = \dfrac{\log_e m}{t'}$

and t' = Rise of temperature required to reduce the insulation resistance to $\dfrac{1}{m}$ of the resistance at initial temperature (*i.e.*, $m = 2$ in this case)

Here $t' = 10°\text{C}$

and the value of k becomes : $k = \dfrac{\log_e 2}{10} = 0.0693$

Now, substituting the various values in eqn. (*i*), we get

$$\log_e R_2 = \log_e 180 - 0.0693\ (36 - 20) = 5.193 - 1.1088$$

i.e., $\mathbf{R_2 = 59.4\ \Omega.}$ **(Ans.)**

(*ii*) **Resistance of copper conductor at 36°C :**

Resistance of conductor for 1600 metres of its length at 20°C,

$$R_{20} = \frac{0.7 \times 1600}{600}$$

(∵ Conductor resistance is *directly* proportional to the *length*)

$$= 1.866\ \Omega$$

Resistance at 36°C, $R_{36} = R_{20}\ [1 + \alpha_{20}\ (360 - 20)]$

$$= 1.866\ [1 + 0.0043 \times 16] = \mathbf{1.99\ \Omega.} \quad \textbf{(Ans.)}$$

2. DEFINITIONS OF IMPORTANT TERMS RELATING NETWORK

1. **Circuit.** A *conducting path* through which an electric current either flows or is intended to flow is called a *circuit*. The various elements of an electric circuit are called *parameters* (*e.g.* resistance, inductance and capacitance). These parameters may be *distributed* or *lumped*.

2. **Linear circuit.** The circuit whose parameters are *constant* (*i.e.*, they do not change with voltage or current) is called a *linear circuit*.

3. **Non-linear circuit.** The circuit whose parameters *change* with voltage or current is called a *non-linear circuit*.

4. **Unilateral circuit.** A unilateral circuit is one whose properties or characteristics change with the direction of its operation (*e.g. diode rectifier*).

5. **Bilateral circuit.** It is that circuit whose properties or characteristics are same in either direction (*e.g. transmission line*).

6. **Electric network.** An electric network arises when a number of parameters or electric elements coexist or combine in any manner or arrangement.

7. **Active network.** An *active network* is one which contains one or more than one sources of *e.m.f.*

8. **Passive network.** A *passive network* is one which does not contain any source of e.m.f.

9. **Node.** A *node* is a junction in a circuit where two or more circuit elements are connected together.

10. **Branch.** The part of a network which lies between two junctions is called **branch.**

3. LIMITATIONS OF OHM'S LAW

In a series circuit or in any branch of a simple parallel circuit the calculation of the current is easily effected by the direct application of Ohm's law. But such a simple calculation is not possible if one of the branches of a parallel circuit contains a source of e.m.f., or if the current is to be calculated in a part of a network in which sources of e.m.f. may be present in several meshes or loops forming the network. The treatment of such cases is effected by the application of fundamental principles of electric circuits. These principles were correlated by Kirchhoff many years ago and enunciated in the form of *two laws*, which can be considered as the foundations of circuit analysis. Other, later, methods have been developed, which when applied to special cases considerably shorten the algebra and arithmetic computation compared with the original Kirchhoff's method.

4. KIRCHHOFF'S LAWS

For complex circuit computations, the following two laws first stated by Gutsav R. Kirchhoff (1824–87) are indispensable.

1. Kirchhoff's Point Law or Current Law (KCL). It states as follows :

The sum of the currents entering a junction is equal to the sum of the currents leaving the junction. Refer Fig. 27.

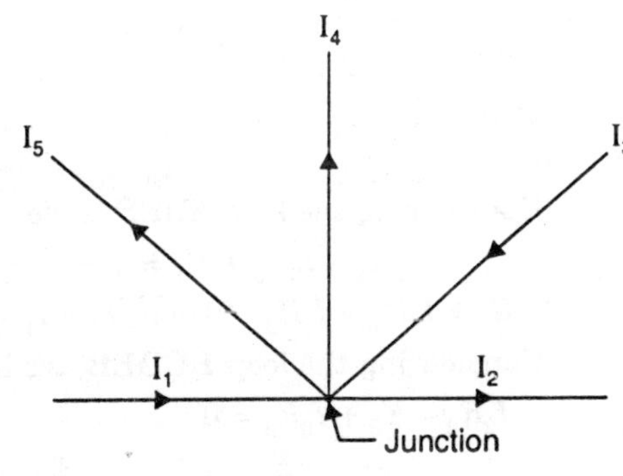

Fig. 27

If the currents *towards* a junction are considered *positive* and those *away* from the same junction *negative*, then this law states that the *algebraic sum of all currents meeting at a common junction is zero.*

i.e., Σ Currents entering = Σ Currents leaving

$$I_1 + I_3 = I_2 + I_4 + I_5 \qquad \text{...[20 (a)]}$$

or

$$I_1 + I_3 - I_2 - I_4 - I_5 = 0 \qquad \text{...[20 (b)]}$$

2. Kirchhoff's Mesh Law or Voltage Law (KVL). It states as follows :

The sum of the e.m.fs (rises of potential) around any closed loop of a circuit equals the sum of the potential drops in that loop.

Considering a rise of potential as positive (+) and a drop of potential as negative (–), the *algebraic sum* of potential differences (voltages) around a closed loop of a circuit is zero :

$$\Sigma E - \Sigma IR \text{ drops} = 0 \text{ (around closed loop)}$$

i.e., $$\Sigma E = \Sigma IR \qquad ...[21\ (a)]$$

or $$\Sigma \text{ Potential rises} = \Sigma \text{ Potential drops} \qquad ...[21\ (b)]$$

To apply this law in practice, assume an arbitrary current direction for each branch current. The end of the resistor through which the current enters, is then positive, with respect to the other end. *If the solution for the current being solved turns out negative, then the direction of that current is opposite to the direction assumed.*

In tracing through any single circuit, whether it is by itself or a part of a network, the following **rules** must be applied :

1. A *voltage drop exists* when tracing through a resistance *with or in the same direction as the current*, or through a battery or generator against their voltage, that is from *positive (+) to negative (–)*. Refer Fig. 28.

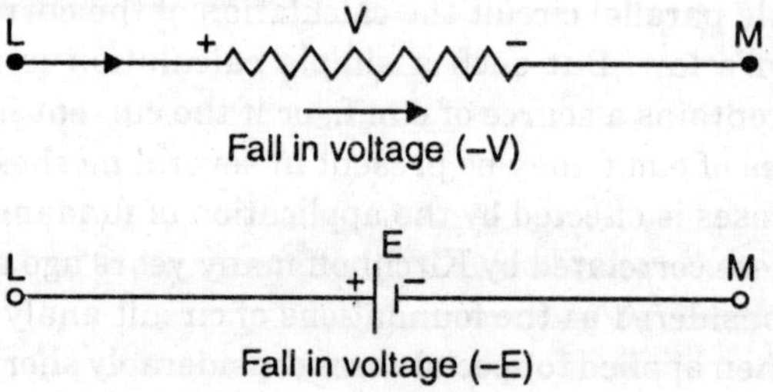

Fig. 28

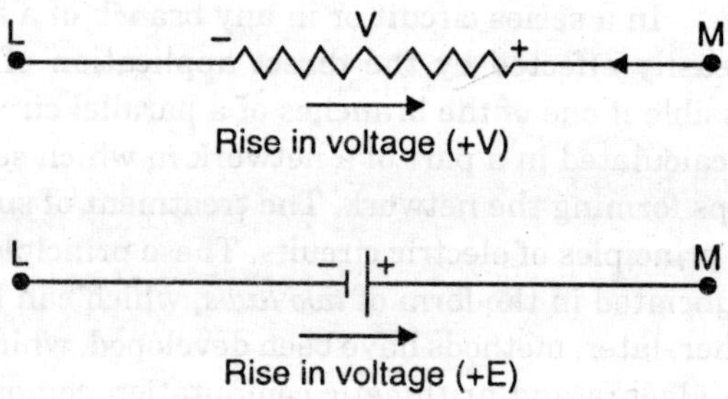

Fig. 29

2. A *voltage rise exists* when tracing through a resistance *against* or *in opposite direction to the current* or through a battery or a generator with their voltage that is from *negative (–) to positive (+)*. Refer Fig. 29.

Illustration. Consider a circuit shown in Fig. 30.

Considering the loop **ABEFA**, we get

$$-I_1R_1 - I_3R_3 + E_1 = 0$$

or $$E_1 = I_1R_1 + I_3R_3 \quad (\text{where } I_3 = I_1 + I_2) \quad ...(i)$$

Considering the loop **BCDEB**, we have

$$I_2R_2 - E_2 + I_3R_3 = 0 \qquad ...(ii)$$

or $$E_2 = I_2R_2 + I_3R_3$$

If E_1, E_2, R_1, R_2 and R_3 are known, then I_1, I_2 and I_3 can be calculated from eqns. (*i*) and (*ii*).

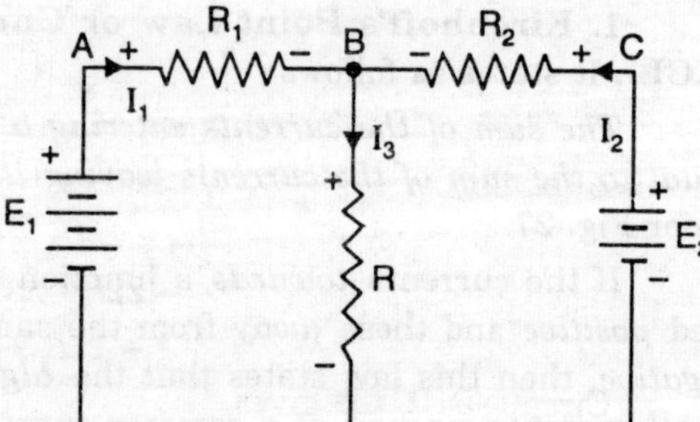

Fig. 30

5. APPLICATIONS OF KIRCHHOFF'S LAWS

Kirchhoff's laws may be employed in the following methods of solving networks :

1. Branch-current method
2. Maxwell's loop (or mesh) current method
3. Nodal voltage method.

5.1. Branch-Current Method

For a multi-loop circuit the following ***procedure*** is adopted for writing equations :

1. Assume currents in different branch of the network.

2. Write down the smallest number of voltage drop loop equations so as to include all circuit elements ; these loop equations are independent.

If there are n nodes of three or more elements in a circuit, then write $(n-1)$ equations as per current law.

3. Solve the above equations simultaneously.

The assumption made about the directions of the currents initially is arbitrary. In case the actual direction is *opposite to the assumed one*, it will be reflected as a negative value for that current in the answer.

The branch-current method (the most primitive one) involves more labour and is not used *except for very simple circuits.*

Example 22. *In the circuit of Fig. 31, find the current through each resistor and voltage drop across each resistor.*

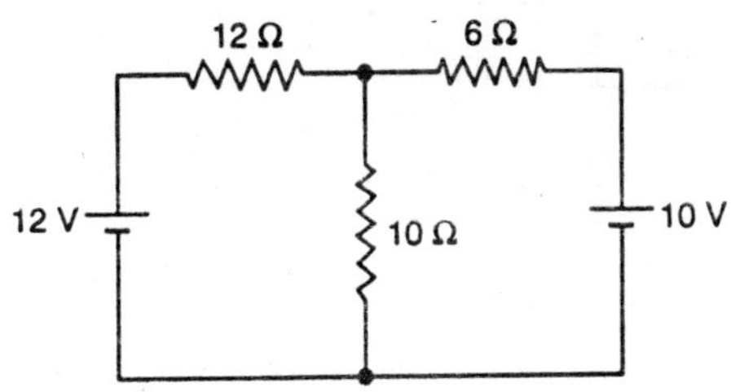

Fig. 31

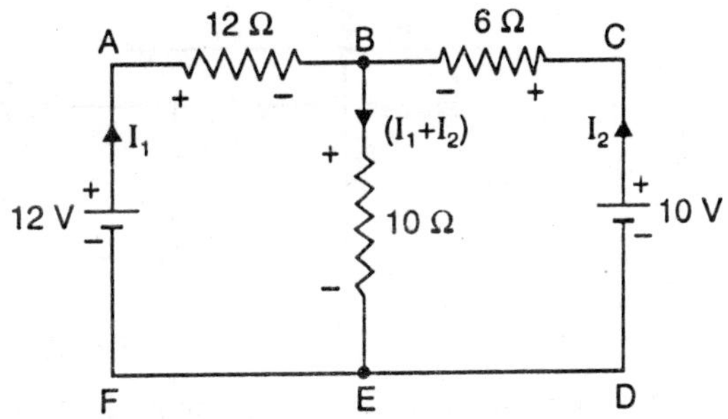

Fig. 32

Solution. Let the currents be as shown in Fig. 32.

Applying Kirchhoff's voltage law to the circuit **ABEFA**, we get

$$-12I_1 - 10(I_1 + I_2) + 12 = 0$$

$$-22I_1 - 10I_2 + 12 = 0$$

or $$+11I_1 + 5I_2 - 6 = 0 \quad \text{...(i)}$$

Circuit **BCDEB** gives,

$$6I_2 - 10 + 10(I_1 + I_2) = 0$$

$$10I_1 + 16I_2 - 10 = 0$$

$$5I_1 + 8I_2 - 5 = 0 \quad \text{...(ii)}$$

Multiplying eqn. (*i*) by 5 and eqn. (*ii*) by 11 and subtracting, we get

$$55I_1 + 25I_2 - 30 = 0$$
$$55I_1 + 88I_2 - 55 = 0$$
$$-\quad -\quad +$$
$$\overline{\quad -63I_2 + 25 = 0 \quad}$$

i.e., $I_2 = 0.397$ A

Substituting this value in eqn. (*i*), we get

$$11I_1 + 5 \times 0.397 - 6 = 0$$

i.e., $I_1 = 0.365$ A

Hence, **Current through 12 Ω resistor,** I_1 **= 0.365 A. (Ans.)**

Current through 6 Ω resistor, I_2 **= 0.397 A. (Ans.)**

Current through 10 Ω resistor, $I_1 + I_2$ **= 0.762 A. (Ans.)**

The voltage drop across :

12 Ω resistor = 0.365 × 12 = **4.38 V. (Ans.)**

6 Ω resistor = 0.397 × 6 = **2.38 V. (Ans.)**

10 Ω resistor = 0.762 × 10 = **7.62 V. (Ans.)**

Example 23. *Find the magnitude and direction of currents in each of the batteries shown in Fig. 33.*

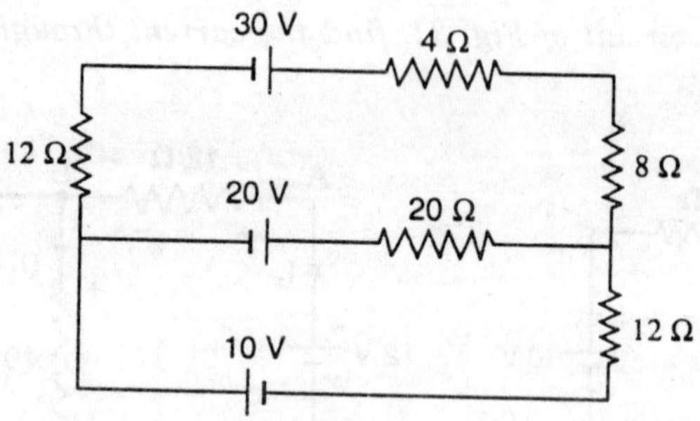

Fig. 33

Solution. Let the directions of currents I_1, I_2 and I_3 in the batteries be as shown in Fig. 34.

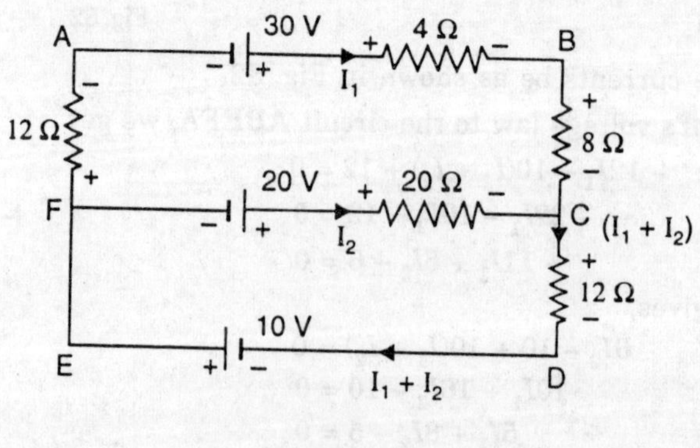

Fig. 34

Applying Kirchhoff's voltage law to the circuit **ABCFA**, we get

$$30 - 4I_1 - 8I_1 + 20I_2 - 20 - 12I_1 = 0$$

$$-24I_1 + 20I_2 + 10 = 0$$

or $$12I_1 - 10I_2 - 5 = 0 \quad ...(i)$$

Circuit **ECDEF** gives,

$$20 - 20I_2 - 12(I_1 + I_2) + 10 = 0$$

$$20 - 20I_2 - 12I_1 - 12I_2 + 10 = 0$$

$$-12I_1 - 32I_2 + 30 = 0$$

or $$6I_1 + 16I_2 - 15 = 0 \quad ...(ii)$$

Multiplying eqn. (*ii*) by 2 and subtracting it from (*i*), we get

$$-42I_2 + 25 = 0$$

i.e., $$I_2 = 0.595 \text{ A}$$

Substituting this value of I_2 in eqn. (*i*), we get

$$12I_1 - 10 \times 0.595 - 5 = 0$$

or $$I_1 = 0.912 \text{ A}$$

Hence current through,

30 V battery, I_1 = **0.912 A.** **(Ans.)**

20 V battery, I_2 = **0.595 A.** **(Ans.)**

10 V battery, $(I_1 + I_2)$ = **1.507 A.** **(Ans.)**

Example 24. *The terminal resistances of batteries A and B are 2.5 Ω and 2 Ω respectively. The battery has an e.m.f. of 20 volts. A resistance of 10 Ω is connected across the battery terminals. Calculate :*

(*i*) *The discharge current of battery A, the discharge current of battery B being 1.75 A.*

(*ii*) *The e.m.f. of battery B.*

(*iii*) *The energy dissipated in 10 Ω resistance in 40 minutes.*

Solution. Refer Fig. 35.

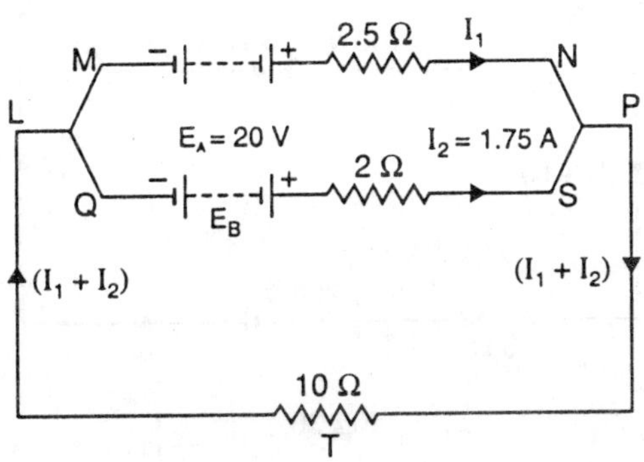

Fig. 35

(*i*) $\mathbf{I_1}$ **:**

Applying Kirchhoff's voltage law to the circuit **LMNPSQL**

$$20 - 2.5I_1 + 1.75 \times 2 - E_B = 0$$

i.e., $$E_B + 2.5I_1 - 23.5 = 0 \quad ...(i)$$

Circuit **LMNPTL** gives,

$$20 - 2.5I_1 - 10(I_1 + I_2) = 0$$

i.e.,
$$20 - 2.5I_1 - 10(I_1 + 1.75) = 0$$

or
$$20 - 12.5I_1 - 17.5 = 0$$

or
$$I_1 = 0.2 \text{ A.}$$

Hence discharge current of battery A = $\mathbf{I_1}$ **= 0.2 A.** **(Ans.)**

(*ii*) $\mathbf{E_B}$ **:**

Substituting the value of I_1 in eqn. (*i*), we get

$$E_B + 2.5 \times 0.2 - 23.5 = 0$$

$$\therefore \quad E_B = 23 \text{ V}$$

Hence e.m.f. of battery B, E_B **= 23 V.** **(Ans.)**

(*iii*) **Energy dissipated in 10 Ω resistor :**

Energy dissipated $= I^2Rt$

$= (I_1 + I_2)^2 \times 10 \times (40 \times 60)$ joules

$= (0.2 + 1.75)^2 \times 10 \times (40 \times 60)$ joules

= 91260 joules. **(Ans.)**

Example 25. *A battery having an e.m.f. of 110 V and an internal resistance of 0.2 Ω is connected in parallel with another battery with e.m.f. of 100 V and a resistance of 0.25 Ω. The two in parallel are placed in series with a regulating resistance of 5 ohms and connected across 220 V mains. Calculate :*

(*i*) *The magnitude and direction of the current in each battery.*

(*ii*) *The total current taken from the mains supply.*

Solution. Refer Fig. 36.

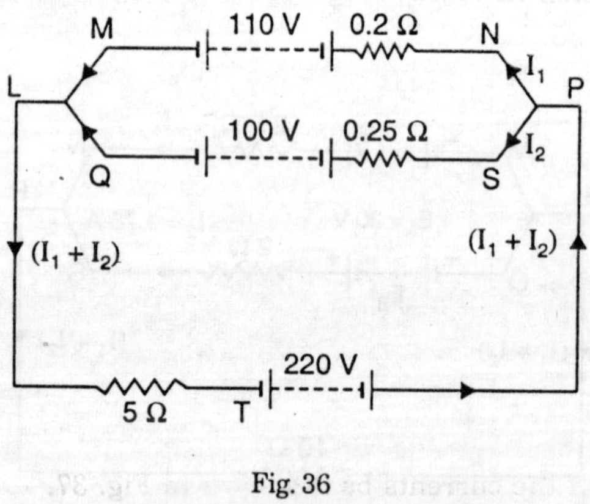

Fig. 36

(*i*) $\mathbf{I_1}$ **:,** $\mathbf{I_2}$ **:**

Let the directions of flow of currents I_1 and I_2 be as shown in Fig. 36.

Applying Kirchhoff's voltage law to **LMNPSQL**, we get

$$110 + 0.2I_1 - 0.25I_2 - 100 = 0$$

$$0.2I_1 - 0.25I_2 = -10$$

or
$$I_1 - 1.25I_2 = -50 \quad ...(i)$$

Circuit **LMNPTL** gives,

$$110 + 0.2I_1 - 220 + 5(I_1 + I_2) = 0$$

$$5.2I_1 + 5I_2 = 110$$

$$I_1 + 0.96I_2 = 21.15 \qquad ...(ii)$$

Subtracting (*ii*) from (*i*), we get

$$-2.21I_2 = -71.15$$

$\therefore$ $\quad$ $\mathbf{I_2 = 32.19}$ **A. (Ans.)**

and $\quad$ $\mathbf{I_1 = -9.75}$ **A. (Ans.)**

Since I_1 turns out to be negative, its actual direction of flow is *opposite* to that shown in Fig. 36. In other words it *is not a charging current but a discharging one*. However, I_2 is a *charging current*.

(*ii*) $(\mathbf{I_1 + I_2})$:

The total current taken from the mains supply,

$$I_1 + I_2 = -9.75 + 32.19 = \mathbf{22.44\ A.\ (Ans.)}$$

Example 26. *In the circuit shown in the Fig. 37 determine :*

(*i*) *All the currents in the network.*

(*ii*) *Voltages between the points.*

Solution. Refer Fig. 37.

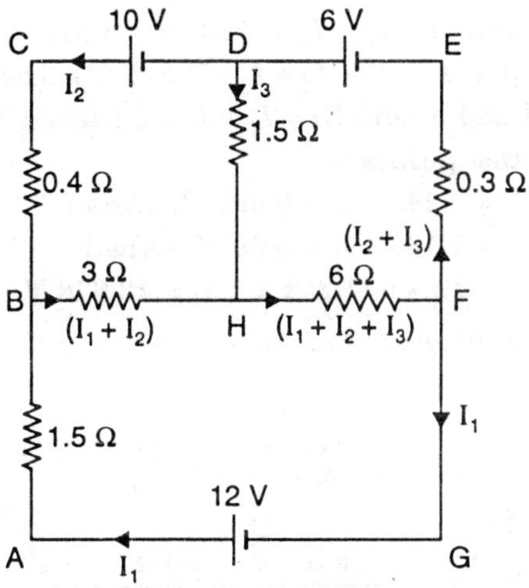

Fig. 37

(*i*) **All the currents in the network :**

Let the directions of the currents be as shown in Fig. 37.

Applying Kirchhoff's voltage law to the circuit **BCDHB**, we get

$$0.4I_2 - 10 - 1.5I_3 + 3(I_1 + I_2) = 0$$

i.e.,
$$3I_1 + 3.4I_2 - 1.5I_3 = 10 \qquad ...(i)$$

Circuit **HDEFH** gives,

$$1.5I_3 - 6 + 0.3(I_2 + I_3) + 6(I_1 + I_2 + I_3) = 0$$

$$6I_1 + 6.3I_2 + 7.8I_3 = 6 \qquad ...(ii)$$

Circuit **ABHFGA** gives,

$$-1.5I_1 - 3(I_1 + I_2) - 6(I_1 + I_2 + I_3) + 12 = 0$$

$$-10.5I_1 - 9I_2 - 6I_3 + 12 = 0$$

i.e., $$10.5I_1 + 9I_2 + 6I_3 = 12 \quad ...(iii)$$

Multiplying eqn. (*ii*) by 2 and subtracting eqn. (*ii*) from eqn. (*i*), we get

$$0.5I_2 - 10.8I_3 = 14 \quad ...(iv)$$

Multiplying eqn. (*ii*) by 10.5 and eqn. (*iii*) by 6 and subtracting eqn. (*iii*) from eqn. (*ii*), we get

$$12.15I_2 + 45.9I_3 = -9 \quad ...(v)$$

Multiplying eqn. (*iv*) by 12.15 and eqn. (*v*) by 0.5 and subtracting eqn. (*v*) from eqn. (*iv*), we get

$$-154.17\ I_3 = 174.6$$

$\therefore$ $$I_3 = -1.132 \text{ A}$$

Substituting the value of I_3 in eqn. (*iv*), we get

$$0.5I_2 - 10.8 \times (-1.132) = 14$$

i.e., $$I_2 = 3.549 \text{ A}$$

Substituting the values of I_2 and I_3 in eqn. (*iii*), we get

$$10.5I_1 + 9 \times 3.549 + 6 \times (-1.132) = 12$$

i.e., $$I_1 = -1.252 \text{ A}$$

Hence the directions of I_1 and I_3 are actually *opposite to the assumed directions.*

The *current between B and H* = $I_1 + I_2 = -1.252 + 3.549 =$ **2.297 A. (Ans.)**

The *current between H and F* = $I_1 + I_2 + I_3 = -1.252 + 3.549 - 1.132 =$ **1.165 A. (Ans.)**

The current between H and F (and E) = $I_1 + I_3 = 3.549 - 1.132 =$ **2.417 A. (Ans.)**

(*ii*) **Voltage between the points :**

Voltage across BH = 2.297 × 3 = **6.891 V. (Ans.)**

Voltage across HF = 1.165 × 6 = **6.99 V. (Ans.)**

Voltage across CE = 12 – (– 1.252 × 1.5) = **13.878 V. (Ans.)**

Example 27. *Determine the magnitude and direction of flow of current in the branch MN for the circuit shown in Fig. 38.*

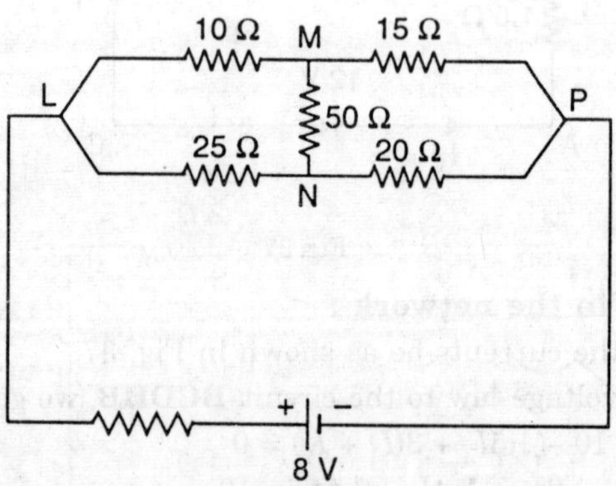

Fig. 38

Solution. Refer Fig. 39.

Let the directions and magnitudes of the currents flowing in the various circuits be as shown in Fig. 39.

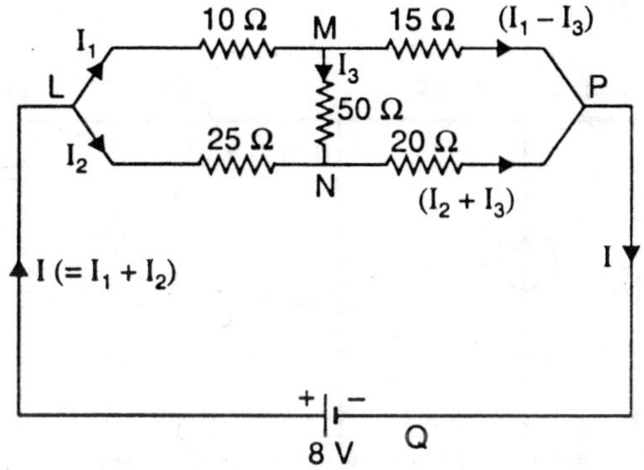

Fig. 39

Applying Kirchhoff's voltage law to the circuit **LMNL**, we get

$$-10I_1 - 50I_3 + 25I_2 = 0$$

or $$I_1 - 2.5I_2 + 5I_3 = 0 \quad ...(i)$$

Circuit **MPNM** gives,

$$-15(I_1 - I_3) + 20(I_2 + I_3) + 50I_3 = 0$$

$$-15I_1 + 20I_2 + 85I_3 = 0$$

or $$I_1 - 1.33I_2 - 5.66I_3 = 0 \quad ...(ii)$$

Circuit **LNPQL** gives,

$$-25I_2 - 20(I_2 + I_3) + 8 = 0$$

$$-45I_2 - 20I_3 + 8 = 0$$

or $$I_2 + 0.44I_3 = 0.177 \quad ...(iii)$$

Subtracting eqn. (*ii*) from eqn. (*i*), we get

$$-1.17I_2 + 10.66I_3 = 0$$

or $$I_2 - 9.11I_3 = 0 \quad ...(iv)$$

Subtracting eqn. (*iv*) from eqn. (*iii*), we get

$$9.55I_3 = 0.177$$

$$I_3 = 0.0185 \text{ A.}$$

Hence magnitude of current (I_3) flowing through the branch MN = 0.0185 A (from M to N). (Ans.)

Example 28. *Determine the current in the 4 Ω resistance of the circuit shown in Fig. 40.*

(Indore University)

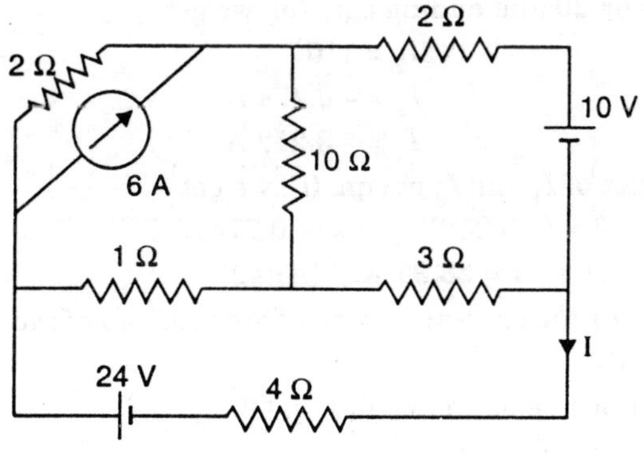

Fig. 40

Solution. Refer Fig. 41.

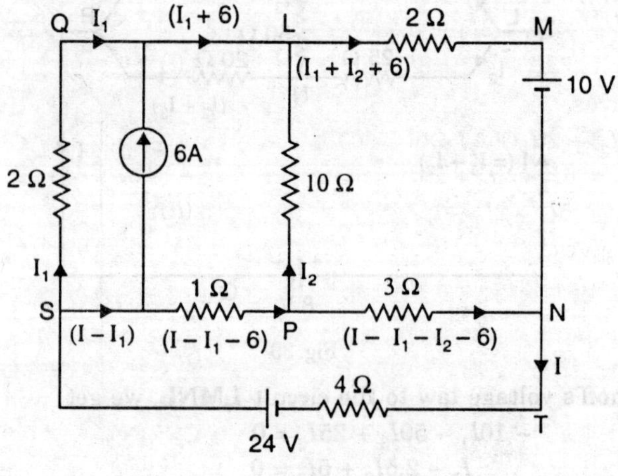

Fig. 41

Let the directions of various currents in different circuits be as shown in Fig. 41.

Applying Kirchhoff's voltage law to the circuit **SQLPS**, we get

$$-2I_1 + 10I_2 + 1(I - I_1 - 6) = 0$$

$$I - 3I_1 + 10I_2 = 6 \quad ...(i)$$

Circuit **LMNPL** gives,

$$-2(I_1 + I_2 + 6) - 10 + 3(I - I_1 - I_2 - 6) - 10I_2 = 0$$

$$3I - 5I_1 - 15I_2 = 40 \quad ...(ii)$$

Circuit **SPNTS** gives,

$$-1(I - I_1 - 6) - 3(I - I_1 - I_2 - 6) - 4I + 24 = 0$$

$$-8I + 4I_1 + 3I_2 = -48$$

or $$8I - 4I_1 - 3I_2 = 48 \quad ...(iii)$$

Multiplying eqn. (*i*) by 3 and subtracting eqn. (*ii*) from eqn. (*i*), we get

$$-4I_1 + 45I_2 = -22$$

or $$I_1 - 11.25I_2 = 5.5 \quad ...(iv)$$

Multiplying eqn. (*i*) by 8 and subtracting eqn. (*iii*) from eqn. (*i*), we get

$$-20I_1 + 83I_2 = 0 \quad ...(v)$$

Multiplying eqn. (*iv*) by 20 and adding eqn. (*v*), we get

$$-142I_2 = 110$$

∴ $$I_2 = -0.774 \text{ A}$$

and $$I_1 = -3.212 \text{ A}$$

Substituting the values of I_1 and I_2 in eqn. (*i*), we get

$$I - 3 \times (-3.212) + 10 \times (-0.774) = 6$$

i.e., **I = 23.37 A. (Ans.)**

Example 29. *Determine the current in each of the resistors of the network shown in Fig. 42.*

Solution. Refer Fig. 42.

Let the current directions be as shown in Fig. 42.

Applying Kirchhoff's voltage law to the circuit **ABDA**, we get

$$-3I_1 - 8I_3 + 4I_2 = 0$$

or $$3I_1 - 4I_2 + 8I_3 = 0 \quad ...(i)$$

Circuit **BCDB** gives,

$$-5(I_1 - I_3) + 2(I_2 + I_3) + 8I_3 = 0$$

$$-5I_1 + 2I_2 + 15I_3 = 0$$

$$5I_1 - 2I_2 - 15I_3 = 0 \quad ...(ii)$$

Circuit **ADCEA** gives,

$$-4I_2 - 2(I_2 + I_3) + 2 = 0$$

$$-6I_2 - 2I_3 = -2$$

or $$3I_2 + I_3 = 1 \quad ...(iii)$$

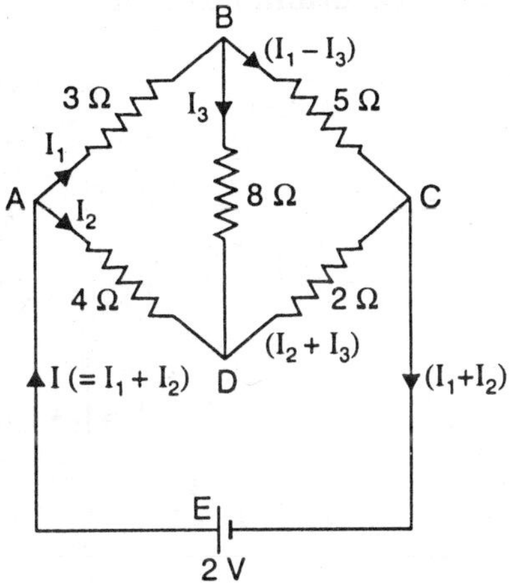

Fig. 42

Multiplying eqn. (*i*) by 5 and eqn. (*ii*) by 3 and subtracting (*ii*) from (*i*), we get

$$15I_1 - 20I_2 + 40I_3 = 0$$

$$15I_1 - \;\; 6I_2 - 45I_3 = 0$$

$$- \qquad + \qquad +$$

$$-14I_2 + 85I_3 = 0$$

or $$14I_2 - 85I_3 = 0 \quad ...(iv)$$

Multiplying eqn. (*iii*) by 14 and eqn. (*iv*) by 3 and subtracting (*iv*) from (*iii*), we get

$$42I_2 + \;\; 14I_3 = 14$$

$$42I_2 - 255I_3 = 0$$

$$- \qquad +$$

$$269I_3 = 14$$

∴ $$I_3 = 0.052 \text{ A}$$

From eqn. (*iv*), $$I_2 = 0.316 \text{ A}$$

From eqn. (*i*), $$I_1 = 0.283 \text{ A}$$

Hence, Current through 3 Ω resistor = $\mathbf{I_1}$ = **0.283 A. (Ans.)**

Current through 4 Ω resistor = $\mathbf{I_2}$ = **0.316 A. (Ans.)**

Current through 8 Ω resistor = $\mathbf{I_3}$ = **0.052 A. (Ans.)**

Current through 5 Ω resistor = $I_1 - I_3$ = **0.231 A. (Ans.)**

Current through 2 Ω resistor = $I_2 + I_3$ = **0.368 A. (Ans.)**

Example 30. *Determine the branch currents in the network of Fig. 43.*

Solution. Refer Fig. 43. Let the current directions be as shown.

Applying Kirchhoff's voltage law to the circuit **ABDA**, we get

$$5 - I_1 \times 1 - I_3 \times 1 + I_2 \times 1 = 0$$

or $$I_1 - I_2 + I_3 = 5 \quad ...(i)$$

Circuit **BCDB** gives,

$$-(I_1 - I_3) \times 1 + 5 + (I_2 + I_3) \times 1 + I_3 \times 1 = 0$$

$$-I_1 + I_2 + 3I_3 = -5$$

or $$I_1 - I_2 - 3I_3 = 5 \quad ...(ii)$$

Circuit **ADCEA** gives,

$$-I_2 \times 1 - (I_2 + I_3) \times 1 + 10 - (I_1 + I_2) \times 1 = 0$$

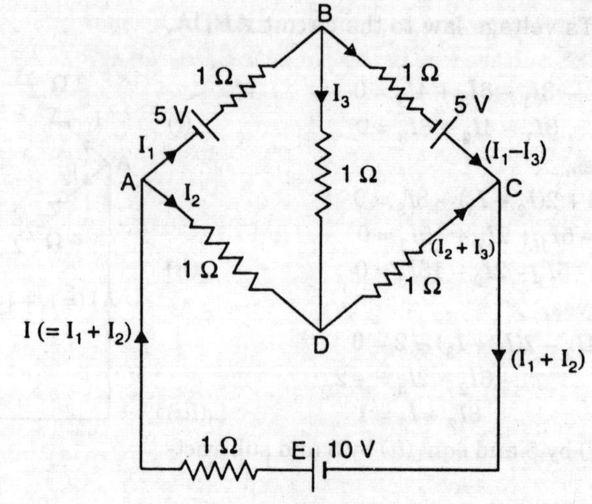

Fig. 43

$$-I_1 - 3I_2 - I_3 = -10$$

or $$I_1 + 3I_2 + I_3 = 10 \quad ...(iii)$$

Subtracting (*ii*) from (*i*), we get $$I_3 = 0$$

From (*i*), we have $$I_1 - I_2 = 5 \quad ...(iv)$$

and from (*iii*), we have $$I_1 + 3I_2 = 10 \quad ...(v)$$

By solving (*iv*) and (*v*), we get $$I_2 = 1.25 \text{ A}$$

and $$I_1 = 6.25 \text{ A}$$

Hence, Current in branch AB = **6.25 A. (Ans.)**

Current in branch BC = **6.25 A. (Ans.)**

Current in branch BD = **0. (Ans.)**

Current in branch AD = **1.25 A. (Ans.)**

Current in branch DC = **1.25 A. (Ans.)**

Current in branch CA = **7.5 A. (Ans.)**

Example 31. *Find the current in the galvanometer arm of the wheatstone bridge shown in Fig. 44.*

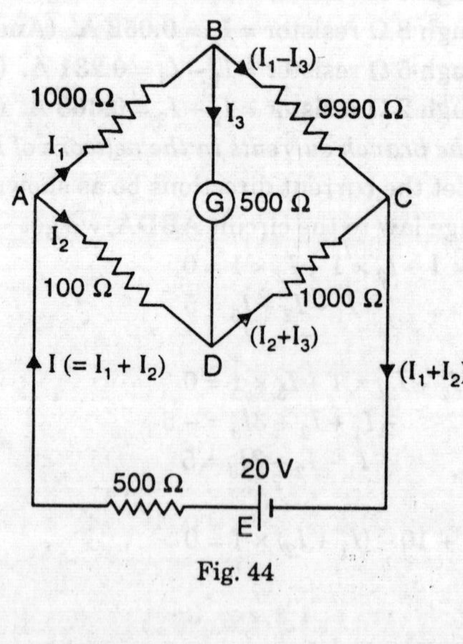

Fig. 44

Solution. Refer Fig. 44. Let the current directions be as shown.

Applying Kirchhoff's voltage law to the circuit **ABDA**, we get

$$-1000I_1 - 500I_3 + 100I_2 = 0$$
$$I_1 - 0.1I_2 + 0.5I_3 = 0 \quad ...(i)$$

Circuit **BCDB** gives,

$$-9990(I_1 - I_3) + 1000(I_2 + I_3) + 500I_3 = 0$$
$$-9990I_1 + 1000I_2 + 11490I_3 = 0$$
$$I_1 - 0.1001I_2 - 1.15I_3 = 0 \quad ...(ii)$$

Circuit **ADCEA** gives,

$$-100I_2 - 1000(I_2 + I_3) + 20 - 500(I_1 + I_2) = 0$$
$$-500I_1 - 1600I_2 - 1000I_3 = -20$$
$$I_1 + 3.2I_2 + 2I_3 = 0.04 \quad ...(iii)$$

Subtracting (*ii*) from (*i*), we get

$$0.0001I_2 + 1.65I_3 = 0 \quad ...(iv)$$

Subtracting (*iii*) from (*ii*), we get

$$+3.3001I_2 + 3.15I_3 = +0.04 \quad ...(v)$$

Solving (*iv*) and (*v*), we get $I_3 = 0.735 \times 10^{-6}$ A

Hence current in the galvanometer arm = 0.735 μA. (Ans.)

Example 32. *Determine the current supplied by the battery in the circuit shown in Fig. 45.*

(Bombay University)

Solution. Refer Fig. 45.

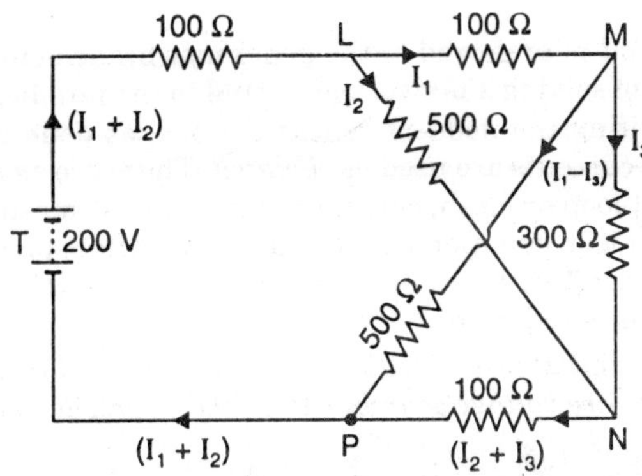

Fig. 45

Applying Kirchhoff's voltage law to the circuit **LMNL**, we get

$$-100I_1 - 300I_3 + 500I_2 = 0$$
$$I_1 - 5I_2 + 3I_3 = 0 \quad ...(i)$$

Circuit **MNPM** gives,

$$-300I_3 - 100(I_2 + I_3) + 500(I_1 - I_3) = 0$$
$$500I_1 - 100I_2 - 900I_3 = 0$$
$$I_1 - 0.2I_2 - 1.8I_3 = 0 \quad ...(ii)$$

Circuit **LMPTL** gives,

$$-100I_1 - 500(I_1 - I_3) + 200 - 100(I_1 + I_2) = 0$$

$$-700I_1 - 100I_2 + 500I_3 = -200$$

$$I_1 + 0.143I_2 - 0.714I_3 = 0.286 \quad ...(iii)$$

Multiplying (*i*) by 1.8 and (*ii*) by 3 and adding, we get

$$1.8I_1 - 9I_2 + 5.4I_3 = 0$$

$$3I_1 - 0.6I_2 - 5.4I_3 = 0$$

$$4.8I_1 - 9.6I_2 = 0$$

or $$I_1 - 2I_2 = 0 \quad ...(iv)$$

Multiplying (*ii*) by 0.714 and (*iii*) by 1.8 and subtracting, we get

$$0.714I_1 - 0.143I_2 - 1.285I_3 = 0$$

$$1.8I_1 + 0.257I_2 - 1.285I_3 = 0.515$$

$$- \quad - \quad + \quad -$$

$$-1.086I_1 - 0.4I_2 = -0.515$$

or $$I_1 + 0.368I_2 = 0.474 \quad ...(v)$$

Subtracting (*v*) from (*iv*), we get $\quad 2.368I_2 = 0.474$

i.e., $$I_2 = 0.2 \text{ A}$$

and $$I_1 = 0.4 \text{ A}$$

$\therefore$ **Current supplied by the battery** $= I_1 + I_2 = 0.2 + 0.46 =$ **0.6 A. (Ans.)**

5.2. Maxwell's Loop (or Mesh) Current Method

The method of *loop* or *mesh* currents is generally used in solving networks having some degree of complexity. Such a degree of complexity already begins for a network of three meshes. It might even be convenient at times to use the method of loop or mesh currents for solving a two-mesh circuit.

The *mesh-current method* is preferred to the general or branch-current method because the unknowns in the initial stage of solving a network are equal to the number of meshes, *i.e.*, the mesh currents. The necessity of writing the node-current equations, as done in the general or branch-current method where branch currents are used, is *obviated*. There are as many mesh-voltage equations as these are independent loop or mesh, currents. Hence, the M-mesh currents are obtained by solving the M-mesh voltages or loop equations for M unknowns. After solving for the mesh currents, only a matter of resolving the confluent mesh currents into the respective branch currents by very simple algebraic manipulations is required.

This method eliminates a great deal of tedious work involved in branch-current method and is best suited when energy sources are voltage sources rather than current sources. This method can be used only for *planar circuits.*

The **procedure** for writing the equations is as follows :

1. Assume the smallest number of mesh currents so that at least one mesh current links every element. As a matter of convenience, all mesh currents are assumed to have a *clockwise direction.*

The *number of mesh currents is equal to the number of meshes in the circuit.*

2. For each mesh write down the Kirchhoff's voltage law equation. Where more than one mesh current flows through an element, the algebraic sum of currents should be used. The algebraic sum of mesh currents may be sum or the difference of the currents flowing through the element depending on the direction of mesh currents.

3. Solve the above equations and from the mesh currents find the branch currents.

Fig. 46 shows two batteries E_1 and E_2 connected in a network consisting of three resistors. Let the loop currents for two meshes be I_1 and I_2 (both clockwise-assumed). It is obvious that current through R_3 (when considered as a part of first loop) is $(I_1 - I_2)$. However, when R_3 is considered part of the second loop, current through it is $(I_2 - I_1)$.

Applying Kirchhoff's voltage law to the *two loops*, we get

$$E_1 - I_1R_1 - R_3(I_1 - I_2) = 0$$

or $$E_1 - I_1(R_1 + R_3) + I_2R_3 = 0 \qquad \text{...Loop 1}$$

Similarly, $$-I_2R_2 - E_2 - R_3(I_2 - I_1) = 0$$

$$-I_2R_2 - E_2 - I_2R_3 + I_1R_3 = 0$$

or $$I_1R_3 - I_2(R_2 + R_3) - E_2 = 0 \qquad \text{...Loop 2}$$

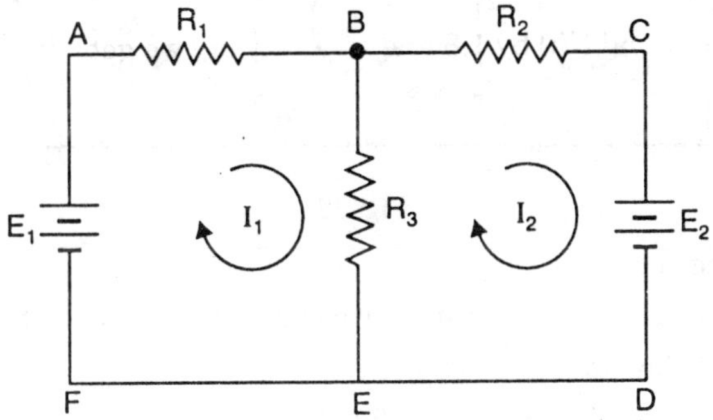

Fig. 46

The above two equations can be solved not only to find loop currents but branch currents as well.

Example 33. *Determine the currents through various resistors of the circuit shown in Fig. 47 using the concept of mesh currents.*

Solution. Refer Fig. 47.

Since there are two meshes, let the loop currents be as shown.

Applying Kirchhoff's law to *loop 1*, we get

$$24 - 4I_1 - 2(I_1 - I_2) = 0$$

$$-6I_1 + 2I_2 + 24 = 0$$

or $$3I_1 - I_2 = 12 \qquad \text{...(i)}$$

For *loop 2*, we have

$$-2(I_2 - I_1) - 6I_2 - 12 = 0$$

$$2I_1 - 8I_2 - 12 = 0$$

$$I_1 - 4I_2 = 6 \qquad \text{...(ii)}$$

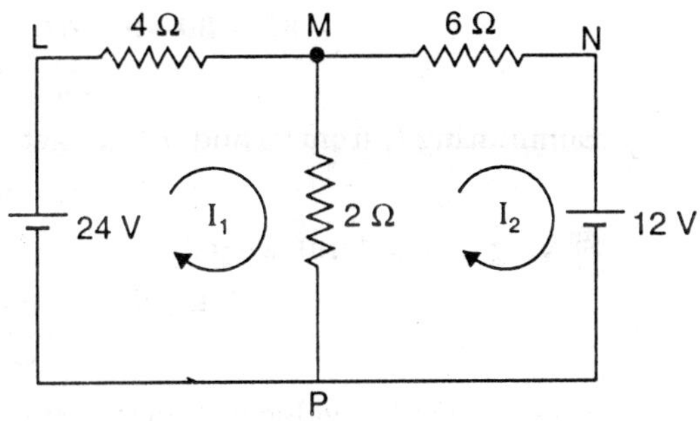

Fig. 47

Solving (*i*) and (*ii*), we get $I_1 = \frac{42}{11}$ A

and $I_2 = -\frac{6}{11}$ A

Hence Current through 4 Ω resistor = $\frac{42}{11}$ **A (from L to M). (Ans.)**

Current through 6 Ω resistor = $\frac{6}{11}$ **A (from N to M). (Ans.)**

Current through 2 Ω resistor = $\frac{42}{11} - \left(-\frac{6}{11}\right) = \frac{48}{11}$ **A (from M to P). (Ans.)**

Example 34. *Determine the current supplied by each battery in the circuit shown in Fig. 48.* **(Aligarh University)**

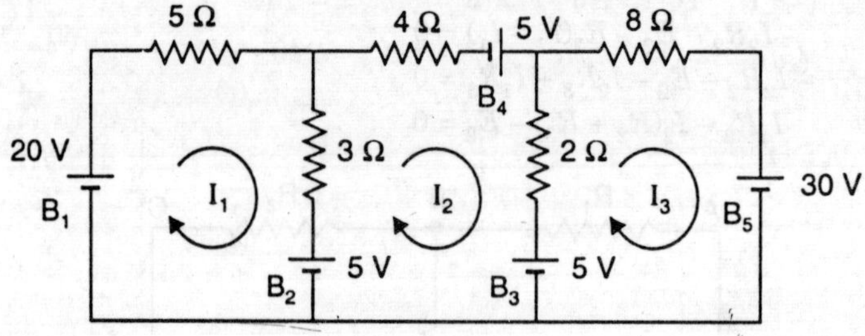

Fig. 48

Solution. Refer Fig. 48.

As there are three meshes, let the three loop currents be as shown.

Applying Kirchhoff's law to *loop 1*, we get

$$20 - 5I_1 - 3(I_1 - I_2) - 5 = 0$$

or $$8I_1 - 3I_2 = 15 \quad ...(i)$$

For *loop 2*, we have

$$-4I_2 + 5 - 2(I_2 - I_3) + 5 + 5 - 3(I_2 - I_1) = 0$$

$$3I_1 - 9I_2 + 2I_3 = -15 \quad ...(ii)$$

For *loop 3*, we have

$$-8I_3 - 30 - 5 - 2(I_3 - I_2) = 0$$

$$2I_2 - 10I_3 = 35 \quad ...(iii)$$

Eliminating I_1 from (*i*) and (*ii*), we get

$$63I_2 - 16I_3 = 165 \quad ...(iv)$$

Solving (*iii*) and (*iv*), we get

$$I_2 = 1.82 \text{ A} \quad \text{and } I_3 = -3.15 \text{ A}$$

(– ve sign means direction of current is *counter-clockwise*)

Substituting the value of I_2 in (*i*), we get

$$I_1 = 2.56 \text{ A}$$

Current through battery B_1 (discharging current) $= I_1 =$ **2.56 A. (Ans.)**

Current through battery B_2 (charging current) $= I_1 - I_2 = 2.56 - 1.82 =$ **0.74 A. (Ans.)**

Current through battery B_3 (discharging current) $= I_2 + I_3 = 1.82 + 3.15 =$ **4.97 A. (Ans.)**

Current through battery B_4 (discharging current) $= I_2 =$ **1.82 A. (Ans.)**

Current through battery B_5 (discharging current) $= I_3 =$ **3.15 A. (Ans.)**

Example 35. *Determine the currents through the different branches of the bridge circuit shown in Fig. 49.*

Solution. Refer Fig. 49.

The three mesh currents are assumed as shown.

The equations for the three meshes are :

For loop 1 : $240 - 20(I_1 - I_2) - 50(I_1 - I_3) = 0$

or $-70I_1 + 20I_2 + 50I_3 = -240$

or $70I_1 - 20I_2 - 50I_3 = 240$ ...(*i*)

For loop 2 : $-30I_2 - 40(I_2 - I_3) - 20(I_2 - I_1) = 0$

or $20I_1 - 90I_2 + 40I_3 = 0$

or $2I_1 - 9I_2 + 4I_3 = 0$ ...(*ii*)

For loop 3 : $-60I_3 - 50(I_3 - I_1) - 40(I_3 - I_2) = 0$

$50I_1 + 40I_2 - 150I_3 = 0$

$5I_1 + 4I_2 - 15I_3 = 0$ ...(*iii*)

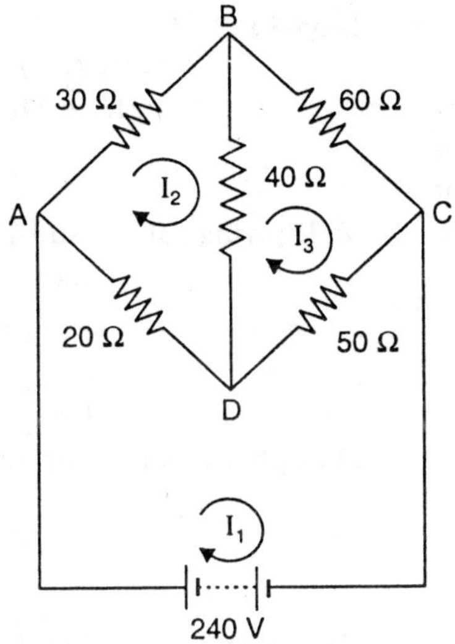

Fig. 49

Solving these equations, we get

$$I_1 = 6.10 \text{ A}, I_2 = 2.56 \text{ A}, I_3 = 2.72 \text{ A}$$

Current through 30 Ω resistor = I_2 = **2.56 A (A to B). (Ans.)**

Current through 60 Ω resistor = I_3 = **2.72 A (B to C). (Ans.)**

Current through 20 Ω resistor = $I_1 - I_2$ = 6.10 – 2.56 = **3.54 A (A to D). (Ans.)**

Current through 50 Ω resistor = $I_1 - I_3$ = 6.10 – 2.72 = **3.38 A (D to C). (Ans.)**

Current through 40 Ω resistor = $I_3 - I_2$ = 2.72 – 2.56 = **0.16 A (D to B). (Ans.)**

Example 36. *Determine the current in the 8 Ω branch in the circuit shown in Fig. 50.*

(Nagpur University)

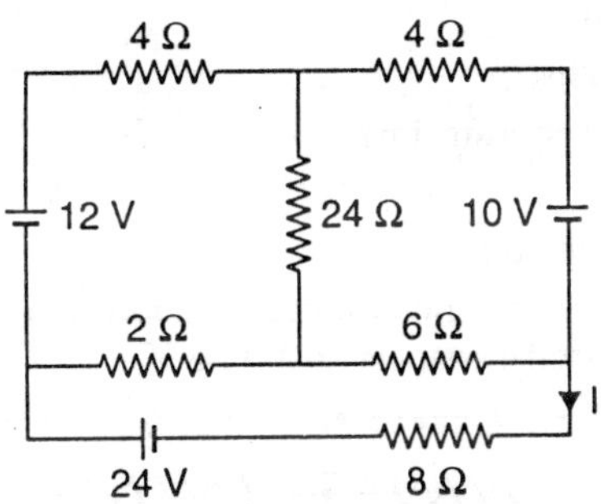

Fig. 50

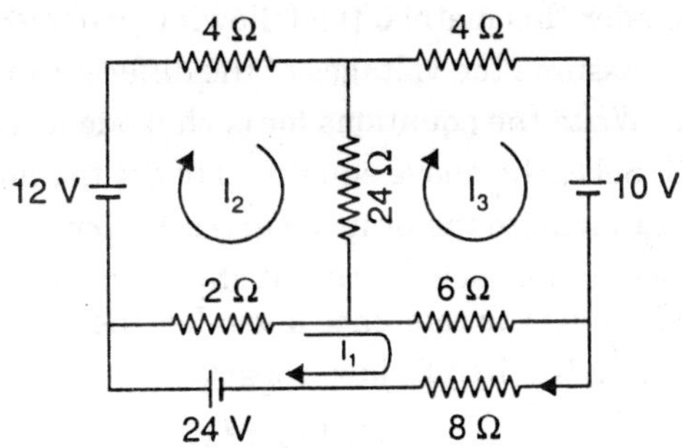

Fig. 51

Solution. The three loop currents are shown in Fig. 51 ; for these loops we have

Loop-1 : $-2(I_1 - I_2) - 6(I_1 - I_3) - 8I_1 + 24 = 0$

$-2I_1 + 2I_2 - 6I_1 + 6I_3 - 8I_1 + 24 = 0$

or $16I_1 - 2I_2 - 6I_3 = 24$

or $8I_1 - I_2 - 3I_3 = 12$ ...(*i*)

Loop-2 : $12 - 4I_2 - 24(I_2 - I_3) - 2(I_2 - I_1) = 0$

or $12 - 4I_2 - 24I_2 + 24I_3 - 2I_2 + 2I_1 = 0$

or $2I_1 - 30I_2 + 24I_3 = -12$

or $I_1 - 15I_2 + 12I_3 = -6$ ...(*ii*)

Loop-3 :

$$-24(I_3 - I_2) - 4I_3 - 10 - 6(I_3 - I_1) = 0$$

or $$-24I_3 + 24I_2 - 4I_3 - 10 - 6I_3 + 6I_1 = 0$$

or $$6I_1 + 24I_2 - 34I_3 = 10$$

or $$3I_1 + 12I_2 - 17I_3 = 5 \quad ...(iii)$$

Multiplying eqn. (*i*) and 15 and subtracting eqn. (*ii*) from (*i*), we get

$$120I_1 - 15I_2 - 45I_3 = 180$$
$$I_1 - 15I_2 + 12I_3 = -6$$
$$- \quad + \quad - \quad +$$
$$119I_1 \qquad - 57I_3 = 186 \quad ...(iv)$$

Multiplying eqn. (*ii*) by 12 and eqn. (*iii*) by 15 and adding eqn. (*iii*) and eqn. (*ii*), we get

$$12I_1 - 180I_2 + 144I_3 = -72$$
$$45I_1 + 180I_2 - 255I_3 = 75$$
$$57I_1 \qquad - 111I_3 = 3 \quad ...(v)$$

Multiplying eqn. (*iv*) by 111 and eqn. (*v*) by 57 and subtracting eqn. (*v*) from eqn. (*iv*), we get

$$13209I_1 - 6327I_3 = 20646$$
$$3249I_1 - 6327I_3 = 171$$
$$- \quad + \quad -$$
$$9960I_1 \qquad = 20475$$

$\therefore$ **I_1** (Current through 8 Ω resistance) $= \frac{20475}{9960} =$ **2.06 A. (Ans.)**

5.3. Nodal Voltage Method

Under this method the following ***procedure*** is adopted :

1. Assume the voltages of the different independent nodes.
2. Write the equations for each mode as per Kirchhoff's current law.
3. Solve the above equations to get the node voltages.
4. Calculate the branch currents from the values of node voltages.

Let us consider the circuit shown in the Fig. 52. L and M are the two independent nodes ; M can be taken as the reference node. Let the voltage of node L (with respect to M) be V_L.

Using Kirchhoff's law, we get

$$I_1 + I_2 = I_3 \quad ...(22)$$

Fig. 52

Ohm's law gives $I_1 = \frac{V_1}{R_1} = \frac{(E_1 - V_L)}{R_1}$

$$I_2 = \frac{V_2}{R_2} = \frac{(E_2 - V_L)}{R_2} \quad ...(23)$$

$$I_3 = \frac{V_3}{R_3} = \frac{V_L}{R_3}$$

$$\frac{E_1 - V_L}{R_1} + \frac{E_2 - V_L}{R_2} = \frac{V_L}{R_3} \quad ...(24)$$

Rearranging the terms, we get

$$V_L\left[\frac{1}{R_1} + \frac{1}{R_2} + \frac{1}{R_3}\right] - \frac{E_1}{R_1} - \frac{E_2}{R_2} = 0 \quad ...(25)$$

It may be noted that the above nodal equation contains the following terms :

(*i*) The node voltage multiplied by the sum of all conductances connected to that anode. This term is *positive*.

(*ii*) The node voltage at the other end of each branch (connected to this node) multiplied by the conductance of branch. These terms are *negative*.

— In this method of solving a network the *number of equations required for the solution is one less than the number of independent nodes in the network.*

— In general the nodal analysis *yields similar solutions.*

— The nodal method is very suitable for *computer work.*

Example 37. *For the circuit shown in Fig. 53, find the currents through the resistances R_3 and R_4.*

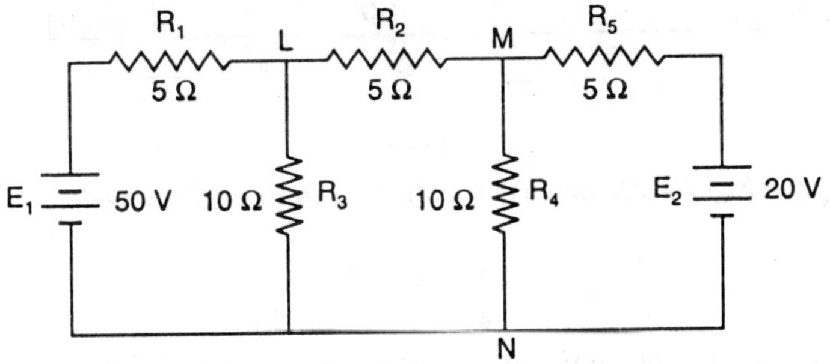

Fig. 53

Solution. Refer Fig. 53.

Let L, M and N = Independent nodes, and

V_L and V_M = Voltages of nodes L and M with respect to node N.

The nodal equations for the nodes L and M are :

$$V_L\left[\frac{1}{R_1}+\frac{1}{R_2}+\frac{1}{R_3}\right]-\frac{E_1}{R_1}-\frac{V_M}{R_2}=0 \quad ...(i)$$

$$V_M\left[\frac{1}{R_2}+\frac{1}{R_4}+\frac{1}{R_5}\right]-\frac{E_2}{R_5}-\frac{V_L}{R_2}=0 \quad ...(ii)$$

Substituting the values in (*i*) and (*ii*) and simplifying, we get

$$V_L\left(\frac{1}{5}+\frac{1}{5}+\frac{1}{10}\right)-\frac{50}{5}-\frac{V_M}{5}=0$$

or $$2.5V_L - V_M - 50 = 0 \quad ...(iii)$$

and $$V_M\left(\frac{1}{5}+\frac{1}{10}+\frac{1}{5}\right)-\frac{20}{5}-\frac{V_L}{5}=0$$

or $$2.5V_M - V_L - 20 = 0$$

or $$-V_L + 2.5V_M - 20 = 0 \quad ...(iv)$$

Solving (*iii*) and (*iv*), we get

$$V_L = 27.6 \text{ V}, V_M = 19.05 \text{ V}$$

Current through R_3 $= \frac{V_L}{R_3} = \frac{27.8}{10} =$ **2.76 A. (Ans.)**

Current through R_4 $= \frac{V_M}{R_4} = \frac{19.05}{10} =$ **1.905 A. (Ans.)**

Example 38. *Using Node voltage method, find the current in 6 Ω resistance for the network shown in Fig. 54.*

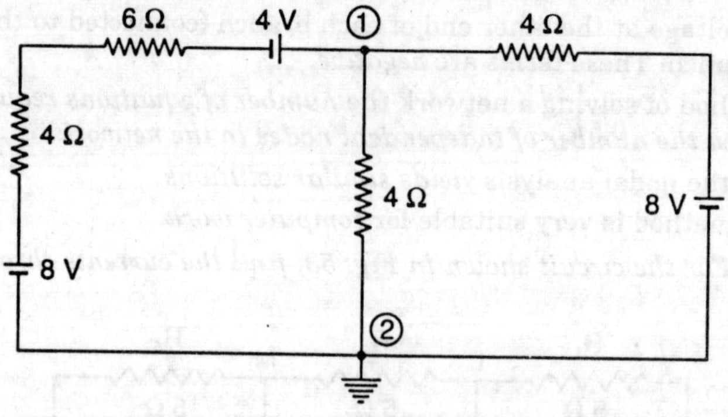

Fig. 54

Solution. Refer Fig. 54. Considering node 2 as the reference node and using node voltage method, we have

$$V_1\left[\frac{1}{(6+4)}+\frac{1}{4}+\frac{1}{4}\right]-\frac{8}{4}-\left(\frac{8+4}{10}\right)=0$$

(The reason for adding the two battery voltages of 4 V and 8 V is because they are connected in additive series).

or $$V_1\,(0.1+0.25+0.25)-2-1.2=0$$

∴ $$V_1=5.33\text{ V}$$

The current flowing through the 6 Ω resistance towards node 1 is

$$=\frac{12-5.33}{6+4}=\mathbf{0.667\ A.\ (Ans.)}$$

Example 39. *Find the branch currents in the circuit of Fig. 55 by nodal analysis.*

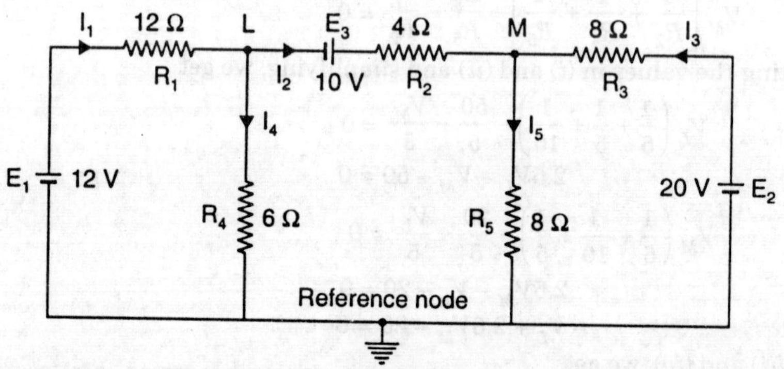

Fig. 55

Solution. Node L :

$$V_L\left[\frac{1}{R_1}+\frac{1}{R_2}+\frac{1}{R_4}\right]-\frac{E_1}{R_1}-\frac{V_M}{R_2}+\frac{E_3}{R_2}=0$$

Substituting the given data, we get

$$V_L\left[\frac{1}{12}+\frac{1}{4}+\frac{1}{6}\right]-\frac{12}{12}-\frac{V_M}{4}+\frac{10}{4}=0$$

$$V_L\left[\frac{1+3+2}{12}\right]-1-\frac{V_M}{4}+\frac{10}{4}=0$$

or $$0.5V_L-1-0.25V_M+2.5=0$$

or $$2V_L-V_M=-6 \qquad ...(i)$$

Node M :

$$V_M\left[\frac{1}{R_2}+\frac{1}{R_3}+\frac{1}{R_5}\right]-\frac{E_2}{R_3}-\frac{V_L}{R_2}-\frac{E_3}{R_2}=0$$

$$V_M\left[\frac{1}{4}+\frac{1}{8}+\frac{1}{8}\right]-\frac{20}{8}-\frac{V_L}{4}-\frac{10}{4}=0$$

or $$V_M\left[\frac{2+1+1}{8}\right]-\frac{20}{8}-\frac{V_L}{4}-\frac{10}{4}=0$$

or $$\frac{V_M}{2}-\frac{20}{8}-\frac{V_L}{4}-\frac{10}{4}=0$$

$$V_M-\frac{V_L}{2}=10 \qquad ...(ii)$$

From eqn. (*i*) and (*ii*), we get

$$V_L=\frac{8}{3}\,V;\quad V_M=\frac{34}{3}\,V$$

$$\mathbf{I_1}=\frac{E_1-V_L}{R_1}=\frac{12-\frac{8}{3}}{12}=\frac{\mathbf{7}}{\mathbf{9}}\ \textbf{A. (Ans.)}$$

$$\mathbf{I_2}=\frac{V_L+E_1-V_M}{R_2}=\frac{\frac{8}{3}+10-\frac{34}{3}}{4}=\frac{\mathbf{1}}{\mathbf{3}}\ \textbf{A. (Ans.)}$$

$$\mathbf{I_3}=\frac{E_2-V_M}{R_3}=\frac{20-\frac{34}{3}}{8}=\frac{\mathbf{13}}{\mathbf{12}}\ \textbf{A. (Ans.)}$$

$$\mathbf{I_4}=\frac{V_L}{6}=\frac{\frac{8}{3}}{6}=\frac{\mathbf{4}}{\mathbf{9}}\ \textbf{A. (Ans.)}$$

$$\mathbf{I_5}=\frac{V_M}{R_5}=\frac{\frac{34}{3}}{8}=\frac{\mathbf{17}}{\mathbf{12}}\ \textbf{A. (Ans.)}$$

6. SOURCE CONVERSION

A given voltage source with a series resistance can be converted into (or replaced by) an equivalent current source with a parallel resistance.

Conversely, a current source with a parallel resistance can be converted into a voltage source with a series resistance.

- Suppose, we want to convert the voltage source of Fig. 56 (*i*) into an equivalent current source. To do so, we shall find the value of current, supplied by the source when a 'short' is put across terminals L and M as shown in Fig. 56 (*ii*) ; this current is $I=\frac{V}{R}$. A current source supplying this current I and having the same resistance R connected in *parallel* with it represents the equivalent source. It is shown in Fig. 56 (*iii*).

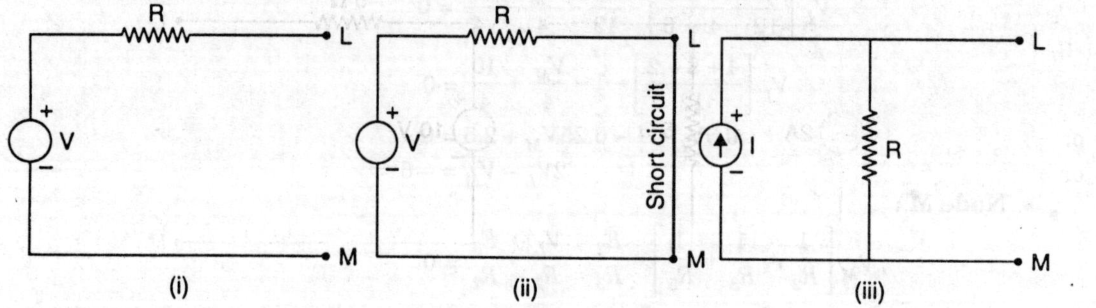

Fig. 56

- Similarly, a current source of I and a parallel resistance R can be converted into a voltage source of voltage $V \equiv IR$ and a resistance R in series with it.

It may be kept in mind that a voltage source-series resistance combination is equivalent to (or replaceable by) a constant source-parallel resistance combination if, and only if their *respective open-circuit voltages are equal, and respective short-circuit currents are equal.*

Example. Refer Fig. 56 (*i*). Voltage across terminals L and M when they are open (*i.e.*, open-circuit voltage V_{OC}) is V itself because there is no drop across R. Short-circuit current across $LM = I = \frac{V}{R}$.

Refer Fig. 56 (*iii*). The open circuit voltage across LM = drop across $R = IR = V$.

If a short is placed across LM, whole of I passes through it because R is completely shorted out.

Example 40. *Convert the voltage source of Fig. 57 (i) into an equivalent current source.*

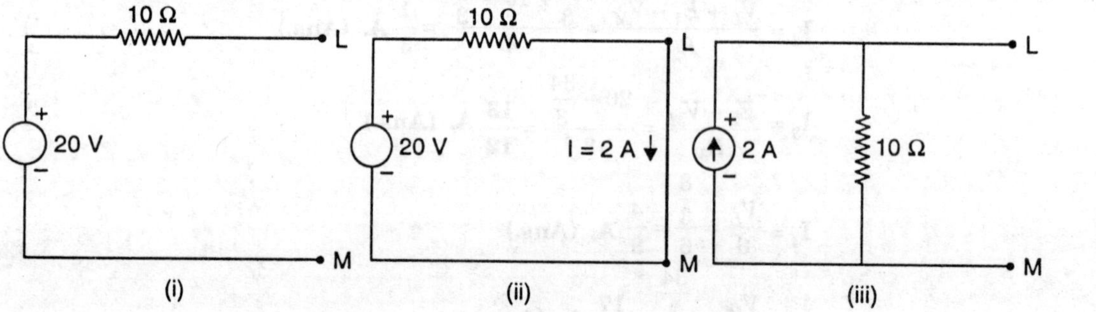

Fig. 57

Solution. As shown in Fig. 57 (*ii*), current obtained by putting a short across terminals L and M is $\frac{20}{10} = 2$ A.

Hence the equivalent current source is as shown in Fig. 57 (*iii*).

Example 41. *Find the equivalent voltage source for the current source in Fig. 58 (i).*

Solution. The open-circuit voltage across terminals A and B in Fig. 58 (*i*) is V_{OC} = drop across $R = 2 \times 5 = 10$ V.

Hence, voltage source has a voltage of 10 V and the same resistance of 5 Ω connected in series [Fig. 58 (*ii*)].

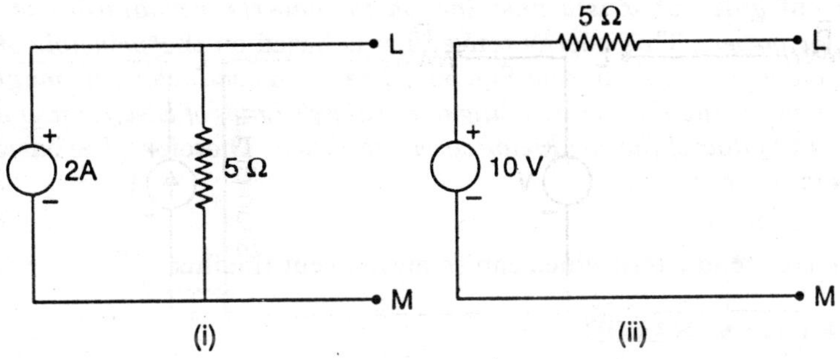

Fig. 58

7. SOLVING EQUATIONS BY DETERMINANTS—CRAMER'S RULE

If the number of equations is more than two, it is easier to get the solution by using determinants.

If I_1, I_2 and I_3 are the three unknowns in a system of three linear equations

$$\begin{aligned} a_{11}I_1 + a_{12}I_2 + a_{13}I_3 &= C_1 \\ a_{21}I_1 + a_{22}I_2 + a_{23}I_3 &= C_2 \\ a_{31}I_1 + a_{32}I_2 + a_{33}I_3 &= C_3 \end{aligned} \quad \text{...(26)}$$

Then, the system can be written in matrix form as follows :

$$\begin{bmatrix} a_{11} & a_{12} & a_{13} \\ a_{21} & a_{22} & a_{23} \\ a_{31} & a_{32} & a_{33} \end{bmatrix} \begin{bmatrix} I_1 \\ I_2 \\ I_3 \end{bmatrix} = \begin{bmatrix} c_1 \\ c_2 \\ c_3 \end{bmatrix} \quad \text{...(27)}$$

The values of I_1, I_2 and I_3 are given by

$$I_1 = \frac{\begin{bmatrix} c_1 & a_{12} & a_{13} \\ c_2 & a_{22} & a_{23} \\ c_3 & a_{32} & a_{33} \end{bmatrix}}{\Delta} \quad \text{...[28 (a)]}$$

$$I_2 = \frac{\begin{bmatrix} a_{11} & c_1 & a_{13} \\ a_{21} & c_2 & a_{23} \\ a_{31} & c_3 & a_{33} \end{bmatrix}}{\Delta} \quad \text{...[28 (b)]}$$

$$I_3 = \frac{\begin{bmatrix} a_{11} & a_{12} & c_1 \\ a_{21} & a_{22} & c_2 \\ a_{31} & a_{32} & c_3 \end{bmatrix}}{\Delta} \quad \text{...[28 (c)]}$$

where $\Delta = \begin{bmatrix} a_{11} & a_{12} & a_{13} \\ a_{21} & a_{22} & a_{23} \\ a_{31} & a_{32} & a_{33} \end{bmatrix}$

This is known as *Cramer's rule* and can be applied to any system of *n* linear equations provided Δ is not zero.

8. PRINCIPLE OF A GENERATOR

An **electrical generator** *is a machine which converts mechanical energy (or power) into electrical energy (or power).* This energy conversion is based on the *principle of the production of dynamically induced e.m.f.* As seen from Fig. 59, *whenever a conductor cuts magnetic flux, dynamically induced e.m.f. is produced in it according to Faraday's Laws of Electromagnetic induction. This e.m.f. causes current to flow if the conductor circuit is closed.* Therefore, the basic essential parts of an electrical generator are :

(*i*) A magnetic field.

(*ii*) A conductor/conductors which can so move to cut the flux.

9. ELEMENTARY GENERATOR

If a single-turn coil (*see* Figs. 59 and 60) is rotated in a uniform magnetic field at a constant speed, as shown in Figs. 59 and 60 (*a*), the e.m.f. induced in a given coil side will vary as the coil moves through various positions from 0° to 360° as shown in Fig. 60.

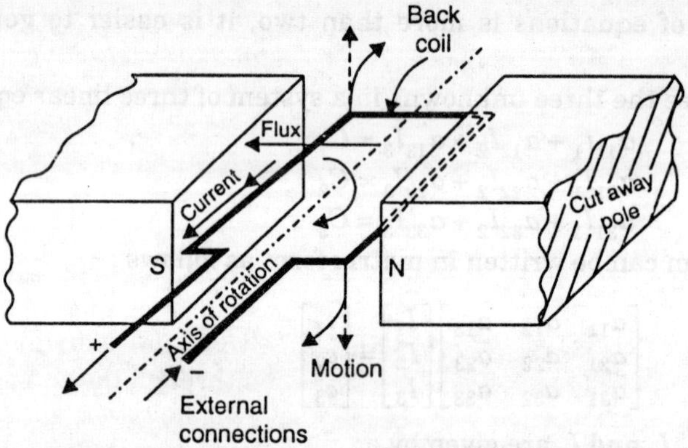

Fig. 59. Elementary generator.

- Using coil side *lm* as a reference, it will be noted that when this coil side is in *position 1* shown in Fig. 60 (*a*), the e.m.f. induced in the coil is *zero,* since-conductor *lm* (and conductor *np*, as well) is *moving parallel to the magnetic field and experiencing no change in flux linkages.*

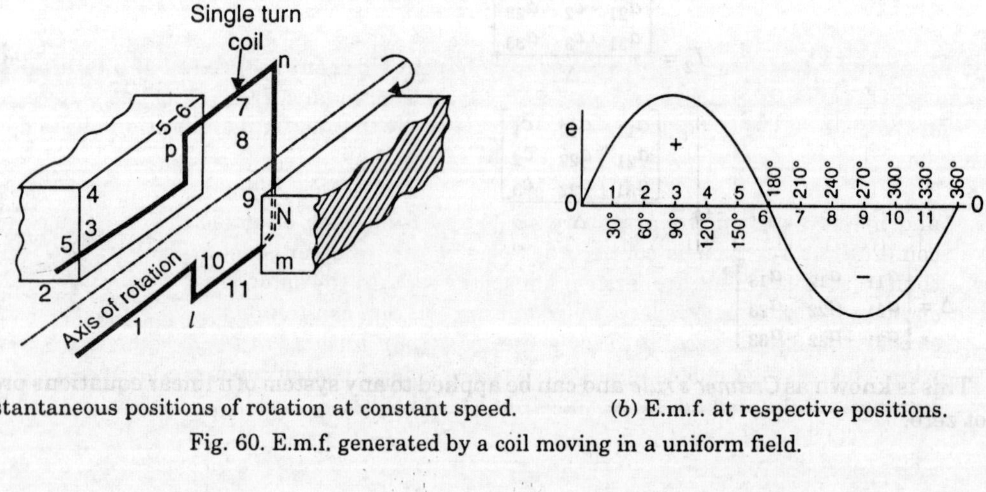

(*a*) Instantaneous positions of rotation at constant speed. (*b*) E.m.f. at respective positions.

Fig. 60. E.m.f. generated by a coil moving in a uniform field.

- When conductor *lm* moves to position 1, rotating in clockwise direction it cuts the uniform magnetic field at an oblique angle of 30°. The e.m.f. induced in this upward-moving conductor with respect to an external load will be positive and will amount to 50 per cent (e.m.f. α sin θ) of the maximum induced voltage. The change in voltage is shown graphically in Fig. 60 (*b*), where the e.m.f. is positive at position 1 and has the value given.
- When the coil moves to position 2, the e.m.f. induced will be positive and will amount to approximately 86.6 per cent of the maximum induced voltage.
- When the coil reaches 90° (*see* Fig. 59), position 3 [*see* Figs. 59 and 60 (*a*)], the conductor *lm* has the *maximum number of flux linkages, since it is moving perpendicularly to the magnetic field, and has the maximum positive value* shown in Fig. 60 (*b*).
- Positions 4 and 5 corresponding to 120° and 150° of rotation respectively, yields an e.m.f. in coil side *lm* identical to those produced at positions 1 and 2 [sin 120° = sin 60°, sin 150° = sin 30°] respectively with a positive polarity since the conductor is still moving upward but the change in flux linkage occurs at a slower rate than a position 3 respectively.
- When the conductor *lm* reaches 180°, position 6, the *induced e.m.f. is again zero,* since no *change in flux linkages occurs when a conductor is moving parallel to a magnetic field.*
- In positions 7 and 8 corresponding to 210° and 240° respectively, the e.m.f. in conductor *lm* is reversed since conductor *lm* is now moving downward in the same uniform magnetic field. The e.m.f. induced increases to a negative maximum at 270°, position 9, and finally decreases through positions 10 and 11 corresponding to 300° and 330° back to zero at position 0.
- It is worth noting that the *e.m.f. in a conductor rotating in a magnetic field is both sinusoidal and alternating.* It may be observed that during this process no e.m.f. is induced in coil-sides *mn* or *pl* since these conductors experience no change in flux linkages. Even if an e.m.f. were produced in these coil sides, they would not contribute to the e.m.f. of the coil because they are moving in the same direction through the same field and hence produce equal e.m.f.'s in opposition. Coil sides *lm* and *np*, however, aid each other, and the total e.m.f. produced by the coil is twice the magnitude represented in Fig. 60 (*b*). It should be noted that no e.m.f. is produced in positions 0 and 6 known as the *interpolar* or neutral zones of the dynamo.
- It may be noted that a *sinusoidal wave shape is produced by a conductor rotating in a theoretically uniform field represented in Fig. 60, in which the air gap is not constant because of the straight pole sides. If the pole tips are curved so that the pole face produces a more uniform gap and flux density except in the inter-polar regions, the wave shape of the induced e.m.f. will tend to be flat-topped, approaching a square wave more nearly than a sine wave.*

Conversion of Alternating Current into Direct Current by means of a commutator : In order to convert the alternating voltage (A.C.) to unidirectional current (D.C.), it is necessary to employ a mechanical switching device which is actuated by the mechanical rotation of the dynamo/generator shaft, called a ***'commutator'*** (*see* Fig. 61).

- The simple split-ring commutator shown consists of two segments (*A* and *B*), secured to and insulated from the armature shaft and from each other as well. Each conducting commutator segment is connected, respectively, to a coil side. Since both coil side and commutator segment are *fastened mechanically* to the same shaft, *the action of the mechanical rotation is to reverse the armature coil and connections to the stationary external circuit at the same instant that the induced e.m.f. reverses in the armature coil side (i.e.,* when the coil side moves under an opposite pole). Fixed brushes are arranged to contact the commutator segments, as shown in Fig. 61.

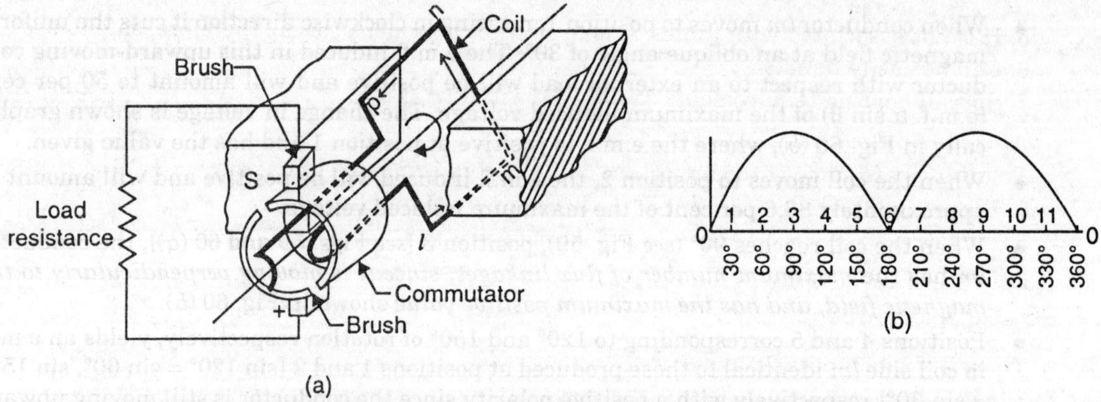

Fig. 61. Biplolar generator with two segment commutator.

- As shown in Figs. 60 and 61, the e.m.f. induced in coil-side *lm* causes a positive polarity for the first 180° of rotation (positions 0 through 6), and a negative polarity for the remaining 180° (positions 6 through 0). But, in Fig. 61, coil side *lm* is connected to commutator segment *A* and the coil side *np* is connected to commutator segment *B*. For the first 180° of rotation, therefore, the positive e.m.f. produced by the coil-side *lm* is connected to the stationary positive brush. For the second 180° of rotation, the negative e.m.f. produced by coil side *lm* is connected to the secondary negative brush. The same effect occurs in reverse order for coil-side *np*.

In effect, therefore, *the action of the commutator is to reverse connections to the external circuit simultaneously and at the same instant the direction of e.m.f. reverses in each of the coil sides. Each brush, positive or negative respectively, is always maintained, therefore, at the same polarity.* Fig. 61 (*b*) shows the e.m.f. (and current) wave form produced as a result of the above process for one full cycle (or 360°) of rotation.

- The pulsating unidirectional current which has a zero value twice each cycle as shown in Fig. 61 (*b*), is hardly suitable for commercial D.C. use. *The output e.m.f. may be made less pulsating by using a large number of coils and commutator segments* (*see* Fig. 62).
- It may be noted that a single turn armature coil generates voltage which is quite small in magnitude. Therefore, the coils employed in commercial generators consist of several turns in series thereby increasing the magnitude of generated e.m.f. in direct proportion to the number of the turns in the coil.

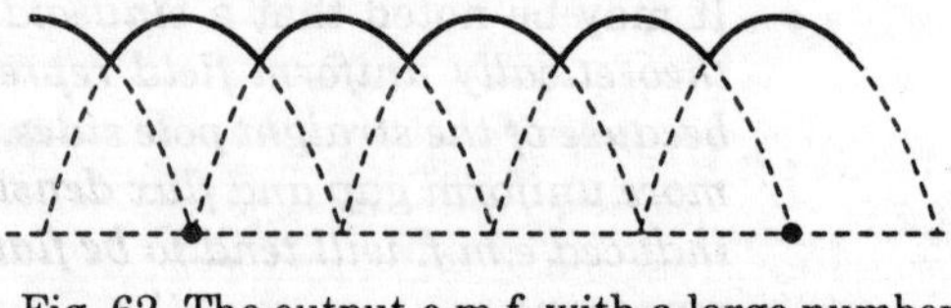

Fig. 62. The output e.m.f. with a large number of coils and cummutator segments.

Note. For details of D.C. generator please refer Chapter 6.

10. ADVANTAGES AND APPLICATIONS OF ELECTRICAL ENERGY

The invisible energy which constitues the flow of electrons in a closed circuit to do work is called **electricity.**

Advantages :

Electrical energy entails the following **advantages** :

1. Very clean type of energy.
2. Cheaper than other forms of energy.
3. Very flexible in use.
4. Easy to operate.

5. It is not associated with fumes, smoke or other poisonous gases ; therefore, safe to use.
6. Can be easily handled.
7. Special heating requirements can be met with.

Applications :

Followings are some of the important *fields of application* of electricity :

1. Domestic
2. Commercial buildings
3. Industry
4. Medical
5. Defence.

11. TYPES OF SUPPLY

There are two types of supply :

1. **Alternating current (A.C.)**

(*The current which changes its magnitude and direction periodically when flows through an electrical circuit is known as* ***alternating current.***)

2. **Direct current (D.C.)**

(*The current which always flows in one direction only in the electrical circuit is known as* ***direct current.***)

A.C. Supplies

The most usual is three-phase A.C. supply at 50 Hz.

Advantages/Features :

1. The generation is simply obtained in a stationary winding on the alternator and the e.m.f. can be readily stepped up, or down, by means of a static piece of apparatus, the *transformer*.
2. The *induction motor will operate on such a supply.*
3. In order to transmit large power economically over long distances, it is necessary to employ high voltages. It is not practicable to wind motors for such high voltages, owing to the *large number of turns of very fine wire which* would be involved. Also the switch and control gear would be extremely expensive *because of the heavy insulation required.* Therefore, a transformer is employed to step down the high transmission voltage to a more suitable value for use with motors.
4. The use of A.C. enables a *large bulk of power to be generated in one unit and stepped up to a higher voltage for transmission*. The transmission line is then tapped at not too frequent intervals to supply the immediately surrounding districts and, at the same time, the voltage is stepped down by means of transformer.
5. The performance of an A.C. motor varies somewhat with the impressed voltage. The voltage variation at the consumer's terminals of a public supply is limited by the Electricity Supply Regulations to a value not exceeding ± 6 per cent.
6. It is not practicable to wind motors below about 50 kW for supplies of 6000 V or higher. These should be arranged for lower pressures, preferably 400 V.
7. A *three-phase supply* requires three conductors or lines. The three phases can either be star connected or delta connected, as shown in Fig. 63. The centre of the star connection is termed the *neutral point*. This point is usually *connected to earth, in the case of the alternator winding*, or *a secondary winding of a step down transformer.*

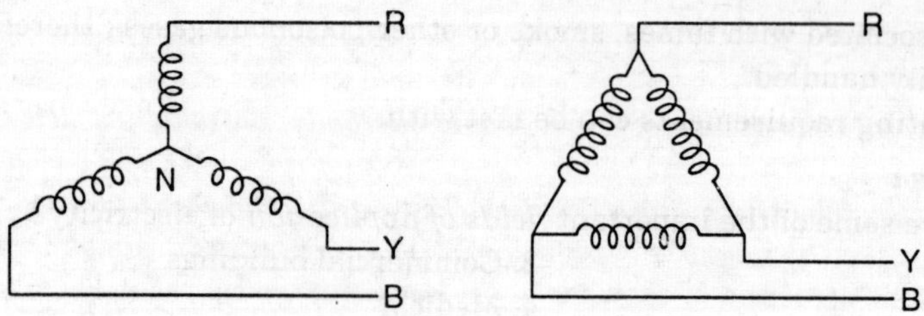

Star connection   Delta connection

Fig. 63. Star and delta connections.

In the star connection, the voltage between one line and the neutral point is $\frac{1}{\sqrt{3}}$ times the voltage between two lines. Thus with 400 V between lines we have 231 V between line and neutral. *Single-phase supplies are taken off between the line and neutral terminals on the transformer secondary*. These are used for *lighting*, and also for supplying *small consumers with power*. In some rural districts, single phase supplies are prevalent even up to 20 kW or more. This *cheapens the cost of installation*. But *single-phase motors are definitely inferior in performance to two or three phase machines.*

D.C. Supplies

- D.C. supply can be generated by using a *rotating armature, commutator*, and *brush gear* (to collect currents). The whole power generated has to pass through the commutator and this restricts the amount which may be generated in one unit.

  A normal commutator could not deal successfully with *more than about 16 V between adjacent segments, so that the diameter of the commutator limits the maximum voltage of the machine.*

- *Standard D.C. voltages* are 230 V and 460 V.

  Voltages on board ship should not exceed :

250 V	for lighting and heating.
500 V	for power.

Advantages/Features :

1. There are several purposes for which a D.C. supply is *preferable* to *A.C.*, and where it may even pay to convert an A.C. supply to D.C. for the particular job. These are :
   (*i*) High-intensity illumination from the D.C. arc light.
   (*ii*) Electric traction tramways, trolley buses, railways.
   (*iii*) Variable-speed motors of large powers, *e.g.*, rolling mill and paper-making drives.
   (*iv*) Variable-speed motors of medium power.
2. Whereas it is possible to obtain the above requirements from an A.C. supply, at times with a very good measure of success, there are certain purposes for which a D.C. supply is essential. These include : *battery charging* and *other electro-chemical processes* such as the production of aluminium by electrolysis of fused bauxite, and electroplating of silver, nickel and chromium etc. The *current required* may readily run into thousands of *amperes in large scale processes.*
3. D.C. is more *readily obtained* than A.C. for small private supplies. A *compound-wound generator will give constant voltage at all loads.*
4. An alternator requires a D.C. generator for excitation purposes ; in addition, an automatic voltage regulator to maintain constant voltage as the load varies.
5. At places where only A.C. mains are available and it is considered necessary to have a D.C. supply, conversion is possible by means of *rotary converter*, rectifier or *motor-generator.*

12. WORK, POWER AND ENERGY

12.1. Work (W)

A *force is said to have done work if it moves the body, on which it acts, through a certain distance.*

Work is *measured by the product of force (P) and displacement (s) both being in the same direction.* Work is positive or negative, according as the force acts in the same direction or in the direction opposite to the direction of displacement. Mathematically,

Work = *Force × displacement in the directions of force.*

Units of work. The units of work depend on the units of force and displacement. In S.I. system, the force is expressed in newtons (N) whereas the distance is expressed in metres (m). Hence the unit of work is *N.m.* When a force of 1 N displaces the body through 1 m, *1 N.m.* work is said to be done. It is also called **1 joule** (briefly written as J). Mathematically,

$$1 \text{ joule} = 1 \text{ N.m.}$$

12.2. Power (P)

"***Power***" *is defined as rate of doing work.*

$$P = \frac{\text{Workdone (W.D.)}}{t}, \text{ where } t \text{ denotes time.}$$

Units of power. The *electric unit of power is joule/second, called watt*(W). Hence another name for the joule is *watt-second.*

The bigger unit is kilowatt,

$$1 \text{ kW} = 1000 \text{ W}$$

Power electrical circuit, $P = \frac{W}{t} = \frac{VQ}{t}$

$= VI$ $\quad (\because \; Q = I \times t)$

$= I^2 R$ $\quad$ (Putting, $V = IR$)

$= \frac{V^2}{R}$ $\quad \left(\text{Putting } I = \frac{V}{R}\right)$

Second unit of power (*Mechanical power*) is called *horse power (H.P.).*

An agent is said to possess one H.P. if it can perform work at the rate 75 m. kg_f/sec. or 4500 m kg_f/min. (*MKS system*).

In *S.I. unit* the unit of power is watt (= 1N m/s).

Generally a bigger unit of power *i.e.*, kW is used, which is equal to 1000 W.

[1 megawatt (MW) = 10^6 W]

In *MKS system* the H.P. developed by the force,

$$\text{H.P.} = \frac{P \times v}{75}, \quad \text{where } P \text{ is in } kg_f \text{ and } v \text{ is in m/sec.}$$

$$(1 \text{ H.P.} = 735.5 \text{ W})$$

In S.I. system, $\quad \text{Power} = \frac{P \times v}{1000} \text{ kW.}$

12.3. Energy

- ***Energy*** *is capacity for doing work.*
- The *units of energy are same as that of work.*
- The energy may exist in the following forms :

  (*i*) Mechanical energy (*ii*) Electrical energy
  (*iii*) Light energy (*iv*) Heat energy

(*v*) Chemical energy (*vi*) Sound energy
(*vii*) Magnetic energy (*viii*) Nuclear energy etc.

Each one of the above energies are convertible to another form.

- The joule or watt-second is very small unit of electrical energy, so for commercial purposes energy is measured in watt hours (Wh) and kilowatts hours (kWh).

$$1 \text{ Wh} = 3600 \text{ J}$$
$$1 \text{ kWh} = 1000 \text{ Wh} = 3600000 \text{ J}$$

12.4. Efficiency

Ordinarily, according to law of conservation of energy when energy is transformed from one or more other forms, the amount of energy input is always equal to the amount of energy (or energies) to which it is transformed or output.

The **efficiency** of *a machine is the ratio of the desired form of energy which may be obtained from the machine to the total energy which is fed to the machine.*

In other words, $$\text{Efficiency} = \frac{\text{Output (useful) energy}}{\text{Total input energy}}$$

The efficiency in % age is expressed as :

$$\% \text{ age efficiency} = \frac{\text{Output energy}}{\text{Input energy}} \times 100$$

Obviously, the *energy lost* during the conversion of energy from one form to another form (or forms), is the *difference between the input and the useful output.* Hence,

$$\text{Energy losses} = \text{Input energy} - \text{output energy}$$

or $$\text{Input energy} = \text{Output energy} + \text{losses}$$

Thus, $$\text{Efficiency} = \frac{\text{Input energy} - \text{losses}}{\text{Input energy}} = \frac{\text{Output energy}}{\text{Output energy} + \text{losses}}.$$

12.5. Heating Effect of Electric Current

When electric current flows through electric conductor, heat is produced which is due to the transfer of electrical energy into heat energy.

Joule's law of electric heating :

The amount of work required to maintain a current of I amperes through a resistance of R ohms for t seconds is

$$\text{Work done, (W.D.)} = I^2 Rt \text{ joules}$$
$$= VIt \text{ joules} \qquad \left(\because\ R = \frac{V}{I}\right)$$
$$= Wt \text{ joules} \qquad (\because\ W = VI)$$
$$= \frac{V^2 t}{R} \text{ joules} \qquad \left(\because\ I = \frac{V}{R}\right)$$

This work is converted into heat and is dissipated away. The amount of heat produced is,

$$H = \frac{\text{Work done}}{\text{Mechanical equivalent of heat}} = \frac{W.D}{J}$$

where, $J = 4186$ joules/kcal = 4200 joules/kcal(approx.) ...in M.K.S. system

$$\therefore \qquad H = \frac{I^2 Rt}{4200} \text{ kcal} = \frac{VIt}{4200} \text{ kcal} = \frac{Wt}{4200} \text{ kcal} = \frac{V^2 t}{4200R} \text{ kcal}$$

In S.I. units $J = 1$.

Thermal efficiency :

It *is defined as the ratio of the heat actually utilised to the total heat produced electrically.*

Let us consider the case of an electric kettle used for boiling water. Out of the total heat produced,

(*i*) some goes to the *heat the apparatus itself,*

(*ii*) some is lost *by radiation and convection etc.* and

(*iii*) rest is *utilised for heating the water* (useful purpose). Hence, *thermal efficiency of this electric apparatus is the ratio of the heat utilised for heating the water to the total heat produced.*

Hence, the relation between heat produced electrically and heat absorbed usefully becomes

$$\frac{VIt}{J} \times \eta = m \times C \times (\theta_2 - \theta_1)$$

where, m = Mass of water (substance)

c = Specific heat of water (substance), and

$\theta_2 - \theta_1$ = Rise of temperature.

Example 41. *An electric installation consists of 15 light points of 60 W each, 8 light points of 40 W lamp, 4 fans of 60 W capacity and a pump motor of $\frac{1}{2}$ H. P. Assuming that 50% of light and fans are used for 4 hours per day and that the water pump works for three hours daily, find out the monthly consumption and cost of electricity bill, based on tariff of Rs. 2.50 per kWh.*

Solution. Total watts of 60 W lamps = 60 × 15 = 900 W

Total watts of 40 W lamps = 40 × 8 = 320 W

Total watts of 60 W fans = 60 × 4 = 240 W

∴ Total watts of lamps and fans = 900 + 320 + 240 = 1460 W

50% of light and fans are used for 4 hours a day,

∴ $$\text{Total watts-hours} = \frac{1460}{2} \times 4 = 2920 \text{ Wh}$$

$$\text{H.P. of pump motor} = \frac{1}{2} \text{ or } \frac{735.5}{2} \simeq 368 \text{ W} \qquad (\because \ 1 \text{ HP} = 735.5 \text{ W})$$

Time for which the motor runs = 3 hours/day

∴ Watt-hours for the motor = 368 × 3 = 1104 Wh/day

Total load = 2920 + 1104 = 4024 Wh/day

∴ Power consumed in 30 day (per month)

= 30 × 4024 = 120720 Wh = 120.72 kWh

Rate per kWh = Rs. 2.50

∴ Total cost = 120.72 × 2.50 = **Rs. 302. (Ans.)**

Example 42. *An electric lift is required to raise a load of 50 kN through a height of 30 metres. One quarter of electrical energy supplied to the lift is lost in motor and gearing. Calculate :*

(i) Energy consumed in kWh,

(ii) If the time required to raise the load is 25 minutes, find the H.P. of the motor and current taken by the motor, the supply voltage being 230 V D.C.

Soltuion. (*i*) **Energy consumed in kWh :**

Load to be lifted = 50 kN

Work to be obtained from the lift = 50 × 30 = 1500 kNm

As 25% of the work is lost in motor and gearing parts etc., so 1500 kNm is 75% of the work which is done by electric motor.

∴ *Energy to be supplied* to the motor

$$= \frac{1500}{0.75} = 2000 \text{ kN m} = 2000 \text{ kWs} \qquad (\because \quad \text{Nm} = \text{W-s})$$

$$= \frac{2000}{3600} = \mathbf{0.556 \text{ kWh.}} \quad \textbf{(Ans.)}$$

(*ii*) H.P. of the motor ; current taken (I) :

Energy to be supplied by the motor = 1500 kNm in 25 minutes

Energy to be supplied by the motor in 1 second

$$= \frac{1500}{25 \times 60} = 1 \text{ kW} = \frac{1}{0.7355} = \mathbf{1.36 \text{ H.P.}} \quad \textbf{(Ans.)} \quad (\because \quad 1 \text{ H.P.} = 0.7355 \text{ kW})$$

$$\text{Energy supplied} = \frac{1 \text{ (kW)}}{0.75} = 1.333 \text{ kW} \quad \text{or} \quad 1333 \text{ W}$$

Now, $V \times I = 1333$ W

or $$I = \frac{1333}{V} = \frac{1333}{230} = \mathbf{5.79 \text{ A.}} \quad \textbf{(Ans.)}$$

Example 43. *A pump is to throw 700 litres of water per minute at a height of 60 metres.*

(*i*) *Calculate the H.P. of the motor when the pump efficiency is 80% and motor efficiency is 85%.*

(*ii*) *If the power factor is 0.8, voltage is 400 V. Find the full load current.*

Solution. Volume of water to be handled per minute = 700 litres = 0.7 m^3

Height upto which water is to be thrown = 60 m

Efficiency of motor, η_{motor} = 80%

Efficiency of pump, η_{pump} = 85%

Power factor = 0.8

Supply voltage = 400 V

(*i*) H.P. of the motor :

Work done by the pump per minute

$$= 0.7 \times 9.81 \times 60 = 412 \text{ kNm } (= 412 \text{ kWs})$$

$$\text{Output of the pump} = \frac{412}{60} = 6.87 \text{ kW} \qquad (\because \quad \text{Weight density of water} = 9.81 \text{ kN/m}^3)$$

$$= \frac{6.87}{0.7355} = 9.34 \text{ H.P.}$$

$$\therefore \quad \text{Input to the pump} = \frac{\text{Pump output}}{\eta_{pump}} = \frac{9.34}{0.8} = 11.67 \text{ H.P.}$$

∴ Input to the pump = Motor output

∴ H.P. of the motor = **11.67 H.P. (Ans.)**

(*ii*) Full load current, I :

$$\text{Motor input} = \frac{\text{Motor output}}{\eta_{motor}}$$

$$= \frac{11.67 \times 0.7355}{0.85} = 10.098 \text{ kW} \quad \text{or} \quad 10098 \text{ W}$$

Also $\sqrt{3}\, VI \cos \phi = 10098$

or $\sqrt{3} \times 400 \times I \times 0.8 = 10098$

$$\therefore \quad I = \frac{10098}{\sqrt{3} \times 400 \times 0.8} = \mathbf{18.22\ A.} \quad \textbf{(Ans.)}$$

Example 44. *A diesel-electric generating set supplies an output of 30 kW. The calorific value of the fuel oil is 52500 kJ/kg. If the overall efficiency of the unit is 35%, calculate*

(*i*) *The mass of the oil required per hour.*

(*ii*) *The electric energy generated per tonne of the fuel.*

Solution. Output of the generating set = 30 kW

Calorific value of the fuel oil, C.V. = 52500 kJ/kg

Overall efficiency of the unit = 35%

(*i*) **The mass of fuel oil required per hour, m_f :**

Input *per hour* $= \frac{30}{0.35} = 85.71$ kWh

∴ Mass of fuel oil required per hour,

$$m_f = \frac{85.71 \times 3600}{\text{C.V.}} = \frac{85.71 \times 3600}{52500} = \mathbf{5.88\ kg/h.} \quad \textbf{(Ans.)}$$

(*ii*) **The electric energy generatred per tonne of the oil :**

Heat content of 1 tonne (*i.e.*, 1000 kg) of oil

$$= 1000 \times 52500 = 52.5 \times 10^6 \text{ kJ} = 52.5 \times 10^6 \text{ kW-s}$$

$$= \frac{52.5 \times 10^6}{3600} = 14583.33 \text{ kWh}$$

∴ Energy output $= 14583.33 \times \frac{35}{100} = \mathbf{5104.2\ kWh.}$ **(Ans.)**

Example 45. *Find the amount of electrical energy expended in raising the temperature of 45 litres of water by 75°C. To what height could a weight of 5 tonnes be raised with the expenditure of the same energy ? Assume efficiencies of the heating equipment and lifting equipment to be 90% and 70% respectively.* **(Elect. Engg. A.M.Ae. S.I.)**

Solution. Mass of water heated, m = 45 litres = $(45 \times 10^{-3}) \times 1000$ = 45 kg

(∵ Mass density of water = 1000 kg/m^3)

Heat required $= m \times c \times (\theta_2 - \theta_1)$

$= 45 \times 4.186 \times 75 = 14127.7$ kJ

(where $\theta_2 - \theta_1$ = rise of temperature = 75°C......Given)

Heat produced electrically $= \frac{14127.7}{0.9} = 15697$ kJ

Hence **electrical energy expended = 15697 kJ (or 4.36 kWh).** **(Ans.)**

Energy available for lifting the load = 0.7 × 15697 = 10988 kJ

If 'h' is the height through which the load of 5 tonnes can be lifted, then potential energy of the load

$= mgh$ joules $= 5000 \times 9.81 \times h$ joules

$$\therefore \quad 5000 \times 9.81 \times h = 10988 \times 10^3$$

or

$$h = \frac{10988 \times 10^3}{5000 \times 9.81} = \mathbf{224\ m.} \quad \textbf{(Ans.)}$$

Example 46. *A pump driven by electric motor lifts 1.5 m^3 of water per minute to a height of 40 m. The pump has an efficiency of 90% and motor has an efficiency of 85%. Determine :*

(i) The power input to the motor.

(ii) The current taken from 480 V supply.

(iii) The electric energy consumed when motor runs at this load for 5 hours.

Assume mass of 1 m^3 of water to be 1000 kg. **(Elect. Engg. Pune University)**

Solution. Weight of water lifted per minute = mg

$$= (1.5 \times 1000) \times 9.81 = 14715 \text{ N or } 14.715 \text{ kN}$$

Height = 40 m ; Time taken = 1 min or 60 s

(*i*) **The power input to the motor :**

Motor output power $= \dfrac{14.715 \times 40}{60} = 9.81 \text{ kW}$

∴ The power input to the motor $= \dfrac{9.81}{\eta_{pump} \times \eta_{motor}}$

$$= \frac{9.81}{0.9 \times 0.85} = \mathbf{12.82\ kW.} \quad \textbf{(Ans.)}$$

(*ii*) **Current drawn, I :**

$$VI = 12.82 \times 10^3$$

$$\therefore \quad I = \frac{12.82 \times 10^3}{480} = \mathbf{26.71\ A.} \quad \textbf{(Ans.)}$$

(*iii*) **Electrical energy consumed by the motor :**

Electrical energy consumed by the motor in 5 hours

$$= 12.82 \times 5 = \mathbf{64.1\ kWh.} \quad \textbf{(Ans.)}$$

Example 47. *The following data relate to a proposed hydro-electric station :*

Available head = 28 m ; Catchment area = 420 sq. km ; rainfall = 140 cm / year ; percentage of total rainfall utilized = 68% ; Penstock efficiency = 94% ; turbine efficiency = 80% ; generator efficiency = 84% and load factor = 44%.

Calculate the power developed.

Solution. Head available, $H = 28$ m

Catchment area, $A = 420$ sq. km $(= 420 \times 10^6 \text{ m}^2)$

Rainfall = 140 cm/year (= 1.4 m)

Rainfall utilized, $h = 68\%$ of the total rainfall

$= (0.68 \times 1.4)$ m per year

Penstock efficiency, $\eta_p = 94\%$

Turbine efficiency, $\eta_t = 80\%$

Generator efficiency, $\eta_g = 84\%$

Load factor = 44%.

Power developed, P :

Quantity of water available *per year*

$$= A \times h = (420 \times 10^6) \times (0.68 \times 1.4) = 399.84 \times 10^6 \text{ m}^3$$

Hence the quantity of water available *per second*,

$$Q = \frac{399.84 \times 10^6}{(365 \times 24) \times 3600} = 12.6 \text{ m}^3$$

$\therefore$ $$P = \eta_0 \times wQH \qquad \text{(where } \eta_0 = \text{overall efficiency} = \eta_p \times \eta_t \times \eta_g)$$

or $$P = \eta_p \times \eta_t \times \eta_g \times wQH$$
$$= 0.94 \times 0.8 \times 0.84 \times 9.81 \times 12.6 \times 28 = 2186 \text{ kW}$$

Hence *average output of generating units* = **2186 kW. (Ans.)**

Example 48. *2.3 kg water in a kettle is to be heated from 20°C to 100°C in 15 minutes. If it is connected to 240 V supply, find the resistance of the kettle. Efficiency of the kettle is 80%.*

Solution. *Given :* $m_w = 2.3$ kg ; $\theta_2 - \theta_1 = 100 - 20 = 80°C$

$t = 15$ minutes, V = 240 V ; $\eta_{kettle} = 80\%$.

Now, Heat given by kettle = Heat gained by water

$$\frac{V^2 t}{R} \times \eta_{kettle} = m_w \times c_w \times (\theta_2 - \theta_1)$$

$$\frac{240^2 \times 15 \times 60}{R} \times 0.8 \text{ (joules)} = 2.3 \times 4186 \times 80 \text{ (joules)}$$

$\therefore$ $$R = \frac{(240)^2 \times 15 \times 60 \times 0.8}{(2.3 \times 4.186 \times 80) \times 10^3} = \mathbf{53.84\ \Omega}. \quad \textbf{(Ans.)}$$

Example 49. *An electric heater is required to dissipate 1 kW, connected to 230 V supply.*

(*i*) *Calculate the resistance of the wire.*

(*ii*) *Find the time required to raise the temperature of 1.5 litres of water at 15°C to boiling point by means of this heater, if efficiency of the heater is 80 per cent.*

Solution. *Given :* Heater wattage, P = 1 kW ; Voltage = 230 V ;

Volume of water to be heated = 1.5 litres ; $\theta_2 - \theta_1 = 100 - 15 = 81°C$, $\eta_{heater} = 80\%$

(*i*) **Resistance of the wire, R :**

$$R = \frac{V^2}{P} = \frac{(230)^2}{1000} = \mathbf{52.9\ \Omega}. \quad \textbf{(Ans.)}$$

(*ii*) **Time required, t :**

Heat given by the heater = Heat gained by water

$$\frac{V^2 t}{R} \times \eta_{water} = m_w \times c_w(\theta_2 - \theta_1)$$

$$\frac{(230)^2 \times t}{52.9} \times 0.8 \text{ joules} = [(1.5 \times 10^{-3} \times 1000) \times 4186 \times 85] \text{ joules}$$

$\therefore$ $$t = 667.14 \text{ or } \mathbf{11.12\ minutes.} \quad \textbf{(Ans.)}$$

HIGHLIGHTS

1. **Electricity** may be defined as a form of energy. It involves making and using energy.
2. The controlled movement of electrons (or drift) through a substance is called *current*. Current is the rate at which electrons move.

3. *Electromotive force* (e.m.f.) is the force that causes a current of electricity to flow. The *volt* is a unit of potential difference and electromotive force.
4. The opposition to flow of electrons (due to bonds between protons and electrons, as well as to collisions) is called *electrical resistance.* Resistance may also be defined as "the property of the electric circuit which opposes the flow of current".
5. The resistance (R) of a conductor is given by :

$$R = \frac{\rho l}{A}$$

where, l = Length of the conductor,
A = Area of cross-section of the conductor, and
ρ = A constant depending on the nature of material of the conductor is known as its specific resistance or resistivity.

6. Conductance (G) is the reciprocal of resistance. Conductivity (σ) is the reciprocal of specific resistance (ρ).
7. The temperature co-efficient of resistance at 0°C may be defined as :
"The change in resistance per ohm for change in temperature of 1°C from 0°C.

Also
$$\alpha_2 = \frac{1}{\frac{1}{\alpha_1} + (t_2 - t_1)}$$

where, α_1 = Temperature co-efficient of resistance at temperature t_1°C.
α_2 = Temperature co-efficient of resistance at temperature t_2°C

8. Ohm's law may be stated as :
"For a fixed metal conductor, the temperature and other conditions remaining constant, the current (I) through it is proportional to the potential difference (V) between its ends".
9. A linear resistor is one which obeys ohm's law. Such elements in which the V/I (volt-ampere) plots are not straight lines but curves are called non-linear resistors or non-linear elements.
10. In a series combination : $R = R_1 + R_2 + R_3$
where R_1, R_2 and R_3 are the resistance in series and R is the equivalent resistance of the combination.
11. In a parallel combination : $\frac{1}{R} = \frac{1}{R_1} + \frac{1}{R_2} + \frac{1}{R_3}$
where R_1, R_2 and R_3 are the resistances connected in parallel and R is the equivalent resistance of the combination.
12. The insulation resistance (R) of a cable is given by :

$$R = \frac{\rho}{2\pi l} \log_e {}^{r_2/r_1}\ \Omega$$

where r_1 = Inner radius of the cable (single core),
r_2 = Outer radius of the cable,
l = Length of the cable, and
ρ = Resistivity of the material.

13. Kirchhoff's law :
First law : Σ Currents entering = Σ Currents leaving
Second law : Σ potential rises = Σ potential drops.
14. Applications of Kirchhoff's laws include the following :
(i) Branch-current method
(ii) Maxwell's loop (or mesh) current method
(iii) Nodal voltage method.

OBJECTIVE TYPE QUESTIONS

Choose the Correct Answers :

1. The flow of current in solids is due to
(a) electrons (b) atoms (c) electrons and ions.
2. An atom is composed of
(a) electrons and protons (b) electrons and neutrons
(c) electrons, protons and neutrons.
3. The proton is a .... particle.
(a) very small and heavy (b) very very large and light
(c) very large and heavy.
4. An electron has a .... charge
(a) negative (b) positive (c) neutral.
5. Those substances whose atoms have their outermost orbit incomplete act as :
(a) good conductors of electricity (b) insulators
(c) semiconductors.
6. The property of the electric circuit which opposes the flow of current is called
(a) resistivity (b) resistance (c) conductance (d) conductivity.
7. The unit of conductivity is
(a) Ω (b) $\mho$ (c) $\mho$/m.
8. If resistances R_1 and R_2 are connected in series then equivalent resistance R will be given by the relation
(a) $R = R_1 + R_2$ (b) $\frac{1}{R} = \frac{1}{R_1} + \frac{1}{R_2}$ (c) $R = \frac{1}{R_1 + R_2}$.
9. The resistivity of insulators .... rapidly with the rise of temperature.
(a) increases (b) decreases (c) first increases then decreases.
10. The insulation resistance varies .... as the length.
(a) directly (b) indirectly (c) three times.
11. Kirchhoff's current law states that
(a) net current flow at the junction is positive
(b) algebraic sum of the currents meeting at the junction is zero
(c) no current can leave the junction without some current entering it
(d) total sum of currents meeting at the junction is zero.
12. According to Kirchhoff's voltage law, the algebraic sum of all IR drops and e.m.fs. in any closed loop of a network is always
(a) negative (b) positive
(c) determined by battery e.m.fs. (d) zero.
13. Kirchhoff's current law is applicable to only
(a) junction in a network (b) closed loops in a network
(c) electric circuits (d) electronic circuits.
14. Kirchhoff's voltage law is related to
(a) junction voltages (b) battery e.m.fs. (c) IR drops (d) both (a) and (b)
(e) none of the above.

Answers

1. (a)	2. (c)	3. (c)	4. (a)	5. (a)
6. (b)	7. (c)	8. (a)	9. (b)	10. (b)
11. (b)	12. (d)	13. (a)	14. (d).	

THEORETICAL QUESTIONS

1. Define the term 'electricity'.
2. Write a short-note on 'Electron theory'?
3. What is an electric current ?
4. What is the difference between electromotive force and potential difference ?
5. What do you mean by the term 'resistance'? On what factors does it depend ?
6. Define the terms 'conductance' and 'conductivity'.
7. What is temperature co-efficient of resistance ?
8. Derive the relation $\alpha_2 = \dfrac{1}{\dfrac{1}{\alpha_1} + (t_2 - t_1)}$ where α_1 and α_2 are the temperature co-efficients of resistance at temperatures t_1 and t_2°C respectively.
9. State and explain Ohm's law.
10. What is the difference between linear and non-linear resistors ?
11. Derive an expression for the equivalent resistance in the following cases :
    (*a*) When the resistances are connected in series.
    (*b*) When the resistances are connected in parallel.
12. How does a conductor differ from an insulator ?
13. Derive an expression for insulation resistance of a cable.
14. Define the following terms :
    Circuit, Electrical network, Active network, Node and Branch.
15. What are the limitations of Ohm's law ?
16. State and explain Kirchhoff's laws.
17. Discuss briefly application of Kirchhoff's laws.
18. Explain the nodal voltage method for solving networks. How are the nodal equations written ?
19. Explain Cramer's rule used for solving equations by determinants.

UNSOLVED EXAMPLES

1. Calculate the resistance of a manganin wire having the following particulars :
   Length of the wire = 100 m
   Uniform cross-sectional area = 0.1 mm²
   Resistivity of the material = 50×10^{-8} Ω m. **[Ans.** 500 Ω]
2. Calculate the area of cross-section and diameter of a copper wire to have a resistance of 0.13 ohm per kilometre ; given that resistivity for copper is 1.7×10^{-8} Ω-m. **[Ans.** 12.9 mm]
3. Find the area of cross-section of a cable of 1 km length to transmit 500 amperes so that total drop in voltage along the cable may not exceed 25 volts.
   Assume : $\rho = 1.7 \times 10^{-8}$ Ω-m. **[Ans.** 3.4 cm²]
   [**Hint.** $R = \dfrac{V}{I} = \dfrac{25}{500} = 0.05\ \Omega$.]
4. Two wires, each of the same length and of the same metal, have resistances of 25 and 36 ohms respectively. What are the relative diameters of the wires ? **[Ans.** 1.2 : 1]
5. A semicircular ring of copper has an inner radius of 6 cm, radial thickness 3 cm and an axial thickness 4 cm. Find the resistance of the ring at 50°C between its two end faces. Assume resistivity of copper at 20°C = 1.724×10^{-8} Ω-m. and temperature co-efficient of resistance of copper at 0°C = 0.0043/°C. **[Ans.** 3.75×10^{-6} Ω]

6. A coil has a resistance of 18 ohms when its mean temperature is 25°C and of 20 ohms when its mean temperature is 55°C. Find its mean temperature rise when its resistance is 21 ohms and the surrounding temperature is 20°C. Assume that temperature co-efficient of resistance is constant. [**Ans.** 50°C]

7. It is required to construct a resistance of 100 ohms having a temperature co-efficient of 0.001/°C. Wires of two materials of suitable cross-sectional area are available. For material *A*, the resistance is 97 ohms per 100 metres and for material *B*, the resistance is 40 ohms per 100 metres. The temperature co-efficient of resistance for material *A* is 0.003/°C and for material *B* is 0.0005/°C. Determine the suitable lengths of wires of materials *A* and *B*. [**Ans.** $l_A = 19.4$ m ; $l_B = 200$ m]

8. The resistance of a copper field coil is 171 ohms at 16° C. What is the average temperature of the coil when the resistance is 128 ohms ? Take the temperature co-efficient of copper as 0.00427 at 0°C. [**Ans.** 55.5°C]

9. The resistance of the field coils with copper conductors of a dynamo is 120 ohms at 25°C. After working for six hours on full load, the resistance of the coil increases to 240 ohms. Calculate the mean temperature rise of the field coils. [**Ans.** 39.8°C]

10. The resistance of the winding of a motor is 3.42 ohms at 20°C. After extended operation at full load the resistance is 4.2 ohms. Find the temperature rise.
$\alpha = 0.00393$ at 20°C. [**Ans.** 59.52°C]

11. Find the resistance between the points *A* and *B* in the series-parallel net-work shown in Fig. 64.

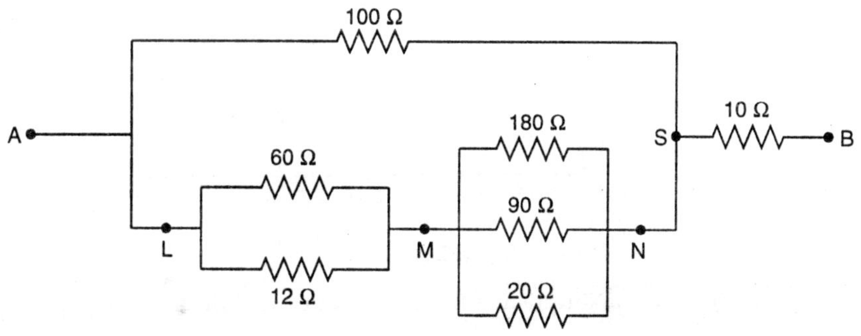

Fig. 64

12. A resistance of 10 ohms is connected in series with a combination of two resistances arranged in parallel each of value 20 ohms. Determine the resistance R_3 which should be shunted across the parallel combination so that current drawn by the circuit is 1.5 A with applied voltage of 20 V. [**Ans.** 5 ohms]

13. A series parallel resistor circuit is shown in Fig. 65. Determine
   (*i*) Equivalent resistance across the battery terminals.
   (*ii*) Current supplied by the battery.
   (*iii*) Current through 4 Ω resistor.
   (*iv*) Equivalent resistance of the circuit with an open circuit at point *L*.
   (*v*) Equivalent resistance of the circuit with an open circuit at point *M*.
   (*vi*) Equivalent resistance of the circuit with points *N* and *P* short circuited.
   (*vii*) Equivalent resistance of the circuit with points *L* and *N* short circuited.

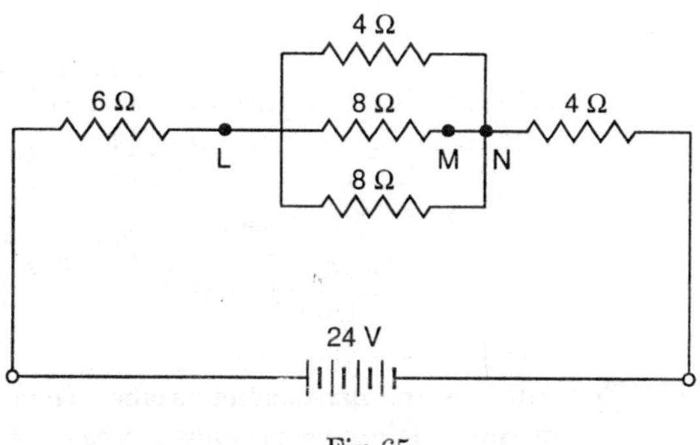

Fig. 65

[**Ans.** (*i*) 12 Ω (*ii*) 2 A (*iii*) 1 A (*iv*) Infinite (*v*) 12.66 Ω (*vi*) 8 Ω (*vii*) 10 Ω]

14. Find the resistance of the circuit shown in Fig. 66. **[Ans. 4.2 Ω]**

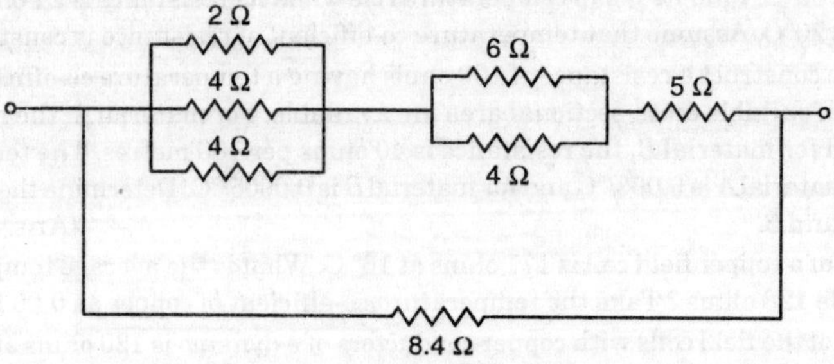

Fig. 66

15. Find the resistance of the circuit shown in Fig. 67. **[Ans. 2.2 Ω]**

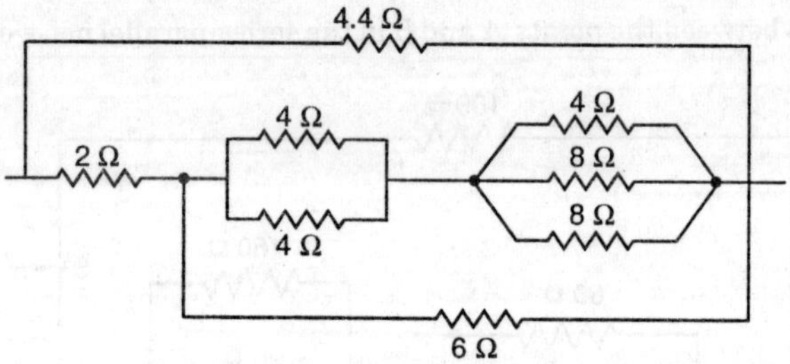

Fig. 67

16. Determine the magnitude and direction of the current in each of the batteries *L*, *M* and *N* shown in the Fig. 68.

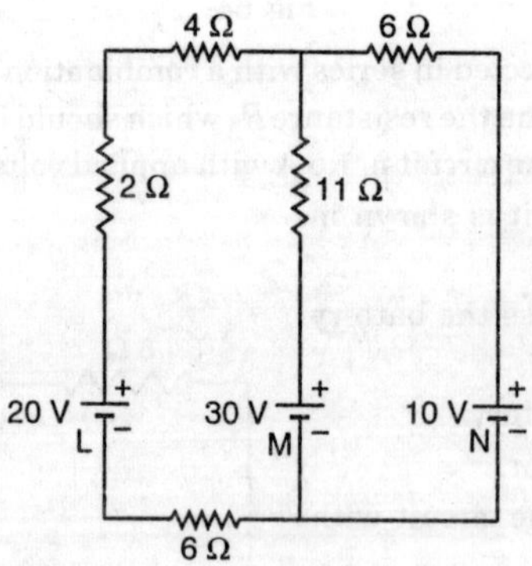

Fig. 68

17. Two batteries are connected in parallel. The e.m.f. and internal resistance of one are 110 V and 6 Ω respectively and the corresponding values for other are 130 V and 4 Ω respectively. A resistance of 20 Ω is connected across the parallel combination. Calculate :

(*i*) The value and direction of the current in each battery.

(*ii*) The terminal voltage. **[Ans.** (*i*) 0.1786 A, 5.2678 A (*ii*) 108.928 V**]**

[Hint. Terminal voltage = $(I_1 + I_2)R$.]

18. Two cells A and B are connected in parallel, unlike poles being joined together. The terminals of the cells are then joined by two resistors of 4 Ω and 2 Ω in parallel. The e.m.f. of A is 2 V, its internal resistance is 1 Ω ; the e.m.f. of B is 1 V, its internal resistance is 2 Ω. Find the current in each of the four branches of the circuit. [**Ans.** 4/3 A, 5/6 A, 1/6 A, 1/3 A]

19. What is the equivalent resistance of the network shown in Fig. 69 ? [**Ans.** 600 Ω]

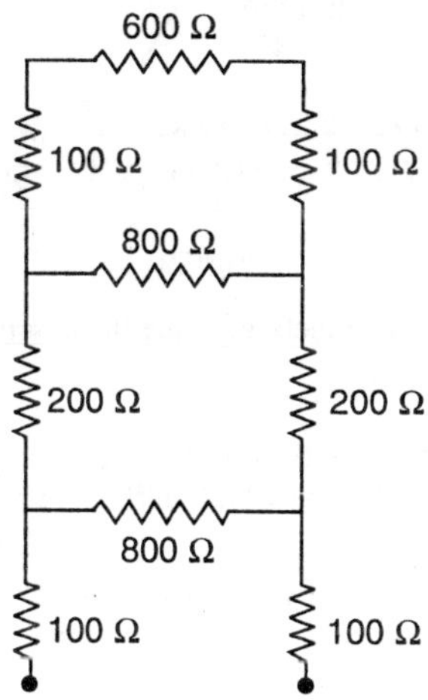

Fig. 69

20. Determine the currents in the three batteries (L, M and N) in the network shown in Fig. 70 and show their values and direction of flow on the diagram. Neglect the internal resistance of batteries. [**Ans.** 2 A, 1 A, 0 A]

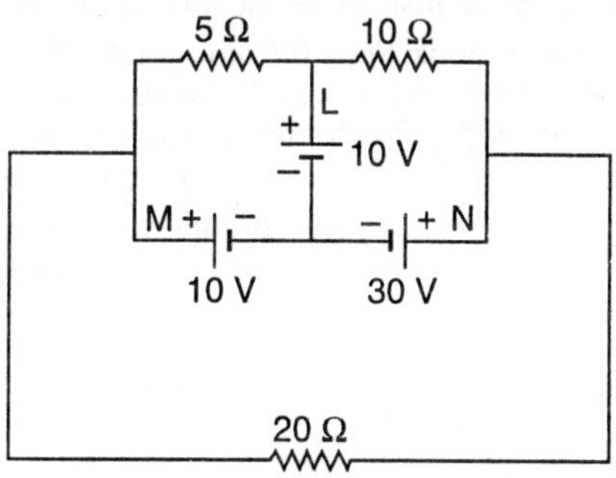

Fig. 70

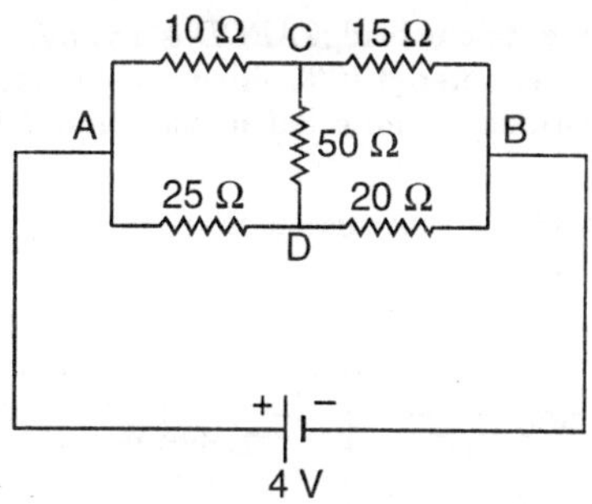

Fig. 71

21. Determine the magnitude and direction of flow of current in the branch CD for the circuit shown in the above Fig. 71. [**Ans.** 7/755 A from C to D]

22. For the lattice-type network shown in Fig. 72, calculate the current in each branch of the network.

(Roorkee University)

$$\left[\textbf{Ans. } I_{ab} = \frac{1}{50}\text{ A} ; I_{bed} = \frac{1}{100}\text{ A} ; I_{bc} = \frac{1}{100}\text{ A};\ I_{ad} = \frac{1}{100}\text{ A} ; I_{dc} = \frac{1}{50}\text{ A} ; I_{cfa} = \frac{3}{100}\text{ A}\right]$$

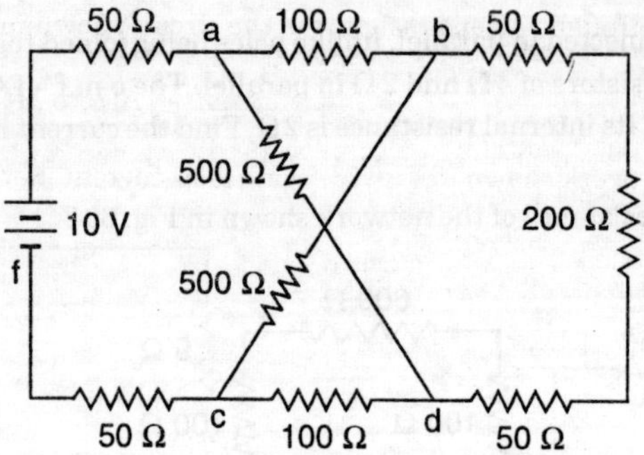

Fig. 72

23. Two storage batteries, *A* and *B*, are connected in parallel to supply a load the resistance of which is 1.2 Ω. Calculate :

(*i*) The current in this load.

(*ii*) The current supplied by each battery if the open-circuit e.m.f. of *A* is 12.5 V and that of *B* is 12.8 V, the internal resistance of *A* being 0.05 Ω and that of *B* 0.08 Ω.

[**Ans.** (*i*) 10.25 A (*ii*) 4 A (*A*), 6.25 A (*B*)]

24. A load having a resistance of 0.1 Ω is fed by two storage batteries connected in parallel. The open circuit e.m.f. of one battery (*A*) is 12.1 V and that of the other battery (*B*) is 11.8 V. The internal resistances are 0.03 Ω and 0.04 Ω respectively. Calculate :

(*i*) The current supplied to the load (*ii*) The current in each battery

(*iii*) The terminal voltage of each battery. [**Ans.** (*i*) 102.2 A, (*ii*) 62.7 A (*A*), 39.5 A (*B*), (*iii*) 10.22 V]

25. A battery having an e.m.f. of 110 V and an internal resistance of 0.2 Ω is connected in parallel with another battery having an e.m.f. of 100 V and internal resistance 0.25 Ω. The two batteries in parallel are placed in series with a regulating resistance of 5 Ω and connected across 200 V mains. Calculate :

(*i*) The magnitude and direction of the current in each battery.

(*ii*) The total current taken from the supply mains. **(Delhi University)**

[**Ans.** (*i*) 11.96 A (discharge) ; 30.43 A (charge) (*ii*) 18.47 A]

26. A wheatstone bridge *ABCD* is arranged as follows : *AB* = 50 ohms, *BC* = 100 ohms and *CD* = 101 ohms. A galvanometer of 1000 ohms resistance is connected between *B* and *D*. A 2-volt battery having negligible resistance is connected across *A* and *C*. Estimate the current flowing through the galvanometer.

[**Ans.** 4.14 μA from *D* to *B*]

27. Determine the current through the galvanometer *G* in the wheatstone bridge network of the given Fig. 73. [**Ans.** 0.52 mA]

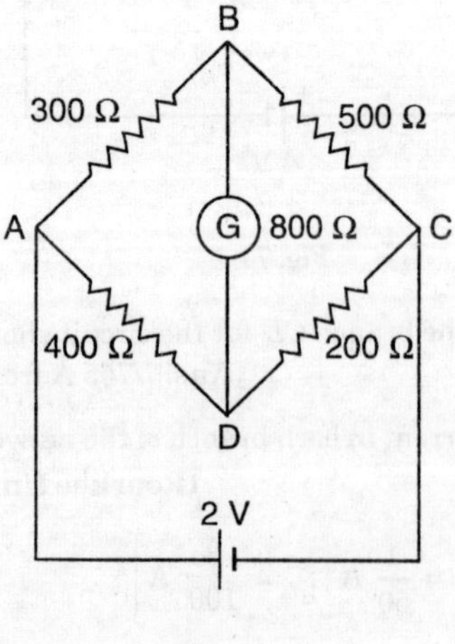

Fig. 73

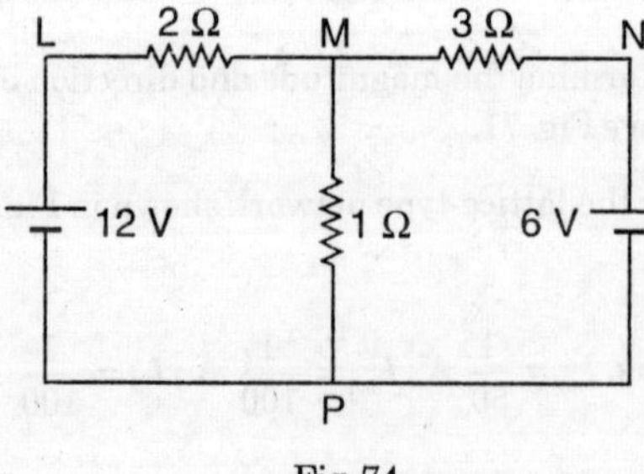

Fig. 74

28. Determine the currents through various resistors of the circuit shown in above Fig. 74.

$$\left[\textbf{Ans. } \frac{42}{11}\text{ A}(L \text{ to } M);\ \frac{6}{11}\text{ A}(N \text{ to } M);\ \frac{48}{11}\text{ A}(M \text{ to } P)\right]$$

29. Find I_1, I_2 and I_3 in the network shown in Fig. 75, using loop-current method. [**Ans.** 1 A, 2 A, 3 A]

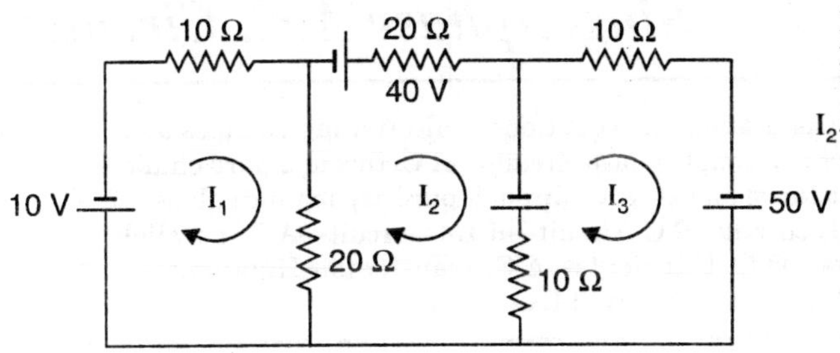

Fig. 75

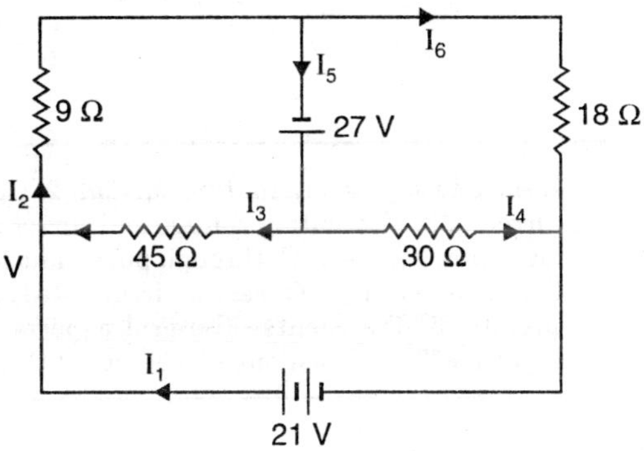

Fig. 76

30. Find the currents in the branches of network of Fig. 76 using nodal voltage method.

[**Ans.** $I_1 = 1.42$ A, $I_2 = 1.68$ A, $I_3 = 0.26$ A, $I_4 = 1.1$ A, $I_5 = 1.36$ A, $I_6 = 0.32$ A]

2

Single-phase A.C. Circuits

1. Introduction to alternating current. 2. Generation and equations of alternating voltages and currents. 3. Alternating voltage and current. 4. Single phase circuits—A.C. through pure ohmic resistance alone—A.C. through pure inductance alone—A.C. through pure capacitance alone—Phasor algebra—A.C. series circuit—R-L circuit—R.C. circuit—R-L-C circuit—A.C. parallel circuits. 5. Transients—General aspects—D.C. transients—A.C. transients—Highlights—Objective Type Questions—Theoretical Questions—Unsolved Examples.

1. INTRODUCTION TO ALTERNATING CURRENT

A.C. means **alternating current**—*The current or voltage which alternates its direction and magnitude every time*. Now a days 95% of the total energy is produced, transmitted and distributed in A.C. supply.

The *reasons* are the following :

(*i*) More voltage can be generated (upto 33000 V) than D.C. (650 V only).

(*ii*) A.C. voltage can be increased and decreased with the help of a static machine called the '*transformer*'.

(*iii*) A.C. transmission and distribution is more economical as line material (say copper) can be saved by transmitting power at higher voltage.

(*iv*) A.C. motors for the same horse power as of D.C. motors are cheaper, lighter in weight, require less space and require lesser attention in operation and maintenance.

(*v*) A.C. can be converted to D.C. (direct current) easily, when and where required but D.C. cannot be converted to A.C. so easily and it will not be economical.

However, D.C. entails the following *merits* and hence finds wide applications.

(*i*) D.C. series motors are most suitable for traction purposes in tramway, railways, crains and lifts.

(*ii*) For electroplating, electrolytic and electrochemical processes (battery charging etc.), D.C. is required.

(*iii*) Arc lamps for search lights and cinema projectors work on D.C.

(*iv*) Arc welding is better than on A.C.

(*v*) Relay and operating time switches, etc., and circuit-breakers, D.C. works more efficiently.

(*vi*) In rolling mills, paper mills, colliery winding, etc., where fine speed control of speeds in both directions is required, D.C. motors are required.

2. GENERATION AND EQUATIONS OF ALTERNATING VOLTAGES AND CURRENTS

Generation of Alternating Voltages and Currents

Alternating voltages may be generated in the following two ways :

1. By *rotating a coil in a stationary magnetic field,* as shown in Fig. 1.

2. By *rotating a magnetic field within a stationary coil,* as shown in Fig. 2.

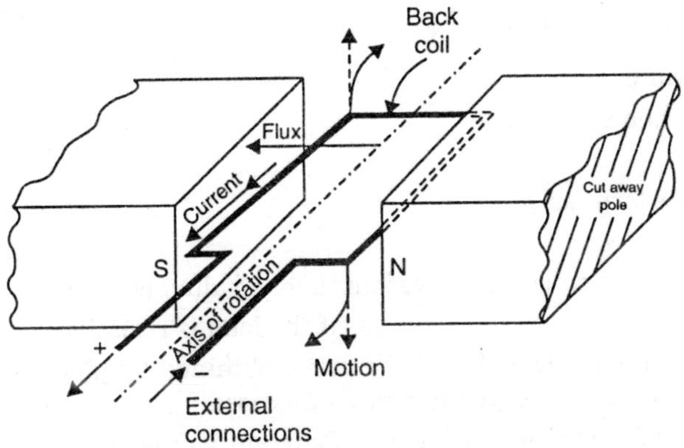

Fig. 1. Rotating a coil in a stationary magnetic field.

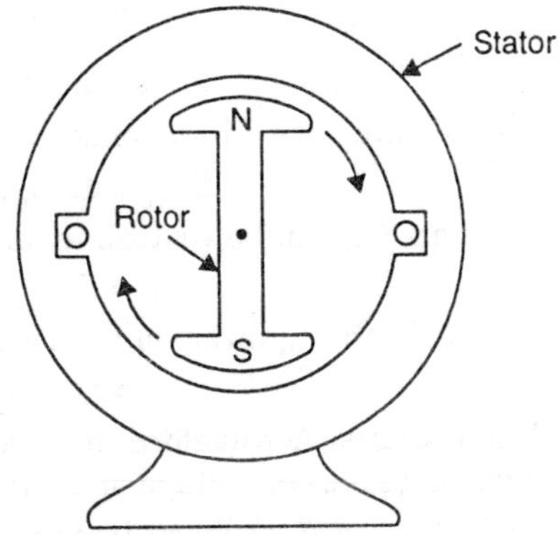

Fig. 2. Rotating a magnetic field within a stationary coil.

The value of the voltage generated in each case depends upon the following factors :

(*i*) The number of turns in the coils ;

(*ii*) The strength of the field ;

(*iii*) The speed at which the coil or magnetic field rotates.

- Out of the above two methods the *rotating-field method is mostly used in practice.*

Equations of Alternating Voltages and Currents

Fig. 3 shows a rectangular coil of N turns rotating clockwise with an angular velocity ω radians per second in a uniform magnetic field.

Since by Faraday's law, the voltage is proportional to the rate at which the conductor its across the magnetic field or to the rate of change of flux linkages, the shape of the wave of voltage applied to the external circuit will be determined by the *flux distribution in the air gap*. For a uniform field between the poles it is evident that *maximum* flux will link with the coil when its plane is in *vertical position i.e.*, perpendicular to the direction of flux between the poles. Also it is obvious that when the plane of coil is *horizontal no flux will link with the coil.*

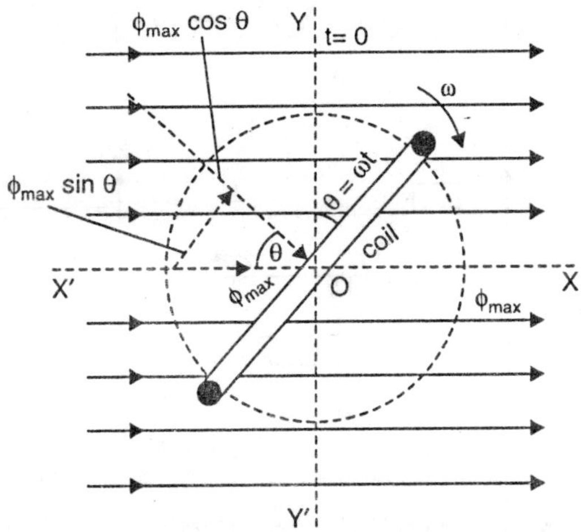

Fig. 3. A coil rotating in a magnet field.

If the position of the coil with reference to the vertical axis be denoted by θ the flux linking with the coil at any instant, as the coil rotates may be determined from the relation,

$$\phi = \phi_{max} \cos \theta$$

$$= \phi_{max} \cos \omega t \qquad ...(i) \quad (\because \theta = \omega t)$$

where, ϕ_{max} = Maximum flux which can link with the coil, and

t = Time taken by the coil to move through an angle θ from vertical position.

Using Faraday's law to eqn. (*i*), in order to determine the voltage equation,

$$e = -N \frac{d\phi}{dt}$$ (where e is the *instantaneous* value of the induced e.m.f.)

$$= -N \frac{d}{dt} (\phi_{max} \cos \omega t) = \omega N \phi_{max} \sin \omega t$$

or $$e = \omega N \phi_{max} \sin \theta \qquad ...(ii)$$

As the value of e will be maximum when $\sin \theta = 1$,

$$\therefore \qquad E_{max} = \omega N \phi_{max}$$

The eqn. (*ii*) can be written in simpler form as

$$e = E_{max} \sin \theta \qquad ...(iii)$$

Similarly the equation of induced alternating current (instantaneous value) is

$$i = I_{max} \sin \theta \qquad \text{(if the load is resistive)} \qquad ...(iv)$$

Waveforms. A waveform (or wave-shape) is the shape of a curve obtained by plotting the instantaneous values of voltage or current as ordinate against time as abscissa.

Fig. 4 (*a, b, c, d, e*) shows irregular waveforms, but each cycle of current/voltage is an exactly replica of the previous one. Alternating e.m.fs and currents produced by machines usually both have positive and negative half waves, the same shape as shown. Fig. 4(*f*) represents a sine wave of A.C. This is the simplest possible waveform, and alternators are designed to give as nearly as possible a sine wave of e.m.f.

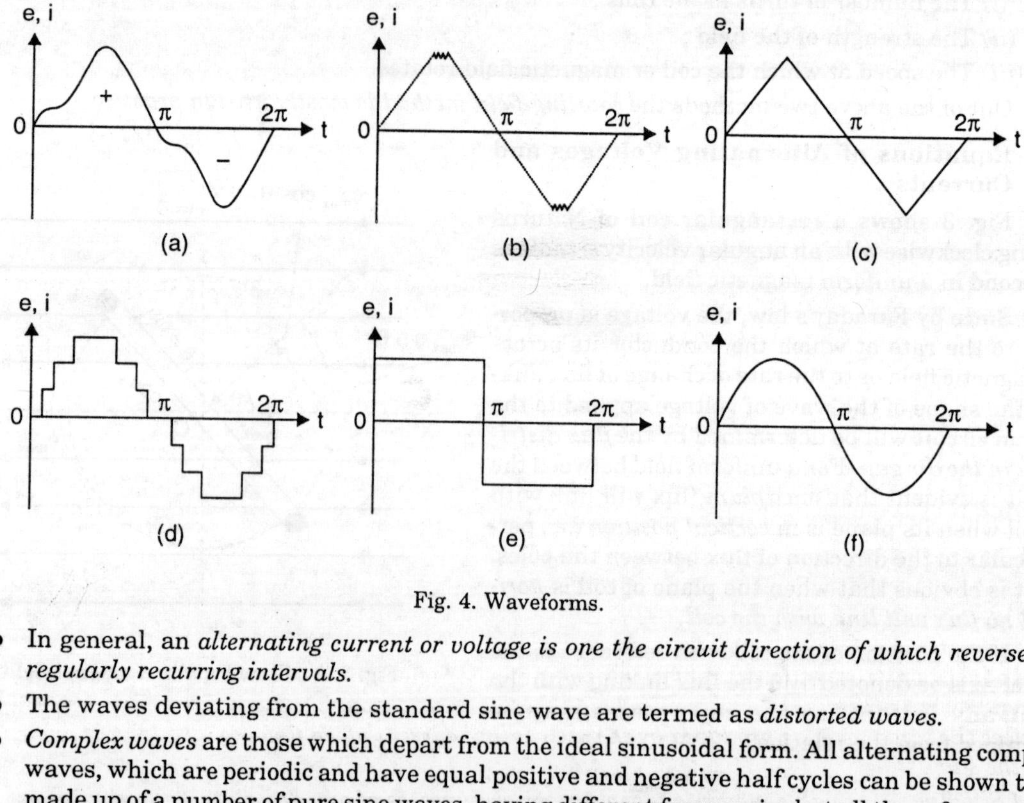

Fig. 4. Waveforms.

- In general, an *alternating current or voltage is one the circuit direction of which reverses at regularly recurring intervals.*
- The waves deviating from the standard sine wave are termed as *distorted waves.*
- *Complex waves* are those which depart from the ideal sinusoidal form. All alternating complex waves, which are periodic and have equal positive and negative half cycles can be shown to be made up of a number of pure sine waves, having different frequencies but all these frequencies are integral multiples of that of the lowest alternating wave, called the *fundamental* (or first harmonic). These waves of higher frequencies are called *harmonics.*

3. ALTERNATING VOLTAGE AND CURRENT

Modern alternators produce an e.m.f. which is for all practical purposes sinusoidal (*i.e.*, a sine curve), the equation between the e.m.f. and time being

$$e = E_{max} \sin \omega t \quad \ldots(1)$$

where, e = Instantaneous voltage ; E_{max} = Maximum voltage ;

ωt = Angle through which the armature has turned from neutral.

Taking the frequency as f hertz (cycles per second), the value of ω will be $2\pi f$, so that the equation reads

$$e = E_{max} \sin (2\pi f)t.$$

The graph of the voltage will be as shown in Fig. 5.

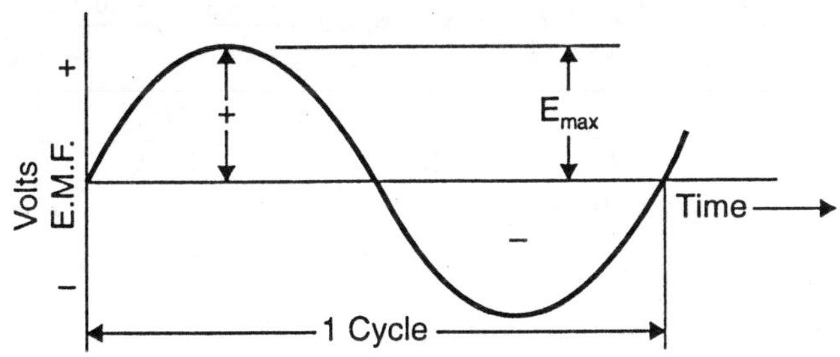

Fig. 5. The graph of the sinusoidal voltage.

1. Cycle. One complete set of positive and negative values of an alternating quantity is known as a *cycle*. A cycle may also sometimes be specified in terms of angular measure. In that case, one complete cycle is said to spread over 360° or 2π radians.

2. Amplitude. The maximum value, positive or negative, of an alternating quantity, is known as its *amplitude*.

3. Frequency (f). The number of cycles/second is called the frequency of the alternating quantity.

Its unit is *hertz* (Hz).

4. Time Period (T). The time taken by an alternating quantity to complete the cycle is called its *time period*. For example, a 50 hertz (Hz) alternating current has a time period of $\frac{1}{50}$ second.

Time period is reciprocal of frequency,

i.e.,
$$T = \frac{1}{f} \left(\text{or } f = \frac{1}{T} \right) \quad \ldots(2)$$

5. Root mean square (R.M.S.) value. The r.m.s. (or effective) value of an alternating current is given by that steady (D.C.) current *which when flowing through a given circuit for a given time produces the same* **heat** *as produced by the alternating current when flowing through the same circuit for the same time.*

R.M.S. value is the value which is taken for power purposes of any description. This value is obtained by finding the square root of the mean value of the squared ordinates for a cycle or half-cycle (See Fig. 5).

This is the value which is used for all power, lighting and heating purposes, as in these cases the power is proportional to the square of the voltage.

Refer Fig. 5.

The equation of sinusoidal alternating current is given as :

$$i = I_{max} \sin \theta$$

The mean of squares of the instantaneous values of current over half cycle is

$$I^2 = \int_0^\pi \frac{i^2 \, d\theta}{(\pi - 0)}$$

$$I^2 = \frac{1}{\pi} \int_0^\pi i^2 \, d\theta = \frac{1}{\pi} \int_0^\pi (I_{max} \sin \theta)^2 \, d\theta$$

$$= \frac{1}{\pi} \int_0^\pi I^2{}_{max} \sin^2 \theta \, d\theta = \frac{I^2{}_{max}}{\pi} \int_0^\pi \left(\frac{1 - \cos 2\theta}{2} \right) d\theta$$

$$= \frac{I^2{}_{max}}{2\pi} \int_0^\pi (1 - \cos 2\theta) \, d\theta = \frac{I^2{}_{max}}{2\pi} \left| \theta - \frac{\sin 2\theta}{2} \right|_0^\pi$$

$$= \frac{I^2{}_{max}}{2\pi} \times \pi = \frac{I^2{}_{max}}{2} \quad \text{or} \quad I = \sqrt{\frac{I^2{}_{max}}{2}} = \frac{I_{max}}{\sqrt{2}}$$

or $\mathbf{I = 0.707\, I_{max}}$. ...(3)

Note. While solving problems, the values of given current and voltage should always be taken as the r.m.s. values, unless indicated otherwise.

6. Average or mean value. The average value of an alternating current is expressed by *that steady current which transfers across any circuit the same* **charge** *as is transferred by that alternating current during the same time.*

The *mean value is only of use in connection with processes where the results depend on the current only, irrespective of the voltage, such as electroplating or battery charging.*

Refer Fig. 6.

The value of instantaneous current is given by

$$i = I_{max} \sin \theta$$

Refer Fig. 6. The value of instantaneous current is given by :

$$i = I_{max} \sin \theta \qquad [\theta = \omega t]$$

$$I_{av} = \frac{1}{(\pi - 0)} \int_0^\pi i d\theta$$

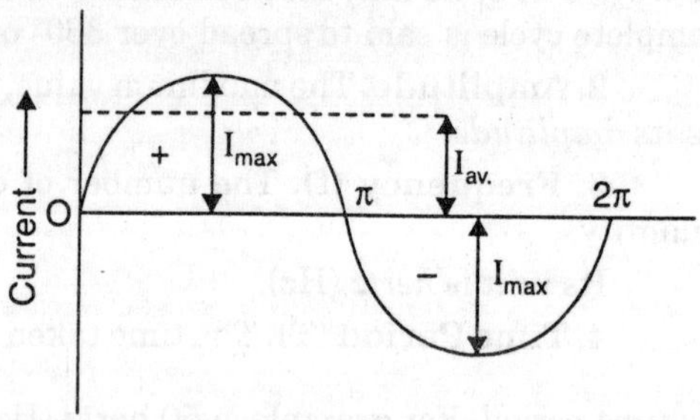

Fig. 6

[Limits are taken from 0 to π, since only first half cycle is considered. For whole cycle, the average value of sine wave is zero.]

$$= \frac{1}{\pi} \cdot \int_0^\pi I_{max} \cdot \sin \theta \, d\theta = \frac{1}{\pi} \cdot I_{max} \left| - \cos \theta \right|_0^\pi$$

$$= \frac{1}{\pi} \cdot I_{max} \, [1 - (-1)] = \frac{2}{\pi} \cdot I_{max}$$

or $\mathbf{I_{av} = 0.637\, I_{max}}$. ...(4)

Note. In case of unsymmetrical alternating current *viz.* half-wave rectified current the average value must always be taken over the whole cycle.

7. Form and Peak Factors

Form factor. *The ratio of r.m.s. (or effective) value to average value is the form factor* (K_f) *of the wave form.* It has use in voltage generation and instrument correction factors.

Peak factor. *The ratio of maximum value to the r.m.s. value is the peak factor (K_p) of the wave form.*

S. No.	*Wave form*	*Form factor* (K_f) $= \frac{\text{r.m.s. value}}{\text{average value}}$	*Peak factor* (K_p) $= \frac{\text{max. value}}{\text{r.m.s. value}}$
1.	**Sine wave :** 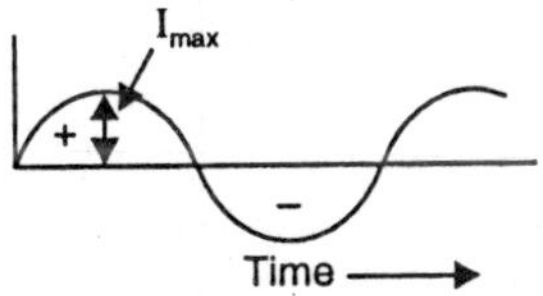 Fig. 7 R.M.S. value $= \frac{I_{max}}{\sqrt{2}} = 0.707\, I_{max}$ Average value $= \frac{2}{\pi} I_{max} = 0.637\, I_{max}$	$K_f = \frac{0.707\, I_{max}}{0.637\, I_{max}} = 1.11$	$K_p = \frac{I_{max}}{0.707\, I_{max}} = 1.41$
2.	**Half wave rectified sine wave :** 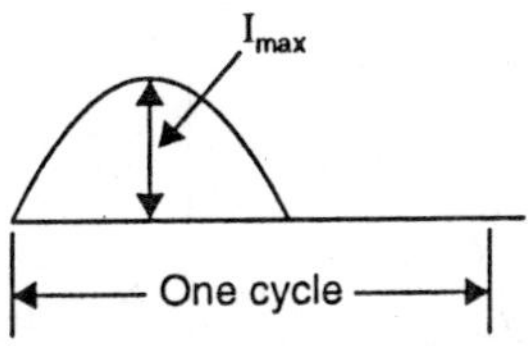 Fig. 8 R.M.S. value $= \frac{I_{max}}{2} = 0.5\, I_{max}$ Average Value $= \frac{1}{\pi} I_{max} = 0.318\, I_{max}$	$K_f = \frac{0.5\, I_{max}}{0.318\, I_{max}} = 1.57$	$K_p = \frac{I_{max}}{0.5\, I_{max}} = 2.0$
3.	**Full wave rectified sine wave :** 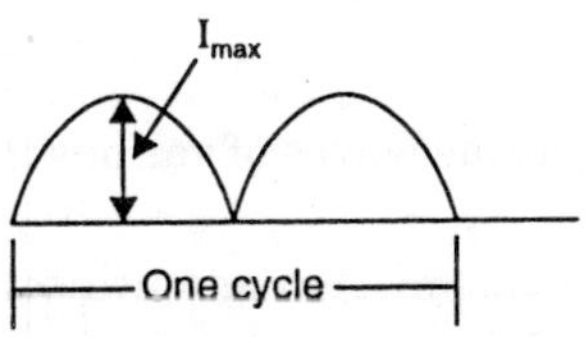 Fig. 9 R.M.S. value $= \frac{I_{max}}{\sqrt{2}} = 0.707\, I_{max}$ Average value $= \frac{2}{\pi} I_{max} = 0.637\, I_{max}$	$K_f = \frac{0.707\, I_{max}}{0.637\, I_{max}} = 1.11$	$K_p = \frac{I_{max}}{0.707\, I_{max}} = 1.41$

S. No.	Wave form	Form factor (K_f) $= \frac{r.m.s.\ value}{average\ value}$	Peak factor (K_p) $= \frac{max.\ value}{r.m.s.\ value}$
4.	**Rectangular wave :** 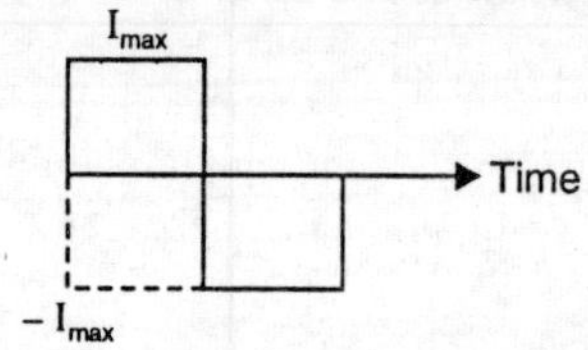 Fig. 10 R.M.S. value = I_{max} Average value = I_{max}	$K_f = 1$	$K_p = 1$
5.	**Triangular wave :** 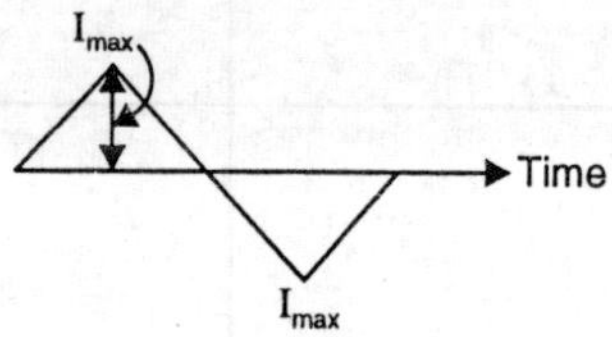 Fig. 11 R.M.S. = $\frac{I_{max}}{\sqrt{3}} = 0.578\, I_{max}$ Average value = $\frac{I_{max}}{2} = 0.5\, I_{max}$	$K_f = \frac{0.578\, I_{max}}{0.5\, I_{max}} = 1.16$	$K_p = \frac{I_{max}}{0.578\, I_{max}} = 1.73$

Reasons for using alternating current (or voltage) of sinusoidal form :

An alternating current (or voltage) of sinusoidal form is normally used because of the following reasons :

1. Mathematically, it is quite simple.
2. Its integrals and differentials both are sinusoidal.
3. It lends itself to vector representation.
4. A complex waveform can be analysed into a series of sine waves of various frequencies, and each such component can be dealt with separately.
5. This waveform is desirable for power generation, transmission and utilisation.

8. Phase and phase angle. The **'phase'** of an A.C. wave may be defined as *its position with respect to a reference axis or reference wave* and **'phase angle'** as *the angle of lead or lag with respect to the reference axis or with respect to another wave.*

Examples. The *phase* of current at point *L* is $\frac{T}{4}$ second where *T* is the time period or expressed in terms of angle θ, it is $\frac{\pi}{2}$ radian (Fig. 12). Similarly

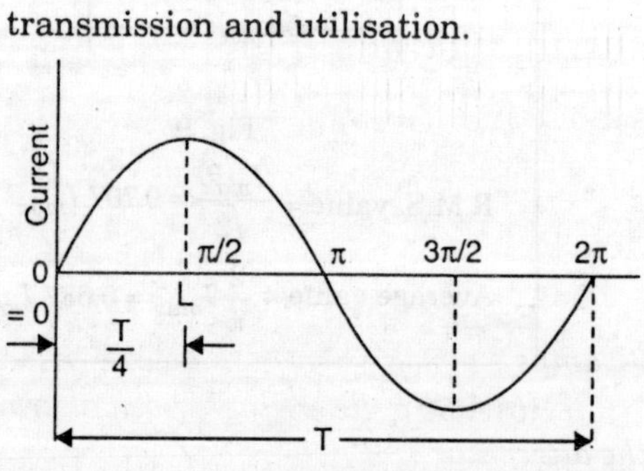

Fig. 12

phase of the rotating coil at the instant shown in Fig. 13 is ωt which is therefore called its *phase angle*.

The e.m.fs. induced in both the coils (Fig. 13) will be of the same frequency and of sinusoidal shape, although the values of instantaneous e.m.f. induced will be different. However, the alternating e.m.fs. would reach their maximum and zero values at the *same time* as shown in Fig. 13 (*b*). Such alternating voltages or curve are said to *in phase* with each other.

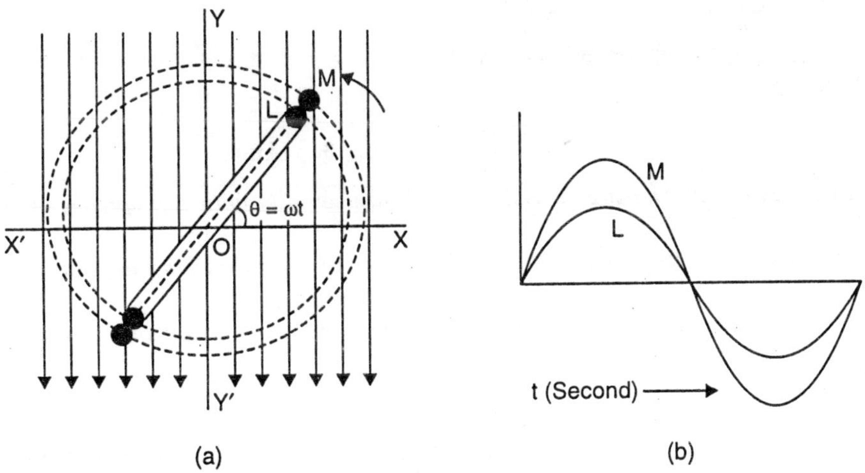

Fig. 13

Refer Fig. 14. *M lags* behind *L* by β and *N lags* behind *L* by (α + β) because they reach their maximum later.

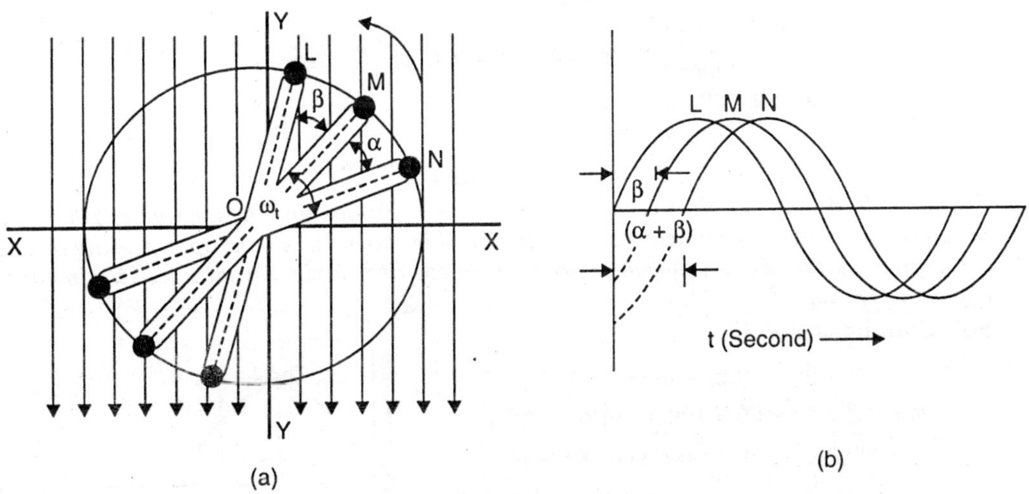

Fig. 14

Example 1. (*a*) *What is the equation of a 25 cycle current sine wave having r.m.s. value of 30 amps ?*

(*b*) *A 60 cycle engine-driven alternator has a speed of 1200 r.p.m. How many poles are there in the alternator ?*

Solution. We know that, $i = I_{max} \sin \omega t$

$= I_{max} \sin 2\pi ft$ $\quad (\because \ \omega = 2\pi f)$

$= 30 \times \sqrt{2} \, . \sin (2\pi \times 25 \times t)$ $\quad \left(\because \dfrac{\text{R.M.S. value}}{\text{Max. value}} = \dfrac{1}{\sqrt{2}}\right)$

$= \mathbf{42.42 \sin 157\, t.}$ **(Ans.)**

Using the relation, $f = \dfrac{Np}{120}$

where, f = frequency, N = speed in r.p.m., and
p = No. of poles

$\therefore \quad 60 = \dfrac{1200\, p}{120}$ or $p = 6.$ **(Ans.)**

Example 2. *An alternating current varying sinusoidally with a frequency of 50 Hz has an r.m.s. value of 40 A. Find :*

(i) The instantaneous value 0.0025 seconds after passing through maximum positive value, and

(ii) The time measured from a maximum value when the instantaneous current is 14.14 A.

Solution. $I_{max} = \sqrt{2} \times 40 = 56.56$ A

$\omega = 2\pi f = 2\pi \times 50 = 100\pi$ radians

(i) $i = I_{max} \cos \omega t$ ... after +ve maximum value

$= 56.56 \cos 100\, \pi t$

$= 56.56 \cos (100\pi \times 0.0025)$ ...(t = 0.0025 s ...given)

$= 56.56 \cos 45°$ (Taking π = 180°)

$= \mathbf{40\ A.}$ **(Ans.)**

(ii) $14.14 = 56.56 \cos (100 \times 180 \times t)$

or $\dfrac{14.14}{56.56} = \cos (100 \times 180 \times t)$

or $\cos^{-1} (0.25) = 100 \times 180 \times t$

$75.5° = 100 \times 180 \times t$

$\therefore \quad t = \mathbf{0.00419\ s.}$ **(Ans.)**

Example 3. *A sinusoidal alternating voltage of 50 Hz has an r.m.s. value of 200 V. Write down the equation for the instantaneous value and find this value 0.0125 sec. after passing through a positive maximum value. At what time measured from a positive maximum value will the instantaneous voltage be 141.4 volts ?*

Solution. Refer Fig. 15.

$V_{max} = \sqrt{2} \times 200 = 282.2$ volts

$\omega = 2\pi f = 2\pi \times 50 = 100\, \pi$ rad/sec.

$\therefore$ Equation for the instantaneous voltage.

$V = V_{max} \sin \omega t$ (with reference to point O)

$= 282.2 \sin 100\, \pi t$ ...*(i)*

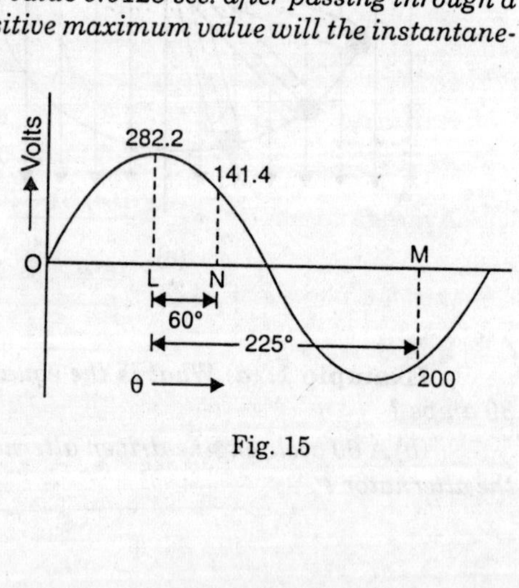

Fig. 15

Since the time (0.0125 sec.) is given from the point L (*i.e.*, from positive maximum value) the equation *(i)* when referred to point L can be written as

$v = 282.2 \sin (90° + 100\, \pi t)$

$= 282.2 \cos 100\, \pi t$

Hence instantaneous value of the voltage 0.0125 sec. after passing through + ve maximum value,

$$v = 282.2 \cos (100 \pi \times 0.0125) \quad \text{...(angle in radians)}$$

$$= 282.2 \cos (100 \times 180 \times 0.0125) \quad \text{...(angle in degrees)}$$

$$= 282.2 \cos 225° = 280.2 \times \left(-\frac{1}{\sqrt{2}}\right)$$

$$= \mathbf{-200\ V\ (point\ M).\ (Ans.)}$$

Also $v = 141.4$ V

$\therefore$ $141.4 = 282.2 \cos (100 \times 180 \times t)$

or $0.5 = \cos (100 \times 180 \times t)$

or $\cos^{-1} (0.5) = 100 \times 180 \times t$

or $60° = 100 \times 180 \times t$

or $t = \frac{1}{300}$ **sec. (point N). (Ans.)**

Example 4. (*a*) *What is the peak value of a sinusoidal alternating current of 4.78 r.m.s. amps ?*

(*b*) *What is the r.m.s. value of a rectangular wave with an amplitude of 9.87 volts ?*

(*c*) *What is the average value over half a cycle of a sinusoidal alternating current whose r.m.s. value is 31 A ?*

Solution. (*a*) Peak value, $I_{max} = \sqrt{2} \times 4.78 =$ **6.76 A. (Ans.)**

(*b*) Refer Fig. 16. If the first half-cycle is divided into n equal parts each of value V, then

$$\text{r.m.s. value} = \left(\frac{V^2 + V^2 + V^2 + \ldots\ldots}{n}\right)^{1/2}$$

$$= V = \mathbf{9.87\ volts.\ (Ans.)}$$

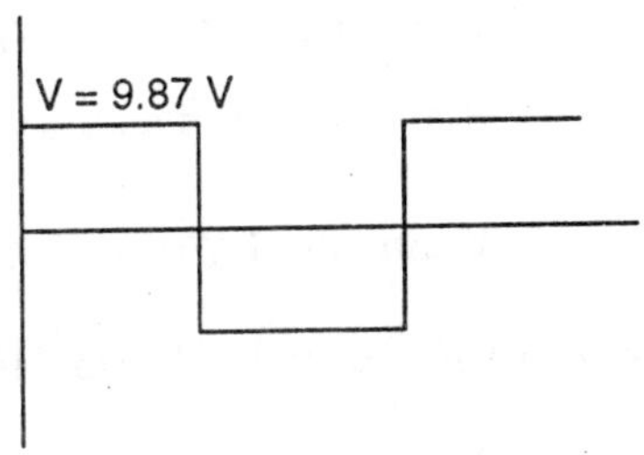

Fig. 16

(*c*) $I_{r.m.s.} = 31$ A

$$I_{av} = \frac{I_{r.m.s.}}{\text{form factor}}$$

$$= \frac{31}{1.11} = \mathbf{27.93\ A.\ (Ans.)}$$

Example 5. *The graph in Fig. 17 shows the variation of voltage with time. Use the graph to calculate the average and r.m.s. value of the voltage. What is the frequency of the voltage ? What would be the r.m.s. value of sine wave having the same peak value ?*

Solution. Refer Fig. 17.

As the graph is symmetrical about time axis, considering only the positive half cycle.

Average value, $V_{av} = \frac{0 + 10 + 20 + 40 + 100 + 120 + 100 + 40 + 20 + 10}{10}$

$= \mathbf{46\ V.\ (Ans.)}$

R.M.S. value, $V = \sqrt{\frac{0^2 + 10^2 + 20^2 + 40^2 + 100^2 + 120^2 + 100^2 + 40^2 + 20^2 + 10^2}{10}}$

$$= \sqrt{\frac{0 + 100 + 400 + 1600 + 10000 + 14400 + 10000 + 1600 + 400 + 100}{10}}$$

$$= \sqrt{\frac{38600}{10}} = \sqrt{3860} = \mathbf{62.1\ V.} \quad \mathbf{(Ans.)}$$

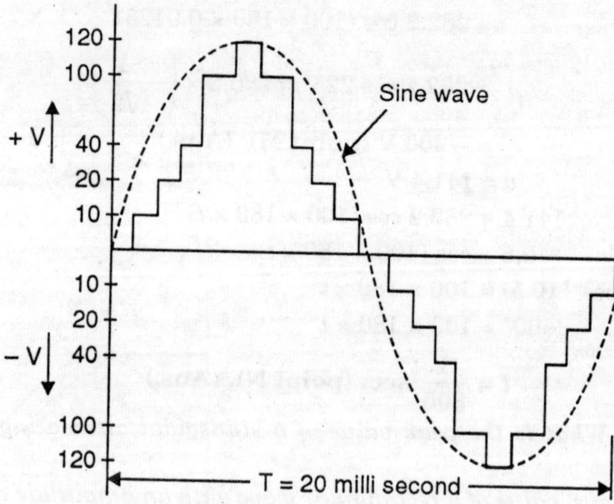

Fig. 17

Since the time period '*T*' is 20 millisec.

$\therefore$ Frequency $\quad 'f' = \frac{1}{T} = \frac{1}{20 \times 10^{-3}} = \mathbf{50\ Hz.}$ **(Ans.)**

R.M.S. value of a sine wave of the same peak value

$$= 0.707 \times 120 = \mathbf{84.84\ V.} \quad \mathbf{(Ans.)}$$

Example 6. *Prove that if a D.C. current of $I_{amps.}$ is superposed in a conductor by an A.C. current of max. value I amps, the r.m.s. value of the resultant is* $\sqrt{\frac{3}{2}}\, I$.

Solution. Let the *A.C. current* be $i = I \sin \theta$ where i is the instantaneous value of the A.C. current and I the *D.C. current.*

The r.m.s. value of $(I + i)$ over one complete cycle is,

$$= \sqrt{\frac{1}{2\pi} \int_0^{2\pi} (I + I \sin \theta)^2 \, d\theta}$$

$$= I \sqrt{\frac{1}{2\pi} \int_0^{2\pi} (I + 2 \sin \theta + \sin^2 \theta)}$$

$$= I \sqrt{\left\{ \frac{1}{2\pi} \int_0^{2\pi} 1 + 2 \sin \theta + \left(\frac{1 - \cos 2\theta}{2} \right) \right\}}$$

$$= I \sqrt{\frac{1}{2\pi} \left| \theta - 2 \cos \theta + \frac{\theta}{2} - \frac{\sin 2\theta}{4} \right|_0^{2\pi}}$$

$$= I \sqrt{\frac{1}{2\pi} (2\pi - 2 + \pi + 2)}$$

$$= I . \sqrt{\frac{3}{2}} . \quad \mathbf{(Ans.)}$$

Example 7. *A resultant current wave is made up of two components : a 4A D.C. component and a 50 Hz A.C. component, which is of sinusoidal waveform and which has a maximum value of 4A.*

(*i*) *Sketch the resultant wave.*

(*ii*) *Write an analytical expression for the current wave, reckoning t = 0 at a point where the A.C. component is at zero value and where di/dt is positive.*

(*iii*) *What is the average value of the resultant current over a cycle ?*

(*iv*) *What is the effective or r.m.s. value of the resultant current ?*

Solution. (*i*) **Sketch of the resultant wave :**

The two current components and the resultant current wave are shown in Fig. 18. **(Ans.)**

(*ii*) **Analytical expression.** The instantaneous value of the resultant current is given by

$$i = (4 + 4 \sin \omega t) = \mathbf{(4 + 4 \sin \theta)}. \quad \textbf{(Ans.)}$$

(*iii*) **Average value.** Since the average value of the alternating current over one complete cycle is zero, hence the *average value of the resultant current is equal to the value of D.C. component i.e.,* **4A (Ans.)**

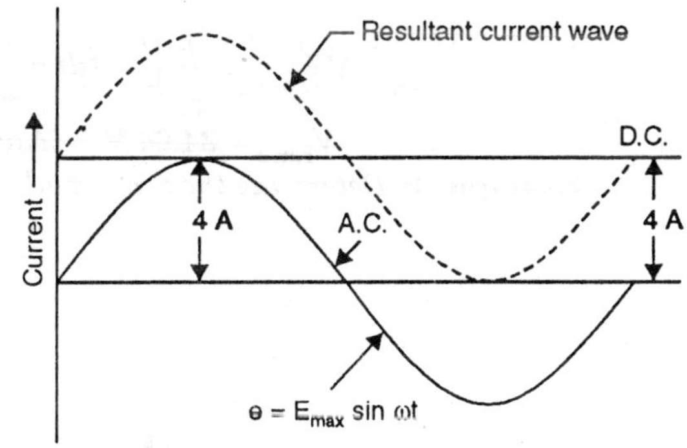

Fig. 18

(*iv*) **Effective or r.m.s. value :**

Mean value of i^2 over complete cycle is

$$= \frac{1}{2\pi} \int_0^{2\pi} i^2 d\theta = \frac{1}{2\pi} \int_0^{2\pi} (4 + 4 \sin \theta)^2 \, d\theta$$

$$= \frac{1}{2\pi} \int_0^{2\pi} (16 + 32 \sin \theta + 16 \sin^2 \theta) \, d\theta$$

$$= \frac{1}{2\pi} \int_0^{2\pi} \left[16 + 32 \sin \theta + 16 \left(\frac{1 - \cos 2\theta}{2} \right) \right] d\theta$$

$$= \frac{1}{2\pi} \int_0^{2\pi} (24 + 32 \sin \theta - 8 \cos 2\theta) \, d\theta$$

$$= \frac{1}{2\pi} \left(24\,\theta - 32 \cos \theta - 8 \times \frac{\sin 2\,\theta}{2} \right)_0^{2\pi}$$

$$= \frac{1}{2\pi} [(48\pi - 32 \cos 2\pi - 4 \sin 4\pi) - (-32)] = \frac{48\pi}{2\pi} = 24 \text{ A}$$

$\therefore$ R.M.S. value, $\quad I = \sqrt{24} = \mathbf{4.9\ A.}$ **(Ans.)**

Example 8. *Determine the average and effective values of the saw-tooth waveform shown in Fig. 19.* **(N.U.)**

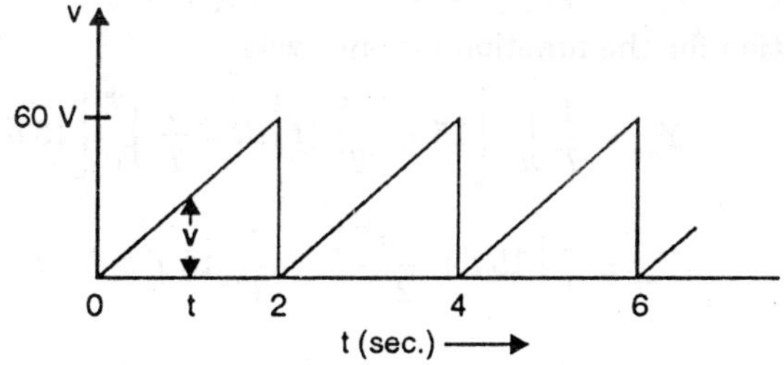

Fig. 19

Solution. Refer Fig. 19.

Since the voltage increases linearly, therefore,

$$V_{av} = \frac{60+0}{2} = \mathbf{30\ V.}\quad \textbf{(Ans.)}$$

The slope, for the internal $0 < t < 2$ is given by :

$$\text{Slope} = \frac{60}{2} = 30$$

$\therefore$ Instantaneous voltage, $v = 30t$ volts

The r.m.s. or effective value of the voltage,

$$V^2_{\text{r.m.s.}} = \frac{1}{T}\int_0^T v^2 dt = \frac{1}{2}\int_0^2 (30t)^2\, dt = 450\int_0^2 t^2\, dt = 450\left|\frac{t^3}{3}\right|_0^2 = 1200$$

or $$V_{\text{r.m.s}} = \mathbf{34.64\ V.}\quad \textbf{(Ans.)}$$

Example. 9. *Determine the r.m.s. and average values of the waveform shown in Fig. 20.*

[Indore University]

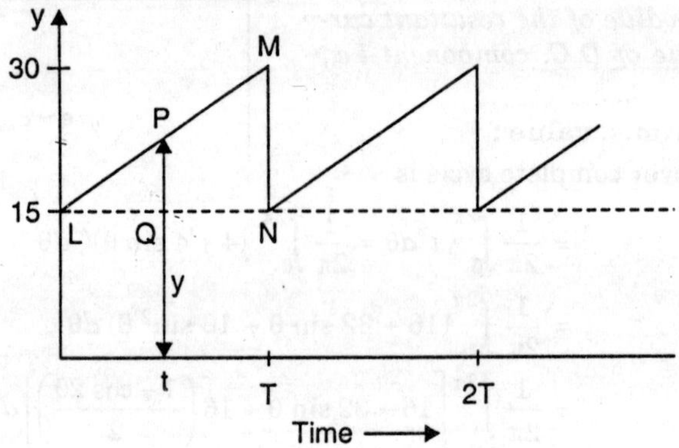

Fig. 20

Solution. Refer Fig. 20.

The slope the curve $\quad LM = \frac{MN}{LN} = \frac{30-15}{T} = \frac{15}{T}$

Now consider the function y at any time 't'. We have,

$$\therefore \quad \frac{PQ}{LQ} = \frac{MN}{LN} = \frac{15}{T}.$$

or $$\frac{y-15}{t} = \frac{15}{T} \quad \text{or} \quad y = 15 + \left(\frac{15}{T}\right)t$$

This gives the equation for the function for one cycle.

$$Y_{av} = \frac{1}{T}\int_0^T \left[15 + \left(\frac{15}{T}\right)t\right] dt = \frac{1}{T}\int_0^T \left[15\, dt + \frac{15}{T}.t.dt\right]$$

or $$Y_{av} = \frac{1}{T}\left[15\,t + \frac{15t^2}{2T}\right]_0^T = \frac{1}{T}[15\,T + 7.5\,T] = \mathbf{22.5.}\quad \textbf{(Ans.)}$$

Mean square value $$= \frac{1}{T}\int_0^T y^2 dt = \frac{1}{T}\int_0^T \left\{15 + \left(\frac{15}{T}\right)t\right\}^2 dt$$

$$= \frac{1}{T}\int_0^T \left(225 + \frac{225t^2}{T^2} + \frac{450}{T}\ t\right) dt$$

$$= \frac{1}{T}\left| 225t + \frac{225t^3}{3T^2} + \frac{450t^2}{2T} \right|_0^T$$

$$= \frac{1}{T}\ |\ 225\ T + 75\ T + 225\ T\ |\ = 525$$

$\therefore$ R.M.S. value $= \sqrt{525}$ = **22.9. (Ans.)**

Example 10. *Find the r.m.s. and average value of the trapezoidal current wave-form shown in the Fig. 21.* **[Bangalore University]**

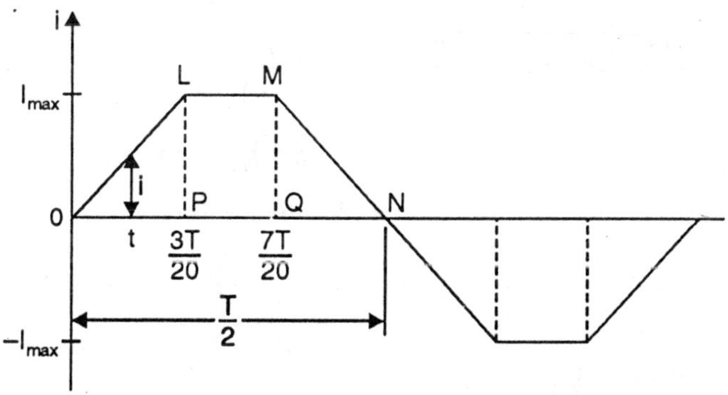

Fig. 21

Solution. Refer Fig. 21. The equation of the current for $0 < t < \frac{3T}{20}$ can be found from the relation,

$$\frac{i}{t} = \frac{I_{max}}{(3T/20)} \quad \text{or} \quad i = \frac{20\ I_{max}}{3T}\ .\ t$$

The equation of the current for $\frac{3T}{20} < t < \frac{7T}{20}$ is given by $i = I_{max}$. Remembering that ΔOLP is identical with MQN,

$$\textit{R.M.S. value of current, } I_{r.m.s.} = \sqrt{\frac{1}{(T/2)}\left[2\int_0^{3T/20} i^2 dt + \int_{3T/20}^{7T/20} I_{max}^2\ dt\right]}$$

$$= \sqrt{\frac{2}{T}\left[2\left\{\left(\frac{20\ I_{max}}{3T}\right)^2 \int_0^{3T/20} t^2\ dt\right\} + I_{max}^2 \int_{3T/20}^{7T/20} dt\right]}$$

$$= \sqrt{\frac{2}{T}\left[2\left\{\left(\frac{20\ I_{max}}{3T}\right)^2 \left|\frac{t^3}{3}\right|_0^{3T/20}\right\} + I_{max}^2\ \Big|\ t\ \Big|_{3T/20}^{7T/20}\right]}$$

$$= \sqrt{\frac{2}{T}\left[2 \times \frac{400\ I_{max}^2}{9T^2} \times \frac{27}{3 \times 8000}\ T^3 + I_{max}^2 \times \left(\frac{7T}{20} - \frac{3T}{20}\right)\right]}$$

$$= \sqrt{\frac{2}{T}\left(0.1\,I_{max}^2\,T + 0.2\,I_{max}^2\,T\right)} = \sqrt{0.6\,I_{max}^2} = \mathbf{0.775\ I_{max}}. \quad \textbf{(Ans.)}$$

Average value of current,

$$I_{av} = \frac{1}{T/2}\left[2\int_0^{3T/20} i\,dt + \int_{3T/20}^{7T/20} I_{max}\,dt\right] = \frac{2}{T}\left[2\int_0^{3T/20}\left(\frac{20\,I_{max}}{3T}\right)t\,dt + I_{max}\int_{3T/20}^{7T/20} dt\right]$$

$$= \frac{2}{T}\left[2\left(\frac{20\,I_{max}}{3T}\right)\left|\frac{t^2}{2}\right|_0^{3T/20} + I_{max}\left|t\right|_{3T/20}^{7T/20}\right]$$

$$= \frac{2}{T}\left[2\left(\frac{20\,I_{max}}{3T}\right)\times\frac{1}{2}\left(\frac{3T}{20}\right)^2 + I_{max}\left(\frac{7T}{20}-\frac{3T}{20}\right)\right]$$

$$= \frac{2}{T}\left[2\left(\frac{20\,I_{max}}{3T}\right)\times\frac{1}{2}\left(\frac{9T}{400}\right)^2 + I_{max}\times\frac{T}{5}\right]$$

$$I_{av} = \frac{2}{T}\left[0.15\,I_{max}\times T + 0.2\,I_{max}\times T\right] = \mathbf{0.7\ I_{max}}. \quad \textbf{(Ans.)}$$

Example 11. *A half wave single anode rectifier has a voltage given by 100 sin ωt applied to it. Estimate the average value on the d.c. side.*

Solution. The wave form on the d.c. side is as shown in Fig. 22.

Mean/Average d.c. voltage will be :

$$I_{av} = \frac{1}{2\pi}\int_0^{\pi} 100\sin\theta\,.\,d\theta$$

$$= \frac{100}{2\pi}\left[-\cos\theta\right]_0^{\pi}$$

$= \mathbf{31.83\ V.}$ **(Ans.)**

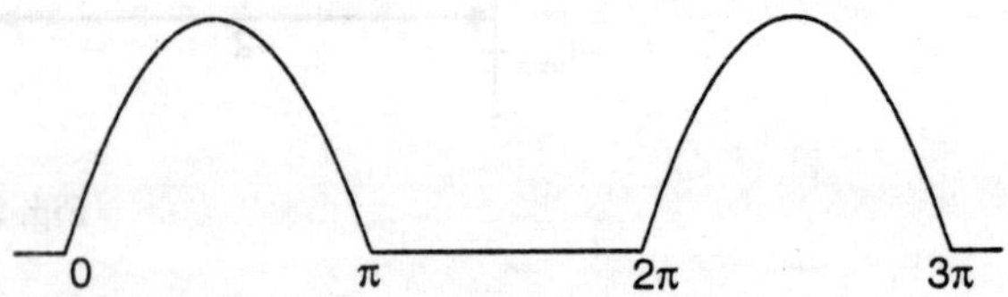

Fig. 22

Example 12. *Calculate from the first principles, the reading which would be indicated by a hot-wire ammeter in a circuit whose current waveform is given by 10 sin ωt + 3 sin 3ωt + 2 sin 5ωt.*

Solution. The expression for instantaneous current is :

$$i = 10\sin\omega t + 3\sin 3\omega t + 2\sin 5\omega t$$

The hot-wire ammeter will read the "r.m.s. value" of the wave form.

Now $i^2 = (10\sin\omega t + 3\sin 3\omega t + 2\sin 5\omega t)^2$

$\therefore$ *R.M.S. value of the current* $= \sqrt{\frac{1}{2\pi}\int_0^{2\pi} i^2\,d(\omega t)}$

$$= \left[\frac{1}{2\pi}\int_0^{2\pi}\left[(10\sin\omega t + 3\sin 3\omega t + 2\sin 5\omega t)^2\,d(\omega t)\right]\right]^{1/2}$$

$$= \left[\frac{1}{2\pi}\int_0^{2\pi}\{10^2\sin^2\omega t + 3^2\sin^2 3\omega t + 2^2\sin^2 5\omega t + 2\times 10\times 3\sin\omega t\sin 3\omega t + 2\times 10\times 2\sin\omega t\sin 5\omega t + 2\times 3\times 2\sin 3\omega t\,.\,\sin 5\omega t\}\,d(\omega t)\right]^{1/2}$$

$$= \left[\frac{1}{2\pi} \int_0^{2\pi} \left\{ \frac{10^2 (1 - \cos 2\omega t)}{2} + \frac{3^2}{2} (1 - \cos 6\omega t) + \frac{2^2}{2} (1 - \cos 10\omega t) + {}^* etc. \right\} d(\omega t) \right]^{1/2}$$

[*All the terms containing the product of two sines when integrated over the range 0 to 2π disappear. This is easily seen by splitting such terms into the difference of two cosines.]

$$= \left[\frac{1}{2\pi} \frac{(10^2 + 3^2 + 2^2)}{2} . 2\pi \right]^{1/2} = \left(\frac{10^2 + 3^2 + 2^2}{2} \right)^{1/2} = \left(\frac{113}{2} \right)^{1/2}$$

$= (56.5)^{1/2} =$ **7.52. (Ans.)**

Example 13. *Four branches A, B, C, D in an A.C. circuit meet at a junction point P. The currents in branches A, B, C flow towards P while the current in branch D flows away from P. The currents in branches A, B and C are*

$$i_A = 20 \sin 628t$$
$$i_B = 15 \sin (628t - \pi/6)$$
$$i_C = 25 \sin (628t + \pi/3)$$

Find an expression for the instantaneous value of current in branch D, and calculate (i) its frequency, and (ii) the heat (watts) that it would produce when flowing in a resistance of 5 ohms.

Solution.

Analytical method. Let the current flowing in the branch D be, $i_D = I_D \sin (628t + \phi)$

$$\Sigma H = I_{Amax} \cos \phi_1 + I_{Bmax} \cos \phi_2 + I_{Cmax} \cos \phi_3 + I_{Dmax} \cos \phi$$
$$= 20 \cos 0° + 15 \cos (-\pi/6) + 25 \cos \pi/3 + I_{Dmax} \cos \phi$$
$$= 20 + 15 \times \frac{\sqrt{3}}{2} + 25 \times \frac{1}{2} + I_{Dmax} \cos \phi$$
$$= 20 + 13 + 12.5 + I_{Dmax} \cos \phi = 45.5 + I_{Dmax} \cos \phi$$
$$\Sigma V = I_{Amax} \sin \phi_1 + I_{Bmax} \sin \phi_2 + I_{Cmax} \sin \phi_3 + I_{Dmax} \sin \phi$$
$$= 20 \sin 0° + 15 \sin (-\pi/6) + 25 \sin \pi/3 + I_{Dmax} \sin \phi$$
$$= 0 - 15 \times \frac{1}{2} + 25 \times \frac{\sqrt{3}}{2} + I_{Dmax} \sin \phi$$
$$= -7.5 + 21.65 + I_{Dmax} \sin \phi = 14.15 + I_{Dmax} \sin \phi$$

Since all the currents are meeting at point P,

$\therefore$ $\Sigma H = 0$

i.e., $45.5 + I_{Dmax} \cos \phi = 0$ or $I_{Dmax} \times \cos \phi = -45.5$ ...(*i*)

and $\Sigma V = 0$

i.e., $14.15 + I_{Dmax} \sin \phi = 0$ or $I_{Dmax} \times \sin \phi = -14.15$ ...(*ii*)

From (*i*) and (*ii*), $I_{Dmax} = (-45.5)^2 + (-14.15)^2 = 47.6$ A.

$$\phi = \tan^{-1} \frac{-14.15}{-45.5} = 197° \text{ or } 3.44°$$

Hence the current in branch D follows the relation,

$i_D = 47.6 \sin (678\, t + 3.44)$. **(Ans.)**

(*i*) **Frequency** $= \frac{\omega}{2\pi} = \frac{628}{2\pi} = 100$ Hz. **(Ans.)**

(*ii*) **Heat produced** $= \left(\frac{I_{Dmax}}{\sqrt{2}} \right)^2 \times R = \left(\frac{47.6}{\sqrt{2}} \right)^2 \times 5 = 5620$ W. **(Ans.)**

4. SINGLE PHASE CIRCUITS

The study of circuits involves three basic types of units (*R, L, C i.e.,* resistance, reactance and capacitance respectively) and four possible series combination of them. The latter, in turn, may be arranged in many kinds of parallel, series-parallel, parallel-series or other complex circuits.

4.1. A.C. Through Pure Ohmic Resistance Alone

The circuit containing a pure resistance *R* is shown in Fig. 23 (*a*). Let the applied voltage be given by the equation,

$$v = V_{max} \sin \theta = V_{max} \sin \omega t \qquad ...(i)$$

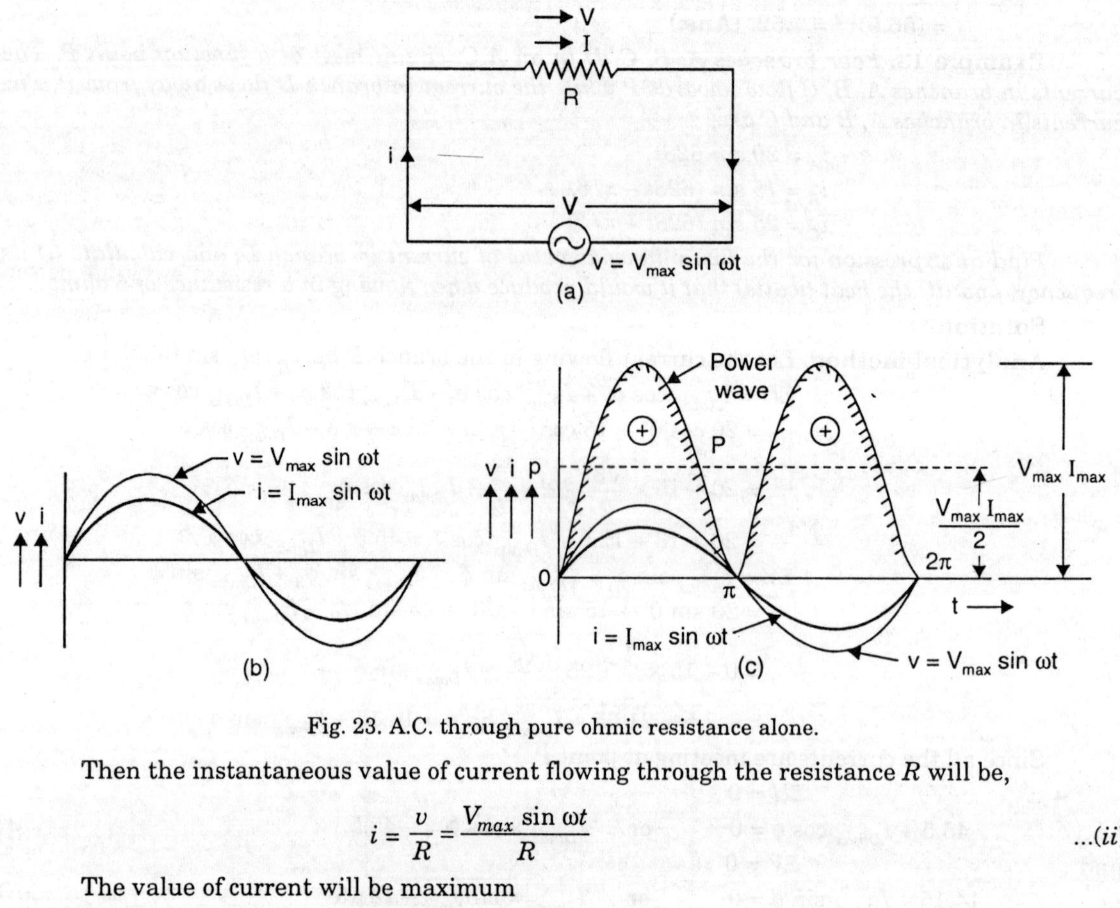

Fig. 23. A.C. through pure ohmic resistance alone.

Then the instantaneous value of current flowing through the resistance *R* will be,

$$i = \frac{v}{R} = \frac{V_{max} \sin \omega t}{R} \qquad ...(ii)$$

The value of current will be maximum

when $\sin \omega t = 1$ or ($\omega t = 90°$)

$$\therefore \quad I_{max} = \frac{V_{max}}{R}$$

Substituting this value in eqn. (*ii*), we get

$$i = I_{max} \sin \omega t \qquad ...(iii)$$

Comparing (*i*) and (*iii*), we find that alternating *voltage and current are in phase with each other* as shown in Fig. 23 (*b*), also shown vectorially in Fig. 23 (*c*).

Power. Refer Fig. 23 (*c*)

Instantaneous power,

$$p = vi = V_{max} \sin \omega t \times I_{max} \sin \omega t = V_{max} I_{max} \sin^2 \omega t$$

$$= \frac{V_{max} I_{max}}{2} \times 2 \sin^2 \omega t = \frac{V_{max} I_{max}}{2} (1 - \cos 2 \omega t)$$

$$= \frac{V_{max}}{\sqrt{2}} \cdot \frac{I_{max}}{\sqrt{2}} - \frac{V_{max}}{\sqrt{2}} \cdot \frac{I_{max}}{\sqrt{2}} \cos 2\omega t$$

(Constant part) (Fluctuating part)

For a complete cycle the average of $\frac{V_{max}}{\sqrt{2}} \cdot \frac{I_{max}}{\sqrt{2}} \cos 2 \omega t$ *is zero.*

Hence, power for the whole cycle,

$$P = \frac{V_{max}}{\sqrt{2}} \cdot \frac{I_{max}}{\sqrt{2}} = V_{r.m.s} \cdot I_{r.m.s.}$$

or $$P = VI \text{ watt}$$

where V = R.M.S. value of applied voltage, and

I = R.M.S. value of the current.

It may be observed from the Fig. 23 (*c*) that no part of the power cycle at any time becomes negative. In other words the power in a purely resistive circuit *never becomes zero*.

Hence in ***pure resistive circuit*** we have :

1. Current is in phase with the voltage.
2. Current $I = \frac{V}{R}$ where I and V are r.m.s. values of current and voltage.
3. Power in the circuit, $P = VI = I^2R$.

4.2. A.C. Through Pure Inductance Alone

Fig. 24 (*a*) shows the circuit containing a pure inductance of L henry.

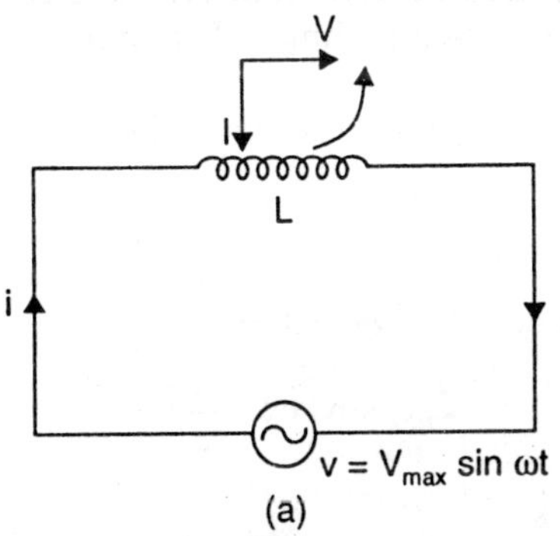

Fig. 24 (*a*)

Let the alternating voltage applied across the circuit be given by the equation,

$$v = V_{max} \sin \omega t \quad ...(i)$$

Whenever an alternating voltage is applied to a purely inductive coil, a back e.m.f. is produced due to the self-inductance of the coil. This back e.m.f. opposes the rise or fall of the current through the coil. Since there is no ohmic drop in this case, therefore, the applied voltage has to overcome this induced e.m.f. only. Thus at every step,

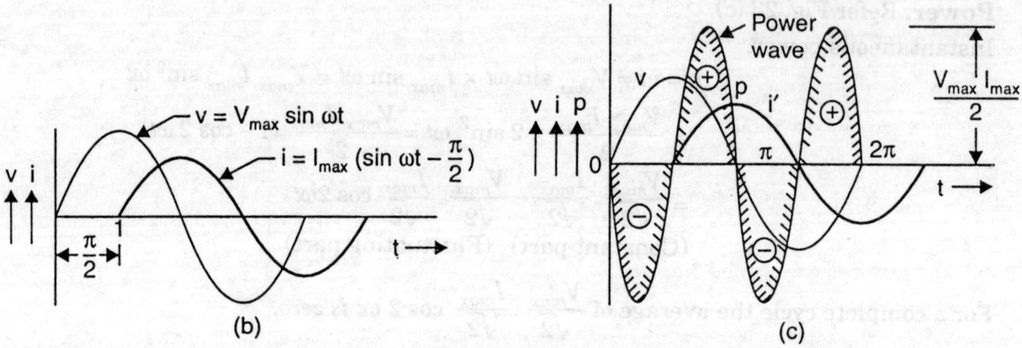

Fig. 24 (*b*), (*c*). A.C. through pure inductance alone. Resultant power zero.

$$v = L\frac{di}{dt}$$

or $$V_{max} \sin \omega t = L\frac{di}{dt}$$

or $$di = \frac{V_{max}}{L} \sin \omega t \, dt$$

Integrating both sides, we get

$$\int di = \int \frac{V_{max}}{L} \sin \omega t \, dt$$

or $$i = \frac{V_{max}}{L}\left(-\frac{\cos \omega t}{\omega}\right) = \frac{V_{max}}{\omega L} \sin\left[\omega t - \frac{\pi}{2}\right]$$

or $$i = \frac{V_{max}}{X_L} \sin\left[\omega t - \frac{\pi}{2}\right] \qquad ...(ii)$$

where $X_L = \omega L$ (opposition offered to the flow of alternating current by a pure inductances) and is called **Inductive reactance.** It is given in ohms if L is in henry and ω is in radian/second.

The value of current will be maximum when $\sin\left(\omega t - \frac{\pi}{2}\right) = 1$

$$\therefore \qquad I_{max} = \frac{V_{max}}{X_L}$$

Substituting this value in eqn. (*ii*), we get

$$i = I_{max} \sin\left(\omega t - \frac{\pi}{2}\right) \qquad ...(iii)$$

Power. Refer Fig. 24 (*c*)

Instantaneous power, $$p = vi = V_{max} \sin \omega t \times I_{max} \sin\left(\omega t - \frac{\pi}{2}\right)$$

$$= -V_{max} I_{max} \sin \omega t \, . \cos \omega t$$

$$= -\frac{V_{max} I_{max}}{2} \times 2 \sin \omega t \cos \omega t$$

$$= -\frac{V_{max}}{\sqrt{2}} \cdot \frac{I_{max}}{\sqrt{2}} \, . \sin 2\omega t$$

$\therefore$ Power for the whole cycle, $P = -\frac{V_{max}}{\sqrt{2}} \frac{I_{max}}{\sqrt{2}} \int_0^{2\pi} \sin 2\omega t = 0$

Hence *average power consumed in a pure inductive circuit is* **zero.**

Hence in a ***pure inductive circuit,*** we have :

1. Current $I = \frac{V}{X_L} = \frac{V}{\omega L} = \frac{V}{2\pi f L}$ amp.
2. Current always *lags* behind the voltage by *90°*.
3. Average power consumed is *zero.*

Variation of X_L and f :

Since $X_L = \omega L = 2\pi f L$, and here if L is constant, then

$$X_L \propto f$$

Fig. 25, shows the variation. As frequency is increased X_L increases and the current taken by the circuit decreases.

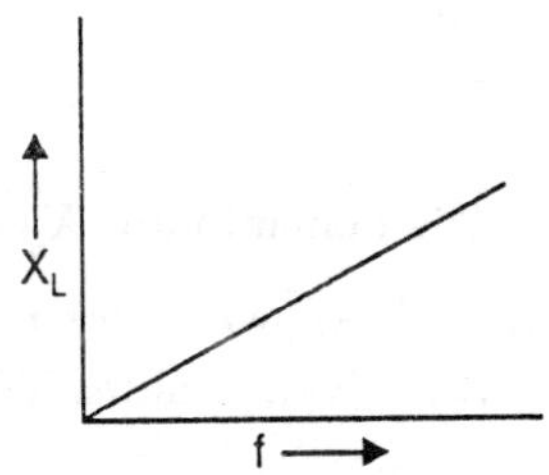

Fig. 25. Variation of X_L with f.

4.3. A.C. Through Pure Capacitance Alone

The circuit containing a pure capacitor of capacitance C farad is shown in Fig. 26 (*a*). Let the alternating voltage applied across the circuit be given by the equation,

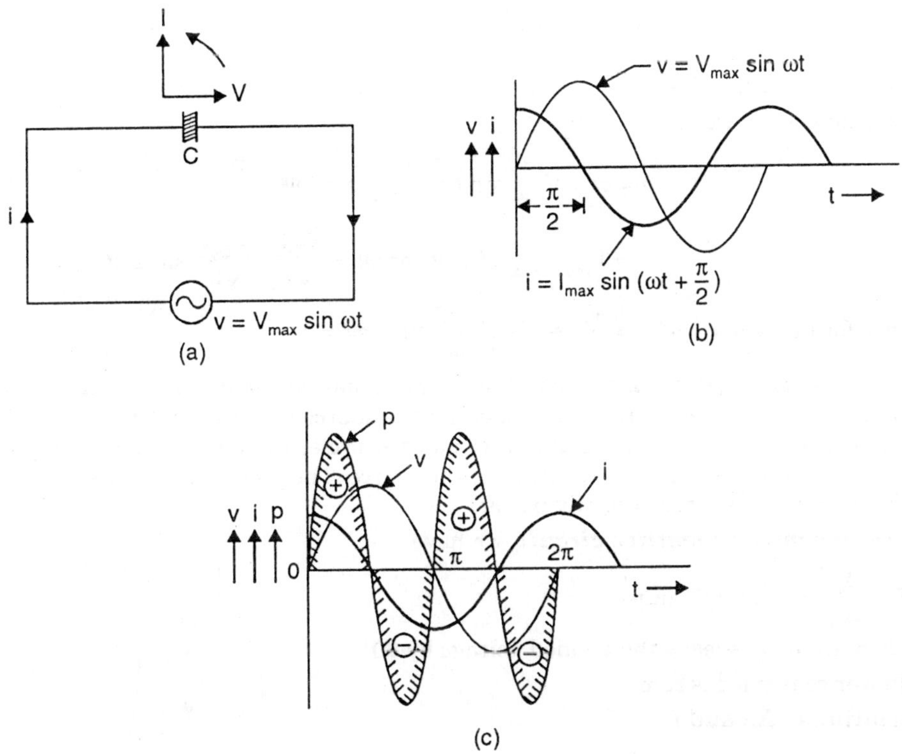

Fig. 26. A.C. through pure capacitance alone. Resultant power is zero.

$$v = V_{max} \sin \omega t \qquad ...(i)$$

Charge on the capacitor at any instant,

$$q = C\,v$$

Current through the circuit,

$$i = \frac{dq}{dt} = \frac{d}{dt}(CV_{max} \sin \omega t) = \omega CV_{max} \cos \omega t$$

or

$$i = \frac{V_{max}}{1/\omega C} \sin\left(\omega t + \frac{\pi}{2}\right)$$

$$\therefore \quad i = \frac{V_{max}}{X_C} \sin\left(\omega t + \frac{\pi}{2}\right) \quad \text{...(ii)}$$

The denominator $X_C = \frac{1}{\omega C}$ (opposition offered to the flow of alternating current by a pure capacitor) is known as *capacitive reactance.*

It is given in ohms if C is in farad and ω in radian/second.

The value of current will be maximum when $\sin\left(\omega t + \frac{\pi}{2}\right) = 1$

$$\therefore \quad I_{max} = \frac{V_{max}}{X_C}$$

Substituting this value in eqn. (*ii*), we get

$$i = I_{max} \sin\left(\omega t + \frac{\pi}{2}\right) \quad \text{...(iii)}$$

Power. Refer Fig. 26 (*c*)

Instantaneous power,

$$p = vi = V_{max} \sin \omega t \times I_{max} \sin\left(\omega t + \frac{\pi}{2}\right)$$

$$= V_{max}\, I_{max} \sin \omega t \cos \omega t = \frac{V_{max}}{\sqrt{2}} \cdot \frac{I_{max}}{\sqrt{2}} \sin 2\omega t$$

$$\text{Power for the whole cycle} = \frac{V_{max}}{\sqrt{2}} \cdot \frac{I_{max}}{\sqrt{2}} \int_0^{2\pi} \sin 2\,\omega t = 0$$

This fact is graphically illustrated in Fig. 26 (*c*). It may be noted that, during the first quarter cycle, what so ever power or energy is supplied by the source is stored in the electric field set-up between the capacitor plates. During the next quarter cycle, the electric field collapses and the power or energy stored in the field is returned to the source. The process is repeated in each alternation and this circuit does not absorb any power.

Hence in a ***pure capacitive circuit,*** we have

1. $I = \frac{V}{X_C} = V \times 2\,\pi f C$ amps.
2. Current always leads the applied voltage by 90°.
3. Power consumed is *zero*.

Variation of X_C and f :

Since $X_C = \frac{1}{2\pi f C}$ and if C is kept constant, than

$$X_C \propto \frac{1}{f}$$

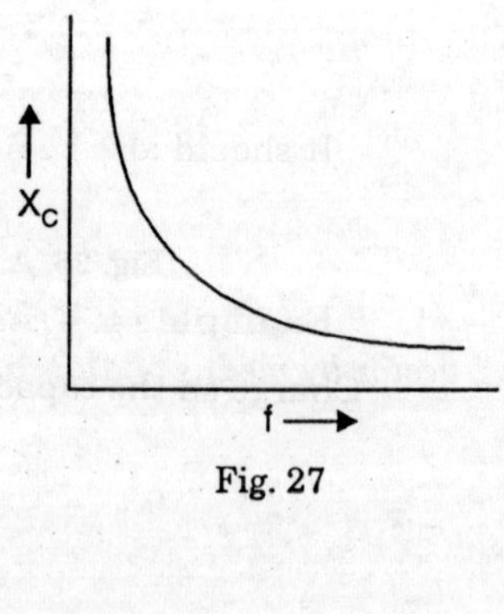

Fig. 27

Fig. 27. shows the variation. As the frequency increases X_C decreases, so the current increases.

4.4. Phasor Algebra

The following are the methods of representing vector quantities :

1. Symbolic notation 2. Trigonometrical form
3. Exponential form 4. Polar form.

A vector as shown in Fig. 28 may be described in the above forms as follows :

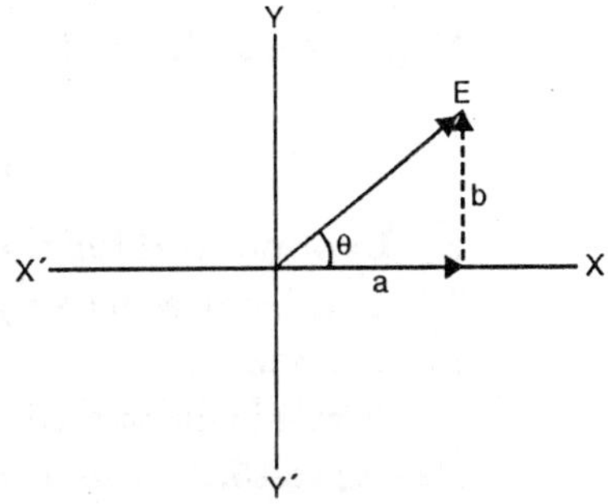

Fig. 28

1. *Symbolic notation* :

$$E = a + jb$$

2. *Trigonometrical form* :

$$E = \sqrt{a^2 + b^2}\ (\cos\theta + j\sin\theta)$$

$$= \sqrt{a^2 + b^2}\ (\cos\theta \pm j\sin\theta) \quad \text{......in general}$$

3. *Exponential form* : $E = \sqrt{a^2 \times b^2}\ e^{+j\theta}$

$$= \sqrt{a^2 + b^2}\ e^{\pm j\theta} \quad \text{......in general}$$

4. *Polar form* : $E = \sqrt{a^2 + b^2}\ \angle\theta$

$$= \sqrt{a^2 + b^2}\ \angle \pm\theta \quad \text{......in general}$$

Significance of operator j. The letter j used in the above expressions is a symbol of an operation. It is used to indicate the counter-clockwise rotation of a vector through 90°. It is assigned a value of $\sqrt{-1}$. The double operation of j on a vector rotates it counter-clockwise (*CCW*) through 180° and hence reverses its sense because, $j \times j = j^2 = \sqrt{(-1)^2} = -1$.

In general, each successive multiplication of j, rotates the phasor further by 90° as given below (Refer Fig. 29)

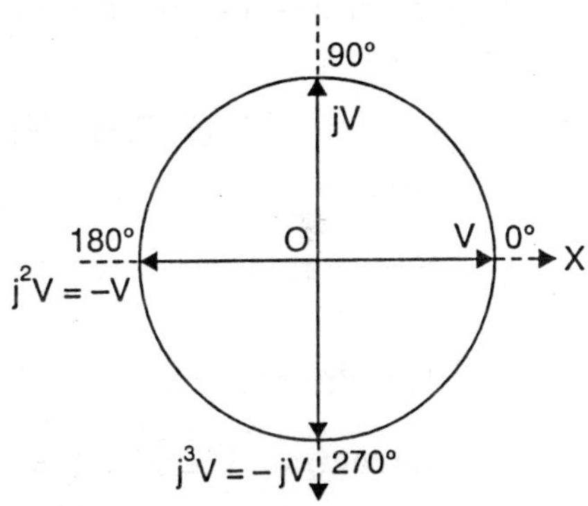

Fig. 29

$$j = \sqrt{-1}$$

...90° CCW rotation from *OX*-axis

$$j^2 = \left(\sqrt{-1}\right)^2 = -1$$

...180° CCW rotation from *OX*-axis

$$j^3 = \left(\sqrt{-1}\right)^3 = -\sqrt{-1} = -j$$

...270° CCW rotation from *OX*-axis

$$j^4 = \left(\sqrt{-1}\right)^4 = +1$$

...360° CCW rotation from *X*-axis

It should also be noted that,

$$\frac{1}{j} = \frac{j}{j^2} = \frac{j}{-1} = -j.$$

Example 14. *Write the equivalent potential and polar forms of vector 6 + j8. Also illustrate the vector by means of diagram.*

Solution. Refer Fig. 30.

Magnitude of the vector $= \sqrt{6^2 + 8^2} = 10$, $\tan \theta = \frac{8}{6}$

$\therefore$ $\theta = \tan^{-1}\left(\frac{8}{6}\right) = 53.1°$

$\therefore$ **Exponential form = 10 $e^{j\,53.1°}$. (Ans.)**

The angle may also be *expressed in radians.*

Polar form = 10 ∠ 53.1°. (Ans.)

The vector is illustrated by means of diagram in Fig. 30.

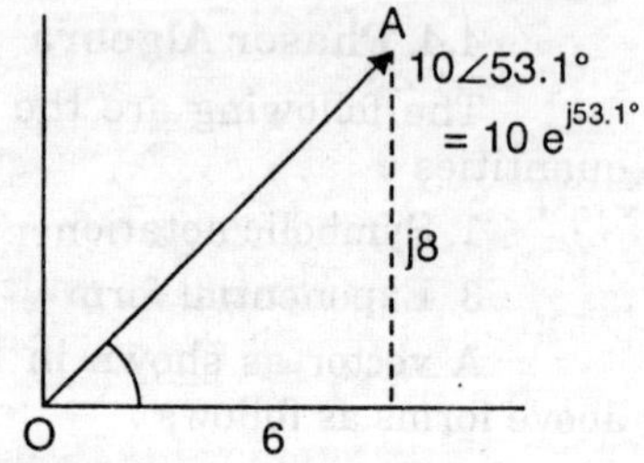

Fig. 30

Example 15. *A vector is represented by* $30e^{-j2\pi/3}$. *Write down the various equivalent forms of the vector and illustrate by means of a vector diagram, the magnitude and position of the above vector.*

Solution. Refer Fig. 31. Draw the vector in a direction making an angle of $\frac{2\pi}{3} = \frac{2 \times 180}{3} = 120°$ in the *clockwise* direction (since the angle is negative).

(*i*) **Rectangular form :**

$$a = 30 \cos(-120°) = -15$$
$$b = 30 \sin(-120°) = -25.98$$

$\therefore$ Expression is **= (– 15 – j 25.98). (Ans.)**

(*ii*) **Polar form = 30 ∠ – 120°. (Ans.)**

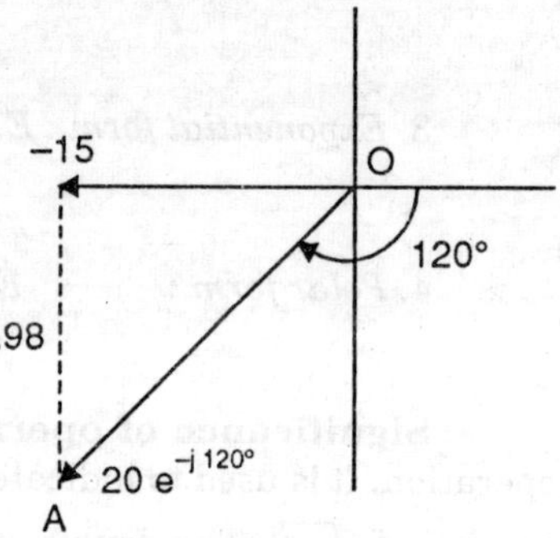

Fig. 31

Addition and subtraction of vector quantities :

For addition and subtraction of vector quantities rectangular form is best suited. Consider two voltage phasors represented as :

$$\overline{V_1} = a_1 + jb_1 \text{ and } \overline{V_2} = a_2 + jb_2$$

Addition. $\overline{V} = \overline{V_1} + \overline{V_2} = (a_1 + jb_1) + (a_2 + jb_2) = (a_1 + a_2) + j(b_1 + b_2)$

The magnitude of the resultant vector $\overline{V} = \sqrt{(a_1 + a_2)^2 + (b_1 + b_2)^2}$

The position of $\overline{V}$ with respect to X-axis is $\theta = \tan^{-1}\left(\frac{b_1 + b_2}{a_1 + a_2}\right)$

Subtraction. $\overline{V} = \overline{V_1} - \overline{V_2} = (a_1 + jb_1) - (a_2 + jb_2) = (a_1 - a_2) + j(b_1 - b_2)$

The magnitude of the resultant vector $\overline{V} = \sqrt{(a_1 - a_2)^2 + (b_1 - b_2)^2}$

The position of $\overline{V}$ with respect to X-axis is $\theta = \tan^{-1}\left(\frac{b_1 - b_2}{a_1 - a_2}\right)$

Multiplication and division of vector quantities :

It the vectors are represented in the polar exponential form, their multiplication and division becomes very easy and simple.

Consider two voltage phasors represented as

$$\overline{V_1} = a_1 + jb_1 = V_1\angle\theta_1, \text{ where } \theta_1 = \tan^{-1}\left(\frac{b_1}{a_1}\right)$$

$$\overline{V_2} = a_2 + jb_2 = V_2\angle\theta_2, \text{ where } \theta_2 = \tan^{-1}\left(\frac{b_2}{a_2}\right)$$

Multiplication. When the phasor quantities are represented in polar form, while multiplying their magnitudes are multiplied and their *angles added algebraically*,

i.e, $$\overline{V_1} \times \overline{V_2} = V_1\angle\theta_1 \times V_2\angle\theta_2 = V_1V_2\angle(\theta_1 + \theta_2)$$

Division. In this case, the magnitudes of phasor quantities (expressed in polar from) are divided and their *angles subtracted algebraically*

i.e. $$\frac{\overline{V_1}}{\overline{V_2}} = \frac{V_1\angle\theta_1}{V_2\angle\theta_2} = \frac{V_1}{V_2}\angle(\theta_1 - \theta_2).$$

Example 16. *Perform the following operation and express the final result in polar form :*

$$10\angle 30° + 16\angle -30°.$$

Solution.
$$10\angle 30° = 10(\cos 30° + j\sin 30°) = 8.66 + j5$$
$$16\angle -30° = 16[\cos(-30°) + j\sin(-30°)] = 13.86 - j8$$
$$\therefore \quad 10\angle 30° + 16\angle -30° = (8.66 + j5) + (13.86 - j8) = 22.52 - j3$$
$$= (\sqrt{22.52^2 + 3^2})\tan^{-1}(-3/22.52)$$
$$= 22.72\tan^{-1}(-3/22.52) = \mathbf{22.72\angle -7.6°}. \quad \textbf{(Ans.)}$$

Example 17. *Subtract the following given vectors from one another.*

$$\overline{A} = 15 + j26 \text{ and } \overline{B} = -19.75 - j7.18.$$

Solution. $\overline{A} - \overline{B} = \overline{C} = (15 + j26) - (-19.75 - j7.18) = 34.75 + j33.18$

$$\therefore \quad \text{Magnitude of } \overline{C} = \sqrt{34.75^2 + 33.18^2} = \mathbf{48}$$
$$\text{Slope of } \overline{C} = \tan^{-1}(33.18/34.75) = 43.68°$$
$$\therefore \quad \overline{C} = \mathbf{48\angle 43.68°}. \quad \textbf{(Ans.)}$$

Example 18. *Perform the operation* $\dfrac{\overline{A}\,\overline{B}}{\overline{C}}$ *and express the final result in polar form for the vectors given below :*

$$\overline{A} = 10 + j10;\ \overline{B} = 15\angle -120°;\ \overline{C} = 5 + j0°.$$

Solution. Rearranging vectors $\overline{A}$ and in polar form, we have

$$\overline{A} = 10 + j10 = \sqrt{10^2 + 10^2}\ \tan^{-1}(10/10) = 14.14\angle 45°$$
$$\overline{C} = 5 + j0 = \sqrt{5^2 + 0^2}\ \tan^{-1}(0/5) = 5\angle 0°$$
$$\therefore \quad \frac{\overline{A}\,\overline{B}}{\overline{C}} = \frac{14.14\angle 45° \times 15\angle -120°}{5\angle 0°} = \frac{14.14 \times 15}{5}\angle(45° - 120° - 0°)$$
$$= \mathbf{42.42\angle -75°}. \quad \textbf{(Ans.)}$$

Example 19. *The instantaneous values of two currents i_1 and i_2 are given as :*

$$i_1 = 5\sin\left(\omega t + \frac{\pi}{4}\right) \text{ and } i_2 = 2.5\cos\left(\omega t - \frac{\pi}{2}\right)$$

Find the r.m.s. value of $i_1 + i_2$ using complex number representation.

(Kerala University)

Solution. *Given :* $i_1 = 5 \sin\left(\omega t + \frac{\pi}{4}\right)$,

$$i_2 = 2.5 \cos\left(\omega t - \frac{\pi}{2}\right) = 2.5 \sin\left[90° + \left(\omega t - \frac{\pi}{2}\right)\right] = 2.5 \sin \omega t$$

$\therefore$ $I_{1(max)} = 5 (\cos 45° + j \sin 45°) = (3.53 + j3.53)$

$I_{2(max)} = 2.5 (\cos 0° + j \sin 0°) = (2.5 + j0)$

The maximum value of resultant current is

$$I_{max} = (3.53 + j3.53) + (2.5 + j0) = 6.03 + j3.53 = 6.987 \angle 30.34°$$

$\therefore$ R.M.S. value $= \frac{6.987}{\sqrt{2}} =$ **4.94 A. (Ans.)**

Conjugate complex numbers

Two numbers are said to be conjugate if they *differ only in the algebraic sign of their quadrature components.* Accordingly, the numbers $(a + jb)$ and $(a - jb)$ are conjugate.

- The *sum* of two conjugate numbers gives in-phase or *active* component and their *difference* gives quadrature or *reactive* component.

i.e. $(a + jb) + (a - jb) = 2a$ (*i.e.,* active component), and

$(a + jb) - (a - jb) = j2b$ (*i.e.,* reactive component).

The resultant is the sum of two vertical components only.

- The resultant arising out of the *multiplication* of two conjugate numbers contains *no quadrature component.*

i.e., $(a + jb) \times (a - jb) = a^2 - j^2b^2 = a^2 + b^2$

The conjugate of a complex number is used to determine the *apparent power* of an A.C. circuit in complex form.

Power and roots of vectors/phasors

The *powers* and *roots* of vectors can be found conveniently in *polar form.* If the vector are not, in polar form, these should be converted into polar form before carrying out the algebraic operations, as mentioned below.

Powers. Consider a vector phasor quantity represented in polar form as $\bar{A} = A\angle\theta$,

Then $(\bar{A})^n = A^n \angle(n \times \theta)$

Example. Suppose it is required to find cube of the vector $4\angle 12°$

Then, $(4\angle 12°)^3 = (4)^3 \angle (3 \times 12°) = 64 \angle 36°$

Roots. Consider a vector (phasor) quantity represented in polar form as $(\bar{A})^{1/n} = (A)^{1/n} \angle \theta/n$.

Example. Suppose it is required to find cube root of $125 \angle 60°$

Then, $(125 \angle 60°)^{1/3} = (125)^{1/3} \angle 60°/3 = 5\angle 20°$.

The 120° operator

In case of 3-phase work, where voltage vectors are displaced by 120° from one another (Fig. 32) it is convenient to use an operator, which rotates a vector/phasor through 120° toward or backwards without altering its length. This operator is '*a*' and any operator which is multiplied by '*a*' remains unaltered in magnitude but is rotated in CCW (counter-clockwise) direction by 120°.

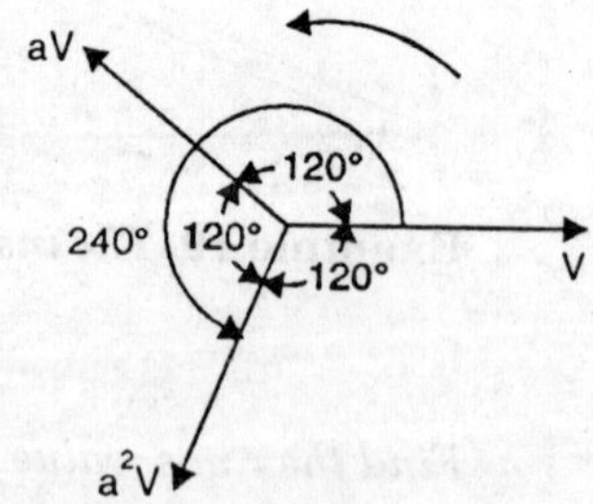

Fig. 32

$\therefore$ $a = 1 \angle 120°$

In cartesian form, $a = \cos 120° + j \sin 120°$

$= -0.5 + j\, 0.866$

Similarly, $a^2 = 1\angle 120° \times 1\angle 120° = 1\angle 240° = \cos 240° + j\sin 240° = -0.5 - j\,0.866$

Hence the operator 'a' will rotate the in CCW by 240° which is the same as rotating the vector in CW(clock-wise) direction by 120°.

$\therefore \qquad a^2 = 1\angle -120°$

Similarly $\qquad a^3 = 1\angle 360° = 1$ (Numerically, a is equivalent to the cube root of unity.)

4.5. A.C. Series Circuits

Under this heading we shall discuss R-L, R-C and R-L-C series circuits.

4.5.1. R-L circuit (Resistance and inductance in series)

Fig. 33 (a) shows a pure resistance R and a pure inductive coil of inductance L connected in series. Such a circuit is known as R-L circuit (usually met a cross in practice).

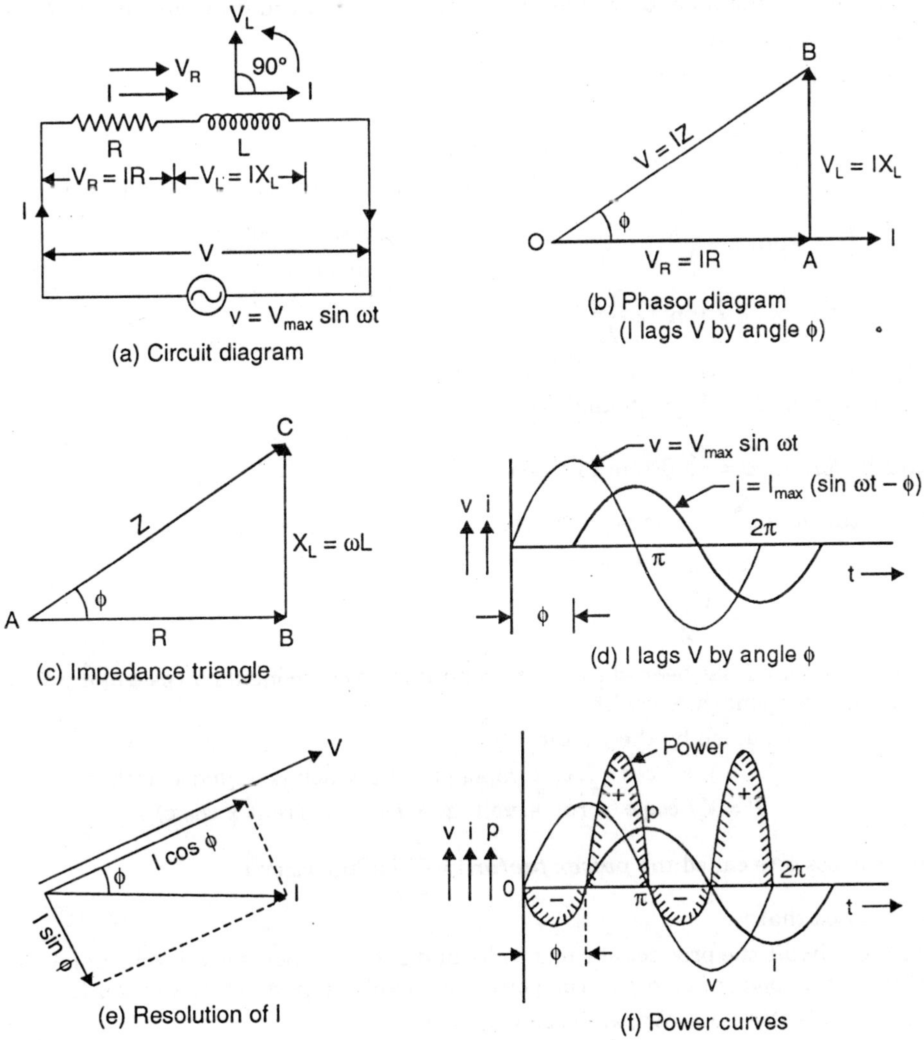

Fig. 33. R-L circuit (Resistance and inductance in series).

Let V = R.M.S. value of the applied voltage,

I = R.M.S. value of the resultant current,

$V_R = IR$ = Voltage drop across R(*in phase* with I), and

$V_L = IX_L$ = Voltage drop across L (coil), *ahead* of I by 90°.

The voltage drop V_R and V_L and shown in voltage triangle OAB in Fig. 33 (*b*), I being taken as the reference vector in the phasor diagram. Vector OA represents ohmic drop V_R and AB represents inductive drop V_L. Vector OB represents the applied voltage V which is the vector sum of the two (*i.e.*, V_R and V_L).

$$\therefore \quad V = \sqrt{V_R^2 + V_L^2} = \sqrt{(IR)^2 + (IX_L)^2} = I\sqrt{(R^2 + X_L^2)}$$

or
$$I = \frac{V}{\sqrt{R^2 + (X_L)^2}} = \frac{V}{Z}$$

where $Z = \sqrt{R^2 + X_L^2}$ (total opposition offered to the flow of alternating current by R-L series circuit) is known as **impedance** of the circuit.

As seen from the "*impedance triangle*" ABC [Fig. 33 (*c*)],

$$Z^2 = R^2 + X_L^2$$

i.e., $(\text{Impedance})^2 = (\text{Resistance})^2 + (\text{Inductive reactance})^2$

From Fig. 33 (*b*) it is evident that voltage V *leads* the current by an angle ϕ such that,

$$\tan\phi = \frac{V_L}{V_R} = \frac{IX_L}{IR} = \frac{X_L}{R} = \frac{\omega L}{R} = \frac{\text{Inductive reactance}}{\text{Resistance}}$$

$$\therefore \quad \phi = \tan^{-1}\left(\frac{X_L}{R}\right)$$

The same is illustrated graphically in Fig. 33 (*d*).

In other words I *lags* V by an angle ϕ.

Power factor, $\cos\phi = \frac{R}{Z}$ [From Fig. 33 (*c*)]

Thus, if the applied voltage is given by $v = V_{max} \sin \omega t$, then current equation is given as,

$$i = I_{max} \sin(\omega t - \phi),$$

where
$$I_{max} = \frac{V_{max}}{Z}$$

In the Fig. 33 (*e*), I has been shown resolved into two components, $I\cos\phi$ along V and $I\sin\phi$ in quadrature (*i.e.*, perpendicular) with V.

Mean power consumed by the circuit

$= V \times I\cos\phi$ (*i.e.*, component of I which is in *phase with V*

i.e., $P = VI\cos\phi$ (= r.m.s. voltage × r.m.s. current × cos ϕ)

The term 'cos ϕ' is called the ***power factor*** $\left(= \frac{R}{Z}\right)$ of the circuit

It may noted that :

— In A.C. circuit the product of r.m.s. volts and r.m.s. amperes gives volt-amperes (*i.e.*, *VA*) and *not true power in watts*. True power (*W*) = volt-amperes (*VA*) × power factor

or Watts = *VA* (Apparent power) × cos ϕ

— The *power consumed is due to ohmic resistance only* since pure inductance consumes no power.

i.e. $$P = VI \cos \phi = VI \times \frac{R}{Z} = \frac{V}{Z} IR = I \times IR = I^2R, \text{ watts}$$

$$\left(\because \quad \cos \phi = R/Z \text{ and } \frac{V}{Z} = I\right)$$

This shows that power is actually consumed in *resistance only*; the *inductor does not consume any power.*

The power consumed in R-L circuit is shown graphically in Fig. 33 (*f*).

Thus in ***R-L circuit*** we have :

1. Impedance, $Z = \sqrt{R^2 + X_L^2}$ (where $X_L = \omega L = 2\pi \times f L$)

2. Current, $I = \frac{V}{Z}$

3. Power factor, $\cos \phi = \frac{R}{Z} \left(= \frac{\text{True power}}{\text{Apparent power}} = \frac{\text{W}}{\text{VA}} \right)$

[or angle of lag, $\phi = \cos^{-1} (R/Z)$]

4. Power consumed, $P = VI \cos \phi \left(= IZ \times I \times \frac{R}{Z} = I^2 R \right)$

Symbolic Notation :

$$Z = R + jX_L$$

The numerical value of impedance vector $= \sqrt{R^2 + X_L^2}$

The phase angle with the reference axis, $\phi = \tan^{-1} (X_L/R)$.

In *polar form :* $\bar{Z} = Z\angle\phi°$.

Apparent, Active (True or real) and Reactive Power :

Every circuit current has two components : (*i*) Active component and (*ii*) Reactive component.

"Active component" consumes power in the circuit while *"reactive component"* is responsible for the field which lags or leads the main current from the voltage.

In Fig. 34. active component is $I_{\text{active}} = I \cos \phi$, and reactive component is $I_{\text{reactive}} = I \sin \phi$

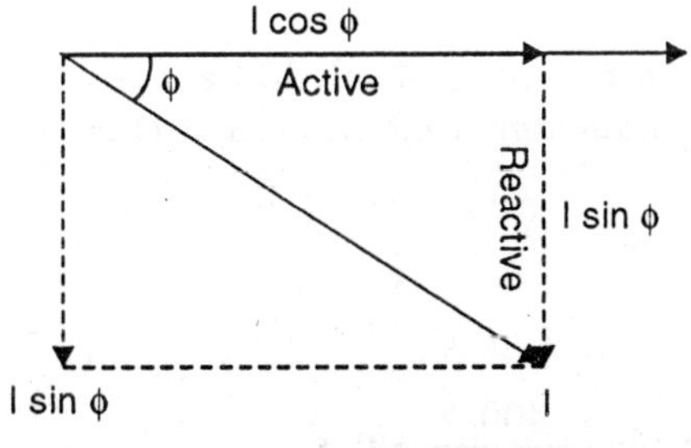

Fig. 34. Active and reactive components of circuit current I.

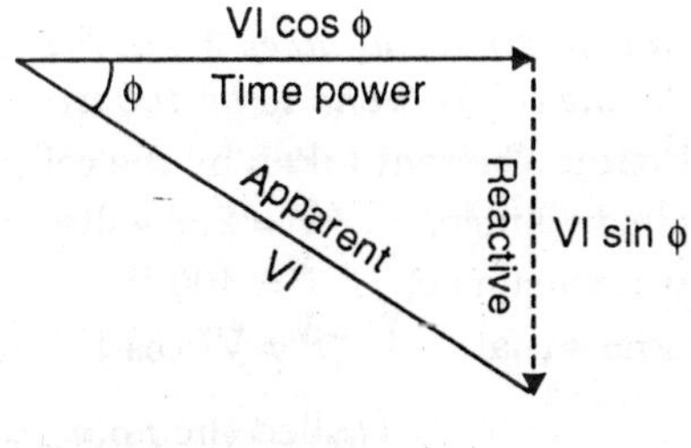

Fig. 35. Apparent, true and reactive power.

So, $$I = \sqrt{(I_{\text{active}})^2 + (I_{\text{reactive}})^2}$$

Refer Fig. 35.

(*i*) **Apparent power (S).** It *is given by the product of r.m.s. values of applied voltage and circuit current.*

$$\therefore \quad S = VI = (I \times Z) \cdot I = I^2Z \text{ volt-amperes } (VA)$$

(*ii*) **Active or true or real power (P or W).** It *is the power which is actually dissipated in the circuit resistance.*

$$P = I^2R = VI \cos \phi \text{ watts}$$

(*iii*) **Reactive power (Q).** A pure inductor and a pure capacitor do not consume any power, since in a half cycle what so ever power is received from the source by these components the same is returned to the source. This power which flows back and forth (*i.e.,* in both directions in the circuit) or *reacts upon itself* is called *"reactive power."*

It may be noted that the *current in phase with the voltage produces active or true or real power while the current 90° out of phase with the voltage contributes to* reactive power.

In a R-L circuit, reactive power which is the power developed in the inductive reactance of the circuit, is given as :

$$Q = I^2 X_L = I^2Z \sin \phi = I \cdot (IZ) \sin \phi$$
$$= VI \sin \phi \text{ volt-amperes-reactive } (VAR)$$

These three powers are shown in Fig. 35

Relation between *VA*, *W* and *VR*

$$W = VA \cos \phi \quad \text{...(i)}$$
$$VAR = VA \sin \phi \quad \text{...(ii)}$$

$$\therefore \quad VA = \frac{W}{\cos \phi} \quad \text{...[From (i)]}$$

$$\text{and,} \quad VA = \frac{VAR}{\sin \phi} \quad \text{...[From (ii)]}$$

$$\text{Power factor } (p.f.) = \frac{W}{VA} = \frac{\text{True power}}{\text{Apparent power}}$$

The larger bigger units of apparent, true and reactive power are *kVA* (or *MVA*), *kW*(or *MW*) and *kVAR* (or *MVAR*) respectively.

The power factor depends on the reactive power component. If it is made equal to the active power component, the power factor becomes unity.

Example 20. *A coil takes 2.5 amps. when connected across 200 volt 50 Hz mains. The power consumed by the coil is found to be 400 watts. Find the inductance and the power factor of the coil.*

Solution. Current taken by the coil, $I = 2.5$ A

Applied voltage, $V = 200$ volts

Power consumed, $P = 400$ W

We know that $P = VI \cos \phi$

$$\text{or} \quad 400 = 200 \times 2.5 \times \cos \phi \text{ or } \cos \phi = \frac{400}{200 \times 2.5} = 0.8$$

Hence power factor of coil is **0.8. (Ans.)**

$$\text{Impedance of the coil,} \quad Z = \frac{V}{I} = \frac{200}{2.5} = 80\ \Omega$$

$$\text{Also} \quad \frac{X_L}{Z} = \sin \phi$$

$$\therefore \quad X_L = Z \sin \phi$$

$$= 80 \sin \phi = 80 \sqrt{1 - \cos^2 \phi}$$

$$= 80 \sqrt{1 - 0.8^2} = 80 \times 0.6 = 48 \, \Omega$$

But $\quad X_L = 2\pi f L$

$\therefore \quad L = \frac{X_L}{2\pi f} = \frac{48}{2\pi \times 50} = $ **0.1529 H (henry). (Ans.)**

Example 21. *A 100 V, 80 W lamp is to be operated on 230 volts, 50 Hz A.C. supply. Calculate the inductance of the choke required to be connected in series with lamp for its operation. The lamp can be taken as equivalent to a non inductive resistance.*

Solution. Current through the lamp when connected across 100 V supply,

$$I = \frac{W}{V} = \frac{80}{100} = 0.8 \text{ A}$$

Resistance of the lamp, $R = \frac{V}{I} = \frac{100}{0.8} = 125 \, \Omega$

If a choke of inductance L henry is connected in series with the lamp to operate it on 230 V, the current through the choke will also be 0.8 A.

The impedance of the circuit when choke is connected in series with the lamp,

$$Z = \frac{V}{I} = \frac{230}{0.8} = 287.5 \, \Omega$$

Reactance of choke coil, $X_L = \sqrt{Z^2 - R^2} = \sqrt{287.5^2 - 125^2} = 258.5 \, \Omega$

But $\quad X_L = 2\pi f L$

or $\quad L = \frac{X_L}{2\pi f} = \frac{258.5}{2\pi \times 50} = 0.825 \text{ H}$

Hence inductance of choke coil, **L = 0.825 H. (Ans.)**

Example 22. *A coil has a resistance of 5 Ω and an inductance of 31.8 mH. Calculate the current taken by the coil and power factor when connected to 200 V, 50 Hz supply.*

Draw the vector diagram.

If a non-inductive resistance of 10 Ω is then connected in series with coil, calculate the new value of current and its power factor.

Solution. $\quad R = 5 \, \Omega$

$L = 31.8 \text{ mH or } 0.0318 \text{ H}$

$\therefore \quad X_L = 2\pi f L$

$= 2\pi \times 50 \times 0.0318 = 10 \, \Omega$

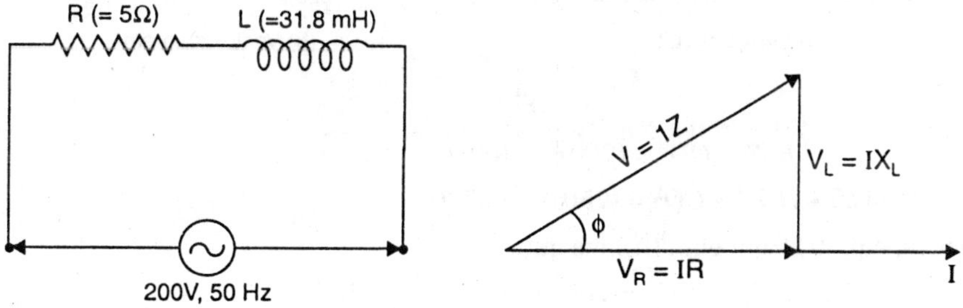

(a) R-L circuit
(b) Vector/phasor diagram

Fig. 36

Impedance of the coil, $Z = \sqrt{R^2 + X_L{}^2}$

$= \sqrt{5^2 + 10^2} = 11.18\ \Omega$

$\therefore$ Current taken by the coil, $I = \frac{V}{Z} = \frac{200}{11.18} =$ **17.9 A. (Ans.)**

Power factor, $\cos\phi = \frac{R}{Z} = \frac{5}{11.18} =$ **0.4475. (Ans.)**

Fig. 36 (*b*) shows the vector diagram.

When non-inductive resistance of 10 Ω is connected in series with the coil :

Total resistance in the circuit, $R' = 5 + 10 = 15\ \Omega$

Reactance in the circuit, $X_L' = X_L = 10\ \Omega$

Impedance of the circuit, $Z' = \sqrt{R'^2 + X'_L{}^2} = \sqrt{15^2 + 10^2} = 18\ \Omega$

Current through the circuit, $I' = \frac{V}{Z'} = \frac{200}{18} =$ **11.11 A. (Ans.)**

Power factor of the circuit, $\cos\phi = \frac{R'}{Z'} = \frac{15}{18} =$ **0.833. (Ans.)**

Example 23. *A current of 5A flows through a non-inductive resistance in series with a choking coil when supplied at 250 V, 50 Hz. If the voltage across the resistance is 125 V and across the coil 200 V, calculate :*

(*i*) *Impedance, reactance and resistance of the coil,*

(*ii*) *The power absorbed by the coil,*

(*iii*) *The total power.*

Draw the vector diagram. **(Elect. Engg. Madras Univ.)**

Solution. Non-inductive resistance connected in series with coil $= \frac{125}{5} = 25\ \Omega$

Refer Fig. 37 (*b*).

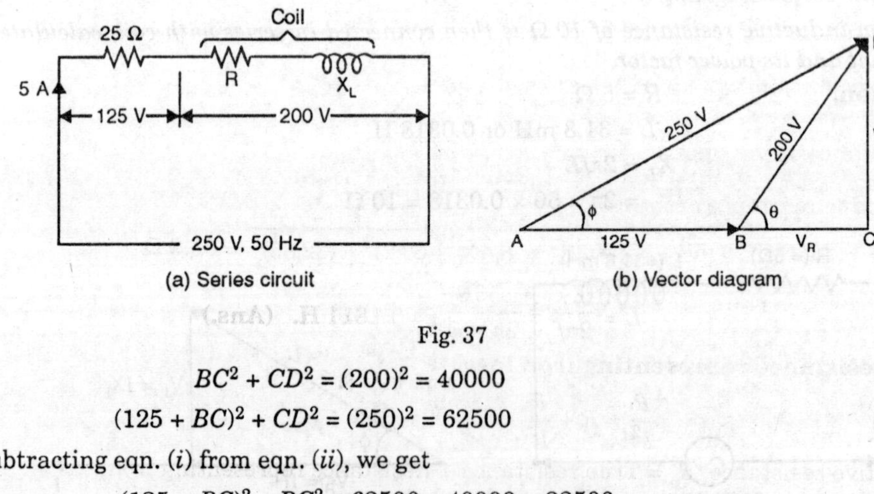

(a) Series circuit (b) Vector diagram

Fig. 37

$$BC^2 + CD^2 = (200)^2 = 40000 \quad ...(i)$$

$$(125 + BC)^2 + CD^2 = (250)^2 = 62500 \quad ...(ii)$$

Subtracting eqn. (*i*) from eqn. (*ii*), we get

$$(125 + BC)^2 - BC^2 = 62500 - 40000 = 22500$$

$$15625 + BC^2 + 250BC - BC^2 = 22500$$

$$\therefore \quad BC = 27.5V\ ;\ CD = \sqrt{(200)^2 - (27.5)^2} = 198.1\ \text{V}$$

(*i*) **Coil impedance,** $Z = \frac{200}{5} = 40\ \Omega$. **(Ans.)**

$$V_R = IR = BC = 27.5 \quad \text{or} \quad 5R = 27.5$$

$$\mathbf{R} = \frac{27.5}{5} = \mathbf{5.5\ \Omega} \quad \textbf{(Ans.)}$$

Also, $V_L = IX_L = CD = 198.1$

$$\therefore \quad \mathbf{X_L} = \frac{198.1}{5} = \mathbf{39.62\ \Omega}. \quad \textbf{(Ans.)}$$

(*ii*) **Power absorbed by the coil,**

$$P = I^2R = 5^2 \times 5.5 = \mathbf{137.5\ W}. \quad \textbf{(Ans.)}$$

(*iii*) **The total power** $= VI \cos \phi = 250 \times 5 \times \frac{AC}{AD}$

$$= 250 \times 5 \times \frac{(125 + 27.5)}{250} = \mathbf{762.5\ W}. \quad \textbf{(Ans.)}$$

The vector diagram is shown in Fig. 37 (*b*). **(Ans.)**

Example 24. *An iron-cored coil has a D.C. resistance of 6 ohms. When it is connected to 230 V, 50 Hz mains, the current taken is 3.5 A at a power factor of 0.5. Determine :*

(*i*) *Effective resistance of the coil.*

(*ii*) *Inductance of the coil.*

(*iii*) *Resistance which represents the effect of the iron loss.*

Solution. *Given :* D.C. resistance (True resistance), R = 6 Ω ; supply voltage = 230 V, $f = 50$ Hz, $I = 3.5$ A ; *p.f.* = 0.5.

(*i*) **Effective resistance of the coil, $\mathbf{R_e}$:**

Total power consumed by the iron-cored choke coil,

$$P = \text{Power loss in ohmic resistance + Iron loss in core} = I^2R + P_i$$

or $\frac{P}{I^2} = R + \frac{P_i}{I^2}$, where $\frac{P}{I^2}$ is known as *effective resistance* of the coil.

$$\therefore \quad \text{Effective resistance, } R_e = \frac{P}{I^2} = \frac{VI \cos \phi}{I^2} = \frac{230 \times 3.5 \times 0.5}{(3.5)^2} = \mathbf{32.86\ \Omega}. \quad \textbf{(Ans.)}$$

(*ii*) **Inductance of the coil, L :**

Impedance of the coil, $Z = \frac{V}{I} = \frac{230}{3.5} = 65.7\ \Omega$

Inductive reactance of the coil,

$$X_L = \sqrt{Z^2 - R_e^2} = \sqrt{(65.7)^2 - (32.86)^2} = 56.9\ \Omega$$

$$\therefore \quad L = \frac{X_L}{2\pi f} = \frac{56.9}{2\pi \times 50} = \mathbf{0.1811\ H}. \quad \textbf{(Ans.)}$$

(*iii*) **Resistance representing iron loss :**

Since $\frac{P}{I^2} = R + \frac{P_i}{I^2}$

Effective resistance, R_e = True resistance + Resistance representing iron loss

32.86 = 6 + Resistance representing iron loss

∴ Resistance representing iron loss = 32.86 – 6 = **26.86 Ω.** **(Ans.)**

Example 25. *Three coils connected in series across a 100 V, 50 Hz supply have the following parameters :*

$$R_1 = 18\ \Omega,\ L_1 = 0.012\ H\ ;\ R_2 = 12\ \Omega,\ L_2 = 0.036\ H\ ;\ R_3 = 3.6\ \Omega,\ L_3 = 0.072\ H$$

Determine the potential drop and phase angle for each coil.

Solution. Fig. 38. shows the circuit diagram.

Total resistance in the circuit, $R = R_1 + R_2 + R_3 = 18 + 12 + 3.6 = 33.6\ \Omega$

Total inductance in the circuit, $L = L_1 + L_2 + L_3 = 0.012 + 0.036 + 0.072 = 0.12$ H

Impedance of coil-1, $Z_1 = \sqrt{R_1^2 + (2\pi f L_1)^2} = \sqrt{(18)^2 + (2\pi \times 50 \times 0.012)^2} = 18.39\ \Omega$

Impedance of coil-2, $Z_2 = \sqrt{R_2^2 + (2\pi f_2 L_2)^2} = \sqrt{(12)^2 + (2\pi \times 50 \times 0.036)^2} = 16.49\ \Omega$

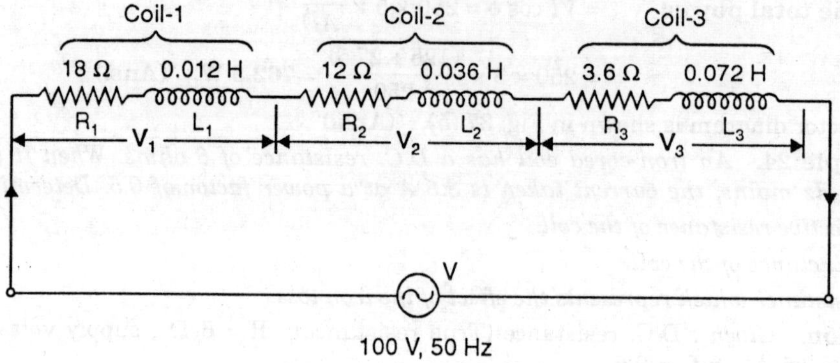

Fig. 38. Circuit diagram.

Impedance of coil–3, $Z_3 = \sqrt{R_3^2 + (2\pi f L_3)^2} = \sqrt{(3.6)^2 + (2\pi \times 50 \times 0.072)^2} = 22.90\ \Omega$

Impedance of the whole circuit, $Z = \sqrt{R^2 + (2\pi f L)^2} = \sqrt{(33.6)^2 + (2\pi \times 50 \times 0.12)^2} = 50.5\ \Omega$

Current through the circuit, $I = \frac{V}{Z} = \frac{100}{50.5} = 1.98\ A$

Potential drop across coil–1, $\mathbf{V_1} = IZ_1 = 1.98 \times 18.39 =$ **36.41 V (Ans.)**

Potential drop across coil–2, $\mathbf{V_2} = IZ_2 = 1.98 \times 16.49 =$ **32.65 V (Ans.)**

Potential drop across coil–3, $\mathbf{V_3} = IZ_3 = 1.98 \times 22.90 =$ **45.34 V (Ans.)**

Phase angle of coil–1, ϕ_1, $\cos^{-1}(R_1/Z_1) = \cos^{-1}(18/18.39) =$ **11.82°. (Ans.)**

Phase angle of coil–2, $\phi_2 = \cos^{-1}(R_2/Z_2) = \cos^{-1}(12/16.49) =$ **43.3°. (Ans.)**

Phase angle of coil–3, $\phi_3 = \cos^{-1}(R_3/Z_3) = \cos^{-1}(3.6/22.90) =$ **80.96°. (Ans.)**

Example 26. *An alternating voltage of (176 + j132) is applied to a circuit and the current in the circuit is given by (6.6 + j 8.8) A. Determine :*

(i) Values of elements of the circuit.

(ii) Power factor of the circuit.

(iii) Power consumed.

Solution. Given : Supply voltage, $\bar{V} = 176 + j132 = 220\ \angle 36.87°$

Circuit current, $\bar{I} = 6.6 + j\,8.8 = 11\ \angle 53.13°$

(*i*) **Values of elements of the circuit, R, C :**

Circuit impedance, $\bar{Z} = \dfrac{\bar{V}}{\bar{I}} = \dfrac{220 \angle 36.87°}{11 \angle 53.13°} = 20 \angle -16.26°$

$= 20\ [\cos(-16.26°) + \sin(-16.26°)]$

$= 20\ (0.96 - j\,0.28) = (19.2 - j5.6)\ \Omega$

$\therefore$ $R = \mathbf{19.2\ \Omega}$ **(Ans.)**

$X_C = 5.6\ \Omega$ or $C = \dfrac{1}{2\pi f X_C} = \dfrac{1}{2\pi \times 50 \times 5.6}$ F = **568.4 μF.** **(Ans.)**

(*ii*) **Power factor of the circuit ; cos φ :**

$\cos\phi = \dfrac{R}{Z} = \dfrac{19.2}{20}$ = **0.96 (leading).** **(Ans.)**

(*iii*) **Power (true) consumed, P :**

Apparent power, $\bar{S} = \bar{V} \times \bar{I}$

$= 220 \angle 36.87° \times 11 \angle -53.13° = 2420 \angle -16.26°$

$= 2323.2 - j\,677.6$

$\therefore$ True power, $P = \mathbf{2323.2\ W.}$ **(Ans.)**

(Alternatively : $P = VI\cos\phi = 220 \times 11 \times 0.96 = 2323.2$ W).

Example 27. *In a circuit, the equations for instantaneous voltage and current are given by,*

$$v = 141.4 \sin\left(\omega t - \frac{2\pi}{3}\right), \text{ volt and}$$

$$i = 7.07 \sin\left(\omega t - \frac{\pi}{2}\right), \text{ amp, where } \omega = 314 \text{ rad/sec.}$$

(*i*) *Sketch a neat phasor diagram for the circuit*

(*ii*) *Use polar notation to calculate impedance with phase angle.*

(*iii*) *Calculate average power.*

(*iv*) *Calculate the instantaneous power at the instant t = 0* **(Pune University)**

Solution. *Given :* $v = 141.4 \sin\left(\omega t - \dfrac{2\pi}{3}\right)$, and $i = 7.07 \sin\left(\omega t - \dfrac{\pi}{2}\right)$, where ω = 314 rad/s.

(*i*) **Phasor diagram :**

From the voltage equation, it is seen that the voltage lags behind the reference quantity by $\dfrac{2\pi}{3}$ rad or $2 \times \dfrac{180}{3} = 120°$. Similarly, current lags behind the reference quantity by $\dfrac{\pi}{2}$ rad or $\dfrac{180}{2}$ = 90°. Between themselves, voltage lags behind the current by (120° – 90°) = 30° as shown in Fig. 39 (*b*).

(*ii*) **Impedance with phase angle (polar notation)**

$$V = \frac{V_{max}}{\sqrt{2}} = \frac{141.4}{\sqrt{2}} = 100\ V\ ;$$

$$I = \frac{I_{max}}{\sqrt{2}} = \frac{7.07}{\sqrt{2}} = 5A.$$

$\therefore$ $V = 100 \angle -120°$ and $I = 5 \angle -90°$

$\therefore$ $Z = \dfrac{V}{I} = \dfrac{100 \angle -120°}{5 \angle -90°} = \mathbf{20 \angle -30°\ \Omega}$ **(Ans.)**

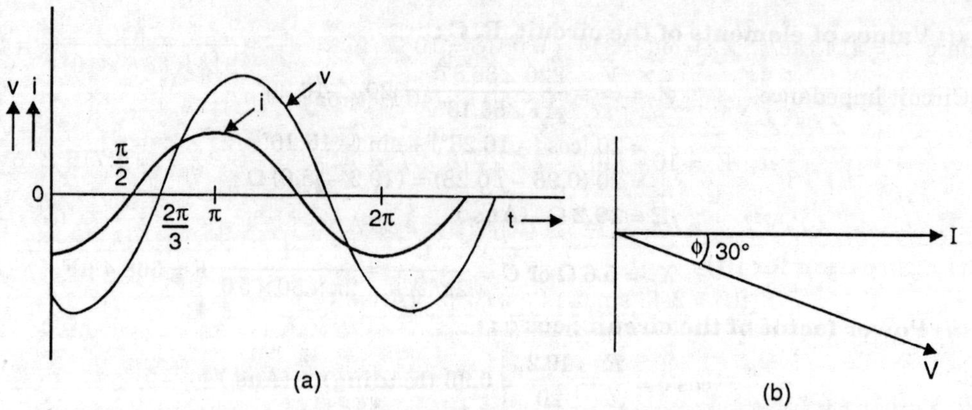

Fig. 39

(*iii*) **Average power :**

Average power = $VI \cos \phi$

$= 100 \times 5 \times \cos 30° =$ **433 W. (Ans.)**

(*iv*) **Instantaneous power at t = 0 :**

At $t = 0$; $v = 141.4 \sin (0 - 120°) = -122.45\ V$

$i = 7.07 \sin (0 - 90°) = -7.07\ A$

∴ Instantaneous power at $t = 0$,

$p = v_i = (-122.45) \times (-7.07) =$ **865.7 W. (Ans.)**

Example 28. *A voltage e(t) = 100 sin 314 t is applied to a series circuit consisting of 10 ohms resistance, 0.0318 henry inductance and a capacitor of 63.6 μF. Determine :*

(*i*) *Expression for i(t).*

(*ii*) *Phase angle between voltage and current.*

(*iii*) *Power factor.*

(*iv*) *Active power consumed.*

(*v*) *Peak value of pulsating energy.* **(Indore University)**

Soution. *Given :* $e(t) = 100 \sin 314\, t$, $R = 10\ \Omega$, $L = 0.0318$ H, $C = 63.6\ \mu F = 63.6 \times 10^{-6}\ F$.

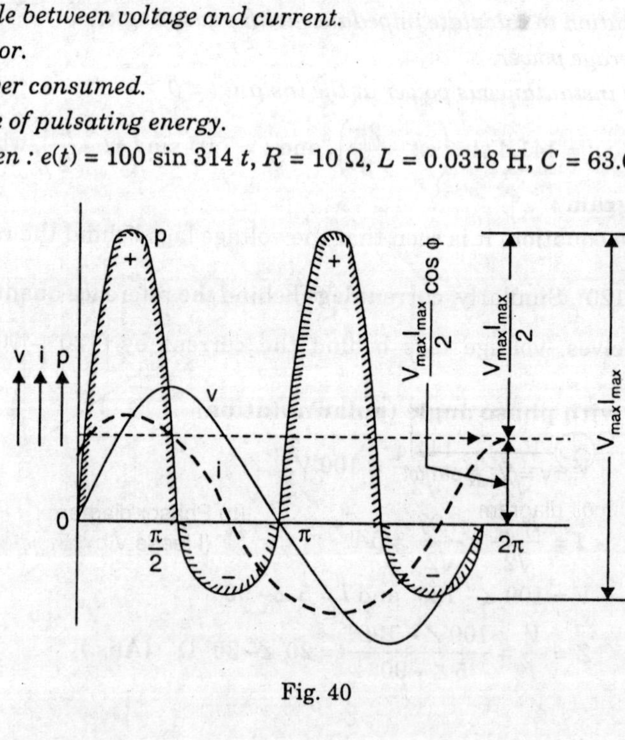

Fig. 40

Here $\omega = 314$ rad/s ; $X_L = \omega L = 314 \times 0.0318 = 10\ \Omega$; $X_C = \dfrac{1}{\omega C} = \dfrac{1}{314 \times 63.6 \times 10^{-6}} = 50\ \Omega$;

$$X = X_L - X_C = 10 - 50 = -40\ \Omega \text{ (capacitive)}$$

$$\bar{Z} = 10 - j40 = 41.2 \angle -76° \ ; \ \bar{I} = \frac{\bar{V}}{\bar{Z}} = \frac{(100/\sqrt{2}) \angle 0°}{41.2 \angle -76°} = 1.716 \angle 76°$$

$$I_{max} = I \times \sqrt{2} = 1.716 \times \sqrt{2} = 2.43 \text{ A}$$

(*i*) **Expression for i(t) :**

$$i(t) = 2.43 \sin (314\, t + 76°). \quad \textbf{(Ans.)}$$

(*ii*) **Phase angle between voltage and current,** ϕ :

$$\phi = \mathbf{76°} \text{ with current leading.}$$

(*ii*) **Power factor, cos** ϕ **:**

$$\cos \phi = \cos 76° = \mathbf{0.24\ (lead).} \quad \textbf{(Ans.)}$$

(*iv*) **Active power consumed, P :**

$$P = VI \cos \phi = (100/\sqrt{2})\,(2.43/\sqrt{2}) \times 0.24 = \mathbf{29.16\ W.} \quad \textbf{(Ans.)}$$

(*iv*) **Peak value of pulsating energy :**

Refer. Fig. 40. The peak value of pulsating energy

$$= \frac{V_{max}\, I_{max}}{2} + \frac{V_{max}\, I_{max}}{2} \cos \phi$$

$$= \frac{V_{max}\, I_{max}}{2} (1 + \cos \phi) = \frac{100 \times 2.43}{2} (1 + 0.24) = \mathbf{150.66\ W.} \quad \textbf{(Ans.)}$$

4.5.2. R-C circuit (Resistance and capacitance in series)

Fig. 41. (*a*) shows a pure resistance R (ohms) and a pure capacitor of capacitance C (farads) connected in series. Such a circuit is known as *R-C* circuit.

Let, V = R.M.S. value of the applied voltage,

I = R.M.S. value of the resultant current,

$V_R = IR$ = Voltage drop across R (in phase with I) and

$V_C = IX_C$ = Voltage drop across C, lagging I by 90°.

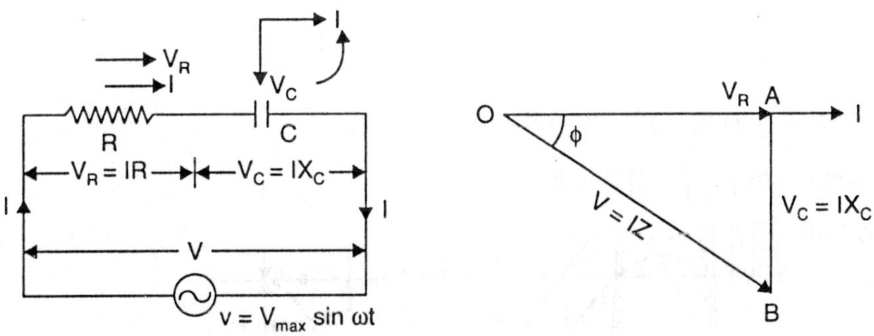

(a) Circuit diagram

(b) Phasor diagram (I leads V by angle ϕ)

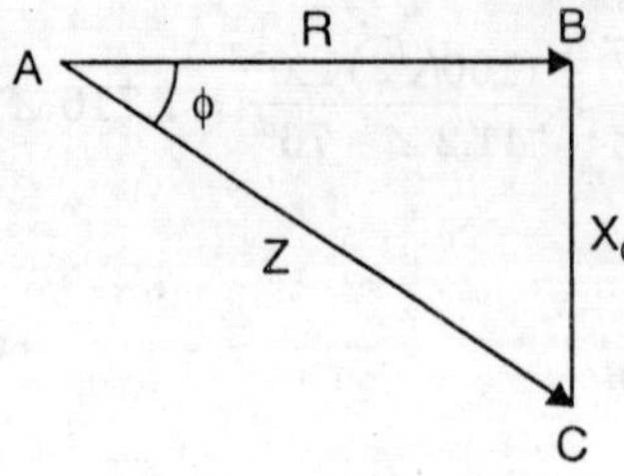

(c) Impedance triangle

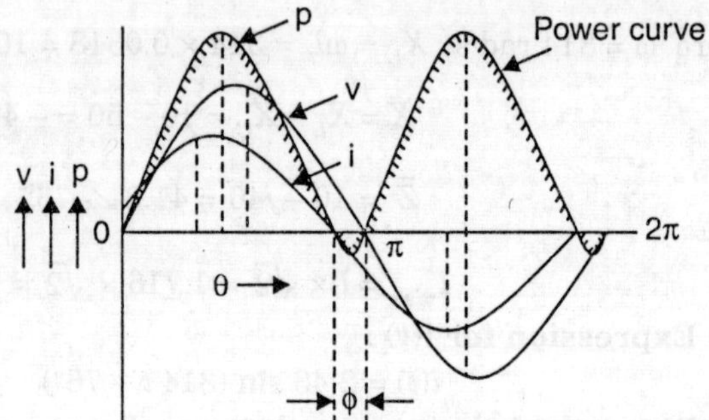

(d) Power curve

Fig. 41. *R-C* circuit (Resistance and capacitance in series).

Voltage drops V_R and V_L are shown in voltage triangle *OAB* in Fig. 41 (*b*) *I* being taken as the reference vector in the phasor diagram. Vector *OA* represents ohmic drop V_R and *AB* represents the capacitive drop V_C. Vector *OB* represents the applied voltage *V*, which is the vector sum of the V_R and V_C)

$$\therefore \quad V = \sqrt{V_R^2 + V_C^2} = \sqrt{(IR)^2 + (IX_C)^2} = I\sqrt{R^2 + X_C^2}$$

or

$$I = \frac{V}{\sqrt{R^2 + X_C^2}} = \frac{V}{Z}$$

where $Z = \sqrt{R^2 + X_C^2}$ (total opposition offered to the flow of alternating current by *R-C* series circuit) is known as the **impedance** of the circuit.

As seen from the *"impedance triangle"* *ABC* [Fig. 41 (*c*)],

$$Z^2 = R^2 + X_C^2$$

i.e., (Impedance)² = (Resistance)² + (Capacitive reactance)²

From Fig. 41 (*b*) it is evident that *I* leads the voltage *V* by an angle ϕ such that,

$$\tan \phi = \frac{V_C}{V_R} = \frac{IX_C}{IR} = \frac{X_C}{R} = \frac{(1/\omega C)}{R} = \frac{\text{Capacitive reactance}}{\text{Resistance}}$$

$$\therefore \quad \phi = \tan^{-1}\left(\frac{X_C}{R}\right)$$

The same is illustrated graphically in Fig. 41 (*d*).

In other words I leads V_R by an angle ϕ.

Power factor, $\cos \phi = \frac{R}{Z}$ [From Fig. 41 (*c*)]

Power. Refer Fig. 41 (*d*),

Instantaneous power,
$$p = vi = V_{max} \sin \omega t \times I_{max} \sin(\omega t + \phi)$$
$$= \frac{V_{max}\, I_{max}}{2} \times 2 \sin(\omega t + \phi) \sin \omega t$$
$$= \frac{V_{max}}{\sqrt{2}} \times \frac{I_{max}}{\sqrt{2}} [\cos \phi - \cos(2\omega t + \phi)]$$

Average power consumed in the circuit over a complete cycle,

$$P = \text{Average of } \frac{V_{max}}{\sqrt{2}} \cdot \frac{I_{max}}{\sqrt{2}} \cos \phi - \text{Average of } \frac{V_{max}}{\sqrt{2}} \cdot \frac{I_{max}}{\sqrt{2}} \cos(2\omega t + \phi)$$

or $$P = \frac{V_{max}}{\sqrt{2}} \cdot \frac{I_{max}}{\sqrt{2}} \cos \phi - \text{zero},$$

or $$P = V_{r.m.s.} \times I_{r.m.s.} \cos \phi = VI \cos \phi$$

where $\cos \phi$ is the *power factor* of the circuit :

Alternatively, $$P = VI \cos \phi = IZ \times I \times \frac{R}{Z} = I^2R$$

This shows that *power is actually consumed in resistance only* ; the *capacitor does not consume any power.*

Thus in ***R-C circuit***, we have :

1. Impedance, $Z = \sqrt{R^2 + X_C{}^2}$ $\left(\text{where } X_C = \frac{1}{\omega C} = \frac{1}{2\pi f, C \text{ being in farad}}\right)$

2. Current, $I = \dfrac{V}{Z}$

3. Power factor, $\cos \phi = \dfrac{R}{Z}\left(= \dfrac{\text{True power}}{\text{Apparent power}} = \dfrac{W}{VA}\right)$

[or angle of lead, $\phi = \cos^{-1} (R/Z)$]

4. Power consumed, $P = VI \cos \phi \; (= I^2R)$.

Example 29. *A capacitance of 20 μF and a resistance of 100 ohms are connected in series across 120 V, 60 Hz mains. Determine the average power expended in the circuit.*

Also draw the vector diagram.

Solution. $R = 100 \, \Omega$

$C = 20 \, \mu F = 20 \times 10^{-6}$ F (farad)

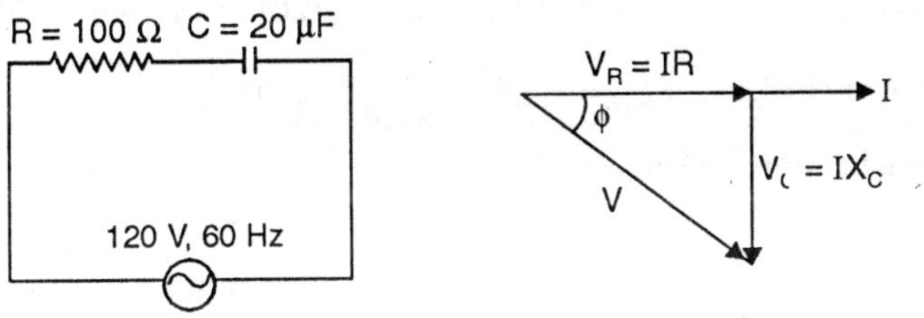

(*a*) R-C circuit (*b*) Vector/phasor diagram

Fig. 42

Capacitive reactance, $$X_C = \frac{1}{2\pi fC} = \frac{1}{2\pi \times 50 \times 20 \times 10^{-6}} = 159 \, \Omega$$

Impedance of the circuit, $$Z = \sqrt{R^2 + X_C{}^2} = \sqrt{100^2 + 150^2} = 188 \, \Omega$$

Current through the circuit $$I = \frac{V}{Z} = \frac{120}{188} = 0.638 \text{ A}$$

Power factor, $$\cos \phi = \frac{R}{Z} = \frac{100}{188} = 0.532$$

Average power expended in the circuit,

$$P_{av} = VI \cos \phi$$
$$= 120 \times 0.638 \times 0.532 = \mathbf{40.75 \text{ W}}. \quad \textbf{(Ans.)}$$

Fig. 42 (*b*) shows the vector/phasor diagram.

Example 30. *A voltage $v = 100 \sin 314t - 50 \cos 314t$, is applied to a circuit having $R = 20\ \Omega$ in series with $C = 100\ \mu F$. Obtain expression for instantaneous current, r.m.s. value of current and the power in the circuit.* **(PTU, 1999)**

Solution. *Given :* $v = 100 \sin 314t - 50 \cos 314t$; $R = 20\ \Omega$; $C = 100\ \mu F$.

The R-C circuit and the phasor diagram for the given instantaneous voltage are shown in Figs. 43 and 44 respectively.

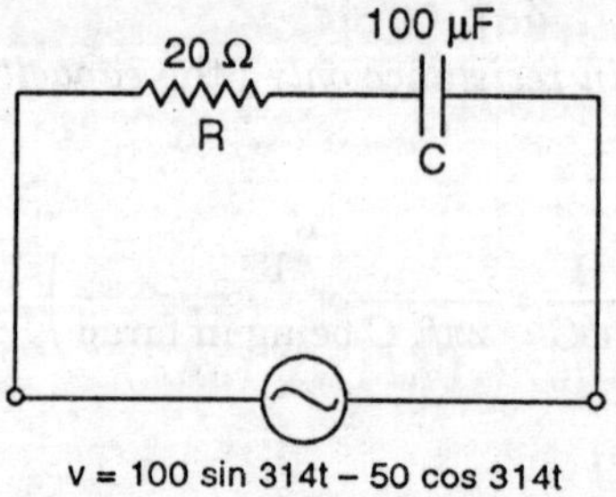

Fig. 43

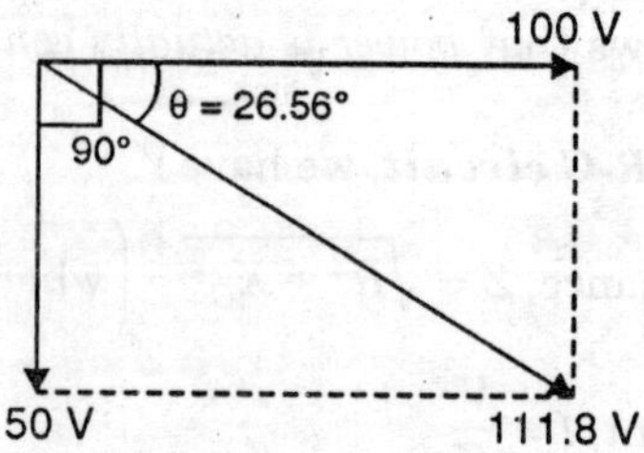

Fig. 44

Resultant voltage, $V_{max(R)} = \sqrt{(100)^2 + (50)^2} = 111.8$ V

Phase angle with the horizontal, $\theta = \tan^{-1}\left(-\dfrac{50}{100}\right) = -26.56°$

$\therefore$ $v = 111.8 \sin (314t - 26.56°)$

Now, $\omega t = 314\, t$ or $\omega = 314$

$\therefore$ Capacitive reactance, $X_C = \dfrac{1}{\omega C} = \dfrac{1}{314 \times 100 \times 10^{-6}} = 31.85\ \Omega$

Circuit impedance, $Z = \sqrt{R^2 + X_C{}^2} = \sqrt{(20)^2 + (31.85)^2} = 37.6\ \Omega$

Maximum value of current, $I_{max} = \dfrac{V_{max}}{Z} = \dfrac{111.8}{37.6} = 2.97$ A

Phase difference between voltage and current,

$$\phi = \cos^{-1}\left(\frac{R}{Z}\right) = \cos^{-1}\left(\frac{20}{37.6}\right) = 57.86° \text{ (leading)}$$

$\therefore$ *Instantaneous value of current*

$$i = 2.97 \sin (314t - 26.56° + 57.86°) = 2.97 \sin (314t + 31.3°)$$

or $$i = 2.97 [\sin 314t \,.\, \cos 31.3° + \cos 314t \,.\, \sin 31.3°]$$

or $$\mathbf{i = 2.54 \sin 314t + 1.54 \cos 314t.} \quad \textbf{(Ans.)}$$

RMS value of the current, $\mathbf{I} = \dfrac{I_{max}}{\sqrt{2}} = \dfrac{2.97}{\sqrt{2}} = \mathbf{2.1\ A.}$ **(Ans.)**

Power in the circuit, $P = VI \cos \phi$

$$= \frac{V_{max(R)}}{\sqrt{2}} \times I \cos \phi = \frac{111.8}{\sqrt{2}} \times 2.1 \times \cos (57.86°) = \mathbf{88.32\ W} \textbf{ (Ans.)}$$

Example 31. *A two element series circuit is connected across an A.C. source $e = 200\sqrt{2} \sin (\omega t + 20°)$ V. The current in the circuit then is found to be $i = 10\sqrt{2} \cos (314\ t - 25°)$ A. Determine parameters of the circuit.* **(Allahabad University)**

Solution. Given : $e = 200\sqrt{2} \sin (\omega t + 20°)$

$i = 10\sqrt{2} \cos (314\, t - 25°)$

Parameters of the circuit, cos φ, R, X_C and C :

The current i can be written as,

$$i = 10\sqrt{2}\ \sin(314\,t - 25° + 90°) = 10\sqrt{2}\ \sin(314\,t + 65°)$$

It is seen that applied voltage leads by 20° and current leads by 65° with regards to the reference quantity, their mutual difference is 65° – 20° = 45°.

Hence, p.f. $\cos\phi = \cos 45° =$ **0.707 (lead). (Ans.)**

Now, $V_{max} = 200\sqrt{2}$ and $I_{max} = 10\sqrt{2}$

$\therefore$ $$Z = \frac{V_{max}}{I_{max}} = \frac{200\sqrt{2}}{10\sqrt{2}} = 20\ \Omega$$

$\therefore$ $R = Z\cos\phi = 20 \times 0.707 =$ **14.14 Ω. (Ans.)**

$X_C = Z\sin\phi = 20 \times 0.707 =$ **14.14 Ω. (Ans.)**

Also, $$X_C = \frac{1}{\omega C} = \frac{1}{2\pi f C} = 14.14$$

where $$f = \frac{314}{2\pi} = 50\ \text{Hz}$$

$\therefore$ $$X_C = 14.14 = \frac{1}{2\pi \times 50 \times C}$$

or $$C = \frac{1}{14.14 \times 2\pi \times 50}\ \text{F} = \mathbf{225.1\ \mu F.}\ \ \textbf{(Ans.)}$$

Hence the *given circuit is an R-C circuit.*

4.5.3. R-L-C circuit (Resistance, inductance and capacitance in series)

Fig. 45 shows a *R-L-C* circuit.

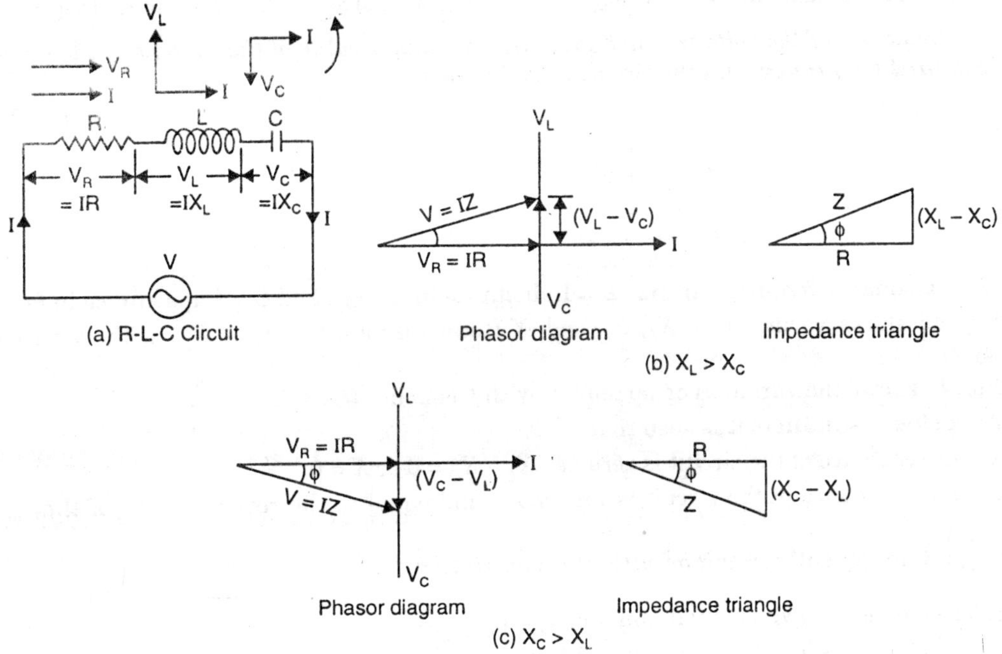

Fig. 45. Resistance, inductance and capacitance in series.

Important formulae :

1. Impedance, $Z = \sqrt{R^2 + (X_L - X_C)^2}$

$\left[\text{where } X_L = 2\pi fL, L \text{ in } henry \text{ and } X_C = \dfrac{1}{2\pi fC}, C \text{ in farad}\right]$

2. Current, $I = \dfrac{V}{Z}$

3. Power factor, $\cos\phi = \dfrac{R}{Z}$

$\left[\text{angle of lag (when } X_L > X_C) \text{ or lead (when } X_C > X_L), \phi = \cos^{-1}\dfrac{R}{Z}\right]$

4. Power consumed $= VI\cos\phi\ (= I^2R)$

Resonance in R-L-C circuits

Refer Fig. 45 (*a*).

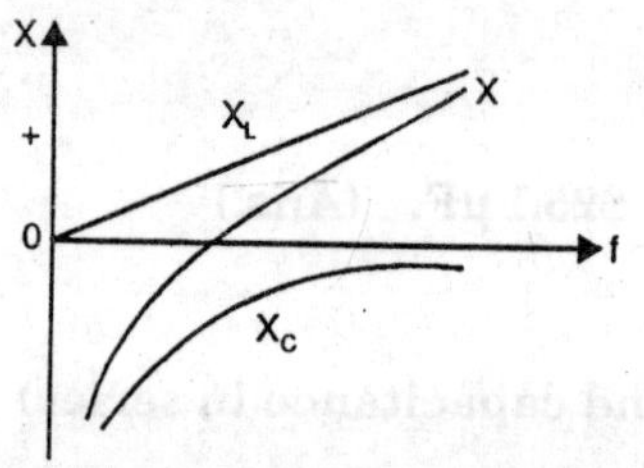

Fig. 46. Reactance (X) v/s frequency (*f*).

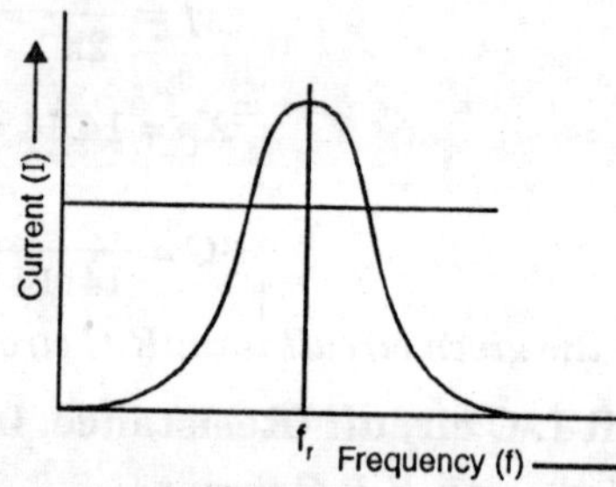

Fig. 47. Current in *R-L-C* circuit v/s frequency.

The frequency of the voltage which gives the maximum value of the current in the circuit is called ***resonant frequency,*** and the circuit is said to be ***resonant.***

At resonance, $X_L = X_C$ (*i.e.*, $Z = R$)

i.e., $$2\pi f_r L = \frac{1}{2\pi f_r C}$$

$$f_r = \frac{1}{2\pi\sqrt{LC}} \quad ...(5)$$

where f_r = Resonance frequency in Hz ; L = Inductance in henry ; and C = Capacitance in farad.

Fig. 46 shows variation of X_L, X_C and X (total reactance = $X_L - X_C$) with variation of frequency f.

Fig. 47 shows the variation of current (I) with frequency (f).

At **series resonance,** it is seen that :

1. Net reactance of the circuit is zero *i.e.,* $X_L - X_C = 0$ or $X = 0$.

2. The impedance of the circuit is *minimum* and equal to the resistance (R) of the circuit $\left(i.e., I = \dfrac{V}{R}\right)$. Consequently *circuit admittance is maximum.*

3. The current drawn is maximum (*i.e.,* $I_r = I_{max}$).

4. The phase angle between the current and voltage is zero ; the *power factor is unity.*

5. The resonant frequency is given by $f_r = \frac{1}{2\pi\sqrt{LC}}$; if the *frequency is below* the resonant frequency the net reactance in the circuit is *capacitive* and if the frequency is above the resonant frequency, the net reactance in the circuit is *inductive*.

6. Although $V_L = V_C$, yet V_{coil} is greater than V_C because of its resistance.

Half power frequencies, Bandwidth and Selectivity

Half power (cut-off) frequencies :

The half power frequencies *are those frequencies at which the power dissipation in the circuit is half of the power dissipation at resonant frequency* f_r. They are the corresponding frequencies f_1 and f_2 at the value of current $I = I_r/\sqrt{2}$; where I_r is the current at resonance in *R-L-C* series circuit (Refer Fig. 48).

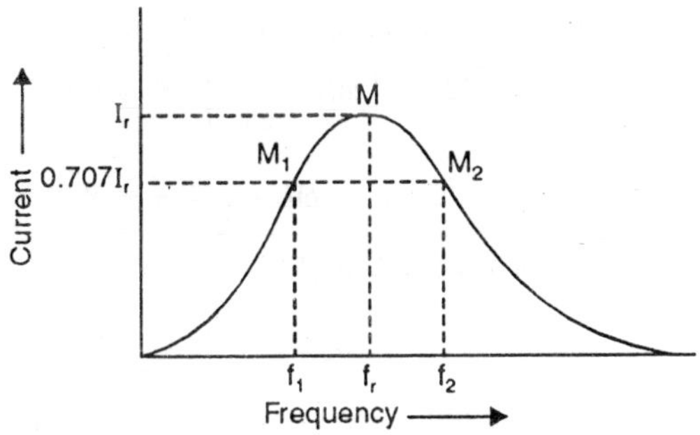

Fig. 48

Hence power, P_r drawn by the circuit at the *resonance* is

$$P_r = I_r^2 R \quad ...(6)$$

Power in the circuit at $f_1 = \left(\frac{I_r}{\sqrt{2}}\right)^2 R = \frac{1}{2} I_r^2 R$

(= half the power at resonance)

Power in the circuit at $f_2 = \left(\frac{I_r}{\sqrt{2}}\right)^2 R = \frac{1}{2} I_r^2 R$ (= half the power at resonance)

Also,
$$f_1 = f_r - \frac{R}{4\pi L}$$
$$f_2 = f_r + \frac{R}{4\pi L}$$
$$f_2 - f_1 = \frac{R}{2\pi L}$$

Bandwidth and Selectivity :

The difference $(f_2 - f_1)$ is called the *bandwidth* (B_{hp}) of the resonant network.

The ratio of the bandwidth to the resonance frequency is defined as the ***selectivity*** of the circuit.

When frequency is varied in *R-L-C* circuit, the selectivity becomes

$$\frac{*(f_2 - f_1)}{f_r} = \frac{1}{Q} \quad ...(7)$$

where Q_r is the *quality factor* of the resonant circuit.

***Relation between bandwidth and quality factor in series resonant conditions :**

A series *R-L-C* circuit is considered. The resonant frequency and angular frequency are expressed by f_r and ω_r respectively. In the above circuit, the current (I) can be described as follows :

$$I = \frac{V}{\sqrt{R^2 + \left(\omega L - \frac{1}{\omega C}\right)^2}}$$

where V, R, L and C are the source voltage, resistance, inductance and capacitance of the circuit respectively.

The current, at a power, half of the maximum power developed at resonant frequency, is $\frac{1}{\sqrt{2}} I_r$, where I_r is the series resonant current *i.e.* $\frac{V}{R}$.

According to the definition of bandwidth,

$$\frac{1}{\sqrt{2}}\frac{V}{R} = \frac{V}{\sqrt{R^2 + \left(\omega L - \frac{1}{\omega C}\right)^2}}$$

or $$\sqrt{R^2 + \left(\omega L - \frac{1}{\omega C}\right)^2} = \sqrt{2}R \quad \text{or} \quad R^2 = \left(\omega L - \frac{1}{\omega C}\right)^2$$

or $$\omega L - \frac{1}{\omega C} = \pm R.$$

According to Fig. 48

$$\omega_2 L - \frac{1}{\omega_2 C} = R \quad \text{...(i)} \qquad \omega_1 L - \frac{1}{\omega_1 C} = -R \quad \text{...(ii)}$$

Adding equations (*i*) and (*ii*), we get

$$(\omega_1 + \omega_2)L - \frac{1}{C}\left(\frac{1}{\omega_1} + \frac{1}{\omega_2}\right) = 0$$

or $$(\omega_1 + \omega_2)L - \frac{1}{C}\left(\frac{\omega_1 + \omega_2}{\omega_1 \omega_2}\right) = 0$$

Since $$\omega_1 + \omega_2 \neq 0,\ \omega_1\omega_2 = \frac{1}{LC}.$$

Again $\omega_r^{\,2} = \frac{1}{LC}$, [$\omega_r$ is the angular frequency at resonant condition.]

Subtracting equations (*i*) and (*ii*), we have

$$L(\omega_2 - \omega_1) + \frac{1}{C}\left(\frac{\omega_2 - \omega_1}{\omega_2 \omega_1}\right) = 2R$$

or $$(\omega_2 - \omega_1) + \frac{1}{LC}\frac{(\omega_2 - \omega_1)}{\omega_2\omega_1} = \frac{2R}{L}$$

or $$(\omega_2 - \omega_1) + \omega_1\omega_2 \cdot \frac{(\omega_2 - \omega_1)}{\omega_1\omega_2} = \frac{2R}{L} \qquad \left[\because \omega_1\omega_2 = \frac{1}{LC}\right]$$

or $$(\omega_2 - \omega_1) = \frac{R}{L}$$

or $$\frac{\omega_2 - \omega_1}{\omega_0} = \frac{R}{\omega_0 L}$$

[ω_0 is the angular frequency at resonant condition]

or $$\omega_2 - \omega_1 = \omega_0 \times \frac{1}{\frac{\omega_0 L}{R}}$$

or $$\omega_2 - \omega_1 = \frac{\omega_0}{Q} \qquad \left[\because Q = \frac{\omega_1 L}{R}\right]$$

or $$f_2 - f_1 = \frac{f_r}{Q} \qquad [\because \omega = 2\pi f]$$

or *Bandwidth* at series resonant condition $= \dfrac{f_r}{Q}$.

$$\left[Q = \frac{\omega_r}{\text{bandwidth}} = \frac{\omega_r}{\omega_2 - \omega_1} = \frac{\omega_r}{R/L} = \frac{\omega_r L}{R} = \frac{L}{R\sqrt{LC}} = \frac{1}{R}\sqrt{\frac{L}{C}}\right]$$

Q-factor of a resonant series circuit :

The Q-factor of an R-L-C series circuit can be defined in the following different ways :

(*i*) ***Q-factor*** is *defined as the voltage magnification in the circuit at the time of resonance.*

Since at resonance current is maximum *i.e.*, $I_r = \dfrac{V}{R}$, the voltage across either coil or capacitor $= I_r X_{Lr}$ or $I_r X_{Cr}$ and supply voltage, $V = I_r R$.

$$\therefore \quad \text{Voltage magnification} = \frac{V_{Lr}}{V} = \frac{I_r X_{Lr}}{I_r R} = \frac{X_{Lr}}{R} = \frac{\omega_r L}{R} = \frac{\text{Reactance}}{\text{Resistance}}$$

or $$\frac{V_{Cr}}{V} = \frac{I_r \times C_r}{I_r R} = \frac{X_{Cr}}{R} = \frac{\text{Reactance}}{\text{Resistance}} = \frac{1}{\omega_r CR}$$

$$\therefore \quad Q\text{-factor} = \frac{\omega_r L}{R} = \frac{2\pi f_r L}{R} = \tan\theta \qquad \text{...[8(a)]}$$

where θ is the circuit *power factor angle of the coil.*

(At resonance, circuit phase angle θ = 0, and $Q = \tan\theta = 0$)

(*ii*) The ***Q-factor*** may also be *defined* as under :

$$Q\text{-factor} = 2\pi \frac{\text{maximum stored energy}}{\text{energy dissipated per cycle}} \quad \text{...... in the circuit}$$

$$= 2\pi \frac{\frac{1}{2}LI_r^2}{I^2RT_r} = 2\pi \frac{\frac{1}{2}L(\sqrt{2}I)^2}{I^2R(1/f_r)} = \frac{I^2 \times 2\pi f_r L}{I^2R} = \frac{\omega_r L}{R}\left(= \frac{1}{\omega_r CR}\right) \qquad \left(\because T_r = \frac{1}{f_r}\right)$$

But resonant frequency, $f_r = \dfrac{1}{2\pi\sqrt{LC}}$ or $2\pi f_r = \dfrac{1}{\sqrt{LC}}$

Putting this value in eqn. 8(*a*), we get

$$Q\text{-factor} = \frac{2\pi f_r L}{R} = \frac{L}{R} \times \frac{1}{\sqrt{LC}} = \frac{1}{R}\sqrt{\frac{L}{C}} \qquad \text{...[8(b)]}$$

In series resonance, higher quality factor *i.e.*, Q-factor means higher voltage magnification as well as higher selectivity of the tuning coil.

Example 32. *A resistance 12 Ω, an inductance of 0.15 H and a capacitance of 100 μF are connected in series across a 100 V, 50 Hz supply. Calculate :*

(*i*) *The current.*

(*ii*) *The phase difference between current and the supply voltage.*

(*iii*) *Power consumed.*

Draw the vector diagram of supply voltage and the line current.

Solution. *Given :* $R = 12\ \Omega$, $L = 0.15$ H or $X_L = 2\pi f L$

$= 2\pi \times 50 \times 0.15 = 47.1\ \Omega$

$C = 100\ \mu F = 100 \times 10^{-6} F$

or $$X_C = \frac{1}{2\pi f C} = \frac{1}{2\pi \times 50 \times 100 \times 10^{-6}} = 31.8\ \Omega$$

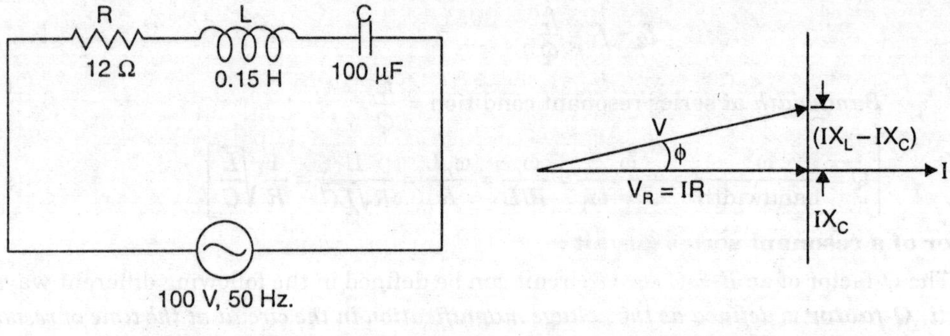

(a) R-L-C circuit (b) Vector/phasor diagram

Fig. 49

(i) **The current, I :** $Z = \sqrt{R^2 + (X_L - X_C)^2}$

$$= \sqrt{12^2 + (47.1 - 31.8)^2} = 19.43\ \Omega$$

Current, $I = \frac{V}{Z} = \frac{100}{19.43} = $ **5.15 A. (Ans.)**

(ii) **Phase difference, ϕ :**

$$\phi = \cos^{-1}\frac{R}{Z}\left[\text{or } \tan^{-1}\frac{X_L - X_C}{R}\right]$$

$$= \cos^{-1}\frac{12}{19.43}\left[\text{or } \tan^{-1}\frac{15.3}{12.0}\right] = 52° \text{ (lag)}$$

Hence current lags **supply voltage by 52°. (Ans.)**

(iii) **Power consumed, P :**

$$P = VI\cos\phi$$

$$= 100 \times 5.15 \times \cos 52° = \mathbf{371.1\ W.} \quad \textbf{(Ans.)}$$

Fig. 49 (a), (b) show the circuit and vector/phasor diagrams respectively.

Example 33. *For the circuit shown in Fig. 50 find the values of (i) current I, (ii) V_1 and V_2 and (iii) p.f.*

Draw the vector diagram. **(Bangalore University)**

Solution. Refer Fig. 50

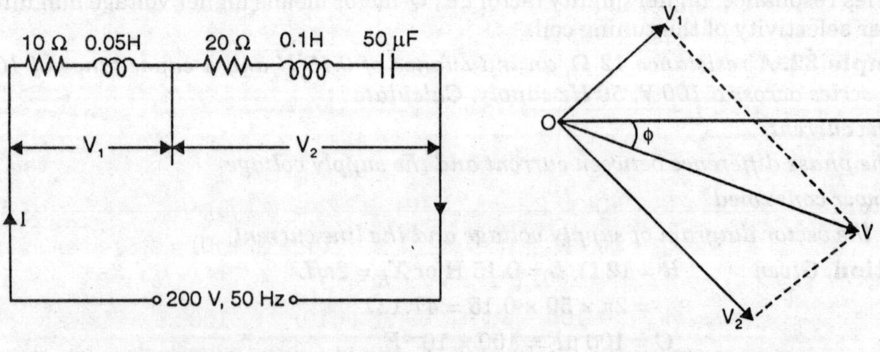

(a) Series circuit (b) Vector diagram

Fig. 50

$$R = 10 + 20 = 30\ \Omega$$
$$L = 0.05 + 0.1 = 0.15\ \text{H}$$
$\therefore$ $$X_L = 2\pi fL = 2\pi \times 50 \times 0.15 = 47.1\ \Omega$$
$$X_C = \frac{1}{2\pi fC} = \frac{1}{2\pi \times 50 \times 10^{-6} \times 50} = 63.7\ \Omega$$
$$X = 47.1 - 63.7 = -16.6\ \Omega$$
$$Z = \sqrt{R^2 + X^2} = \sqrt{(30)^2 + (-16.6)^2} = 34.3\ \Omega$$
(*i*) $$I = \frac{V}{Z} = \frac{200}{34.3} = \mathbf{5.83\ A.\ (Ans.)}$$
(*ii*) $$X_{L1} = 2\pi \times 50 \times 0.05 = 15.7\ \Omega$$
$$Z_1 = \sqrt{10^2 + 15.7^2} = 18.6\ \Omega$$
$\therefore$ $$V_1 = IZ_1 = 5.83 \times 18.6 = \mathbf{108.4\ V.\ (Ans.)}$$
$$\phi_1 = \cos^{-1}(10/18.6) = 57.5° \text{ (lag)}$$
$$X_{L2} = 2\pi \times 50 \times 0.1 = 31.4\ \Omega$$
$$X = 31.4 - 63.7 = -32.3\ \Omega$$
$$Z_2 = \sqrt{20^2 + (-32.3)^2} = 38\ \Omega$$
$\therefore$ $$V_2 = IZ_2 = 5.83 \times 38 = \mathbf{221.5\ V.\ (Ans.)}$$
$$\phi_2 = \cos^{-1}(20/38) = 58.2° \text{ (lead)}$$
(*iii*) Combined p.f. $$= \cos\phi = \frac{R}{Z} = \frac{30}{34.3} = \mathbf{0.875\ (lead)}$$

Vector diagram is shown in Fig. 50 (*b*).

Example 34. *For the circuit shown in Fig. 51. Calculate :*

(*i*) *Current ;* (*ii*) *Voltage drops* V_1, V_2 *and* V_3 *;*

(*iii*) *Power absorbed by each importance ;* (*iv*) *Total power absorbed by the circuit.*

Take voltage vector along the reference axis.

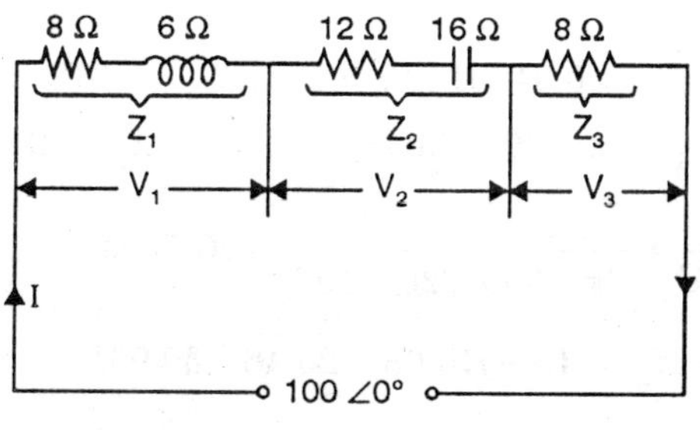

Fig. 51 **(Pune University)**

Solution. $Z_1 = (8 + j6)\ \Omega$; $Z_2 = (12 - j16)\ \Omega$; $Z_3 = (8 + j0)$

$$Z = Z_1 + Z_2 + Z_3 = (8 + j6) + (12 - j16) + (8 + j0) = (28 - j10)\ \Omega$$

Taking $V = V\angle 0° = 200\angle 0° = (200 + j0)$

$\therefore$ $$I = \frac{V}{Z} = \frac{100}{(28 - j10)} = \frac{200(28 + j10)}{(28 - j10)(28 + j10)} = \frac{100(28 + j10)}{(28)^2 + (10)^2}$$
$$= \frac{200(28 + j10)}{884} = 3.17 + j1.13$$

(*i*) Magnitude of current $= \sqrt{(3.17)^2 + (1.13)^2}$ = **3.36 A. (Ans.)**

(*ii*)
$$V_1 = IZ_1 = (3.17 + j1.13)(8 + j6)$$
$$= 3.17 \times 8 + 3.17 \times j6 + 8 \times j1.13 + j1.13 \times j6$$
$$= 25.36 + j19.02 + j9.04 - 6.78 = \mathbf{18.58 + j28.06.\ (Ans.)}$$
$$V_2 = IZ_2 = (3.17 + j1.13)(12 - j16)$$
$$= 38.04 - j50.72 + j13.56 + 18.08 = \mathbf{56.12 - j37.16.\ (Ans.)}$$
$$V_3 = IZ_3 = (3.17 + j1.13)(8 + j0) = \mathbf{25.36 + j9.04.\ (Ans.)}$$
$$[V = V_1 + V_2 + V_3 = (18.58 + j28.06) + (56.12 - j37.16) + (25.36 + j9.04) = 100 + j0 \text{ (check)}]$$

Example 35. *Fig. 52 shows a circuit connected to a 230 V, 50 Hz supply. Determine the following :*

(*i*) *Current drawn* (*ii*) *Voltages V_1 and V_2*

(*iii*) *Power factor.*

Draw also the phasor diagram.

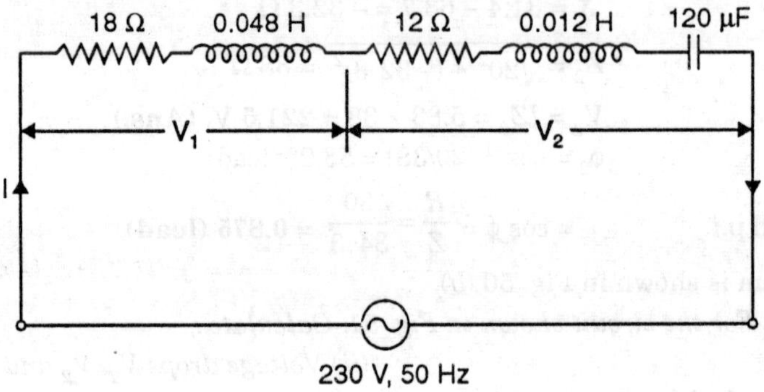

Fig. 52 **(Pune University)**

Solution. Refer Fig. 52.

Given : $R_1 = 18\ \Omega, L_1 = 0.048\ \text{H}\ ;\ R_2 = 12\ \Omega, L_2 = 0.012\ H\ ;\ C = 120\ \mu F = 120 \times 10^{-6}$ F.

$\therefore$ $X_{L1} = 2\pi f L_1 = 2\pi \times 50 \times 0.048 = 15.08\ \Omega\ ;\ X_{L2} = 2\pi f L_2 = 2\pi \times 50 \times 0.012 = 3.77\ \Omega$

$$X_{C2} = \frac{1}{2\pi f C} = \frac{1}{2\pi \times 50 \times 120 \times 10^{-6}} = 26.53\ \Omega$$

Impedance, $\overline{Z}_1 = R_1 + jX_{L1} = 18 + j15.08 = 23.48 \angle 39.96°$

Impedance, $\overline{Z}_2 = R_2 + jX_{L2} - jX_C = 12 + j3.77 - j26.53 = 12 - j22.76 = 25.73 \angle -62.2°$

Total impedance, $\overline{Z} = \overline{Z}_1 + \overline{Z}_2 = (18 + j15.08) + (12 - j22.76) = 30 - j7.68 = 30.97 \angle -14.36°$

Current drawn, I :

Taking supply voltage as reference vector, $\overline{V} = V\angle 0° = 230\angle 0°$.

Current, $\overline{I} = \frac{\overline{V}}{\overline{Z}} = \frac{230\angle 0°}{30.97\angle -14.36°}$ = **7.43 ∠14.36°A. (Ans.)**

Voltages V_1 and V_2 :

Voltage, $\bar{V_1} = \bar{I}\bar{Z_1} = 7.43 \angle 14.36° \times 23.48 \angle 39.96°$

$= 7.43 \times 23.48 \angle(14.36° + 39.96°)$

$= \mathbf{174.46 \angle 54.32° \text{ V.}}$ **(Ans.)**

Voltage $\bar{V_2} = \bar{I}\bar{Z_2} = 7.43 \angle 14.36° \times 25.73 \angle -62.2°$

$= 7.43 \times 25.73 \angle(14.36° - 62.2°)$

$= \mathbf{191.2 \angle -47.84°}$. **(Ans.)**

Phase angle between supply voltage and current *i.e.*, V and I, $\phi = 14.36°$ (lead)

Power factor, cos ϕ :

$\cos \phi = \cos (14.36°)$

$=$ **0.9687 (leading). (Ans.)**

$[\bar{V} = \bar{V_1} + \bar{V_2} = 174.46 \angle 54.32° + 191.2 \angle -47.84° = 230 \angle 0°$ (check)]

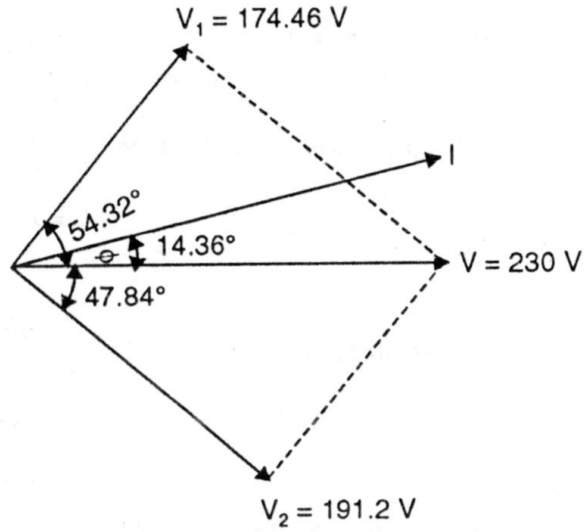

Fig. 53

Resonance, Q-Factor and Bandwidth

Example 36. *A circuit consisting of a coil having an inductance of 0.25 H and a resistance of 3 Ω is arranged in series with a capacitor of capacitance 20 μF. Calculate at what frequency resonance will take place and current flow if an alternating voltage of 40 V at the resonant frequency is applied to the circuit. Find also the voltage across the capacitor.*

Solution. Resonant frequency, $f_r = \dfrac{1}{2\pi\sqrt{LC}}$ [L in henry, C in farad]

$$= \frac{1}{2\pi\sqrt{0.25 \times 20 \times 10^{-6}}} = \mathbf{71.2 \text{ Hz. (Ans.)}}$$

At the resonant frequency, I_r (or I_{max}) $= \dfrac{V}{R} = \dfrac{40}{3} =$ **13.33 A. (Ans.)**

Voltage across the capacitor, $V_c = IX_c$

$$= \frac{13.33}{2\pi fC} \text{ (where } C \text{ is in farad)}$$

$$= \frac{13.33}{2\pi \times 71.2 \times 20 \times 10^{-6}} = \mathbf{1489.8 \text{ V. (Ans.)}}$$

Example 37. *A coil of inductance 0.64 H and resistance 40 Ω is connected in series with a capacitor of capacitance 12 μF.*

Estimate :

(*i*) *The frequency at which resonance will occur.*

(*ii*) *The voltage across the coil and capacitor, respectively and also the supply voltage when a current of 1.5 A at the resonant frequency is flowing.*

(*iii*) *The three voltages in (ii) with a current of 1.5 A flowing at a frequency of 50 Hz.*

Solution. (*i*) $f_r = \dfrac{1}{2\pi\sqrt{LC}}$ [L is in henry, C is in farad]

$$= \frac{1}{2\pi\sqrt{0.64 \times 12 \times 10^{-6}}} = \mathbf{57.4 \text{ Hz. (Ans.)}}$$

(*ii*) At resonance the supply voltage $= IR$

$= 1.5 \times 40 =$ **60 V. (Ans.)**

Voltage across the coil $= I\sqrt{R^2 + X_L{}^2}$

$= 1.5\sqrt{40^2 + (2\pi \times 57.4 \times 0.64)^2}$ = **351.4 V. (Ans.)**

Voltage across the capacitor $= IX_C$

$= I \times \dfrac{1}{2\pi fC}$ [C in μF]

$= \dfrac{1.5}{2\pi \times 57.4 \times 12 \times 10^{-6}}$ = **346.6 V. (Ans.)**

(*iii*) **At 50 Hz :**

Voltage across the coil $= I\sqrt{R^2 + X_L{}^2}$

$= 1.5\sqrt{40^2 + (2\pi \times 50 \times 0.64)^2}$ = **307.5 V. (Ans.)**

Voltage across the capacitor $= IX_C$

$= \dfrac{1.5}{2\pi \times 50 \times 12 \times 10^{-6}} \simeq$ **398 V. (Ans.)**

Voltage across the entire circuit,

$$V = I\sqrt{R^2 + (X_L - X_C)^2}$$

$$= 1.5\sqrt{(40)^2 + \left\{(2\pi \times 50 \times 0.64) - \left(\frac{1}{2\pi \times 50 \times 12 \times 10^{-6}}\right)\right\}^2}$$

$= 1.5\sqrt{(40)^2 + (-64.2)^2}$ = **113.5 V. (Ans.)**

Example 38. *A coil having an inductance of 50 mH and resistance 10 Ω is connected in series with a 25 μF capacitor across a 200 V A.C. supply. Calculate :*

(*i*) *Resonance frequency of the circuit ;*

(*ii*) *Current flowing at resonance ;*

(*iii*) *Value of Q by using different data.* **(Bombay University)**

Solution. (*i*) $f_r = \dfrac{1}{2\pi\sqrt{LC}} = \dfrac{1}{2\pi\sqrt{50 \times 10^{-3} \times 25 \times 10^{-6}}}$ = **142.3 Hz. (Ans.)**

(*ii*) $I_{max} = \dfrac{V}{R} = \dfrac{200}{10}$ = **20 A. (Ans.)**

(*iii*) $Q = \dfrac{1}{R}\sqrt{\dfrac{L}{C}} = \dfrac{1}{10}\sqrt{\dfrac{50 \times 10^{-3}}{25 \times 10^{-6}}}$ = **4.47. (Ans.)**

Example 39. *A series R-L-C series circuit consists of R = 800 Ω, L = 80 mH and C = 8 picofarad. The applied voltage across the circuit is 100 V. Determine :*

(*i*) *Resonant frequency of the circuit.*

(*ii*) *Q-factor of the circuit at the resonant frequency.*

(*iii*) *Bandwidth of the resonant circuit.*

(*iv*) *Frequencies at which the half power points occur.*

(*v*) *Bandwidth of the circuit.* **(Delhi University)**

Solution. *Given :* $R = 800\ \Omega$; $L = 80$ mH $= 0.08$ H ; $C = 8 \times 10^{-12}$ F ; $V = 100$ volts.

(*i*) **Resonant frequency of the circuit, f_r :**

$f_r = \dfrac{1}{2\pi\sqrt{LC}} = \dfrac{1}{2\pi\sqrt{0.08 \times 8 \times 10^{-12}}}$ = **198.94 kHz. (Ans.)**

(*ii*) **Q-factor at the resonant frequency :**

$$(Q\text{-factor})_{\text{resonant frequency}} = \frac{1}{R}\sqrt{\frac{L}{C}} = \frac{1}{800}\sqrt{\frac{0.08}{8\times10^{-12}}} = \mathbf{125.} \quad \textbf{(Ans.)}$$

(*iii*) **Bandwidth of the resonant circuit ; B_{hp} ;**

$$B_{hp} = \frac{R}{2\pi L} = \frac{800}{2\pi\times0.08} = \mathbf{1591.5\ Hz.} \quad \textbf{(Ans.)}$$

Also,
$$B_{hp} = \frac{f_r}{Q} = \frac{198.94\text{ kHz}}{125} = 1.5915\text{ kHz} = 1591.5\text{ Hz} \quad \text{...as above}$$

(*iv*) **Frequencies at which the half power points occur, ω_1, ω_2 :**

$$f_1 = f_r - \frac{R}{4\pi L} = 198.94\text{ (kHz)} - \frac{800}{4\pi\times0.08}\times\frac{1}{1000}\text{ (kHz)} = \mathbf{198.14\ kHz.} \quad \textbf{(Ans.)}$$

$$f_2 = f_r + \frac{R}{4\pi L} = 198.94 + \frac{800}{4\pi\times0.08}\times\frac{1}{1000} = \mathbf{199.74\ kHz.} \quad \textbf{(Ans.)}$$

(*v*) **Bandwidth of the resonant circuit :**

Bandwidth $= f_2 - f_1 = 199.74 - 198.14 =$ **1.6 kHz. (Ans.)**

Example 40. *A series R-L-C circuit consists of R = 20 Ω, L = 20 mH and C = 0.5 μF. If the circuit is connected to a 20 V variable frequency supply calculate the following :*

(*i*) *Resonant frequency f_r.*

(*ii*) *Resonance circuit Q-factor using L/C ratio.*

(*iii*) *Half-power bandwidth, using f_r and Q-factor.*

(*iv*) *Half-power bandwidth using the general formula for any bandwidth.*

(*v*) *Half-power bandwidth using the given component values.*

(*vi*) *Maximum power dissipated at f_r.* **(Delhi University)**

Solution. *Given :* $R = 20\ \Omega$; $L = 20\text{ mH} = 0.02\text{ H}$; $C = 0.5\times10^{-6}\text{ F}$; V = 20 volts.

(*i*) **Resonant frequency,** $\mathbf{f_r} = \dfrac{1}{2\pi\sqrt{LC}}$

$$= \frac{1}{2\pi\sqrt{0.02\times0.5\times10^{-6}}} = \mathbf{1591\ Hz.} \quad \textbf{(Ans.)}$$

(*ii*) **Q-factor**
$$= \frac{1}{R}\sqrt{\frac{L}{C}} = \frac{1}{20}\sqrt{\frac{0.02}{0.5\times10^{-6}}} = \mathbf{10.} \quad \textbf{(Ans.)}$$

(*iii*) **Half-power bandwidth** $\left(\textit{using } \dfrac{L}{C} \textit{ ratio}\right)$, $\mathbf{B_{hp}} = \dfrac{f_r}{Q\text{-factor}} = \dfrac{1591}{10} = \mathbf{159.1\ Hz.}$ **(Ans.)**

(*iv*) **Half-power bandwidth** *(using the general formula)*,

$$B_{hp} = \frac{f_r Q}{(Q)_{\text{resonance}}} = \frac{1591\times\tan 45^\circ}{10} = \mathbf{159.1\ Hz.} \quad \textbf{(Ans.)}$$

($\because$ At half power points, $Q = \tan\theta = \tan 45^\circ = 1$)

(*v*) **Half-power bandwidth** *(using component values)*,

$$B_{hp} = \frac{R}{2\pi L} = \frac{20}{2\pi\times0.02} = \mathbf{1591.5.} \quad \textbf{(Ans.)}$$

(*vi*) **Maximum power dissipated at $\mathbf{f_r}$,**

$$P_r = I_r^2 R = \left(\frac{V}{R}\right)^2 R = \frac{V^2}{R} = \frac{(20)^2}{20} = \mathbf{20\ W.} \quad \textbf{(Ans.)}$$

4.6. A.C. Parallel Circuits

4.6.1. Introduction

Now-a-days, owing to multiple system of transmission and distribution, we come across parallel circuits (*i.e.*, impedances joined in parallel) more often. Practically all lighting and power circuits are constant voltage circuits with the loads connected in parallel. In a parallel A.C. circuit (like parallel D.C. circuit) the *voltage is the same across each branch.*

4.6.2. Methods for solving A.C. parallel circuits

The following *three* methods are available to solve such circuits :

1. Phasor or vector method
2. Admittance method
3. Vector algebra (symbolic method or *j*-method)

1. *Vector or phasor method :*

Consider a parallel circuit consisting of two branches of impedances $Z_1(R_1, L)$ and $Z_2(R_2, C)$ respectively, and connected in parallel across an alternating voltage *V* volts (r.m.s.), as shown in Fig. 54 (*a*). Since the two branches are connected in parallel therefore, the voltage across each branch is the same and equal to supply voltage *V* but currents through them will be different.

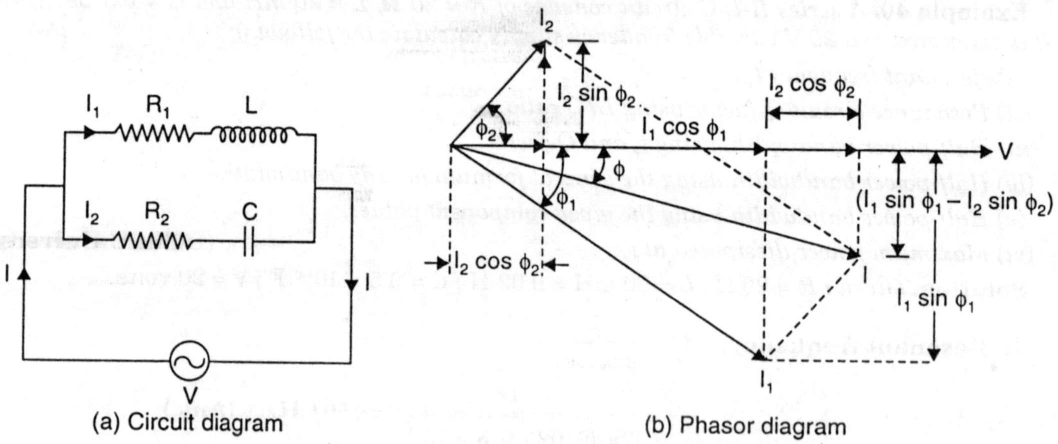

Fig. 54. Single-phase parallel circuit—Phasor method.

Branch-1 Impedance, $Z_1 = \sqrt{R_1^2 + X_L^2}$

Current, $I_1 = \dfrac{V}{Z_1}$;

Power factor, $\cos \phi_1 = \dfrac{R_1}{Z_1}$ or $\phi_1 = \cos^{-1}\left(\dfrac{R_1}{Z_1}\right)$;

Current, I_1 lags behind the applied voltage by ϕ_1.

Branch-2 Impedance, $Z_2 = \sqrt{R_2^2 + X_C^2}$;

Current, $I_2 = \dfrac{V}{Z_2}$;

Power factor, $\cos \phi_2 = \dfrac{R_2}{Z_2}$ or $\phi_2 = \cos^{-1}\left(\dfrac{R_2}{Z_2}\right)$

Current I_2 leads V by ϕ_2 [Fig. 54 (*b*)]

Resultant current I, which is phasor sum of I_1 and I_2, can be determined either by using parallelogram law of phasors, as shown in Fig. 54 (*b*) or by resolving branch currents I_1 and I_2 along X-axis and Y-axis and then determining the resultant of these components analytically.

Component of resultant current I along X-axis

= Sum of components of branch currents I_1 and I_2 along X-axis

or $$I \cos \phi = I_1 \cos \phi_2 + I_2 \cos \phi_2$$

Similarly, component of resultant current I along Y-axis

= Sum of components of branch currents I_1 and I_2 along Y-axis

or $$I \sin \phi = -I_1 \sin \phi_1 + I_2 \sin \phi_2$$

$\therefore$ $$I = \sqrt{(I_1 \cos \phi_1 + I_2 \cos \phi_2)^2 + (I_2 \sin \phi_2 - I_1 \sin \phi_1)^2} \quad ...(9)$$

and $$\tan \phi = \frac{I_2 \sin \phi_2 - I_1 \sin \phi_1}{I_1 \cos \phi_1 + I_2 \cos \phi_2} = \frac{Y\text{-component}}{X\text{-component}} \quad ...[9(a)]$$

If tan ϕ is +ve, then current I will lead the applied voltage V, if ϕ is –ve current I will lag behind the applied voltage V.

Power factor of the whole circuit is given by,

$$\cos \phi = \frac{I_1 \cos \phi_1 + I_2 \cos \phi_2}{I} = \frac{X\text{-component}}{I} \quad ...[9(b)]$$

2. *Admittance method :*

Admittance (denoted by symbol Y) of a circuit is defined as the *reciprocal of its impedance.*

$\therefore$ $$Y = \frac{1}{Z} = \frac{\text{I}}{V} \quad \text{or} \quad Y = \frac{\text{r.m.s. amperes}}{\text{r.m.s. volts}}$$

The unit of admittance is *siemens* (S). The old unit was mho (℧).

As the impedance Z of a circuit has two components R and X (See Fig. 55), similarly, shown in Fig. 56, admittance Y also has two components G (*conductance*-X-component) and B (*susceptance*-Y-component).

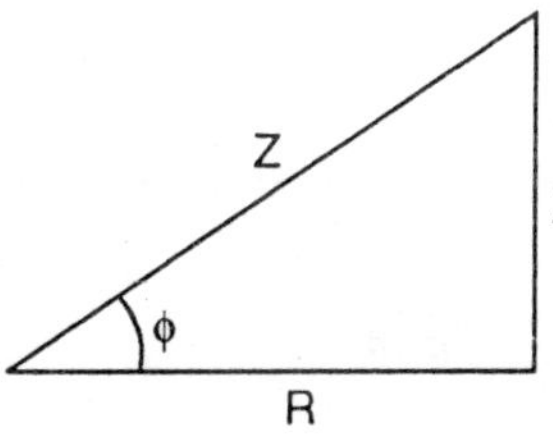

Fig. 55

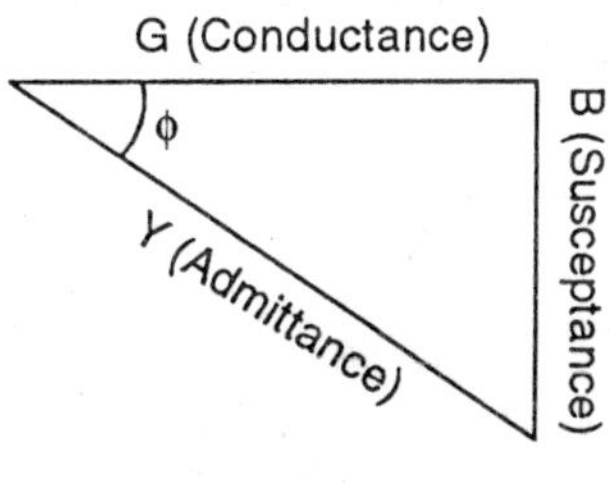

Fig. 56

Obviously, $$G = Y \cos \phi = \frac{1}{Z} \cdot \frac{R}{Z} = \frac{R}{Z^2} = \frac{R}{R^2 + X^2}$$

Similarly, $$B = Y \sin \phi = \frac{1}{Z} \cdot \frac{X}{Z} = \frac{X}{Z^2} = \frac{X}{R^2 + X^2}$$

$\therefore$ Admittance, $Y = \sqrt{G^2 + B^2}$ just as $Z = \sqrt{R^2 + X^2}$

The units of G, B and Y are in Siemens. Here, we shall consider capacitive susceptance as +ve and inductive capacitance as –ve.

Application of admittance method in solution of single-phase parallel circuits : Refer Fig. 57. Determine conductance and susceptance of individual branches from the relations

$$G = \frac{R}{Z^2} \text{ and } B = \frac{X}{Z^2}$$

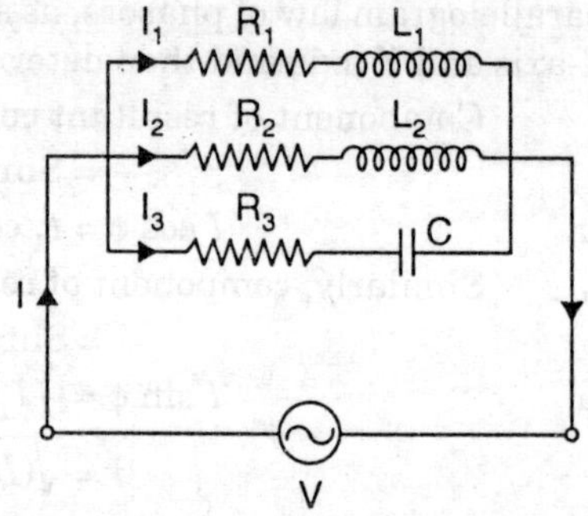

Fig. 57. Admittance method.

Taking B as +ve if X is capacitive and as –ve if X is inductive. Let the conductances of the three branches of circuit shown in Fig. 57 be G_1, G_2 and G_3 respectively and susceptances be B_1, B_2 and B_3 respectively. Total conductance is found by merely adding the conductances of three branches. Similarly, total susceptance is found by algebraically adding the individual susceptances of different branches.

∴ Total conductance, $G = G_1 + G_2 + G_3$

and, total susceptance, $B = B_1 + B_2 + B_3$

∴ Total admittance $Y = \sqrt{G^2 + B^2}$ ...(10)

Total current, $I = VY$ ...(11)

Power factor, $\cos \phi = \frac{G}{Y}$ ...(12)

3. *Complex or Phasor algebra :*

Consider the parallel circuit shown in Fig. 58 in which two impedances $\overline{Z}_1$ and $\overline{Z}_2$, being in parallel, have the same potential difference across them

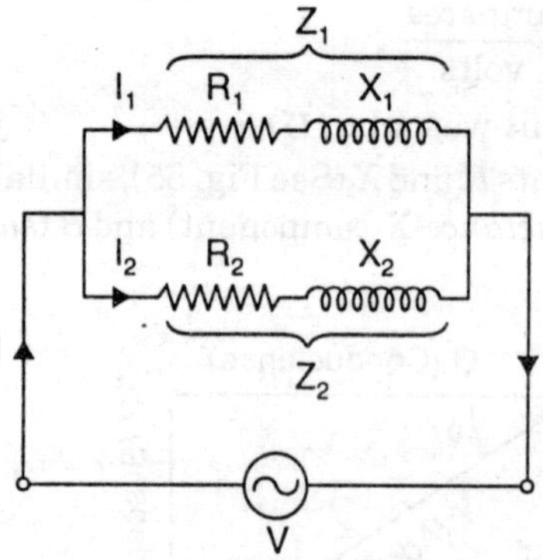

Fig. 58

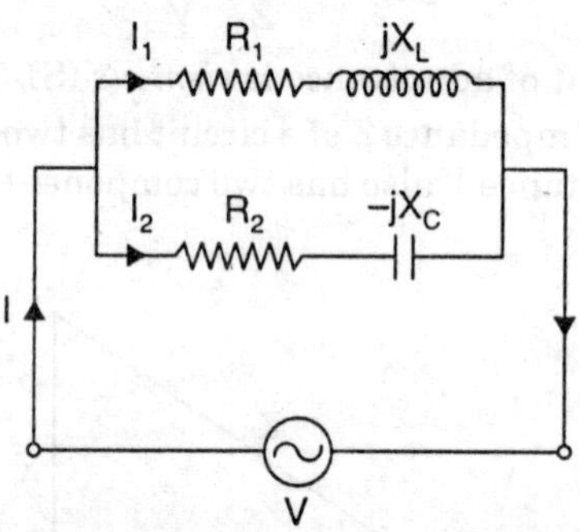

Fig. 59

Now, $\overline{I}_1 = \frac{\overline{V}}{\overline{Z}_1}$ and $\overline{I}_2 = \frac{\overline{V}}{\overline{Z}_2}$

and current $\overline{I} = \overline{I}_1 + \overline{I}_2 = \frac{\overline{V}}{\overline{Z}_1} + \frac{\overline{V}}{\overline{Z}_2}$

$$= \overline{V}\left(\frac{1}{\overline{Z}_1} + \frac{1}{\overline{Z}_2}\right) = \overline{V}(\overline{Y} + \overline{Y}_2) = \overline{V}\overline{Y} \quad ...(13)$$

where $\overline{Y}$ (= total admittance) = $\overline{Y}_1 + \overline{Y}_2$ ...(14)

It may be noted that the *admittances* are *added in parallel branches*, whereas *impedances are added for branches in series.*

It is most important to remember that *admittances and impedances being complex quantities must be added in complex form.*

Let us now consider the two parallel branches shown in Fig. 59, we have

$$\bar{Y}_1 = \frac{1}{\bar{Z}_1} = \frac{1}{R_1 + jX_L} = \frac{R_1 - jX_L}{(R_1 + jX_L)(R_1 - jX_L)}$$

$$= \frac{R_1 - jX_L}{R_1^2 + X_L^2} = \frac{R_1}{R_1^2 + X_L^2} - j\frac{X_L}{R_1^2 + X_L^2} = G_1 - jB_1$$

where $G_1 = \frac{R_1}{R^2 + X_L^2}$ ...Conductance of upper branch,

$B_1 = -\frac{X_L}{R_1^2 + X_L^2}$ ...Susceptance of upper branch.

Similarly, $\bar{Y}_2 = \frac{1}{\bar{Z}_2} = \frac{1}{R_2 - jX_C} = \frac{(R_2 + jX_C)}{(R_2 - jX_C)(R_2 + jX_C)}$

$$= \frac{R_2 + jX_C}{R_2^2 + X_C^2} = \frac{R_2}{R_2^2 + X_C^2} + j\frac{X_C}{R_2^2 + X_C^2} = G_2 + jB_2$$

Total admittance $Y = \bar{Y}_1 + \bar{Y}_2 = (G \ \ jB_1) + (G_2 + jB_2) = (G_1 + G_2) - j(B_1 - B_2) = G - jB$

$$\bar{Y} = \sqrt{(G_1 + G_2)^2 + (B_1 - B_2)^2} \quad ...(15)$$

and $$\phi = \tan^{-1}\left(\frac{B_1 - B_2}{G_1 + G_2}\right) \quad ...(16)$$

For admittance the *polar form* is :

$\bar{Y} = Y\angle\phi°$, where ϕ is as given above

$$Y = \sqrt{G^2 + B^2}\ \angle\tan^{-1}\left(\frac{B}{G}\right)$$

Total current $\bar{I} = \bar{V}\,\bar{Y}$; $I_1 = \bar{V}\,\bar{Y}_1$ and $I_2 = \bar{V}\,\bar{Y}_2$

If $\bar{V} = V\angle 0°$ and $\bar{Y} = Y\angle\phi$ then $\bar{I} = \bar{V}\,\bar{Y} = V\angle 0° \times Y\angle\phi = VY\angle\phi$

In general, *i.e.*, $\bar{V} = V\angle\alpha$ and $\bar{Y} = Y\angle\beta$, then

$$\bar{I} = \bar{V}\,\bar{Y} = V\angle\alpha \times Y\angle\beta = VY\angle(\alpha + \beta) \quad ...(17)$$

Thus, it is worth noting that when vector algebra is multiplied by admittance either in complex (rectangular) or polar form, the result is vector current in its proper phase relationship with respect to the voltage, irrespective of the axis to which the voltage may have been referred to.

Example 41. *A resistance of 60 Ω, an inductance of 0.18 H and a capacitance of 120 μF are connected in parallel across a 100 V, 50 Hz supply. Calculate :*

(*i*) *Current in each path.*

(*ii*) *Resultant current.*

(*iii*) *Phase angle between the resultant current and the supply voltage.*

(*iv*) *Power factor of the circuit.*

Solution. *Given :* $R = 60\ \Omega$; $L = 0.18$ H ; $C = 120\ \mu F = 120 \times 10^{-6}$ F, $V = 100$ volts, 50 Hz.

∴ Inductance reactance, $X_L = 2\pi fL = 2\pi \times 50 \times 0.18 = 56.55\ \Omega$, and

capacitance reactance, $X_C = \dfrac{1}{2\pi fC} = \dfrac{1}{2\pi \times 50 \times 120 \times 10^{-6}} = 26.53\ \Omega$

(*i*) **Current in each path :**

Current through reactance,

$$\mathbf{I_1} = \frac{V}{R} = \frac{100}{60} = \textbf{1.67 A in phase with voltage V.} \quad \textbf{(Ans.)}$$

Current through inductance,

$$\mathbf{I_2} = \frac{V}{X_L} = \frac{100}{56.55} = \textbf{1.77 A lagging behind voltage V by 90°.} \quad \textbf{(Ans.)}$$

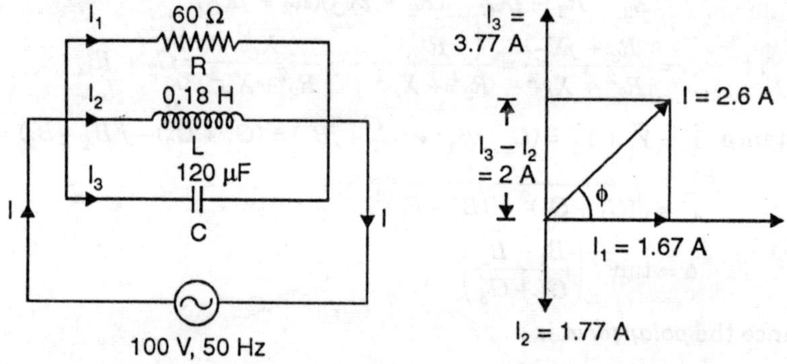

(a) Circuit diagram (b) Phasor diagram

Fig. 60

Current through capacitance,

$$\mathbf{I_3} = \frac{V}{X_C} = \frac{100}{26.53} = \textbf{3.77 A leading the voltage V by 90°.} \quad \textbf{(Ans.)}$$

The circuit and phasor diagrams are shown in Fig. 60 (*a*) and (*b*) respectively.

(*ii*) **Resultant current, I :**

Resultant current, $I = \sqrt{(I_1)^2 + (I_3 - I_2)^2} = \sqrt{(1.67)^2 + (3.77 - 1.77)^2} = \mathbf{6\ A.}$ **(Ans.)**

(*iii*) **Phase angle between the resultant current and the supply voltage, φ :**

$$\phi = \tan^{-1}\left(\frac{I_3 - I_2}{I_1}\right) = \tan^{-1}\left(\frac{2}{1.67}\right) = \textbf{50.14° (lead).} \quad \textbf{(Ans.)}$$

(*iv*) **Power factor of the circuit, cos φ :**

$$\cos\phi = \cos 50.14° = \textbf{0.641 (lead).} \quad \textbf{(Ans.)}$$

Example 42. *Determine the r.m.s. value of current in each branch and total current of the circuit shown in Fig. 61. Draw the phasor diagram.* **(PTU, Jan. 2000)**

Solution. Refer Fig. 61 (*a*).

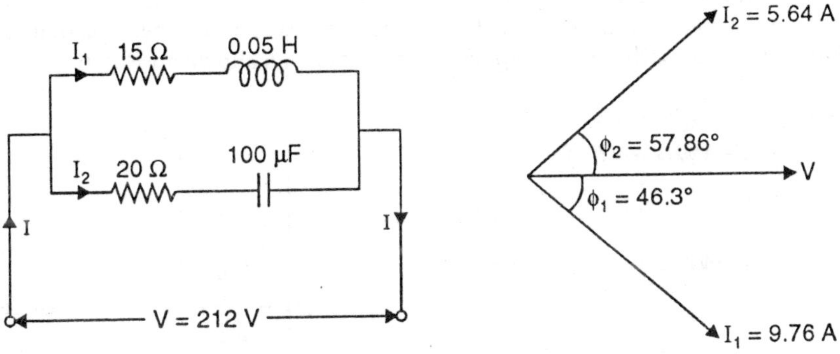

(*a*) Parallel circuit (*b*) Phasor diagram

Fig. 61

Branch No. 1 :

$$R_1 = 15\ \Omega, X_1 = 2\pi \times f \times L = 2\pi \times 50 \times 0.05 = 15.7\ \Omega$$

$$Z_1 = \sqrt{R_1^2 + X_1^2} = \sqrt{15^2 + 15.7^2} = 21.71\ \Omega$$

$$I_1 = \frac{V}{Z_1} = \frac{212}{21.71} = 9.76 \text{ A}$$

$$\cos \phi_1 = \frac{R_1}{Z_1} = \frac{15}{21.71} = 0.691 \text{ [or } \phi_1 = \cos^{-1}(0.691) = 46.3° \textit{ (lagging)}]$$

$$\sin \phi_1 = \sin (46.3°) = 0.723$$

Branch No. 2 :

$$R_2 = 20\ \Omega, X_2 = \frac{1}{2\pi fC} = \frac{1}{2\pi \times 50 \times 100 \times 10^{-6}} = 31.83\ \Omega$$

$$Z_2 = \sqrt{R_2^2 + X_2^2} = \sqrt{20^2 + 31.83^2} = 37.59\ \Omega$$

$$I_2 = \frac{V}{Z_2} = \frac{212}{37.59} = 5.64 \text{ A}$$

$$\cos \phi_2 = \frac{R_2}{Z_2} = \frac{20}{37.59} = 0.532 \qquad \text{[or } \phi_2 = \cos^{-1}(0.532) = 57.86° \textit{ (leading)}]$$

$$\sin \phi_2 = \sin (57.86°) = 0.847$$

The phasor diagram is shown in Fig. 60 (*b*)

X-components of I_1 and $I_2 = I_1 \cos \phi_1 + I_2 \cos \phi_2$

$$= 9.76 \times 0.691 + 5.64 \times 0.532 = 9.74 \text{ A}$$

Y-components of I_1 and $I_2 = -I_1 \sin \phi_1 + I_2 \sin \phi_2$

$$= -9.76 \times 0.723 + 5.64 \times 0.847 = -2.28 \text{ A}$$

Total current, $I = \sqrt{(9.74)^2 + (2.28)^2}$ = **10 A. (Ans.)**

Example 43. *The currents in each branch of a two-branched parallel circuit are given by the expressions $i_A = 7.07 \sin (314t - \pi/4)$ and $i_B = 21.2 \sin (314t + \pi/3)$. The supply voltage is given by the expression $v = 354 \sin 314t$. Derive a similar expression for the supply current and calculate the ohmic value of the components, assuming two pure components in each branch. State whether the reactive components are inductive or capacitive.* **(Elect. Engg. Calcutta University)**

Solution. From the given expressions of currents, we find that :

- i_A lags the voltage by $\pi/4$ radian or 45° and i_B leads it by $\pi/3$ radian or 60°. Hence branch A consists of a resistance in series with a pure inductive reactance. Branch B consists of a resistance in series with a pure capacitive reactance as shown in Fig. 62

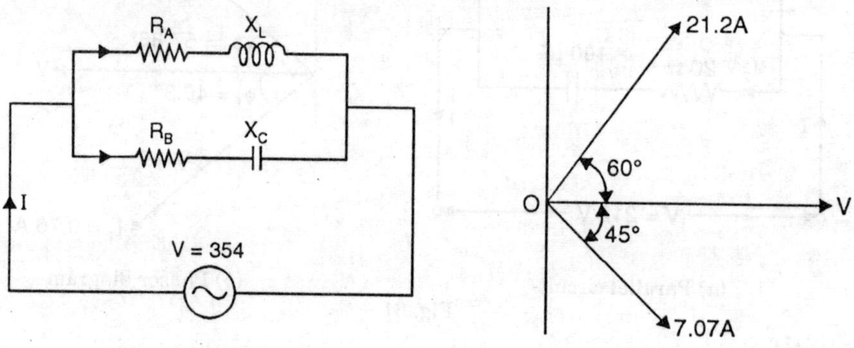

(*a*) Parallel circuit (*b*) Phasor diagram

Fig. 62

- Maximum value of current in branch A is 7.07 A and in branch B is 21.2 A.

The resultant current can be found as follows :

As seen from the vector diagram [Fig. 62 (*b*)],

X-component = 21.2 cos 60° + 7.07 cos 45° = 15.6 A

Y-component = 21.2 sin 60° – 7.07 sin 45° = 13.36 A

Maximum value of the resultant current

$$= \sqrt{(15.6)^2 + (13.36)^2} = 20.54 \text{ A}$$

$$\phi = \tan^{-1}(13.36/15.6) = 40.6° \text{ (lead)}$$

Hence the expression for the supply current is :

$$\mathbf{i = 20.54 \sin (314t + 40.6°)} \textbf{ (Ans.)}$$

$$Z_A = \frac{354}{7.07} = \mathbf{50\ \Omega}\ ; \cos\phi_A = \cos 45° = 0.707, \sin\phi_A = \sin 45° = 0.707$$

$$R_A = Z_A \cos\phi_A = 50 \times 0.707 = \mathbf{35.4\ \Omega}$$

$$X_L = Z_A \sin\phi_A = 50 \times 0.707 = \mathbf{35.4\ \Omega}.$$

$$Z_B = \frac{354}{21.2} = \mathbf{16.7\ \Omega.\ (Ans.)}$$

$$R_B = Z_B \cos 60° = 16.7 \times \cos 60° = \mathbf{8.35\ \Omega}$$

$$X_B = Z_B \sin 60° = 16.7 \times \sin 60° = \mathbf{14.46\ \Omega.\ (Ans.)}$$

Example 44. *Two impedances given by $Z_1 = (10 + j5)$ and $Z_2 = (8 + j6)$ are joined in parallel and connected across a voltage of $v = 200 + j0$. Calculate the circuit current, its phase and the branch currents. Draw the phasor diagram.* **(M.S. University Baroda)**

Solution. Refer Fig. 63 (*a*).

Branch A,
$$Y_1 = \frac{1}{Z_1} = \frac{1}{(10 + j5)}$$

$$= \frac{1}{(10 + j5)} \times \frac{10 - j5}{(10 - j5)} = \frac{10 - j5}{100 + 25} = \frac{10 - j5}{125}$$

$$= (0.08 - j0.04) \text{ siemens}$$

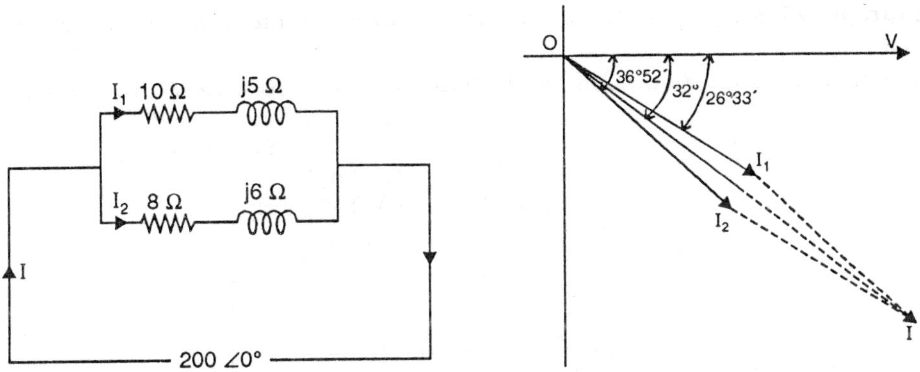

(a) Parallel circuit (b) Phasor diagram (not to scale)

Fig. 63

Branch B, $Y_2 = \frac{1}{Z_2} = \frac{1}{(8 + j6)}$

$= \frac{1}{(8 + j6)} \times \frac{(8 - j6)}{(8 - j6)} = \frac{8 - j6}{64 + 36} = \frac{8 - j6}{100}$

$= (0.08 - j0.06)$ siemens

$\therefore$ $Y = Y_1 + Y_2 = (0.08 - j0.04) + (0.08 - j0.06)$

$= (0.16 - j0.1)$ siemens

Alternatively :

$$\frac{1}{Z} = \frac{1}{Z_1} + \frac{1}{Z_2} = \frac{Z_1 + Z_2}{Z_1 Z_2}$$

$$\therefore \quad Y = \frac{Z_1 + Z_2}{Z_1 Z_2} = \frac{(10 + j5) + (8 + j6)}{(10 + j5)(8 + j6)} = \frac{(18 + j11)}{50 + j100}$$

Rationalising the above expression, we get

$$Y = \frac{(18 + j11)(50 - j100)}{(50 + j100)(50 - j100)} = \frac{2000 - j1250}{12500}$$

$= 0.16 - j0.1$ (same as before)

Now $V = 200\angle 0°$

$\therefore$ $I = VY = (200 + j0)(0.16 - j0.1)$

$= 32 - j20 =$ **37.74 ∠– 32°** ...polar form. **(Ans.)**

It lags behind the applied voltage by 32°.

Power factor $= \cos 32° =$ **0.848. (Ans.)**

$I_1 = VY_1 = (200 + j0)(0.08 - j0.04)$

$= 16 - j8 =$ **17.89 ∠ – 26° 33′. (Ans.)**

It lags behind the applied voltage by 26° 33′.

$I_2 = VY_2 = (200 + j0)(0.08 - j0.06)$

$= 16 - j12 = 20\angle - 36° 52'$

It lags behind the applied voltage by 36° 52′.

The phasor diagram is shown in Fig. 63 (b).

Example 45. *Fig. 64 shows a parallel circuit in which the values of the parameters are as given below :*

$R_1 = 70\ \Omega$ *(non-inductive) ; Coil :* $R_C = 30\ \Omega$, $L_C = 0.5\ H$; $R_2 = 100\ \Omega$; $X_C = 157\ \Omega$ *(at 50 Hz).*

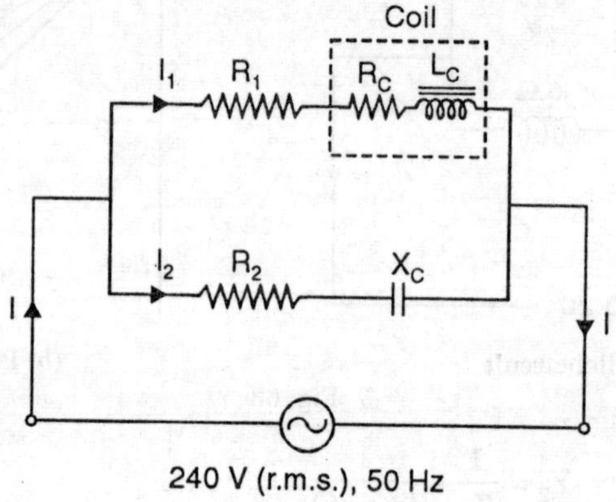

Fig. 64

(*i*) *Determine the branch currents and the total current.*

(*ii*) *Draw the phasor diagram indicating the currents and voltages across coil and condenser.*

(*iii*) *If the A.C. source is replaced by an equivalent D.C. source, what current would be drawn by the circuit ?*

(Pune University)

Solution. *Given :* $R_1 = 70\ \Omega$; $R_C = 30\ \Omega$, $L_C = 0.5$ H ; $R_2 = 100\ \Omega$; $X_C = 157\ \Omega$,

Applied voltage = 240 V (r.m.s.), 50 Hz

(*i*) **The branch currents (I_1, I_2) and the total current (I) :**

Resistance of the inductive branch $= R_1 + R_C = 70 + 30 = 100\ \Omega$

Reactance of the inductive branch $= 2\pi f L_C = 2\pi \times 50 \times 0.5 = 157\ \Omega$

Conductance of inductive branch,

$$G_1 = \frac{R_1 + R_C}{(R_1 + R_C)^2 + (2\pi f L_C)^2} = \frac{100}{(100)^2 + (157)^2} = 0.00288 \text{ S (siemens)}$$

Susceptance of inductive branch,

$$B_1 = \frac{2\pi f L_C}{(R_1 + R_C)^2 + (2\pi f L_C)^2}$$

$$= \frac{157}{(100)^2 + (157)^2} = -\ 0.00453 \text{ S (Being inductive)}$$

Conductance of capacitive branch,

$$G_2 = \frac{R_2}{{R_2}^2 + {X_C}^2} = \frac{100}{(100)^2 + (157)^2} = 0.00288 \text{ S}$$

Susceptance of capacitive branch,

$$B_2 = \frac{X_C}{{R_2}^2 + {X_C}^2} = \frac{157}{(100)^2 + (157)^2} = 0.00453 \text{ S (Being capacitive)}$$

Total conductance of the circuit,

$$G = G_1 + G_2 = 0.00288 + 0.00288 = 0.00576 \text{ S}$$

Total susceptance of the circuit,

$$B = B_1 + B_2 = -\ 0.00453 + 0.00453 = 0$$

Total admittance of the circuit,

$$Y = \sqrt{G^2 + B^2} = \sqrt{(0.00576)^2 + 0^2} = 0.00576 \text{ S}$$

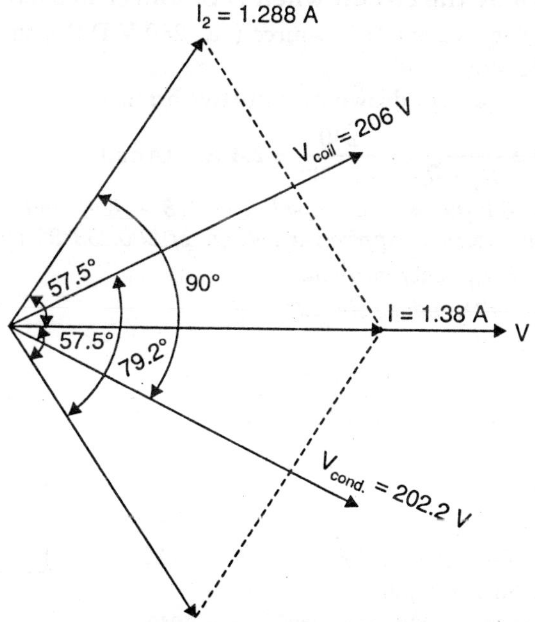

Fig. 65. Phasor diagram.

Current in inductive branch, $\mathbf{I_1} = V \times Y_1 = V\sqrt{G_1^2 + B_1^2}$

$$= 240\sqrt{(0.00288)^2 + (-0.00453)^2} = \mathbf{1.288\ A.} \quad \textbf{(Ans.)}$$

Phase angle, $\phi_1 = \tan^{-1}\left(\dfrac{B_1}{G_1}\right) = \tan^{-1}\left(\dfrac{-0.00453}{0.00288}\right) = \mathbf{-57.5°.}$ **(Ans.)**

Current in capacitive branch, $\mathbf{I_2} = V \times Y_2 = V\sqrt{G_2^2 + B_2^2}$

$$= 240\sqrt{(0.00288)^2 + (0.00453)^2} = \mathbf{1.288\ A.} \quad \textbf{(Ans.)}$$

Phase angle, $\phi_2 = \tan^{-1}\left(\dfrac{B_2}{G_2}\right) = \tan^{-1}\left(\dfrac{0.00453}{0.00288}\right) = \mathbf{57.5°.}$ **(Ans.)**

Total current, $\mathbf{I} = V \times Y = 240 \times 0.00576 = \mathbf{1.38\ A.}$ **(Ans.)**

Phase angle, $\phi = \tan^{-1}\left(\dfrac{B}{G}\right) = \tan^{-1}\left(\dfrac{0}{0.00576}\right) = 0°$

Voltage across condenser

$= I_2X_C = 1.288 \times 157 =$ **202.2 V** *lagging behind I_2 by 90°.* **(Ans.)**

Voltage across coil $= I_1 \times \sqrt{(R_C)^2 + (2\pi f L_C)^2} = 1.288\sqrt{(30)^2 + (157)^2} = 206$ V

Phase angle with current $I_1 = \tan^{-1}\left(\dfrac{157}{30}\right) = \mathbf{79.2°.}$ **(Ans.)**

(*ii*) **Phasor diagram :**

The phasor diagram indicating the currents and voltages across coil and condenser is shown in Fig. 65.

(*iii*) **Current drawn by the circuit when D.C. source is used :**

When energised by equivalent D.C. source (*i.e.*, 240 V D.C.), the capacitive branch will be open and current drawn by the circuit

= Current drawn by inductive branch

$$= \frac{V}{R_1 + R_C} = \frac{240}{70 + 30} = \mathbf{2.4\ A.}\quad \textbf{(Ans.)}$$

Example 46. *Fig. 66 shows two impedances (18 + j24) and (15 – j30) Ω connected in parallel ; across the combination is applied a voltage of 200∠53°8′. Determine :*

(i) kVA, kVAR and kW in each branch.

(ii) The power factor of the whole circuit.

Solution. Refer Fig. 66

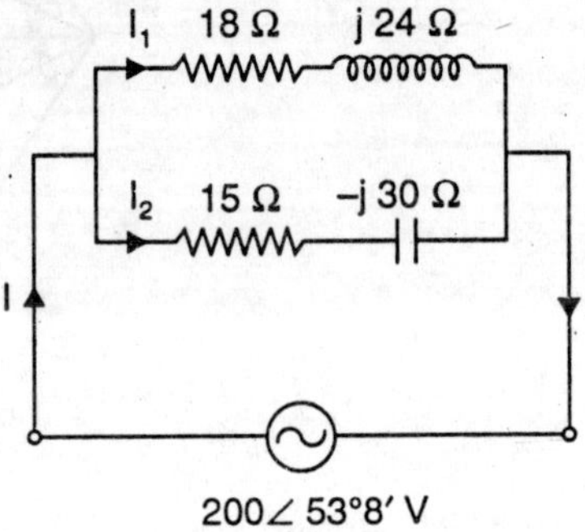

Fig. 66

$$\overline{Y}_1 = \frac{1}{18 + j24} = \frac{18 - j24}{(18 + j24)(18 - j24)}$$

$$= \frac{18 - j24}{900} = (0.02 - j0.0266)\ \text{S}$$

$$\overline{Y}_2 = \frac{1}{15 - j30} = \frac{15 + j30}{(15 + j30)(15 - j30)} = \frac{15 + j30}{1125}$$

$$= (0.0133 + j0.0266)\ \text{S}$$

Now, $V = 200\angle 53°8' = 200\,(\cos 53°8' + j \sin 53°8')$

$= 200(0.6 + j0.8) = (120 + j160)$ volts

$$\overline{I}_1 = \overline{V}\,\overline{Y}_1 = (120 + j160)(0.02 - j0.0266)$$

$= 2.4 + j3.2 - j3.2 + 4.26 = (6.66 + j0)$ ...(along the reference axis)

$\therefore$ $\overline{I}_2 = \overline{V}\,\overline{Y}_2 = (120 + j160)(0.0133 + j0.0266)$

$= 1.6 + j3.2 + j2.13 - 4.26 = -2.66 + j5.33$ (leading)

(*i*) **kVA, kVAR and kW in each branch :**

Power calculations can be calculated by the method of conjugates.

Branch 1 (*Inductive branch*)

The current conjugate of $(6.66 + j0)$ is $(6.66 - j0)$

$$\therefore\quad \overline{V}\,\overline{I}_1 = (120 + j160)(6.66 - j0) = 800 + j1066 \quad \therefore\quad \textbf{kW} = \frac{800}{1000} = \mathbf{0.8.}\quad \textbf{(Ans.)}$$

$$\therefore\quad \textbf{kVAR} = \frac{1066}{1000} = \mathbf{1.066.}\quad \textbf{(Ans.)}$$

The fact that it is positive merely shows that reactive *VA* (volt-amps) are due to lagging current

$$\textbf{kVA} = \sqrt{(kW)^2 + (kVAR)^2} = \sqrt{(0.8)^2 + (1.066)^2} = \mathbf{1.33.}\quad \textbf{(Ans.)}$$

Branch 2 (*Capacitive branch*) :

The current conjugate of $(-2.66 + j5.33)$ is $(-2.66 - j5.33)$

$$\therefore\quad \overline{V}\,\overline{I}_2 = (120 + j160)(-2.66 - j5.33) = 533.6 - j1065.2$$

$$\therefore\quad \textbf{kW} = \frac{533.6}{1000} = \mathbf{0.5336.}\quad \textbf{(Ans.)}\ ;\quad \textbf{kVAR} = \frac{-1065.2}{1000} = \mathbf{-1.065.}\quad \textbf{(Ans.)}$$

The negative sign merely indicates the reactive volt-amps are due to leading current.

$$\therefore \qquad \textbf{kVA} = \sqrt{(0.5336)^2 + (-1.065)^2} = \mathbf{1.191.} \quad \textbf{(Ans.)}$$

(*ii*) **The power factor of the whole circuit, cos φ :**

$$\overline{Y} = \overline{Y}_1 + \overline{Y}_2 = (0.02 - j0.0266) + (0.0133 + j0.0266) = 0.0333 + j0$$

$$\overline{I} = \overline{V}\,\overline{Y} = (120 + j160)(0.0333 + j0) = 4 - j5.33 = 6.66 \angle 53°8'$$

or

$$\overline{I} = \overline{I}_1 + \overline{I}_2 = (6.66 + j0) + (-2.66 + j5.33) = 4 - j5.33 \quad \text{(Same as above)}$$

Power factor of the circuit, $\cos\phi = \cos 0° = \mathbf{1.}$ **(Ans.)**

($\because$ Current is in phase with voltage)

Example 47. *For the circuit of Fig. 67, calculate the current supplied by the voltage source and the voltage across the current source.* **(PTU June 2000 ; May 2001)**

Solution. Refer Figs. 67 and 68.

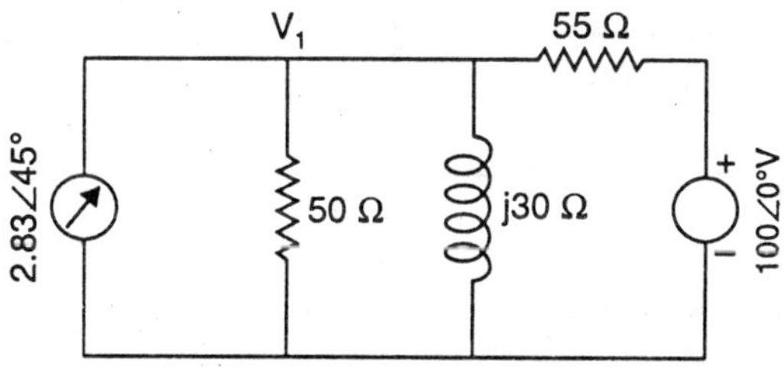

Fig. 67

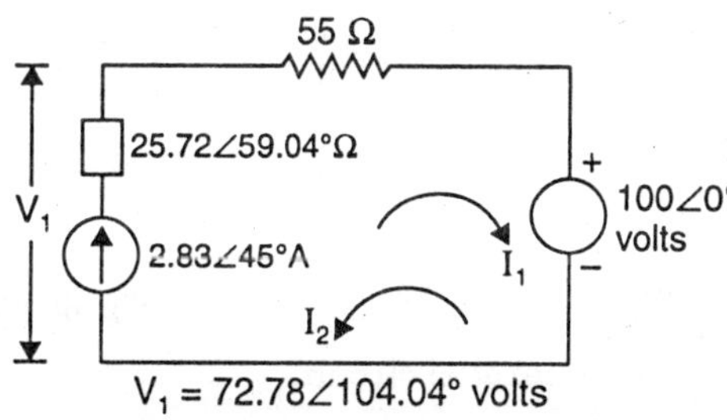

Fig. 68

Let the impedance of the current source be Z_i, then

$$\frac{1}{Z_i} = \frac{1}{50\angle 0°} + \frac{1}{30\angle 90°} = \frac{30\angle 90° + 50\angle 0°}{1500\angle 90°} = \frac{50 + j30}{1500\angle 90°} = \frac{58.31\angle 30.96°}{1500\angle 90°}$$

$$Z_i = \frac{1500\angle 90°}{58.31\angle 30.96°} = 25.72 \angle 59.04° = (13.23 + j22.05)\ \Omega$$

Voltage across current source,

$$V_1 = IZ_i = 2.83 \angle 45° \times 25.72 \angle 59.04° = \mathbf{72.78 \angle 104.04° \text{ volts.}} \quad \textbf{(Ans.)}$$

Current supplied by one source,

$$I_1 = \frac{V_1}{Z_i + R} = \frac{72.78\angle 104.04°}{13.25 + j22.05 + 55} = \frac{72.78\angle 104.04°}{68.25 + j22.05}$$

$$= \frac{72.78\angle 104.04°}{71.72\angle 17.91°} = 1.015 \angle 86.13° = (0.0685 + j1.01)\ \text{A}$$

Current supplied by the other source,

$$I_2 \frac{V_2}{Z_i + R} = \frac{100\angle 0°}{71.72\angle 17.91°} = 1.394 \angle(-17.91°) = 1.326 - j0.428$$

Current supplied by the voltage source,

$$I = I_2 - I_1 = 1.326 - j0.428 - 0.0685 - j1.01 = 1.2575 - j1.438$$

$$= \mathbf{1.91 \angle -48.83 \text{ amp.}} \quad \textbf{(Ans.)}$$

4.6.3. Series-Parallel Circuits

Series-parallel circuits may be solved by the following methods :

1. Admittance method.
2. Symbolic method.

1. *Admittance method :*

In series-parallel circuits, the parallel circuit is first reduced to an equivalent series circuit and then combined with the rest of the circuit as usual.

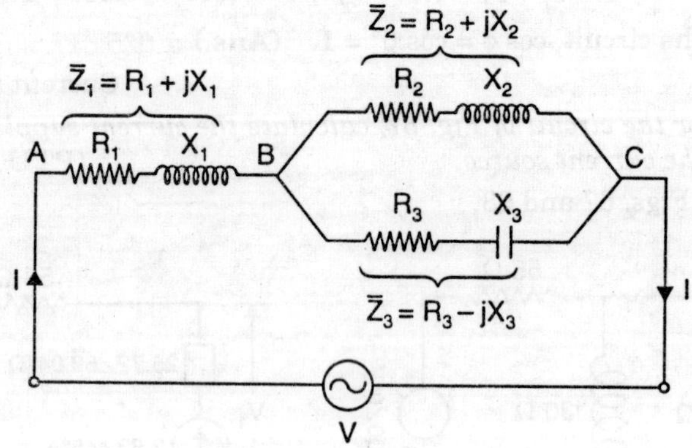

Fig. 69. Series-parallel circuit.

For a parallel circuit,

Equivalent series resistance, $R_{eq} = Z \cos \phi = \frac{1}{Y} \cdot \frac{G}{Y} = \frac{G}{Y^2}$

Equivalent series reactance, $X_{eq} = Z \sin \phi = \frac{1}{Y} \cdot \frac{B}{Y} = \frac{B}{Y^2}$

2. *Symbolic method :*

Refer series-parallel circuit shown in Fig. 69. First calculate the equivalent impedance of parallel branches and then add it to the series impedance to get the total impedance of the circuit. Then current flowing through the circuit is found as follows :

$$\bar{Y}_2 = \frac{1}{R_2 + jX_2} \quad \text{and} \quad \bar{Y}_3 = \frac{1}{R_3 - jX_3}$$

$$\therefore \quad \bar{Y}_{23} = \frac{1}{R_2 + jX_2} + \frac{1}{R_3 - jX_3}$$

$$\therefore \quad \bar{Z}_{23} = \frac{1}{\bar{Y}_{23}}; \; \bar{Z}_1 = R_1 + jX_1$$

$$\bar{Z} = \bar{Z}_{23} + \bar{Z}_1$$

$$\therefore \quad \bar{I} = \frac{\bar{V}}{\bar{Z}}.$$

Example 48. *For the circuit shown in Fig. 70 calculate :*

(i) Currents I_A, I_B and I_C.

(ii) Total power factor for the whole circuit.

Draw the complete phasor diagram.

Solution. Refer Fig. 70.

Given : $\overline{Z}_A = 2 + j1.5 = 2.5\angle 36.9°$; $\overline{Z}_B = 5 - j3.5 = 6.1\angle -35°$, $\overline{Z}_C = 3 + j2.5 = 3.9\angle 39.8°$

(*i*) **Currents I_A, I_B and I_C :**

$$\frac{1}{\overline{Z}_{AB}} = \frac{1}{\overline{Z}_A} + \frac{1}{\overline{Z}_B} \quad \text{or} \quad \overline{Z}_{AB} = \frac{\overline{Z}_A \overline{Z}_B}{\overline{Z}_A + \overline{Z}_B} = \frac{2.5\angle 36.9° \times 6.1\angle -35°}{(2 + j1.5) + (5 - j3.5)}$$

$$= \frac{15.25\angle 1.9°}{7 - j2} = \frac{15.25\angle 1.9°}{7.28\angle -16°} = 2.095\angle 17.9° \simeq 2 + j0.65$$

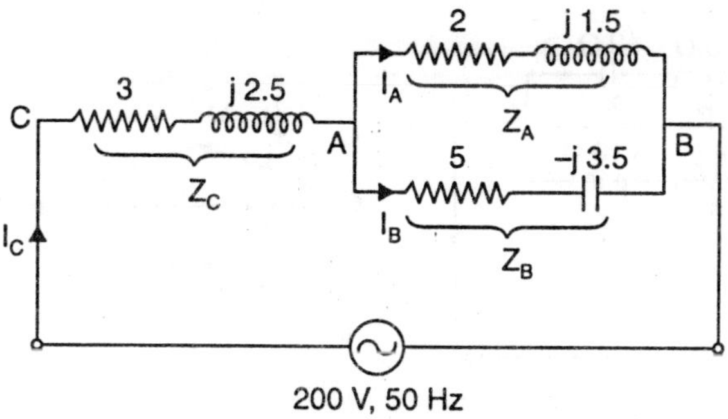

Fig. 70. Circuit diagram.

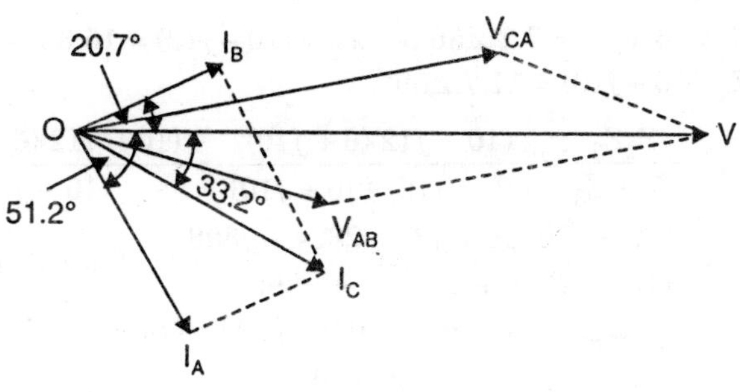

Fig. 71. Phasor diagram.

$$\overline{Z} = \overline{Z}_C + \overline{Z}_{AB} = (3 + j2.5) + (2 + j0.65) = 5 + j3.15 = 5.9\angle 32.2°$$

Let $\quad \overline{V} = 200\angle 0°$,

Then, $\quad \mathbf{I_C} = \frac{\overline{V}}{\overline{Z}} = \frac{200\angle 0°}{5.9\angle 32.2°} = \mathbf{33.9\angle -32.2°}$. **(Ans.)**

$$\mathbf{I_A} = \overline{I}_C \times \frac{\overline{Z}_B}{\overline{Z}_A + \overline{Z}_B} = 33.9\angle -32.2° \times \frac{6.1\angle -35°}{7.28\angle -16°} = \mathbf{28.4\angle -51.2°}. \quad \textbf{(Ans.)}$$

$$\mathbf{I_B} = \overline{I}_C \times \frac{\overline{Z}_A}{\overline{Z}_A + \overline{Z}_B} = 33.9\angle -32.2° \times \frac{2.5\angle 36.9°}{7.28\angle -16°} = \mathbf{11.64\angle 20.7°}. \quad \textbf{(Ans.)}$$

(*ii*) **Total power factor in the whole circuit :**

The phase angle between V and total circuit current I_C is 32.2°.

∴ Power factor for the whole circuit ; cos φ = cos 32.2° = **0.846 (lag). (Ans.)**

For drawing the phasor diagram of Fig. 71, V_{CA} and V_B have to be calculated.

$$\bar{V}_{CA} = \bar{I}_C \bar{Z}_C = 33.9\angle -32.2° \times 3.9\angle 39.8° = 132.2\angle 7.6°$$

$$\bar{V}_{AB} = \bar{I}_C \bar{Z}_{AB} = 33.9\angle -32.2° \times 2.095\angle 17.9° = 71\angle -14.3°$$

Fig. 71 shows the complete phasor diagram.

Example 49. *Determine the current drawn by the following circuit (Fig. 72) when a voltage of 200 V is applied across the same. Draw the phasor diagram.*

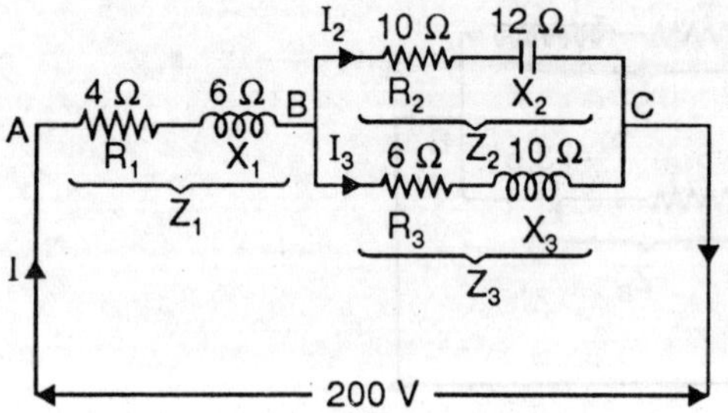

Fig. 72. Series parallel circuit

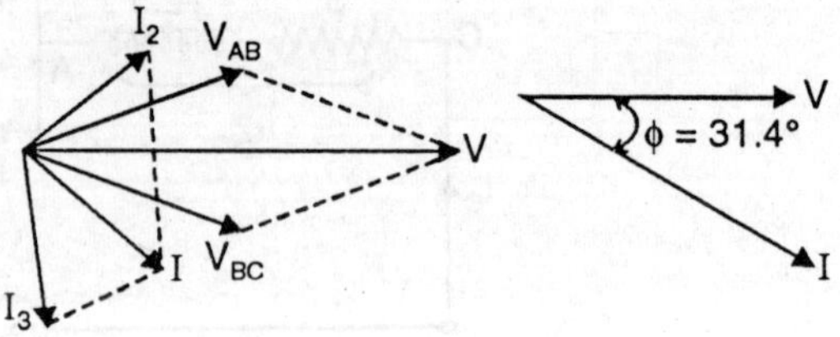

Fig. 73. Phasor diagram

(Elect. Engg. Bangalore University)

Solution. Refer Fig. 72.

$$Z_1 = (4 + j6) = 7.2\angle 56.3° \; ; \; Z_2 = (10 - j12) = 15.6\angle -50.2°$$

$$Z_3 = (6 + j10) = 11.7\angle 59°$$

$$Z_{BC} = \frac{Z_2 Z_3}{Z_2 + Z_3} = \frac{(10 - j12)(6 + j10)}{(10 - j12) + (6 + j10)} = \frac{(10 - j12)(6 + j10)}{(16 - j2)} = \frac{(180 + j28)}{(16 - j2)}$$

$$= \frac{(180 + j28)(16 + j2)}{(16 - j2)(16 + j2)} = \frac{2824 + j808}{260} = 10.9 + j3.1 = 11.3\angle 15.9°$$

$$Z = Z_1 + Z_{BC} = (4 + j6) + (10.9 + j3.1) = (14.9 + j9.1) = 17.5\angle 31.4°$$

Assuming $$V = 200\angle 0° \; ; \; I = \frac{V}{Z} = \frac{200\angle 0°}{17.5\angle 31.4°} = 11.4\angle -31.4°$$

For drawing the phasor diagram, let us find the following quantities :

$$V_{AB} = IZ_1 = 11.4\angle -31.4° \times 7.2\angle 56.3° = 82.08\angle 24.9°$$

$$V_{BC} = IZ_{BC} = 11.4\angle -31.4° \times 11.3\angle 15.9° = 128.8\angle -15.5°$$

$$I_2 = \frac{V_{BC}}{Z_L} = \frac{128.8\angle -15.5°}{15.6\angle -50.2°} = 8.26\angle 34.7°$$

$$I_3 = \frac{V_{BC}}{Z_3} = \frac{128.8\angle -15.5°}{11.7\angle 59°} = 11\angle -74.5°$$

Various currents and voltages are shown in their phase relationship in Fig. 73.

4.6.4. Resonance in parallel circuits

In case of a series circuit consisting of R (resistance), L (inductance) and C (capacitance), resonance takes place when $V_L = V_C$ *i.e.*, when $X_L = X_C$. In other words, resonance takes place when the power factor of the circuit approaches *unity*. The basic condition for resonance, *i.e.*, power factor of the entire circuit being unity, remains the same for parallel circuits also. Thus, *resonance in a parallel circuit will occur, when the power factor of the entire circuit becomes unity.*

A parallel circuit consisting of an inductive coil in parallel with a capacitor is shown in Fig. 74 (*a*). The phasor diagram of this circuit with applied voltage as the reference phasor is shown in Fig. 74 (*b*). The current drawn by the inductive branch lags the applied voltage by an angle ϕ_L. The current drawn by the capacitive branch leads the applied voltage by 90°.

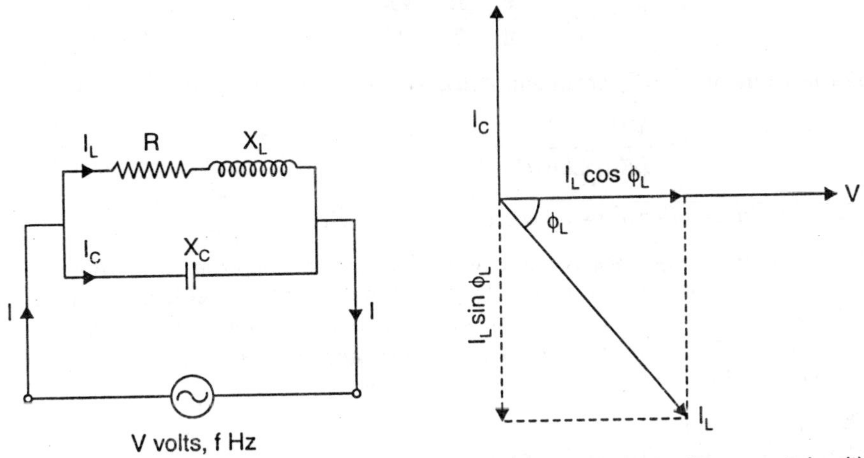

(a) Circuit diagram (Resonant circuit) (b) Phasor diagram (Resonant circuit)

Fig. 74. Resonance in a parallel circuit.

The power factor of the current becomes unity when the total current drawn by the entire circuit is in phase with the applied voltage. This will happen only when the current drawn by the capacitive branch I_C equals the active component of the current of the inductive branch I_L [Fig. 74 (*b*)].

Hence for *resonance* in parallel circuit.

$$I_C = I_L \sin \phi_L \quad \text{...(18)}$$

Now, $$I_L = \frac{V}{Z} \; ; \sin \phi_L = \frac{X_L}{Z} \text{ and } I_C = \frac{V}{X_C}$$

Hence, condition for resonance becomes

$$\frac{V}{X_C} = \frac{V}{Z} \times \frac{X_L}{Z} \quad \text{or} \quad Z^2 = X_L \times X_C$$

Now, $$X_L = \omega L, \; X_C = \frac{1}{\omega C}$$

∴ $$Z^2 = \frac{\omega L}{\omega C} = \frac{L}{C} \quad \text{...(19)}$$

or $$R^2 + X_L{}^2 = R^2 + (2\pi f_r L)^2 = \frac{L}{C}$$

or $$(2\pi f_r L)^2 = \frac{L}{C} - R^2 \quad \text{or} \quad 2\pi f_r = \sqrt{\frac{1}{LC} - \frac{R^2}{L^2}}$$

or $$f_r = \frac{1}{2\pi}\sqrt{\frac{1}{LC} - \frac{R^2}{L^2}} \quad \text{...(20)}$$

This is the resonant frequency and is given in Hz if R is in ohm, L is in henry and C in farad. If R is negligible, then

$$f_r = \frac{1}{2\pi\sqrt{LC}} \quad \text{...same as for series resonance.}$$

Current at resonance

Refer Fig. 74 (*b*). Since wattless component is zero, the circuit current is given as :

$$I = I_L \cos \phi_L = \frac{V}{Z} \times \frac{R}{Z} = \frac{VR}{Z^2}$$

Putting the value of $Z^2 = \frac{L}{C}$ from eqn. (19), we get

$$I = \frac{VR}{L/C} = \frac{V}{L/CR} \qquad \text{...(21)}$$

Thus, the *impedance* offered by a resonant parallel circuit $= \frac{L}{CR}$.

This impedance is purely resistive and generally termed as *equivalent* or *dynamic impedance* of the circuit. As the resultant current drawn by a resonant parallel circuit is minimum, the circuit is normally called ***rejector*** circuit. Such a type of circuit is quite useful in *radio work*.

The phenomenon of resonance in *parallel circuits* is normally termed as ***"current resonance"*** whereas it is termed ***"voltage resonance"*** in *series circuit*.

Resonance characteristics :

Fig. 75 shows the characteristics of the parallel circuit consisting of an inductance L and capacitance C in parallel plotted against frequency, the voltage applied being constant.

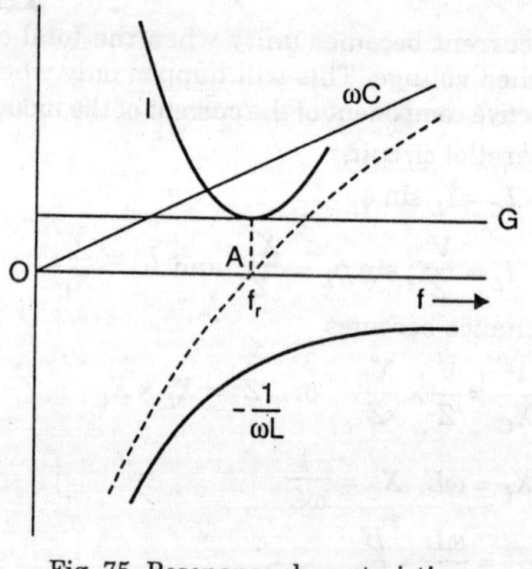

Fig. 75. Resonance characteristics.

Inductive susceptance, $\quad b = -\frac{1}{X_L} = -\frac{1}{\omega L} = -\frac{1}{2\pi f L}$

Thus inductive susceptance is *inversely proportional* to the *frequency* and is represented by rectangular hyperbola in the fourth quadrant (because it is assumed –ve).

Capacitive susceptance, $\quad b = \frac{1}{X_C} \ \omega C = 2\pi f C$

Thus capacitive susceptance is *directly proportional to the frequency* and is represented by a straight line passing through the origin.

Net susceptance, B is the difference of the two susceptances and is represented by the dotted hyperbola. The net susceptance is zero at point A, hence *admittance is minimum* and is equal to G. Thus, at point A, *line current is minimum.* The frequency at which the total current becomes minimum is the resonance frequency f_r.

Evidently, below the resonant frequency, the inductive susceptance predominates, thus making the circuit current to be lagging, whereas beyond f_r capacitive susceptance predominates and the current leads the applied voltage. At resonant frequency f_r, the current is in phase with the applied voltage.

Hence at ***parallel resonance*** it is seen that :

1. The admittance of the circuit is *minimum* and is *equal to* the conductance of the circuit.
2. The *current drawn is minimum.*
3. The phase angle between the current and voltage is zero, the *power factor is unity.*
4. The resonant frequency is given by $f_r = \frac{1}{2\pi\sqrt{LC}}$ if the *resistance* in the inductance and capacitance branches is *negligible.*

4.6.5. Comparison of series and parallel resonant circuits

S. No.	*Aspects*	*Series circuit (R-L-C)*	*Parallel circuit (R-L and C)*
1.	*Impedance at resonance*	Minimum	Maximum
2.	*Current at resonance*	Maximum = $\frac{V}{R}$	Minimum = $V/(L/CR)$
3.	*Effective impedance*	R	L/CR
4.	*Power factor at resonance*	Unity	Unity
5.	*Resonant frequency*	$\frac{1}{2\pi\sqrt{LC}}$	$\frac{1}{2\pi}\sqrt{\left(\frac{1}{LC}-\frac{R^2}{L^2}\right)}$
6.	*It magnifies*	Voltage	Current
7.	*Magnification is*	$\frac{\omega L}{R}$	$\frac{\omega L}{R}$

4.6.6. Q-factor of a parallel circuit

It is defined as the *ratio of the current circulating between its two branches to the line current drawn from the supply or simply, as the current magnification.*

$$\text{Q-factor} = \frac{1}{R}\sqrt{\frac{L}{C}} \qquad \text{...(22)}$$

- It may be noted that in *series circuits, Q-factor gives the voltage magnification, whereas in parallel circuits it gives the current magnification.*

Bandwidth of a parallel resonant circuit :

The bandwidth of a parallel circuit is defined in the same way as that for a series circuit. This circuit also has upper and lower half-power frequencies where power dissipated is half of that at resonant frequency.

The net susceptance B, at bandwidth frequencies, equals conductance. Hence,

At f_2 : $\quad B = B_{C2} - B_{L2}$

At f_1 : $\quad B = B_{L1} - B_{C1}$

Hence $Y = \sqrt{G^2 + B^2} = \sqrt{2}G$ and $\phi = \tan^{-1}\left(\frac{B}{G}\right) = \tan^{-1}(1) = 45°$

However, at off-resonant frequencies, $Y > G$ and $B_C \neq B_L$ and phase angle is greater than zero.

Example 50. *An inductive circuit of resistance 2 ohms and inductance 0.01 H is connected to a 250 V, 50 Hz supply.*

(*i*) *What capacitance placed in parallel will produce resonance ?*

(*ii*) *Determine also the total current taken from the supply and the currents in the branch circuit.* **(Kerala University)**

Solution. *Given* : $R = 2\ \Omega$; $L = 0.01$ H, Supply voltage = 250 V, 50 Hz.

(*i*) **Value of capacitance which will produce resonance, C :**

Now, $$X_L = 2\pi f L = 2\pi \times 50 \times 0.01 = 3.14\ \Omega$$

$$Z = \sqrt{R^2 + X_L^2} = \sqrt{2^2 + 3.14^2} = 3.72\ \Omega$$

We know that, $$Z^2 = \frac{L}{C} \quad \text{or} \quad C = \frac{L}{Z^2}$$

$$\therefore \quad C = \frac{0.01}{(3.72)^2} = 722.6 \times 10^{-6}\text{ F} \quad \text{or} \quad \mathbf{722.6\ \mu F.} \quad \textbf{(Ans.)}$$

(*ii*) **Total current and currents in the branch circuits, I, I_L, I_C :**

$$I_{R\text{-}L} = \frac{V}{Z} = \frac{250}{3.72} = \mathbf{67.2\ A.} \quad \textbf{(Ans.)}$$

$$\tan \phi_L = \frac{3.14}{2} = 1.57 \quad \text{or} \quad \phi_L = \tan^{-1}(1.57) = \mathbf{57.5°} \quad \textbf{(Ans.)}$$

Hence, current in *R-L* branch lags the applied voltage by 57.5°.

$$I_C = \frac{V}{X_C} = \frac{V}{1/\omega C} = \omega VC = 2\pi \times 50 \times 250 \times 722.6 \times 10^{-6} = \mathbf{56.75} \quad \textbf{(Ans.)}$$

This current leads the applied voltage by 90°.

Total current, $$I = I_{R\text{-}L} \cos \phi = 67.2 \cos 57.5° = \mathbf{36.1\ A.} \quad \textbf{(Ans.)}$$

$$\left[\text{or } I = \frac{V}{L/CR} = \frac{VCR}{L} = \frac{250 \times 722.6 \times 10^{-6} \times 2}{0.01} = 36.1\text{ A} \right]$$

5. TRANSIENTS

5.1. General Aspects

If a circuit is switched from one condition to another either by a change in the applied voltage or change in a circuit parameter, there *exists a transitional period during which the branch currents and voltage drops change from their former values to new ones. After transition period, the circuit becomes steady.*

The ***transient disturbances*** in the electrical circuits are *disturbances caused by sudden switching off and on or short circuit of the circuit and sudden change in the applied voltage.* The current developed in the circuit due to this disturbance is called the "*transient current*". The "*resultant current*" in the circuit is the *steady state current with a transient current superimposed.* The *transient currents* are found to be *associated with the changes in stored energy in capacitors and inductors.* Hence in a *purely resistive circuit no transient current is developed since there is no stored energy in a resistor.*

Single energy and double energy transients :

- Single *energy transient* is the transient disturbance where only one form of energy, *either electromagnetic or electrostatic* is involved *e.g.*, transient disturbance in a circuit consisting of resistor and inductor *i.e.*, *R-L* circuit or a circuit consisting of resistor and capacitor *i.e.*, *R-C* circuit.
- *Double energy transient* is the transient disturbance where *both electromagnetic and electrostatic energies* are involved *e.g.*, transient disturbance in a circuit consisting of resistor, inductor and capacitor *i.e.*, *R-L-C* circuit.

5.2. D.C. Transients

(*i*) **R-L transients :**

In the *R-L* circuit shown in Fig. 76

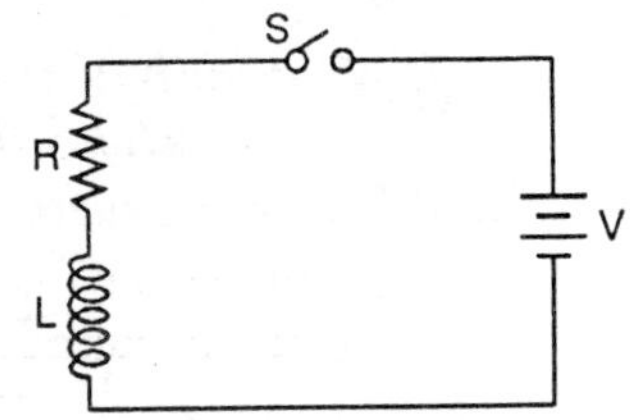

Fig. 76. *R-L* circuit.

$$i = \frac{V}{R}[1 - e^{-(R/L)t}] \qquad ...(23)$$

The plot of i (exponential rise equation) versus time is shown in Fig. 77.

The time constant (λ) for the above function is the time at which the exponent of e is unity. Thus in this case time constant (λ) is L/R. At one time constant, the value of i will be

$$i = (1 - e^{-1}) = 1 - 0.368 = 0.632$$

At this time current will be *63.2% of its final value.*

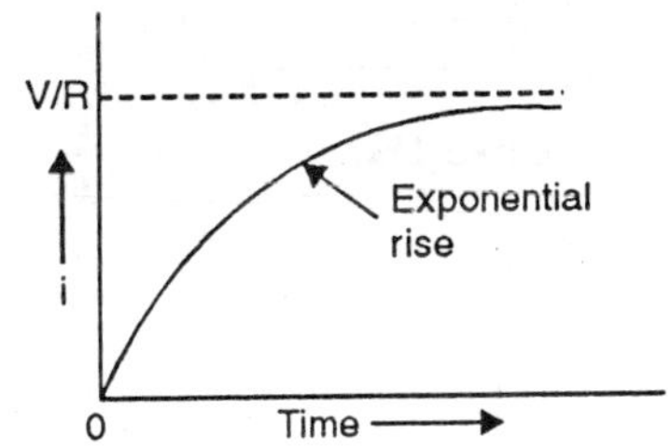

Fig. 77

The voltage across inductance,

$$v_L = L\frac{di}{dt} = Ve^{-(R/L)t} \qquad ...(24)$$

and voltage across resistor,

$$v_R = V[1 - e^{-(R/L)t}] \qquad ...(25)$$

The exponential rise of resistor voltage and exponential decay of inductor voltage are shown in Fig. 78.

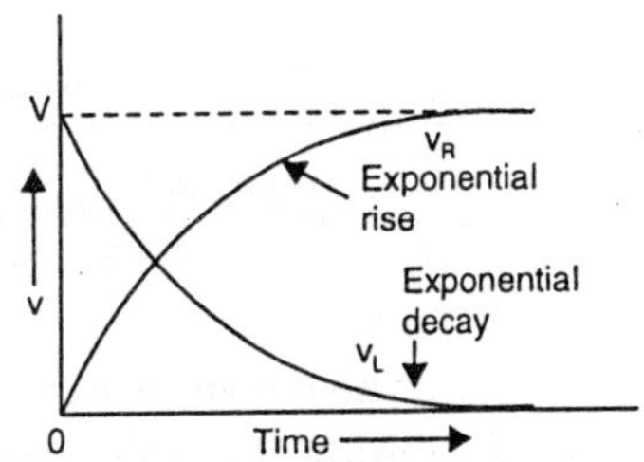

Fig. 78

Also, $\qquad v_R + v_L = V[1 - e^{-(R/L)t}] + Ve^{-(R/L)t} = V$

Power in the circuit elements is given by

$$P_R = \frac{V^2}{R}[1 - 2e^{-(R/L)t} + e^{-2(R/L)t}] \qquad ...(26)$$

$$P_L = \frac{V^2}{R}[e^{-(R/L)t} - e^{-2(R/L)t}] \qquad ...(27)$$

Total power,

$$P = P_R + P_L = \frac{V^2}{R}[1 - e^{-(R/L)t}] \qquad ...(28)$$

(*ii*) **R-C transients :**

In the *R-C* circuit shown in Fig. 79,

$$i = \frac{V}{R}e^{-t/RC}$$

Transients voltages across *R* and *C* are given by

$$v_R = Ve^{-t/RC} \qquad ...(29)$$

$$v_C = V(1 - e^{-t/RC}) \qquad ...(30)$$

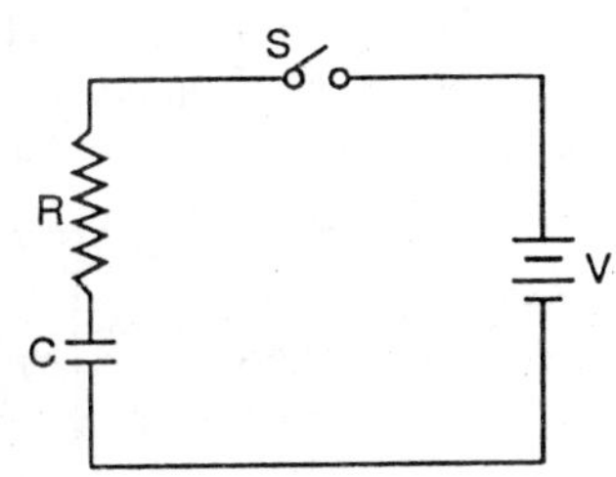

Fig. 79. *R-C* circuit.

Also, the power in circuit elements is given by

$$P_R = \frac{V^2}{R}e^{-2t/RC} \qquad ...(31)$$

$$P_L = \frac{V^2}{R}(e^{-t/RC} - e^{-2t/RC}) \qquad ...(32)$$

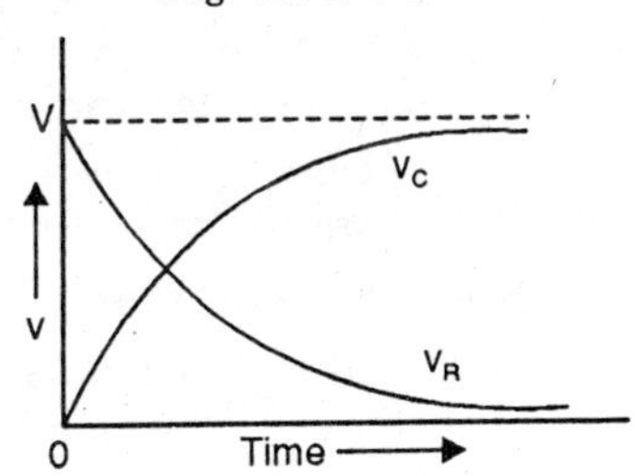

Fig. 80

(*iii*) **R-L-C transients :**

For *R-L-C* circuit shown in Fig. 81 the following integro-differential equation can be written as follows :

$$R_i + L\frac{di}{dt} + \frac{1}{C}\int idt = V \qquad ...(33)$$

While solving for i, the following three cases are considered :

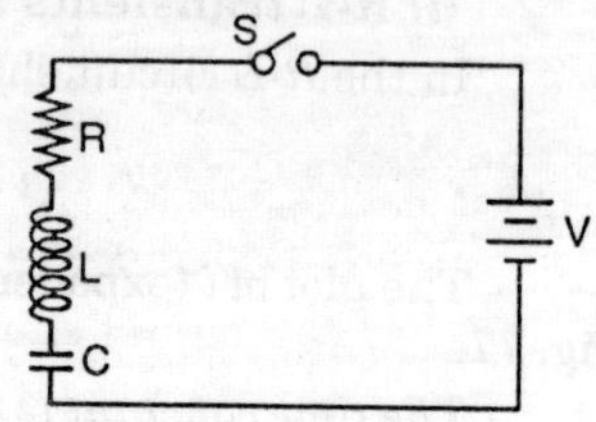

Fig. 81. *R-L-C* circuit.

Case I. $\left(\frac{R}{2L}\right)^2 > \frac{1}{LC}$

In this case the current is given by

$$i = C_1 e^{\alpha t}\, C_1(e^{\beta t} + C_2 e^{-\beta t}) \quad ...(34)$$

Case II. $\left(\frac{R}{2L}\right)^2 = \frac{1}{LC}$

Here, $i = e^{\alpha t}\,(C_1 + C_2 t)$ ...(35)

Case III. $\left(\frac{R}{2L}\right)^2 < \frac{1}{LC}$

Here, $i = e^{\alpha t}\,(C_1 \cos \beta t + C_2 \sin \beta t)$ ...(36)

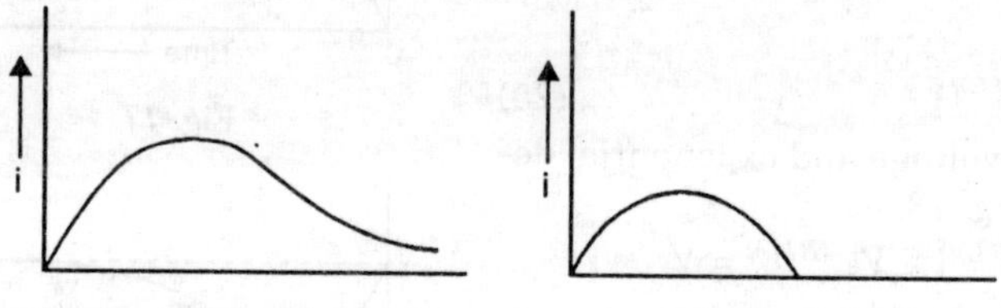

Fig. 82

In all the above cases the current contains the factor $e^{\alpha t}$ and since $\alpha = -R/2L$ the final value is zero, assuming that the complimentary function decays in a relatively short time. Fig. 82 shows the value of i for initial values zero and initial slope positive.

5.3. A.C. Transients

(*i*) **R-L sinusoidal transient :** Refer Fig. 83

Here the voltage function could be at any point in the period at the instant of closing the switch and therefore the phase angle ϕ can take any values from 0 to 2 π rad/sec.

In this case, the current (i) is given by

$$i = e^{-(R/L)t}\left[\frac{-V_{max}}{\sqrt{R^2+\omega^2L^2}}\sin\left(\psi - \tan^{-1}\frac{\omega L}{R}\right)\right] + \frac{V_{max}}{\sqrt{R^2+\omega^2L^2}}\sin\left(\omega t + \psi - \tan^{-1}\frac{\omega L}{R}\right) \quad ...(37)$$

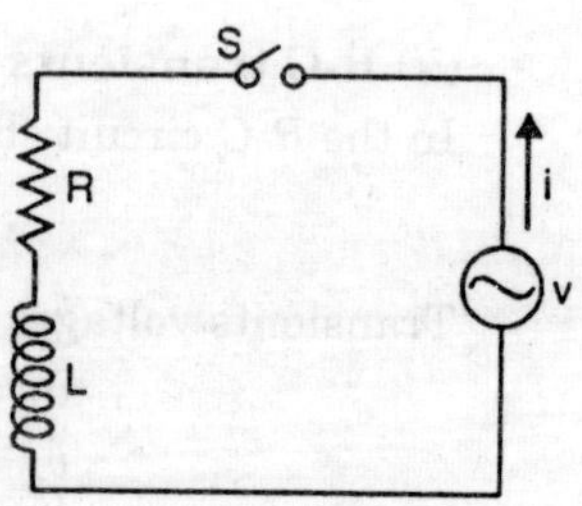

$v = V_{max} \sin(\omega t + \psi)$

Fig. 83. *R-L* circuit.

where $\tan^{-1}\left(\frac{\omega L}{R}\right) = \phi$.

It may be noted that :

— The *first part (transient component of current,* i_t, of the above equation contains the factor $e^{-(R/L)t}$ which has a *value of zero in a relatively short time.*

— The *second part* of the above equation is the *steady current* (i_s) which lags the applied voltage by $\tan^{-1}\frac{\omega L}{R}$.

Here, $$\frac{V_{max}}{R^2+\omega^2L^2} = I_{max},\ \tan^{-1}\left(\frac{\omega L}{R}\right) = \phi$$

(*ii*) **R-C sinusoidal transient :**

For *R-C* circuit shown in Fig. 84 the basic equation is :

$$Ri + \frac{1}{C}\int v\,dt = V_{max}\sin(\omega t + \psi) \quad ...(38)$$

Here the current *i* is given by,

$$i = e^{-(t/RC)}\left[\frac{V_{max}}{R}\sin\psi - \frac{V_{max}}{\sqrt{R^2+\left(\frac{1}{\omega C}\right)^2}}\sin\left(\psi + \tan^{-1}\frac{1}{\omega CR}\right)\right]$$
$$+ \frac{V_{max}}{\sqrt{R^2+\left(\frac{1}{\omega C}\right)^2}}\sin\left(\omega t + \psi + \tan^{-1}\frac{1}{\omega CR}\right) \quad ...(39)$$

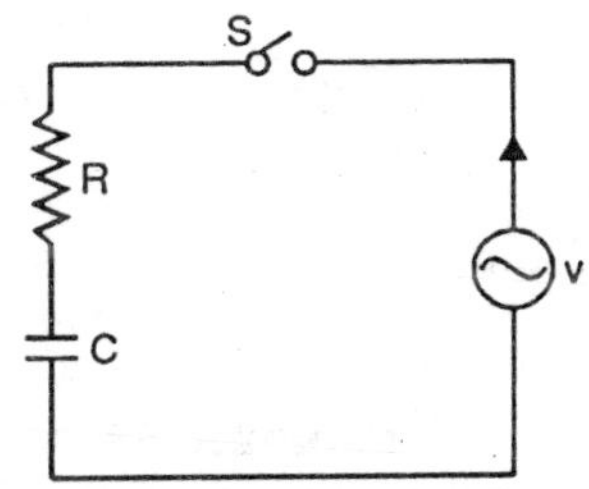

$v = V_{max}\sin(\omega t + \psi)$
Fig. 84. *R-C* circuit.

It may be noted that :

—The *first part* of the above equation is *the transient with decay factor* $e^{-t/RC}$

—The *second part is the steady current which leads the applied voltage by* $\tan^{-1}\frac{1}{\omega CR}$.

Example 51. *A series circuit has R = 10 Ω and L = 0.1 H. A direct voltage of 200 V is suddenly applied to it. Calculate the following :*

(*i*) *The voltage drop across the inductance at the instance of switching on and at 0.01 second, and*

(*ii*) *The flux linkages at these instants.* **(Bombay University)**

Solution. *Given :* $R = 10\ \Omega$; $L = 0.1$ H ; $V = 200$ volts

(*i*) **The voltage drop across the inductance :**

(*a*) *Switching instant.* At the instant of switching on, $i = 0$, so that $iR = 0$ hence all the applied voltage must drop across the inductance only. Therefore, voltage drop across inductance **= 200 V. (Ans.)**

(*b*) *When t = 0.01 second*

$$i = \frac{V}{R}[1 - e^{-(R/L)t}] \quad ...[\text{Eqn. 23}]$$
$$= \frac{200}{10}[1 - e^{-(10/0.1)\times 0.01}] = 20(1 - e^{-1}) = 12.64 \text{ A}$$

and, $$iR = 12.64 \times 10 = 126.4 \text{ V}$$

∴ The voltage drop across the inductance = $\sqrt{(200)^2 - (126.4)^2}$ = **155 V. (Ans.)**

(*ii*) **The flux linkages :**

Now, $$L = \frac{N\phi}{i} \text{ or } N\phi = Li$$

∴ Flux linkages ($N\phi$) = $Li = 0.1 \times 12.64$ = **1.264 Wb-turns. (Ans.)**

Example 52. *A choke has a resistance of 50 Ω and inductance of 1.0 H. It is supplied with an A.C. voltage given by 141 sin 314t. Find the expression for transient component of the current flowing through the choke after the voltage is suddenly switched on.* **(Jadavpur University)**

Solution. *Given* : $R = 50\ \Omega$; $L = 1.0$ H ; $e = 141 \sin 314t$

Expression for transient component of the current :

The equation of the transient component of the current is given by :

$$i_t = I_{max} \sin \phi e^{-(R/L)t} \qquad \text{...[Eqn. 37]}$$

Here
$$I_{max} = \frac{V_{max}}{Z} = \frac{V_{max}}{\sqrt{R^2 + \omega^2 L^2}} = \frac{141}{\sqrt{(50)^2 + (314)^2 (1.0)^2}} = 0.443 \text{ A}$$

and,
$$\phi = \tan^{-1}\left(\frac{\omega L}{R}\right) = \tan^{-1}\left(\frac{314 \times 1.0}{50}\right) = 80.95°$$

$$\therefore \quad i_t = 0.443 \sin 80.95° \, e^{-(50/1.0)t} = \mathbf{0.437\, e^{-50t}}. \quad \textbf{(Ans.)}$$

Example 53. *A series circuit has R = 10 Ω and L = 0.1 H. A 50 Hz sinusoidal voltage of maximum value of 400 V is applied across this circuit. Find an expression for the value of current at any instant after the voltage is applied, assuming that the voltage is zero at the instant of application. Calculate its value 0.02 second after switching on.* **(Punjab University)**

Solution. *Given* : $R = 10\ \Omega$; $L = 0.1$ H ; $f = 50$ Hz ; $V_{max} = 400$ V ; $t = 0.02$ s.

The current consists of transient component and steady-state component. The equation of the resultant current is given by :

$$i = e^{-(R/L)t}\left[\frac{-V_{max}}{\sqrt{R^2 + \omega^2 L^2}} \sin\left(\psi - \tan^{-1}\frac{\omega L}{R}\right)\right]$$

(Transient current)

$$+ \frac{V_{max}}{\sqrt{R^2 + \omega^2 L^2}} \sin\left(\omega t + \psi - \tan^{-1}\frac{\omega L}{R}\right) \qquad \text{...[Eqn. 37]}$$

(Steady-state current)

where
$$\frac{V_{max}}{\sqrt{R^2 + \omega^2 L^2}} = I_{max}, \quad \tan^{-1}\left(\frac{\omega L}{R}\right) = \phi$$

Here, $\psi = 0$ as per given condition, then

$$i = e^{-(R/L)t}\left[-\frac{V_{max}}{\sqrt{R^2 + \omega^2 L^2}} \sin\left(-\tan^{-1}\frac{\omega L}{R}\right)\right]$$

$$+ \frac{V_{max}}{\sqrt{R^2 + \omega^2 L^2}} \sin\left(\omega t - \tan^{-1}\frac{\omega L}{R}\right) \qquad \text{...(i)}$$

Now,
$$\frac{V_{max}}{R^2 + \omega^2 L^2} = I_{max} = \frac{400}{\sqrt{10^2 + (314)^2 (0.1)^2}} = 12.14 \text{ A}$$

$$\tan^{-1}\frac{\omega L}{R} = \phi = \tan^{-1}\left(\frac{314 \times 0.1}{10}\right) = 72.3° = 1.262 \text{ rad.}$$

Substituting the value in eqn. (*i*), we get

$$i = e^{-(10/0.1) \times 0.02}\,[-12.14 \sin(-72.3°)] + 12.14 \sin(314 \times 0.02 - 1.262)$$

$$= e^{-2}\,[12.14 \sin(72.3°)] + 12.14 \underset{\text{(rad)}}{\sin(5.018)}$$

$$= 0.1353(12.14 \times 0.9527) + 12.14 \underset{\text{(deg)}}{\sin(287.5°)}$$

$$= 1.56 - 11.58 = \mathbf{-10.02\ A.} \quad \textbf{(Ans.)}$$

Example 54. *In a simple saw-tooth generator circuit with the thyratron switches on at 150 V and switches off at 10 V. If this circuit is supplied with 250 V D.C. source ; find the time period of saw-tooth wave. The resistance and capacitance have the values of 10 kΩ and 1 μF respectively.*

(Jadavpur University)

Solution. *Given :* $R = 10\ k\Omega$; $\mu = 1\ \mu F = 1 \times 10^{-6}$ F ; Circuit voltage = 250 V ; Switching on voltage = 150 V ; Switching off voltage = 10 V.

Time period of saw-tooth wave : Refer Fig. 84

Let, V = Applied voltage

v_{C1} = Switching-off voltage of thyratron = 10 V

v_{C2} = Switching-on voltage of thyratron = 150 V

Now, $v = V(1 - e^{-t/RC})$ ...[Eqn. 30]

$$\therefore \quad v_{C1} = V(1 - e^{-t_1/RC}) \quad ...(i)$$

$$v_{C2} = V[1 - e^{-(t_1 + T)/RC}] \quad ...(ii)$$

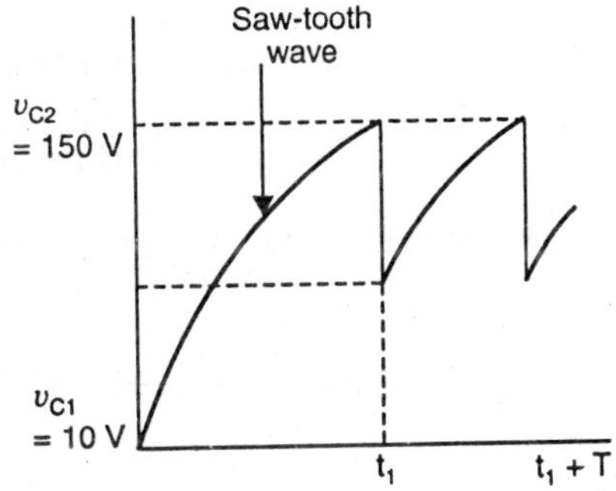

Fig. 84

where, T is the time period of the saw-tooth wave.

Eqns. (*i*) and (*ii*) can be written as :

$$V - v_{C1} = e^{-t_1/RC} \quad ...(iii)$$

$$V - v_{C2} = e^{-(t_1 + T)/RC} \quad ...(iv)$$

Dividing (*i*) by (*ii*), we get

$$\frac{V - v_{C1}}{V - v_{C2}} = \frac{e^{-t_1/RC}}{e^{-(t_1+T)/RC}} = e^{[-t_1/RC - \{-(t_1+T)/RC\}]} = e^{T/RC}$$

or

$$T/RC = \ln\left(\frac{V - v_{C1}}{V - v_{C2}}\right) \quad \text{or} \quad T = \ln\left(\frac{V - v_{C1}}{V - v_{C2}}\right) \times RC$$

Here, $v_{C1} = 10$ V ; $v_{C2} = 150$ V

Substituting the values, we get

$$T = \ln\left(\frac{250 - 10}{250 - 150}\right) \times 10 \times 10^3 \times 1 \times 10^{-6} = 0.00875 \text{ s.} \quad \textbf{(Ans.)}$$

HIGHLIGHTS

1. Modern alternators produce an e.m.f. which is for all practical purposes sinusoidal (*i.e.* a sine curve), the equation between the e.m.f. and time being

$$e = E_{max} \sin \omega t$$

where e = instantaneous voltage ; E_{max} = maximum voltage;

ωt = angle through which the armature has turned from neutral.

2. The *r.m.s. value* of an alternating current is given by that steady (D.C.) current which when flowing through a given circuit produces the same *heat* as is produced by the alternating current when flowing through the same circuit for the same time.

$$I_{rms} = 0.707\ I_{max}.$$

3. The *average or mean value* of an alternating current is expressed by that steady current which transfers across any circuit the same **charge** as is transferred by that alternating current during the same time.

$$I_{av} = 0.637\, I_{max}$$

Form factor is the ratio of r.m.s. value to average value of the wave form.

Peak factor is the ratio of maximum value to the r.m.s. value of the wave form.

4. The frequency (series circuit) of the voltage which gives the maximum value of the current in the circuit is called *resonant frequency* (f_r) and the circuit is said to be *resonant*.

$$f_r = \frac{1}{2\pi\sqrt{LC}}.$$

5. Q-factor of a series circuit is defined as equal to the voltage magnification in the circuit at resonance.

$$\text{Q-factor} = \frac{1}{R}\sqrt{\frac{L}{C}}.$$

6. Q-factor of a parallel circuit is defined as the ratio of the current circulating between its two branches to the line current drawn from the supply or simply, as the current magnification.

$$\text{Q-factor} = \frac{1}{R}\sqrt{\frac{L}{C}}.$$

OBJECTIVE TYPE QUESTIONS

Choose the Correct Answer :

1. A sine wave has a frequency of 50 Hz. Its angular frequency is ...... radian/second.
   (*a*) 100 π (*b*) 50 π
   (*c*) 25 π (*d*) 5 π.
2. The reactance offered by a capacitor to alternating current of frequency 50 Hz is 20 Ω. If frequency is increased to 100 Hz, reactance becomes ...... ohms.
   (*a*) 2.5 (*b*) 5
   (*c*) 10 (*d*) 15.
3. The period of a wave is
   (*a*) the same as frequency (*b*) time required to complete one cycle
   (*c*) expressed in amperes (*d*) none of the above.
4. The form factor is the ratio of
   (*a*) peak value to r.m.s. value (*b*) r.m.s. value to average value
   (*c*) average value to r.m.s. value (*d*) none of the above.
5. The period of a sine wave is $\frac{1}{50}$ seconds. Its frequency is
   (*a*) 20 Hz (*b*) 30 Hz
   (*c*) 40 Hz (*d*) 50 Hz.
6. An A.C. current is given by $i = 200 \sin 100\,\pi t$. It will achieve a value of 100 A after ...... second.
   (*a*) $\frac{1}{900}$ (*b*) $\frac{1}{800}$
   (*c*) $\frac{1}{700}$ (*d*) $\frac{1}{600}$.
7. A heater is rated as 230 V, 10 kW, A.C. The value 230 V refers to
   (*a*) average voltage (*b*) r.m.s. voltage
   (*c*) peak voltage (*d*) none of the above.
8. If two sinusoids of the same frequency but of different amplitudes and phase angles are subtracted, the resultant is
   (*a*) a sinusoid of the same frequency (*b*) a sinusoid of half the original frequency
   (*c*) a sinusoid of double the frequency (*d*) not a sinusoid.

9. The peak value of a sine wave is 200 V. Its average value is
(*a*) 127.4 V (*b*) 141.4 V
(*c*) 282.8 V (*d*) 200 V.

10. If two sine waves of the same frequency have a phase difference of π radians, then
(*a*) both will reach their minimum values at the same instant
(*b*) both will reach their maximum values at the same instant
(*c*) when one wave reaches its maximum value, the other will reach its minimum value
(*d*) none of the above.

11. The r.m.s. value of a sine wave is 100 A. Its peak value is
(*a*) 70.7 A (*b*) 141.4 A
(*c*) 150 A (*d*) 282.8 A.

12. If two waves are expressed as $e_1 = E_{m_1} \sin(\omega t + \alpha_1)$ and $e_2 = E_{m_2} \sin(\omega t + \alpha_2)$, then
(*a*) e_1 is leading e_2 by $\angle(\alpha_2 - \alpha_1)$ (*b*) e_2 is leading e_1 by $\angle(\alpha_2 - \alpha_1)$
(*c*) e_2 is leading e_1 by $\angle(\alpha_1 - \alpha_2)$ (*d*) e_1 is in phase with e_2.

13. The voltage of domestic supply is 220 V. This figure represents
(*a*) mean value (*b*) r.m.s. value
(*c*) peak value (*d*) average value.

14. Two waves of the same frequency have opposite phase when the phase angle between them is
(*a*) 360° (*b*) 180°
(*c*) 90° (*d*) 0°.

15. The power consumed in a circuit element will be least when the phase difference between the current and voltage is
(*a*) 180° (*b*) 90°
(*c*) 60° (*d*) 0°.

16. The r.m.s. value and mean value is the same in the case of
(*a*) triangular wave (*b*) sine wave
(*c*) square wave (*d*) half wave rectified sine wave.

17. For the same peak value which of the following wave will have the highest r.m.s. value ?
(*a*) square wave (*b*) half wave rectified sine wave
(*c*) triangular wave (*d*) sine wave.

18. For the same peak value, which of the following wave has the *least* mean value ?
(*a*) half wave rectified sine wave (*b*) triangular wave
(*c*) sine wave (*d*) square wave.

19. For a sine wave with peak value I_{max} the r.m.s. value is
(*a*) $0.5\ I_{max}$ (*b*) $0.707\ I_{max}$
(*c*) $0.9\ I_{max}$ (*d*) $1.414\ I_{max}$.

20. Form Factor is the ratio of
(*a*) average value/r.m.s. value (*b*) average value/peak value
(*c*) r.m.s. value/average value (*d*) r.m.s. value/peak value.

21. Form factor for a sine wave is
(*a*) 1.414 (*b*) 0.707
(*c*) 1.11 (*d*) 0.637.

22. For a sine wave with peak value E_{max} the average value is
(*a*) $0.636\ E_{max}$ (*b*) $0.707\ E_{max}$
(*c*) $0.434\ E_{max}$ (*d*) $1.414\ E_{max}$.

23. The current in a circuit is given by : $i = 100 \sin 314\, t$ amperes
The maximum value and frequency of current are
(*a*) $50\sqrt{2}$ A, 100 Hz (*b*) $100\sqrt{2}$ A, 100 Hz
(*c*) 100 A, 50 Hz (*d*) 70.7A, 50 Hz.

24. For a frequency of 200 Hz, the time period will be
(*a*) 0.05 s (*b*) 0.005 s
(*c*) 0.0005 s (*d*) 0.5 s.

25. The phase difference between voltage and current wave through a circuit element is given as 30°. The essential condition is that
(*a*) both waves must have same frequency
(*b*) both waves must have identical peak values
(*c*) both waves must have zero value at the same time
(*d*) none of the above.

26. An A.C. voltage of 50 Hz has a maximum value of 50 V. Its value after 1/600 second after the instant the current is zero, will be
(*a*) 5 V (*b*) 12.5 V
(*c*) 25 V (*d*) 43.3 V.

27. When two waves are in phase they have peak values at an interval of
(*a*) 180° (*b*) 120°
(*c*) 90° (*d*) none of the above.

28. For 200 V r.m.s. value triangular wave, the peak voltage will be
(*a*) 200 V (*b*) 222 V
(*c*) 282 V (*d*) 346 V.

29. A sine wave of voltage varies from zero to maximum of 200 V. How much is the voltage at the instant of 30° of the cycle ?
(*a*) 50 V (*b*) 82.8 V
(*c*) 100 V (*d*) 173.2 V.

30. How much r.m.s. current does a 300 W 200 V bulb take from the 200 V, 50 Hz power line ?
(*a*) 0.5 A (*b*) 1.5 A
(*c*) 2 A (*d*) 3 A.

31. Two sinusoidal currents are given by $i_1 = 100 \sin (\omega t + \pi/3)$, and $i_2 = 150 \sin (\omega t - \pi/4)$
The phase difference between them is ...... degrees.
(*a*) 15 (*b*) 50
(*c*) 60 (*d*) 105.

32. The r.m.s. value of a half-wave rectified current is 100 A. Its value for full-wave rectification would be ...... amperes.
(*a*) 141.4 (*b*) 200
(*c*) $200/\pi$ (*d*) $40/\pi$.

33. From the two voltages equations $e_1 = E_{max} \sin 100\pi t$, and $e_2 = E_{max} \sin (100\pi t + \pi/6)$, it is obvious that
(*a*) 1 leads 2 by 30° (*b*) 2 lags behind 1
(*c*) 2 achieves its maximum value $\frac{1}{600}$ second before 1 does
(*d*) 1 achieves its zero value $\frac{1}{600}$ second before 2.

34. The r.m.s. value of a sinusoidal A.C. current is equal to its value at an angle of ...... degrees.
(*a*) 90 (*b*) 60
(*c*) 45 (*d*) 30.

35. Capacitive reactance is more when
(*a*) capacitance is less and frequency of supply is less
(*b*) capacitance is less and frequency of supply is more
(*c*) capacitance is more and frequency of supply is less
(*d*) capacitance is more and frequency of supply is more.

36. Time constant of a capacitive circuit increases with the
(*a*) increase of capacitance and decrease of resistance
(*b*) increase of capacitance and increase of resistance
(*c*) decrease of capacitance and decrease of resistance
(*d*) decrease of capacitance and increase of resistance.

37. In a series circuit on resonance, following will occur
(*a*) $V = V_R$ (*b*) $X_L = X_C$
(*c*) $Z = R$ (*d*) $V_L = V_C$
(*e*) all above.

38. In a series resonant circuit, the impedance of the circuit is
(*a*) minimum (*b*) maximum
(*c*) zero (*d*) none of the above.

39. Power factor of an electrical circuit is equal to
(*a*) R/Z
(*b*) cosine of phase angle difference between current and voltage
(*c*) kW/kVA (*d*) ratio of useful current to total current I_W/I
(*e*) all above.

40. The best place to install a capacitor is
(*a*) very near to inductive load (*b*) across the terminals of the inductive load
(*c*) far away from the inductive load (*d*) any where.

41. Poor power factor
(*a*) reduces load handling capability of electrical system
(*b*) results in more power losses in the electrical system
(*c*) overloads alternators, transformers and distribution lines
(*d*) results in more voltage drop in the line (*e*) results in all above.

42. Capacitors for power factor correction are rated in
(*a*) kW (*b*) kVA
(*c*) kV (*d*) kVAR.

43. In series resonant circuit, increasing inductance to its twice value and reducing capacitance to its half value
(*a*) will change the maximum value of current at resonance
(*b*) will change the resonance frequency
(*c*) will change the impedance at resonance frequency
(*d*) will increase the selectivity of the circuit.

44. Pure inductive circuit
(*a*) consumes some power on average
(*b*) does not take power at all from a line
(*c*) takes power from the line during some part of the cycle and then returns back to it during other part of the cycle
(*d*) none of the above.

45. Inductance affects the direct current flow
(*a*) only at the time of turning off (*b*) only at the time of turning on
(*c*) at the time of turning on and off (*d*) at all the time of operation.

46. Inductance of a coil varies
(*a*) directly as the cross-sectional area of magnetic core
(*b*) directly as square of number of turns (*c*) directly as the permeability of the core
(*d*) inversely as the length of the iron path (*e*) as (*a*) to (*d*).

47. All the rules and laws of D.C. circuit also apply to A.C. circuit containing
(*a*) capacitance only (*b*) inductance only
(*c*) resistance only (*d*) all above.

48. Time constant of an inductive circuit
(*a*) increases with increase of inductance and decrease of resistance
(*b*) increases with the increase of inductance and the increase of resistance
(*c*) increases with decrease of inductance and decrease of resistance
(*d*) increases with decrease of inductance and increase of resistance.

49. Power factor of an inductive circuit is usually improved by connecting capacitor to it in
(*a*) parallel (*b*) series
(*c*) either (*a*) or (*b*) (*d*) none of the above.

50. In a highly capacitive circuit the
(*a*) apparent power is equal to the actual power (*b*) reactive power is more than the apparent power
(*c*) reactive power is more than the actual power (*d*) actual power is more than its reactive power.

51. Power factor of the following circuit will be zero
(*a*) resistance (*b*) inductance
(*c*) capacitance (*d*) both (*b*) and (*c*).

52. Power factor of the following circuit will be unity
(*a*) inductance (*b*) capacitance
(*c*) resistance (*d*) both (*a*) and (*b*).

53. Power factor of the system is kept high
(*a*) to reduce line losses
(*b*) to maximise the utilization of the capacities of generators, lines and transformers
(*c*) to reduce voltage regulation of the line
(*d*) due to all above reasons.

54. The time constant of the capacitance circuit is defined as the time during which voltage
(*a*) falls to 36.8% of its final steady value (*b*) rises to 38.6% of its final steady value
(*c*) rises to 63.2% of its final steady value (*d*) none of the above.

55. In the *R-L-C* containing $R = 4.5\ \Omega$, $L = 0.06$ H, $C = 0.6\ \mu$F the power factor will be
(*a*) zero (*b*) lagging
(*c*) leading (*d*) unity.

ANSWERS

A. Choose the Correct Answer :

1. (*a*)	2. (*c*)	3. (*b*)	4. (*b*)	5. (*d*)	6. (*d*)
7. (*b*)	8. (*a*)	9. (*a*)	10. (*c*)	11. (*b*)	12. (*b*)
13. (*b*)	14. (*b*)	15. (*b*)	16. (*c*)	17. (*a*)	18. (*a*)
19. (*b*)	20. (*c*)	21. (*c*)	22. (*a*)	23. (*c*)	24. (*b*)
25. (*a*)	26. (*c*)	27. (*d*)	28. (*d*)	29. (*c*)	30. (*b*)
31. (*d*)	32. (*a*)	33. (*c*)	34. (*c*)	35. (*a*)	36. (*b*)
37. (*e*)	38. (*a*)	39. (*e*)	40. (*b*)	41. (*e*)	42. (*d*)
43. (*d*)	44. (*c*)	45. (*c*)	46. (*e*)	47. (*c*)	48. (*a*)
49. (*a*)	50. (*c*)	51. (*d*)	52. (*c*)	53. (*d*)	54. (*c*)
55. (*c*).					

THEORETICAL QUESTIONS

1. Define the following terms :
Circuit, Electrical network, Active network, Node and Branch.
2. What are the limitations of ohm's law ?
3. State and explain Kirchhoff's laws.
4. Discuss briefly application of Kirchhoff's laws.

5. Explain the nodal voltage method for solving networks. How are the nodal equations written ?
6. Explain Cramer's rule used for solving equations by determinants.
7. State and explain Superposition theorem.
8. State Norton's theorem. List the steps for finding the current in a branch of a network with the help of this theorem.
9. State Thevenin's theorem.
10. State the maximum power transfer theorem and explain its importance.
11. Define the following terms as applied to an alternating current :
    Cycle, frequency, time period, amplitude.
12. What do you mean by the term "Phase difference" ?
13. Explain the following terms relating alternating current :
    (*i*) R.M.S. value (*ii*) Average value
    (*iii*) Form factor (*iv*) Peak factor.
14. Explain briefly the following as applied to A.C. series and parallel circuits :
    (*i*) Resonance frequency (*ii*) Q-factor.
15. What do you mean by transient disturbances ?
16. Define single energy and double energy transients

UNSOLVED EXAMPLES

1. An alternating current of frequency 60 Hz has a maximum value of 120 A. Write down the equation for its instantaneous value. Reckoning time from the instant the current is zero and is becoming positive, find :
   (*i*) The instantaneous value after $\frac{1}{360}$ second ;
   (*ii*) The time taken to reach 96 A for the first time. [**Ans.** 103.9 A, 0.00245 second]
2. An alternating current of frequency 50 Hz has a maximum value of 100 A. Calculate :
   (*i*) Its value 1/600 second after the instant the current is zero and its value decreasing thereafter wards.
   (*ii*) How many seconds after the instant the current is zero (increasing thereafter wards) will the current attain the value of 86.6 A ? [**Ans.** – 50 A, 1/300 s]
3. Calculate the r.m.s. value, the form factor of a periodic voltage having the following values for equal time intervals changing suddenly from one value to the next : 0, 5, 10, 20, 50, 60, 50, 20, 10, 5, 0, 5, 10 V etc. What would be the r.m.s. value of sine wave having the same peak value ?
   [**Ans.** 31 V ; 23 V ; 1.35 ; (app.) ; 42.2 V]
4. A sinusoidally varying alternating current has an average value of 127.4 A. When its value is zero, then its rate of change is 62,800 A/s. Find the analytical expression for the sine wave.
   [**Ans.** $i = 200 \sin 100\,\pi t$]
5. A coil of resistance 10 Ω and inductance 0.1 H is connected in series with a 150 μF capacitor across a 200 V, 50 Hz supply. Calculate (*i*) the inductive reactance, (*ii*) the capacitive reactance, (*iii*) the impedance (*iv*) the current, (*v*) the power factor, (*vi*) the voltage across the coil and the capacitor respectively.
   [**Ans.** (*i*) 31.4 Ω, (*ii*) 21.2 Ω, (*iii*) 14.3 Ω, (*iv*) 14 A, (*v*) 0.7 lag (*vi*) 460 V, 297 V]
6. A circuit is made up of 10 Ω resistance, 12 mH inductance and 281.5 μF capacitance in series. The supply voltage is 100 V (constant). Calculate the value of the current when the supply frequency is (*i*) 50 Hz and (*ii*) 150 Hz. [**Ans.** 8 A leading ; 8 A lagging]
7. A coil having a resistance of 10 Ω and an inductance of 0.2 H is connected in series with a capacitor of 50.7 μF. The circuit is connected across a 100 V, 50 Hz A.C. supply. Calculate (*i*) the current flowing (*ii*) the voltage across the capacitor (*iii*) the voltage across the coil. Draw a vector diagram to scale.
   [**Ans.** (*i*) 10 A (*ii*) 628 V (*iii*) 635 V]
8. A coil is in series with a 20 μF capacitor across a 230 V, 50 Hz supply. The current taken by the circuit is 8 A and the power consumed is 200 W. Calculate the inductance of the coil if the power factor of the circuit is (*i*) leading and (*ii*) lagging.

Sketch a vector diagram for each condition and calculate the coil power factor in each case.

[**Ans.** 0.415 H ; 0.597 H ; 0.0238 ; 0.0166]

9. An A.C. series circuit has a resistance of 10 Ω, an inductance of 0.2 H and a capacitance of 60 μF. Calculate :

(*i*) the resonant frequency (*ii*) the current and

(*iii*) the power at resonance.

Given that the applied voltage is 200 V. [**Ans.** 46 Hz ; 20 A ; 4 kW]

10. A circuit consists of an inductor which has a resistance of 10 Ω and an inductance of 0.3 H, in series with a capacitor of 30 μF capacitance. Calculate :

(*i*) The impedance of the circuit to currents of 40 Hz ;

(*ii*) The resonant frequency ;

(*iii*) The peak value of stored energy in joules when the applied voltage is 200 V at the resonant frequency.

[**Ans.** 58.31 Ω ; 53 Hz ; 120 J]

11. A resistor and a capacitor are connected in series with a variable inductor. When the circuit is connected to a 240 V, 50 Hz supply, the maximum current given by varying the inductance is 0.5 A. At this current, the voltage across the capacitor is 250 V. Calculate the values of the following :

(*i*) The resistance ; (*ii*) The capacitance ;

(*iii*) The inductance.

Neglect the resistance of the inductor. [**Ans.** 480 Ω, 6.36 μF ; 1.59 H]

12. A resistance, a capacitor and a variable inductance are connected in series across a 200 V, 50 Hz supply. The maximum current which can be obtained by varying the inductance is 314 mA and the voltage across the capacitor is then 300 V. Calculate the capacitance of the capacitor and the values of the inductance and resistance. [**Ans.** 3.33 μF, 3.04 H, 637 Ω]

13. A circuit consisting of a coil of resistance 12 Ω and inductance 0.15 H in series with a capacitor of 12 μF is connected to a variable frequency supply which has a constant voltage of 24 V.

Calculate : (*i*) The resonant frequency, (*ii*) The current in the circuit at resonance, (*iii*) The voltage across the capacitor and the coil at resonance. [**Ans.** (*i*) 153 Hz, (*ii*) 2 A, (*iii*) 224 V]

14. A resistance of 24 Ω, a capacitance of 150 μF and an inductance of 0.16 H are connected in series with each other. A supply at 240 V, 50 Hz is applied to the ends of the combination. Calculate (*i*) the current in the circuit (*ii*) the potential differences across each element of the circuit (*iii*) the frequency to which the supply would need to be changed so that the current would be at unity power-factor and find the current at this frequency. [**Ans.** (*i*) 6.37 A (*ii*) V_R = 152.8 V, V_C = 320 V, V_L = 123.3 V (*iii*) 32 Hz ; 10 A]

15. A coil-A of inductance 80 mH and resistance 120 Ω is connected to a 230 V, 50 Hz single-phase supply. In parallel with it is a 16 μF capacitor is series with a 40 Ω non-inductive resistor *B*. Determine

(*i*) The power factor of the combined circuit,

(*ii*) The total power taken from the supply. [**Ans.** (*i*) 0.945 lead (*ii*) 473 W]

16. A choking coil of inductance 0.08 H and resistance 12 ohm, is connected in parallel with a capacitor of 120 μF. The combination is connected to a supply at 240 V, 50 Hz. Determine the total current from the supply and its power factor. Illustrate your answers with a phasor diagram. [**Ans.** 3.94 A, 0.943 lag]

17. A choking coil having a resistance of 20 Ω and an inductance of 0.07 henry is connected with a capacitor of 60 μF capacitance which is in series with a resistor of 50 Ω. Calculate the total current and the phase angle when this arrangement is connected to 200 V, 50 Hz mains. [**Ans.** 7.15 A, 24° 39′ lag]

18. A coil of resistance of 15 Ω and inductance 0.05 H is connected in parallel with a non-inductive resistance of 20 Ω. Find (*i*) the current in each branch ; (*ii*) the total current (*iii*) the phase angle of whole arrangement for an applied voltage of 200 V at 50 Hz. [**Ans.** 9.22 A ; 10 A ; 22.1°]

19. A sinusoidal 50 Hz voltage of 200 V (r.m.s.) supplies the following three circuits which are in parallel : (*i*) a coil of inductance 0.03 H and resistance 3 Ω ; (*ii*) a capacitor of 400 μF in series with a resistance of 100 Ω ; (*iii*) a coil of inductance 0.02 H and resistance 7 Ω in series with a 300 μF capacitor. Find the total current supplied and draw a complete vector diagram. [**Ans.** 29.4 A]

20. In a series-parallel circuit, the two parallel branches *A* and *B* are in series with *C*. The impedances are $Z_A = (10 - j8)$ Ω, $Z_B = (9 - j6)$ Ω and $Z_C = (100 + j0)$. Find the currents I_A and I_B and the phase difference between them. Draw the phasor diagram. [**Ans.** $I_A = 12.71 \angle -30° 58'$ $I_B = 15 \angle -35° 56'$; 4° 58′]

3

Three-phase Circuits

1. INTRODUCTION

- Generation, transmission and heavy-power utilisation of A.C. electric energy almost invariably involve a type of system or circuit called a *polyphase system* or *polyphase circuit*. In such a system, each voltage source consists of a group of voltages having relative magnitudes and phase angles. Thus, a *m-phase system will employ voltage sources which, conventionally, consist of m voltages substantially equal in magnitude and successively displaced by a phase angle of 360° / m*.
- A *3-phase system* will employ voltage sources which, conventionally, consist of *three voltages substantially equal in magnitude* and *displaced by phase angles of 120°*. Because it *possesses definite economic and operating advantages, the 3-phase system is by far the most common,* and consequently emphasis is placed on 3-phase *circuits*.

2. ADVANTAGES OF POLYPHASE SYSTEMS

The advantages of polyphase systems *over single-phase systems* are :

1. A polyphase transmission line requires less conductor material than a single-phase line for transmitting the same amount power at the same voltage.

2. For a given frame size a polyphase machine gives a higher output than a single-phase machine. For example, output of a 3-phase motor is 1.5 times the output of single-phase motor of same size.

3. Polyphase motors have a uniform torque where most of the single-phase motors have a pulsating torque.

4. Polyphase induction motors are self-starting and are more efficient. On the other hand single-phase induction motors are not self-starting and are less efficient.

5. Per unit of output, the polyphase machine is very much cheaper.

6. Power factor of a single-phase motor is lower than that of polyphase motor of the same rating.

7. Rotating field can be set up by passing polyphase current through stationary coils.

8. Parallel operation of polyphase alternators is simple as compared to that of single-phase alternators because of pulsating reaction in single-phase alternator.

It has been found that the above advantages are *best realised in the case of three-phase systems*. Consequently, the electric power is generated and transmitted in the form of three-phase system.

3. GENERATION OF THREE-PHASE VOLTAGES

- Let us consider an elementary 3-phase 2-pole generator as shown in Fig. 1. On the armature are three coils, *ll'*, *mm'*, and *nn'* whose axes are displaced 120° in space from each other.

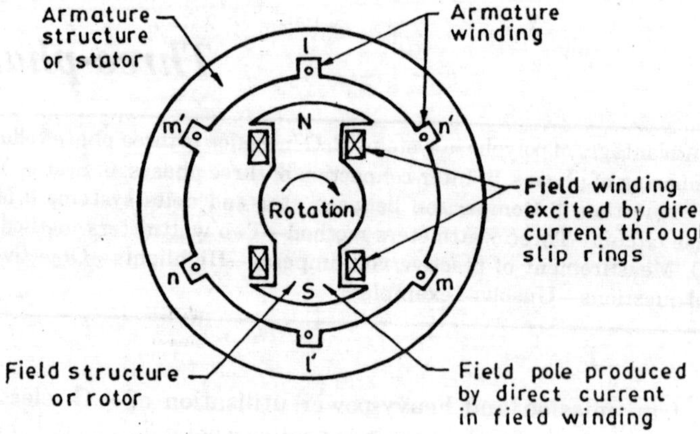

Fig. 1. Elementary 3-phase 2-pole generator.

- When the field is excited and rotated, voltages will be generated in the three phases in accordance with Faraday's law. If the field structure is so designed that the flux is distributed sinusoidally over the poles, the flux linking any phase will vary sinusoidally with time and sinusoidal voltages will be induced in three-phases. These three waves will be displaced 120 electrical degrees (Fig. 2) in time as a result of the phases being displaced

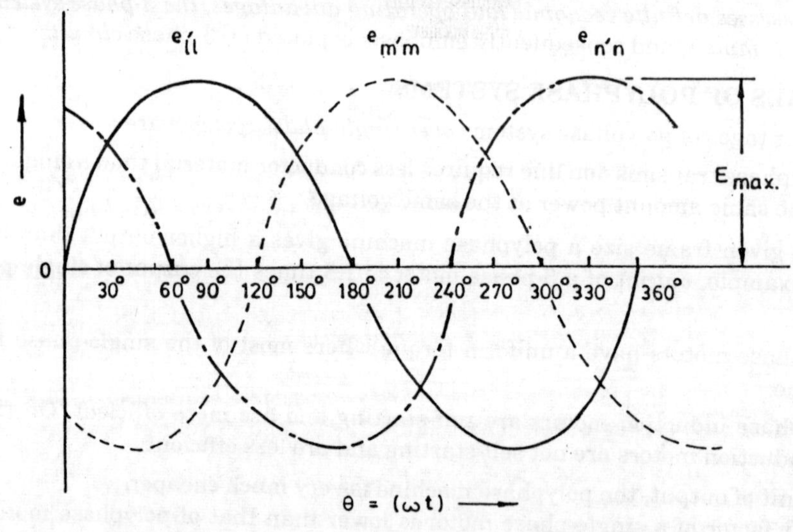

Fig. 2. Voltage waves generated in windings of Fig. 1.

120° in space. The corresponding phasor diagram is shown in Fig. 3. The equations of the instantaneous values of the three voltages (given by Fig. 2) are :

$$e_{l'l} = E_{max.} \sin \omega t$$
$$e_{m'm} = E_{max.} \sin (\omega t - 120°)$$
$$e_{n'n} = E_{max.} \sin (\omega t - 240°)$$

The sum of the above three e.m.fs. is always zero as shown below :

Resultant instantaneous e.m.f.

$= e_{l'l} + e_{m'm} + e_{n'n}$

$= E_{max.} \sin \omega t + E_{max.} \sin (\omega t - 120°) + E_{max.} \sin (\omega t - 240°)$

$= E_{max.} [\sin \omega t + (\sin \omega t \cos 120° - \cos \omega t \sin 120° + \sin \omega t \cos 240° - \cos \omega t \sin 240°)]$

$= E_{max.} [\sin \omega t + (- \sin \omega t \cos 60° - \cos \omega t \sin 60° - \sin \omega t \cos 60° + \cos \omega t \sin 60°)]$

$= E_{max.} (\sin \omega t - 2 \sin \omega t \cos 60°)$

$= E_{max.} (\sin \omega t - \sin \omega t) = 0.$

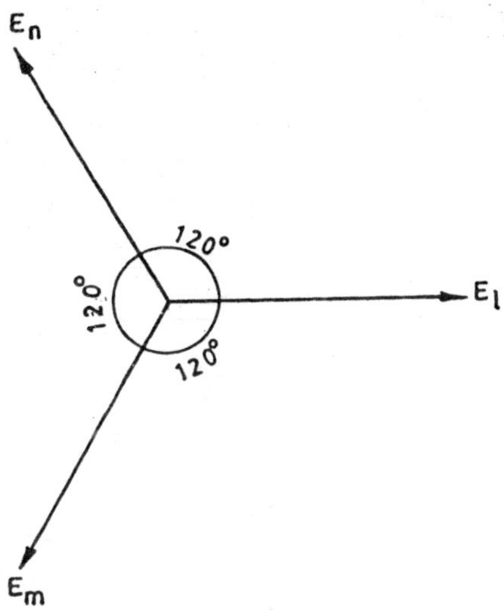

Fig. 3. Phasor diagram of generated voltages.

4. PHASE SEQUENCE AND NUMBERING OF PHASES

- *By phase sequence is meant the order in which the three phases attain their peak or maximum.*

In the generation of three-phase e.m.fs. in Fig. 2 clockwise rotation of the field system in Fig. 1 was assumed. This assumption made the e.m.f. of phase '**m**' lag behind that of '**l**' by 120° and in a similar way, made that of '**n**' lag behind that of '**m**' by 120° (or that of **l** by 240°). Hence, the order in which the e.m.fs. of phases **l, m** and **n** attain their maximum value is **lmn**. It is called the *phase order* or phase sequence **l → m → n**. If now the rotation of field structure of Fig. 1 is reversed *i.e.* made counter-clockwise, then the order in which three phases would attain their corresponding maximum voltages would also be reversed. The phase sequence would become **l → n → m**. This means that e.m.f. of phase '**n**' would now lag behind that of phase '**l**' by 120° instead of 240° as in the previous case.

The phase sequence of the voltages applied to a load, in general, is determined by the order in which the 3-phase lines are connected. The *phase sequence can be reversed* by *interchanging any pair of lines*. (In the case of an induction motor, reversal of sequence results in the reversed direction of motor rotation).

- The three-phases may be *numbered* **l, m, n** or 1, 2, 3 or they may be given three *colours* (as is customary).

The colours used commercially are *red, yellow* (or sometimes white) and *blue*. In this case sequence is RYB.

Evidently in any three-phase system, there are two possible sequences, in which three coils or phase voltages may pass through their maximum value *i.e.* red → yellow → blue (RYB) or red → blue → yellow (RBY).

By convention :

RYB ...... taken as *positive*.

RBY ...... taken as *negative*.

5. INTER-CONNECTION OF THREE PHASES

Each coil of three phases has two terminals [one 'start' (S) and another 'finish' (F)] and if individual phase is connected to a separate load circuit, as shown in Fig. 4, we get a non-interlinked

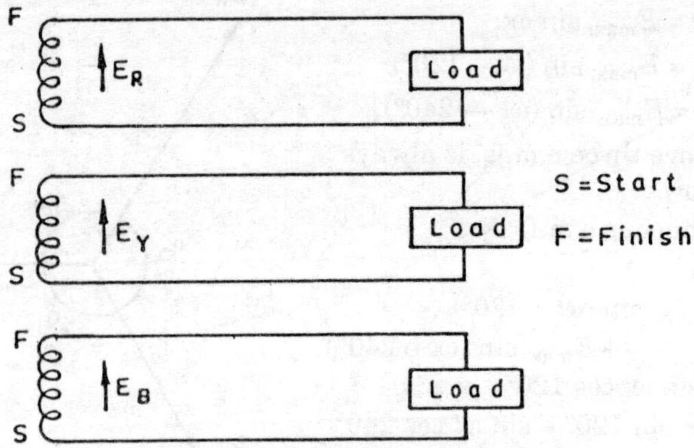

Fig. 4. Non-interlinked 3-phase system.

3-phase system. In such a system each circuit will require two conductors, therefore, 6 conductors in all. This makes the whole system *complicated* and *expensive*. Hence the *three phases are generally interconnected which results in substantial saving of copper*.

The general method of inter-connections are :

1. Star or Wye (Y) connection.
2. Mesh or delta (Δ) connection.

6. STAR OR WYE (Y) CONNECTION

- In this method of inter-connection the similar ends either the 'start' or 'finish' are joined together at point N. This common point N [Fig. 5 (a)] is called *star point* or *neutral point*. Ordinarily only three wires are carried to the external circuit giving 3-phase, 3-wire start connected system but sometimes a *fourth-wire*, known as *neutral wire* is carried to the neutral point of the external load circuit giving 3-phase, 4 wire star connected system.
- The *voltage between any line and the neutral point* (*i.e.* voltage across the phase winding) is called the **'phase voltage' (E_{ph})** ; while the *voltage available between any pair of terminals* (or outers) is called the **'line voltage' (E_L)**.

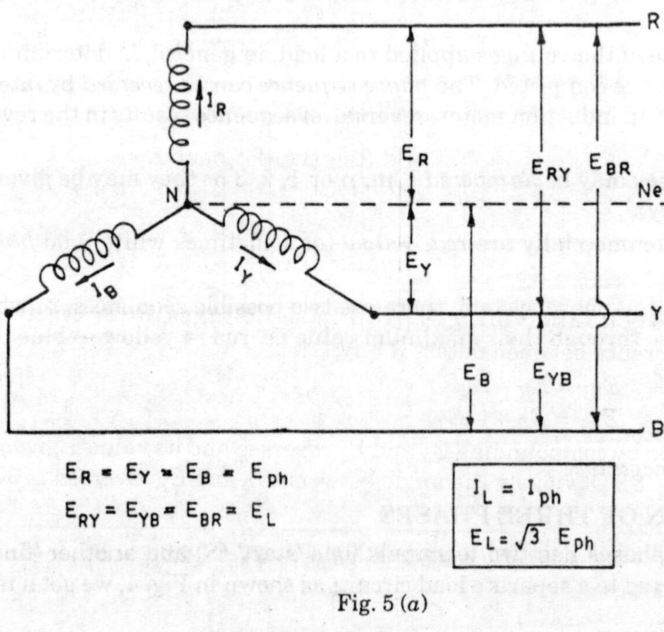

Fig. 5 (a)

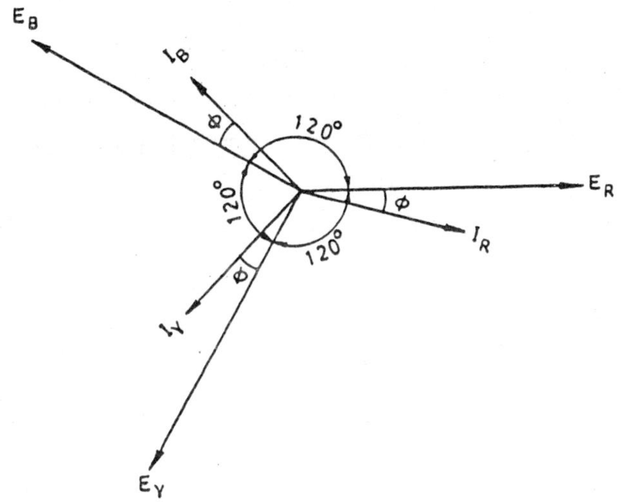

Fig. 5. (*b*) Star-connected three-phase net work.

- In star connection, as is evident in Fig. 5 (*a*) there are two phase windings between each pair of terminals, but since their *similar* ends have been joined together, they are in *opposition*. Obviously, the instantaneous value of potential difference between any two terminals is the *arithmetic difference* of the two-phase e.m.fs. concerned. However, the r.m.s. value of this potential difference is given by the *vector difference* of the two-phase e.m.fs.

- Fig. 5 (*b*) shows the vector diagram for phase voltages and currents in a star connection where a *balanced system has been assumed*. [A balanced system is one in which (*i*) the voltages in all phases are equal in magnitude and differ in phase from one another by equal angles, in this case, the angle = 360/3 = 120°, (*ii*) the currents in the three phases are equal in magnitude and also differ in phase from one another by equal angles. A three-phase balanced load is that in which the loads connected across the three-phases are *identical*]. Thus, we have

$$E_R = E_Y = E_B = E_{ph} \text{ (phase e.m.f.)}$$

Line voltage, $E_{RY}(= E_L)$ = Vector difference of E_R and E_Y

$$= E_R - E_Y$$

Line voltage, $E_{YB} = E_Y - E_B$

Line voltage, $E_{BR} = E_B - E_R$.

(*a*) **Relation Between Line Voltages and Phase Voltages.** Refer Fig. 6.

The potential difference between outers R any Y is

$$E_{RY} = E_R - E_Y \qquad \text{[vector difference]}$$

or

$$E_{RY} = E_R + (-E_Y) \qquad \text{[vector sum]}$$

Hence, E_{RY} is found by compounding E_R and E_Y reversed and its value is given by the diagonal of the parallelogram (Fig. 6). Obviously the angle between E_R and E_Y reversed is 60° and the value of

$$E_{RY} \text{ (or } E_L) = \sqrt{E_R^2 + E_Y^2 + 2E_R E_Y \cos 60°} = \sqrt{E_{ph}^2 + E_{ph}^2 + 2E_{ph} \times E_{ph} \times \tfrac{1}{2}} = \sqrt{3}E_{ph} \qquad ...(1)$$

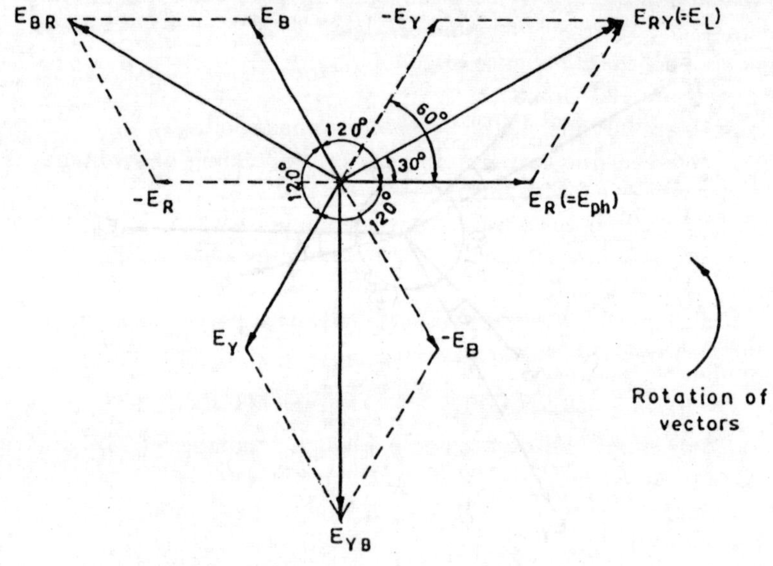

Fig. 6. Vector diagram for star connected net-work.

Similarly $\quad E_{YB}(=E_L) = E_Y - E_B = \sqrt{3}E_{ph}$

and $\quad E_{BR}(=E_L) = E_B - E_R = \sqrt{3}E_{ph}$

i.e., $\quad E_{RY} = E_{YB} = E_{BR} = E_L = \sqrt{3}E_{ph}$

Hence, $\quad E_L = \sqrt{3}E_{ph}$ ...(2)

(*i.e., Line voltage* = $\sqrt{3}$ *phase voltage*).

(*b*) **Relation between Line Currents and Phase Currents.** Since in star-connected system each line conductor is connected to separate phase, so the current flowing through the line and phase are same.

Current in outer (or line) $\quad R = I_R$

Current in outer $\quad Y = I_Y$

Current in outer $\quad B = I_B$

Since $\quad I_R = I_Y = I_B$ = say, I_{ph}—the phase current

∴ Line current, $\quad I_L = I_{ph}$ ...(3)

(*c*) **Power.** If the phase current has a phase difference of ϕ with phase voltage,

Power per phase $\quad = E_{ph}I_{ph}\cos\phi$

Total power (true), $\quad P = 3 \times$ power per phase

$\quad P = 3 \times E_{ph}I_{ph}\cos\phi$ ...[4 (*a*)]

Now $\quad E_{ph} = \dfrac{E_L}{\sqrt{3}}$ and $I_{ph} = I_L$

Hence in terms of line values, the above expression becomes

$$P = 3 \times \frac{E_L}{\sqrt{3}} I_L \cos\phi$$

or $\quad P = \sqrt{3}E_LI_L\cos\phi$ ...[4 (*b*)]

(Apparent power $= \sqrt{3}E_LI_L$).

In a *balanced star-connected net work* the following points are worthnoting :

(*i*) *Line voltages are* $\sqrt{3}$ *times the phase voltages.*

(*ii*) *Line currents are equal to phase currents.*

(*iii*) Line voltages are 120° apart.

(*iv*) Line voltages are 30° ahead of the respective phase voltages.

(*v*) The angle between line currents and the corresponding line voltages is $(30° \pm \phi)$ + for lagging currents – ve for leading currents.

(*vi*) True power = $\sqrt{3}\ E_L I_L \cos \phi$, where ϕ is the angle between respective phase current and phase voltage, not between the line current and line voltage.

(*vii*) Apparent power = $\sqrt{3} E_L I_L$.

(*viii*) In balance system, the potential of the neutral or star point is zero.

∴ Potential at neutral (or star) point $= E_{NR} + E_{NY} + E_{NB} = 0$.

STAR OR WYE CONNECTION

Example 1. *Three equal impedances each having a resistance of 25 Ω and reactance of 40 Ω are connected in star to a 400 V, 3-phase, 50 Hz system. Calculate :*

(*i*) *The line current* (*ii*) *Power factor, and*

(*iii*) *Power consumed.*

Solution. Resistance per phase, $R_{ph} = 25\ \Omega$

Reactance per phase, $X_{ph} = 40\ \Omega$

Line voltage, $E_L = 400$ V

Line current, I_L **:**

Power factor, $\cos \phi$ **:**

Power consumed, P **:**

Refer Fig. 7.

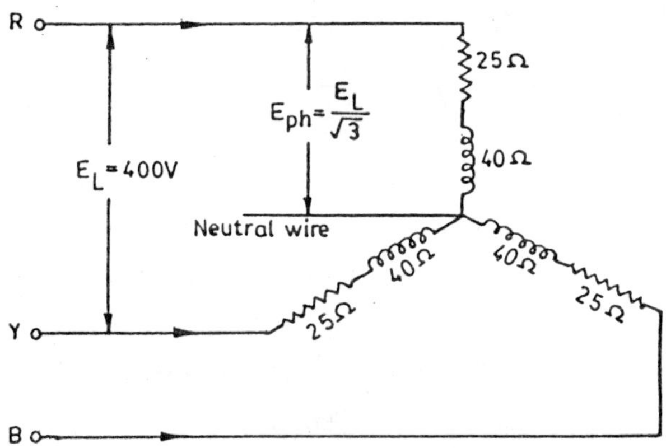

Fig. 7

Impedance per phase, $Z_{ph} = \sqrt{R_{ph}^2 + X_{ph}^2}$

∴ $Z_{ph} = \sqrt{25^2 + 40^2} = 47.17\ \Omega$

Phase voltage, $E_{ph} = \frac{E_L}{\sqrt{3}} = \frac{400}{\sqrt{3}} = 231$ V

Phase current, $I_{ph} = \frac{E_{ph}}{Z_{ph}} = \frac{231}{47.17} = 4.9$ A (app.)

(*i*) **Line current,** I_L = phase current, I_{ph}

$\therefore$ $\mathbf{I_L = 4.9\ A.}$ **(Ans.)**

(*ii*) **Power factor,** $\cos\phi = \dfrac{R_{ph}}{Z_{ph}} = \dfrac{95}{47.17} = \mathbf{0.53\ (lag).}$ **(Ans.)**

(*iii*) **Power consumed,** $P = \sqrt{3}E_L I_L \cos\phi = \sqrt{3} \times 400 \times 4.9 \times 0.53$

$= \mathbf{1800\ W\ (app.).}$ **(Ans.)**

[or $P = 3I_{ph}{}^2 R_{ph} = 3 \times 4.9^2 \times 25 = 1800$ W].

Example 2. *Three identical coils are connected in star to a 400 V (line voltage), 3-phase A.C. supply and each coil takes 300 W. If the power factor is 0.8 (lagging). Calculate :*

(*i*) *The line current,* (*ii*) *Impedance, and*

(*iii*) *Resistance and inductance of each coil.*

Solution. Line voltage, $E_L = 400$ V

Power taken by each coil, $P_{ph} = 300$ W

Power factor, $\cos\phi = 0.8$ (lagging)

$\mathbf{I_L\ ;\ Z\ ;\ R_{ph}\ ;\ L_{ph}}$ **:**

Phase voltage, $E_{ph} = \dfrac{E_L}{\sqrt{3}} = \dfrac{400}{\sqrt{3}}$ V

Also $P_{ph} = E_{ph} I_{ph} \cos\phi$

$300 = \dfrac{400}{\sqrt{3}} \times I_{ph} \times 0.8$

$\therefore$ $I_{ph} = \dfrac{300 \times \sqrt{3}}{400 \times 0.8} = 1.62$ A.

(*i*) **Line current,** I_L = phase current, I_{ph}

$\therefore$ $\mathbf{I_L = 1.62\ A.}$ **(Ans.)**

(*ii*) **Coil impedance,** $Z_{ph} = \dfrac{E_{ph}}{I_{ph}} = \dfrac{\frac{400}{\sqrt{3}}}{1.62} = 142.5\ \Omega$

$\therefore$ $\mathbf{Z_{ph} = 142.5\ \Omega.}$ **(Ans.)**

(*iii*) $R_{ph} = Z_{ph} \cos\phi = 142.5 \times 0.8 = 114\ \Omega$

Coil reactance, $X_{ph} = Z_{ph} \sin\phi = 142.5 \times 0.6 = \mathbf{85.5\ \Omega}$ **(Ans.)**

But $X_{ph} = 2\pi f L_{ph}$

$\therefore$ $L_{ph} = \dfrac{X_{ph}}{2\pi f} = \dfrac{85.5}{2\pi \times 50} = 0.272$ H.

Hence, $\mathbf{R_{ph} = 114\ \Omega.}$ **(Ans.)**

and $\mathbf{L_{ph} = 0.272\ H.}$ **(Ans.)**

Example 3. *In a 3-phase, 3-wire system with star-connected load the impedance of each phase is (3 + j4) Ω. If the line voltage is 230 V, calculate :*

(*i*) *The line current, and* (*ii*) *The power absorbed by each phase.*

Solution. Line voltage, $E_L = 230$ V

Resistance per phase, $R_{ph} = 3\ \Omega$

Reactance per phase, $X_{ph} = 4\ \Omega$

$\mathbf{I_L\ ;\ P_{ph}}$ **:**

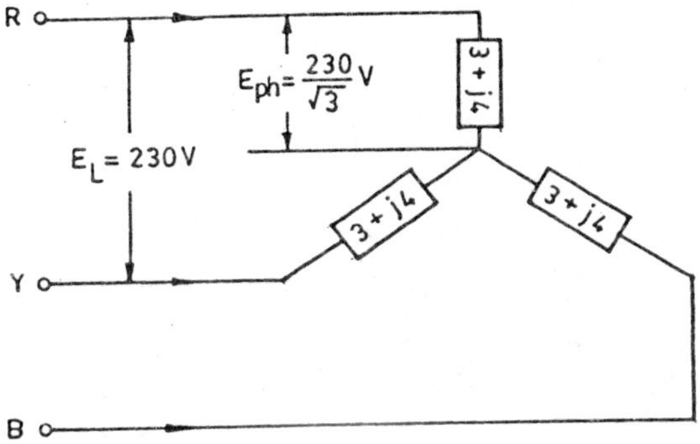

Fig. 8

Phase voltage, $$E_{ph} = \frac{E_L}{\sqrt{3}} = \frac{230}{\sqrt{3}} \text{ V}$$

Impedance per phase, $$Z_{ph} = \sqrt{R_{ph}^2 + X_{ph}^2} = \sqrt{3^2 + 4^2} = 5 \ \Omega$$

Power factor, $$\cos \phi = \frac{R}{Z} = \frac{3}{5} = 0.6$$

Phase current, $$I_{ph} = \frac{E_{ph}}{Z_{ph}} = \frac{230}{\sqrt{3} \times 5} = 26.56 \text{ A}$$

Line current, $\mathbf{I_L = I_{ph} = 26.56}$ **A. (Ans.)**

Power absorbed by each phase,

$$\mathbf{P_{ph}} = E_{ph} I_{ph} \cos \phi = \frac{230}{\sqrt{3}} \times 26.56 \times 0.6 = \mathbf{2116 \text{ W (Ans.)}}$$

$$[P_{ph} = I_{ph}^2 R_{ph} = 26.56^2 \times 3 = 2116 \text{ W}]$$

Solution by Symbolic Notation. In Fig. 9 E_R, E_Y and E_B are the phase voltages whereas I_R, I_Y and I_B are phase currents.

Taking E_R *as the reference vector,* we get

$$E_R = \frac{230}{\sqrt{3}} \angle 0° = 133 \angle 0° = 133 + j0 \text{ volt}$$

$$E_Y = 133 \angle -120° = 133 (-0.5 - j0.866) = (-66.5 - j115) \text{ volts}$$

$$E_B = 133 \angle 120° = 133 (-0.5 + j0.866) = (-66.5 + j115) \text{ volts}$$

$$Z = 3 + j4 = 5 \angle 53° \ 8'$$

$$I_R = \frac{E_R}{Z} = \frac{133 \angle 0°}{5 \angle 53° \ 8'} = 26.6 \angle -53° \ 8'$$

This current lags behind the reference voltage (E_R) by 53° 8′ (Fig. 9).

$$I_Y = \frac{E_Y}{Z} = \frac{133 \angle -120°}{5 \angle 53° \ 8'} = 26.6 \angle -173° \ 8'$$

It lags the reference vector *i.e.*, E_R by 173° 8′ which amounts to lagging behind its phase voltage E_Y by 53° 8′,

$$I_B = \frac{E_B}{Z} = \frac{133 \angle 120°}{5 \angle 53° \ 8'} = 26.6 \angle 66° \ 52'$$

This current leads E_R by 66° 52′ which is the same as *lagging* behind its phase voltage by 53° 8′.

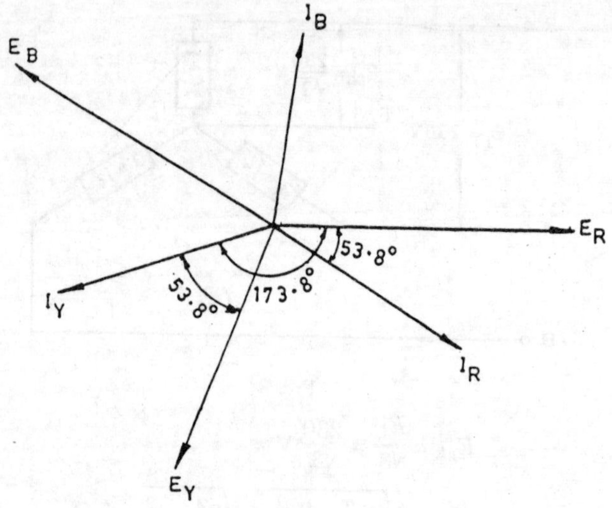

Fig. 9

Let us consider R-phase for calculation of power

$$E_R = (133 + j0)\ ;\ I_R = 26.6\ (0.6 - j0.8) = (15.96 - j21.28)$$

Using method of conjugates, we get

$$P_{VA} = (133 - j0)\ (15.96 - j21.28) = 2116 - j2830$$

$\therefore$ Real power absorbed/phase = 2116 W ...(As before)

Example 4. *A star-connected, 6000 V, 3-phase alternator is supplying 4000 kW at a power factor of 0.8. Calculate the active and reactive components of the current in each phase.*

Solution. Line voltage, $E_L = 6000$ V

Power supplied, $P = 4000$ kW

Power factor, $\cos \phi = 0.8$

Active and reactive components of current :

We know that, $P = \sqrt{3}\ E_L I_L \cos \phi$

$$4000 \times 1000 = \sqrt{3} \times 6000 \times I_L \times 0.8$$

i.e., $$I_L = \frac{4000 \times 1000}{\sqrt{3} \times 6000 \times 0.8} = 481 \text{ A}$$

$\therefore$ $I_{ph} = I_L = 481$ A

Active component $= I_{ph} \cos \phi = 481 \times 0.8 =$ **384.8 A. (Ans.)**

Reactive component $= I_{ph} \sin \phi = 481 \times 0.6 =$ **288.6 A. (Ans.)**

Example 5. *In a 3-phase, 4-wire system, two phases have currents of 20 A and 12 A in lagging power factors of 0.8 and 0.6 respectively, while the third phase is open-circuited. Calculate the current in the neutral and draw the vector diagram.*

Solution. Refer Figs. 10 and 11.

$$\cos \phi_1 = 0.8$$

$\therefore$ $$\phi_1 = \cos^{-1} 0.8 = 36°\ 52'\ \text{(lag)}$$

$$\cos \phi_2 = 0.6$$

$\therefore$ $$\phi_2 = \cos^{-1} 0.6 = 53°\ 8'\ \text{(lag)}$$

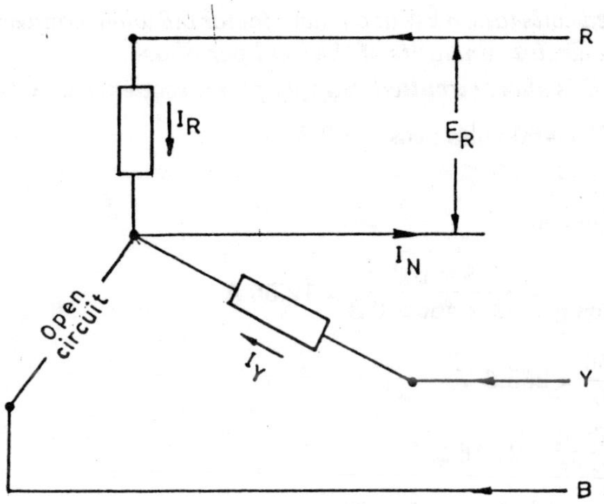

Fig. 10

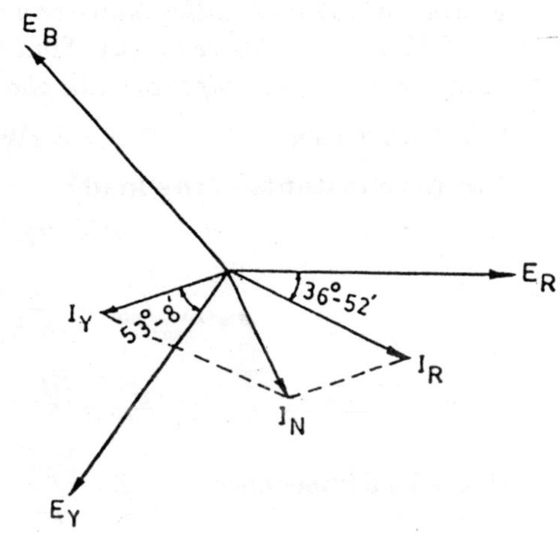

Fig. 11

Let E_R be the reference vector.

Then $\quad I_R = 20 \angle -36° \, 52' = 20\,(0.8 - j0.6) = (16 - j12)$

and $\quad I_Y = 12 \angle -173° \, 8' = 12(-1 - j0.12) = (-12 - j1.44)$

The current through the neutral

$$I_N = I_R + I_Y$$
$$= (16 - j12) + (-12 - j1.44) = 4 - j13.44 = 14 \angle -73° \, 24'$$

Hence, **current in the neutral = 14 ∠– 73° 24′. (Ans.)**

Example 6. *A balanced 3-phase star connected load of 100 kW takes a leading current of 80 A, when connected across a 3-phase, 1100 V, 50 Hz supply. Find the circuit constants of the load per phase.*

Solution. *Given* : $P = 100$ kW ; I_{ph} $(= I_L) = 80$ A ; $E_L = 1100$ V ; $f = 50$ Hz

Circuit constants of the load per phase, R, C :

As the 3-ϕ load is balanced and star connected, line or phase current,

$$I_L = (I_{ph}) = \frac{P}{\sqrt{3}\, E_L \cos \phi} = \frac{100 \times 10^3}{\sqrt{3} \times 1100 \times \cos \phi}$$

or $$\cos \phi = \frac{100 \times 10^3}{\sqrt{3} \times 1100 \times 80} = 0.656$$

Load impedance, $$Z = \frac{E_{ph}}{I_{ph}} = \frac{(1100/\sqrt{3})}{80} = 7.94\ \Omega$$

∴ $$\mathbf{R} = Z \cos \phi = 7.94 \times 0.656 = \mathbf{5.2\ \Omega}. \text{ (Ans.)}$$

Now $X_C = \dfrac{1}{2\pi fC}$ as the current given is *leading* current.

∴ $$C = \frac{1}{2\pi f X_C}$$

But $$X_C = Z \sin \phi = 7.94 \times 0.755 = 5.99\ \Omega$$

∴ $$\mathbf{C} = \frac{1}{2\pi \times 50 \times 5.99}\ \text{F}$$

$$= \frac{10^6}{2\pi \times 50 \times 5.99}\ \mu\text{F} = \mathbf{531.4\ \mu F}. \text{ (Ans.)}$$

Example 7. *Three identical star-connected coils take 8 kW at a power factor 0.8 when connected across a 460 V, 3-phase, 3-wire supply. Find the circuit constants of the load per phase.*

Find the line currents if the coil in phase C is short circuited. Supply phase sequence is A-B-C.

Solution. *Given :* $P = 8$ kW ; $V = 460$ volts ; $\cos \phi = 0.8$

Circuit constants of the load :

$$P = \sqrt{3} E_L I_L \cos \phi$$

$$I_L (= I_{ph}) = \frac{P}{\sqrt{3}\, E_L \cos \phi} = \frac{8 \times 10^3}{\sqrt{3} \times 460 \times 0.8} = 12.55 \text{ A}$$

Phase voltage, $E_{ph} = \dfrac{E_L}{\sqrt{3}} = \dfrac{460}{\sqrt{3}} = 265.6$ V

Hence load impedance, $Z = \dfrac{E_{ph}}{I_{ph}} = \dfrac{265.6}{12.55} = 21.16\ \Omega$

Load resistance, $\mathbf{R} = Z \cos \phi = 21.16 \times 0.8 = \mathbf{16.93\ \Omega}$. **(Ans.)**

Load reactance, $\mathbf{X} = Z \sin \phi = 21.16 \times 0.6 = \mathbf{12.7\ \Omega}$. **(Ans.)**

Line currents if the coil in phase C is short circuited :

The connection arrangement is shown in Fig. 12 when coil C is shorted.

Given that phase sequence is A-B-C.

Hence, $E_{AB} = 460 \angle 0°$

$E_{BC} = 460 \angle -120°$

$E_{CA} = 460 \angle +120°$

Also, $Z = 21.6 \angle 36.8°$

$$I_{AB} = \frac{E_{AB}}{2Z} = \frac{460 \angle 0°}{2 \times 21.16 \angle 36.8°} = 10.87 \angle -36.8°$$

$$I_{BC} = \frac{E_{BC}}{Z} = \frac{460 \angle -120°}{21.16 \angle 36.8°} = 21.74 \angle -156.8°$$

$$I_{CA} = \frac{E_{CA}}{Z} = \frac{460 \angle 120°}{21.16 \angle 36.8°} = 21.74 \angle 83.2°$$

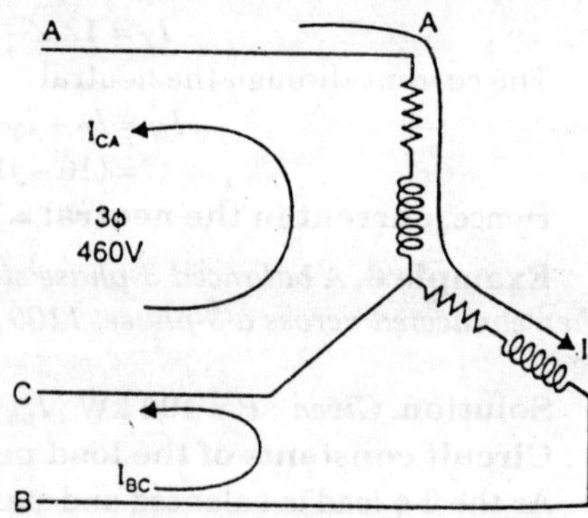

Fig. 12

Line current in line A $= I_{AB} - I_{CA}$

$= 10.87 \angle -36.8° - 21.74 \angle 83.2°$

$= 10.87(0.8 - j0.6) - 21.74(0.118 + j0.993)$

$= 8.70 - j6.52 - 2.56 - j21.59 = \mathbf{(6.14 - j28.11)A.}$ **(Ans.)**

Line current in line B $= I_{BC} - I_{AB}$

$= 21.74 \angle -156.8° - 10.87 \angle -36.8°$

$= 21.74(-0.92 - j0.394) - (8.70 - j6.52)$

$= -20.0 - j8.56 - 8.70 + j6.52 = \mathbf{(-28.70 - j2.04)A.}$ **(Ans.)**

Line current in line $C = I_{CA} - I_{BC}$

$= (-2.56 - j21.59) - (-20 - j8.56) = -2.56 - j21.59 + 20 + j8.56$

$= \mathbf{(17.44 - j13.03)A.}$ **(Ans.)**

Example 8. *In a star-connected load each phase consists of a resistance of 50 Ω in parallel with a capacitor of capacitance 16 μF. When it is connected to a 400 V, 3-phase, 5 Hz supply, calculate :*

(i) The line current, *(ii) The power factor,*

(iii) The power absorbed, and *(iv) The total kVA.*

Solution. Resistance per phase, $R = 50\ \Omega$

Capacitance, $C = 16\ \mu F = 16 \times 10^{-6}$ F

The circuit is shown in Fig. 13.

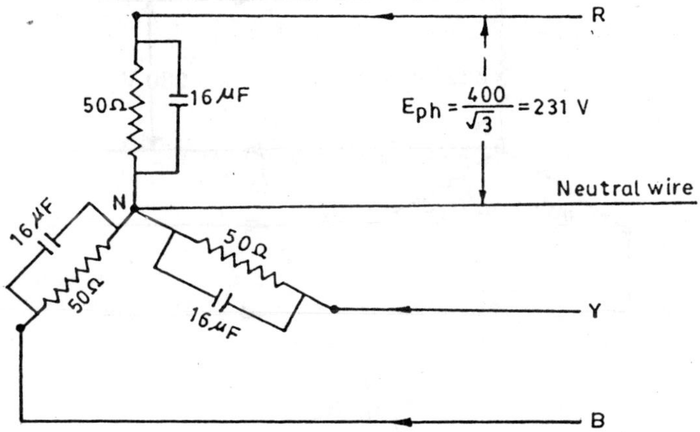

Fig. 13

(i) $$E_{ph} = \left(\frac{400}{\sqrt{3}}\right)\angle 0° = 231 \angle 0° = (231 + j0)$$

Admittance of each phase,

$$Y_{ph}\left(=\frac{1}{Z_{ph}}\right) = \frac{1}{R} + j\omega C$$

$$= \frac{1}{50} + j(2\pi \times 50 \times 16 \times 10^{-6}) = 0.02 + j0.005$$

$\therefore$ $$I_{ph} = E_{ph} \cdot Y_{ph} = 231\,(0.02 + j0.005) = 4.62 + j1.155 = 4.76 \angle 14°$$

For a star-connection, $I_{ph} = I_L$

$\therefore$ $$I_L = 4.76 \text{ A}$$

Hence, **line current = 4.76 A. (Ans.)**

(ii) **Power factor** = cos 14° = **0.97 (leading). (Ans.)**

(iii) Now $E_{ph} = (231 + j0)$; $I_{ph} = (4.62 + j1.155)$

$\therefore$ $$P_{VA} = (231 + j0)\,(4.62 - j1.155)$$

$$= 231 \times 4.62 - j1.155 \times 231 = 1067.22 - j266.8$$

$$= 1100 \angle -14°$$ (per phase)

$\therefore$ **Total power absorbed** = 3 × 1067.22 = 3201.66 W = **3.201 kW. (Ans.)**

(iv) **Total volt-amperes** = 3 × 1100 = 3300 VA = **3.3 kVA. (Ans.)**

Example 9. *A 3-phase, star-connected system with 230 V between each phase and neutral has resistances of 8, 10 and 20 Ω respectively in three phases, calculate :*

(i) The current flowing in each phase, *(ii) The neutral current, and*

(iii) The total power absorbed.

Solution. Refer Figs. 14 (*a*) and (*b*).

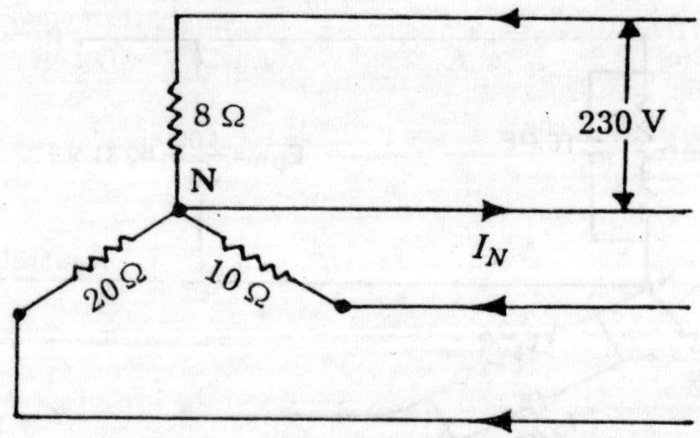

(*a*)

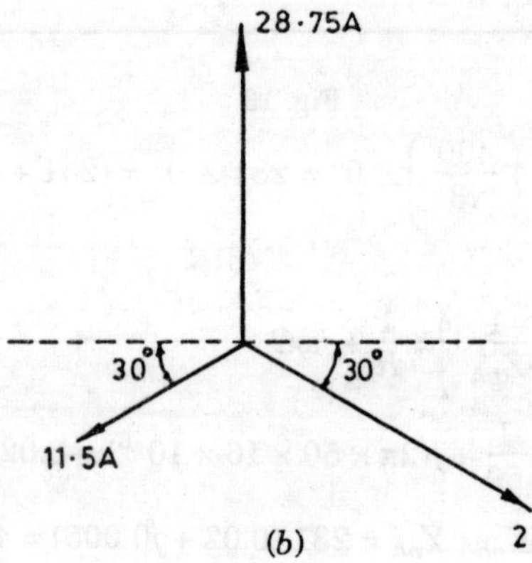

(*b*)

Fig. 14

Phase voltage, $E_{ph} = 230$ V

(*i*) Current in 8 Ω resistor $= \dfrac{230}{8} =$ **28.75 A. (Ans.)**

Current in 10 Ω resistor $= \dfrac{230}{10} =$ **23 A. (Ans.)**

Current in 12 Ω resistor $= \dfrac{230}{20} =$ **11.5 A. (Ans.)**

(*ii*) The above currents are mutually displaced by 120°. The neutral current I_N is the vector sum of these three currents.

I_N can be found by splitting up these three-phase currents into their *X*-components and *Y*-components and then by combining them together.

ΣX-components $= 23 \cos 30° - 11.5 \cos 30° = 11.5 \cos 30° = 9.96$ A

ΣY-components $= 28.75 - 23 \sin 30° - 11.5 \sin 30° = 28.75 - 34.5 \sin 30° = 11.5$ A

∴ **Neutral current,** $I_N = \sqrt{(9.96)^2 + (11.5)^2} =$ **15.21 A. (Ans.)**

(*iii*) **Total power absorbed,**

$P = 230\,(28.75 + 23 + 11.5) =$ **14547.5 W. (Ans.)**

Example 10. *In a 3-phase, 4-wire system, there is a balanced 3-phase motor load taking 10 kW at a power factor of 0.8 lagging while lamps connected between phase conductors and the neutral are taking 2.5 kW, 2 kW and 5 kW respectively. Voltage between line conductors is 430 V. Calculate :*

(*i*) *The current in the neutral wire.* (*ii*) *The current in each conductor.*

Solution. Motor load = 10 kW or 10000 W

Power factor of motor load, $\cos \phi = 0.8$

Lamp loads : 2.5 kW, 2 kW and 5 kW

Voltage between line conductors, $E_L = 430$ V.

Refer Figs. 15 and 16.

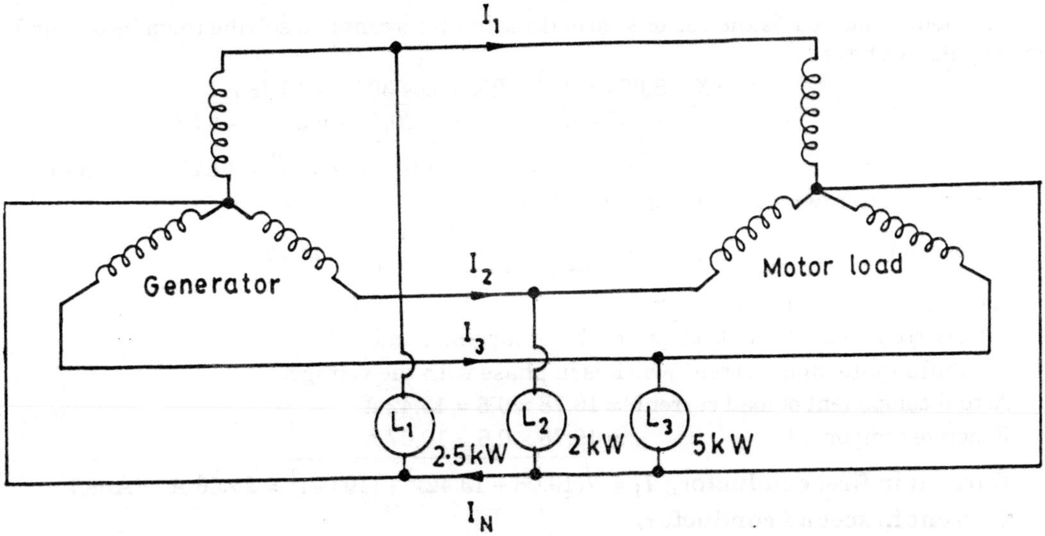

Fig. 15

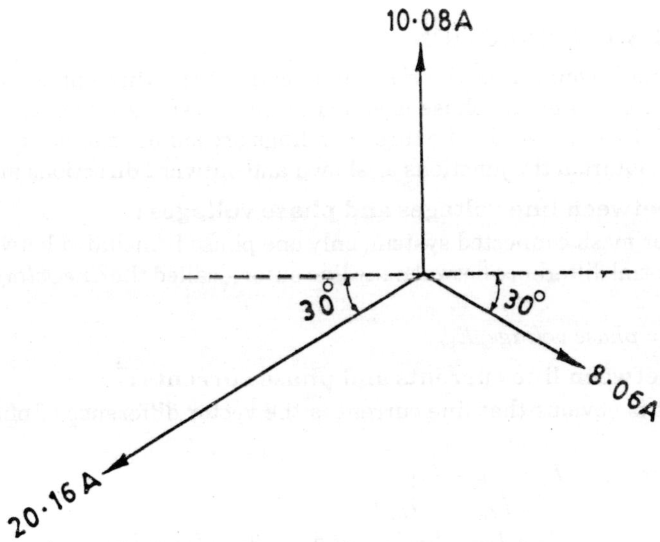

Fig. 16

Let us first find the current in the neutral wire due to lamp loads L_1 (2.5 kW), L_2 (2 kW) and L_3 (5 kW) respectively.

Phase voltage, $E_{ph} = \frac{E_L}{\sqrt{3}} = \frac{430}{\sqrt{3}} = 248$ V

Current through lamp L_1 $= \frac{2.5 \times 1000}{248} = 10.08$ A

Current through lamp L_2 $= \frac{2 \times 1000}{248} = 8.06$ A

Current through lamp L_3 $= \frac{5 \times 1000}{248} = 20.16$ A.

The neutral current is the vector sum of three lamp currents. Resolving them into their X and Y components, we have

$$X = 8.06 \cos 30° - 20.16 \cos 30° = -10.48 \text{ A}$$

$$Y = 10.08 - 8.06 \sin 30° - 20.16 \sin 30° = -4.03 \text{ A}$$

$$\therefore \quad I_N = \sqrt{X^2 + Y^2} = \sqrt{(-10.48)^2 + (-4.03)^2} = \mathbf{11.23\ A.}\ \ \textbf{(Ans.)}$$

Line current drawn by the balanced 3-phase motor load, $I_L = \frac{P}{\sqrt{3} E_L \cos \phi} = \frac{10000}{\sqrt{3} \times 430 \times 0.8} = 16.78$ A

Each conductor carries two currents :

(*i*) a current of 16.78 A at 0.8 power factor lagging ; and

(*ii*) appropriate lamp current which is in phase with the voltage.

Actual component of load current $= 16.78 \times 0.8 = 13.42$ A

Reactive component $= 16.78 \times 0.6 = 10.07$ A

Current in first conductor, $I_1 = \sqrt{(10.08 + 13.42)^2 + (10.07)^2} = \mathbf{25.56\ A.}$ **(Ans.)**

Current in second conductor,

$$I_2 = \sqrt{(8.06 + 13.42)^2 + (10.07)^2} = \mathbf{23.72\ A}\ \ \textbf{(Ans.)}$$

Current in third conductor, $I_3 = \sqrt{(20.16 + 13.42)^2 + (10.07)^2} = \mathbf{35.06\ A.}$ **(Ans.)**

7. DELTA (Δ) OR MESH CONNECTION

In a delta or mesh connection the *dissimilar* ends of the three-phase windings are joined together *i.e.* the '*starting*' end of one phase is joined to the '*finishing*' end of the other phase and so on as shown in Fig. 17. In other words, the three windings are joined in series to form a closed mesh. Three leads are taken out from the junctions as shown and outward directions are taken as positive.

(*a*) **Relation between line voltages and phase voltages :**

Since in delta or mesh connected system, only one phase is included between any pair of line outers, therefore potential difference between the line outers, called the *line voltage*, is equal to phase voltage.

i.e., *Line voltage*, E_L = *phase voltage*, E_{ph}.

(*b*) **Relation between line currents and phase currents :**

From Fig. 17 it is obvious that line current is the vector difference of phase currents of two phases concerned.

Thus, line current, $I_R = I_{YR} - I_{RB}$ (Vector difference)

$= I_{YR} + (-I_{RB})$ (Vector sum)

Similarly, $I_Y = I_{BY} - I_{YR}$ and $I_B = I_{RB} - I_{BY}$

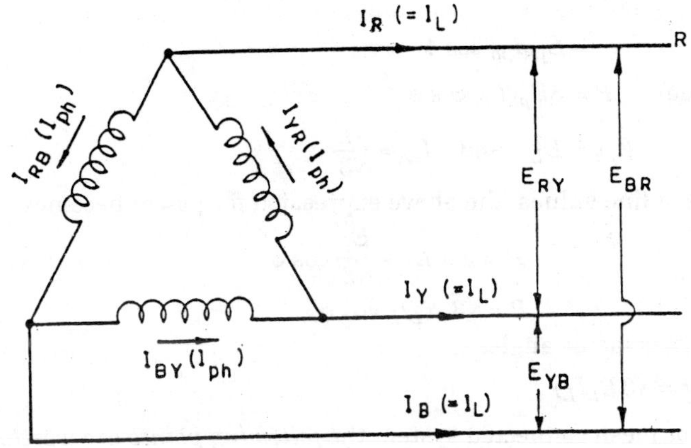

$E_{RY} = E_{YB} = E_{BR} = E_{ph} = E_L$

$I_R = I_Y = I_B = I_L \; (= \sqrt{3}\, I_{ph})$

Fig. 17. Delta or mesh connected diagram.

Refer Fig. 18. Since phase angle between phase current I_{YR} and $-I_{RB}$ is 60°,

$$\therefore \qquad I_R = \sqrt{I_{YR}^2 + I_{RB}^2 + 2I_{YR}I_{RB}\cos 60^\circ}$$

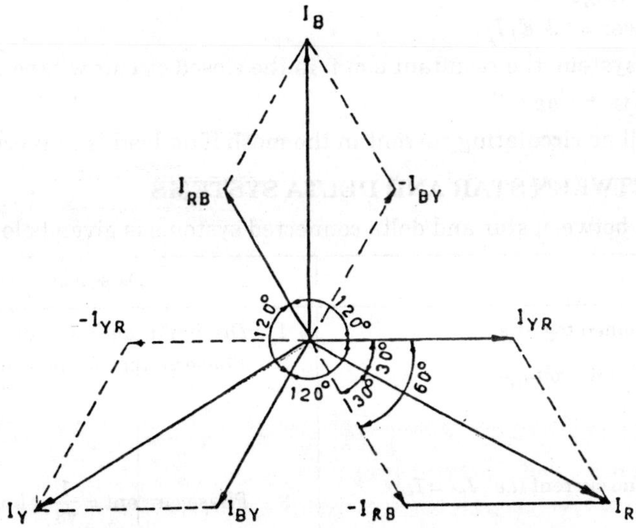

Fig. 18. Vector diagram for delta connected net work.

Assuming the delta connected system or network be balanced, the phase current in each winding is equal and let each be equal to I_{ph} (*i.e.*, $I_{YR} = I_{BY} = I_{RB} = I_{ph}$)

$$\therefore \qquad I_R\,(= I_L) = \sqrt{I_{ph}^2 + I_{ph}^2 + 2I_{ph}I_{ph}\cos 60^\circ} = \sqrt{2I_{ph}^2 + 2I_{ph}^2 \times \frac{1}{2}} = \sqrt{3}I_{ph}$$

Similarly, $\qquad I_Y = I_B = \sqrt{3}I_{ph}$

Hence, $\qquad I_L = \sqrt{3}\, I_{ph} \qquad ...(5)$

(*i.e.*, *line current* = $\sqrt{3}$ *phase current*)

(c) **Power :**

Power/phase $= E_{ph}I_{ph} \cos \phi$

Total power (true) $P = 3E_{ph}I_{ph} \cos \phi$

But $E_{ph} = E_L$ and $I_{ph} = \frac{I_L}{\sqrt{3}}$

Hence, in terms of line values, the above expression for power becomes

$$P = 3 \times E_L \times \frac{E_L}{\sqrt{3}} \cos \phi$$

or

$$P = \sqrt{3}\, E_L I_L \cos \phi \quad \text{...(6)}$$

where ϕ = the phase power factor angle.

(*Apparent power* = $\sqrt{3} E_L I_L$)

In case of delta or mesh connected system the *following points are worthnoting* :

(*i*) *Line voltages are equal to phase voltages.*

(*ii*) Line currents are $\sqrt{3}$ times phase currents.

(*iii*) Line currents are 120° apart.

(*iv*) Line currents are 30° behind their respective phase currents.

(*v*) The angle between line currents and corresponding line voltages is $(30° \pm \phi)$ as in the star system.

(*vi*) True power = $\sqrt{3}\, E_L I_L \cos \phi$, where ϕ is the phase angle between respective phase current and phase voltage.

(*vii*) Apparent power = $\sqrt{3}\, E_L I_L$.

(*viii*) In balanced system, the resultant e.m.f. in the closed circuit will be zero.

i.e., $E_{RY} + E_{YB} + E_{BR} = 0$.

Hence, there will no circulating current in the mesh if no-load is connected to the lines.

8. COMPARISON BETWEEN STAR AND DELTA SYSTEMS

The comparison between star and delta connected systems is given below :

Star connected system	*Delta connected system*
1. *Similar ends* are joined together.	1. *Dissimilar ends* are joined.
2. Phase voltage = $\frac{1}{\sqrt{3}}$ line voltage $\left(i.e., E_{ph} = \frac{E_L}{\sqrt{3}}\right)$.	2. Phase voltage = line voltage (*i.e.*, $E_{ph} = E_L$).
3. Phase current = line current (*i.e.*, $I_{ph} = I_L$).	3. Phase current = $\frac{1}{\sqrt{3}} \times$ line current $\left(i.e., I_{ph} = \frac{I_L}{\sqrt{3}}\right)$.
4. Possible to carry neutral to the load.	4. Neutral wire not available.
5. Provides 3-phase 4-wire arrangement.	5. Provides 3-phase 3-wire arrangement.
6. Can be used for lighting as well as power load.	6. Can be used for power loads only.
7. Neutral wire of a star connected alternator can be connected to earth, so relays and protective devices can be provided in the star connected alternators for safety.	7. Not possible. Delta connected system is mostly used in transformer for running of small low voltage 3-phase motors and *best suited for rotary converters*.

DELTA OR MESH CONNECTION

Example 11. *Three identical coils connected in delta across 400 V, 50 Hz, 3-phase supply take a line current of 15 A at a power factor 0.8 lagging. Calculate :*

(*i*) *The phase current, and*

(*ii*) *The impedance, resistance and inductance of each winding.*

Solution. Line voltage, $E_L = 400$ V

Line current, $I_L = 15$ A

Power factor, $\cos\phi = 0.8$ lagging

$\mathbf{I_{ph}}$; $\mathbf{Z_{ph}}$; $\mathbf{R_{ph}}$; **L :**

Phase voltage, $E_{ph} = E_L = 400$ V

(*i*) **Phase current,** $I_{ph} = \frac{I_L}{\sqrt{3}} = \frac{15}{\sqrt{3}} = \mathbf{8.66\ A.}$ **(Ans.)**

(*ii*) **Impedance of each phase,** $\mathbf{Z_{ph}} = \frac{E_{ph}}{I_{ph}} = \frac{400}{8.66} = \mathbf{46.19\ \Omega.}$ **(Ans.)**

Resistance of each phase, $\mathbf{R_{ph}} = Z_{ph}\cos\phi = 46.19 \times 0.8 = \mathbf{36.95\ \Omega.}$ **(Ans.)**

Reactance of each phase, $X_{ph} = Z_{ph}\sin\phi = 46.19\sqrt{(1-\cos^2\phi)}$

$$= 46.19\sqrt{1-(0.8)^2} = 27.71\Omega$$

$\therefore$ **Inductance,** $\mathbf{L} = \frac{X_{ph}}{2\pi f} = \frac{27.71}{2\pi \times 50} = \mathbf{0.088\ H.}$ **(Ans.)**

Example 12. *A 220 V, 3-phase voltage is applied to a balanced delta-connected 3-phase load of phase impedance (6 + j8).*

(*i*) *Find the phasor current in each line.*

(*ii*) *What is the power consumed per phase ?*

(*iii*) *What is the phasor sum of the three line currents ? What does it have this value ?*

Solution. Refer Fig. 19.

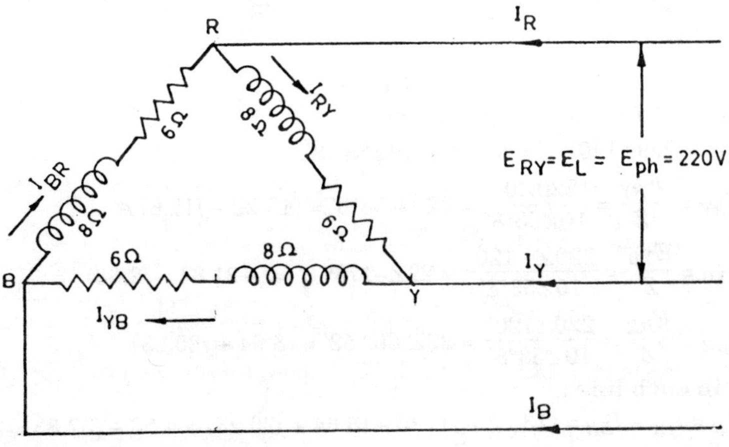

Fig. 19

Resistance per phase, $R_{ph} = 6\ \Omega$

Reactance per phase, $X_{ph} = 8\ \Omega$

$E_L = E_{ph} = 220$ V

Impedance per phase, $Z_{ph} = \sqrt{R_{ph}^2 + X_{ph}^2} = \sqrt{6^2 + 8^2} = 10\ \Omega$

(*i*) Phase current, $I_{ph} = \frac{E_{ph}}{Z_{ph}} = \frac{220}{10} = 22\ \text{A}$

$\therefore$ **Line current,** $I_L = \sqrt{3} \times 22 =$ **38.1 A. (Ans.)**

(*ii*) **Power consumed per phase,**

$$P_{ph} = I_{ph}^2 \times R_{ph} = 22^2 \times 6 = \mathbf{2904\ W.}\ \ \textbf{(Ans.)}$$

(*iii*) Phasor sum would be zero because the three currents are equal in magnitudes and have a mutual phase difference of 120°.

Solution by Symbolic Notation. Let E_{RY} is taken as a reference vector (Fig. 20).

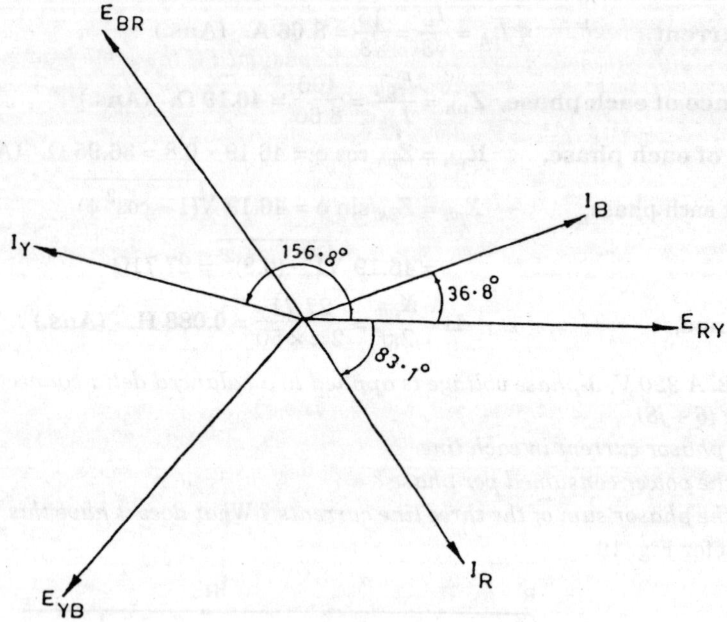

Fig. 20

$E_{RY} = 220\angle 0°,\ E_{YB} = 220\angle -120°$

$E_{BR} = 220\angle 120°,\ Z = 6 + j8 = 10\angle 53°\ 8'$,

$$I_{RY} = \frac{E_{RY}}{Z} = \frac{220\angle 0°}{10\angle 53°8'} = 22\angle -53°8' = (13.22 - j17.6)\ \text{A}$$

$$I_{YB} = \frac{E_{YB}}{Z} = \frac{220\angle -120°}{10\angle 53°8'} = 22\angle -173°\ 8' = (-21.84 - j2.63)$$

$$I_{BR} = \frac{E_{BR}}{Z} = \frac{220\angle 120°}{10\angle 53°8'} = 22\angle\ 66°\ 52' = (8.64 + j20.23)$$

(*i*) **Current in each line :**

$$I_R = I_{RY} - I_{BR} = (13.22 - j17.6) - (8.64 + j20.23) = 4.58 - j37.83 = \mathbf{38.1\angle 83.1°.}\ \ \textbf{(Ans.)}$$

$$I_Y = I_{YB} - I_{RY} = (-21.84 - j2.63) - (13.22 - j17.6)$$

$$= -21.84 - j2.63 - 13.22 + j17.60 = -35.06 + j14.97 = \mathbf{38.12\angle 156.8°.}\ \ \textbf{(Ans.)}$$

$$I_B = I_{BR} - I_{YB} = (8.64 + j20.23) - (-21.84 + j2.63) = 8.64 + j20.23 + 21.84 + j2.63$$

$$= 30.48 + j22.86 = \mathbf{38.1\angle 36.8°.}\ \ \textbf{(Ans.)}$$

(*ii*) **Power consumed per phase :**

Using conjugate of voltage, we get for R-phase

$$P_{VA} = E_{RY} \cdot I_{RY} = (220 - j0)(13.22 - j17.6) = (2908.4 - j3872) \text{ volt ampere}$$

True power per phase = **2.908 kW. (Ans.)**

(*iii*) **Phase sum of the three line currents**

$$= I_R + I_Y + I_B$$
$$= (4.58 - j37.83) + (-35.06 + j14.96) + (30.48 + j22.86) = 0$$

Hence, **the phasor sum of three line currents drawn by a 'balanced load' is zero. (Ans.)**

Example 13. *A delta-connected balanced 3-phase load is supplied from a 3-phase, 400 V supply. The line current is 30 A and the power taken by the load is 12 kW. Find :*

(i) Impedance in each branch ; and

(ii) The line current, power factor and power consumed if the same load is connected is star.

Solution. Delta-connection :

$$E_{ph} = E_L = 400 \text{ V}$$
$$I_L = 30 \text{ A}$$
$$\therefore \quad I_{ph} = \frac{I_L}{\sqrt{3}} = \frac{30}{\sqrt{3}} = 17.32 \text{ A.}$$

(*i*) **Impedance per phase**

$$\mathbf{Z_{ph}} = \frac{E_{ph}}{I_{ph}} = \frac{400}{17.32} = \mathbf{23.09\ \Omega.} \quad \textbf{(Ans.)}$$

Now
$$P = \sqrt{3} E_L I_L \cos \phi$$
$$12000 = \sqrt{3} \times 400 \times 30 \times \cos \phi.$$

or
$$\cos \phi \text{ (power factor)} = \frac{12000}{\sqrt{3} \times 400 \times 30} = 0.577$$

(*ii*) **Star-connection**

$$E_{ph} = \frac{E_L}{\sqrt{3}} = \frac{400}{\sqrt{3}} = 231 \text{ V}$$
$$\mathbf{I_L} = I_{ph} = \frac{E_{ph}}{Z_{ph}} = \frac{231}{23.09} = \mathbf{10\ A} \quad \textbf{(Ans.)}$$

Power factor, $\cos \phi = 0.577$ (since impedance is same)

Power consumed $= \sqrt{3}\, E_L I_L \cos \phi = \sqrt{3} \times 400 \times 10 \times 0.577 =$ **3997.6 W (Ans.)**

Example 14. *Three 50 Ω non-inductive resistances are connected in (i) star, (ii) delta across a 400 V, 50 Hz., 3-phase mains. Calculate the power taken from the supply system in each case. In the event of one of the three resistances getting opened, what would be the value of the total power taken from the mains in each of the two cases.*

Solution. Star connection :

Phase voltage, $$E_{ph} = \frac{E_L}{\sqrt{3}} = \frac{400}{\sqrt{3}} = 231 \text{ V}$$

Phase current, $$I_{ph} = \frac{E_{ph}}{R_{ph}} = \frac{231}{50} = 4.62 \text{ A}$$

Power consumed, $$P = 3 I_{ph}^2 R_{ph} = 3 \times 4.62^2 \times 50 = \mathbf{3200\ W.} \quad \textbf{(Ans.)}$$

$$[\text{or } P = \sqrt{3} E_L I_L \cos \phi = \sqrt{3} \times 400 \times 4.62 \times 1 = 3200 \text{ W}]$$

Delta connection :

Phase voltage, $E_{ph} = E_L = 400$ V

Phase current, $I_{ph} = \frac{E_{ph}}{R_{ph}} = \frac{400}{50} = 8$ A

Power consumed, $P = 3I_{ph}{}^2 R_{ph} = 3 \times 8^2 \times 50 =$ **9600 W. (Ans.)**

When one of the resistances is disconnected :

(*i*) **Star connection.** Refer Fig. 21.

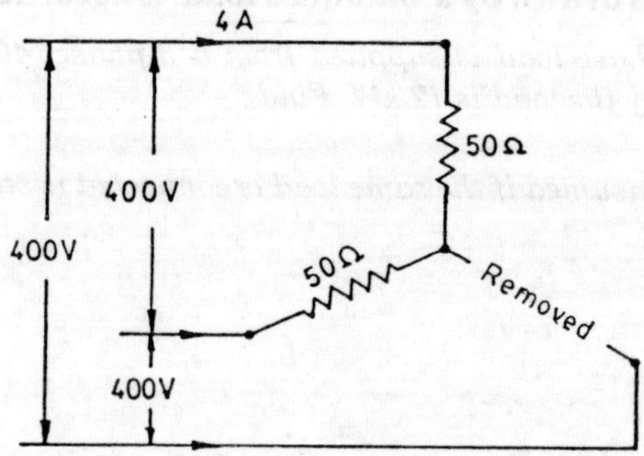

Fig. 21

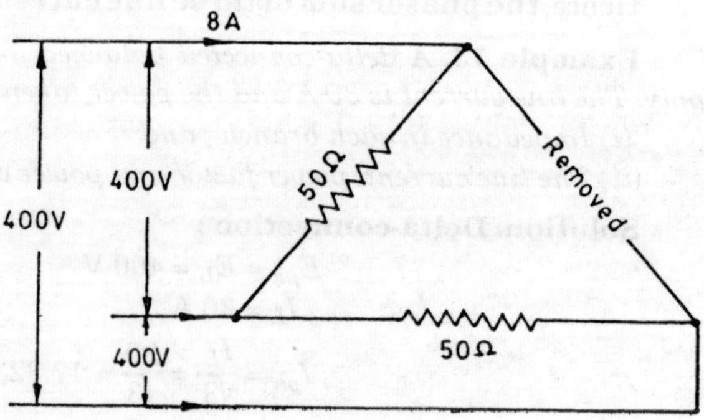

Fig. 22

When one of the resistances is disconnected, the circuit is no longer 3-phase but converted into single-phase circuit, having two resistances each of 50 ohm connected in series across supply of 400 V.

Hence line current, $I_L = \frac{E_L}{2R_{ph}} = \frac{400}{2 \times 50} = 4$ A

Power consumed, $P = 4^2 (50 + 50) =$ **1600 W. (Ans.)**

[or $P = VI \cos \phi = 400 \times 4 \times 1 = 1600$ W].

(*ii*) **Delta connection.** Refer Fig. 22.

Potential difference across each resistance, $E_L = 400$ V

Current in each resistance $= \frac{400}{50} = 8$ A

Power consumed in both resistances $= 2 \times 8^2 \times 50 =$ **6400 W. (Ans.)**

[or $P = 2 \times E_{ph} I_{ph} \cos \phi = 2 \times 400 \times 8 \times 1 = 6400$ W].

Example 15. *The secondary of a 3-phase star-connected transformer, which has a phase voltage of 230 V feeds a 3-phase delta connected load ; each phase of which has a resistance of 30 Ω and an inductive reactance of 40 Ω. Draw the circuit diagram of the system and calculate :*

(*i*) *The voltage across each phase of load,*

(*ii*) *The current in each phase of load,*

(*iii*) *The current in the transformer secondary windings, and*

(*iv*) *The total power taken from the supply and its power factor.*

Solution. Refer Fig. 23.

Resistance per phase, $R_{ph} = 20$ Ω

Reactance per phase, $X_{ph} = 40$ Ω

Phase voltage across transformer secondary,

$E_{ph} = 230$ V

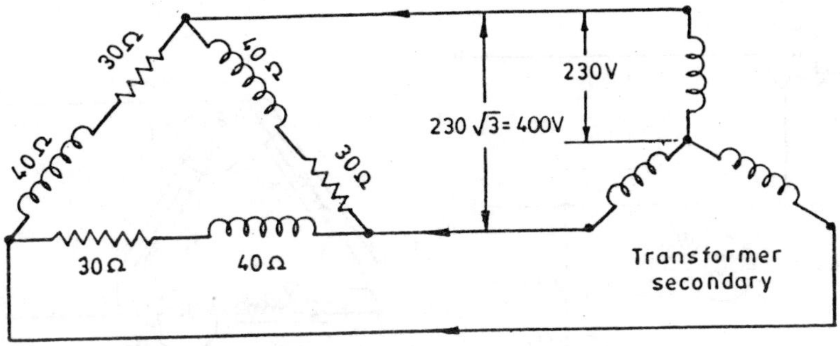

Fig. 23

Line voltage across delta connected load = line voltage across transformer secondary

$$= \sqrt{3} \times 230 = 400 \text{ V (app.)}$$

(*i*) **Voltage across each phase of the load,**

$$E_{ph} = E_L = \mathbf{400\ V.}\ \ \textbf{(Ans.)}$$

(*ii*) **Current in each phase of the load,**

$$I_{ph} = \frac{E_{ph}}{Z_{ph}} = \frac{400}{\sqrt{R_{ph}^2 + X_{ph}^2}} = \frac{400}{\sqrt{30^2 + 40^2}} = \frac{400}{50} = \mathbf{8\ A.}\ \ \textbf{(Ans.)}$$

(*iii*) **Current in the transformer secondary**

$$= \text{line current of load} = \sqrt{3} \times I_{ph} = \sqrt{3} \times 8 = \mathbf{13.86\ A.}\ \textbf{(Ans.)}$$

(*iv*) **Power factor,** $\cos \phi = \frac{R_{ph}}{Z_{ph}} = \frac{30}{50} = \mathbf{0.6.}\ \ \textbf{(Ans.)}$

Total power consumed, $P = \sqrt{3} E_L I_L \cos \phi = \sqrt{3} \times 400 \times 13.86 \times 0.6 = \mathbf{5761.5\ W.}\ \ \textbf{(Ans.)}$

9. MEASUREMENT OF POWER IN 3-PHASE CIRCUIT

- The power in 3-phase load can be measured by using the following methods :
  1. Three wattmeters method.
  2. Two wattmeters method.
  3. One wattmeter method.
- A wattmeter consists of two coils : Refer Fig. 24.
  1. *Current coil*—possesses a *low resistance*.
  2. *Pressure or potential coil*—possesses a *high resistance*.

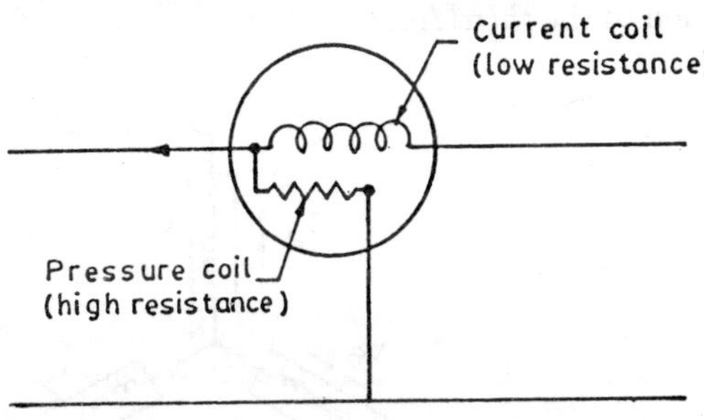

Fig. 24. Connections of a wattmeter.

The *'current coil'* is connected in *series with the line carrying the current* and the *'pressure coil' is connected across the two points whose potential difference is to be measured.* A wattmeter shows a reading which is proportional to the *product* of the *current* through its current coil, the *potential difference* across its pressure coil and *cosine of the angle* between this voltage and current.

9.1. Three-Wattmeters Method. Figs. 25 and 26 show the connection diagram for star-connected and delta-connected loads respectively. As indicated in the figures three wattmeters are connected in each of the three phases of the load whether star or delta connected. The current coil

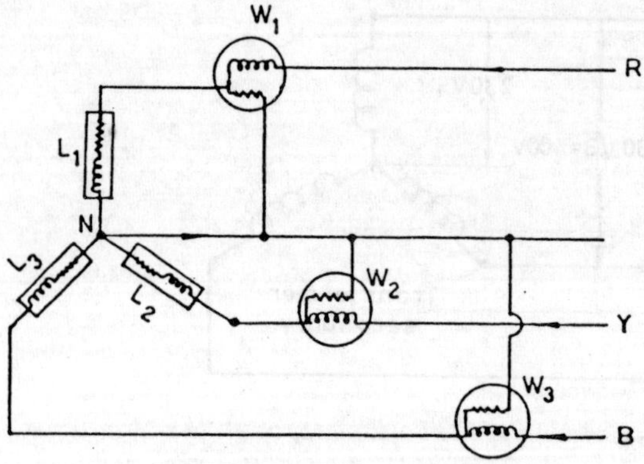

Fig. 25. Star-connected load.

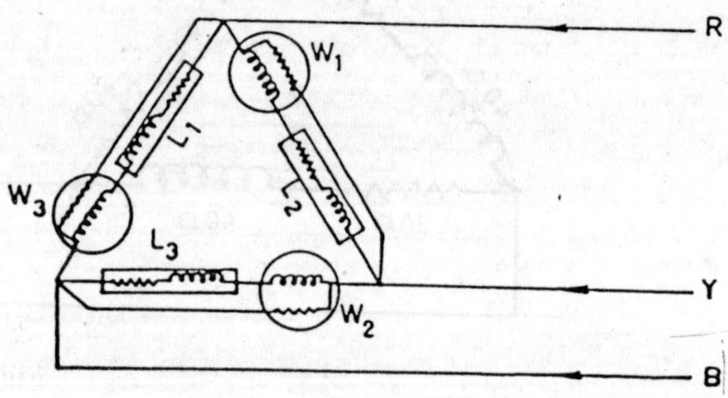

Fig. 26. Delta-connected load.

of each wattmeter carries the current of one phase only and the pressure coil measures the phase-voltage of the phase. Hence, *each wattmeter measures the power in a single phase. The total power in the load is given by the algebraic sum of the readings of the three wattmeters.*

While using this method following *difficulty* is met with :

- In case of star-connected load it is not always possible to get at neutral point which is required for connections (Fig. 25).
- In case of delta-connected load, under ordinary conditions it is not generally feasible to break into the phases of the load.

To measure power it is not necessary to use three wattmeters, two wattmeters can be used for the purpose as explained in the Article 9.2.

9.2. Two-Wattmeters method. Balanced or unbalanced load, Figs. 27 and 28 show connection diagrams for star-connected and delta-connected loads respectively. In this method the current coils of the two wattmeters are inserted in *any two* lines and the pressure (or potential) coil of each joined to the *third line*.

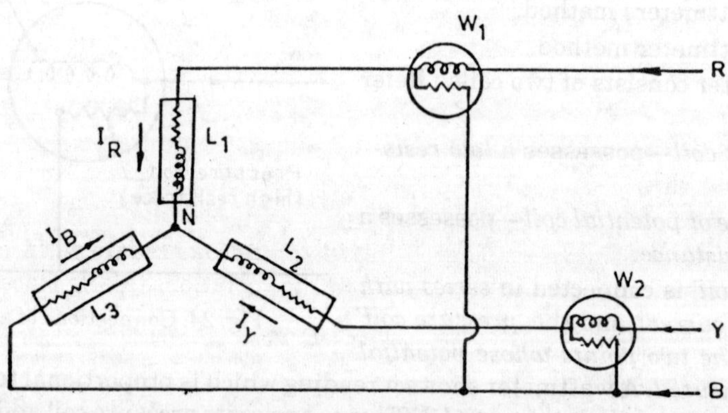

Fig. 27. Star-connected load.

It can be proved that the sum of the instantaneous powers indicated by W_1 and W_2 gives the instantaneous power absorbed by the three loads L_1, L_2 and L_3. Let us consider a star-connected load (although it can be equally applied to a delta-connected load which can always be replaced by an equivalent star-connected load).

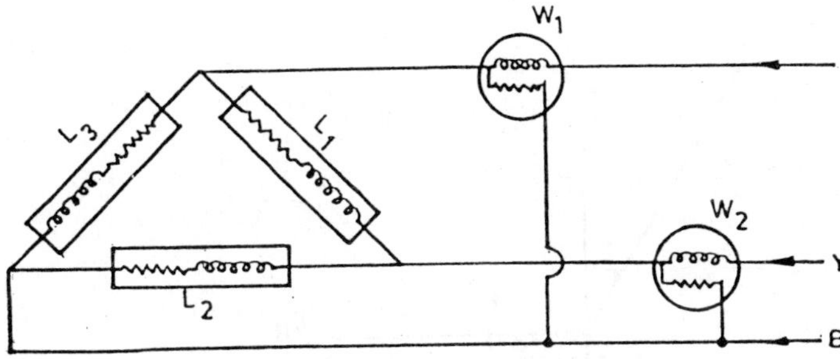

Fig. 28. Delta-connected load.

Keeping in mind that it is important to take *the direction of the voltage through the circuit as the same as that taken for the current when establishing the readings of the two wattmeters.*

Instantaneous current through $W_1 = i_R$

Instantaneous potential difference across $W_1 = e_{RB} = e_R - e_B$

Instantaneous power read by $W_1 = i_R(e_R - e_B)$

Instantaneous current through $W_2 = i_Y$

Instantaneous potential difference across $W_2 = e_{YB} = e_Y - e_B$

Instantaneous power read by $W_2 = i_Y(e_Y - e_B)$

$$\therefore \quad W_1 + W_2 = i_R(e_R - e_B) + i_Y(e_Y - e_B) = i_Re_R + i_Ye_Y - e_B(i_R + i_Y)$$

Now according to Kirchhoff's point law

$$i_R + i_Y + i_B = 0$$

$$\therefore \quad i_R + i_Y = -i_B$$

$$\therefore \quad W_1 + W_2 = i_Re_R + i_Ye_Y + i_Be_B = p_1 + p_2 + p_3$$

where p_1 = power absorbed by load L_1, p_2 = power absorbed by load L_2, and
p_3 = power absorbed by load L_3.

$$\therefore \quad W_1 + W_2 = \text{total power absorbed.}$$

This proof is true whether the load is *balanced* or *unbalanced.*

In case the load is star-connected, then it should have no neutral connection (*i.e.*, 3-phase, 3-wire connected) and if it has a neutral connection (*i.e.*, 3-phase, 4-wire connected) that it should be exactly balanced so that in each case there is no neutral current i_N otherwise Kirchhoff's point law will give $i_R + i_Y + i_B + i_N = 0$.

In the above derivation we have considered the *instantaneous* readings. In fact the moving system wattmeter, due to its inertia, cannot quickly follow the variations taking place in cycle, hence it indicates the *average power.*

$$\therefore \text{ Total power} = W_1 + W_2$$

$$= \frac{1}{T}\int_0^T i_Re_{RB}\,dt + \frac{1}{T}\int_0^T i_Ye_{YB}\,dt. \qquad \text{...(7)}$$

Two-Wattmeter Method—Balanced load. The total power consumed by a *balanced load* can be found by using two wattmeters (Figs. 27 and 28). When load is assumed inductive in Fig. 25, the vector diagram for such a balanced star-connected load is shown in Fig. 29.

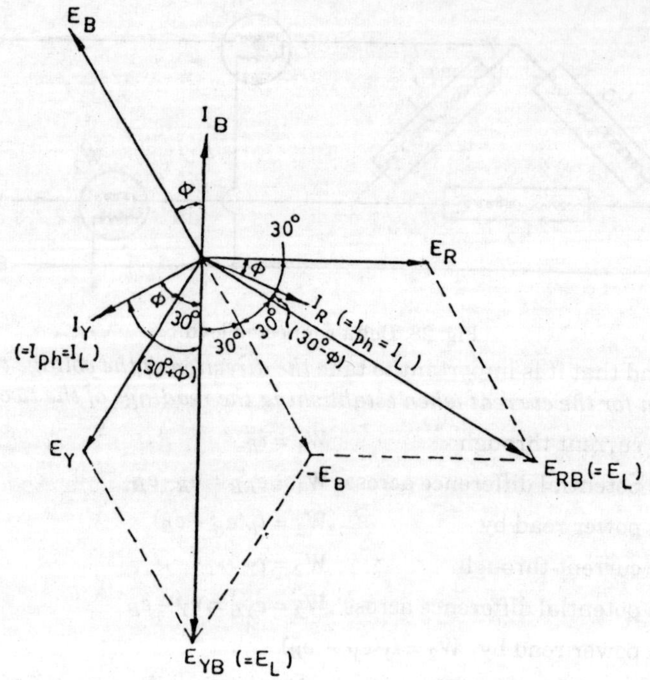

Fig. 29. Vector diagram—two wattmeters method.

Let us consider the problem in terms of r.m.s. values (instead of instantaneous values).

Let E_R, E_Y, E_B = r.m.s. values of the three phase voltages,

and I_R, I_Y, I_B = r.m.s. values of the currents.

Since these voltages and currents are assumed sinusoidal, they can be represented by vectors, the currents lagging behind their respective phase voltages by ϕ.

Refer Fig. 29.

Current through wattmeter $W_1 = I_R$

Potential difference across pressure coil of wattmeter

$$W_1 = E_{RB} = E_R - E_B \quad \text{(Vectorially)}$$

The value of E_{RB} is found by compounding E_R and E_B reversed as shown in Fig. 29. It may be observed that phase difference between E_{RB} and $I_R = (30° - \phi)$.

$\therefore$ Reading of wattmeter $W_1 = E_{RB} I_R \cos (30° - \phi)$ ...(i)

Similarly, current through wattmeter $W_2 = I_Y$

Potential difference across pressure coil of wattmeter

$$W_2 = E_{YB} = E_Y - E_B \quad \text{(Vectorially)}$$

The value of E_{YB} is found by compounding E_Y and E_B reversed as shown in Fig. 29. The phase difference between E_{YB} and $I_Y = (30° + \phi)$.

$\therefore$ Reading of wattmeter $W_2 = E_{YB} I_Y \cos (30° + \phi)$ ...(ii)

Since the load is balanced, $E_{RB} = E_{YB} = E_L$ (Line voltage)

and $I_R = I_Y = I_L$ (Line current)

$\therefore$ $$W_1 = E_L I_L \cos (30° - \phi)$$

and $$W_2 = E_L I_L \cos (30° + \phi)$$

$\therefore$ Total power,
$$\begin{aligned} P &= W_1 + W_2 \\ &= E_L I_L \cos (30° - \phi) + E_L I_L \cos (30° + \phi) \\ &= E_L I_L [\cos (30° - \phi) + \cos (30° + \phi)] \\ &= E_L I_L [\cos 30° \cos \phi + \sin 30° \sin \phi + \cos 30° \cos \phi - \sin 30° \sin \phi] \\ &= E_L I_L (2 \cos 30° \cos \phi) = E_L I_L \times 2 \times \frac{\sqrt{3}}{2} \cos \phi = \sqrt{3} E_L I_L \cos \phi \end{aligned}$$

i.e., $$P = \sqrt{3} E_L I_L \cos \phi$$

Hence the *sum of the readings of the two wattmeters give the total power consumption in the 3-phase load.*

It is worth noting that in the above case the phase sequence of *RYB* has been assumed, the readings of the two wattmeters will change if the phase sequence is reversed.

Variations in wattmeter readings. As shown above that for a *lagging power factor*

$$W_1 = E_L I_L \cos (30° - \phi)$$

and $$W_2 = E_L I_L \cos (30° + \phi)$$

From above it is evident that individual readings of the wattmeters *not only depend on the load but also upon its power factor*. Let us take up the following cases :

(*i*) When $\phi = 0$

i.e., power factor is *unity* (*i.e.,* load is *resistive*)

Then $$W_1 = W_2 = E_L I_L \cos 30°$$

The reading of each wattmeter will be equal and opposite (*i.e.,* up-scale reading).

(*ii*) When $\phi = 60°$

i.e., power factor = 0.5 (lagging)

Then $$W_2 = E_L I_L \cos (30° + 60°) = 0$$

Hence, the power is measured by W_1 above.

(*iii*) When $90° > \phi > 60°$

i.e., $0.5 > \text{p.f.} > 0$

Then W_1 is still *positive* but reading of W_2 is *reversed*. For a leading p.f., conditions are just the opposite of this. In that case, W_1 will read *negative* because the phase angle between the current and voltage is *more than* 90°. For getting the total power, the reading of W_2 is to be *subtracted* from that of W_1. Under this condition, W_2 will read *'down scale' i.e., backwards*. Hence, *to obtain a reading on W_2, it is necessary to reverse either its pressure coil or current coil, usually the former.*

All readings taken after reversal of pressure coil are to be taken as negative.

(*iv*) When $\phi = 90°$

(*i.e.,* p.f. = 0 *i.e.,* pure inductive or capacitive load)

Then $$W_1 = E_L I_L \cos (30° - 90°) = E_L I_L \sin 30°$$

and $$W_2 = E_L I_L \cos (30° + 90°) = - E_L I_L \sin 30°$$

These two readings are equal in magnitude but opposite in sign

$\therefore$ $$W_1 + W_2 = 0.$$

So far we have considered *lagging angles* (taken as *positive*). Now let us discuss how the readings of wattmeters change when the power factor is *leading* one.

— For $\phi = +60°$ (lag) : $W_2 = 0$

— For $\phi = -60°$ (lead) : $W_1 = 0$

Thus we find that for angles of lead the readings of the two wattmeters are interchanged. Hence, when the power is *leading* :

$$W_1 = E_L I_L \cos(30° + \phi)$$
$$W_2 = E_L I_L \cos(30° - \phi).$$

Power Factor—When the load is 'balanced'. When load is balanced with a *lagging power factor* and the voltage and currents are sinusoidal :

$$W_1 + W_2 = E_L I_L \cos(30° - \phi) + E_L I_L \cos(30° + \phi) = \sqrt{3} E_L I_L \cos\phi \quad ...(i)$$

Similarly,
$$W_1 - W_2 = E_L I_L \cos(30° - \phi) - E_L I_L \cos(30° + \phi)$$
$$= E_L I_L (2 \sin\phi \sin 30°) = E_L I_L \sin\phi \quad ...(ii)$$

Dividing (*ii*) by (*i*), we get

$$\tan\phi = \frac{\sqrt{3}(W_1 - W_2)}{(W_1 + W_2)} \quad ...(8)$$

For a *leading power*, this expression becomes

$$\tan\phi = -\frac{\sqrt{3}(W_1 - W_2)}{(W_1 + W_2)} \quad ...(9)$$

After finding $\tan\phi$, hence ϕ, the value of power factor $\cos\phi$ can be found (from trigonometrical tables).

One important point which must be kept in mind is that if W_2 reading has been taken after reversing the pressure coil *i.e.*, if W_2 is *negative*, then the eqn. (8) becomes

$$\tan\phi = \sqrt{3}\,\frac{W_1 - (-W_2)}{W_1 + (-W_2)}$$

or
$$\tan\phi = \sqrt{3}\,\frac{W_1 + W_2}{W_1 - W_2} \quad ...(10)$$

The power factor may also be expressed in terms of ratio of the readings of the two wattmeters.

Let
$$\frac{\text{Smaller reading}}{\text{Larger reading}} = \frac{W_2}{W_1} = \alpha$$

Then from eqn. (8) above, we have

$$\tan\phi = \frac{\sqrt{3}\left[1 - \left(\frac{W_2}{W_1}\right)\right]}{\left[1 + \left(\frac{W_2}{W_1}\right)\right]} = \frac{\sqrt{3}(1 - \alpha)}{(1 + \alpha)}$$

We know that,
$$\sec^2\phi = 1 + \tan^2\phi$$

or
$$\frac{1}{\cos^2\phi} = 1 + \tan^2\phi \quad \text{or} \quad \cos^2\phi = \frac{1}{1 + \tan^2\phi}$$

$$\cos\phi = \frac{1}{\sqrt{1 + \tan^2\phi}} = \frac{1}{\sqrt{1 + \left[\frac{\sqrt{3}(1 - \alpha)}{1 + \alpha}\right]^2}}$$

or $$= \frac{1}{\sqrt{1+3\left(\frac{1-\alpha}{1+\alpha}\right)^2}} = \frac{1+\alpha}{\sqrt{(1+\alpha)^2+3\,(1-\alpha)^2}}$$

$$= \frac{1+\alpha}{\sqrt{1+\alpha^2+2\alpha+3\,(1+\alpha^2-2\alpha)}} = \frac{1+\alpha}{\sqrt{4+4\alpha^2-4\alpha}}$$

$$= \frac{1+\alpha}{2\sqrt{1-\alpha+\alpha^2}}$$

i.e., $$\cos\phi = \frac{1+\alpha}{2\sqrt{1-\alpha+\alpha^2}} \qquad \text{...(11)}$$

If a curve is plotted between α and $\cos\phi$, then the curve obtained will be as shown in Fig. 30, this curve is called **watt-ratio curve.**

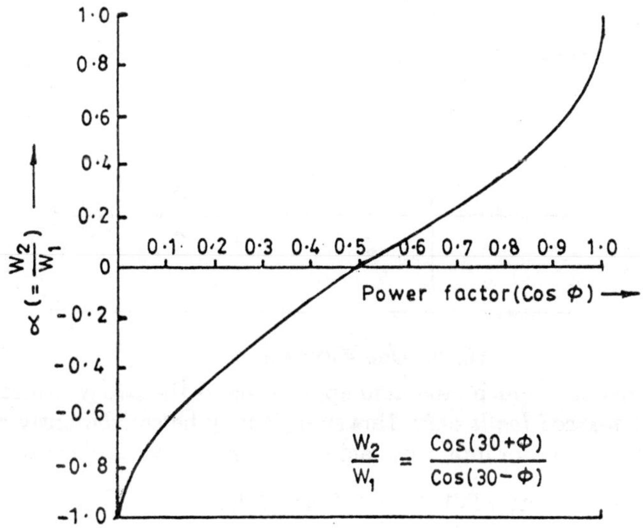

Fig. 30. Watt-ratio curve.

Reactive volt amperes (*with two wattmeters*)

We know that $$\tan\phi = \frac{\sqrt{3}(W_1 - W_2)}{W_1 + W_2}$$

As the tangent of the angle of lag between phase current and phase voltage of a circuit is always equal to the ratio of reactive power to the true power (Fig. 31). Hence, in case of a balanced load, the reactive power is given by $\sqrt{3}$ *times the difference of the readings of the two wattmeters* used to measure the power of a 3-phase circuit by two wattmeter method.

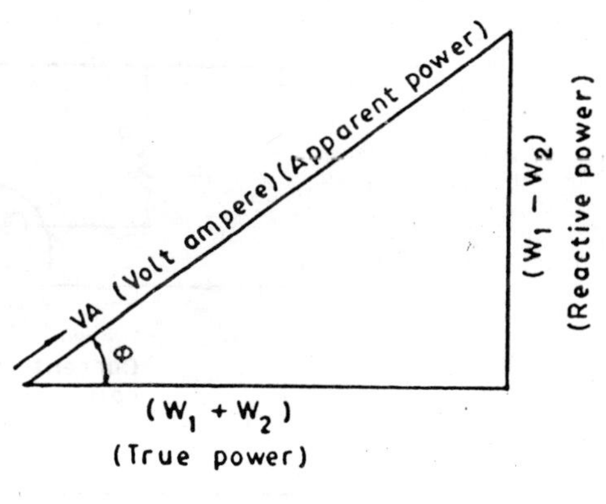

Fig. 31

Mathematical proof is as follows :

$$\sqrt{3}\,(W_1 - W_2) = \sqrt{3}\,[(E_L I_L \cos (30° - \phi) - E_L I_L \cos (30° + \phi)]$$
$$= \sqrt{3}\,E_L I_L\,[(\cos 30° \cos \phi + \sin 30° \sin \phi) - (\cos 30° \cos \phi - \sin 30° \sin \phi)]$$
$$= \sqrt{3}\,E_L I_L\,[\cos 30° \cos \phi + \sin 30° \sin \phi - \cos 30° \cos \phi + \sin 30° \sin \phi]$$
$$= \sqrt{3}\,E_L I_L\,(2 \sin 30° \sin \phi) = \sqrt{3}\,E_L I_L \sin \phi$$

i.e., $\sqrt{3}\,(W_1 - W_2) = \sqrt{3}\,E_L I_L \sin \phi.$

9.3. One-Wattmeter Method. In this method the current coil is connected in any one line and the pressure coil is connected alternately between this and the other two lines as shown in Fig. 32. The two readings so obtained, for a balanced load, correspond to those obtained by normal two wattmeter method.

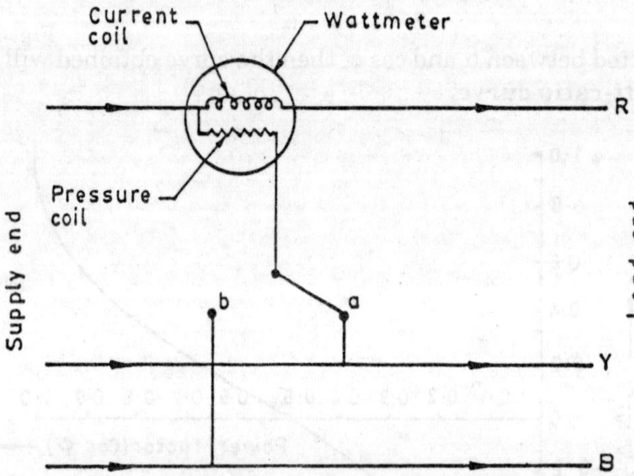

Fig. 32. One wattmeter method.

This method is not of as much universal application as the two wattmeter method because it is restricted to *fairly balanced loads only*. However, it may be conveniently applied, for instance, when it is desired to find the power input to a factory motor *in order to check the load upon the motor*.

10. MEASUREMENT OF REACTIVE VOLT AMPERES

In order to measure reactive power in a single phase circuit a *compensated wattmeter* is used. In this wattmeter the voltage applied to the pressure coil is *90° out of phase with the actual voltage* and hence it will read *VI* cos (90° – ϕ) *i.e., VI* sin ϕ, reactive power.

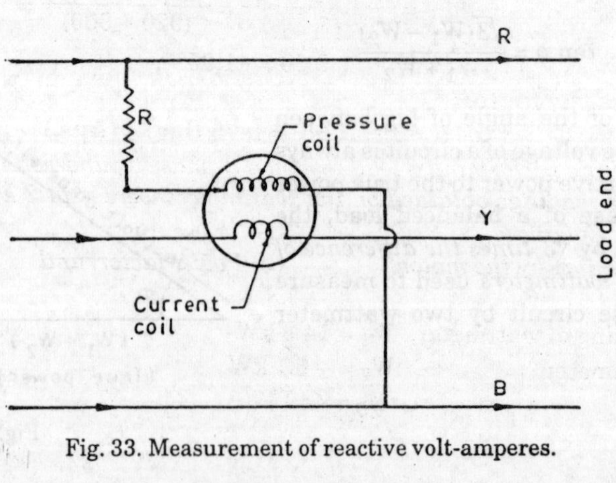

Fig. 33. Measurement of reactive volt-amperes.

In balanced three-phase circuit the reactive power can be determined by using one wattmeter. The necessary connections are shown in Fig. 33. The current coil is inserted in one line and the pressure coil is connected across the other two lines.

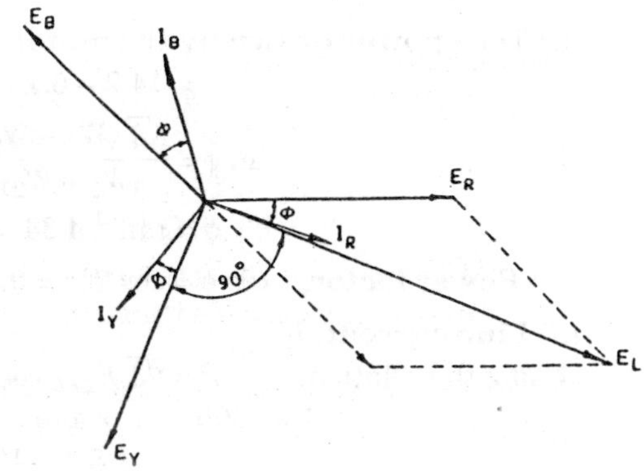

Fig. 34

The current following through the current coil of the wattmeter

$$= I_Y = I_{ph} \text{ (say)}$$

The potential difference across the potential coil of wattmeter,

$$E = E_R - E_B$$
$$= \sqrt{3}\, E_{ph} \text{ leading vector } E_Y \text{ by } 90°$$
$$= \sqrt{3}\, E_{ph} \text{ leading the vector of current } I_Y \text{ by } (90° + \phi)$$

∴ Reading of the wattmeter (Fig. 34)

$$= \sqrt{3}\, E_{ph} I_{ph} \cos (90° + \phi)$$
$$= -\sqrt{3}\, E_{ph} I_{ph} \sin \phi = -W_r$$

Total reactive power of the circuit,

$$= 3\, E_{ph} I_{ph} \sin \phi = -\sqrt{3}\, W_r$$

Reactive power (as earlier started) can also be determined from two wattmeter readings connected for measurement of power.

$$W_1 - W_2 = E_L I_L \cos (30° - \phi) - E_L I_L \cos (30° + \phi)$$

or $$W_1 - W_2 = E_L I_L \times 2 \sin 30° \sin \phi$$

or $$E_L I_L \sin \phi = W_1 - W_2$$

Reactive power of load circuit.

$$W_r = \sqrt{3}\, E_L I_L \sin \phi = \sqrt{3}\,(W_1 - W_2).$$

POWER MEASUREMENT IN 3-PHASE CIRCUITS

Example 18. *The power input to a 3-phase induction motor is read by two wattmeters. The readings are 920 W and 300 W. Calculate the power factor of the motor.*

Solution. Reading of wattmeter, $W_1 = 920$ W

Reading of wattmeter, $W_2 = 300$ W

Power factor of the motor, cos ϕ :

Using the relation, $$\tan \phi = \frac{\sqrt{3}\,(W_1 - W_2)}{(W_1 + W_2)} = \frac{\sqrt{3}\,(920 - 300)}{(920 + 300)} = \frac{\sqrt{3} \times 620}{1220} = 0.88$$

∴ $$\phi = \tan^{-1} 0.88 = 41.35°$$

Power factor of the motor,

$$\cos \phi = \cos 41.35° = \mathbf{0.75 \text{ (lag)}}. \quad \textbf{(Ans.)}$$

Example 19. *While performing a load test on a 3-phase wound-rotor induction motor by two wattmeters method, the readings obtained on two wattmeters were + 14.2 kW and – 6.1 kW and the line voltage was 440 V. Calculate :*

(i) True power drawn by the motor *(ii) Power factor, and*

(iii) Line current.

Solution. Reading of wattmeter, $W_1 = 14.2$ kW

Reading of wattmeter, $W_2 = -6.1$ kW

Line voltage, $E_L = 440$ V

(*i*) **True power drawn by the motor**

$$= 14.2 - 6.1 = \mathbf{8.1\ kW.}\ \textbf{(Ans.)}$$

(*ii*) $$\tan \phi = \frac{\sqrt{3}\,(W_1 - W_2)}{(W_1 + W_2)} = \frac{\sqrt{3}\,[14.2 - (-6.1)]}{[14.2 + (-6.1)]} = \frac{\sqrt{3} \times 20.3}{8.1} = 4.34$$

$$\therefore \quad \phi = \tan^{-1} 4.34 = 77°$$

∴ **Power factor,** $\cos \phi = \cos 77° = \mathbf{0.2249\ (lag).}$ **(Ans.)**

(*iii*) **Line current, I_L :**

Using the relation, $P = \sqrt{3}\, E_L I_L \cos \phi$

$$8.1 \times 1000 = \sqrt{3} \times 440 \times I_L \times 0.2249$$

$$\therefore \quad I_L = \frac{8.1 \times 1000}{\sqrt{3} \times 440 \times 0.2249} = 47.26\ \text{A}$$

Hence, **line current = 47.26 A. (Ans.)**

Example 20. *A 3-phase, 440 V motor load has a power factor of 0.6. Two wattmeters connected to measure the power show the input to be 25 kW. Find the reading on each instrument.*

Solution. Input power = 25 kW

Line voltage, $E_L = 440$ V

Power factor of the motor load, $\cos \phi = 0.6$

$\mathbf{W_1}$; $\mathbf{W_2}$:

Using the relation : $$\tan \phi = \frac{\sqrt{3}\,(W_1 - W_2)}{W_1 + W_2} \quad ...(i)$$

Now, $\cos \phi = 0.6, \quad \therefore \quad \phi = 53.13°$

$$\therefore \quad \tan \phi = \tan 53.13° = 1.333$$

Also $W_1 + W_2 = 25$ kW (*given*) ...(*ii*)

Substituting these values in (*i*), we get

$$1.333 = \frac{\sqrt{3}\,(W_1 - W_2)}{25}$$

$$\therefore \quad W_1 - W_2 = \frac{1.333 \times 25}{\sqrt{3}} = 19.24\ \text{kW} \quad ...(iii)$$

From (*ii*) and (*iii*), we get $\mathbf{W_1 = 22.12\ kW}$

and $\mathbf{W_2 = 2.88\ kW.}$ **(Ans.)**

☞**Example 21.** *In a 3-phase circuit two wattmeters used to measure power indicate 1200 W and 600 W respectively. Find the power factor of the circuit :*

(*i*) *When both wattmeter readings are positive.*

(*ii*) *When the latter is obtained by reversing the current coil connections.*

(AMIE Summer, 1999)

Solution. (*i*) *When both wattmeter readings are positive :*

Reading of wattmeter, $W_1 = 1200$ W

Reading of wattmeter, $W_2 = 600$ W

We know that, $$\tan \phi = \frac{\sqrt{3}\,(W_1 - W_2)}{W_1 + W_2} = \frac{\sqrt{3}\,(1200 - 600)}{(1200 + 600)} = 0.577$$

$$\phi = \tan^{-1} 0.577 = 30°$$

Power factor, $\mathbf{\cos \phi = \cos 30° = 0.866\ (lag).}$ **(Ans.)**

(*ii*) *When the reading of wattmeter W_2 is obtained by reversing the coil connection* :

Reading of wattmeter, $W_1 = 1200$ W

Reading of wattmeter, $W_2 = -600$ W

We know that, $\tan \phi = \dfrac{\sqrt{3}\,(W_1 - W_2)}{(W_1 + W_2)} = \dfrac{\sqrt{3}\,[1200 - (-600)]}{[1200 + (-600)]} = \dfrac{\sqrt{3} \times 1800}{600} = 5.196$

or $\phi = \tan^{-1} 5.196 = 79.1°$

Hence, **power factor** $= \cos \phi = \cos 79.1° = 0.1889$. **(Ans.)**

Example 22. *In order to measure the power input and the power factor of an over-excited synchronous motor two wattmeters are used. If the meters indicate (– 3.5 kW) and (+ 8.0 kW) respectively. Calculate :*

(*i*) *Power factor of the motor.* (*ii*) *Power input to the motor.*

Solution. (*i*) *Since an over-excited synchronous motor runs with a leading power factor,* we should use the relation,

$$\tan \phi = -\frac{\sqrt{3}\,(W_1 - W_2)}{(W_1 + W_2)}$$

Moreover it is W_1 that gives negative reading and not W_2. Hence $W_1 = -3.5$ kW.

$$\therefore \quad \tan \phi = -\frac{\sqrt{3}\,(-3.5 - 8)}{(-3.5 + 8)} = -\frac{\sqrt{3} \times 11.5}{5.5} = 3.62$$

$$\therefore \quad \phi = \tan^{-1} 3.62 = 74.6° \text{ (lead)}$$

$\therefore$ **Power factor,** $\cos \phi = \cos 74.6° = 0.2655$ **(lead). (Ans.)**

(*ii*) **Power input** $= W_1 + W_2 = -3.5 + 8 = 4.5$. **W. (Ans.)**

Example 23. *Two wattmeters are used to measure power input to a 1.5 kV, 50 Hz, 3-phase motor running on full-load at an efficiency of 85 per cent. Their readings are 250 kW and 80 kW respectively. Calculate :*

(*i*) *Input,* (*ii*) *Power factor,*

(*iii*) *Line current, and* (*iv*) *Output.*

Solution. *Since the motor is running at full-load, its power factor must be greater than 0.5. Hence W_2 reading is positive*

$\therefore$ $W_1 = +250$ kW and $W_2 = +80$ kW

(*i*) **Input** $= W_1 + W_2 = 250 + 80 = 330$ **kW. (Ans.)**

(*ii*) $\tan \phi = \dfrac{\sqrt{3}\,(W_1 - W_2)}{W_1 + W_2} = \dfrac{\sqrt{3}\,(250 - 80)}{(250 + 80)} = 0.892$

$\therefore$ $\phi = \tan^{-1} 0.892 = 41.74°$

and **power factor** $= \cos \phi = \cos 41.74° = 0.746$ **(lag). (Ans.)**

(*iii*) **Power,** $P = \sqrt{3}\, E_L I_L \cos \phi$

$\therefore$ $I_L = \dfrac{P}{\sqrt{3}\, E_L \cos \phi} = \dfrac{330 \times 1000}{\sqrt{3} \times 1.5 \times 1000 \times 0.746} = 170.27$

Hence, **line current = 170.27 A. (Ans.)**

(*iv*) **Output** = input × efficiency = 330 × 0.85 = **280.5 kW. (Ans.)**

☞**Example 24.** *Two wattmeters are used to measure power input to a synchronous motor. Each of them indicates 60 kW. If the power factor be changed to 0.866 leading, determine the readings of the two wattmeters, the total input power remaining the same. Draw the vector diagram for the second condition of the load.*

Solution. Reading of wattmeter, $W_1 = 60$ kW

Reading of wattmeter, $W_2 = 60$ kW.

First case. In the first case, *both wattmeters read equal and positive. Hence, motor must be running at unity power factor.*

Second case. *Power factor is 0.866 leading.*

In this case : $W_1 = E_L I_L \cos (30° + \phi)$

$W_2 = E_L I_L \cos (30° - \phi)$

$\therefore \quad W_1 + W_2 = \sqrt{3}\, E_L I_L \cos \phi$

$W_1 - W_2 = - E_L I_L \sin \phi$

$$\therefore \quad \tan \phi = - \frac{\sqrt{3}\,(W_1 - W_2)}{(W_1 + W_2)}$$

Since $\cos \phi = 0.866$ *(given)*

$\therefore \quad \phi = \cos^{-1} 0.866 = 30°$

and $\tan \phi = \tan 30° = \dfrac{1}{\sqrt{3}}$

$$\therefore \quad \frac{1}{\sqrt{3}} = - \frac{\sqrt{3}\,(W_1 - W_2)}{120}$$

[$\because$ $W_1 + W_2 = 120$ kW *(given)*]

$$\therefore \quad W_1 - W_2 = - \frac{120}{3} = - 40 \quad ...(i)$$

But $W_1 + W_2 = 120$ ...(ii)

From (i) and (ii), we get $\mathbf{W_1 = 40}$ **kW**, $\mathbf{W_2 = 80}$ **kW. (Ans.)**

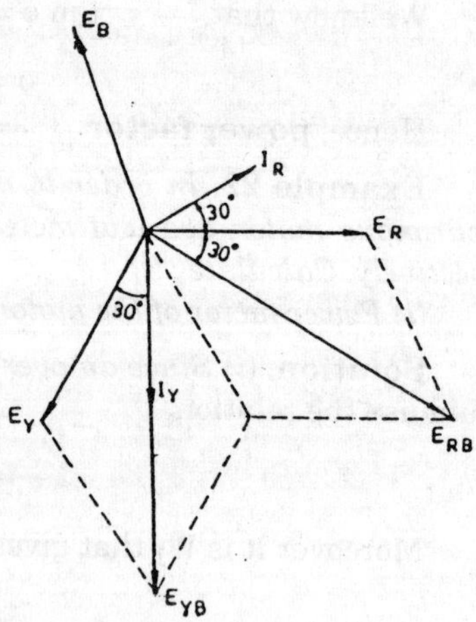

Fig. 35

For connection diagram, please refer to Fig. 28. The vector or phasor diagram is shown in Fig. 35.

Example 25. *Two wattmeters connected to read the total power in a 3-phase system supplying a balanced load read 10.5 kW and – 2.5 kW respectively. Calculate the total power and power factor.*

Also, explain the significance of (i) *equal wattmeter readings and* (ii) *a zero reading on one wattmeter.*

Solution. *Given :* $W_1 = 10.5$ kW ; $W_2 = - 2.5$ kW.

For two wattmeter method,

total power = $W_1 + W_2 = 10.5 + (- 2.5) =$ **8.0 kW. (Ans.)**

$$\text{We know that,} \quad \tan \phi = \frac{\sqrt{3}\,(W_1 - W_2)}{W_1 + W_2} = \frac{\sqrt{3}\,[10.5 - (-2.5)]}{[10.5 + (-2.5)]} = 2.8145$$

or $\phi = 70.44°$

$\therefore$ **Power factor,** $\cos \phi = \cos 70.44° =$ **0.335. (Ans.)**

(i) For readings of the two wattmeters to be equal,

$W_1 = W_2$

or $\sqrt{3}\, E_L I_L \cos (30° - \phi) = \sqrt{3}\, E_L I_L \cos (30° + \phi)$

or $\cos (30° - \phi) = \cos (30° + \phi)$

$\therefore \quad \phi = 0°$ or $\cos \phi = 1$

i.e., for *unity power factor, the readings of two wattmeters are equal*. **(Ans.)**

(ii) As readings of wattmeters are

$W_1 = \sqrt{3} E_L I_L \cos (30° - \phi)$, and

$W_2 = \sqrt{3}\, E_L I_L \cos (30° + \phi)$,

for *reading of one of the wattmeters to be zero* ϕ *must be 60°, which makes* $W_2 = \sqrt{3}\, E_L I_L \cos 90° = \mathbf{0}$.

(Ans.)

Example 26. *A star connected 3-phase load consists of three similar impedances. When load is connected to a 3-phase 400 V, 50 Hz supply it takes 30A line current at a power factor of 0.8 lagging.*

(*i*) *Calculate the total power taken by the load.*

(*ii*) *Calculate the resistance and reactance of each phase of the load.*

(*iii*) *If power is measured by two wattmeters method, then calculate the reading of each wattmeter.*

Solution. *Given* : $E_L = 400$ V ; $f = 50$ Hz ; $I_L = (I_{ph}) = 30$ A, $\cos\phi = 0.8$ lagging.

(*i*) **Total power taken by the load, P :**

$$P = \sqrt{3}E_L I_L \cos\phi$$
$$= \sqrt{3} \times 400 \times 30 \times 0.8 = 16627 \text{ W or } \mathbf{16.627\ kW.}\ \textbf{(Ans.)}$$

(*ii*) **Resistance and reactance of each phase of the load, R, X :**

$$E_{ph} = \frac{E_L}{\sqrt{3}} = \frac{400}{\sqrt{3}} = 230.95 \text{ V}$$

$\therefore$ Load impedance, $Z = \dfrac{E_{ph}}{I_{ph}} = \dfrac{230.95}{30} = 7.7\ \Omega$

$\therefore$ Load resistance, $\mathbf{R} = Z\cos\phi = 7.7 \times 0.8 = \mathbf{6.16\ \Omega}$. **(Ans.)**

and, load reactance, $\mathbf{X} = Z\sin\phi = 7.7 \times 0.6 = \mathbf{4.62\ \Omega}$. **(Ans.)**

(*iii*) **Reading of each wattmeter, W_1, W_2 :**

Reading of wattmeter 1, $W_1 = E_L I_L \cos(30° + \phi)$

Reading of wattmeter 2, $W_2 = E_L I_L \cos(30° - \phi)$

where $\phi = \cos^{-1} 0.8 = 36.9°$

$\therefore$ $\mathbf{W_1} = 400 \times 30 \cos(30° + 36.9°) = 4708$ W or **4.708 kW. (Ans.)**

and $\mathbf{W_2} = 400 \times 30 \cos(30° - 36.9°) = 11913$ W or **11.913 kW. (Ans.)**

Example 27. *A 3-phase, 3-wire, 415 V system supplies a balanced load of 20 A at a power factor 0.8 lag. The current coil of wattmeter (1) is in phase R and of wattmeter 2 in phase B. Calculate :*

(*i*) *The reading on 1 when its voltage coil is across R and Y ;*

(*ii*) *The reading on 2 when its voltage coil is across B and Y ;*

(*iii*) *The reading on 1 when its voltage coil is across Y and B.*

Solution. *Given* : $E_L = 415$ V, $I_L = 20$ A, $\cos\phi = 0.8$ (lag) or $\phi = \cos^{-1} 0.8 = 36.87°$ (lag)

(*i*) *The reading on 1 when its voltage coil is across R and Y and current coil is in R phase*

$$= E_L I_L \cos(30° - \phi) = 415 \times 20 \times \cos(30° - 36.87°)$$
$$= 8240 \text{ W or } \mathbf{8.24\ kW.}\ \textbf{(Ans.)}$$

(*ii*) *Reading on wattmeter 2 when its voltage coil is across B and Y and current coil is in phase B*

$$= E_L I_L \cos(30° + \phi) = 415 \times 20 \cos(30° + 36.87°)$$
$$= 3260 \text{ W or } \mathbf{3.26\ kW.}\ \textbf{(Ans.)}$$

(*iii*) *When the current coil is in one phase (phasor R) and the voltage coil is across other two phases (Y and R),* the wattmeter reads $\dfrac{\text{total } kVAR}{\sqrt{3}}$ and

Total $kVAR = \sqrt{3}(W_1 - W_2)$

$\therefore$ Wattmeter reading $= \dfrac{\text{Total } kVAR}{\sqrt{3}} = \dfrac{\sqrt{3}(W_1 - W_2)}{\sqrt{3}}$

$$= W_1 - W_2 = 8240 - 3260 = \mathbf{4980\ VAR.}\ \textbf{(Ans.)}$$

Example 28. *The ratio of the readings of two wattmeters connected to measure power in a balanced 3 – φ, 3 wire load is 5 : 3. The load is known to be inductive with a lagging power factor. Calculate the power factor of the load.* **(Punjab University)**

Solution. Let the ratio of wattmeter readings be *r i.e.*, $\frac{W_2}{W_1} = r$

Now,
$$\tan \phi = \sqrt{3}\left(\frac{W_1 - W_2}{W_1 + W_2}\right) = \sqrt{3}\left[\frac{1 - \frac{W_2}{W_1}}{1 + \frac{W_2}{W_1}}\right] = \sqrt{3}\left(\frac{1-r}{1+r}\right)$$

and power factor,
$$\cos \phi = \frac{1}{\sec \phi} = \frac{1}{\sqrt{1 + \tan^2 \phi}} = \frac{1}{\sqrt{1 + 3\left(\frac{1-r}{1+r}\right)^2}}$$

Substituting $r = \frac{W_2}{W_1} = \frac{3}{5} = 0.6$ in the above expression, we get

$$\cos \phi = \frac{1}{\sqrt{1 + 3\left(\frac{1 - 0.6}{1 + 0.6}\right)^2}} = \mathbf{0.918\ (lag).}\quad \textbf{(Ans.)}$$

Example 29. *A symmetrical star-connected three-phase load is made of three identical coils having an internal resistance of 12 Ω and inductance of 142 mH. The power taken by this load when connected to a balanced 3-phase, three wire supply of 400 V, 50 Hz is measured by using the two wattmeters method. The wattmeters used are similar and are of type of 400 V, 5 A, UPF, class 1 accuracy.*

(*i*) *Determine expected readings of W_1 and W_2.*

(*ii*) *Compute the total power.* **(GATE, 1995)**

Solution. Line voltage, $E_L = 400$ V

Phase voltage, $E_{ph} = \frac{400}{\sqrt{3}} = 231$ V

Coil resistance, $R_{ph} = 12\ \Omega$

Coil inductive reactance, $X_{ph} = 2\pi f L_{ph} = 2\pi \times 50 \times 142 \times 10^{-3} = 44.61\ \Omega$

Coil impedance, $Z_{ph} = \sqrt{R_{ph}^2 + X_{ph}^2} = \sqrt{12^2 + 44.61^2} = 46.2\ \Omega$

Coil phase angle, $\phi = \cos^{-1} \frac{R_{ph}}{Z_{ph}} = \cos^{-1}\left(\frac{12}{46.2}\right) = 74.94°$ (lagging)

Phase current, $I_{ph} = \frac{E_{ph}}{Z_{ph}} = \frac{231}{46.2} = 5$ A

(*i*) **Readings of wattmeters W_1, W_2 :**

Readings of wattmeter, $W_1 = E_L I_L \cos (30° - \phi)$

$= 400 \times 5.0 \cos (30° - 74.94°) =$ **1415.7 W. (Ans.)**

($\because\ I_L = I_{ph} = 5.0$ A)

Reading of wattmeter, $W_2 = E_L I_L \cos (30° + \phi)$

$= 400 \times 5.0 \times \cos (30° + 74.24°) = -$ **515.6 W. (Ans.)**

(*ii*) **Total power :**

Total power $= W_1 + W_2$

$= 1415.7 + (-515.6) =$ **900.1 W. (Ans.)**

Example 30. *Two wattmeters are being used to measure power of a balanced load of 30 A at power factor 0.8 lagging being supplied by a 3-phase, 3-wire, 440 V supply. Calculate :*

(*i*) *Power consumed.* (*ii*) *Reading of wattmeter No. 1*

(*iii*) *Reading of wattmeter No. 2.*

Solution. Line voltage, $E_L = 440$ V

Line current, $I_L = 30$ A

Power factor, $\cos\phi = 0.8$

Phase angle, $\phi = \cos^{-1} 0.8 = 36.87°$

(*i*) **Power consumed,** $P = \sqrt{3}\, E_L I_L \cos\phi = \sqrt{3} \times 440 \times 30 \times 0.8$

$= 18290$ W or **18.29 kW. (Ans).**

(*ii*) **Reading of wattmeter No. 1**

$= E_L I_L \cos(30° - \phi) = 440 \times 30 \cos(30° - 36.87°) =$ **13105 W. (Ans.)**

(*iii*) **Reading of wattmeter No. 2**

$= E_L I_L \cos(30° + \phi) = 440 \times 30 \times \cos(30° + 36.87°)$

= 5185 W (app.). (Ans.)

☞**Example 31.** *A balanced star-connected load, each having a resistance of 10 Ω and inductive reactance of 20 Ω is connected to a 400 V, 50 Hz supply. The phase sequence is RYB. Two wattmeters connected to read total power have their current coils connected in the red and blue lines respectively. Calculate the reading of each wattmeter.*

Draw the circuit and vector diagram.

Solution. Resistance per phase, $R_{ph} = 10\ \Omega$

Reactance per phase, $X_{ph} = 20\ \Omega$

Line voltage, $E_L = 400$ V

Impedance per phase, $Z_{ph} = \sqrt{R_{ph}^2 + X_{ph}^2} = \sqrt{10^2 + 20^2} = 22.36\ \Omega$

Phase voltage, $E_{ph} = \dfrac{E_L}{\sqrt{3}} = \dfrac{400}{\sqrt{3}} = 231$ V

$$I_{ph}\ (= I_L) = \frac{E_{ph}}{Z_{ph}} = \frac{231}{22.36} = 10.33 \text{ A}$$

Power factor, $\cos\phi = \dfrac{R_{ph}}{Z_{ph}} = \dfrac{10}{22.36} = 0.447$

$\therefore$ $\phi = 63.43°$

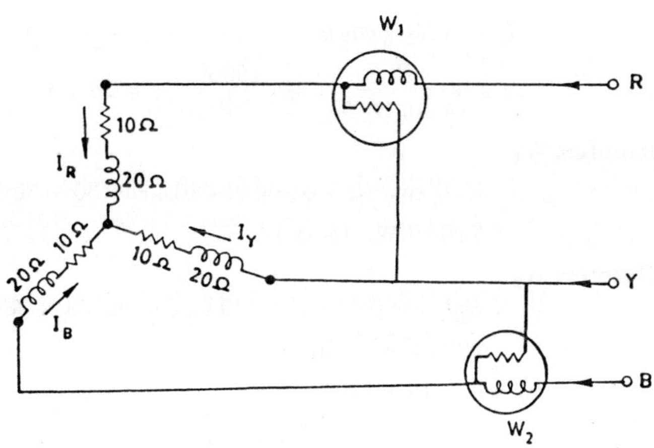

Fig. 36

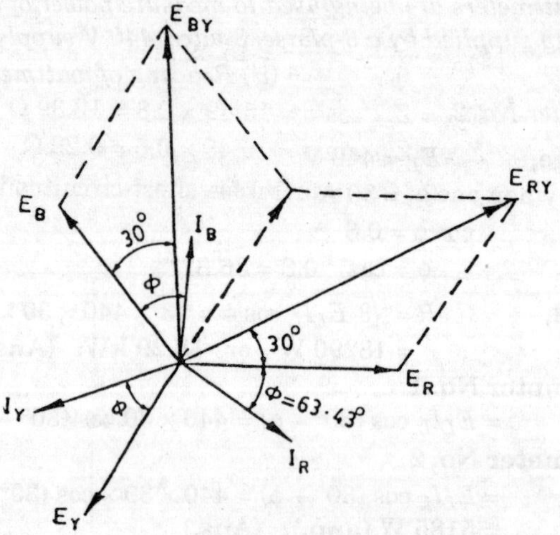

Fig. 37

The circuit and vector diagrams are shown in Figs. 36 and 37 respectively.

Reading of wattmeter, $W_1 = E_{RY} \times I_R \times \cos(30° + \phi) = E_L \times I_L \times \cos(30° + \phi)$

$= 400 \times 10.33 \cos(30° + 63.43°) = -$ **247 W (app.). (Ans.)**

Reading of wattmeter, $W_2 = E_{BY} \times I_B \times \cos(30° - \phi) = E_L \times I_L \times \cos(30° - 63.43°)$

$= 400 \times 10.33 \times \cos(30° - 63.43°) =$ **3448 W (app.). (Ans.)**

☞**Example 32.** *Three equal star-connected inductors take 10 kW at power factor 0.8 when connected to a 440 V, 3 φ, 3-wire supply. Give the connection diagram of two single-phase wattmeters to measure the power in the circuit. Allocate the reading of each wattmeter. Find the line currents if one of the inductors is short-circuited. Draw the vector diagrams of the currents and voltages under the condition.*

Solution. Power, $P = 10 \text{ kW} = 10000 \text{ W}$

Line voltage, $E_L = 440 \text{ V}$

Power factor, $\cos\phi = 0.8$ (lag)

$\therefore$ $\phi = 36.9°$ (lag)

The connection diagram of two single-phase wattmeters to measure the power in the circuit is shown in Figs. 38 and 39.

$$P = \sqrt{3}\, E_L I_L \cos\phi$$

$\therefore$ Line current, $I_L = \dfrac{P}{\sqrt{3}\, E_L \cos\phi} = \dfrac{10000}{\sqrt{3} \times 440 \times 0.8} = 16.4 \text{ A}.$

Reading of wattmeter, W_1

$= E_L I_L \cos(30° - \phi) = 440 \times 16.4 \cos(30° - 36.9°)$

= 7163.7 W. (Ans.)

Reading of wattmeter, W_2

$= E_L I_L \cos(30° + \phi) = 440 \times 16.4 \cos(30° + 36.9°)$

= 2831.1 W. (Ans.)

Phase current, $I_{ph} = I_L = 16.4 \text{ A}$

Phase impedance, $Z_{ph} = \dfrac{E_{ph}}{I_{ph}} = \dfrac{\frac{440}{\sqrt{3}}}{16.4} = 15.49\ \Omega$

Phase resistance, $R_{ph} = Z_{ph} \cos\phi = 15.49 \times 0.8 = 12.39\ \Omega$

Phase reactance, $X_{ph} = Z_{ph} \sin\phi = 15.49 \times 0.6 = 9.29\ \Omega$

The circuit diagram when one of the inductors is short-circuited is shown in Fig. 38.

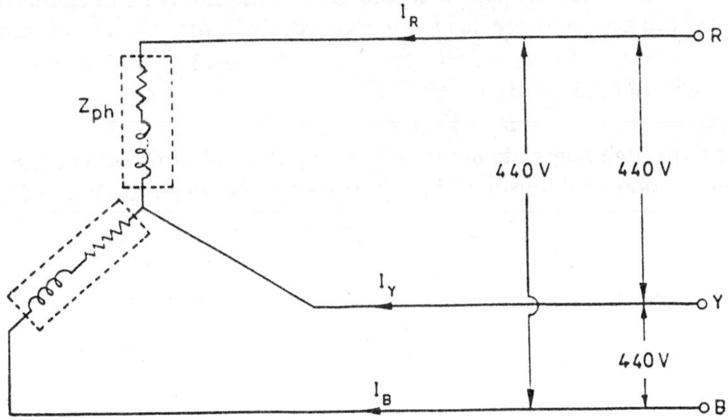

Fig. 38

Current in line R, $I_R = \dfrac{E_{RY}}{Z_{ph}} = \dfrac{440}{15.49} = 28.4\ \text{A}$

Current in line B, $I_B = \dfrac{E_{BY}}{Z_{ph}} = \dfrac{440}{15.49} = 28.4\ \text{A}$

The currents in line R and line B are equal in magnitude but differ in phase by 60°, as shown in Fig. 39.

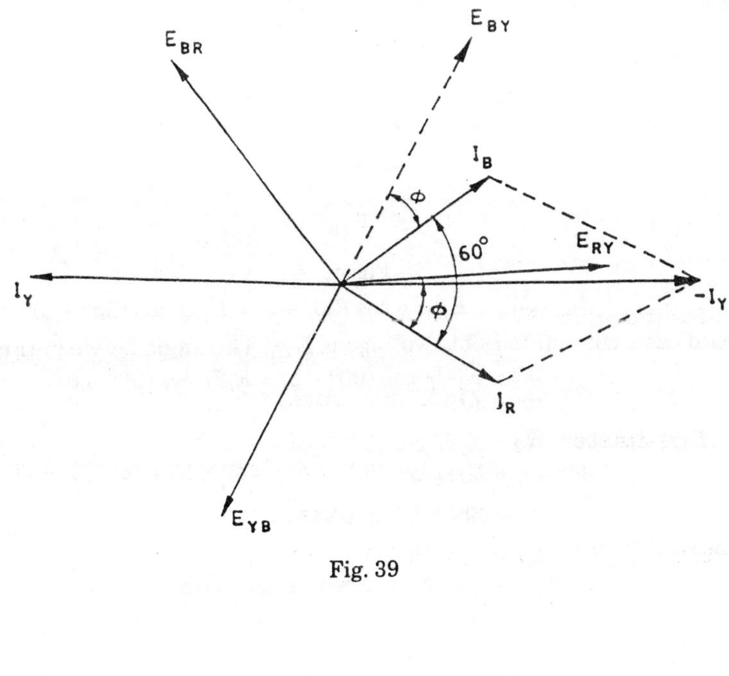

Fig. 39

Current in line Y, $I_Y = -(I_R + I_B)$

Magnitude of current flowing in line Y,

$$I_Y = 2I_R \cos 30° = 2 \times 28.4 \times \cos 30° = \mathbf{49.19\ A.}\quad \textbf{(Ans.)}$$

Example 33. *A 3-phase, 3-wire balanced load with a lagging power factor is supplied at 440 V (between lines). A single phase wattmeter (scaled in kW) when connected with its current coil in the R-line and voltage coil between R and Y lines gives a reading of 8 kW. When the same terminals of the voltage coil are switched over to Y-line and B-line, the current coil connections remaining the same, the reading of the wattmeter remains unchanged. Calculate the line current and power factor of the load.*

Solution. Line voltage, $E_L = 440$ V

Phase sequence $R \to Y \to B$

The current through the wattmeter is I_R and potential difference across its pressure coil is E_{RY}. As seen from the phasor diagram of Fig. 40 the angle between the two is $(30° + \phi)$.

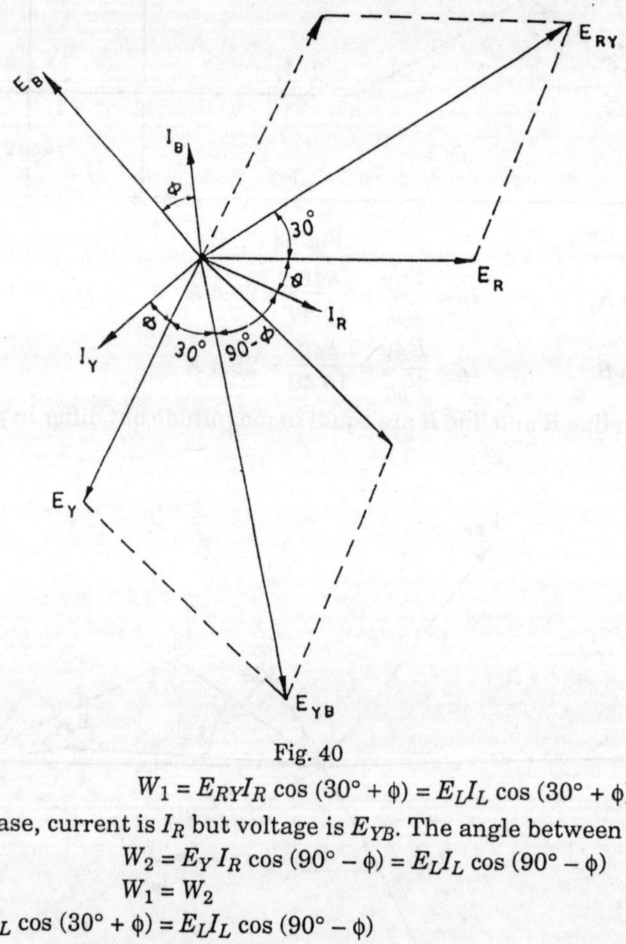

Fig. 40

$$\therefore \qquad W_1 = E_{RY} I_R \cos (30° + \phi) = E_L I_L \cos (30° + \phi) \qquad ...(i)$$

In the second case, current is I_R but voltage is E_{YB}. The angle between the two is $(90° - \phi)$.

$$\therefore \qquad W_2 = E_Y I_R \cos (90° - \phi) = E_L I_L \cos (90° - \phi) \qquad ...(ii)$$

Since $W_1 = W_2$ *(given)*

$$E_L I_L \cos (30° + \phi) = E_L I_L \cos (90° - \phi)$$

$$\therefore \qquad 30° + \phi = 90° - \phi$$

or $2\phi = 60° \quad \therefore \quad \phi = 30°$

$\therefore$ **Load power factor,**

$$\cos \phi = \cos 30° = \mathbf{0.866\ (lag).}\quad \textbf{(Ans.)}$$

Now, $W_1 = W_2 = 8 \text{ kW} = 8000 \text{ W}$

From (*ii*), we get $8000 = E_L I_L \cos(90° - \phi) = 440 \times I_L \cos(90° - 30°)$

$$\therefore \quad I_L = \frac{8000}{440 \times \cos 60°} = 36.36 \text{ A}$$

Hence, **line current = 36.36 A. (Ans.)**

Example 34. *A wattmeter indicates 6 kW when its current coil is connected in red phase and its potential coil is connected between the neutral and red phase of a symmetrical 3-phase system supplying a balanced load of 40 A at 440 V. What will be reading of the instrument if the current coil connections remain unchanged, the potential coil be connected between blue and yellow phases ?*

Solution. When the wattmeter current coil is connected in red phase and its pressure coil is connected between red phase and neutral, the wattmeter reads per phase $E_{ph} I_{ph} \cos \phi$

or $$E_{ph} I_{ph} \cos \phi = 6 \text{ kW} = 6000 \text{ W}$$

But $$E_{ph} = \frac{440}{\sqrt{3}} \text{ V and } I_{ph} = 40 \text{ A} \quad \textit{(given)}$$

$\therefore$ Power factor, $$\cos \phi = \frac{6000}{E_{ph} I_{ph}} = \frac{6000}{\frac{440}{\sqrt{3}} \times 40} = 0.59$$

and $$\sin \phi = \sqrt{1 - (0.59)^2} = 0.807$$

When the pressure coil is connected between blue and yellow phases and current coil is connected in red phase, the instrument will read $\frac{\text{Total kVAR}}{\sqrt{3}}$ or $\sqrt{3}$ kVAR per phase.

$\therefore$ **Reading of instrument** $= \sqrt{3} E_{ph} I_{ph} \sin \phi = \sqrt{3} \times \frac{440}{\sqrt{3}} \times 40 \times 0.807$

$= 14203$ or **14.203 kVAR. (Ans.)**

Example 35. *A wattmeter reads 5.54 kW when its current coil is connected in R phase and its voltage coil is connected between the neutral and R phase of a symmetrical 3-phase system supplying a balanced load of 30 A at 400 V. What will be the reading on the instrument if the connections to the current coil remain unchanged and the voltage coil is connected between B and Y phases ? Take phase sequence RYB. Draw corresponding phasor diagram.* **(Panjab University)**

Solution. Phase current, $I_{ph} = I_L = 30 \text{ A}$

Phase voltage, $$E_{ph} = \frac{E_L}{\sqrt{3}} = \frac{400}{\sqrt{3}} = 231 \text{ V}$$

Power supplied per phase, P = reading of wattmeter $= 5.54$ kW

Load power factor, $$\cos \phi = \frac{P}{E_{ph} \cdot I_{ph}} = \frac{5.54 \times 10^3}{231 \times 30} = 0.8$$

or $$\phi = \cos^{-1} 0.8 = 36.87°$$

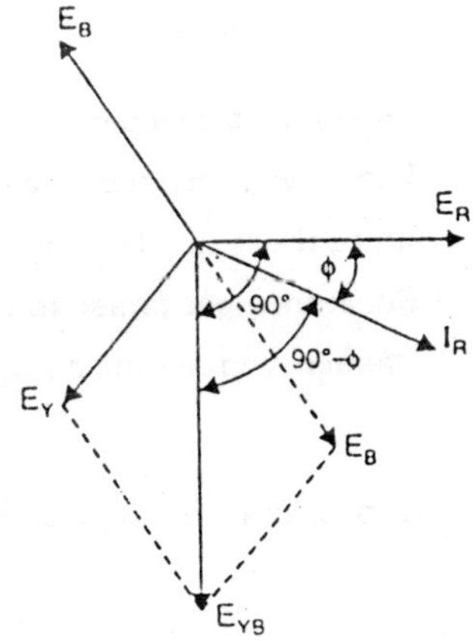

Fig. 41

In the *second case* when pressure coil is connected across phases *B* and *Y*, the phase angle between current in phase *R* and voltage across voltage coil is (90° – ϕ) as obvious from the phasor diagram, shown in Fig. 41.

$\therefore$ Wattmeter reading $= E_{YB} I_R \cos (90° - \phi)$

$= 400 \times 30 \times \cos (90° - 36.87°)$

$= 7200$ W or **7.2 kW. (Ans.)**

Example 36. *Two wattmeters A and B give readings as 5000 W and 1000 W respectively during the power measurement of 3 – φ, 3 wire balanced load system. (a) Calculate the power and power factor (i) when both meters read direct and (ii) when one of them reads in reverse (b) if the voltage of the circuit is 400 V, what is the value of the capacitance which must be introduced in each phase to cause the whole of the power to appear on A ? The frequency of supply is 50 Hz.*

(Nagpur University, 1998)

Solution. *Given* : $W_1 = 5000$ W ; $W_2 = 1000$ W ; $V = 400$ volts ; $f = 50$ Hz.

(*a*) **Power (P) and power factor (cos φ) :**

(*i*) *When both meters read direct :*

Total power, $P = W_1 + W_2 = 5000 + 1000 = 6000$ W or **6 kW. (Ans.)**

We know that, $\tan \phi = \dfrac{\sqrt{3}\,(W_1 - W_2)}{W_1 + W_2} = \dfrac{\sqrt{3}\,(5000 - 1000)}{(5000 + 1000)} = 1.1547$

or $\phi = \tan^{-1} (1.1547) = 49.1°$

$\therefore$ Power factor, $\cos \phi = \cos 49.1° =$ **0.655 (lagging). (Ans.)**

(*ii*) *When one of the meters reads in reverse :*

Here, $W_2 = -\,1000$ W

Total power, $P = W_1 + W_2 = 5000 + (-\,1000) = 4000$ W or **4 kW. (Ans.)**

Again, $\tan \phi = \dfrac{\sqrt{3}\,(W_1 - W_2)}{W_1 + W_2} = \dfrac{\sqrt{3}\,[5000 - (-\,1000)]}{(5000 - 1000)} = 2.598$

or $\phi = \tan^{-1} 2.598 = 68.95°$

$\therefore$ $\cos \phi = \cos 68.95° =$ **0.359 (lagging). (Ans.)**

(*b*) **Capacitance to be introduced in each phase, C :**

Reading of wattmeter, *B will be zero and whole of the power would appear on wattmeter A when power factor cos φ′ = 0.5 or φ′ = 60°*. It means power factor is to be improved from 0.359 to 0.5.

Load cement per phase, $I_{ph} (= I_L) = \dfrac{P}{\sqrt{3}\, E_L \cos \phi} = \dfrac{4000}{\sqrt{3} \times 400 \times 0.359} = 16.08$ A

Load impedance per phase, $Z_{ph} = \dfrac{E_{ph}}{I_{ph}} = \dfrac{400/\sqrt{3}}{16.08} = 14.36\ \Omega$

Load resistance per phase, $R_{ph} = Z_{ph} \cos \phi = 14.36 \times 0.359 = 5.156\ \Omega$

Load reactance per phase, $X_{ph} = Z_{ph} \sin \phi = 14.36 \sin \cos^{-1} (0.359) = 13.4\ \Omega$

Since there is no change in resistance, therefore,

Reactance per phase in the circuit $X'_{ph} = R_{ph} \,.\, \tan \phi' = 5.156 \tan 60° = 8.93\ \Omega$

Capacitive reactance introduced in each phase,

$$X_c = X_{ph} - X'_{ph} = 13.4 - 8.93 = 4.47\ \Omega$$

Capacitance introduced in each phase, $\mathbf{C} = \dfrac{1}{2\pi f X_L}$

$= \dfrac{1}{2\pi \times 50 \times 4.47} \times 10^6\ \mu\text{F} =$ **712 μF. (Ans.)**

HIGHLIGHTS

1. *In a star-connected system*

$$E_{ph} = \frac{E_L}{\sqrt{3}}$$
$$I_{ph} = I_L$$
$$P = \sqrt{3} E_L I_L \cos \phi$$

2. *In a delta-connected system*

$$E_{ph} = E_L$$
$$I_{ph} = \frac{I_L}{\sqrt{3}}$$
$$P = \sqrt{3} E_L I_L \cos \phi.$$

3. *Two-wattmeter method is generally used to measure 3-phase power. In this method, the current coils of the two wattmeters are connected in any two lines and their potential coils to the remaining third line. The sum of the two wattmeters readings gives the total power in the circuit. If the load is balanced, then its power factor can also be calculated from these two readings. The readings of the two wattmeters are :*

(i)
$$\left.\begin{aligned} W_1 &= E_L I_L \cos (30° - \phi) \\ W_2 &= E_L I_L \cos (30° + \phi) \\ \tan \phi &= \frac{\sqrt{3}(W_1 - W_2)}{W_1 + W_2} \end{aligned}\right] \text{Lagging power factor}$$

(ii)
$$\left.\begin{aligned} W_1 &= E_L I_L \cos (30° + \phi) \\ W_2 &= E_L I_L \cos (30° - \phi) \\ \tan \phi &= -\frac{\sqrt{3}(W_1 - W_2)}{W_1 + W_2} \end{aligned}\right] \text{Leading power factor}$$

4. *There is a 'floating neutral' or 'star' point in an unbalanced star-connected load without neutral wire. The potential of the load star point is different from that of the generator star point.*

Under such conditions, the phase voltage of load is **not** $\frac{1}{\sqrt{3}}$ *of the line voltage.*

OBJECTIVE TYPE QUESTIONS

A. Choose the Correct Answer :

1. In two wattmeters method of power measurement, one of the wattmeters will show negative reading when the load power factor angle is strictly
   (*a*) less than 30° (*b*) less than 60°
   (*c*) greater than 30° (*d*) greater than 60°. **[U.P.S.C., 1996]**
2. Three identical impedances are connected in delta to a 3-phase supply of 400 V. The line-current is 34.65 A and the total power taken is 14.4 kW. The resistance of the load in each phase (in ohm) is
   (*a*) 20 (*b*) 16
   (*c*) 12 (*d*) 10. **[U.P.S.C., 1993]**
3. The power in a 3-phase system is given by $\sqrt{3} V_L I_L \cos \phi$ where ϕ is the phase angle between
   (*a*) line-voltage and line current (*b*) phase voltage and phase current
   (*c*) line voltage and phase current (*d*) phase voltage and line current.
4. Which of the following statements associated with 3-phase delta connected circuits is *true* ?
   (*a*) Line voltage is equal to phase voltage (*b*) Line current is equal to phase current
   (*c*) Line voltage is $\sqrt{3}$ time of phase voltage (*d*) Line currents are 60° apart.

5. In a 3-phase supply, floating neutral is undesirable because it may result in ...... across the load.
(a) unequal line voltages (b) high voltage
(c) low voltage.

6. Phase reversal in a 4-wire unbalanced load supplied from a balanced 3-phase supply causes change in
(a) the power consumed (b) magnitude of phase currents
(c) only the magnitude of the neutral current
(d) magnitude as well as phase angle of the neutral current.

7. A 3-phase star connected symmetrical load consumes P watts of power from a balanced supply. If the same load is connected in delta to the same supply, the power consumption will be
(a) P (b) $\sqrt{3}P$
(c) $3P$ (d) not determined from the given data.
[U.P.S.C., 1995]

8. Three unequal impedances are connected in star to a 3-phase system. The sum of three line currents will be
(a) equal to the each line current (b) zero
(c) none of these.

9. Three equal impedances are first connected in delta across a 3 – φ balanced supply. If the same impedances are connected in star across the same supply
(a) phase currents will be one-third (b) line currents will be one-third
(c) power consumed will be one-third (d) none of the above.

10. Which of the following is a four wire system ?
(a) Delta (b) Star
(c) Both delta and star (d) Neither delta nor star.

11. In a 3-phase balanced star-connected load, neutral current is equal to
(a) zero (b) I_P
(c) I_L (d) unpredictable.

12. Which of the following equations is valid for a 3-phase 4-wire balanced star-connected load ?
(a) $I_R + I_Y + I_B = I_N = 0$ (b) $I_R + I_Y - I_B = I_N$
(c) $I_R - I_Y + I_B = I_N$ (d) $\frac{V_B + V_R + V_Y}{Z} = I_N$.

13. Three unequal impedances are connected in delta to a 3-phase, 3-wire system.
(a) The voltages across the three phases will be different
(b) Both of the phase currents and line currents will be unbalanced
(c) Phase currents will be unbalanced but the line currents will be balanced
(d) None of the above.

14. The relationship between the line and phase voltages of a delta-connected circuit is given by
(a) $V_L = V_P$ (b) $V_L = \sqrt{3}V_P$
(c) $V_L = \frac{V_P}{\sqrt{2}}$ (d) $V_L = \frac{2}{\pi}V_P$. **[B.T.E. M.P. 1996]**

15. In case of a delta connected load, if one resistor is removed, the power will become
(a) zero (b) one-third
(c) two-third (d) none of the above.

16. Readings of 1154 and 577 watts are obtained when two wattmeters method was used on a balanced load. The delta connected load impedance for a system of 100 V will be
(a) $15 \angle\pm 30°$ (b) $15 \angle+ 30°$
(c) $15 \angle- 30°$ (d) $15 \angle+ 90°$. **[U.P.S.C., 1993]**

17. Which of the following statements is true about two-wattmeters method for power measurement in 3-phase circuit ?

(*a*) Power can be measured using two wattmeters method only for star-connected 3-phase circuits

(*b*) When two wattmeters show identical readings, the power factor is 0.5

(*c*) When power factor is unity, one of the wattmeter reads zero

(*d*) When the readings of the two wattmeters are equal but of opposite sign, the power factor is zero.

[U.P.S.C., 1992]

18. The minimum number of wattmeters to measure power in a 3-phase unbalanced star-connected load is

(*a*) one (*b*) two

(*c*) three (*d*) four.

19. While measuring power in a three phase load by two-wattmeters method, the readings of two wattmeters will be equal and opposite when

(*a*) *pf* is unity (*b*) load is balanced

(*c*) phase angle is between 60° and 90° (*d*) the load is purely inductive. **[U.P.S.C., 1995]**

20. In the measurement of 3-phase power by two-wattmeters method, if the two wattmeter readings are equal the power factor of the circuit is

(*a*) 0.8 lagging (*b*) 0.8 leading

(*c*) zero (*d*) unity. **[A.M.I.E., 1998]**

21. Which of the following are the necessary conditions for an entire 3-ϕ system to be balanced ?

1. The line voltages are equal in magnitude
2. The phase differences between successive line voltages are equal
3. The impedances in each of the phases are identical.

Select the correct answer using the codes given below :

Codes :

(*a*) 1, 2 and 3 (*b*) 1 and 3

(*c*) 1 and 2 (*d*) 2 and 3. **[U.P.S.C., 1996]**

22. The power delivered to a 3-phase load can be measured by the use of 2 wattmeters only when the

(*a*) load is balanced (*b*) load is unbalanced

(*c*) 3-phase load is connected to the source through 3-wires

(*d*) 3-phase load is connected to the source through 4-wires. **[U.P.S.C., 1995]**

23. The ratio of the readings of two wattmeters connected to measure power in a balanced 3-phase load is 5 : 3 and the load is inductive. The power factor of the load is

(*a*) 0.917 lead (*b*) 0.917 lag

(*c*) 0.6 lead (*d*) 0.6 lag. **[U.P.S.C., 1994]**

24. In two phase system phase voltages differ by

(*a*) 60° (*b*) 90°

(*c*) 120° (*d*) 180°.

25. W_1 and W_2 are the readings of two wattmeters used to measure power of a 3-phase balanced load. The reactive power drawn by the load is

(*a*) $W_1 + W_2$ (*b*) $W_1 - W_2$

(*c*) $\sqrt{3}(W_1 + W_2)$ (*d*) $\sqrt{3}(W_1 - W_2)$. **[A.M.I.E., 1997]**

26. In a two wattmeters method of measuring power in a 3-phase system one of the wattmeters reads negative implying

(*a*) wattmeter connection is faulty (*b*) load is unbalanced

(*c*) power flow is in the reverse direction (*d*) power factor is less than 0.5. **[A.M.I.E., 1998]**

27. v_{RN}, v_{YN} and v_{BN} are the instantaneous line to neutral voltages and i_R, i_Y and i_B are instantaneous line currents in a balanced three phase circuit, the computation $v_{RN}(i_Y - i_B) - (v_{YN} - v_{BN})i_R$ will yield a quantity proportional to the

(a) active power (b) power factor
(c) reactive power (d) complex power. **[GATE, 1993]**

28. If the R-phase of a 3-phase star connected alternator becomes reversely connected by mistake, it will affect
(a) E_{RY} and E_{YB} (b) E_{YB} and E_{BR}
(c) E_{RY} and E_{BR} (d) E_{RY}, E_{YB} and E_{BR}.

29. When phase sequence at the 3-phase load is reversed
(a) phase currents are changed in magnitude
(b) phase currents change in phase angle but not in magnitude
(c) total power consumed is changed
(d) phase powers are changed.

30. Lamp load is being supplied from a 3-phase, 4-wire, 230/400 V ac supply system. If a 3-phase motor is now switched on across the same supply, then neutral current will
(a) increase (b) decrease
(c) remain unchanged (d) unpredictable.

31. Lamps of 40 W, 100 W and 200 W ratings are connected in different phases of a 3-phase, 4-wire supply. If the neutral wire breaks, then the lamp likely to fuse first will be
(a) 200 W (b) 100 W
(c) 40 W (d) all lamps.

32. The instantaneous values of currents in phases R and Y of a 3-phase system are 25 A each. For a phase sequence of BRY, the instantaneous value of current in phase B is
(a) 25 A (b) 50 A
(c) 12.5 A (d) 43.3 A.

33. The power measurement in balanced 3-ϕ circuit can be done by
(a) one wattmeter method only (b) two wattmeters method only
(c) three wattmeters method only (d) any one of the above. **[B.T.E. M.P., 1998]**

ANSWERS

1. (d)	2. (c)	3. (b)	4. (a)	5. (a)	6. (d)	7. (c)
8. (b)	9. (c)	10. (b)	11. (a)	12. (a)	13. (b)	14. (a)
15. (c)	16. (b)	17. (d)	18. (b)	19. (d)	20. (d)	21. (a)
22. (c)	23. (b)	24. (b)	25. (d)	26. (d)	27. (c)	28. (c)
29. (b)	30. (c)	31. (c)	32. (b)	33. (d).		

B. Say 'Yes' or 'No' :

1. Polyphase system means the system which consists of numerous windings or circuits.
2. Electrical displacement = $\dfrac{\text{180 electrical degrees}}{\text{number of phases}}$.
3. A load circuit is said to be balanced when the loads (impedances) connected in various phases are same in magnitude as well as in phase.
4. Rotating magnetic field can be set up by passing polyphase current through stationary coils.
5. Polyphase system is less capable and reliable than single phase system.
6. Parallel operation of polyphase alternators is simple as compared to that of single phase alternators because of pulsating reaction in single-phase alternators.
7. Double script notation is usually preferred in A.C. electrical circuits.
8. Phase sequence is the order or sequence in which the currents or voltages in different phases attain their maximum values one after the other.
9. If the phase sequence is reversed by interchanging any two terminals of the 3-phase supply, the motor would rotate in the same direction.

10. Star-connected system is obtained by joining together similar ends either the start or the finish, the other ends are joined to the line wires.
11. The common point at which similar (start or finish) ends, are connected is called the neutral or star point.
12. In a balanced star-connected system line voltage = phase voltage.
13. In a balanced delta-connected system line current is equal to phase current.
14. Delta-connected system is best suited for rotary converters.
15. Three wattmeters method is useful for measuring power in 3-phase, 3-wire load circuits.
16. Power in a 3-phase circuit = $\sqrt{3}\, E_L I_L \cos \phi$.
17. In the case of balanced three-phase circuit the reactive power can be determined by use of one wattmeter.
18. A load having a low lagging power factor gives a poor voltage regulation.
19. Static capacitors are installed to improve the power factor of a group of A.C. motors.
20. Static capacitors are practically loss free.

ANSWERS

1. Yes	2. No	3. Yes	4. Yes	5. No	6. Yes	7. Yes
8. Yes	9. No	10. Yes	11. Yes	12. No	13. No	14. Yes
15. No	16. Yes	17. Yes	18. Yes	19. Yes	20. Yes.	

THEORETICAL QUESTIONS

1. State the advantages of A.C. polyphase supply system over single-phase system.
2. Why is the number of phases in a polyphase system always three rather than any other number ?
3. Explain clearly what is meant by 'phase sequence' of 3-phase voltages.
4. What are the two systems in which three-phases can be connected ? What are the advantages and disadvantages of each system ?
5. What are the advantages of inter-connecting the 3-phases ?
6. Derive an expression for power in a 3-phase star-connected system in terms of (*i*) phase values and (*ii*) line values of voltages and currents.
7. Derive the numerical relationship between line and phase currents for a balanced 3-phase delta-connected load.
8. Deduce an expression for power in a 3-phase balanced load circuit. Show that it is the same irrespective of the load being connected in star or delta.
9. Compare the star and delta connections in a 3-phase system.
10. Enumerate various methods for 3-phase power measurement, and describe in detail two wattmeters method for 3-phase power measurement.
11. Prove that the power in a balanced 3-phase circuit can be deduced from the readings of two wattmeters. Draw the relevant connection diagram and vector diagram.
12. Discuss two-wattmeters method for power measurement in 3-phase system and obtain a relation for the power factor.
13. Derive an expression for star-connected impedances equivalent to three delta-connected impedances.
14. What does the following signify in case of measurement of balanced 3-phase power by two-wattmeters method :
    (*i*) When the reading of one wattmeter is zero.
    (*ii*) When the readings of two wattmeters are equal.
    (*iii*) When the readings of two wattmeters are equal and opposite.
    In each case give reasons to your answer.

15. Explain what will be the consequences of disconnecting the neutral in a 3-phase, 4-wire system.
16. What are the disadvantages of having a low lagging power factor in a system.
17. Enumerate the various equipments which can be employed to improve the power factor.

UNSOLVED EXAMPLES

Star/Delta Connections

1. Three equal impedances each having a resistance of 20 Ω and reactance of 15 Ω are connected in star to a 400 V, 3-phase, 50 Hz system. Calculate :
   (*i*) The line current, (*ii*) The power factor, and
   (*iii*) The power consumed. [**Ans.** 9.24 A ; 0.8 (lag) ; 5120 W]
2. Three equal impedances are star-connected to a 3-phase, 50 Hz supply. If the resistance and reactance of each branch are 25 Ω and 38 Ω respectively, calculate :
   (*i*) The line current, and (*ii*) The power consumed. [**Ans.** 5.28 A ; 2086 W]
3. Three resistances of 20 Ω each are connected in star across the 400 V, 3-phase A.C. supply. Calculate :
   (*i*) The line and phase currents, (*ii*) Phase voltages, and
   (*iii*) Total power taken. [**Ans.** (*i*) 11.55 A ; (*ii*) 231 V ; (*iii*) 8.0 kW]
4. A star-connected, 3-phase load consists of three similar impedances. When the load is connected to a 3-phase, 500 V, 50 Hz supply, the line current is 28.85 A and the power factor is 0.8 lagging, calculate :
   (*i*) The total power taken by the load, and (*ii*) The resistance of each phase of the load.
   [**Ans.** 20 kW ; 8 Ω]
5. Three non-inductive resistances each of 50 Ω are connected in star across 400 V, 3-phase A.C. supply. Calculate the current through each. Calculate the current if they were connected in delta across the same supply. [**Ans.** 4.62 A ; 8 A]
6. Three identical coils are connected in star to a 200 V (line voltage), 3-phase, A.C. supply and each coil takes 200 W. The power factor is 0.8 (lagging), calculate :
   (*i*) The line current, (*ii*) Impedance, and
   (*iii*) Resistance and inductance of each coil. [**Ans.** 2.165 A ; 53.334 Ω, 42.662 Ω, 0.102 H]
7. In a three-phase, 3-wire system with star-connected load the impedance of each phase is $(6 + j8)$ Ω. If the line voltage is 230 V, calculate :
   (*i*) The line current, and (*ii*) The power absorbed by each phase.
   [**Ans.** 13.3 A ; 1067 W]
8. A balanced star-connected load of $(8 + j6)$ Ω per phase is connected to 3-phase, 230 V supply. Find :
   (*i*) Line current, (*ii*) Power factor,
   (*iii*) Power, (*iv*) Reactive volt-amperes, and
   (*v*) Total volt-amperes. [**Ans.** 13.28 A ; 0.8 (lag) ; 4.232 W ; 3174 ; 5290]
9. A star-connected, 5000 V, 3-phase alternator is supplying 3,000 kW at power factor of 0.8. Calculate the active and reactive components of the current in each phase. [**Ans.** 346.2 A ; 260 A]
10. In a 3-phase, 4-wire system, two phases have currents of 10 A and 6 A in lagging power factors of 0.8 and 0.6 respectively, while the third phase is open-circuited. Calculate the current in the neutral and sketch the vector diagram. [**Ans.** $7 \angle -73° 24'$]
11. In a star-connected load each phase consists of a resistance of 100 Ω in parallel with a capacitor of capacitance 31.8 μF. When it is connected to a 416 V, 3-phase, 50 Hz supply, calculate :
    (*i*) The line current, (*ii*) The power factor,
    (*iii*) The power absorbed, and (*iv*) The total kVA.
    [**Ans.** 3.39 A ; 0.707 (leading) ; 1.728 kW ; 2.443 kVA]

12. Three identical coils connected in delta across 400 V, 50 Hz, 3-phase supply takes a line current of 17.32 A at power factor 0.5 lagging. Determine :
   (*i*) The current in each phase, and
   (*ii*) Resistance, inductance and impedance of each phase winding. [**Ans.** 10 A ; 20 Ω ; 0.11 H, 40 Ω]

13. A 220 V, 3-phase voltage is applied to a balanced delta-connected 3-phase load of phase impedance $(15 + j20)$ Ω. Find :
   (*i*) The phasor current in each line. (*ii*) What is the power consumed per phase ?
   (*iii*) What is the phasor sum of the three line currents ? Why does it have this value ?
   [**Ans.** 15.24 A ; 1161.6 W ; zero]

14. Calculate (*i*) line current and (*ii*) the total power absorbed when three coils each having a resistance of 10 Ω and reactance of 7 Ω are connected (*a*) in star and (*b*) in delta across a 400 V, 3-phase supply.
   [**Ans.** 18.93 A, 10750 W ; 56.7 A, 32250 W]

15. A delta-connected balanced 3-phase load is supplied from a 3-phase 400 V supply. The line current is 20 A and power taken by the load is 10000 W. Find :
   (*i*) Impedance in each branch ; and
   (*ii*) The line current, power factor and power consumed if the same load is connected in star.
   [**Ans.** 34.6 Ω, 20/3 A, 0.7217, 3330 W]

16. A balanced 3-phase load consists of resistances of 4 Ω each. Determine the total power when the resistances connected to 400 V supply are :
   (*i*) Star connected (*ii*) Delta connected. [**Ans.** 40 kW ; 120 kW]

17. Three non-inductive resistances, each of 100 Ω are connected in star across 400 V supply. Calculate the current through each. What would be the current through each if they are connected in delta across the same supply. [**Ans.** 2.31 A ; 4 A]

18. Three 100 Ω non-inductive resistances are connected in (*i*) star (*ii*) delta across a 400 V, 50 Hz, 3-phase mains. Calculate the power taken from the supply system in each case. In the event of one of the three resistances opened, what would be the value of the total power taken from the mains in each of the two cases. [**Ans.** 1600 W ; 4800 W]

19. A 3-phase delta connected load ; each phase of which has an inductive reactance of 40 Ω and a resistance of 25 Ω, is fed from the secondary of a 3-phase star-connected transformer, which has phase voltage of 240 V. Draw the circuit diagram of the system and calculate :
   (*i*) The current in each phase of the load, (*ii*) The voltage across each phase of the load,
   (*iii*) The current in the transformer secondary windings, and
   (*iv*) The total power taken from the supply and its power factor. [**Ans.** 8.8 A ; 415.7 V ; 15.3 A ; 5820 W]

20. Three similar resistors are connected in star across 400 V, 3-phase supply. The line current is 5 A. Calculate the value of each resistor. To what value line voltage be changed to obtain the same current with the resistors connected in delta ? [**Ans.** 46.2 Ω ; 133.3 V]

21. A 3-phase, star-connected system with 230 V between each phase and neutral has a resistance of 4, 5 and 6 Ω respectively in three phases. Calculate :
   (*i*) The current flowing in each phase, (*ii*) The neutral current, and
   (*iii*) The total power absorbed. [**Ans.** 57.5 A, 46 A, 38.3 A ; 16.71 A ; 32610 W]

22. In a 3-phase, 4 wire system there is a balanced 3-phase motor load taking 20 kW at a power factor of 0.8 lagging while lamps connected between phase conductors and the neutral are taking 5, 4 and 10 kW respectively. Voltage between line conductors is 430 V. Calculate the current in each conductor and in the neutral wire of the feeder supplying these loads. [**Ans.** 51.2 A ; 47.5 A ; 80 A ; 22.4 A]

Power Measurement in 3-phase Circuits

23. The power input to a 3-phase induction motor is read by two wattmeters. The readings are 860 W and 240 W. What is the power factor of the motor ? [**Ans.** 0.7155 (lag)]

24. While performing a load test on a 3-phase wound rotor induction motor by two wattmeters method, the readings obtained on two wattmeters were + 12.5 kW and – 4.8 kW and the line voltage was 440 V. Calculate :
(*i*) Power drawn by the motor, (*ii*) Power factor, and
(*iii*) Line current. **[Ans.** 7.7 kW ; 0.2334 ; 43.3 A]

25. The input power to a 3-phase delta-connected motor was measured by two wattmeters method. The readings were 20.8 kW and – 6.8 kW and the line voltage was 400 V. Calculate :
(*i*) Input power, (*ii*) Power factor, and
(*iii*) Line current. **[Ans.** 14 kW ; 0.281 ; 72 A]

26. A 3-phase, 500 V motor load has a power factor of 0.4. Two wattmeters connected to measure the power show the input to be 30 kW. Find the reading on each instrument. **[Ans.** 35 kW ; – 5 kW]

27. The power in a 3-phase circuit is measured by two wattmeters. If the total power is 100 kW and power factor is 0.66 leading, what will be the reading of each wattmeter ? For what power factor will one of the wattmeters read zero ? **[Ans.** 17.15 kW ; 82.85 kW ; p.f. = 0.5]

28. Two wattmeters used to measure the power input in a 3-phase circuit indicate 1000 W and 500 W respectively. Find the power factor of the circuit
(*i*) when both wattmeter readings are positive ; and
(*ii*) when the latter is obtained by reversing the current coil connections.
[Ans. 0.866 (lag) ; 0.1889 (lag)]

29. Two wattmeters connected to read the total power in a 3-phase system supplying a balanced load read 10.5 kW and – 2.5 kW respectively. Calculate the total power and the power factor. Draw suitable phasor diagram, explain the significance of
(*i*) equal wattmeter readings ; and (*ii*) a zero reading on one wattmeter.
[Ans. 8 kW ; 0.3348 (lag) (*i*) unity p.f., (*ii*) 0.5 p.f.]

30. The power input to a 1000 V, 3-phase induction motor running on full-load is measured by two wattmeters which indicate 300 kW and 100 kW respectively. Determine :
(*i*) Input, (*ii*) Line current,
(*iii*) Power factor, and (*iv*) Output if the efficiency of the motor is 92%.
[Ans. 400 kW ; 303.5 A ; 0.756 (lag) ; 368 kW]

31. Two wattmeters are used for measuring the power input and the power factor of an over-excited synchronous motor. If the readings of the meters are – 2.0 kW and + 7.0 kW respectively, calculate the input and power factor of the motor. **[Ans.** 5 kW ; 0.3057 (leading)]

32. The power input to a 2-kV, 50 Hz, 3-phase motor running on full-load at an efficiency of 90 per cent is measured by two wattmeters which indicate 300 kW and 100 kW respectively. Calculate :
(*i*) Input, (*ii*) Power factor,
(*iii*) Line current, and (*iv*) Output. **[Ans.** 400 kW ; 0.756 (lag) ; 153 A ; 360 kW]

33. In a balanced three-phase system power is measured by two-wattmeters method and the ratio of the two wattmeter readings is 2 : 1. Find the power factor of the system. **[Ans.** 0.866 (log)]

34. The power input of a synchronous motor is measured by two wattmeters both of which indicate 50 kW. If the power factor of the motor be changed to 0.866 leading, determine the readings of the two wattmeters, the total input power remaining the same. Draw the vector diagram for the second condition of load. **[Ans.** 33.33 kW ; 66.67 kW]

35. Each phase of a 3-phase delta-connected load consists of an impedance $Z = 20 \angle 60° \ \Omega$. The line voltage is 440 V at 50 Hz. Calculate the power consumed by each phase impedance and the total power. What will be the readings of the two wattmeters connected ? **[Ans.** 4840 W ; 14520 W ; 14520 W. zero]

36. A 3-phase 3-wire, 415 V supplies a balanced load of 20 A at power factor 0.8 lagging. Two wattmeters are used to measure power. Calculate :
(*i*) Power, (*ii*) Reading of wattmeter No 1, and
(*iii*) Reading of wattmeter No. 2. **[Ans.** 11.5 kW ; 8240 W ; 3260 W]

37. A balanced load is supplied from a 3-phase, 400 V, 3-wire system whose power is measured by two wattmeters. If the total power supplied is 26 kW at 0.75 power factor lagging, find the readings of each of the two wattmeters. [**Ans.** 19.62 kW ; 6.38 kW]

38. A 3-phase 4-wire, star-connected system supplies only non-inductive load. The current in line R is 8 A, the current in line Y is 10 A and the current in line B is 6 A. The voltage from each line to neutral is 120 V. Find :

(*i*) Wattage shown by each of the three wattmeters, and

(*ii*) Power taken by the lighting load. [**Ans.** 960 W, 1200 W, 720 W ; 2880 W]

39. Three identical coils, each having a resistance of 20 Ω and reactance of 20 Ω are connected in (*i*) star ; and (*ii*) delta across 440 V, 3-phase supply. Calculate for each method of connection the line current and reading on each of the two wattmeters connected to measure power.

[**Ans.** 8.98 A ; 3817.5 W, 1022.5 W ; 26.95 A ; 12452.5 W ; 3067.5 W]

40. A balanced star-connected load, each having a resistance of 10 Ω and inductive reactance of 30 Ω is connected to a 400 V, 50 Hz supply. The phase sequence is RYB. Two wattmeters connected to read total power have their current coils connected in the red and blue lines respectively. Calculate the reading of each wattmeter. Draw the circuit and vector diagrams. [**Ans.** $W_1 = -585$ W ; $W_2 = 2185$ W]

41. Three equal star-connected inductors take 8 kW at power factor 0.8 when connected to a 460 V, 3ϕ, 3-wire supply. Give the connection diagram of 2 single-phase wattmeters to measure the power in the circuit. Allocate the reading of each wattmeter. Find the line currents if one of the inductor is short-circuited. Draw the vector diagrams of the currents and voltages under the condition.

[**Ans.** 5732 W, 2268 W ; 21.74 A, 21.74 A]

42. A 3-phase, 3-wire balanced load with a lagging power factor is supplied at 400 V (between lines). A single-phase wattmeter (scaled in kW) when connected with its current coil in the R-line and the voltage coil between R and Y lines gives a reading of 6 kW. When the same terminals of the same voltage coil are switched over to Y-line and B-line, the current-coil connections remaining the same, the reading of the wattmeter remains unchanged. Calculate the line current and power factor of the load. Phase sequence is $R \rightarrow Y \rightarrow B$. [**Ans.** 30 A ; 0.866 (lag)]

43. A wattmeter reads 5.54 kW when its current coil is connected in red phase and its potential coil is connected between the neutral and red phase of a symmetrical 3-phase system supplying a balanced load of 30 A at 400 V. What will be the reading of the instrument if the current coil connections remains unchanged but the potential coil be connected between blue and yellow phase ? What does this figure represent ? [**Ans.** 7 kVAR]

44. A 3-phase, 434 V, 50 Hz supply is connected to a 3-phase star-connected induction motor and synchronous motor. Impedance of each phase of induction motor is $1.25 + j2.17$ Ω. The 3-phase synchronous motor is over-excited and it draws a current of 120 A at 0.87 leading power factor. Two wattmeters are connected in usual manner to measure power drawn by the two motors. Calculate :

(*i*) Reading on each wattmeter, and (*ii*) Combined power factor.

[**Ans.** 63970 W, 52050 W ; 0.9845 (lag)]

45. A star-connected balanced load is supplied from a 3-phase balanced supply with a line voltage of 416 V at a frequency of 50 Hz. Each phase of load consists of a resistance and a capacitor joined in series and the readings on two wattmeters connected to measure the total power supplied are 782 W and 1980 W, both positive. Calculate :

(*i*) The power factor of the circuit, (*ii*) The line current, and

(*iii*) The capacitance of each capacitor. [**Ans.** 0.8 (leading) ; 4.8 A ; 106 μF]

UNIT-II : MAGNETIC CIRCUITS

Chapter :

4. Magnetic Circuits

4

Magnetic Circuits

1. Magnetic field. 2. Terms connected with magnetic materials. 3. Classification of magnetic materials. 4. Magnetically soft materials. 5. Magnetically hard materials. 6. Laws of magnetic force. 7. Magnetic field due to a current carrying conductor. 8. Magnetic flux and flux density. 9. Magnetomotive force and magnetic field strength. 10. Force on a current-carrying conductor lying in a magnetic field. 11. Magnetising force of a long straight conductor. 12. Magnetising force of a long solenoid. 13. Force between two parallel conductors-Ampere's law. 14. Terms connected with magnetic circuit. 15. Comparison of electric and magnetic circuits. 16. Magnetic leakage and fringing. 17. Magnetisation (B-H) curves. 18. Losses in magnetic materials—hysteresis loss—eddy current losses. 19. Faraday's laws of electromagnetic induction. 20. Induced e.m.f.—Dynamically induced e.m.f.—Statically induced e.m.f. 21. Inductances in series. 22. Inductances in parallel—Highlights—Objective Type Questions—Theoretical Questions—Unsolved Examples.

1. MAGNETIC FIELD

- Substances that are magnetic have a north-seeking pole and south-seeking pole, corresponding to the earth's magnetic poles. In effect, *each magnet has a magnetic field around* it just as the earth does.
- The magnetic field is *strongest at the ends of the magnet*. In the *centre* of the magnet the *strength is negligible*. A magnetic field can be seen by placing a magnet under a piece of paper and then sprinkling iron filings onto the paper. The iron filings line up to show the magnetic field and its concentration, as shown in Fig. 1 (*a*).

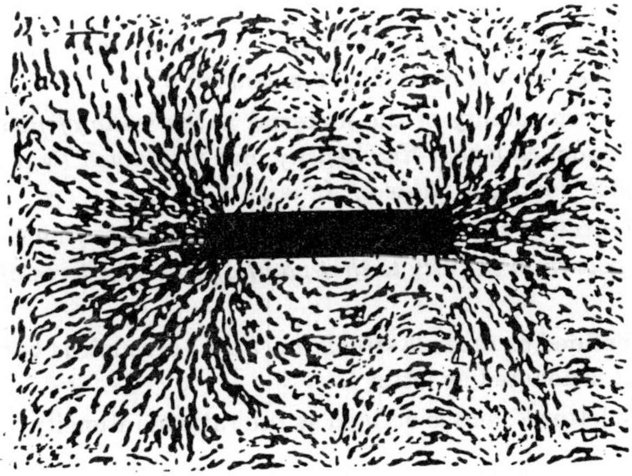

(*a*) Illustrating a magnetic field by the use of iron filings.

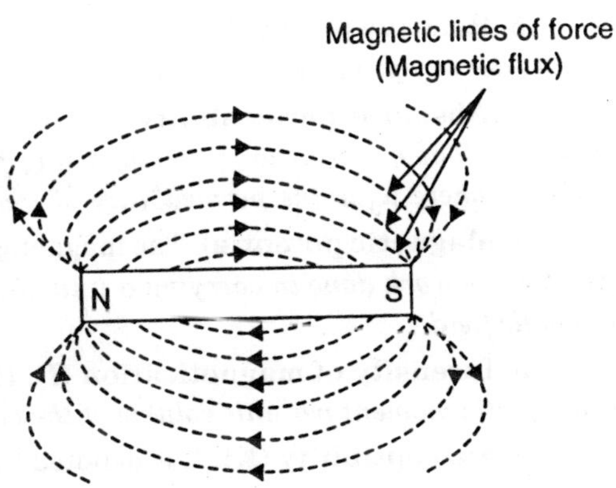

(*b*)

Fig. 1. Magnetic field around a magnet.

- The unlike poles of two magnets attract each other while the like poles repel each other (Fig. 2).

The following points are worth noting about magnets and magnetic lines of force.

1. Each magnet has two poles called north pole (N) and south pole (S).
2. Unlike poles attract, like poles repel.
3. A force field exists around a magnet which is concentrated at the poles of the magnet. The strength of the magnetic field is indicated by the density or concentration of lines of force.
4. Lines of force repel each other.
5. Lines of force cannot intersect.
6. Lines of force form closed loops.

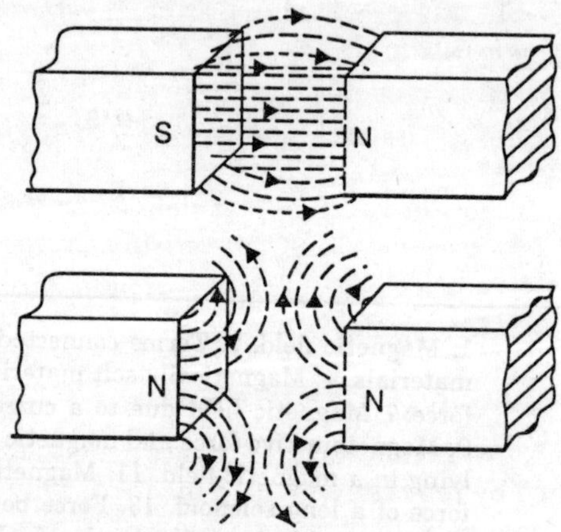

Fig. 2. Unlike poles attract each other ; like poles repel each other.

2. TERMS CONNECTED WITH MAGNETIC MATERIALS

1. **Magnetic force.** It is the *force exerted by one magnet on another either to attract it or repel it.*

2. **Unit pole strength.** It is defined as the *strength of that pole which when placed in vacuum at a distance of one metre from a similar and equal pole, repels it with a force of one newton.*

3. **Magnetic flux density (B).** It is usually represented by letter B and is defined as *the flux or lines of force passing per unit area through any substance through a plane at right angles to the direction* of magnetic flux ; it is measured in Wb/m^2 (or T *i.e.*, Tesla).

4. **Magnetic field strength.** It may be defined in the following two ways :

(*i*) Field strength at any point within a magnetic field is *the number of lines of force passing through a unit area round the point considered and held perpendicular to the lines.*

(*ii*) Field strength at any point within a magnetic field *is the force exerted by a unit North pole at that point.*

5. **Relative permeability** (μ_r). It is *the ratio of flux density (B) produced in that material to the flux density produced in vacuum by the same magnetising force (H). It is denoted by* μ_r.

6. **Absolute permeability** (μ). Absolute permeability of the material is the *ratio of flux density in that material to the magnetising force producing that flux density and is denoted by* μ ; $\mu = \mu_0\mu_r$ *where* μ_0 *is the permeability of free space having a value of* $4\pi \times 10^{-7}$ *H/m.*

7. **Magnetic potential.** The magnetic potential at any point within a magnetic field is measured by the *work done in carrying a unit North pole from infinity to that point against the force of magnetic field.*

8. **Intensity of magnetisation (I).** It is defined as *the pole strength per unit area of the bar or magnetic moment per unit volume of the bar.* It is denoted by letter I.

9. **Susceptibility (K).** It is denoted by the letter K, and is defined as the ratio of intensity of magnetisation (I) to magnetising force (H). In other words $K = \frac{I}{H}$.

10. **Magnetomotive force.** It is *that force which drives or tends to drive the flux through a magnetic circuit.* In short it is written as m.m.f. It is the product of number of turns (N) and current (I) in amperes in these turns *i.e.*, m.m.f. = NI.

11. **Magnetic reluctance.** It is *that property of the material which opposes the production of magnetic flux in it.*

12. **Co-ercive force.** It may be defined as the *demagnetising force which is necessary to neutralise completely the magnetism in an electromagnet after the value of magnetising force becomes zero.*

13. **Resonance.** It is defined as the *magnetic flux density which still persists in magnetic material even when the magnetising force is completely removed.* It is expressed in Wb/m^2.

14. **Retentivity.** It is *that property of magnetic material which is measured by its maximum value of the residual induction.*

3. CLASSIFICATION OF MAGNETIC MATERIALS

All materials possess magnetic properties to a greater or lesser degree and these are determined by the facts that (1) a magnetic field exert forces and torques on the bodies, (2) a body placed in a magnetic field distorts the field. The magnetic properties of the materials are characterised by their relative permeabilities. In accordance with the value of *relative permeability* the materials may be classified in the following three ways :

1. **Ferromagnetic materials.** The relative permeabilities of these materials are *much greater than unity* and are dependent on the field strengths. They attract the lines of force strongly [Fig. 3 (*a*)]. The principal ferromagnetic elements are *iron, cobalt, nickle*. Gadolinium however also comes under this classification. These have high susceptibility.

2. **Paramagnetic materials.** These have relative permeability *slightly greater than unity* and are magnetised slightly. They attract the lines of forces weakly [Fig. 3 (*b*)]. *Aluminium, platinum* and *oxygen* belong to this category.

3. **Diamagnetic materials.** The relative permeability of these materials is *slightly less than unity*. They repel the lines of force slightly [Fig. 3 (*c*)]. The examples are *Bismuth, silver, copper* and *hydrogen*.

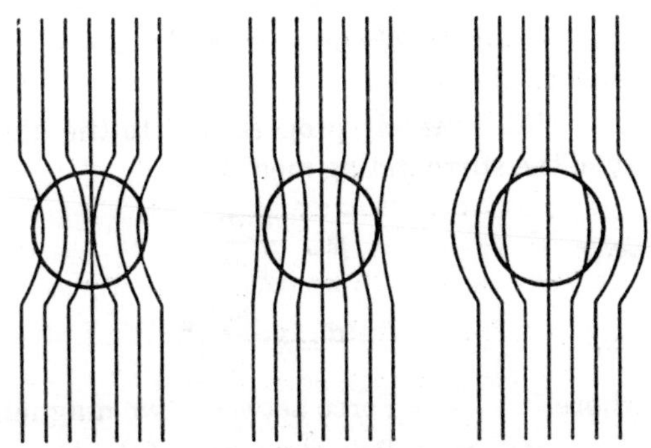

(*a*) Ferromagnetic (*b*) Paramagnetic (*c*) Diamagnetic

Fig. 3

4. MAGNETICALLY SOFT MATERIALS

The magnetically soft materials (*suitable for making electromagnets*) are characterised as follows :

1. They have high permeability.
2. The magnetic energy stored is not high.
3. They have negligible co-ercive force (due to which these are **not** *suitable for making permanent magnets*).
4. They have low resonance.

The various magnetically soft materials are enumerated below :

(*i*) Pure or ingot iron (*ii*) Cast iron
(*iii*) Carbon steels (*iv*) Silicon steels
(*v*) Manganese and nickle steels (*vi*) Permalloy
(*vii*) Mumetal (*viii*) Peminvar.

5. MAGNETICALLY HARD MATERIALS

Magnetically hard materials *are used for making* **permanent magnets** and have the following characteristics :

1. They possess high value of *BH* product.
2. High retentivity.
3. High co-ercivity.
4. Strong magnetic reluctance.
5. Hysteresis loop is more rectangular in shape.

Important magnetically hard materials are enumerated below :

(*i*) Tungsten steel (*ii*) Cobalt steel
(*iii*) Chromium steel (*iv*) Alnico
(*v*) Cunife (*vi*) Hypernic.

6. LAWS OF MAGNETIC FORCE

Coulomb, through his experiments, found that the force between two magnetic poles placed in a medium is

(*i*) directly proportional to their pole strengths,

(*ii*) inversely proportional to the square of the distance between them, and

(*iii*) inversely proportional to the absolute permeability of the surrounding medium

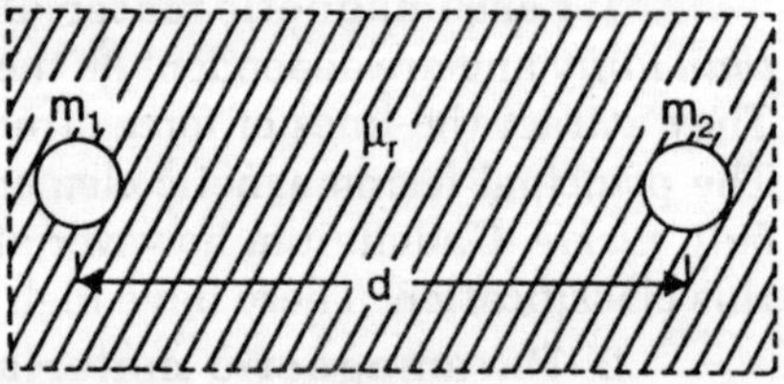

Fig. 4

i.e., $$F \propto \frac{m_1 m_2}{\mu d^2}$$

i.e., $$F = k\frac{m_1 m_2}{\mu d^2}$$

where F = Force between two magnetic poles,

m_1, m_2 = Pole strengths of the two poles respectively,

d = Distance between the poles,

μ = Absolute permeability, and

k = Constant.

In the S.I. system, the value of the constant, $k = \dfrac{1}{4\pi}$

$$\therefore \quad F = \frac{m_1 m_2}{4\pi\mu d^2}\ \text{N} \quad \text{...(1)}$$

or $$F = \frac{m_1 m_2}{4\pi\mu_0\mu_r d^2} \quad \text{...... in a medium} \quad \text{...[1 (a)]}$$

$$= \frac{m_1 m_2}{4\pi\mu_0 d^2} \quad \text{...... in air} \quad \text{...[1 (b)]}$$

7. MAGNETIC FIELD DUE TO A CURRENT CARRYING CONDUCTOR

When an electric current flows through a wire, a magnetic field is built up around the wire itself. This can be seen using a cardboard, iron filings, and a current-carrying wire. When the wire passed through the cardboard and the current flowing, iron filings are sprinkled onto the card board. They can be seen arranging themselves in a magnetic field (Fig. 5). The magnetic lines of force are referred to as flux. Just as in a natural magnet, the field is strongest near the wire and diminishes as the distance from the wire increases.

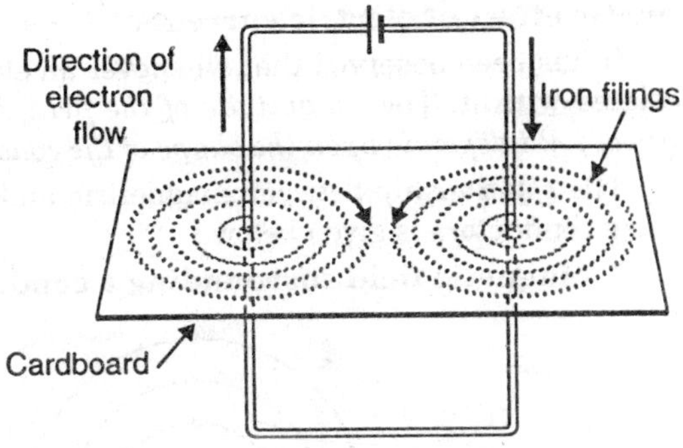

Fig. 5. Magnetic field around a wire that is carrying electric current.

- Flux around a wire does have direction. Flux direction is determined by the direction of electron flow within the wire. As shown in Fig. 6, the *North pole of the compass needle indicates the direction of flux or magnetic field around the wire.* The **dot** *in the centre of the wire on the left indicated the point of the current-direction arrow coming* **toward the observer**; the × at the right represents the tail of the current arrow **pointing away from the observer.** If the direction of electron flow within the wire is reversed, the compass needles will reverse themselves, indicating a change in flux direction.

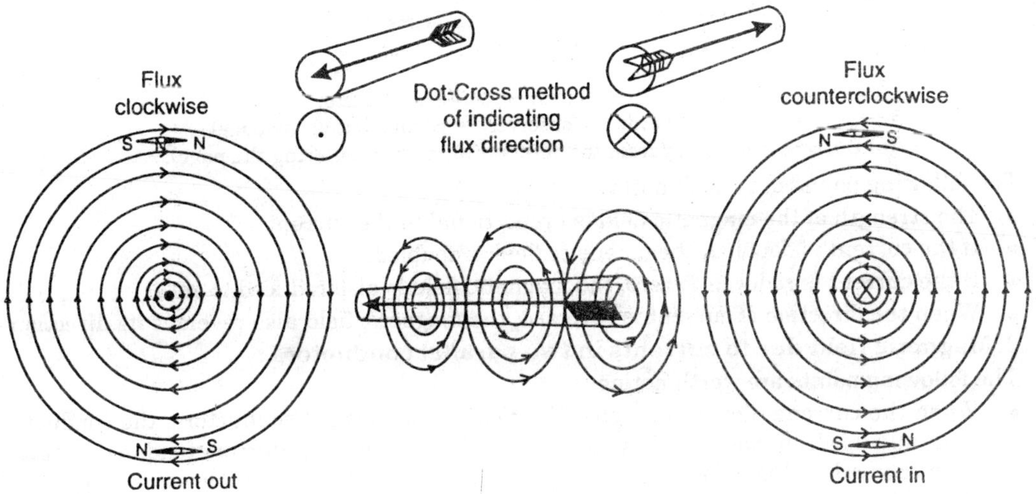

Fig. 6. Compasses indicate the direction of flux around a wire.

Right hand rule (or right hand screw rule)

The direction of the magnetic field can be found by using *right hand rule or the right hand screw rule.* The right hand rule states as follows :

"Grasp the wire in the right hand, with the thumb pointing in the direction of the current. The fingers will curl around the wire in the direction of the magnetic field".

Fig. 7 illustrates the rule.

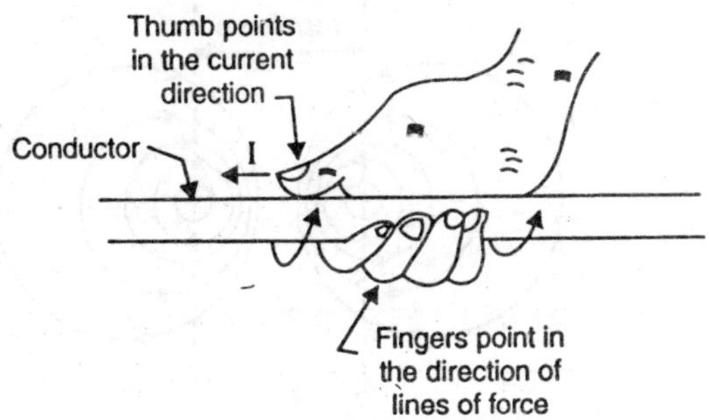

Fig. 7. Right hand rule (or right hand screw rule).

The right hand screw rule can be explained as follows :

As a wood screw is turned clockwise it moves (or progresses) into the wood. The *horizontal direction of the screw is analogous to the direction of current in a conductor. The circular motion of the screw shows the direction of the magnetic flux around the conductor.*

Magnetic effect of electric currents

It has been observed that whenever an electric current flows, a magnetic field is invariably associated with it. The *magnitude of the field depends on the magnitude of the current, and the nature of field depends upon the shape of the conductor.*

The relationship between magnetism and the electric current for various shapes of current carrying conductors is given below :

1. Magnetic field surrounding a conductor :

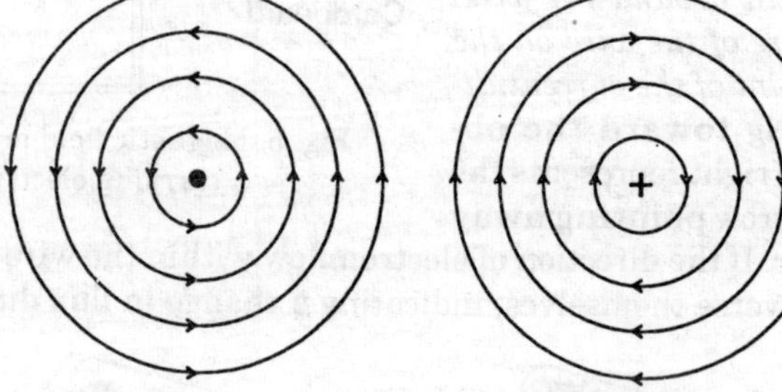

Fig. 8. Magnetic field surrounding a conductor.
(*a*) Current coming towards the observer (or coming out of the page)
(*b*) Current going away from the observer (or current entering the page).

The following points are worth noting :

- The strength of the magnetic field is proportional to the current.
- If the current is doubled, the magnetic field is doubled.
- If the current is reduced to zero, the magnetic field is also reduced to zero.
- When the direction of current is reversed, the magnetic field also reverses its direction.

2. Magnetic field due to currents in two parallel conductors :

The following points are worth noting :

- When the currents flow in *opposite direction* in the parallel conductors, the conductors mutually *repel*, so that in such a case, the current carrying conductors tend to *move away from each other* [Fig. 9 (*a*)].
- When the currents flow in the *same direction* in parallel conductors, the conductors mutually *attract*, so that in such a case, the current carrying conductors tend to move *towards each other* [Fig. 9 (*b*)].

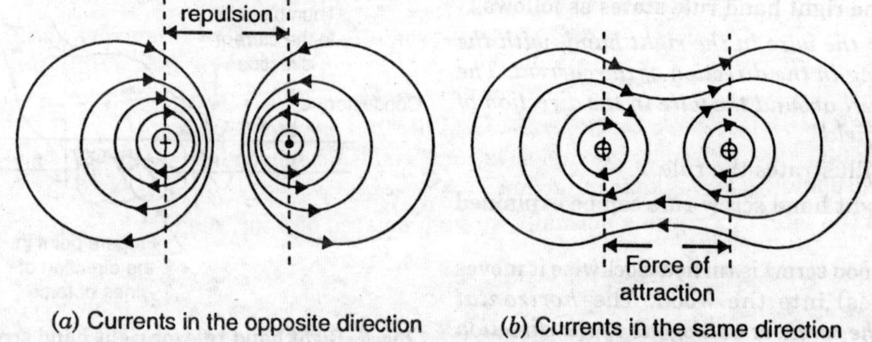

(*a*) Currents in the opposite direction (*b*) Currents in the same direction

Fig. 9. Force between parallel current carrying conductors.

3. Magnetic field due to a circular loop :

When a single conductor is looped to form a single turn of coil, the magnetic fields are set up as small loops around the coil shown as *AA* is Fig. 10 (*a*). The magnetic lines enter the inside of the coil on one face and leave it on the other. The *direction of the field in the centre of coil will be perpendicular to the plane of the loop* as shown in Fig. 10.

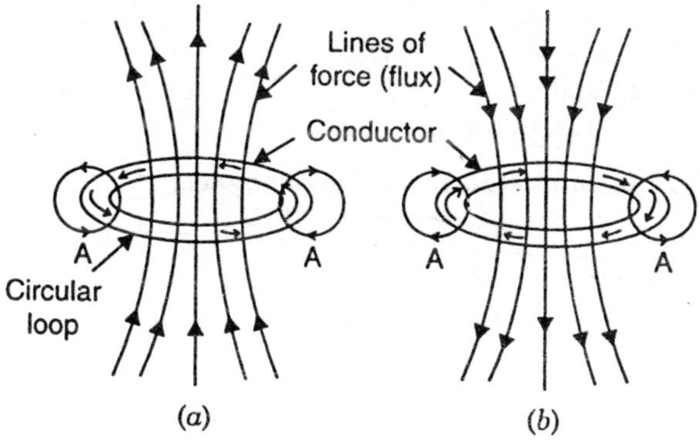

Fig. 10. Magnetic field due to circular loop.

4. Magnetic field due to solenoid :

A *solenoid is a coil of wire wound on a long straight core*. When a current passes through a solenoid a magnetic field is produced. The lines are approximately parallel to the axis of the solenoid until they reach the ends where they begin to diverge (magnetic field pattern long similar to bar magnet).

The solenoid produces strong magnetic fields for such applications as *relay coils, transformer coils and circuit breakers*.

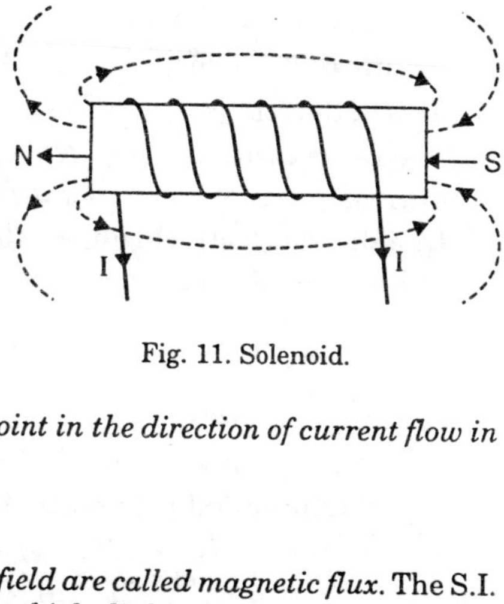

Fig. 11. Solenoid.

The right hand rule for determining the direction of flux from a solenoid states as follows :

"Hold the solenoid in right hand such that the fingers point in the direction of current flow in the coils. The thumb will point towards N pole of the field".

8. MAGNETIC FLUX AND FLUX DENSITY

Magnetic flux (ϕ). *The total lines of force in a magnetic field are called magnetic flux*. The S.I. unit of magnetic flux is *weber (Wb)*. It is defined as *the flux which, linking in a single turn coil, induces an e.m.f. of one volt when the flux is reduced to zero at a uniform rate in one second.*

Flux density (B). It is given by the *flux passing per unit area through a plane at right angles to the flux*. It is usually designated by the capital letter *B* and is measured in weber/metre^2 (Wb/m^2). Flux density is also known as *magnetic induction*.

If ϕ Wb is the total flux passing normally through an area of *A* m^2, then

$$B = \frac{\phi}{A} \text{ Wb/m}^2 \text{ or tesla } (T) \qquad \text{...(2)}$$

9. MAGNETOMOTIVE FORCE AND MAGNETIC FIELD STRENGTH

Magnetomotive force (m.m.f.). Magnetomotive force drives or tends to drive flux through a magnetic circuit and corresponds to electromotive force (e.m.f.) in an electric circuit. The m.m.f. of a coil is the *product of current in the coil and the number of turns of the coil* and has the unit of ampere turns (*AT*).

Magnetic field strength (H). The magnetic field strength (*H*) is defined as the *m.m.f. per metre length of magnetic circuit i.e.,*

$$H = \frac{NI}{l} \text{ AT/m}$$

where, N = Number of turns of a coil,
I = Nurrent (amperes), and
l = Length of the magnetic circuit (*m*).

Example 1. *The following data relate to an electromagnet :*

Total flux $= 8 \times 10^{-4}$ *Wb*
Cross-sectional area of the core = *200 mm²*
Number of turns = *100*
Magnitude of current = *2 A*
Length of the magnetic circuit = *400 mm*

Calculate the following :

(i) Flux density in the coil,
(ii) Magnetomotive force, and
(iii) Magnetic field strength.

Solution. Total flux, $\phi = 8 \times 10^{-4}$ Wb
Area of the coil, $A = 200 \text{ mm}^2 = 2 \times 10^{-4} \text{ m}^2$
Number of turns, $N = 100$
Current, $I = 2$ A
Length of magnetic circuit, $l = 400$ mm = 0.4 m.

(*i*) **Flux density in the coil, B :**

$$B = \frac{\phi}{A} = \frac{8 \times 10^{-4}}{2 \times 10^{-4}} = \textbf{4 Wb/m}^2 \textbf{ or T. (Ans.)}$$

(*ii*) **Magnetomotive force, m.m.f. :**

m.m.f. = NI = 100 × 2 = **200 AT. (Ans.)**

(*iii*) **Magnetic field strength, H :**

$$H = \frac{NI}{l} = \frac{200}{0.4} = \textbf{500 AT/m. (Ans.)}$$

10. FORCE ON A CURRENT-CARRYING CONDUCTOR LYING IN A MAGNETIC FIELD

It has been found that *whenever a current-carrying conductor is placed in a magnetic field, it experiences a force which acts in a direction perpendicular both to the direction of the current and the field.*

— Fig. 12 (*a*) shows the field set up by the poles.
— Fig. 12 (*b*) shows the conductor field due to flow of current in the conductor.
— Fig. 12 (*c*) shows the resultant field produced when the current-carrying conductor of [Fig. 12 (*b*) (*i*)] is inserted in the pole-field, *with the axis of the conductor at right angles to the direction of the flux.*

On the *upper side* of the conductor in Fig. 12 (*c*) the magnetizing forces of the field and of the current in the conductor are *additive*, while on the *lower side* these are subtractive. This explains why the resultant field is strengthened above and weakened below the conductor. Thus the conductor has a force on it which tends to move it *downwards*.

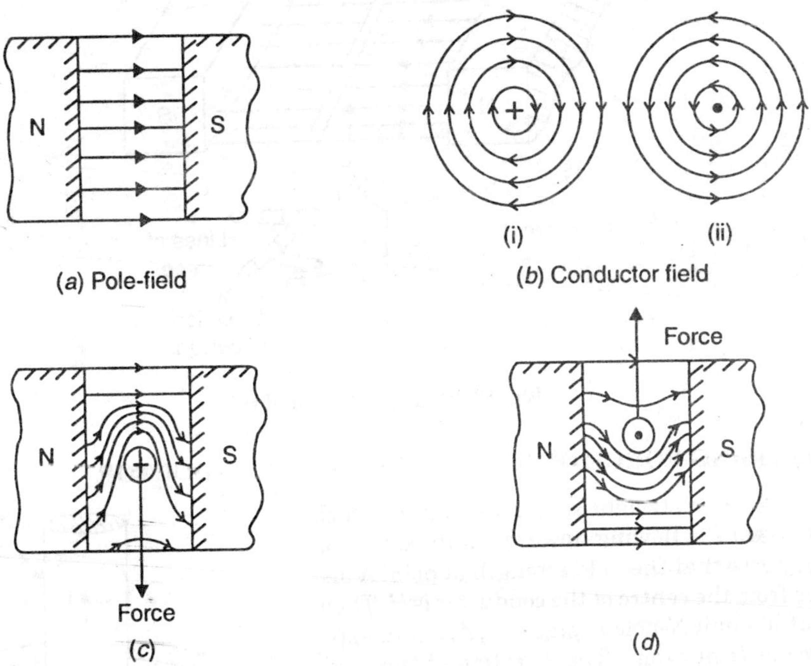

Fig. 12. Force on a current-carrying conductor lying in a magnetic field.

— Fig. 12 (*d*) shows that when the current in the conductor is *reversed*, the direction of the force is also *reversed*.

The force developed in the conductor is given by the relation :

$$F = BIl \text{ newtons} \quad ...(3)$$

$$(= \mu_0\mu_r HIl \text{ newtons}) \quad ...[3\,(a)]$$

where F = Force developed in the conductor,

B = Flux density, T (Wb/m²),

I = Current in the conductor, A, and

l = Exposed length of the conductor, m

μ_0 = Absolute permeability

μ_r = Relative permeability

H = Magnetising force.

The *direction of this force may be easily found by Fleming's left hand rule* which states as follows :

"Hold your left hand with index finger, middle finger and thumb at right angles. If the index finger points in the direction of the flux from north to south and middle finger points in the direction of the imposed voltage and its resulting conventional current flow, then the thumb will point in the direction of the force that is developed".

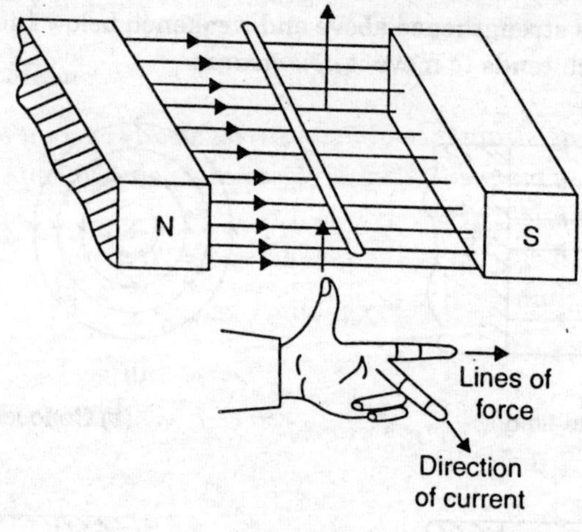

Fig. 13. Fleming's left hand rule.

11. MAGNETISING FORCE OF A LONG STRAIGHT CONDUCTOR

Fig. 14 shows a straight solid conductor in which the steady current I is flowing upwards in the direction indicated. Suppose that the field strength at point A distant r metres from the centre of the conductor is H. Then, it means that if a unit N-pole is placed at A, it will experience a force of H newtons. The direction of this force will would be *tangential* to the circular line of force passing through A. If this unit N-pole is moved once round the conductor against this force, then work done is

$$= \text{force} \times \text{distance}$$

$$= H \times 2\pi r \text{ joules.}$$

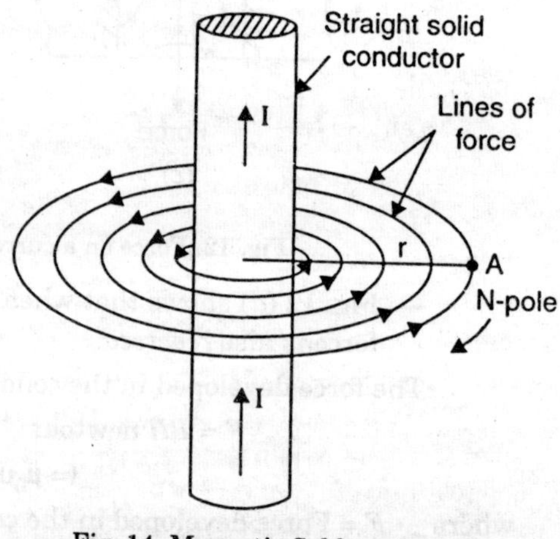

Fig. 14. Magnetic field produced by current in a straight solid conductor.

According to *work law* (It states that the *net work in joules done on or by a unit N-pole in moving around any single completed path in a magnetic field is numerically equal to the ampere-turns linked with its path*), this must be equal to the ampere-turns included within the path.

Ampere-turns linked $= NI$

$$= 1 \times I = I$$

$$\therefore \quad H \times 2\pi r = I$$

or
$$H = \frac{1}{2\pi r} \text{ AT/m} \quad \text{...(4)}$$

Also
$$B = \mu H = \frac{\mu I}{2\pi r} \text{ Wb} \quad \text{...(5)}$$

(where $\mu = \mu_0 \mu_r$)

For N conductors,
$$H = \frac{NI}{2\pi r} \text{ AT/m} \quad \text{...(6)}$$

and $$B = \frac{\mu_0 \mu_r NI}{2\pi r} \text{ Wb/m}^2 \quad \text{...... in a medium} \quad ...(7)$$

$$= \frac{\mu_0 NI}{2\pi r} \text{ Wb/m}^2 \quad \text{...... in air} \quad ...[7\,(a)]$$

Example 2. *A long straight conductor carries steady current of 20 A. Find :*

(i) The intensity of magnetic field produced at a point 300 mm from the axis of the conductor.

(ii) Flux density of field at that point.

Solution. Current carried by the conductor, $I = 20$ A

Distance, $r = 300$ mm $= 0.3$ m

(*i*) **Magnetic field intensity, H :**

$$H = \frac{I}{2\pi r} = \frac{20}{2\pi \times 0.3} = 10.61 \text{ A/m. (Ans.)}$$

(*ii*) **Flux density, B :**

$$B = \mu_0 \mu_r H$$
$$= 4\pi \times 10^{-7} \times 1 \times 10.61$$
$$= \mathbf{133.3 \times 10^{-7}\ T\ (or\ Wb/m^2).\ (Ans.)}$$

12. MAGNETISING FORCE OF A LONG SOLENOID

Fig. 15 shows magnetic field produced by current in a solenoid.

Let H = Magnetising force along the axis of the solenoid.

Assumptions :

H remains constant through the length l of the solenoid.

The value of H outside the solenoid is negligible.

Let a unit N-pole is placed at a point A (outside the solenoid) and is taken once round the completed path (shown dotted in Fig. 15) in a direction opposite to H.

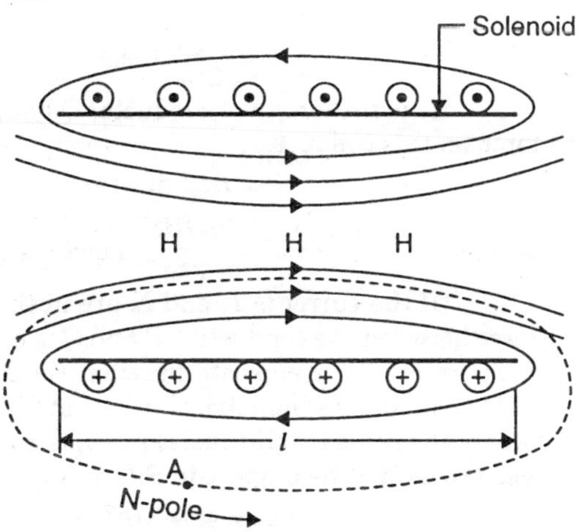

Fig. 15. Magnetic field produced by current in a solenoid.

Now work done in one round = $H \times l$ joules

The ampere-turns linked with this path = NI

[where N = Number of turns of solenoid, and
I = Current (A) passing through the solenoid.]

As per work law, $H \times l = N \times I$

or $$H = \frac{NI}{l} \text{ AT/m}$$

Also $$B = \frac{\mu_0 \mu_r NI}{l} \quad \text{......in a medium} \quad ...(8)$$

$$= \frac{\mu_0 NI}{l} \quad \text{...... in air} \quad ...[8\,(a)]$$

Example 3. *A solenoid of 300 turns is wound on a continuous ring of magnetic material of relative permeability 1000. If the flux density in the magnetic material of the core in the solenoid is 1.3 T, what is the current in the solenoid ? Mean diameter of the solenoid is 100 mm.*

Solution. Number of turns of the solenoid, $N = 300$

Relative permeability, $\mu_r = 1000$

Flux density, $B = 1.3$ T

Mean diameter of the solenoid, $d = 100$ mm $= 0.1$ m

$\therefore$ Mean circumference of the ring, $l = 2\pi r = 2\pi \times \frac{0.1}{2} = 0.314$ m

Current in the solenoid, I :

$$B = \frac{\mu_0 \mu_r NI}{l}$$

$$1.3 = \frac{4\pi \times 10^{-7} \times 1000 \times 300 \times I}{0.314}$$

$$\therefore \quad I = \frac{1.3 \times 0.314}{4\pi \times 10^{-7} \times 1000 \times 300} = \textbf{1.08 A. (Ans.)}$$

13. FORCE BETWEEN TWO PARALLEL CONDUCTORS—AMPERE'S LAW

Let P and Q be the two conductors, d metres apart, carrying currents I_1 and I_2 amperes as shown in Fig. 16.

The flux density at Q due to P,

$$B = \frac{\mu_0 I_1}{2\pi d} \text{ tesla}$$

If l is the length of conductor Q lying in this flux density, then force on this conductor,

$$F = BI_2 l \text{ newtons}$$

or
$$F = \frac{\mu_0 I_1 I_2 l}{2\pi d} \text{ newtons} \quad ...(9)$$

(If the currents I_1 and I_2 are in the *same* direction, then the force between the conductors P and Q is one of *attraction* ; whereas if the currents are in opposite direction, the conductors *repel* each other). Since eqn. (9) is symmetrical in I_1 and I_2, the force on conductor P is also of the same magnitude. Eqn. (9) is known as *Ampere's law* and is used to define the ampere in S.I. units.

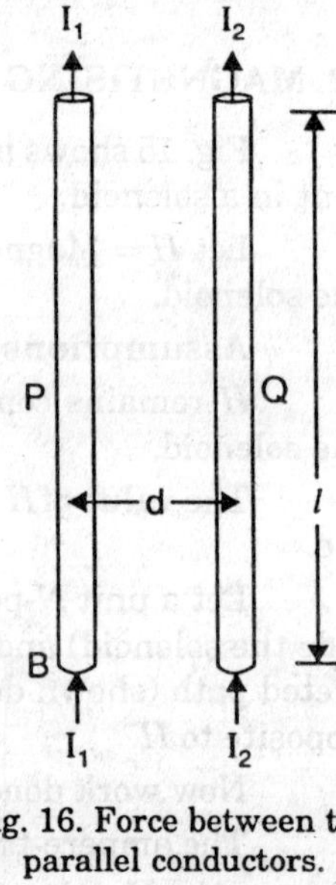

Fig. 16. Force between two parallel conductors.

$$\mu_0 = 4\pi \times 10^{-7} \text{ H/m}$$

If, $l = d = 1$ m,

and, $I_1 = I_2 = 1$ A,

then $F = 2 \times 10^{-7}$ N

Hence, one ampere current is defined as follows :

"An ampere is that current which when flowing in each of the two infinitely long parallel conductors situated in vacuum and separated 1 metre between centres, produces on each conductor a force of 2×10^{-7} N per metre length."

Example 4. *Two long parallel conductors carrying current, in the same direction, of 2500 A and 5000 A are placed with their axis 100 mm apart. Calculate the force between them in N/m.*

Solution. Current, $I_1 = 2500$ A

Current, $I_2 = 5000$ A

Distance between the conductors, $d = 100$ mm $= 0.1$ m.

Force between the conductors per metre, F :

$$F = \frac{\mu_0 I_1 I_2 l}{2\pi d}$$

or
$$\frac{F}{l} = \frac{\mu_0 I_1 I_2}{2\pi d} = \frac{4\pi \times 10^{-7} \times 2500 \times 5000}{2\pi \times 0.1}$$

$$= \mathbf{25\ N/m.\ (Ans.)}$$

Example 5. *Two long straight parallel wires, standing in air 2 m apart, carry currents I_1 and I_2 in the same direction. The magnetic intensity at a point midway between the wires is 7.95 AT/m. If the force on each wire per unit length is 2.4×10^{-4} N, evaluate I_1 and I_2.* **(Madras University)**

Solution. Distance between the conductors, $d = 2$ m

Magnetic intensity midway between the wires = 7.95 AT/m

Force on each wire per unit length = 2.4×10^{-4} N

I_1, I_2 :

We know that when the two currents flow in the same direction, net field strength midway between the two conductors is the difference of the two field strengths.

Now,
$$H_1 = \frac{I_1}{2\pi r} = \frac{I_1}{2\pi \times \frac{2}{2}} = \frac{I_1}{2\pi} \qquad \left(\because\ r = \frac{d}{2} = \frac{2}{2} = 1\ \text{m}\right)$$

and
$$H_2 = \frac{I_2}{2\pi r} = \frac{I_2}{2\pi \times \frac{2}{2}} = \frac{I_2}{2\pi}$$

$\therefore$
$$H_1 - H_2 = \frac{I_1}{2\pi} - \frac{I_2}{2\pi} = 7.95$$

or
$$I_1 - I_2 = 2\pi \times 7.95 = 50 \qquad ...(i)$$

Also using the relation,

$$F = \frac{\mu_0 I_1 I_2 l}{2\pi d}$$

or
$$\frac{F}{l} = \frac{\mu_0 I_1 I_2}{2\pi d}$$

or
$$2.4 \times 10^{-4} = \frac{4\pi \times 10^{-7} \times I_1 I_2}{2\pi \times 2} \qquad \left[\because\ \frac{F}{l} = 2.4 \times 10^{-4}\ \text{N ...... Given}\right]$$

or
$$I_1 I_2 = \frac{2.4 \times 10^{-4} \times 2\pi \times 2}{4\pi \times 10^{-7}} = 2400 \qquad ...(ii)$$

Substituting the value of I_1 from (*i*) in (*ii*), we get

$$(50 + I_2) \times I_2 = 2400$$

or
$$I_2^2 + 50 I_2 - 2400 = 0$$

or
$$I_2 = \frac{-50 \pm \sqrt{2500 + 9600}}{2}$$

$$= \frac{-50 \pm 110}{2} = 30\ \text{A} \qquad \textbf{(taking + ve value only)}$$

$\therefore$
$$I_1 = 80\ \text{A}.$$

Hence **I_1 = 80 A and I_2 = 30 A. (Ans.)**

14. TERMS CONNECTED WITH MAGNETIC CIRCUIT

A *magnetic circuit* is defined as *the route or path which is followed by a magnetic flux.*

Some important terms connected with magnetic circuits are discussed below :

1. **Permeability (μ).** Permeability of any material is a measure of ease with which the atoms can be arranged. The S.I. unit for permeability is henry/metre (H/m).

Mathematically, $\mu = \mu_0 \mu_r$ H/m ...(10)

where μ_0 = Permeability of free space
$= 4\pi \times 10^{-7}$ H/m (in S.I. units)
μ_r = Relative permeability (= 1 for air and non-magnetic materials)

Relative permeability (μ_r) is simply a numeric which expresses the degree to which the material is a better conductor of magnetic flux as compared to free space.

> **Note.** μ_r for *air* (and non-magnetic materials) = 1
> μ_r for *diamagnetic materials* = slightly less than one
> μ_r for *paramagnetic materials* = slightly higher than one
> μ_r for *ferromagnetic materials* = in the hundreds or thousands
> (μ_r for ferromagnetic materials changes with the magnetic field strength)

2. **Magnetomotive force (m.m.f.).** Magnetomotive force *drives or tends to drive flux through a magnetic circuit* and corresponds to electromotive force (e.m.f.) in a circuit. M.m.f. is equal to the work done in joules in carrying a unit magnetic pole once through the entire magnetic circuit.

M.m.f. is measured in **ampere-turns (AT).** Ampere-turns are found by *multiplying number of turns (N) of a magnetic circuit and the current in amperes (I)* in those turns.

Mathematically, $AT = NI$.

3. **Reluctance.** Just as the current in an electric circuit is limited by resistance, the flux in the magnetic circuit is limited by the *reluctance* (symbol S) of the magnetic circuit. Thus *reluctance is a measure of opposition offered by a magnetic circuit to the setting up of the flux.*

The reluctance of a magnetic circuit is given by,

$$S = \frac{l}{\mu A} = \frac{l}{\mu_0 \mu_r A} \quad ...(11)$$

where, l = Length of the magnetic circuit,
A = Cross-sectional area of the magnetic circuit,
μ_0 = Absolute permeability, and
μ_r = Relative permeability.

Reluctance *of a magnetic circuit is the ratio of m.m.f. and flux.*

i.e.,
$$\text{Reluctance} = \frac{\text{m.m.f.}}{\text{flux}}$$

or
$$S = \frac{NI}{\phi} \quad ...(12)$$

In other words the reluctance of a magnetic circuit is the number of ampere-turns required per weber of magnetic flux in the circuit. The *unit of reluctance* is *AT/Wb*. Since 1 AT/Wb = 1/henry, the unit of reluctance is *"reciprocal henry"*.

4. **Permeance.** The *reciprocal of reluctance* is known as *permeance.*

i.e.,
$$\text{Permeance} = \frac{1}{\text{Reluctance}} = \frac{1}{S}$$

It is measured in *Wb/AT* or *henry.*

5. **Reluctivity.** It is the *specific reluctance* and corresponds to resistivity which is specific resistance.

Relation between flux density (B) and magnetic field strength (H) :

Comparing eqns. (12) and (13), we get

$$\frac{l}{A\mu_0\mu_r} = \frac{NI}{\phi}$$

or $$\frac{\phi}{A\mu_0\mu_r} = \frac{NI}{l} \qquad \text{(By rearranging)}$$

But $$\frac{\phi}{A} = B,$$

and $$\frac{NI}{l} = H$$

$\therefore$ $$\frac{B}{\mu_0\mu_r} = H$$

or $$B = \mu_0\mu_r H \qquad ...(13)$$

or $$B = \mu H$$

[where μ (permeability) = $\mu_0\mu_r$]

Thus, *permeability is the ratio of flux density to magnetic field strength.*

15. COMPARISON OF ELECTRIC AND MAGNETIC CIRCUITS

The analogy between electric and magnetic circuits is given in the Table 1.

Table 1. Analogy between Electric and Magnetic Circuits (Similarities)

Aspect	Electric circuit	Magnetic circuit
	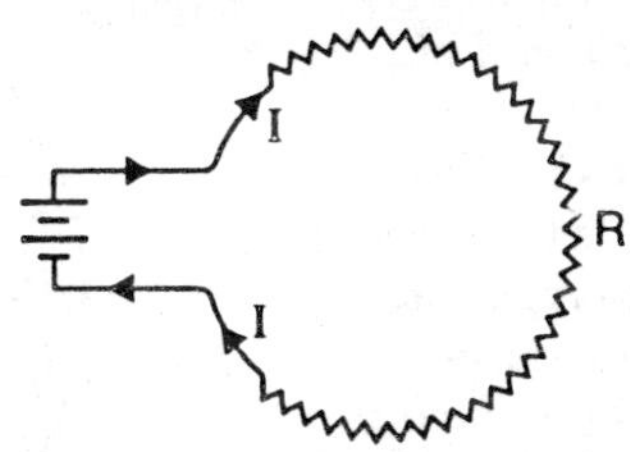 Fig. 17	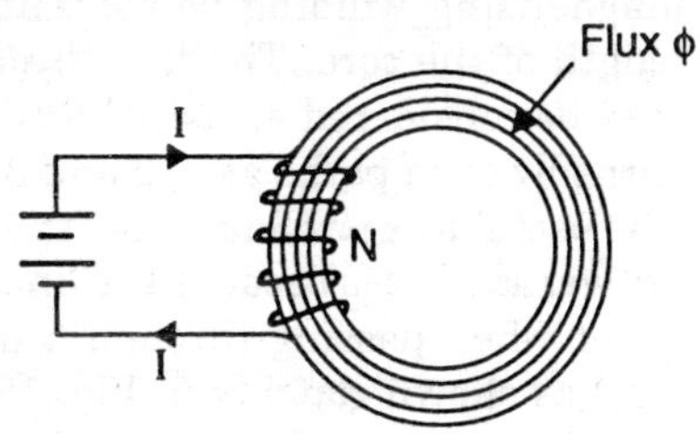 Fig. 18
1. *Equivalent circuit*	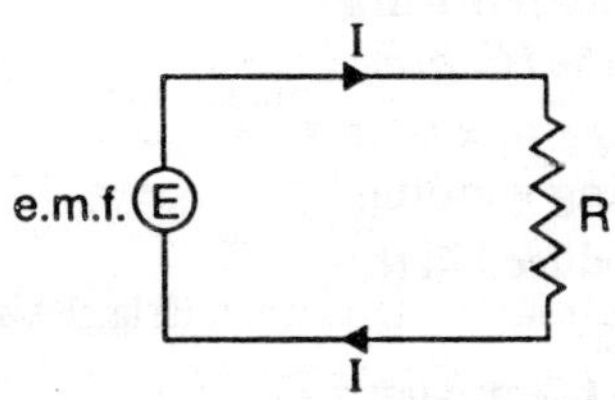	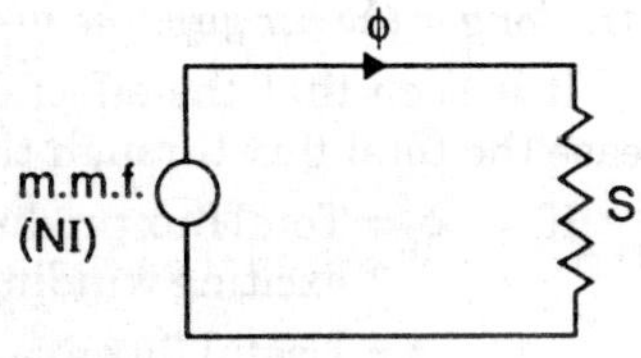
2. *Exciting force*	Battery voltage (E)	Ampere-turns (AT)
3. *Response*	Current (I)	Flux (ϕ)
4. *Ohm's law*	$I = \frac{E}{R}$	$\phi = \frac{\text{m.m.f.}}{\text{reluctance}} = \frac{NI}{S}$

5. *By dimensions*	$R = \rho \dfrac{l}{a}$ (Conductance = 1/R)	$S = \dfrac{l}{A\mu_0\mu_r}$ (Permeance = 1/S)
6. *Proportionality*	ρ	$\dfrac{1}{\mu} = \left(\dfrac{l}{\mu_0\mu_r}\right)$
7. *Field intensity*	Electric field intensity = $\dfrac{E}{l}$ V/m	Magnetic field intensity = $\dfrac{NI}{l}$ AT/m
8. *Density*	Current density (A/m²)	Flux density (Wb/m²)

Differences :

1. The *resistivity* of conductors is more or less *constant* but the *permeability* of the ferromagnetic materials *varies* greatly with magnetic field strength.

2. Strictly speaking, flux does not actually 'flow' in the sense in which an electric current flows.

3. In an electric circuit energy must be supplied to *maintain* the flow of electricity in the circuit, whereas (in a magnetic circuit) the magnetic flux once, it is set up, does not require any further supply of energy.

(**Example.** Once the flux produced by a current in the solenoid has attained its steady value, the energy subsequently absorbed by that solenoid is all dissipated as heat due to the resistance of the winding).

4. In ferromagnetic materials if the magnetic field strength is increased the flux density increases till a state of saturation is reached after which there is no increase in flux density for further increase in field strength. There is no such phenomenon in electric circuits.

16. MAGNETIC LEAKAGE AND FRINGING

Fig. 19 shows a metal ring symmetrically situated relative to the air gap in the iron ring. Let the magnetising winding be concentrated over a short length of the core. The flux through the *metal ring* may be considered as *useful flux* and that which returns by such paths as 1, 2 and 3 is the *leakage flux*. The useful flux passing across the gap tends to buldge outwards (this is because the lines of force repel each other when passing through a non-magnetic material) as shown roughly in Fig. 19, thereby *reducing the effective area of the gap and reducing the flux density in the gap*. This effect is referred to as **fringing** and the *longer the air gap, the greater is the fringing*.

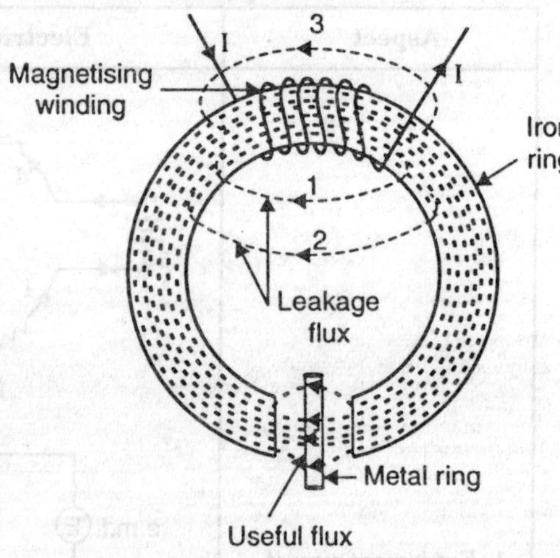

Fig. 19. Magnetic leakage and fringing.

It is seen that the effect of leakage flux is to increase the total flux through the exciting winding.

If ϕ_t = Total flux produced (produced in the exciting winding), and

ϕ = Useful flux available in the air-gap,

Then, leakage coefficient (or factor) $\lambda = \dfrac{\phi_t}{\phi}$

The value of λ for electrical machines is usually about 1.15 to 1.25.

[**Note.** Magnetic leakage can be minimised by placing the exciting coils or windings as close as possible to the air gap or to the points in the magnetic circuit where the flux is to be utilised.]

17. MAGNETISATION (B-H) CURVES

Consider a toroidal solenoid wound on a non-magnetic core, such as shown in Fig. 20. If the flux density is measured on the centre line of the toroid, the relationship between B and H is given by the straight line OA in Fig. 21. If now the space within the toroid is filled with an unmagnetised ferromagnetic material, the well known magnetisation curve $OBCDE$, is obtained. The magnetisation has many names being referred to as :

B-H curve, the magnetic saturation c... ...irgin curve or simply the saturation curve.

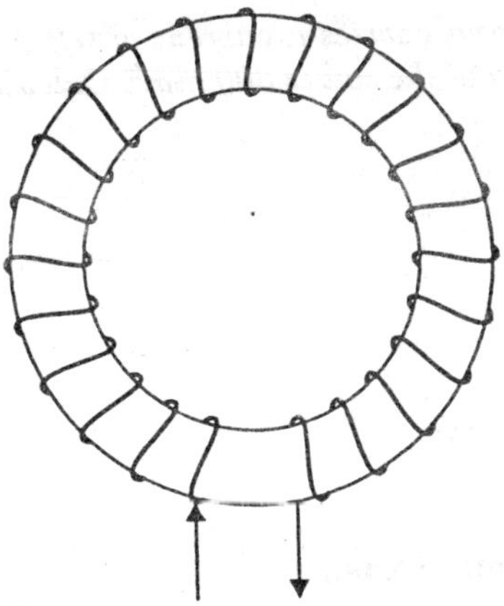

Fig. 20. Toroid.

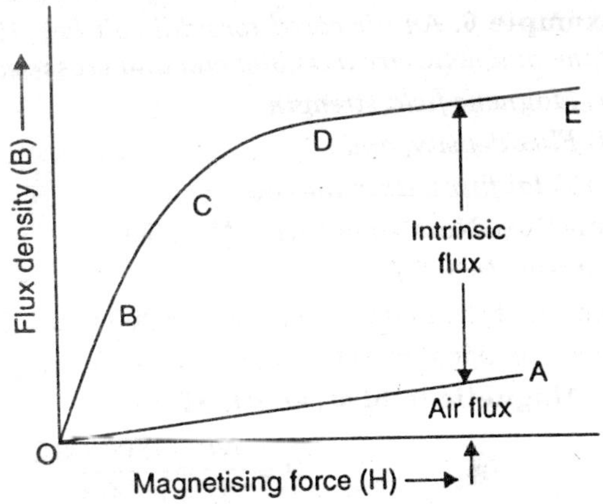

Fig. 21. Magnetising curve for a ferromagnetic material.

The difference in flux between the saturation curve and the air line, OA at any magnetising force, is due to the contribution of the magnetic material. This flux is known as the *intrinsic flux* and gives a true measure of magnetic properties than does the total flux, especially at very high m.m.fs. The point at which the intrinsic flux density curve becomes horizontal gives the intrinsic saturation.

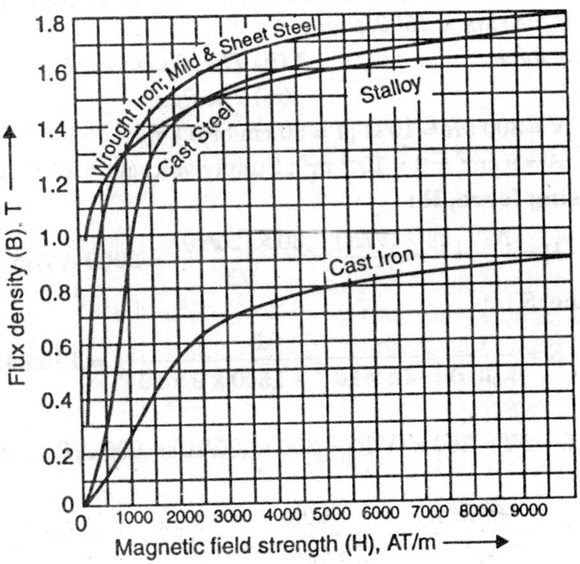

Fig. 22. *B-H* curves.

The magnetisation curves can be determined by the following methods provided the materials are in the form of a ring :

(*i*) By means of a ballistic galvanometer

(*ii*) By means of a fluxmeter.

The graphs in Fig. 22 show the relationship between the flux density (B) and the magnetic field strength obtained for different qualities of iron. The data for mild steel, wrought iron and sheet steel are so similar that they can be represented by a common graph. *Stalloy* is an alloy of iron and silicon commonly used in the construction of transformers and A.C. machines.

Example 6. *An air cored toroidal coil has 1500 turns and carries a current of 0.2 A. The length of the magnetic circuit is 300 mm and cross-sectional area of the coil is 800 mm². Calculate :*

(*i*) *Magnetic field strength,*

(*ii*) *Flux density, and*

(*iii*) *Total flux within the coil.*

Solution. Number of turns, $N = 1500$

Current, $I = 0.2$ A

Length of the magnetic circuit, $l = 300$ mm $= 0.3$ m

Cross-sectional area of the coil, $A = 800\text{ mm}^2 = 800 \times 10^{-6}\text{ m}^2$

(*i*) **Magnetic field strength, H :**

$$H = \frac{NI}{l} = \frac{1500 \times 0.2}{0.3} = \mathbf{1000\ AT/m.}\quad \textbf{(Ans.)}$$

(*ii*) **Flux density, B :**

$$B = \mu_0 \mu_r H$$

$$= \mu_0 H \qquad (\because \text{ For air cored coil, } \mu_r = 1)$$

$$= 4\pi \times 10^{-7} \times 1000 = \mathbf{1.25 \times 10^{-3}\ T.}\quad \textbf{(Ans.)}$$

(*iii*) **Total flux within the coil, ϕ :**

$$\phi = B \times A$$

$$= 1.256 \times 10^{-3} \times 800 \times 10^{-6} = \mathbf{1.005 \times 10^{-6}\ Wb.}\quad \textbf{(Ans.)}$$

Example 7. *A coil of 300 turns and of resistance 10 Ω is wound uniformly over a steel ring of mean circumference 30 cm and cross-sectional area 9 cm². It is connected to a supply at 20 V D.C. If the relative permeability of the ring is 1500, find :*

(*i*) *The magnetising force,* (*ii*) *The reluctance,*

(*iii*) *The m.m.f., and* (*iv*) *The flux.* **(VTU, 1999)**

Solution. *Given* : $N = 300$; $R = 10\ \Omega$; $l = 30$ cm or 0.3 m ;

$A = 9\text{ cm}^2 = 9 \times 10^{-4}\text{ m}^2$; Supply voltage, $= 20$ V ; $\mu_r = 1500$.

(*i*) **The magnetising force, H :**

$$H = \frac{NI}{l} = \frac{N \times (V/R)}{l} = \frac{30 \times (20/10)}{0.3} = \mathbf{2000\ AT/m.}\quad \textbf{(Ans.)}$$

(*ii*) **The reluctance, S :**

$$S = \frac{l}{\mu_0 \mu_r A} = \frac{0.3}{4\pi \times 10^{-7} \times 1500 \times 9 \times 10^{-4}} = \mathbf{176.84 \times 10^3\ AT/Wb.}\quad \textbf{(Ans.)}$$

(*iii*) **The m.m.f. :**

$$\text{m.m.f.} = NI = 300 \times (V/R) = 300 \times (20/10) = \mathbf{600\ AT.}\quad \textbf{(Ans.)}$$

(*iv*) **The flux, ϕ :**

$$\phi = \frac{\text{m.m.f.}}{\text{reluctance}} = \frac{600}{176.84 \times 10^3} = 3.39 \times 10^{-3}\text{ Wb} = \mathbf{3.39\ mWb.}\quad \textbf{(Ans.)}$$

Example 8. *A magnetic circuit consists of an iron ring of mean circumference 80 cm with cross-sectional area of 12 cm² throughout. A current of 1 A in the magnetising coil of 200 turns produces a total flux of 1.2 mWb in the iron. Calculate :*

(*i*) *The flux density in the iron.*

(*ii*) *The absolute and relative permeability of iron.*

(*iii*) *The reluctance of the circuit.* **(P.U., 2000)**

Solution. *Given :* $l = 80$ cm $= 0.8$ m ; $a = 12$ cm² $= 12 \times 10^{-4}$ m² ;

$I = 1$ A ; $N = 200$; $\phi = 1.2$ mWb $= 0.0012$ Wb.

(*i*) **The flux density in the iron, B :**

$$B = \frac{\phi}{A} = \frac{0.0012}{12 \times 10^{-4}} = \mathbf{1\ Wb/m^2.\ (Ans.)}$$

(*ii*) **The absolute and relative permeability of iron, μ, μ_r :**

The magnetising force, $H = \dfrac{NI}{l} = \dfrac{200 \times 1}{0.8} = 250$ AT/m

Absolute permeability, $\mu = \mu_0\mu_r = \dfrac{B}{H}$ or $\mu = \dfrac{1}{250} = \mathbf{4 \times 10^{-3}}$ **unit. (Ans.)**

and $$\mu_r = \frac{\mu}{\mu_0} = \frac{4 \times 10^{-3}}{4\pi \times 10^{-7}} = \mathbf{3183.\ (Ans.)}$$

(*iii*) **The reluctance of the circuit, S :**

$$S = \frac{l}{\mu_0\mu_r A} = \frac{0.8}{4\pi \times 10^{-7} \times 3183 \times 12 \times 10^{-4}}$$

$$= \mathbf{166.67 \times 10^3\ AT/Wb.\ (Ans.)}$$

Example 9. *An iron ring 200 mm mean diameter is made of 30 mm round iron of permeability 900, has an air gap 10 mm wide. It has 800 turns. If the current flowing through this winding is 6.8 amperes, determine :*

(*i*) *The magnetomotive force.*

(*ii*) *The total reluctance of the circuit.*

(*iii*) *The flux in the ring.*

(*iv*) *The flux density in the ring.*

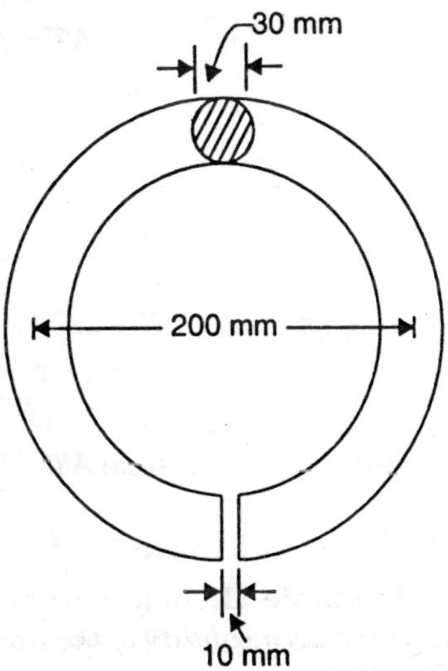

Fig. 23

Solution. Length of air gap, $l_g = 10$ mm $= 0.01$ m

Mean dia. of iron ring = 200 mm = 0.2 m

$\therefore$ Mean length of iron path,

$$l_i = (\pi \times 0.2 - 0.01)$$
$$= 0.618 \text{ m}$$

Area of cross-section, $A = \dfrac{\pi}{4} \times 0.03^2$

$= 7.068 \times 10^{-4}$ m²

Number of turns, $N = 800$

Permeability of iron, $\mu_r = 900$

Current through the winding, $I = 68$ A.

(*i*) **Magnetomotive force, m.m.f. :**

m.m.f. $= NI$ ampere-turns

$= 800 \times 6.8 =$ **5440 AT. (Ans.)**

(ii) Total reluctance of the circuit, S :

$$S = \frac{l_i}{\mu_0 \mu_r A} + \frac{l_g}{\mu_0 A} \qquad (\because \mu_r \text{ for air} = 1)$$

$$= \frac{0.618}{4\pi \times 10^{-7} \times 900 \times 7.068 \times 10^{-4}} + \frac{0.01}{4\pi \times 10^{-7} \times 7.068 \times 10^{-4}}$$

$$= \frac{1}{4\pi \times 10^{-7} \times 7.068 \times 10^{-4}} \left(\frac{0.618}{900} + 0.01 \right)$$

$$= \frac{1}{88.82 \times 10^{-11}} \times 0.0107 = \mathbf{0.12 \times 10^8 \ AT/Wb.} \ \textbf{(Ans.)}$$

(iii) Flux in the ring, ϕ :

$$\phi = \frac{\text{m.m.f.}}{\text{reluctance}} = \frac{5440}{0.12 \times 10^8} = \mathbf{4.533 \times 10^{-4} \ Wb.} \ \textbf{(Ans.)}$$

(iv) Flux density in the ring, B :

$$B = \frac{\phi}{A} = \frac{4.533 \times 10^{-4}}{7.068 \times 10^{-4}} = \mathbf{0.64 \ T.} \ \textbf{(Ans.)}$$

Example 10. *An iron of 15 cm diameter and 10 cm² cross-section with a saw cut 2 mm wide is wound with 200 turns of wire. The flux density is 1 Wb/m². The relative permeability of iron is 600. Find the exciting current. Neglect leakage.* **(K.U.)**

Solution. *Given :* $l_i = \pi \times 15 - \frac{2}{10} = 46.72$ cm $= 0.4692$ m ;

$l_g = 2$ mm $= 0.002$ m ; $N = 200$; $B = 1$ Wb/m², $\mu_r = 600$.

Exciting current, I :

Total AT (ampere-turns) required

$= AT$ required for iron ring (AT_i) + AT required for air gap (AT_g)

i.e.,

$$AT = AT_i + AT_g$$

$$= H_i l_i + H_g l_g$$

$$= \frac{B_i}{\mu_0 \mu_r} l_i + \frac{B_g}{\mu_0} l_g$$

$$= \frac{1}{4\pi \times 10^{-7} \times 600} \times 0.4692 + \frac{1}{4\pi \times 10^{-7}} \times 0.002 \qquad (\because B_i = B_g = 1 \text{ Wb/m}^2)$$

$$= \frac{1}{4\pi \times 10^{-7}} \left(\frac{0.4692}{600} + 0.002 \right) = 2213.84$$

But total $AT = NI$

$$\therefore \quad I = \frac{2213.84}{200} = \mathbf{11.07.} \ \textbf{(Ans.)}$$

Example 11. *An iron ring of mean length 1 metre has an air gap of 2 mm and a winding 400 turns. If the permeability of the iron is 400 when a current of 2.5 amperes flow through the coil, find the flux density.* **(Indore University)**

Solution. Mean length of iron ring (or iron path), $l_i = 1$ m (neglecting air gap length)

Length of air gap, $l_g = 2$ mm $= 0.002$ m

Number of turns, $N = 400$

Permeability of iron, $\mu_r = 400$
Current through the coil, $I = 2.5$ A

Flux density, B :

Total AT required = AT required for iron ring (AT_i) + AT required for air gap (AT_g)

i.e.,
$$AT = AT_i + AT_g$$
$$= H_i l_i + H_g l_g$$
$$= \frac{B}{\mu_0 \mu_r} l_i + \frac{B}{\mu_0} l_g$$

i.e.,
$$AT = \frac{B}{\mu_0 \times 400} \times 1 + \frac{B}{\mu_0} \times 0.002$$
$$= \frac{B}{\mu_0}\left(\frac{1}{400} + 0.002\right) = 0.0045 \frac{B}{\mu_0}$$

No. of AT supplied = $NI = 400 \times 2.5 = 1000$

$$\therefore \quad 0.0045 \frac{B}{\mu_0} = 1000$$

and
$$B = \frac{1000\,\mu_0}{0.0045} = \frac{1000 \times 4\pi \times 10^{-7}}{0.0045} = \mathbf{0.279\ T.\ (Ans.)}$$

Example 12. *An iron ring of mean circumference of 200 mm and cross-sectional area of 250 mm² has flux of 500 μWb. Calculate the m.m.f. acting on the iron ring. An air gap of 2 mm is now cut in the ring. Find the flux in the ring if the m.m.f. remains same. A relative permeability of 1200 may be assumed.* **(Roorkee University)**

Solution. *Given :* $l_i = 900 - 2 = 898$ mm $= 0.898$ m ; $A = 250$ mm² $= 250 \times 10^{-6}$ m²
$\phi = 500$ μWb $= 500 \times 10^{-6}$ Wb ; $l_g = 2$ mm $= 0.002$ m ; $\mu_r = 1200$.

(*i*) **Before the air gap is cut :**

$$\phi = \frac{\text{m.m.f.}}{\text{reluctance}} = \frac{NI}{l/\mu_0 \mu_r A}$$

$$\therefore \quad \textbf{M.m.f.,}\ NI\ (i.e., AT) = \frac{\phi l}{\mu_0 \mu_r A} = \frac{500 \times 10^{-6} \times 0.9}{4\pi \times 10^{-7} \times 1200 \times 250 \times 10^{-6}} = \mathbf{1193.7\ AT.\ (Ans.)}$$

(*ii*) **After the air gap is cut :**

$$\text{Total } AT = AT_i + AT_g$$

or
$$1193.7 = \frac{B_i}{\mu_0 \mu_r} l_i + \frac{B_g}{\mu_0} l_g$$
$$= \frac{B}{\mu_0}\left(\frac{l_i}{\mu_r} + l_g\right) \qquad (\because\ B_i = B_g = B)$$
$$= \frac{\phi}{\mu_0 A}\left(\frac{l_i}{\mu_r} + l_g\right)$$

or
$$1193.7 = \frac{\phi}{4\pi \times 10^{-7} \times 250 \times 10^{-6}}\left(\frac{0.898}{1200} + 0.002\right)$$
$$= \frac{\phi}{3141.6 \times 10^{-13}} \times 0.00275$$

$$\therefore \qquad \phi = \frac{1193.7 \times 3141.6 \times 10^{-13}}{0.00275} = 1.36 \times 10^{-4} \text{ or } 136 \times 10^{-6} \text{ Wb}$$

$$= \mathbf{136\ \mu Wb.\ (Ans.)}$$

Example 13. *A ring has a diameter of 24 cm and a cross-sectional area of 1000 mm². The ring is made up of semicircular sections of cast iron and cast steel with each joint having a reluctance equal to an air gap of 0.2 mm. Find the ampere turns required to produce a flux of 8 × 10⁻⁴ Wb. The relative permeability of cast-steel and cast iron are 900 and 170 respectively. Neglect fringing and leakage effects.*

(Gujrat University)

Solution. *Given :* $\phi = 8 \times 10^{-4}$ Wb ; $A = 1000 \text{ mm}^2 = 1000 \times 10^{-6} = 0.001 \text{ m}^3$;

Total length of air gap, $l_g = 2 \times 0.2 = 0.4 \text{ mm} = 0.0004 \text{ m}$; $(\mu_r)_{c.i.} = 900$; $(\mu_r)_{c.s.} = 170$.

$$AT = (AT)_{c.i} + (AT)_g + (AT)_{c.s.}$$

$$= \frac{B}{\mu_0(\mu_r)_{c.i.}} l_{c.i.} + \frac{B}{\mu_0} . l_g + \frac{B}{\mu_0(\mu_r)_{c.s.}} l_{c.s.}$$

$$(\because \quad B_{c.i.} = B_g = B_{i.s.} = B)$$

or
$$AT = \frac{B}{\mu_0}\left[\frac{l_{c.i.}}{(\mu_r)_{c.i.}} + l_g + \frac{l_{c.s.}}{(\mu_r)_{c.s.}}\right] \qquad ...(i)$$

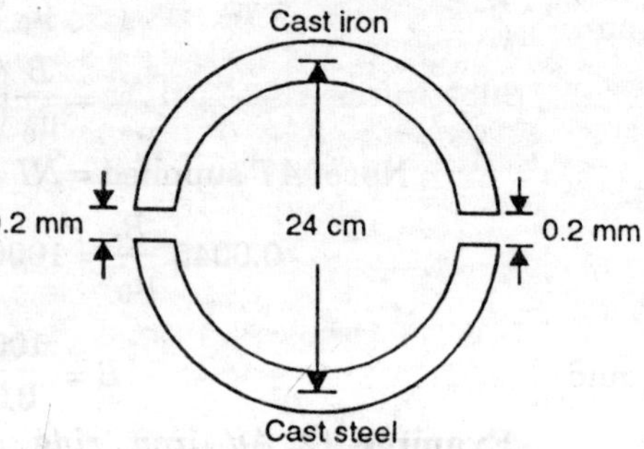

Fig. 24

Here, $\quad B = \frac{\phi}{A} = \frac{8 \times 10^{-4}}{0.001} = 0.8 \text{ Wb/m}^2$

$$l_{c.i.} = l_{c.s.} = \frac{\pi D}{2} = \frac{\pi \times 0.24}{2} = 0.377 \text{ m}$$

Substituting the values in eqn. (*i*), we get

$$AT = \frac{0.8}{4\pi \times 10^{-7}}\left[\frac{0.377}{900} + 0.0004 + \frac{0.377}{170}\right] = \mathbf{1933.\ (Ans.)}$$

Example 14. *A magnetic ring has a mean circumference of 1.8 m and is of 0.012 m² in cross-section, and is wound with 210 turns. A saw cut of 4.8 mm width is made in the ring. Calculate the magnetising current required to produce a flux of 0.8 milliweber in the air gap.*

Assume leakage factor as 1.25 and permeability of iron as 400.

Solution. Mean circumference of the ring = 1.8 m

Length of air gap, $\quad l_g = 4.8 \text{ mm} = 0.0048 \text{ m}$

$\therefore$ Length of iron path $\quad = 1.8 - l_g = 1.8 - 0.0048 = 1.795 \text{ m}$

Area of cross-section, $\quad A = 0.012 \text{ m}^2$

Flux in the air gap, $\quad \phi_g = 0.8 \text{ m Wb}$

Number of turns, $\quad N = 210$

Permeability of iron, $\quad \mu_r = 400$

Leakage factor, $\quad \lambda = 1.25$

Magnetising current I :

Total ampere-turns required (AT) = AT required for iron ring + AT required for air gap

i.e.,
$$AT = AT_i + AT_g$$

$$= H_i l_i + H_g l_g$$

$$= \frac{B_i}{\mu_0 \mu_r} l_i + \frac{B_g}{\mu_0} . l_g \qquad ...(i)$$

Now, leakage factor $\lambda = \frac{\phi_i}{\phi_g} = 1.25$ (given)

[where ϕ_i = flux in the iron path, and ϕ_g = flux in the air gap]

$$\therefore \quad \frac{\phi_i}{0.8 \times 10^{-3}} = 1.25$$

$$\phi_i = 0.8 \times 10^{-3} \times 1.25 = 1 \times 10^{-3} \text{ Wb}$$

Hence, $$B_i = \frac{\phi_i}{A} = \frac{1 \times 10^{-3}}{0.012} = 0.0833 \text{ T}$$

and $$B_g = \frac{\phi_g}{A} = \frac{0.8 \times 10^{-3}}{0.012} = 0.0666 \text{ T}$$

Substituting these values in eqn. (*i*), we get

$$AT = \frac{0.0833}{4\pi \times 10^{-7} \times 400} \times 1.795 + \frac{0.0666}{4\pi \times 10^{-7}} \times 0.0048$$

$$= 297.5 + 254.4 = 551.9$$

i.e., $$NI = 551.9$$

$$\therefore \quad \mathbf{I} = \frac{551.9}{N} = \frac{551.9}{210} = \mathbf{2.63\ A.\ (Ans.)}$$

Example 15. *A rectangular core shown in Fig. 25 has a mean length of magnetic path of 120 cm, cross-section of (3 cm × 3 cm), relative permeability of 1400 and an air gap of 4 mm cut in the core. The three coils carried by the core have number of turns $N_1 = 350$, $N_2 = 500$ and $N_3 = 500$; and the respective currents are 1.5 A, 3.5 A and 2.5 A. The directions of the currents are as shown. Find the flux in the air gap.* **(Pune University)**

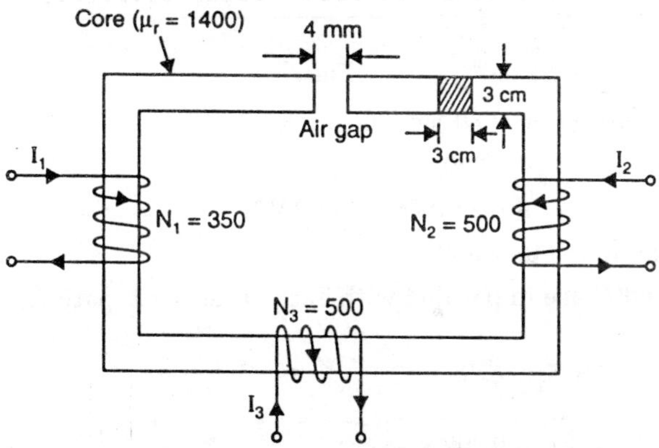

Fig. 25

Solution. *Given* : $l_i = 1.2 - 0.004 = 1.196$ m ; $\mu_r = 1400$; $l_g = 4$ mm $= 0.004$ m ;

$A = 3 \times 3 = 9 \text{ cm}^2 = 9 \times 10^{-4} \text{ m}^2$

$N_1 = 350$; $N_2 = 500$; $N_3 = 500$; $I_1 = 1.5$ A ; $I_2 = 3.5$ A ; $I_3 = 2.5$ A.

Flux in the air gap, ϕ_g :

By applying the Right-hand thumb rule, it is found that fluxes produced by the currents I_1 and I_2 are directed in the *clockwise* direction through the iron core whereas that produced by current I_3 is directed in the *anti-clockwise* direction through the core.

$\therefore$ Total m.m.f. $= N_1I_1 + N_2I_2 - N_3I_3$

$= 350 \times 1.5 + 500 \times 3.5 - 500 \times 2.5 = 1025\ AT$

Reluctance of the air gap $= \dfrac{l_g}{\mu_0 A} = \dfrac{0.004}{4\pi \times 10^{-7} \times 9 \times 10^{-4}} = 3.536 \times 10^6\ AT/Wb$

Reluctance of the iron path $= \dfrac{l_i}{\mu_0 \mu_r A} = \dfrac{1.196}{4\pi \times 10^{-7} \times 1400 \times 9 \times 10^{-4}} = 0.755 \times 10^6\ AT/Wb$

Total reluctance $= (3.536 + 0.755) \times 10^6 = 4.291 \times 10^6\ AT/Wb$

The flux in the air gap is the same as in the iron core.

Air gap flux, $\phi_g = \dfrac{\text{m.m.f.}}{\text{reluctance}} = \dfrac{1025}{4.291 \times 10^6} = 238.9 \times 10^{-6} =$ **238.9 μWb. (Ans.)**

Example 16. *A cast steel magnetic structure made from a bar of section 2 cm × 2 cm is shown in Fig. 26. Determine the current that the 600 turn magnetising coil on the left limb should carry so that a flux of 2 mWb is produced in the right limb. Take μ_r = 600 and neglect leakage.*

(Allahabad University)

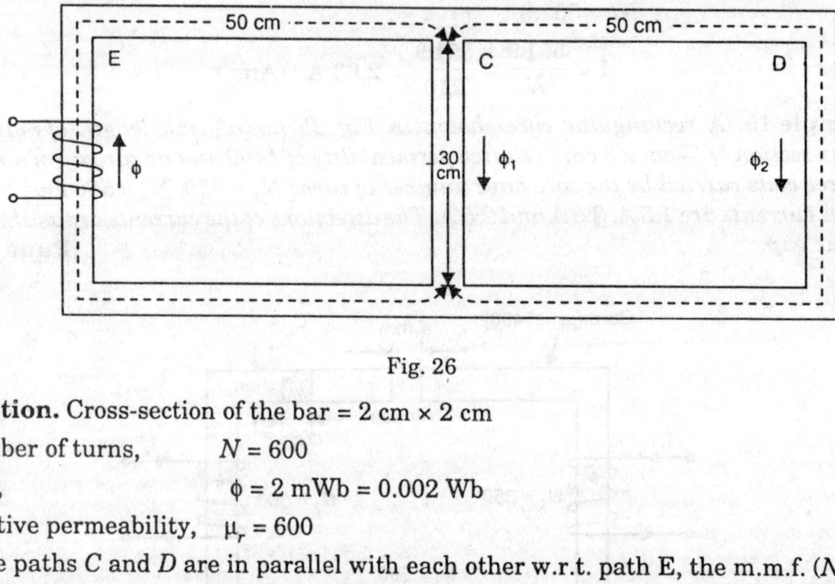

Fig. 26

Solution. Cross-section of the bar = 2 cm × 2 cm

Number of turns, $N = 600$

Flux, $\phi = 2\ \text{mWb} = 0.002\ \text{Wb}$

Relative permeability, $\mu_r = 600$

Since paths C and D are in parallel with each other w.r.t. path E, the m.m.f. (NI) across the two is the same.

i.e., $\phi_1 S_1 = \phi_2 S_2$

$$\phi_1 \times \frac{30}{\mu A} = 0.002 \times \frac{50}{\mu A} \qquad \left(\because\ \phi = \frac{NI}{S}\right)$$

$\therefore$ $\phi_1 = 0.00333$ Wb

$\therefore$ $\phi = \phi_1 + \phi_2 = 0.00333 + 0.002 = 0.00533$ Wb.

Total AT required $= AT$ required for path E

$+ AT$ required for either of the two paths C or D

$$= \frac{Bl}{\mu_0 \mu_r} + \frac{B_2 l_2}{\mu_0 \mu_r} = \frac{1}{\mu_0 \mu_r A}(\phi l + \phi_2 l_2) \qquad \left(\because\ B = \frac{\phi}{A}\right)$$

$$= \frac{1}{4\pi \times 10^{-7} \times 600 \times (2 \times 2 \times 10^{-4})} (0.00533 \times 0.5 + 0.002 \times 0.5)$$

$$= 3.316 \times 10^6 \times 3.665 \times 10^{-3} = 12153$$

$\therefore$ Current required, $I = \frac{12153}{600}$ = **20.26 A. (Ans.)**

18. LOSSES IN MAGNETIC MATERIALS

When magnetic materials are subjected to an alternating flux, two types of losses occur which are :

1. Hysteresis loss
2. Eddy current loss.

The above losses comprise the *total core loss*.

18.1. Hysteresis Loss

- Below *curie temperature* (it is the rising temperature at which the given material ceases to be ferromagnetic, or the falling temperature at which it becomes ferromagnetic) all ferromagnetic materials exhibit the phenomenon called **hysteresis** which is defined as *the lagging of magnetisation or induction flux density (B) behind the magnetising force (H) or it is that quality of a magnetic substance due to which energy is dissipated in it on the reversal of its magnetism*.
- Fig. 27 shows a typical hysteresis loop. It is a curve plotted between *B* and *H* for various values of *H* from a maximum value in the positive direction to maximum value in the negative direction and back again.

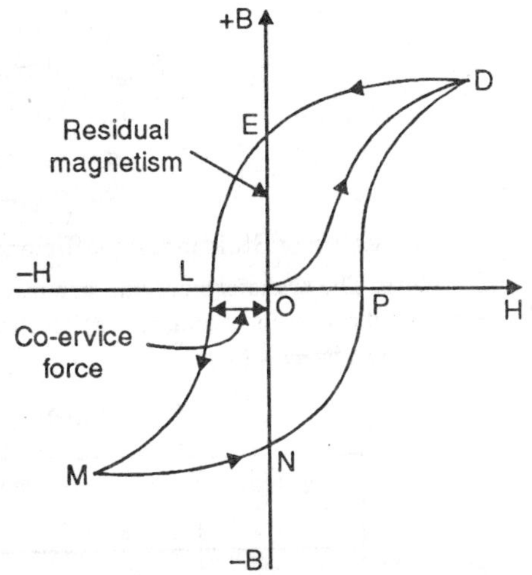

Fig. 27. Hysteresis loop for a ferromagnetic material.

— Starting at zero with a coil wound round a toroid of unmagnetised iron, the magnetisation curve follows curve *OD*. If the m.m.f. is gradually reduced, the flux curve follows the line *DE*. As the m.m.f. is reversed the flux falls into the point *L* and thence to *M*. As the m.m.f. is returned to zero, the flux traces out the path *MN*. Then, as m.m.f. is again increased, curve NPD is followed. *The area within the closed loop is a measure of energy lost during the cycle*. This energy is, in effect, a frictional loss and shows up as heat in the material. *The distance OE is a measure of the residual flux* left in the closed magnetic circuit when the current is zero. It is known as B_r, the *residual induction*. B_r describes circuits in which there are no air gaps, *e.g.* the iron toroid, should not be confused with *remanance*, a more general term which refers to magnetic induction remaining in the magnetic circuit (usually in the air gap when one is present) after magnetising force has been removed.

— The distance *OL* is known as $-H_c$, the *co-ercive force*, and is *the value of the demagnetising m.m.f. required to bring the residual or remanent magnetic induction to zero when such a loop is being traced out*.

- If a ferromagnetic substance is subjected to an alternating m.m.f., the first hysteresis loops traced out do not necessarily fall upon each other. When successive loops retrace preceding ones, the material is said to be in a cyclically magnetised condition. *For electro-magnet core materials, values of B_r and $-H_c$ are determined from a hysteresis loop taken when material is cyclically magnetised. Permanent magnet values, however are taken from the first hysteresis loop, since permanent magnets need be magnetised only once.*
- The *hysteresis loop equals the work which is necessary to reverse the direction of magnetisation*. The actual shape and area of loop depend on the internal structure and composition of the ferromagnetic substance.

— Thus *work done* (W) = (*area of B-H loop*) joules/m³/cycle.

It may be noted that while calculating the actual area, scales of B and H should be taken into consideration.

For example, if scales are :

$1 \text{ cm} = x \text{ AT/m}$ ... for H

$1 \text{ cm} = y \text{ Wb/m}^2$ ... for B

then $W = x \,.\, y$ (area of B-H loop) joules/m³/cycle.

— *Steinmetz* developed an empirical relationship to express this loss in following terms.

$$P_h = K_h f B_m^{\ k} \quad ...(14)$$

In this expression

P_h = Hysteresis loss in watt per m³ or per kg,

B_{max} = Maximum flux density, Wb/m²,

K_h = Hysteresis coefficient,

k = Steinmetz coefficient, and

f = Frequency of magnetization, Hz.

The value of Steinmetz coefficient k is approximately 2 *for all modern magnetic materials*.

Note. The transformers and generators cores and armatures of the electric motors etc. which are subjected to rapid reversals of magnetisation should be made of such substances which have *low coefficient in order to reduce the hysteresis loss.*

Table 2. Hysteresis Coefficients

S. No.	Material	Hysteresis coefficient K_h (joules/m³) × 10^2
1.	Cast iron	27.63 to 40.2
2.	Sheet iron	10.05
3.	Cast steel	7.54 to 30.14
4.	Hard cast steel	63 to 70.34
5.	Silicon steel (4.8 % Si)	1.91
6.	Hard tungsten steel	145.7
7.	Good dynamo sheet steel	5.02
8.	Mild steel castings	7.54 to 22.61
9.	Nickel	32.66 to 100.5
10.	Permalloy	0.25

18.2. Eddy Current Losses

- The term "eddy currents" is applied to *those electric currents which circulate within a mass of conducting material when the latter is situated in a varying magnetic field.* The conducting material may be considered as consisting of large number of closed conducting paths, each of which behaves like a short circuited winding. The varying magnetic field induces eddy e.m.fs. in these closed elemental paths giving rise to eddy currents. These eddy currents produce loss in power resulting in heating of materials. This loss is of considerable importance as it effects the efficiency and heating of electrical machines.
- *The eddy currents produce a magnetic field of their own which opposes the main magnetic field. As the effect of eddy currents is not uniform over the cross section of the material, this results in a flux distribution which is not uniform, the flux density in the outer portions being greater than that at the centre. Thus there is a reduction in effective cross-section. The effect of eddy currents upon flux distribution is chiefly of importance in transformers where the material otherwise would be worked at a uniform flux density.*
- The magnetic materials used for varying magnetic fields are laminated (made up of thin sheets insulated from each other) so as to reduce eddy currents and associated losses, as by laminating, the *area of path of eddy currents is reduced giving rise to a large value of resistance.*

Eddy Current Loss in Thin Sheets :

Fig. 28 shows a thin plate of thickness t and width b, thickness being *considerably smaller than the width*. Let us suppose that this sheet carries a flux, $B = B_{max} \sin \omega t$, the field running parallel to the axis of the sheet. Eddy currents would flow in the sheet in the elemental paths as shown in Fig. 28.

Let us consider 1 metre axial length and also 1 metre width of sheet. Take an elemental path of thickness dx at a distance x from the axis. Flux enclosed by 1 metre length of this path,

$$\phi = B \text{ (flux density)} \times 2x \times 1 \text{ (area)} = 2xB$$
$$= 2xB \sin \omega t$$

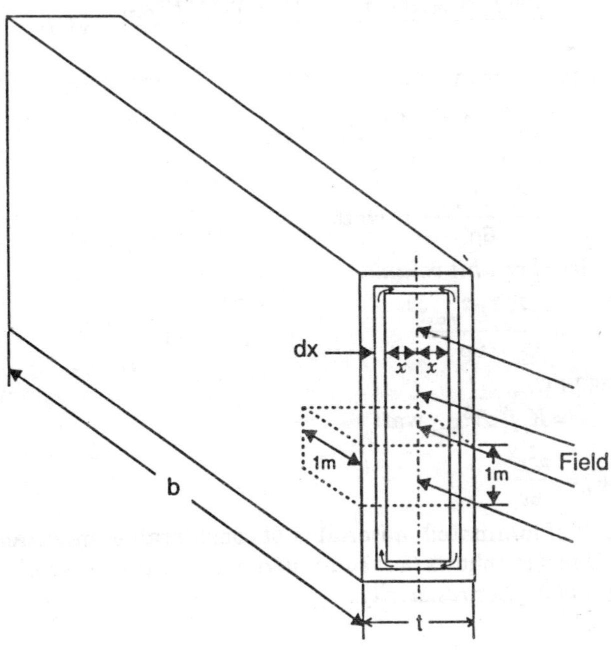

Fig. 28. Eddy currents in lamination.

Instantaneous value of eddy e.m.f. induced in the elemental path,

$$e_x = \frac{d\phi}{dt} = \frac{d}{dt}(2x\,B_{max}\sin\omega t)$$
$$= 2x\,B_{max}\cos\omega t$$
$$= 2x\,\omega\,B_{max}\cos\omega t.$$

R.m.s. value of eddy e.m.f. in the elemental path,

$$E_{r.m.s.} = \frac{1}{\sqrt{2}} \times E_{max}$$
$$= \frac{1}{\sqrt{2}} \times 2x\,\omega\,B_{max} \qquad [\because \quad E_{max} = 2x\omega\,B_{max} \text{ i.e., when } \cos\omega t = 1]$$
$$= \sqrt{2}x\,\omega\,B_{max} = \sqrt{2}x \times 2\pi f\,B_{max}$$
$$= 2\sqrt{2}x\,\pi f\,B_{max}$$

If ρ is the resistivity of material, the resistance of each elemental path

$$R_{xe} = \frac{2\times\rho}{1\times dx} = \frac{2\rho}{dx}$$

(Neglecting portions of path parallel to edges of plates).

$\therefore$ Eddy current in elemental path,

$$I_{xe} = \frac{E_{xe}}{R_{xe}} = \frac{2\sqrt{2}x\,\pi f\,B_{max}}{2\rho/dx}$$
$$= \frac{\sqrt{2}\,\pi f\,B_{max}x\,dx}{\rho}$$

Eddy current loss in this path with 1 metre length and 1 metre width is

$$P_{xe} = I_{xe}^{\,2}\,R_{xe}$$
$$= \frac{2\pi^2 f^2 B^2{}_{max}(x\,dx)^2}{\rho^2} \times \frac{2\rho}{dx} = \frac{4\pi^2 f^2 B^2{}_{max}}{\rho}\,.\,x^2\,dx$$

$\therefore$ Total eddy current loss in a plate 1 metre wide, 1 metre long and t metre thick is

$$= \int_{x=0}^{x=t/2} \frac{4\pi^2 f^2 B^2{}_{max}}{\rho}\,.\,x^2\,dx$$
$$= \frac{\pi^2 f^2 B^2{}_{max}}{6\rho}\,t^3 \text{ watt} \qquad ...(15)$$

$\therefore$ Eddy current loss *per unit volume*

$$= \frac{\pi^2 f^2 B^2{}_{max} t^2}{6\rho} \qquad ...(16)$$

(the volume being $t\ m^3$)

$$= K_e f^2\,B^2{}_{max} \text{ watt} \qquad ...(17)$$

where
$$K_e = \frac{\pi^2 t^2}{6\rho} \qquad ...(18)$$

— The thickness t of laminated material is of considerable importance as regards to eddy current loss. Eqn. (16) shows *that eddy current losses are reduced by using thinner plates and a material of higher resistivity.*

— Eqn. (16) is derived by *assuming that magnetic field is uniform throughout and parallel to the axis and also the sheets are very thin*. Therefore, as such, it *should not be applied to cases where frequency is abnormal, the sheets thick or the variation of flux density is not sinusoidal with time.*

The *eddy current loss can be* **reduced** in the following ways :

1. It has been found that this loss can be minimised by building up the required cross-section for the flux path by stacking thin pieces known as "*laminations*" as shown in Fig. 29. Since the e.m.fs. set up in the material by the varying flux are usually of small magnitude, the natural oxide on the surface of the sheet iron or steel from which the laminations are punched will effectively insulate the laminations from one another, and thus limit each eddy current path to a single lamination.

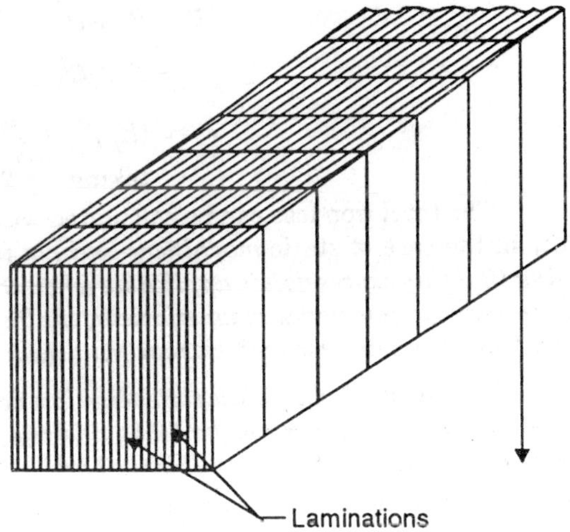

Fig. 29

2. This loss may also be reduced by grinding the ferromagnetic material to a powder and mixing it with a binder that effectively insulates the particles from one another. This mixture is then formed under pressure into the desired shape and heat treated. Magnetic cores for use in communication equipment are frequently made by this process.

- It may be noted that *iron losses* (hysteresis and eddy current losses) *if allowed to take place unchecked, not only reduce the efficiency of electrical equipment but also raise the temperature of the core*. Hence these losses should be kept *as small as is economically possible*.

Other Factors Affecting Core Loss

1. *Rivets and bolts*. Laminations must be held together. A common method is to use rivets or bolts, but they *tend to short circuit lamination insulation*. Rivets and bolts very often form part of a closed magnetic loop which, under A.C. operation, may *cause a serious loss of electrical energy*.

2. *Burrs*. As dies become worn, the burrs that they form on punched edges of lamination stock become larger and larger. Thus burrs are not always removed by deburring operation, and they *cut through sheet insulation forming contacts between laminations and allow excessive eddy currents to flow*. These are commonly referred to as *intersheet eddy currents*. Machining operations that are performed on stacked laminations, such as the grinding of induction motor stators or broaching of solenoid faces, often smear the lamination together, forming good connections between sheets. Many schemes, such as treating the faces with acid or alkali, have been tried in an attempt to eat away the metal particles after laminations have been assembled into apparatus, but this attempted cure often causes other troubles, such as rusting. Burning of the burrs with a flame has also been attempted, with varying success.

3. *Pressure*. Mechanical strains set up in magnetic materials either from *cold working during rolling or punching* or *distortion during final assembly operations*, affect magnetic properties in several ways. The hysteresis and eddy current losses as well as the magnetising current are increased. *The increased hysteresis loss due to punching depends upon the material, its thickness, and the configuration of the grains*. The loss may be as high as 300 per cent of the loss in the unpunched sheet. The strains produced by punching operations increase the loss primarily in the narrow section next to the punched edge. The per cent increase in hysteresis loss is much lower for high flux densities than for low ones.

The *effect of tension on silicon steel is to increase the permeability until high flux densities are reached. The effect of compression, however, is much more marked than that of tension and reduces the permeability considerably.*

Total iron losses

Total iron loss is the sum of hysteresis and eddy current loss and is given by the relation :

$$P_i = P_h + P_e$$

$$= K_h f B_{max}^k + K_e f^2 B_{max}^2 \text{ watts per m}^3 \text{ or per kg} \quad ...(19)$$

$$= (K_h f + K_e f^2) B_{max}^2 \text{ watt per m}^3 \text{ or per kg} \quad ...(20)$$

taking $k = 2$.

The total iron loss can be calculated with the aid of above expression but it gives good results only in the case of static *machinery and not in the case of dynamo electric machinery. The reason being that this expression is applicable when the field is simply alternating, but when the reversal of flux are due to rotational magnetic field, the losses do not follow the same laws. Therefore in practice, iron loss curves are used which represent the loss in watt per kg as a function of flux density.*

Example 17. *Calculate the energy lost per hour in a specimen of iron subjected to magnetisation at 50 c/s. The specimen weighs 50 kg and the hysteresis loop is equivalent in area to 250 joules per cubic metre. Density of iron is 7500 kg/m³.*

Solution. Loss = 250 joules/m³/cycle.

Volume of 1 kg of iron $= \dfrac{1}{\text{Density of iron}} = \dfrac{1}{7500}$

∴ Loss per kg $= 250 \times \dfrac{1}{7500}$ joules/cycle

No. of reversal/sec = 50

Weight of specimen = 50 kg

∴ Hysteresis loss/sec. $= 50 \times 250 \times \dfrac{1}{7500} \times 50 = \dfrac{250}{3}$ joules sec.

and Loss per hour $= \dfrac{250}{3} \times 60 \times 60 =$ **3×10^5 joules. (Ans.)**

Example 18. *The hysteresis loop of a specimen of iron having 10 kg mass is equivalent in area to 250 joules/m³ of iron. Find the loss of energy per hour at the rate of 50 c/s. Assume the density of iron as 7500 kg/m³.*

Solution. Energy loss/m³/cycle = 250 joules

Density of iron = 7500 kg/m³

∴ Volume of iron/kg $= \dfrac{1}{7500}$ m³

Mass of iron specimen = 10 kg

∴ Total volume of iron $= 10 \times \dfrac{1}{7500}$ m³

No. of cycle per second = 50

Total No. of cycles per hour = 50 × 60 × 60

∴ Total loss $= \dfrac{10}{7500} \times 250 \times 50 \times 60 \times 60 =$ **60000 joules. (Ans.)**

Example 19. *The hysteresis loop for a certain magnetic material is drawn to the following scale : 1 cm = 4 AT / cm and 1 cm = 0.05 Wb / m². The loop area is 90 cm² and density of given material is 7800 kg / m³. Calculate the hysteresis loss in watts / kg at 50 c / s.*

Solution. Hysteresis loss = xy × (area of B-H loop) joules/m³/cycle

Also $\quad 1 \text{ cm} = 4 \text{ AT/cm} = 400 \text{ AT/m}$

and $\quad 1 \text{ cm} = 0.05 \text{ Wb/m}^2$

i.e., $\quad x = 400$

and $\quad y = 0.05$

$\therefore \quad \text{loss} = 400 \times 0.05 \times 90 = 1800 \text{ joules/m}^3\text{/cycle}$

Volume of 1 kg of material $= \dfrac{1}{7800} \text{ m}^3$

$\therefore$ Loss/cycle $= 1800 \times \dfrac{1}{7800}$ joules

Number of reversals/sec. $= 50$

$\therefore$ Loss/kg $= 1800 \times \dfrac{1}{7800} \times 50$ joules/sec. or watts.

$= \textbf{11.54 watts. (Ans.)}$

Example 20. *In a certain transformer, the hysteresis loss is 300 W when the maximum flux density is 0.9 Wb / m² and the frequency 50 c / s. What would be the hysteresis loss if the maximum flux density were increased to 1.1 Wb / m² and the frequency reduced to 40 c / s. Assume the hysteresis loss over this range to be proportional to* $B_{max}^{1.7}$.

Solution. We know that,

Hysteresis loss, $\quad P_h = \eta \, B_{max}^{1.7} f V$ joules/sec.

$\therefore \quad P_h \propto B_{max}^{1.7} f$

In the *first case*, $\quad P_h = 300 \text{ W} \,;\, B_{max} = 0.9 \text{ Wb/m}^2 \,;\, f = 50 \text{ c/s}.$

In the *second case*, $P_h = ? \,;\, B_{max} = 1.1 \text{ Wb/m}^2 \,;\, f = 40 \text{ c/s}.$

or
$$\frac{300}{P_h} = \frac{(0.9)^{1.7} \times 50}{(1.1)^{1.7} \times 40}$$

or
$$P_h = 300 \times \left(\frac{1.1}{0.9}\right)^{1.7} \times \frac{4}{5} = \textbf{337.7 watts. (Ans.)}$$

Example 21. *Calculate the energy loss per hour by hysteresis in 50 kg iron core which is subjected to a sinusoidal flux alternating at 50 c / s. The hysteresis loop for the core has an area of 150 cm², when plotted to a scale of 1 cm = 0.008 Wb / m², 1 cm = 20 AT / cm.*

Solution. Here $\quad 1 \text{ cm} = 20 \text{ AT/cm} = 2000 \text{ AT/m}$

$1 \text{ cm} = 0.008 \text{ Wb/m}^2$

Area of hysteresis loop $= 150 \text{ cm}^2$

$\therefore$ Loss $= xy \times$ (area of B / H loop) joules/m³/cycle

$= 2000 \times 0.008 \times 150 = 2400 \text{ joules/m}^3\text{/cycle}$

Taking density of iron 7500 kg/m³.

Volume of core $= \dfrac{50}{7500} \text{ m}^3 = \dfrac{1}{150} \text{ m}^2$

Loss for total volume $= 2400 \times \frac{1}{150} = 16$ joules/cycle

No. of reversals per second $= 50$

$\therefore$ Loss/sec $= 16 \times 50 = 800$ joules

or Loss/hour $= 800 \times 3600 =$ **2880000 joules. (Ans.)**

Example 22. *Calculate the loss per kg in a specimen of alloy steel for a maximum flux density of 1.1 Wb/m² and a frequency of 50 Hz, using 0.5 mm thick sheets. The resistivity of alloy steel is 30 × 10⁻⁸ Ω m. The density is 7800 kg/m³. Hysteresis loss in each cycle is 380 W-s/m³.*

Solution. Maximum flux density, $B_{max} = 1.1$ Wb/m²

Frequency $= 50$ Hz

Thickness of sheet $= 0.5$ mm

Resistivity of alloy steel, $\rho = 30 \times 10^{-8}\ \Omega$ m

Density $= 7800$ kg/m³

Hysteresis loss/cycle $= 380$ W-s/m³

Total iron loss :

Eddy current loss is given by,

$$P_e = \frac{\pi^2 f^2 B_{max}^2 t^2}{6\rho} \text{ watt per m}^3$$

$$= \frac{\pi^2 f^2 B_{max}^2 t^2}{6\rho \times 7800} \text{ watt/kg}$$

$$= \frac{\pi^2 \times 50^2 \times 1.1^2 \times (0.5/1000)^2}{6 \times 30 \times 10^{-8} \times 7800} = 0.5316 \text{ W/kg}$$

Hysteresis loss, $P_h = \frac{380 \times 50}{7800} = 2.4359$ W/kg at 50 Hz.

$\therefore$ Total iron loss $= 0.5316 + 2.4359 =$ **2.9675 W/kg. (Ans.)**

Example 23. *A specimen of cold rolled grain oriented 0.3 mm thick stampings has a resistivity of 4.8 × 10⁻⁷ Ω m. The published hysteresis loop is essentially rectangular in form, with a co-ercive force, of 12 AT/m for all values of peak flux densities upto 1.6 Wb/m². The manufacturer quotes the iron loss in material as 1.2 W/kg, with a sinusoidal flux density 1.0 Wb/m² (peak) at 100 Hz. Calculate the loss in the material from its properties and compare it with the quoted value. The density is 7650 kg/m³.*

Solution. Thickness of stamping $= 0.3$ mm

Resistivity, $\rho = 4.8 \times 10^{-7}\ \Omega$ m

Density $= 7650$ kg/m³

The material is subjected to a peak flux density of 1 Wb/m² and therefore the value of coercive force is 12 AT/m for all densities. The hysteresis loop is thus a rectangle with sides $2 \times 12 = 24$ AT/m and $2 \times 1 = 2$ Wb/m².

Loss per cycle = Area of hysteresis loop $= 24 \times 2 = 48$ W-s/m³

Hysteresis loss/kg at 100 Hz $= \frac{48 \times 100}{7650} = 0.627$ W/kg

Eddy current loss, $P_e = \frac{\pi^2 f^2 B_{max}^2 t^2}{6\rho}$ W/m³ $= \frac{\pi^2 f^2 B_{max}^2 t^2}{6\rho \times 7650}$ W/kg

$$= \frac{\pi^2 \times (100)^2 \times (1.0)^2 \times (0.3/1000)^2}{6 \times 4.8 \times 10^{-7} \times 7650} = 0.403 \text{ W/kg}$$

Total iron loss, $P_i = P_h + P_e = 0.627 + 0.403$
$= \mathbf{1.03\ W/kg.}$ **(Ans.)**

The calculated loss *is smaller* than the quoted. **(Ans.)**

19. FARADAY'S LAWS OF ELECTROMAGNETIC INDUCTION

"The phenomenon whereby an e.m.f. and hence current is induced in any conductor which is cut across or is cut by a magnetic flux is known as ***electromagnetic induction***".

Faraday's First Law. It states as follows :

"Whenever the magnetic flux linked with a circuit changes, an e.m.f. is always induced in it".

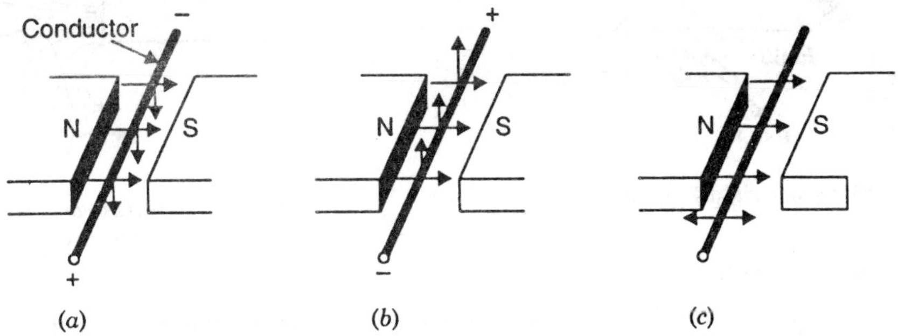

(*a*) Voltage induced across a wire moving downward.

(*b*) Voltage induced across a wire moving upward.

(*c*) No voltage is induced in a wire moving parallel to the field.

Fig. 30. When a conductor is moved across a magnetic field a voltage is induced in the conductor.

Or

"Whenever a conductor cuts magnetic flux, an e.m.f. is induced in that conductor".

Faraday's Second Law. It states as follows :

"The magnitude of induced e.m.f. is equal to the rate of change of flux-linkages".

Mathematically, $$e = -N \frac{d\phi}{dt} \text{ volts} \qquad \text{...(21)}$$

where e = Induced e.m.f.,

$\frac{d\phi}{dt}$ = Rate of change of flux, and

N = Number of turns of the coil.

[Usually, a *minus sign* is given to the right-hand side expression to signify the fact that the *induced e.m.f. sets up current in such a direction that magnetic effect produced by it opposes the very cause producing it.*]

Direction of induced e.m.f. and current :

The direction of the induced current may be found easily by applying either Fleming's Right-hand Rule (Fig. 31) or Lenz's Law. *Fleming's rule is used where induced e.m.f. is due to flux-cutting (i.e., dynamically induced e.m.f.) and Lenz's Law when it is due to change by flux-linkages (i.e., statically induced e.m.f.).*

— **Lenz's law.** Fig. 32 shows induction of an e.m.f. in a simple circuit. The direction of the induced e.m.f. is determined by Lenz's law, *which states that the current produced by the induced e.m.f. opposes the change in flux.*

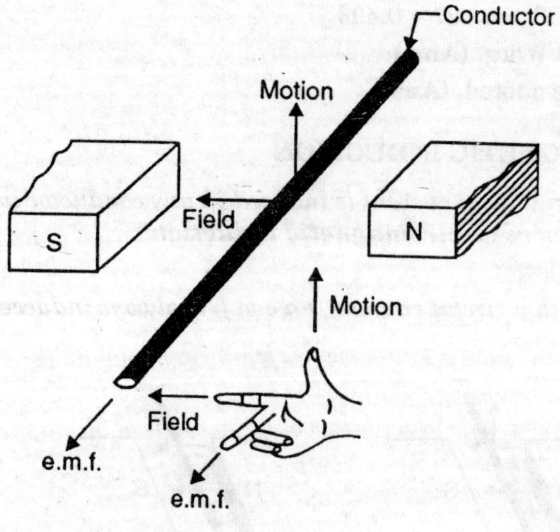

Fig. 31. Fleming's right hand rule.

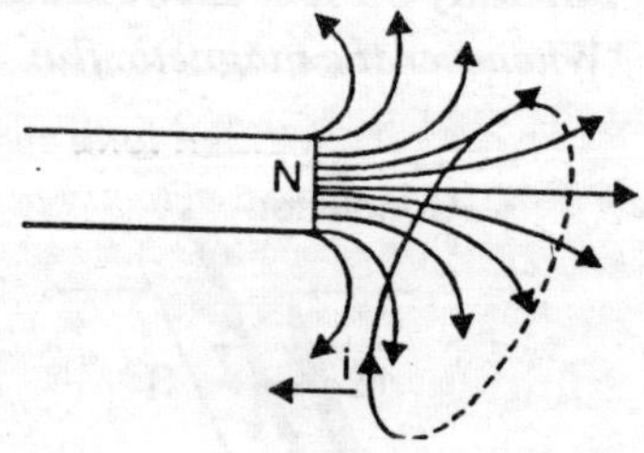

Fig. 32. Induction of e.m.f. in a simple circuit.

Lenz's law may also be stated as follows :

"In all cases of electromagnetic induction, an induced voltage will cause a current to flow in a closed circuit in such a direction that the magnetic field which is caused by that current will oppose the change that produced the current".

Example 24. *A coil has 480 turns and flux in it charges from 0.24 m Wb to 1.2 m Wb in 0.24 second. Calculate the voltage induced in the coil.*

Solution. *Given* : $N = 480$; $d\phi = 0.002 - 0.00024 = 0.00096$ Wb ; $dt = 0.24$ s

Induced e.m.f., e : $$e = N\frac{d\phi}{dt} = \frac{480 \times 0.00096}{0.24} = \mathbf{1.9\ V.\ (Ans.)}$$

Example 25. *The field coils of a six-poles D.C. generator each having 600 turns, are connected in series. On the excitation of the field, there is a magnetic flux of 0.024 Wb / pole. If the field circuit is opened in 0.24 second and residual magnetism is 0.0024 Wb / pole, calculate the average voltage which is induced across field terminals. In which direction is this voltage directed relative to the direction of the current.*

Solution. Total number of turns $= 6 \times 600 = 3600$

Total initial flux $= 6 \times 0.024 = 0.144$ **Wb**

Total residual flux $= 6 \times 0.0024 = 0.0144$ **Wb**

Change in flux, $d\phi = 0.144 - 0.0144 = 0.1296$ **Wb**

Time of opening the circuit, $dt = 0.024$ second

$$\therefore \quad \text{Induced e.m.f., } \mathbf{e} = N\frac{d\phi}{dt} = 3600 \times \frac{0.1296}{0.024} = \mathbf{19440\ V.\ (Ans.)}$$

The direction of this induced e.m.f. is the same as the initial direction of exciting current. **(Ans.)**

20. INDUCED E.M.F.

Induced e.m.f. may be of the following two types :

1. Dynamically induced e.m.f.
2. Statically induced e.m.f.

20.1. Dynamically Induced e.m.f. :

Refer Fig. 30. The e.m.f. induced (e) in the conductor is given by :

$$e = Blv \text{ volt} \quad \text{...(22)}$$

where, B = Flux density of the magnetic field in *tesla*,
l = Length of the conductor in *metres*, and
v = Velocity of the conductor in *m/s*.

If the conductor moves at an angle θ with the direction of flux then the induced e.m.f.

$$e = Blv \sin \theta \text{ volts} \quad \text{...(23)}$$

The direction of the induced e.m.f. is given by Fleming's Right hand rule.

Example 26. *A 1.2 m long conductor moves at right angles to a uniform magnetic field of flux density 1.8 Wb / m² with a velocity of 60 metres / second. Calculate the e.m.f. induced in it.*

Find also the value of induced e.m.f. when the conductor moves at an angle of 30° to the direction of the field.

Solution. *Given* : l = 1.2 m ; B = 1.8 Wb/m^2 ; v = 60 m/s

(*i*) Induced e.m.f., $\mathbf{e} = Bl\,v$ .... [See Fig. 33 (*a*)]

$= 1.8 \times 1.2 \times 60 =$ **129.6 V. (Ans.)**

(*ii*) Induced e.m.f., $\mathbf{e} = Bl\,v \sin \theta$ .... [See Fig. 33 (*b*)]

$= 1.8 \times 1.2 \times 60 \times \sin 30° =$ **64.8 V. (Ans.)**

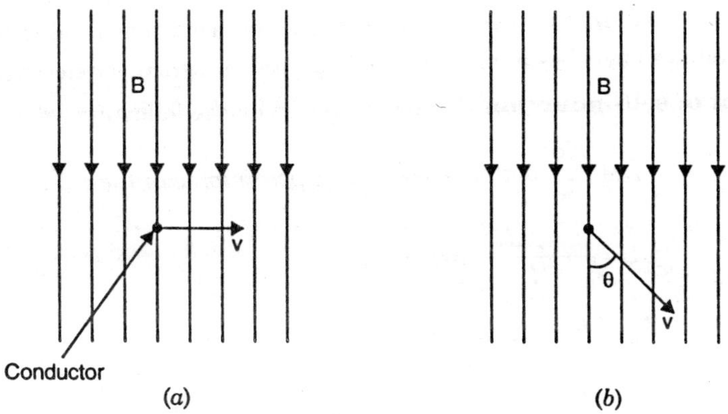

Fig. 33

Example 27. *In a four-pole dynamo, the flux/pole is 18 m Wb. If the armature is driven at 600 r.p.m., what is the average e.m.f. induced in one of the armature conductors ?*

Solution. Flux/pole = 18 m Wb ;

∴ Flux cut in one revolution $= 18 \times 4 = 72 \text{ m Wb} = 72 \times 10^{-3} \text{ Wb}$

Speed of the armature $= 600 \text{ r.p.m.} = \dfrac{600}{60} = 10 \text{ r.p.s.}$

∴ Time taken for one revolution $= \dfrac{1}{10} = 0.1 \text{ s}$

Average e.m.f. induced, $\mathbf{e} = N \dfrac{d\phi}{dt}$

$N = 1$; $d\phi = 72 \times 10^{-3}$ Wb ; $dt = 0.1$ s

∴ $\mathbf{e} = 1 \times \dfrac{72 \times 10^{-3}}{0.1} =$ **0.72 V. (Ans.)**

20.2. Statically Induced e.m.f. :

The e.m.f. induced by *variation of flux is termed as "statically induced e.m.f."*.

Statically induced e.m.f. can be further subdivided as follows :

(*i*) Self-induced e.m.f.

(*ii*) Mutually induced e.m.f.

20.2.1. Self-induced e.m.f.

Self-induced e.m.f. is the *e.m.f. induced in a coil due to the change of its own flux linked with it. If the current through the coil (Fig. 34) is changed* then the flux linked with its own turns will also change which will produce in it, what its called *self-induced e.m.f.* $\left(e = -N\dfrac{d\phi}{dt}\right)$.

The *direction of this e.m.f.* is given by *Lenz's law* (and would be such as *to oppose any change of flux which is, in fact, the very cause of its production*).

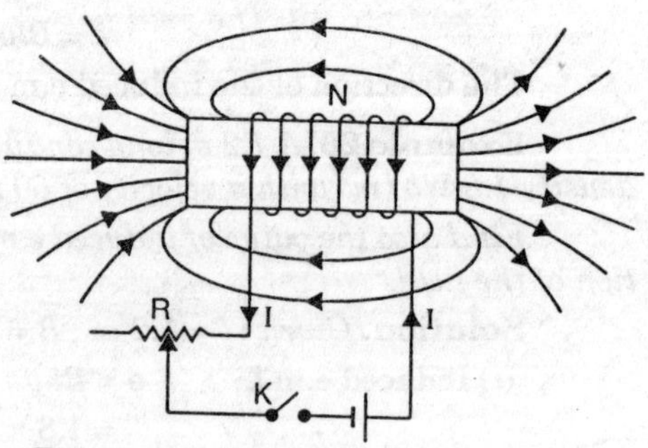

Fig. 34. Self-induced e.m.f. and self-induction.

Self-inductance :

The *property of the coil due to which it opposes any increase or decrease of current or flux through it, is known as* **self-inductance.** It is measured in terms of self-induction L (in *henry*).

Self-induction is sometimes analogously called *electromagnetic or electrical inertia*.

Co-efficient of self-induction (L) may be found by the following relations :

1. $$L = \frac{N\phi}{i} \text{ henry (Weber-turns per ampere in the coil)} \quad ...(24)$$

2. $$L = \frac{\mu_0 \mu_r A N^2}{l} \text{ henry (H)} \quad ...(25)$$

3. $$L = \frac{e_L}{\frac{di}{dt}} \text{ henry (H)} \quad ...(26)$$

where N = Number of turns of the solenoid,

A = Area of cross-section,

e_L = Induced e.m.f., and

$\dfrac{di}{dt}$ = Rate of change of current.

Energy in electromagnetic field :

If the current through an inductor is made to increase from 0 to i in a time 't', an e.m.f. e is induced in the coil and the total energy W absorbed by the inductor in time interval 0 to t is given as :

$$W = \int_0^t ei\, dt = \int_0^t \left(L\frac{di}{dt}\right) i\, dt$$

or

$$W = \int_0^i Li\, di = \frac{1}{2} L i^2 \text{ joules} \quad ...(27)$$

Thus the inductor has absorbed an energy which is proportional to inductance and the square of current. This energy is stored in the magnetic field. When the current changes from i to zero, this energy is returned to the source. Eqn. (27) can be written as :

$$L = \frac{2W}{i^2} \text{ henrys} \qquad ...(28)$$

Eqn. (28) is *an energy description of inductance.*

Example 28. *The field winding of a D.C. electromagnets is wound with 800 turns and has resistance of 40 Ω when exciting voltage is 230 V, the magnetic flux around the coil is 0.004 Wb. Calculate :*

(*i*) *The self inductance of the coil*

(*ii*) *The energy stored in the magnetic field.*

Solution. *Given* : $N = 800$; $R = 40\ \Omega$; $V = 230$ V ; $\phi = 0.004$ Wb,

(*i*) **The self inductance of the coil, L :**

$$\mathbf{L} = \frac{N\phi}{i} \qquad ...(\text{Eqn. } 24)$$

$$= \frac{N\phi}{(V/R)} = \frac{800 \times 0.004}{(230/40)} = \mathbf{0.556\ H.\ (Ans.)}$$

(*ii*) **The energy stored in the magnetic field, W :**

$$\mathbf{W} = \frac{1}{2} L\, i^2 \text{ joules} \qquad ...(\text{Eqn. } 27)$$

$$= \frac{1}{2} \times 0.556 \times \left(\frac{230}{40}\right)^2 = \mathbf{9.19\ J.\ (Ans.)}$$

Example 29. *An iron ring 30 cm mean diameter is made of square iron of 2 cm × 2 cm cross-section and is uniformly wound with 400 turns of wire of 2 mm² cross-section. Calculate the value of the self-inductance of the coil. Assume $\mu_r = 800$.* **(Gwalior University)**

Solution. *Given* : $A = 2 \times 2 = 4 \text{ cm}^2 = 4 \times 10^{-4} \text{ m}^2$; $N = 400$, $l = \pi d = \pi \times 0.3 = 0.3\,\pi\, m$;

μ_r (relative permeability) = 800

Self inductance of coil, L :

$$L = \frac{\mu_0 \mu_r A N^2}{l} \text{ henry} \qquad [\text{Eqn. } (25)]$$

$$= \frac{4\pi \times 10^{-7} \times 800 \times 4 \times 10^{-4} \times 400^2}{0.3\pi} = \mathbf{0.0683\ H.\ (Ans.)}$$

Example 30. *An air-cored solenoid has a length of 60 cm and a diameter of 4 cm. Calculate its inductance if it has 1000 turns and also find the energy stored if the current rises from zero to 6 A.* **(VTU August, 2000)**

Solution. *Given* : $l = 60$ cm = 0.6 m ; $d = 4$ cm = 0.04 m ; $N = 1000$; $i = 5 - 0 = 5$ A

Inductance, L :

$$L = \frac{\mu_0 \mu_r A N^2}{l} \qquad [\text{Eqn. } (25)]$$

$$= \frac{4\pi \times 10^{-7} \times 1 \times \left(\frac{\pi}{4} \times 0.04^2\right)^2 \times (1000)^2}{0.6}$$

($\because$ $\mu_r = 1$, since solenoid is air-cored)

$= 2.632 \times 10^{-3}$ H or 2.632 mH

Energy stored :

$$\text{Energy stored} = \frac{1}{2} Li^2$$

$$= \frac{1}{2} \times 2.632 \times 10^{-3} \times 6^2 = \mathbf{0.0474\ joules.\ (Ans.)}$$

Example 31. *If a coil of 200 turns is linked with a flux of 0.01 Wb when carrying current of 10 A, calculate the inductance of the coil. If this current is uniformly reversed in 0.01 second, calculate the induced electromotive force.*

Solution. *Given* : $N = 200$; $\phi = 0.01$ Wb, $i = 10\ A$; $dt = 0.01$ s

$$L = \frac{N\phi}{i} = \frac{200 \times 0.01}{10} = 0.2 \text{ H}$$

Induced electromotive force, e_L :

$$e_L = L\frac{di}{dt} \qquad \text{[Eqn. (26)]}$$

$$= 0.2 \times \frac{[10 - (-10)]}{0.01} = \mathbf{400\ V.\ (Ans.)}$$

Example 32. *A coil of 1500 turns carrying a current of 5 A produces a flux of 2.5 mWb. If the current is reversed in 0.2 second, find the average vlaue of e.m.f. induced in the coil. Also find the self inductance of the coil.* **(VTU August, 2001)**

Solution. *Given* : $N = 1500$; $i = 5$ A ; $\phi = 2.5$ mWb ; $dt = 0.2$ s.

We know that, $$\mathbf{L} = \frac{N\phi}{i} = \frac{1500 \times (2.5 \times 10^{-3})}{5} = 0.75 \text{ H}$$

$\therefore$ E.m.f. induced, $$e_L = L\frac{di}{dt} = 0.75 \times \frac{[5 - (-5)]}{0.2} = \mathbf{37.5\ V.\ (Ans.)}$$

Example 33. *A direct current of one ampere is passed through a coil of 6000 turns ; it produces a flux of 0.1 mWb. Assuming that whole of this flux threads all the turns, what is the inductance of the coil ? What would be the voltage developed across the coil if the current were interrupted in 10^{-3} second ? What is the energy stored in the coil.*

Solution. *Given* : $i = 1$ A ; $N = 6000$; $\phi = 0.1$ mWb $= 0.1 \times 10^{-3}$ Wb ; $dt = 10^{-3}$ s.

$$L = \frac{N\phi}{i} = \frac{6000 \times (0.1 \times 10^{-3})}{1} = \mathbf{0.6\ H.\ (Ans.)}$$

Inducted e.m.f. $$e_L = L \cdot \frac{di}{dt} = \frac{0.6 \times 1}{10^{-3}} = 600 \text{ V}$$

Energy stored in the coil $$= \frac{1}{2} Li^2 = \frac{1}{2} \times 0.6 \times 1^2 = \mathbf{0.3\ J.\ (Ans.)}$$

20.2.2. Mutually induced e.m.f. Refer Fig. 35. *Production of e.m.f. in coil B due to change in current A is called mutually induced e.m.f.*

- Consider two coils *A* and *B* lying close to each other (Fig. 35). Coil A is joined to a *battery*, a *switch* (*K*) and a *variable resistance* (*R*) whereas coil *B* is connected to a *sensitive voltmeter* V.
- When current through *A* is established by closing the switch, its *magnetic field is set up and partly links with or threads through the coil B.* As the current through *A* is changed, the flux linked with *B* is also changed. Hence mutually induced *e.m.f.* is produced in *B* where *magnitude is given by Faraday's law and direction by Lenz's law.*

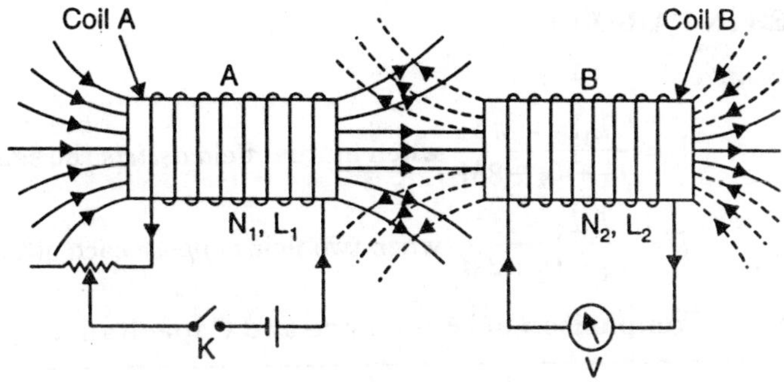

Fig. 35. Mutually induced e.m.f. and mutual inductance.

- Now, if battery is connected to *B* and the voltmeter across *A*, then the situation is reversed and now a change of current in *B* will produce mutually-induced *e.m.f. A*.

— From the examples considered above, it is evident that there is no movement of any conductor, the flux variations being brought about by variations in current strength only. Such an e.m.f. induced in one coil by the influence of the other coil is called (statically but) mutually induced e.m.f.

Mutual inductance :

Mutual inductance may be defined *as the ability of one coil (or circuit) to produce e.m.f. in a nearly coil by induction when the current in the first coil changes.* This action being reciprocal, the second coil can also induce an e.m.f. in the first coil when current in the second coil changes. This ability of reciprocal induction is measured in terms of the coefficient of mutual induction *M*.

Co-efficient of mutual induction (M) may be found by the following relations :

1. $$M = \frac{N_2\phi_1}{i_1} \text{ henry (H)} \qquad ...(29)$$

2. $$M = \frac{\mu_0\mu_r A N_1 N_2}{l} \text{ henry (H)} \qquad ...(30)$$

3. $$M = -\frac{e_M}{\frac{di_1}{dt}} \qquad ...(31)$$

Co-efficient of coupling (k) :

It is defined as the *ratio of mutual inductance between the coils and the square root of product of self-inductance of each coil.*

In other words, $$k = \frac{M}{\sqrt{L_1L_2}} \qquad ...(32)$$

21. INDUCTANCES IN SERIES

In general we have,

$$L = L_1 + L_2 + 2M \text{ ...... if } \textit{m.m.fs. are additive} \qquad ...(33)$$

$$L = L_1 + L_2 - 2M \text{ ...... if } \textit{m.m.fs. are subtractive} \qquad ...(34)$$

22. INDUCTANCES IN PARALLEL

In general we have,

$$L = \frac{L_1L_2 - M^2}{L_1 + L_2 - 2M} \text{ when mutual field } \textit{assists} \text{ the separate fields} \quad ...(35)$$

$$L = \frac{L_1L_2 - M^2}{L_1 + L_2 + 2M} \text{ when two fields } \textit{oppose} \text{ each other} \quad ...(36)$$

Comparison of Inductance and Capacitance

Inductance	Capacitance
1. Conducts current (I)	1. Stores charge (Q)
2. $L = L_1 + L_2 + L_3 + ...$ *in series*	2. $\frac{1}{C} = \frac{1}{C_1} + \frac{1}{C_2} + \frac{1}{C_3} + ...$ *in series*
3. $\frac{1}{L} = \frac{1}{L_1} + \frac{1}{L_2} + \frac{1}{L_3} + ...$ *in parallel*	3. $C = C_1 + C_2 + C_3 + ...$ *in parallel*
4. Requires wire conductor.	4. Requires dielectric as insulator.

Example 34. *If a coil of 150 turns is linked with a flux of 0.01 Wb when carrying a current of 10 A ; calculate the inductance of the coil. If this current is uniformly reversed in 0.1 second, calculate the induced e.m.f. If a second coil of 100 turns is uniformly wound over the first coil, find the mutual inductance between the coils.* **(Pune University)**

Solution. *Given :* $N_1 = 150$; $\phi_1 = 0.01$ Wb ; $i_1 = 10$ A ; $di = 10 - (-10) = 20$ A ;

$dt = 0.1$ s ; $N_2 = 100$

Induced e.m.f. L_1 :

$$L_1 = \frac{N_1\phi_1}{i_1} = \frac{150 \times 0.01}{10} = \mathbf{0.15\ H.\ (Ans.)}$$

Mutual inductance, M :

$$M = \frac{N_2\phi_1}{i_1} = \frac{100 \times 0.01}{10} = \mathbf{0.1\ H.\ (Ans.)}$$

Example 35. *Two coils having 30 and 600 turns respectively are wound side-by-side on a closed iron circuit of cross-section 100 sq. cm. and mean length 200 cm.*

(*i*) *Estimate the mutual inductance between the coils if the relative permeability of the iron is 2000.*

(*ii*) *If a current of zero ampere grows to 20 A in a time of 0.02 second in the first coil, find the e.m.f. induced in the second coil.* **(JNT University Warrangal)**

Solution. *Given :* $N_1 = 30$; $N_2 = 600$; $A = 100 \times 10^{-4}\ m^2$; $l = 200$ cm $= 2$ m

$\mu_r = 2000$; $di = 20 - 0 = 20$ A ; $dt = 0.02$ s.

(*i*) Mutual inductance, M :

$$M = \frac{\mu_0\mu_r AN_1N_2}{l} \text{ H} \quad ...(37)$$

$$= \frac{4\pi \times 10^{-7} \times 2000 \times 100 \times 10^{-4} \times 30 \times 600}{2} = 0.226 \text{ H}$$

(*ii*) **Mutually induced e.m.f., e_M :**

$$e_M = M \cdot \frac{di_1}{dt} \qquad ...(38)$$

$$= 0.226 \times \frac{20}{0.02} = \mathbf{226\ V.}\ \textbf{(Ans.)}$$

Example 36. *Two coils A and B are wound on the same ions core. There are 600 turns on A and 3600 turns on B. The current of 4 amperes through coil A produces a flux of 500 × 10^{-6} Wb in the core. If this current is reversed in 0.02 second, calculate average e.m.f. induced in coils A and B.*

Solution. *Given* : $N_1 = 600$; $N_2 = 3600$; $i_1 = 4$ A ; $\phi_1 = 500 \times 10^{-6}$ Wb ; $dt = 0.02$ s.

E.m.f. induced in coils A and B, e_1, e_2 :

Self inductance of coil A, $L_1 = \frac{N_1\phi_1}{i_1} = \frac{600 \times 500 \times 10^{-6}}{4} = 0.075$ H

Mutual inductance between the coils, $M = \frac{N_2\phi_1}{i_1} = \frac{3600 \times 500 \times 10^{-6}}{4} = 0.45$ H

(Assuming flux created in coil A links completely with the coil B, *i.e.*, $\phi_2 = \phi_1$).

Average e.m.f. induced in coil A, $\quad e_1 = L_1 \frac{di_1}{dt} = 0.075 \times \frac{[4-(-4)]}{0.02} = \mathbf{30\ V.}$ **(Ans.)**

Average e.m.f. induced in coil B, $\quad e_2 = M \frac{di_1}{dt} = 0.45 \times \frac{[4-(-4)]}{0.02} = \mathbf{1.80\ V.}$ **(Ans.)**

Example 37. *This coils A of 11450 turns and B of 14500 turns lie in parallel planes so that 65 per cent of flux produced in A links coil B. It is found that a current of 6 A in A produces a flux of 0.7 m Wb while the same current in B produces 0.9 m Wb. Determine :*

(*i*) *Mutual inductance* ; (*ii*) *Co-efficient of coupling.*

Solution. *Given* : $N_1 = 11450$; $N_2 = 14500$; $\phi_2 = 0.65\ \phi_1$; $i_1 = 6$ A ;

$\phi_1 = 0.7$ m Wb $= 0.7 \times 10^{-3}$ Wb ; $\phi_2 = 0.9$ m Wb $= 0.9 \times 10^{-3}$ Wb.

(*i*) **Mutual inductance M :**

$$\mathbf{M} = \frac{N_2\phi_2}{i_1} = \frac{14500 \times 0.65 \times 0.7 \times 10^{-3}}{6} = \mathbf{1.1.\ H.}\ \textbf{(Ans.)}$$

(*ii*) **Co-efficient coupling *k* :**

$$L_1 = \frac{N_1\phi_1}{i_1} = \frac{11450 \times 0.7 \times 10^{-3}}{6} = 1.336 \text{ H}$$

$$L_2 = \frac{N_2\phi_2}{i_1} = \frac{14500 \times 0.9 \times 10^{-3}}{6} = 2.175 \text{ H}$$

$$\therefore \quad \mathbf{k} = \frac{M}{\sqrt{L_1 L_2}} \qquad ...[\text{Eqn. (32)}]$$

$$= \frac{1.1}{\sqrt{1.336 \times 2.175}} = \mathbf{0.645.}\ \textbf{(Ans.)}$$

Example 38. *The combined inductance of two coils connected in series is 0.6 H or 0.1 H depending on the relative directions of the currents in the coils. If one of the coils when isolated has a self inductance of 0.2 H. Calculate :*

(*i*) *Mutual inductance* ; (*ii*) *Coupling coefficient.* **(Indore University)**

Solution. (*i*) $L = L_1 + L_2 + 2M$

or $0.6 = L_1 + L_2 + 2M$ ...(*i*)

and $0.1 = L_1 + L_2 - 2M$ ...(*ii*)

From (*i*) and (*ii*), we get **M = 0.125 H. (Ans.)**

(*ii*) Let $L_1 = 0.2$ H, then substituting this value in (*i*) above, we get $L_2 = 0.15$ H

∴ Coupling coefficients, $k = \frac{M}{\sqrt{L_1 L_2}} = \frac{0.125}{\sqrt{0.2 \times 0.15}}$ = **0.72. (Ans.)**

HIGHLIGHTS

1. Laws of magnetic force :
$$F = \frac{m_1 m_2}{4\pi\mu_0\mu_r d^2}.$$
2. Flux density (*B*).
$$B = \frac{\phi}{A}$$
3. Magnetic field strength (*H*).
$$H = \frac{NI}{l} \text{ AT/m.}$$
4. Force on a current carrying conductor lying in a magnetic field (*F*) :
$$F = BIl \text{ newtons}$$
$$(= \mu_o\mu_r HIl \text{ newtons}).$$
5. Magnetising force of a long straight conductor
$$H = \frac{NI}{2\pi r} \text{ AT/m and } B = \frac{\mu_0\mu_r NI}{2\pi r}.$$
6. Magnetising force of a long solenoid
$$H = \frac{NI}{l} \text{ AT/m and } B = \frac{\mu_0\mu_r NI}{l}.$$
7. Force between two parallel conductors
$$F = \frac{\mu_0 I_1 I_2 l}{2\pi d} \text{ newtons.}$$
8. **Permeability** is the ratio of flux density to magnetic field strength.
9. **Hysteresis** is defined as the lagging of magnetisation or induction flux density (*B*) behind the magnetising force (*H*).
10. *"The phenomenon whereby an e.m.f. and hence current is induced in any conductor which is cut across or is cut by a magnetic flux is known as* **electromagnetic induction.**
11. **Faraday's first law.** It states that : *"Whenever the magnetic flux linked with a circuit changes, an e.m.f. is always induced in it".*

    Or

    "Whenever a conductor cuts magnetic flux, an e.m.f. is induced in that conductor".
12. **Lenz's law.** It states that : *"In all cases of electromagnetic induction, an induced voltage will cause a current to flow in a closed circuit in such a direction that magnetic field which is caused by that current will oppose the change that produced the current".*

13. The e.m.f. induced by variation of flux is termed as "statically induced e.m.f.." It is further subdivided as :

(i) Self-induced e.m.f. ; (ii) Mutually induced e.m.f.

14. *The property of the coil due to which it opposes any increase or decrease of the current or flux through it, is known as* **self-inductance.**

15. *Mutual inductance* is the ability of one coil (or circuit) to produce e.m.f. in a near by coil by induction when the current in the first coil changes.

16. *Co-efficient of coupling* is defined as the ratio of mutual inductance between the coils and the square root of product of self-inductance of each coil.

OBJECTIVE TYPE QUESTIONS

A. Choose the Correct Answer :

1. An air gap is usually inserted in magnetic circuits to
(a) increase m.m.f (b) increase the flux (c) prevent saturation (d) none of the above.

2. The relative permeability of a ferromagnetic material is
(a) less than one (b) more than one
(c) more than 10 (d) more than 100 or 1000.

3. The unit of magnetic flux is
(a) henry (b) weber (c) ampere turn/metre (d) ampere/meter.

4. Permeability in a magnetic circuit corresponds to .... in an electric circuit.
(a) resistance (b) resistivity (c) conductivity (d) conductance.

5. Point out the *wrong* statement. Magnetic leakage is undesirable in electric machines because it
(a) lowers their power efficiency (b) increases their cost of manufacture
(c) leads to their increased weight (d) produces fringing.

6. Relative permeability of vacuum is
(a) 1 (b) 1 H/m (c) $1/4\pi$ (d) $4\pi \times 10^{-7}$ H/m.

7. The magnetising force (H) and magnetic flux density (B) are connected by the relation
(a) $B = \mu_r H/\mu_0$ (b) $B = \mu H$ (c) $B = H/\mu_0\mu_r$ (d) $B = \mu_0 H/\mu_r$.

8. Permanent magnets are normally made of
(a) alnico alloys (b) aluminium (c) cast iron (d) wrought iron.

9. Energy stored by a coil is doubled when its current is increased by ...... percent.
(a) 25 (b) 50 (c) 41.4 (d) 100.

10. Those magnetic materials are best suited for making armature and transformer cores which have ...... permeability and ..... hysteresis loss.
(a) high, high (b) low, high (c) high, low (d) low, low.

11. The rate of rise of current through an inductive coil is maximum
(a) at 63.2% of its maximum steady value (b) at the start of the current flow
(c) after one time constant (d) near the final maximum value of current.

12. When both the inductance and resistance of a coil are doubled the value of
(a) time constant remains unchanged (b) initial rate of rise of current is doubled
(c) final steady current is doubled (d) time constant is halved.

13. The initial rate of rise of current through a coil of inductance 10 H when suddenly connected to a D.C. supply of 200 V is ...... A/s
(a) 50 (b) 20 (c) 0.05 (d) 500.

14. A material for good magnetic memory should have
(*a*) low hysteresis loss (*b*) high permeability (*c*) low retentivity (*d*) high retentivity.
15. Conductivity is analogous to
(*a*) retentivity (*b*) resistivity (*c*) permeability (*d*) inductance.
16. In a magnetic material hysteresis loss takes place primarily due to
(*a*) rapid reversals of its magnetisation (*b*) flux density lagging behind magnetising force
(*c*) molecular friction (*d*) its high retentivity.
17. Those materials are well suited for making permanent magnets which have ...... retentivity and ...... coercivity
(*a*) low, high (*b*) high, high (*c*) high, low (*d*) low, low.
18. If the area of hysteresis loop of a material is large, the hysteresis loss in this material will be
(*a*) zero (*b*) small (*c*) large (*d*) none of the above.
19. Hard steel is suitable for making permanent magnets because
(*a*) it has good residual magnetism (*b*) its hysteresis loop has large area
(*c*) its mechanical strength is high (*d*) its mechanical strength is low.
20. Silicon steel is used in electrical machines because it has
(*a*) low co-ercivity (*b*) low retentivity
(*c*) low hysteresis loss (*d*) high co-ercivity.
21. Conductance is analogous to
(*a*) permeance (*b*) reluctance (*c*) flux (*d*) inductance.
22. The property of a material which opposes the creation of magnetic flux in it is known as
(*a*) reluctivity (*b*) magnetomotive force (*c*) permeance (*d*) reluctance.
23. The unit of retentivity is
(*a*) weber (*b*) weber/sq. m (*c*) ampere turn/metre (*d*) ampere turn.
24. Reciprocal of reluctance is
(*a*) reluctivity (*b*) permeance (*c*) permeability (*d*) susceptibility.
25. While comparing magnetic and electric circuits, the flux of magnetic circuit is compared with which parameter of electrical circuit ?
(*a*) E.m.f. (*b*) Current (*c*) Current density (*d*) Conductivity.
26. The unit of reluctance is
(*a*) metre/henry (*b*) henry/metre (*c*) henry (*d*) 1/henry.
27. A ferrite core has less eddy current loss than an iron core because
(*a*) ferrites have high resistance (*b*) ferrites are magnetic
(*c*) ferrites have low permeability (*d*) ferrites have high hysteresis.
28. A ferromagnetic core subjected to cycles of magnetisation will exhibit hysteresis when the cycle is
(*a*) rotating (*b*) alternating (*c*) pulsating (*d*) any of the above.
29. In order to minimise loss due to hysteresis, the magnetic material should have
(*a*) high resistivity (*b*) low hysteresis co-efficient
(*c*) large *B-H* loop area (*d*) high retentivity.
30. Hysteresis loss least depends on
(*a*) volume of material (*b*) frequency
(*c*) steinmetz co-efficient of material (*d*) ambient temperature.
31. Laminated cores, in electrical machines, are used to reduce
(*a*) copper loss (*b*) eddy current loss (*c*) hysteresis loss (*d*) all of the above.
32. The area of hysteresis loss is a measure of
(*a*) permittivity (*b*) permeance (*c*) energy loss per cycle (*d*) magnetic flux.

33. Permeance is to reluctance as conductance is to
(*a*) ampere turns (*b*) inductance (*c*) capacitance (*d*) resistance.

34. According to steinmetz hysteresis law, hysteresis loss in a material is proportional to
(*a*) $B^{3.6}$ (*b*) $B^{1.6}$ (*c*) $B^{1.2}$ (*d*) $B^{2.6}$.

35. The transformer cores operating at microwave frequency range, are usually made of
(*a*) carbon (*b*) copper (*c*) silicon steels (*d*) ferrites.

36. The property of coil by which a counter e.m.f. is induced in it when the current through the coil changes is known as
(*a*) self-inductance (*b*) mutual inductance (*c*) series aiding inductance (*d*) capacitance.

37. As per Faraday's laws of electromagnetic induction, an e.m.f. is induced in a conductor whenever it
(*a*) lies perpendicular to the magnetic flux (*b*) lies in a magnetic field
(*c*) cuts magnetic flux
(*d*) moves parallel to the direction of the magnetic field.

38. Which of the following circuit element stores energy in the electromagnetic field ?
(*a*) Inductance (*b*) Condenser (*c*) Variable resistor (*d*) Resistance.

39. The inductance of a coil will increase under all the following conditions *except*
(*a*) when more length for the same number of turns is provided
(*b*) when the number of turns of the coil increase
(*c*) when more area for each turn is provided
(*d*) when permeability of the core increases.

40. Higher the self-inductance of a coil,
(*a*) lesser its weber-turns (*b*) lower the e.m.f. induced
(*c*) greater the flux produced by it
(*d*) longer the delay in establishing steady current through it.

41. In an iron cored coil the iron core is removed so that the coil becomes an air cored coil. The inductance of the coil will
(*a*) increase (*b*) decrease
(*c*) remain the same (*d*) initially increase and then decrease.

42. An open coil has
(*a*) zero resistance and inductance (*b*) infinite resistance and zero inductance
(*c*) infinite resistance and normal inductance (*d*) zero resistance and high inductance.

43. Both the number of turns and the core length of an inductive coil are doubled. Its self-inductance will be
(*a*) unaffected (*b*) doubled (*c*) halved (*d*) quadrupled.

44. If current in a conductor increases then according to Lenz's law self-induced voltage will
(*a*) air the increasing current (*b*) tend to decrease the amount of current
(*c*) produce current opposite to the increasing current
(*d*) aid the applied voltage.

45. The direction of induced e.m.f. can be found by
(*a*) Laplace's law (*b*) Lenz's law
(*c*) Fleming's right hand rule (*d*) Kirchhoff's voltage law.

46. Air-core coils are practically free from
(*a*) hysteresis losses (*b*) eddy current losses (*c*) both (*a*) and (*b*) (*d*) none of the above.

47. The magnitude of the induced e.m.f. in a conductor depends on the
(*a*) flux density of the magnetic field (*b*) amount of flux cut
(*c*) amount of flux linkages (*d*) rate of change of flux-linkages.

48. Mutually inductance between two magnetically-coupled coils depends on
(*a*) permeability of the core (*b*) the number of their turns
(*c*) cross-sectional area of their common core (*d*) all of the above.

49. A laminated iron core has reduced eddy-current losses because
(*a*) more wire can be used with less D.C. resistance in coil
(*b*) the laminations are insulated from each other
(*c*) the magnetic flux is concentrated in the air gap of the core
(*d*) the laminations are stacked vertically.

50. The co-efficient of self-inductance for a coil is given as
(*a*) $\frac{NI}{\phi}$ (*b*) $\frac{N\phi}{I}$ (*c*) $\frac{NI^2}{\phi}$ (*d*) $\frac{N\phi}{I^2}$.

51. The law that the induced e.m.f. and current always oppose the cause producing them is due to
(*a*) Faraday (*b*) Lenz (*c*) Newton (*b*) Coulomb.

52. Which of the following is not a unit of inductance ?
(*a*) Henry (*b*) Coulomb/volt ampere
(*c*) Volt second per ampere (*d*) All of the above.

53. In case of an inductance, current is proportional to
(*a*) voltage across the inductance (*b*) magnetic field
(*c*) both (*a*) and (*b*) (*d*) neither (*a*) nor (*b*).

54. Which of the following circuit elements will oppose the change in circuit current ?
(*a*) Capacitance (*b*) Inductance (*c*) Resistance (*d*) All of the above.

55. For a purely inductive circuit which of the following is *true* ?
(*a*) Apparent power is zero (*b*) Relative power is zero
(*c*) Actual power of the circuit is zero
(*d*) Any capacitance even if present in the circuit will not be charged.

56. Which of the following is unit of inductance ?
(*a*) Ohm (*b*) Henry (*c*) Ampere turns (*d*) Webers/metre.

57. An e.m.f. of 16 volts is induced in a coil of inductance 4H. The rate of change of current must be
(*a*) 64 A/s (*b*) 32 A/s (*c*) 16 A/s (*d*) 4 A/s.

58. The core of a coil has a length of 200 mm. The inductance of coil is 6 mH. If the core length is doubled, all other quantities, remaining the same, the inductance will be
(*a*) 3 mH (*b*) 12 mH (*c*) 24 mH (*d*) 48 mH.

59. The self inductances of two coils are 8 mH and 18 mH. If the co-efficients of coupling is 0.5, the mutual inductance of the coils is
(*a*) 4 mH (*b*) 5 mH (*c*) 6 mH (*d*) 12 mH.

60. Two coils have inductances of 8 mH and 18 mH and a co-efficient of coupling of 0.5. If the two coils are connected in series aiding, the total inductance will be
(*a*) 32 mH (*b*) 38 mH (*c*) 40 mH (*d*) 48 mH.

61. A 200 turns coil has an inductance of 12 mH. If the number of turns is increased to 400 turns, all other quantities (area, length etc.) remaining the same, the inductance will be
(*a*) 6 mH (*b*) 14 mH (*c*) 24 mH (*d*) 48 mH.

62. Two coils have self-inductances of 10 H and 2 H, the mutual inductance being zero. If the two coils are connected in series, the total inductance will be
(*a*) 6 H (*b*) 8 H (*c*) 12 H (*d*) 24 H

63. In case all the flux from the current in coil 1 links with coil 2, the co-efficient of coupling will be
(*a*) 2.0 (*b*) 1.0 (*c*) 0.5 (*d*) zero.

64. The energy in joules stored in the magnetic field of 0.15 H inductance with a 180 mA current will be
(*a*) 2.43 (*b*) 2.43×10^{-3} (*c*) 2.43×10^{-6} (*d*) 2.43×10^{-9}.

65. A coil with negligible resistance has 50 V across it with 10 mA. The inductive reactance is
(*a*) 50 ohms (*b*) 500 ohms (*c*) 1000 ohms (*d*) 5000 ohms.

Answers

1. (*c*)	**2.** (*d*)	**3.** (*b*)	**4.** (*c*)	**5.** (*a*)	**6.** (*a*)	**7.** (*b*)
8. (*a*)	**9.** (*c*)	**10.** (*c*)	**11.** (*b*)	**12.** (*a*)	**13.** (*b*)	**14.** (*d*)
15. (*c*)	**16.** (*d*)	**17.** (*b*)	**18.** (*c*)	**19.** (*a*)	**20.** (*c*)	**21.** (*a*)
22. (*d*)	**23.** (*b*)	**24.** (*b*)	**25.** (*b*)	**26.** (*d*)	**27.** (*d*)	**28.** (*d*)
29. (*b*)	**30.** (*d*)	**31.** (*b*)	**32.** (*c*)	**33.** (*d*)	**34.** (*b*)	**35.** (*d*)
36. (*a*)	**37.** (*c*)	**38.** (*a*)	**39.** (*a*)	**40.** (*d*)	**41.** (*b*)	**42.** (*b*)
43. (*b*)	**44.** (*c*)	**45.** (*b*)	**46.** (*c*)	**47.** (*d*)	**48.** (*d*)	**49.** (*b*)
50. (*b*)	**51.** (*b*)	**52.** (*b*)	**53.** (*b*)	**54.** (*b*)	**55.** (*c*)	**56.** (*b*)
57. (*d*)	**58.** (*a*)	**59.** (*c*)	**60.** (*b*)	**61.** (*d*)	**62.** (*c*)	**63.** (*b*)
64. (*b*)	**65.** (*d*).					

THEORETICAL QUESTIONS

1. Define the following terms :
   (*i*) Magnetic flux density (*ii*) Relative permeability
   (*iii*) Absolute permeability (*iv*) Magnetomotive force.
2. State the differences between magnetically soft and magnetically hard materials.
3. Sketch the shape of magnetic field around a current-carrying conductor. How can the direction of the field be determined ?
4. What do you understand by the ferromagnetic, paramagnetic and diamagnetic materials ?
5. Sketch the magnetic field that occurs around two adjacent current carrying conductors (*i*) for currents in opposite direction (*ii*) for currents in the same direction.
6. Derive an expression for magnetising force of a long straight conductor.
7. Derive an expression for magnetising force of a long solenoid.
8. Derive an expression for force between two parallel conductors.
9. Define the following terms as applied to a magnetic circuit :
   Permeability ; Magnetomotive force (m.m.f.) ; Reluctance ; Permeance ; Reluctivity.
10. Give the analogy between electric and magnetic circuits (similarities).
11. What do you mean by magnetic leakage and fringing ?
12. Draw and explain a magnetising curve for a ferromagnetic material.
13. Explain briefly the following two types of losses which occur when magnetic materials are subjected to an alternating flux :
    (*i*) Hysteresis loss
    (*ii*) Eddy current loss.
14. What do you understand by "electromagnetic induction".
15. Explain briefly Faraday's first and second laws of electromagnetic induction.
16. State and explain Fleming's right hand rule.
17. What is "Lenz's law" ?
18. What is difference between the dynamically and statically induced e.m.fs. ?

19. Define the term "Self-inductance".

20. Derive an expression for energy stored in an electromagnetic field.

21. Explain briefly "Mutually induced e.m.f.".

22. What is "Mutual inductance" ?

23. What do you understand by "Co-efficient of coupling" ?

UNSOLVED EXAMPLES

MAGNETISM

1. The total flux of an electromagnet is 8×10^{-4} Wb. (*i*) If the cross-sectional area of the core is 100 mm^2, find the flux density in the core. (*ii*) The electromagnet has 50 turns and a current of 2 A flows through the coil. If the length of the magnetic circuit is 200 mm, find the m.m.f. and magnetic field strength. **[Ans.** (*i*) 8 T, (*ii*) 100 AT, 500 AT/m]

2. A long straight conductor carries a steady current of 10 A. Find (*i*) the intensity of magnetic field produced at a point 150 mm from the axis of the conductor and (*ii*) flux density of the field at that point. **[Ans.** 10.61 A/m, 133.3×10^{-7} T]

3. A solenoid of 400 turns is wound on a continuous ring of magnetic material of relative permeability 1200. If the flux density of the magnetic material of the core in the solenoid is 1.3 T, what is the current in the solenoid ? Mean diameter of the solenoid is 200 mm. **[Ans.** 1.35 A]

4. Two wires carrying current, in the same direction, of 5000 A and 10000 A are placed with their axis 50 mm apart. Calculate the force between them in N/m. **[Ans.** 200 N/m]

5. Two long parallel linear conductors carry 50 A. If the conductors are separated by 25 mm find the force per metre length on the conductor if the current flow is (*i*) in the opposite direction (*ii*) in the same direction. **[Ans.** (*i*) 0.02 N (repulsive) (*ii*) 0.02 (alternative)]

6. An air cored toroidal coil has 3000 turns and carries a current of 0.1 A. The length of the magnetic circuit is 150 mm and the cross-sectional area of the coil is 400 mm^2. Calculate : (*i*) Magnetic field strength, (*ii*) flux density and (*iii*) total flux within the coil. **[Ans.** (*i*) 2000 AT/m, (*ii*) 2.51×10^{-3} T, (*iii*) 1.004×10^{-6} Wb]

7. A cast steel ring of 200 mm mean diameter and 300 mm^2 cross-section has a winding of 200 turns. If the relative permeability of steel is 1000, what current will be required to produce a flux of 3.75×10^{-4} Wb. **[Ans.** 3.17 A]

8. An iron ring of mean length 500 mm has an air gap of 1 mm and a winding of 200 turns. If the permeability of iron is 400 when a current of 1.25 A flows through the coil, find the flux density. **[Ans.** 0.134 T]

9. An iron ring 100 mm mean diameter is made of 15 mm round iron of permeability 900, has an air gap 5 mm wide. It has 400 turns. If the current through this winding is 3.4 A, determine : (*i*) magnetomotive force, (*ii*) the total reluctance of the circuit, (*iii*) the flux in the ring and (*iv*) the flux density in the ring. **[Ans.** (*i*) 1360 AT, (*ii*) 0.2394×10^{8} AT/Wb, (*iii*) 5680×10^{-8} Wb, (*iv*) 0.321 T]

10. A magnetic ring has a mean circumference of 1.5 m and is of 0.01 m^2 in cross-section, and is wound with 175 turns. A saw cut of 4 mm width is made in the ring. Calculate the magnetising current required to produce a flux of 0.8 m Wb in the air gap.
Assume leakage factor as 1.25 and permeability of iron as 400. **[Ans.** 3.16 A]

ELECTROMAGNETISM

11. A flux of 4×10^{-4} Wb is created by a current of 10 A flowing through a 100 turns coil. Calculate the inductance of coil corresponding to a complete reversal of the current in 0.01 second. Find the magnitude of induced e.m.f. **[Ans.** 4×10^{-3} H ; 8 V]

12. The field winding of a D.C. electromagnet is wound with 960 turns and has resistance of 50 Ω when exciting voltage is 230 V, the magnetic flux linking with the coil is 0.005 Wb. Calculate : (*i*) The self inductance of the coil ; (*ii*) The energy stored in the magnetic field. **[Ans.** (*i*) 1.0435 H ; (*ii*) 11.04 J]

13. A solenoid has 1200 turns and carries a current of 2 A. The iron core has a length of 0.4 m and cross-section of 80 cm^2, the relative permeability is 1000. Calculate the self induced e.m.f. in the solenoid, if the current is switched off in 0.01 second. **[Ans.** 7238 V]

14. A coil of 300 turns, wound on a core of non-magnetic material has an inductance of 10 mH. Calculate ;

    (*i*) The flux produced by a current of 5 A ;

    (*ii*) The average value of e.m.f. induced when a current of 5 A is reversed in 8 milli-seconds.

    [Ans. (*i*) 0.1667 m Wb, (*ii*) 12.5 V]

15. Two coils A and B are wound on the same iron core. There are 500 turns on A and 3000 turns on B. A current of 5 amperes through coil A produces a flux of 600 × 10^6 Wb in the core. If this current is reversed in 0.02 s, calculate the average e.m.f. induced in coils A and B. **[Ans.** 30 V ; 180 V]

16. Two coils, A of 12500 turns and B of 16000 turns, lie in parallel planes so that 60% of the flux produced in A links coil B. It is found that a current of 5 A in A produces a flux of 0.6 × 10^{-3} Wb while the same current in B produces 0.8 × 10^{-3} Wb. Determine :

    (*i*) Mutual inductance ;

    (*ii*) Coefficient of coupling. **[Ans.** (*i*) 1.15 H ; (*ii*) 0.586]

UNIT-III : D.C. MACHINES AND TRANSFORMERS

Chapters :

5. Direct Current (D.C.) Machines
6. Transformers

5

Direct Current (D.C.) Machines

1. CONSTRUCTION OF D.C. MACHINES

A D.C. machine consists of *two* main parts :

(*i*) **Stationary part.** It is designed mainly for *producing a magnetic flux*.

(*ii*) **Rotating part.** It is called the *armature*, where mechanical energy is converted into electrical (electrical generator), or conversely, electrical energy into mechanical (electric motor).

The stationary and rotating parts are separated from each other by an *air gap*.

- The *stationary part* of a D.C. machine consists of *main poles*, designed to create the magnetic flux, *commutating poles* interposed between the main poles and designed *to*

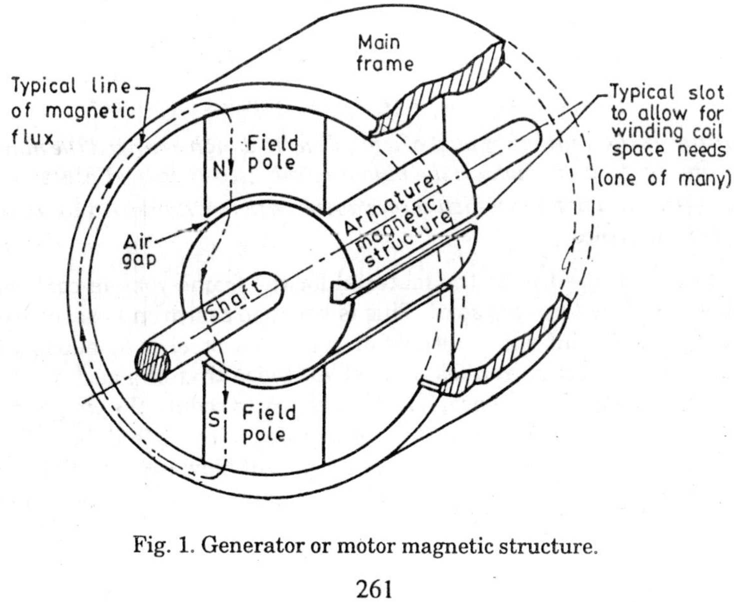

Fig. 1. Generator or motor magnetic structure.

ensure sparkless operation of the brushes at the commutator (in very small machines with a lack of space commutating poles are not used) ; and a *frame*/yoke.

- The *armature* is a cylindrical body rotating in the space between the poles and comprising a *slotted armature core,* a *winding* inserted in the armature core slots, a *commutator,* and *brush gear*.

Fig. 1 shows generator or motor magnetic structure.

Description of Parts of D.C. Machines :

1.1. Frame. Fig. 2 shows the sectional view of four pole D.C. machine.

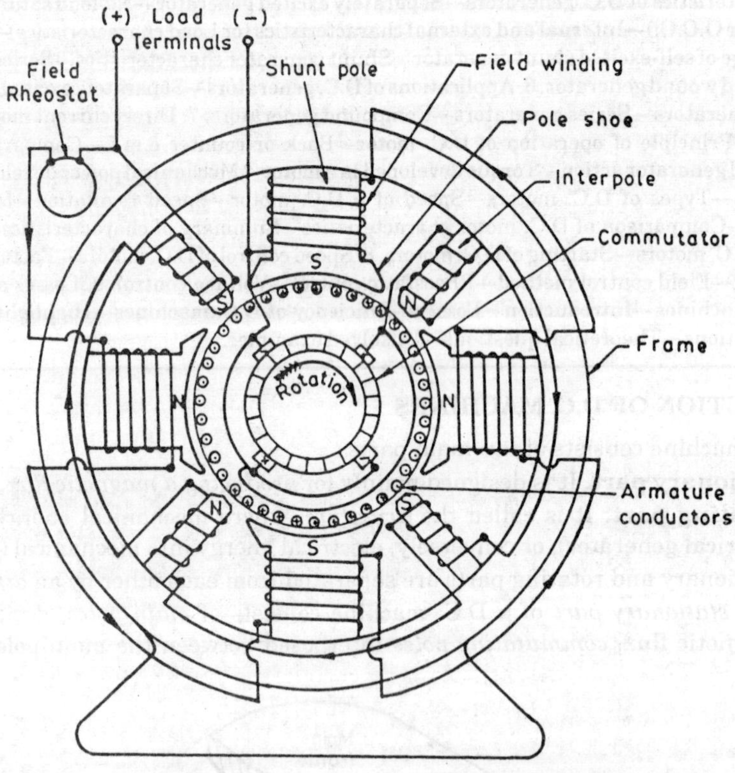

Fig. 2. Sectional view of a four pole D.C. machine.

- *The frame is the stationary part of a machine to which are fixed the main and commutating poles and by means of which the machine is bolted to its bed plate.*
- *The ring-shaped portion which serves as the path for the main and commutating pole fluxes is called the* '**yoke**'.

Cast iron used to be the material for the frame/yoke in *early machines* but now it has been *replaced by cast steel.* This is because cast iron is saturated by a flux density of about 0.8 Wb/m^2 while saturation with cast steel is at about 1.5 Wb/m^2. Thus the cross-section of a cast iron frame is about twice that of a cast steel frame for the same value of magnetic flux. Hence, if it is necessary to reduce the weight of machine, cast steel is used. Another disadvantage with the use of cast iron is that its mechanical and magnetic properties are uncertain due to the presence of blow holes in the casting. Lately, rolled steel yokes have been developed with the improvements in the welding techniques. The *advantages of fabricated yokes are that there are no pattern charges and the magnetic and mechanical properties of the frame are absolutely consistent.*

It may be advantageous to use cast iron for frames but for medium and large sizes usually rolled steel is used.

- If the armature diameter does not exceed 35 to 45 cm, then, in addition to the poles, end shields or frame-heads which carry the bearings are also attached to the frame. When the armature diameter exceeds 1 m, it is common practice to use pedestal-type bearings, mounted separately, on the machine bed plate outside the frame.
- The end shield bearings, and sometimes the pedestal bearings, are of ball or roller type. However, more frequently plain pedestal bearings are used.
- In machines with large diameter armatures a brush-holder yoke is frequently fixed to the frame.

1.2. Field Poles

- Formerly the poles were cast integral with the yoke. This practice is still being followed for small machines. But in present day machines *it is usual to use either a completely laminated pole, or solid steel poles with laminated pole shoes.*
- Laminated construction is necessary because of the pulsations of field strength that result when the notched armature rotor magnetic structure passes the pole shoe. Variations in field strength result in internal eddy currents being generated in a magnetic structure. These eddy currents cause losses ; they may be largely prevented by having laminated magnetic structures. Laminated structures allow magnetic flux to pass along the length of the laminations, but do not allow electric eddy currents to pass across the structure from one lamination to another. The assembled stack of laminations is held together as a unit by appropriately placed rivets. *The outer end of the laminated pole is curved to fit very closely into the inner surface of the main frame.*
- Fig. 3 shows the constructional details of a field pole. *The pole shoe acts as a support to the field coils and spreads out the flux in the air gap and also being of larger cross-section reduces the reluctance of the magnetic path.*

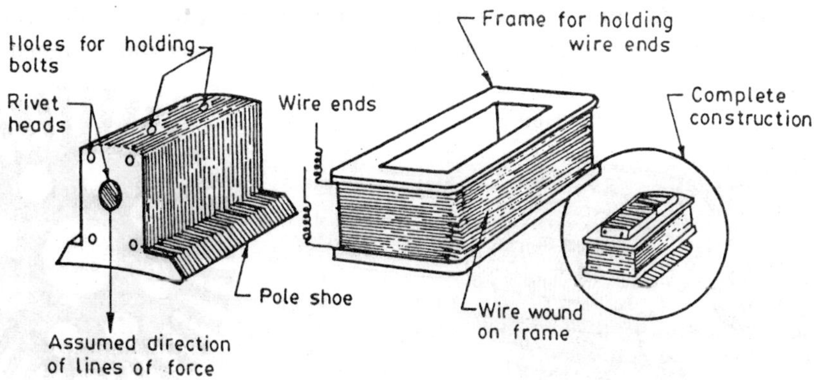

Fig. 3. Constructional details of a field pole.

- Different methods are used for attaching poles to the yoke. In case of *smaller sizes,* the back of the pole is drilled and tapped to receive pole bolts (*see* Fig. 4). In *larger sizes*, a circular or a rectangular pole bar is fitted to the pole. This pole bar is drilled and tapped and the pole bolts passing through laminations screw into the tapped bar (*see* Fig. 5).

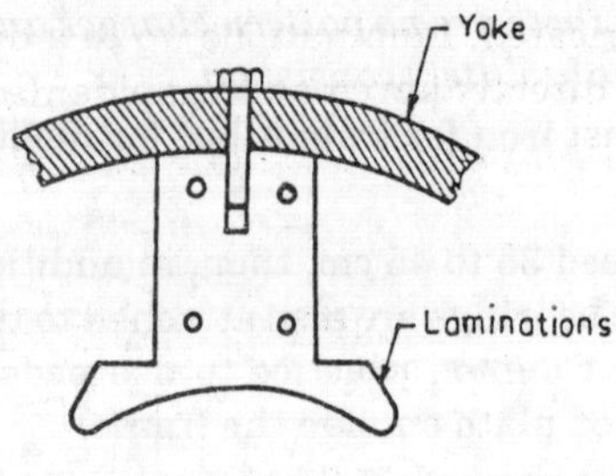

Fig. 4. Fixing pole to the yoke.

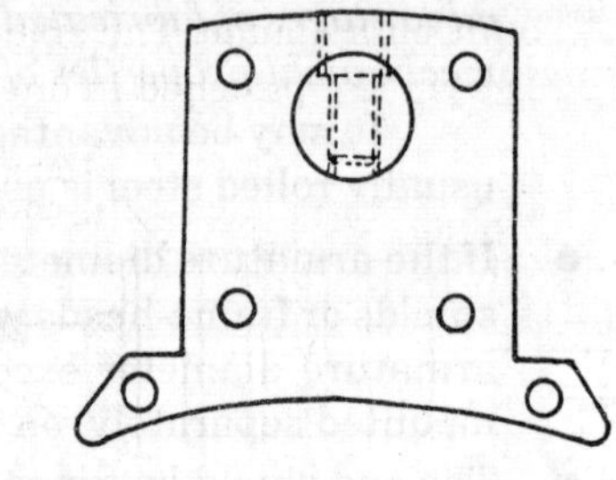

Fig. 5

1.3. Commutating Poles

- A commutating pole (also called *interpole*) is similar to a main pole and consists of core terminating in a pole shoe, which may have various shapes, and coil mounted on the core.
- The commutating poles *are arranged strictly midway between the main poles* and are bolted to the yoke.
- *Commutating poles are usually made of solid steel, but for machines operating on sharply varying loads they are made of sheet steel.*

1.4. Armature

- The armature consists of core and winding. *Iron being the magnetic material is used for armature core.* However, iron is also a good conductor of electricity. The rotation of solid iron core in the magnetic field results in eddy currents. The flow of eddy currents in the core leads to wastage of energy and creates the problem of heat dissipation. *To reduce the eddy currents the core is made of thin laminations.*
- *The armature of D.C. machines (see Fig. 6) is built up of thin laminations of low loss silicon steel.* The laminations are usually 0.4 to 0.5 mm thick and are insulated with varnish.
- The armature **laminations**, in *small machines,* are fitted directly on to the shaft and are clamped tightly between the flanges which also act as supports for the armature winding. One end flange rests against a shoulder on the shaft, the laminations are fitted and other end is pressed on the shaft and retained by a key.

The core (except in small size) is divided into number of packets by radial ventilation spacers. The spacers are usually I sections welded to thick steel laminations and arranged to pass centrally down each tooth.

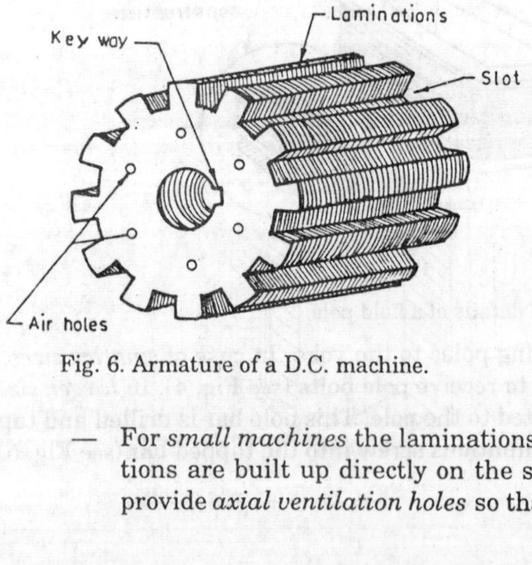

Fig. 6. Armature of a D.C. machine.

Fig. 7. Drum armature stamping with axial flow ventilation system.

— For *small machines* the laminations are punched in one piece (*see* Fig. 7). These laminations are built up directly on the shaft. With such an arrangement, it is necessary to provide *axial ventilation holes* so that air can pass into ventilating ducts.

— The armature laminations of *medium size machines* (having more than four poles) are built on a spider. The spider may be fabricated. Laminations up to a diameter of about 100 cm are punched in one piece and are directly keyed on the spider (*see* Fig. 8).

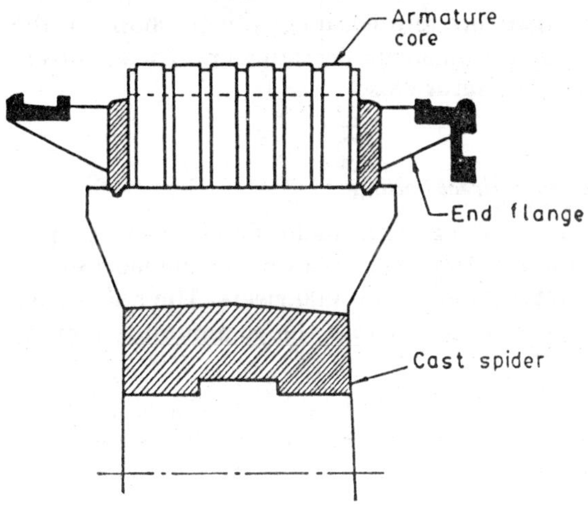

Fig. 8. Clamping of an armature core.

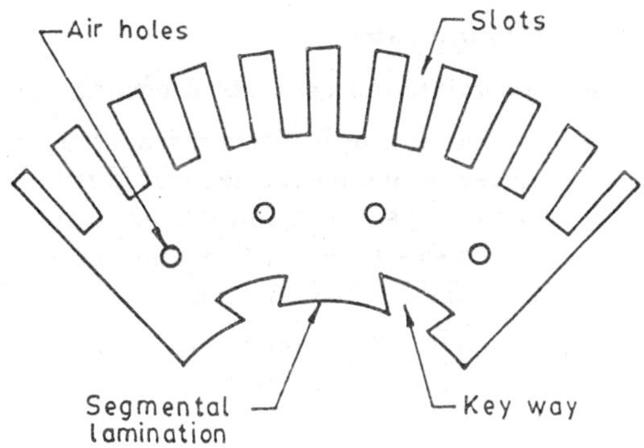

Fig. 9. Segmental stampings.

— In case of *large machines,* the laminations of such thin sections are difficult to handle because they tend to distort and become wavy when assembled together. Hence circular laminations instead of being cut in one piece are cut in a number of suitable sections or *segments* which form part of a complete ring (*see* Fig. 9). A complete circular limination is made up of four or six or even eight segmental laminations. Usually two keyways are notched in each segment and are dove-tailed or wedge shaped to make the laminations self-locking in position.

- The armature winding is housed in slots on the surface of the armature. The conductors of each coil are so spaced that when one side of the coil is under a north pole, the opposite is under a south pole.

Fig. 10 shows the arrangement of conductors and insulation in a slot.

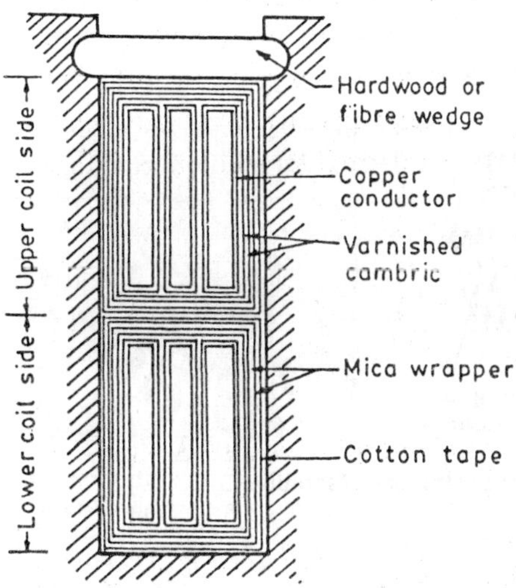

Fig. 10. Cross-section of an armature slot.

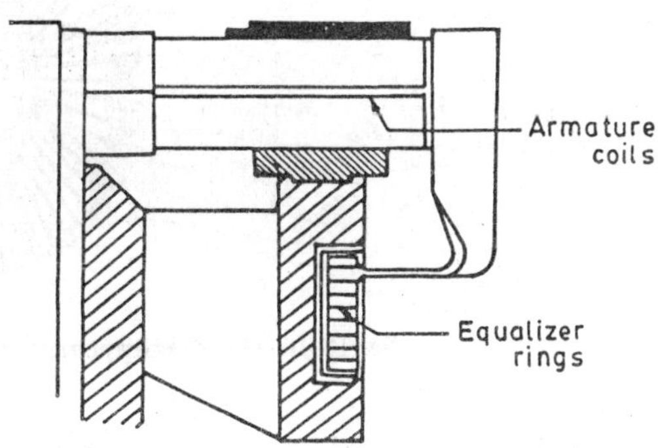

Fig. 11. Ring type equilizers.

- *In D.C. machines two layer winding with diamond shaped coils is used.* The coils are usually former wound. In *small machines,* the coils are held in position by band of steel wire, wound under tension along the core length. In *large machines*, it is useful to employ wedges of fibre or wood to hold coils in place in the slots. Wire bands are employed for holding the overhang. The *equilizer connections* are located under the overhang on the side of the commutator. Fig. 11 shows a typical arrangement for equilizers. The equilizers can be accommodated on the other end of the armature also.

1.5. Commutator

- *A commutator converts alternating voltage to a direct voltage.*
- A commutator is a cylindrical structure built up of segments made of hard drawn copper. These segments are separated from one another and from the frame of the machine by *mica strips.* The segments are connected to the winding through risers. The risers have air spaces between one another so that air is drawn across the commutator thereby keeping the commutator cool.

Fig. 12 shows the components of a commutator. The general appearance of a commutator when completed is as shown in Fig. 13 (*a*). The commutator and armature assembly is shown in Fig. 13 (*b*).

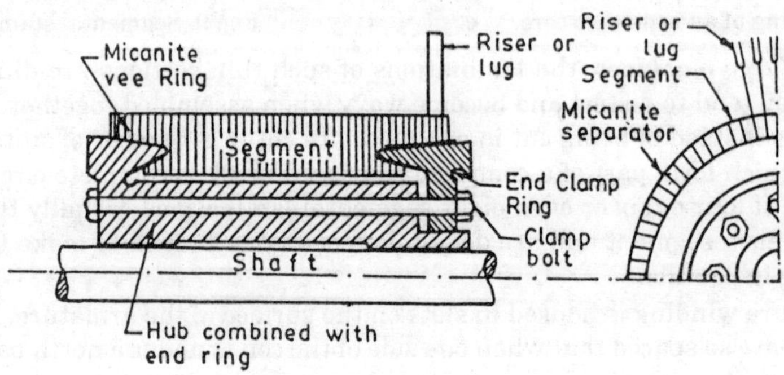

Fig. 12. Commutator components.

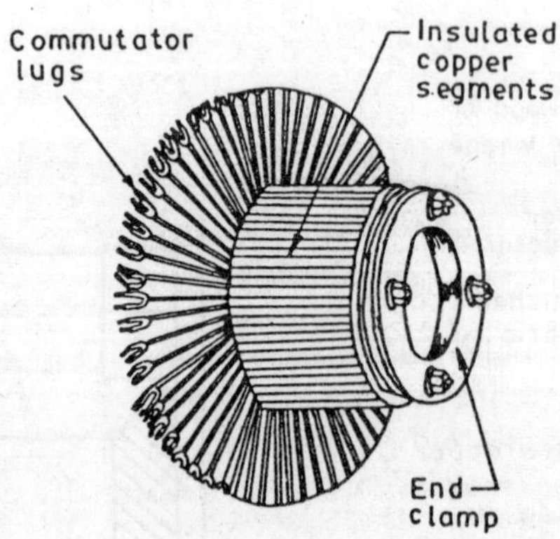

Fig. 13. (*a*) General appearance of a commutator after assembly.

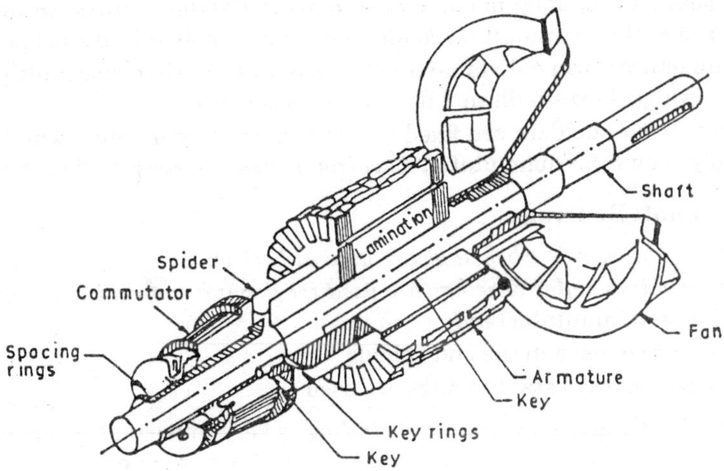

Fig. 13. (*b*) Commutator and armature assembly.

1.6. Brush Gear. To collect current from a rotating commutator or to feed current to it use is made of *brush-gear* which consists of :

(*i*) Brushes
(*ii*) Brush holders
(*iii*) Brush studs or brush-holder arms
(*iv*) Brush rocker
(*v*) Current-collecting busbars.

Brushes. The brushes used for D.C. machines are divided into five classes :

(*i*) Metal graphite
(*ii*) Carbon graphite
(*iii*) Graphite
(*iv*) Electro-graphite
(*v*) Copper.

- The allowable *current density* at the brush contact varies from 5 A/cm^2 in case of carbon to 23 A/cm^2 in case of copper.
- The use of *copper brushes* is made for machines designed for *large currents at low voltages.* Unless, very carefully lubricated, they cut the commutator very quickly and in any case, the wear is rapid. *Graphite and carbon graphite brushes are self-lubricating and, are, therefore, widely used.* Even with the softest brushes, however, there is a gradual wearing away of the commutator, and if mica between the commutator segments does not wear down so rapidly as the segments do, the high mica will cause the brushes to make poor contact with segments, and sparking will result and consequent damage to commutator. So to prevent this, the mica is frequently '*undercut*' to a level below the commutator surface by means of a narrow milling cutter.

Brush holders. *Box type brush holders* are used in all ordinary D.C. machines. A box type brush holder is shown in Fig. 14. At the outer end of the arm, a brush box, open at top and bottom is attached. The brush is pressed on to the commutator by a *clock spring.* The *pressure can be adjusted by a lever arrangement provided with the spring.* The brush is connected to a flexible conductor called *pig tail.* The flexible conductor may be attached to the brush by a screw or may be soldered.

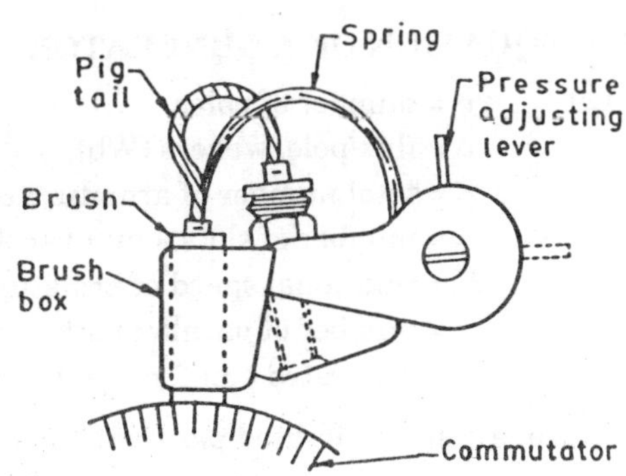

Fig. 14. Box type brush holder.

- The bush boxes are usually made of *bronze casting* or *sheet brass*. In low voltage D.C. machines where the commutation conditions are easy galvanised steel box may be used.
- Some manufacturers use individual brush holders while others use multiple holders, *i.e.*, a number of single boxes built up into one long assembly.

Brush rockers. Brush holders are fixed to brush rockers with bolts. The brush rocker is arranged concentrically round the commutator. *Cast iron is usually, used for brush rockers.*

1.7. Armature Shaft Bearings

- *With small machines roller bearings are used at both ends.*
- For *larger machines roller bearings are used for driving end and ball bearings are used for non-driving* (commutator) *end*.
- The bearings are housed in the end shields.
- For large machines pedestal bearings are used.

1.8. Armature Windings. The *armature winding* is very important element of a machine, as it directly takes part in the conversion of energy from one form into another. The requirements which a winding must meet are diverse and often of a conflicting nature. Among these requirements the following are of major importance.

- The winding must be designed with the *most advantageous utilisation of the material in respect to weight and efficiency*.
- The winding should *provide the necessary mechanical, thermal and electrical strength of the machine* to ensure the usual service life of 16-20 years.
- For D.C. machines proper current collection at the commutator (*i.e.*, absence of detrimental sparking) must be ensured.
- According to the degree of closure produced by winding, armature windings are of the following two types :

  1. Open coil winding   2. Closed coil winding.

  The closed armature windings are of two types :

  (*i*) Ring winding   (*ii*) Drum winding

  In general there are two types of drum armature windings :

  (*i*) Lap winding   (*ii*) Wave winding.

  "Lap winding" is *suitable for comparatively low voltage* but high current generators whereas **"wave of winding"** is used for high voltage, low current machines.

— In '*lap winding*' the finish of each coil is connected to the start of the next coil so that winding or commutator pitch is unity.

— In '*wave winding*' the finish of coil is connected to the start of another coil well away from the fixed coil.

2. E.M.F. EQUATION OF A GENERATOR

Let p = number of poles,

ϕ = flux/pole, webers (Wb),

Z = total number of armature conductors,

= number of slots × number of conductors/slot,

N = rotational speed of armature, r.p.m.,

a = number of parallel paths in armature, and

E_g = generated e.m.f. *per parallel path in armature*.

Average e.m.f. generated per conductor $= \frac{d\phi}{dt}$ volt.

Now, flux cut per conductor in one revolution, $d\phi = p\phi$ Wb.

Number of revolutions/second $= \frac{N}{60}$

$\therefore$ Time for one revolution, $dt = \frac{60}{N}$ seconds

Hence, according to Faraday's laws of electromagnetic induction,

E.m.f. generated per conductor $= \frac{p\phi N}{60}$ volts.

For a lap wound generator :

Number of parallel paths, $a = p$

Number of conductor (*in series*) in one path $= \frac{Z}{p}$

$\therefore$ E.m.f. generated *per path* $= \frac{\phi p N}{60} \times \frac{Z}{p} = \frac{\phi Z N}{60}$ volt.

For a wave wound generator :

Number of parallel paths, $a = p$

Number of conductor (in series) in one path $= \frac{Z}{2}$

$\therefore$ E.m.f. generator per path $= \frac{p\phi N}{60} \times \frac{Z}{2} = \frac{p\phi Z N}{120}$ volt.

In general, generated e.m.f.

$$E_g = \frac{\phi Z N}{60} \times \left(\frac{p}{a}\right) \text{volt} = \frac{p\phi Z N}{60a} \qquad ...(1)$$

where $a = p$ ...... for lap winding

$= 2$ ...... for wave winding.

Example 1. *A six pole lap wound D.C. generator has 720 conductors, a flux of 40 m Wb per pole is driven at 400 r.p.m. Find the generated e.m.f.*

Solution. Number of poles, $p = 6$

Total number of conductors, $Z = 720$

Flux per pole, $\phi = 40$ m Wb $= 40 \times 10^{-3}$ Wb

Speed of rotation, $N = 400$ r.p.m.

Number of parallel paths, $a = p = 6$ [Since the generator is *lap wound.*]

Generated e.m.f. E_g :

Using the relation, $E_g = \frac{p\phi Z N}{60a} = \frac{6 \times 40 \times 10^{-3} \times 720 \times 400}{60 \times 6} = 192$ V.

Hence, **generated e.m.f. E_g = 192 V. (Ans.)**

Example 2. *A six-pole lap connected generator has a useful flux/pole of 0.045 Wb. If the no load voltage at 400 r.p.m. is 300 V, find the conductors on the armature periphery.*

Solution. Number of poles, $p = 6$

Useful flux/pole, $\phi = 0.045$ Wb

No load voltage, $E_g = 300$ V

Number of conductors, Z :

Number of parallel paths, $a = p = 6$ [Since the generator is *lap wound.*]

We know that, $E_g = \frac{p\phi Z N}{60a}$

$$300 = \frac{6 \times 0.045 \times Z \times 400}{60 \times 6}$$

$$\therefore \quad Z = \frac{300 \times 60 \times 6}{6 \times 0.045 \times 400} \quad i.e. \quad Z = 1000.$$

Hence, **total number of armature conductors = 1000. (Ans.)**

Example 3. *An 8-pole wave connected D.C. generator has 1000 armature conductors and flux/pole 0.035 Wb. At what speed must it be driven to generate 500 V?*

Solution. Number of poles, $p = 8$

Total number of armature conductor, $Z = 1000$

Flux/pole, $\phi = 0.035$ Wb

Generated voltage, $E_g = 500$ V

Number of parallel paths, $a = 2$ [Since the generator is wave wound.]

Speed of rotation, N :

Using the relation, $E_g = \frac{p\phi ZN}{60a}$

$$500 = \frac{8 \times 0.035 \times 1000 \times N}{60 \times 2}$$

$$\therefore \quad N = \frac{500 \times 60 \times 2}{8 \times 0.035 \times 1000} = 214.3 \text{ r.p.m.}$$

Hence, **speed of generator = 214.3 r.p.m. (Ans.)**

Example 4. *The armature of a 6-pole D.C. generator has a wave winding containing 650 conductors. Calculate the generated e.m.f. when the flux per pole is 0.055 Wb and the speed is 300 r.p.m.*

Calculate speed at which the armature must be driven to generate an e.m.f. of 550 V if the flux per pole is reduced to 0.05 Wb.

Solution. Number of poles, $p = 6$

Total number of conductors, $Z = 650$

Flux per pole, $\phi = 0.055$ Wb

Speed of rotation, $N = 300$ r.p.m.

E.m.f. generated, $E_g = ?$

Generated e.m.f. (2nd case) $= 550$ V

Flux per pole (2nd case) $= 0.05$ Wb

Speed of rotation, $N = ?$

Case I. E.m.f. generated, E_g :

Using the relation, $E_g = \frac{p\phi ZN}{60a}$

$$= \frac{6 \times 0.055 \times 650 \times 300}{60 \times 2} \quad [\because a = 2, \text{ as the generator is wave wound}]$$

$$= 536.25 \text{ V.}$$

Hence, **e.m.f. generated = 536.25 V. (Ans.)**

Case II. Speed of rotation, N :

$$E_g = \frac{p\phi ZN}{60a}$$

$$550 = \frac{6 \times 0.05 \times 650 \times N}{60 \times 2}$$

$$N = \frac{550 \times 60 \times 2}{6 \times 0.05 \times 650} = 338.46 \text{ r.p.m.}$$

Hence, **speed of rotation = 338.46 r.p.m. (Ans.)**

Example 5. *A six pole lap wound D.C. armature has 70 slots with 20 conductors/slot. The ratio of pole arc to pole pitch is 0.68. The diameter of bore of the pole shoe is 0.46 m. The length of pole shoe is 0.3 m. If the air gap flux density is 0.3 Wb/m² and the e.m.f. induced in the armature is 500 V, find the speed at which it runs.*

Solution. Number of poles, $p = 6$

Number of slots $= 70$

Conductors/slot $= 20$

$\therefore$ Total number of conductors, $Z = 70 \times 20 = 1400$

Ratio of pole arc to pole pitch $= 0.68$

Diameter of bore of the pole shoe $= 0.46$ m

Length of the pole shoe $= 0.3$ m

Air gap flux density, $B = 0.3$ Wb/m^2

E.m.f. induced, $E_g = 500$ V

Speed of rotation, N :

$$\frac{\text{Pole arc}}{\text{Pole pitch}} = 0.68$$

$\therefore$ Pole arc

$$= 0.68 \times \text{pole pitch}$$

$$= 0.68 \times \frac{\pi D}{p} = \frac{0.68 \times \pi \times 0.46}{6} = 0.1638 \text{ m}$$

Area of pole shoe, A = pole arc × length of pole shoe

$$= 0.1638 \times 0.3 = 0.04914 \text{ m}^2$$

Now, $\phi = B \times A = 0.3 \times 0.04914 = 0.01474$ Wb

Using the relation, $E_g = \dfrac{p\phi ZN}{60a}$

$$500 = \frac{6 \times 0.01474 \times 1400 \times N}{60 \times 6} \quad [\because \; a = p = 6, \text{ generator being lap wound}]$$

$$\therefore \quad N = \frac{500 \times 60 \times 6}{6 \times 0.01474 \times 1400} = 1453.7 \text{ r.p.m.}$$

Hence, **speed of rotation = 1453.7 r.p.m. (Ans.)**

3. TYPES OF D.C. GENERATORS

- The power stations of modern design generate practically only three-phase alternating current. A large part of this power is used in the form of alternating current in industry, for *lighting* and *domestic needs*. When industrial needs make it necessary too or when it is of greater advantage to *use direct current* (for *chemical* and *metallurgical plants, electric traction*, etc.) *it is generally obtained by converting A.C. to D.C. with the help of converters of ionic or machine types.* In the latter case wide use is made of such installations as motor generator sets in which A.C. motor is coupled to a D.C. generator on a common shaft.
- As primary sources of power, D.C. generators are *mainly used in self-contained plants such as automobiles and air planes, for electric arc welding, train car lighting, in sub-marines, etc.*

3.1. Classification

According to *method of excitation* D.C. generators are classified as follows :

1. Separately excited generators, 2. Self-excited generators.

3.1.1. Separately excited generators

These are those generators whose field magnets are *energised* from an independent *external source* of D.C. current. Such a generator is shown in Fig. 15.

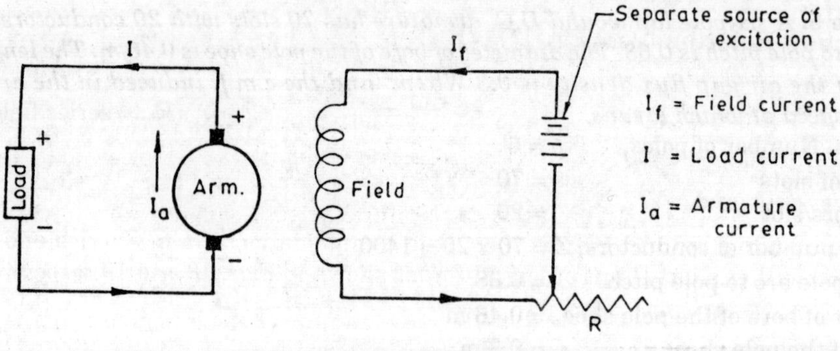

Fig. 15. Separately excited generator.

3.1.2. Self-excited generators

These are those generators whose field magnets are *energised by the current produced by the generators themselves.* Due to residual magnetism, there is always present some flux in the poles. When the armature is rotated, some e.m.f. and hence some induced current produced which is partly or fully passed through the field coils thereby *strengthening the pole flux.*

Self excited generators can be divided, in accordance with how the *field winding is connected into generators*, as follows :

(*i*) Shunt wound generators

(*ii*) Series wound generators

(*iii*) Compound wound generators :

(*a*) Short shunt

(*b*) Long shunt

(*i*) **Shunt wound generators :** Refer Fig. 16. In these generators the field windings are connected *across or in parallel with the armature conductors,* and have the full voltage of the generator across them.

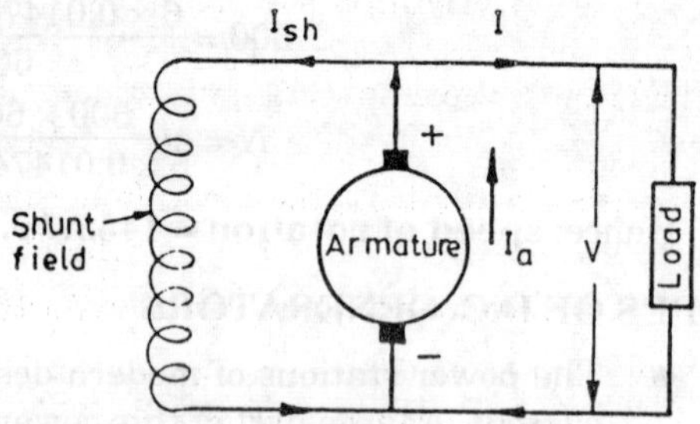

Fig. 16. Shunt wound generator.

Important relations : Refer Fig. 16.

(*i*) $I_{sh} = \dfrac{V}{R_{sh}}$

(*ii*) $I_a = I_{sh} + I$

(*iii*) $V = E_g - I_a R_a$

(*iv*) Power developed $= E_g I_a$

(*v*) Power delivered $= VI$

where I_{sh} = shunt field current, I_a = armature current,

I (or I_l) = load current, R_a = armature resistance,

R_{sh} = shunt field resistance, E_g = generated e.m.f., and

V = terminal voltage.

(*ii*) **Series wound generators :** Refer Fig. 17. In this case, the field windings are joined in series with armature conductors. *As they carry full load current, they consist of relatively few turns of thick wire or strip.* The use of such generators is limited to special purposes (as *boosters* etc.).

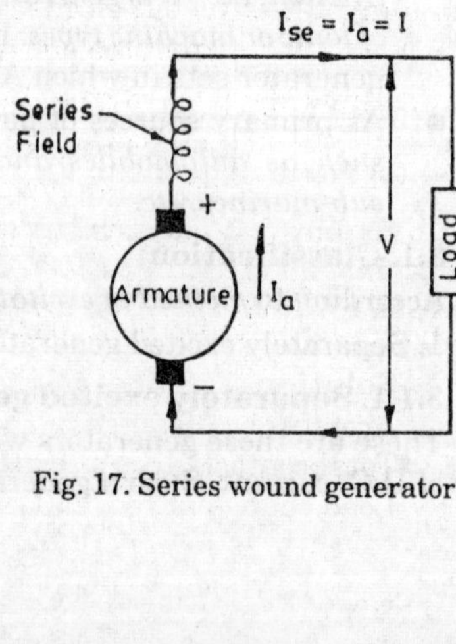

Fig. 17. Series wound generator.

Important Relations. (*see* Fig. 17) :

(*i*) $I_a = I_{se} = I$ (I_{se} = series field current)

(*ii*) $V = E_g - I(R_a + R_{se})$ (R_{se} = series field resistance)

(*iii*) Power developed = $E_g I$

(*iv*) Power delivered = VI.

(*iii*) **Compound wound generators.** It is a combination of a few series and a few shunt windings and be either short shunt or long shunt as shown in Figs. 18 and 19 respectively.

Important Relations :

(*a*) **Short shunt compound wound.** (*See* Fig. 18) :

(*i*) $I_{se} = I$

(*ii*) $I_{sh} = \dfrac{V + I_{se}R_{se}}{R_{sh}}$

(*iii*) $I_a = I + I_{sh}$

(*iv*) $V = E_g - I_aR_a - I_{se}R_{se}$

(*v*) Power developed = E_gI_a

(*vi*) Power delivered = VI.

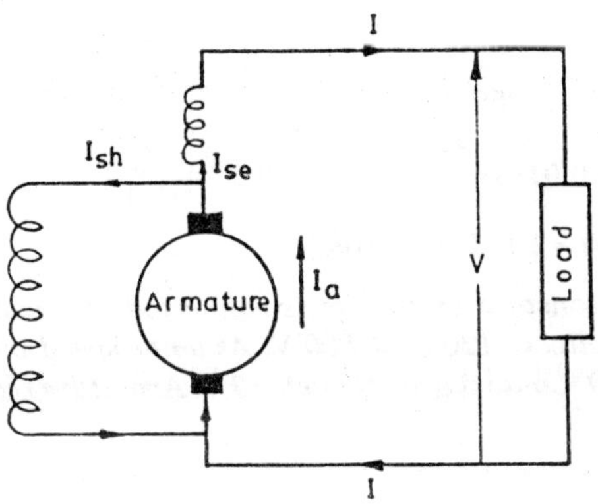

Fig. 18. Short shunt compound wound generator.

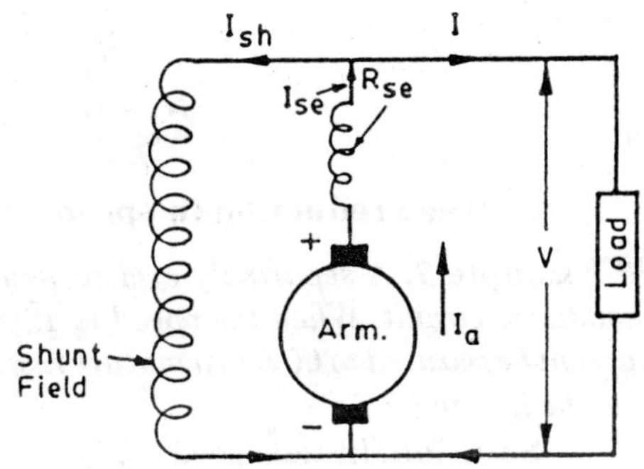

Fig. 19. Long shunt compound wound generator.

(*b*) **Long shunt compound wound.** (*See* Fig. 19) :

(*i*) $I_{sh} = \dfrac{V}{R_{sh}}$

(*ii*) $I_a = I_{se} = I + I_{sh}$

(*iii*) $V = E_g - I_aR_a - I_{se}R_{se} = E_g - I_a(R_a + R_{se})$

(*iv*) Power developed = E_gI_a

(*v*) Power delivered = VI.

Example 6. *The terminal voltage of a separately excited D.C. generator with constant excitation is constant and is equal to 250 V. Determine the percentage reduction in speed when the load changes from 250 kW to 150 kW. The armature resistance is 0.012 Ω and total contact drop at brushes = 2 V. Neglect armature reaction.*

Solution. Terminal voltage, $V = 250$ volts

Armature resistance, $R_a = 0.012\ \Omega$

Total contact drop at brushes = 2 V

Load change : From 250 kW to 150 kW

Percentage reduction in speed :

We know that generated e.m.f.,

$$E_g = \frac{p\phi ZN}{60a} \quad i.e., \quad E_g \propto \phi N$$

Since ϕ is constant, $E_g \propto N$

— When the load is 250 kW, the armature current

$$I_{a1} = \frac{250 \times 1000}{250} = 1000 \text{ A}$$

and $E_{g1} = V + I_{a1} R_{a1}$ + drop in brushes

$= 250 + 1000 \times 0.012 + 2 = 250 + 12 + 2 = 264$ V

— When the load is 150 kW, the armature current

$$I_{a2} = \frac{150 \times 1000}{250} = 600 \text{ A}$$

and $E_{g2} = V + I_{a2} R_a$ + drop in brushes

$= 250 + 600 \times 0.012 + 2 = 250 + 7.2 + 2 = 259.2$ V

Now since $E_g \propto N$

$$\therefore \quad \frac{E_{g2}}{E_{g1}} = \frac{N_2}{N_1}$$

i.e., $\frac{N_2}{N_1} = \frac{259.2}{264}$ or $\frac{N_1}{N_2} = \frac{264}{259.2}$

or $$\frac{N_1 - N_2}{N_1} = \frac{264 - 259.2}{264} = 0.01818$$

$\therefore$ **%age reduction in speed = 0.01818 × 100 = 1.818%. (Ans.)**

☞**Example 7.** *A separately excited generator with constant excitation is connected to a constant resistance circuit. When the speed is 1200 r.p.m. it delivers 120 A at 500 V. At what speed will the current be reduced to 60 A ? Armature resistance = 0.1 Ω. Contact drop / brush = 1 V. Armature reaction may be ignored.*

Solution. Initial speed, $N_1 = 1200$ r.p.m.

Initial load current, $I_{a1} = 120$ A

Initial terminal voltage, $V_1 = 500$ V

Final load current, $I_{a2} = 60$ A

Armature resistance, $R_a = 0.1\ \Omega$

Contact drop/brush = 1 V

Speed N_2 :

Refer Fig. 20.

We know that e.m.f. generated (at load current of 120 A)

$E_{g1} = V_1 + I_{a1} R_a$ + drop in brushes

$= 500 + 120 \times 0.1 + 2 \times 1 = 514$ V

Load resistance (constant),

$$R = \frac{500}{120} = 4.167\ \Omega$$

Let E_{g2} be the generated e.m.f. when current delivered is 60 A.

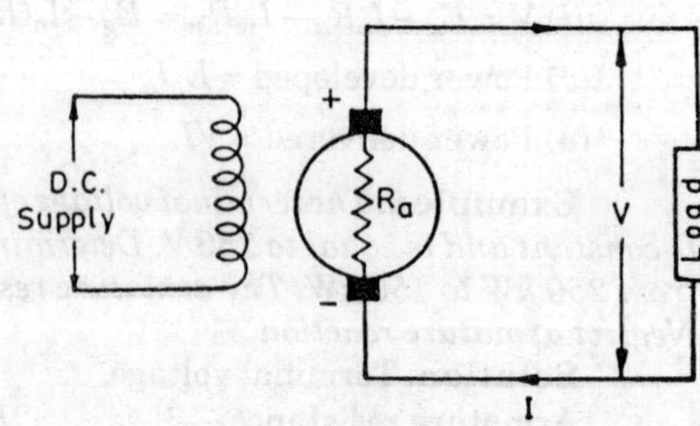

Fig. 20

$$V_2 = I_{a2}R = 60 \times 4.167 = 250 \text{ V}$$
$$E_{g2} = V_2 + I_{a2}R_a + 2 \times 1$$
$$= 250 + 60 \times 0.1 + 2 \times 1 = 258 \text{ V}$$

Now using the relation, $\dfrac{N_2}{N_1} = \dfrac{E_{g2}}{E_{g1}}$

$$\frac{N_2}{1200} = \frac{258}{514}$$
$$N_2 = \frac{1200 \times 258}{514} = 602.33 \text{ r.p.m.}$$

Hence, **speed at a load of 60 A = 602.33 r.p.m. (Ans.)**

Example 8. *A shunt generator supplied 500 A at 500 V. Calculate its generated e.m.f. if its armature and shunt field resistances are 0.02 Ω and 125 Ω respectively.*

Solution. Load current, $I = 500$ A

Terminal voltage, $V = 500$ volts

Armature resistance, $R_a = 0.2\ \Omega$

Shunt field resistance, $R_{sh} = 125\ \Omega$

Refer Fig. 21.

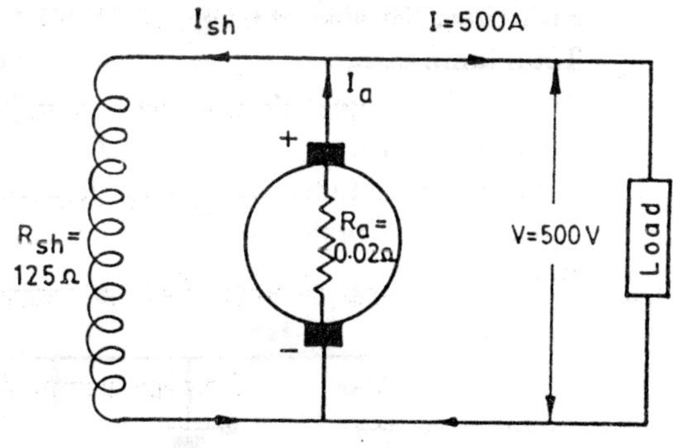

Fig. 21

Shunt field current,

$$I_{sh} = \frac{V}{R_{sh}} = \frac{500}{125} = 4 \text{ A}$$

Armature current,

$$I_a = I + I_{sh} = 500 + 4 = 504 \text{ A}$$

Generated e.m.f.,

$$E_g = V + I_aR_a$$
$$= 500 + 504 \times 0.02 = 510.08 \text{ V}$$

Hence, **generated e.m.f. = 510.08 V. (Ans.)**

Example 9. *A 4-pole, D.C. shunt generator, with a shunt field resistance of 100 ohms and an armature resistance of 1 ohm, has 378 wave-connected conductors in its armature. The flux per pole is 0.02 Wb. If a load resistance of 10 ohms is connected across the armature terminals and the generator is driven at 1000 r.p.m., calculate power absorbed by load.*

Solution. *Given* : $p = 4$; $R_{sh} = 100\ \Omega$; $R_a = 1\ \Omega$, $Z = 378$; $a = 2$; $\phi = 0.02$ Wb ; $R_L = 10\ \Omega$; $N = 1000$ r.p.m.

Power absorbed by the load :

The generator arrangement is shown in the Fig. 22.

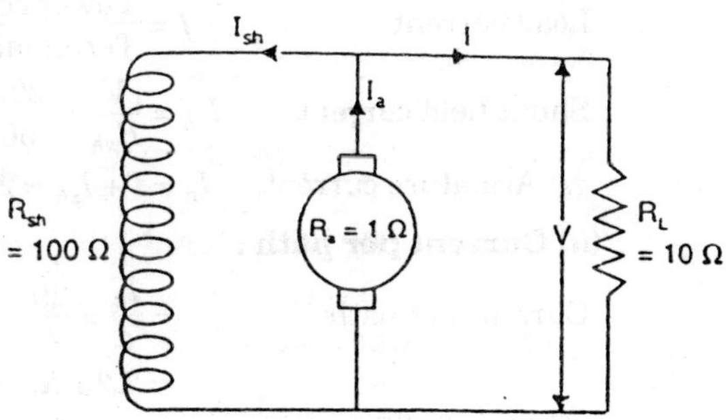

Fig. 22

For the generator,

$$E_g = \frac{p\phi ZN}{60\,a}$$
$$= \frac{4 \times 0.02 \times 378 \times 1000}{60 \times 2} = 252 \text{ V}$$
$$V = E_g - I_aR_a$$
$$\therefore \quad I_a = \frac{E_g - V}{R_a} = \frac{252 - V}{1} = 252 - V$$

Now $\quad I = \dfrac{V}{R_L} = I_a - I_{sh} = (252 - V) - \dfrac{V}{100}$

$$\therefore \quad \frac{V}{10} = (252 - V) - \frac{V}{100} \quad \text{or} \quad 252 - V = \frac{11\,V}{100}$$

$$\text{or} \quad 25200 - 100\,V = 11\,V \quad \text{or} \quad V = \frac{25200}{111} = 227 \text{ V}$$

$$I = \frac{227}{10} = 22.7 \text{ A}$$

$\therefore$ **Power absorbed by the load** $= I^2R_L$

$= 22.7^2 \times 10 = 5153$ W or **5.153 kW. (Ans.)**

Example 10. *A 4-pole lap wound shunt generator supplies to 50 lamps of 100 watts, 200 V each. The field and armature resistances are 50 Ω and 0.2 Ω respectively. Allowing a brush drop of 1 V each brush, calculate the following :*

(i) Armature current. *(ii) Current per path.*
(iii) Generated e.m.f. *(iv) Power output of D.C. armature.*

Solution. Number of poles, $p = 4$
Total lamp load, $P = 50 \times 100 = 5000$ W
Terminal voltage, $V = 200$ Volts
Field resistance, $R_{sh} = 50\ \Omega$
Armature resistance, $R_a = 0.2\ \Omega$
Voltage drop/brush $= 1$ V
Refer Fig. 23.

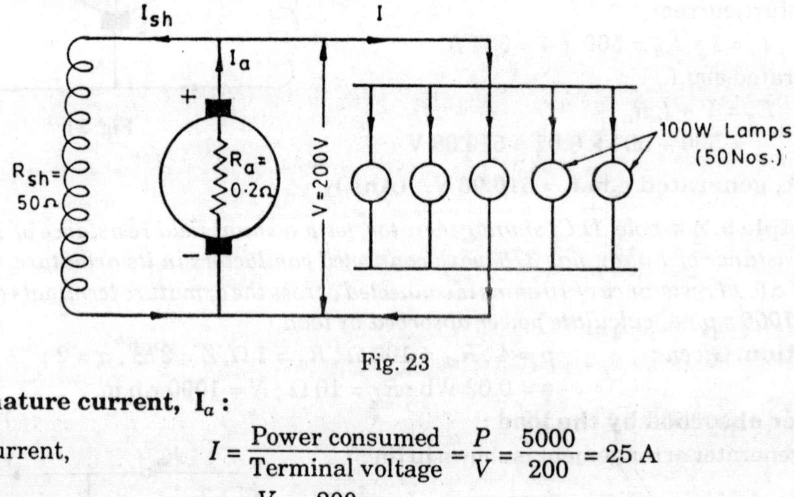

Fig. 23

(i) **Armature current, I_a :**

Load current, $I = \dfrac{\text{Power consumed}}{\text{Terminal voltage}} = \dfrac{P}{V} = \dfrac{5000}{200} = 25$ A

Shunt field current, $I_{sh} = \dfrac{V}{R_{sh}} = \dfrac{200}{50} = 4$ A

$\therefore$ Armature current, $I_a = I + I_{sh} =$ **25 + 4 = 29 A. (Ans.)**

(ii) **Current per path :**

Current per path $= \dfrac{I_a}{a} = \dfrac{29}{4}$ [$\because$ $a = p = 4$, generator being lap wound]

= 7.25 A. (Ans.)

(iii) **Generated e.m.f., E_g :**

$$E_g = V + I_aR_a + \text{brush drop} = 200 + 29 \times 0.2 + 2 \times 1 = 207.8 \text{ V}$$

Hence, **generated e.m.f. = 207.8 V. (Ans.)**

(*iv*) **Power output of D.C. armature :**

Power output of D.C. armature

$$= \frac{E_g I_a}{1000} = \frac{207.8 \times 29}{1000} \text{ kW} = \mathbf{6.026\ kW.}\quad \textbf{(Ans.)}$$

☞**Example 11.** *A 4-pole, 500 V wave-wound shunt generator delivers a load current of 140 A. It has 65 slots with 12 conductors / slot and runs at 800 r.p.m. The shunt field and armature resistances are 250 Ω and 0.2 Ω respectively. The diameter of the bore of the pole shoe is 45 cm, the angle subtended by the pole shoe is 70° and it is 25 cm in length. Assuming contact drop / brush as 1 V, calculate the flux density in the air gap.*

Solution. Refer Fig. 24.

Number of poles, $p = 4$

Number of parallel paths, $a = 2$

[generator being wave wound]

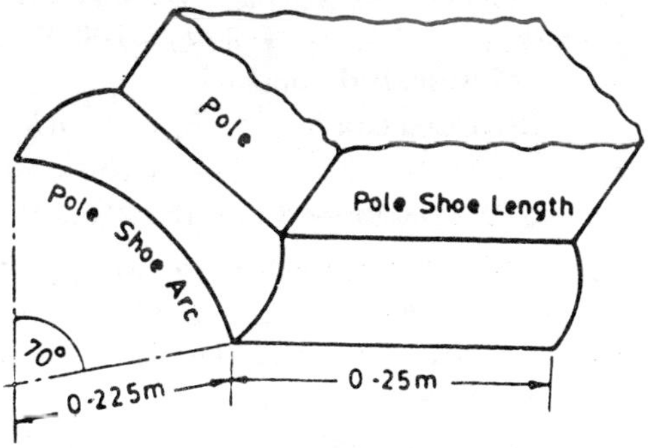

Fig. 24

Terminal voltage, $V = 500$ Volts

Load current, $I = 140$ A

Speed of rotation, $N = 800$ r.p.m.

Number of slots = 65

Number of conductors/slot = 12

∴ Total number of conductors,

$$Z = 12 \times 65 = 780$$

Shunt field resistance, $R_{sh} = 250\ \Omega$

Armature resistance, $R_a = 0.2\ \Omega$

Diameter of the bore of pole shoe, $D = 45$ cm (= 0.45 m)

Angle subtended by the pole shoe, $\theta = 70°$

Length of pole shoe = 25 cm (= 0.25 m)

Contact drop/brush = 1 V

Flux density in the air gap, B :

Shunt field current, $I_{sh} = \frac{V}{R_{sh}} = \frac{500}{250} = 2$ A

Armature current, $I_a = I + I_{sh} = 140 + 2 = 142$ A

Generated e.m.f., $E_g = V + I_a R_a$ + brush drop

$$= 500 + 142 \times 0.2 + 2 \times 1 = 530.4 \text{ V}$$

Also, $E_g = \frac{p\phi ZN}{60a}$

$$530.4 = \frac{4 \times \phi \times 780 \times 800}{60 \times 2}$$

or $$\phi = \frac{530.4 \times 60 \times 2}{4 \times 780 \times 800} = 0.0255 \text{ Wb}$$

Arc length of pole shoe $= \frac{\pi D\theta}{360} = \frac{\pi \times 0.45 \times 70}{360} = 0.275$ m

Area of pole shoe, A = Arc length × length of pole shoe

$$= 0.275 \times 0.25 = 0.06875 \text{ m}^2$$

Flux density in the air gap, $B = \frac{\phi}{A} = \frac{0.0255}{0.06875} = 0.371$ T (or Wb/m^2)

Hence, **flux density in the air gap = 0.371 T. (Ans.)**

Example 12. *A series generator delivers a current of 100 A at 250 V. Its armature and series field resistances are 0.1 Ω and 0.055 Ω respectively. Find :*

(*i*) *Armature current* (*ii*) *Generated e.m.f.*

Solution. *See* Fig. 25.

Load current, $I = 100$ A

Terminal voltage, $V = 250$ Volts

Armature resistance, $R_a = 0.1\ \Omega$

Series field resistance, $R_{se} = 0.055\ \Omega$

(*i*) **Armature current, I_a :**

Armature current (I_a) = load current (I)

$\therefore$ I_a = **100 A. (Ans.)**

(*ii*) **Generated e.m.f. :**

Generated e.m.f. $E_g = V + I(R_a + R_{se})$

$= 250 + 100(0.1 + 0.055) = 265.5$ V

Hence, **generated e.m.f. = 265.5 V. (Ans.)**

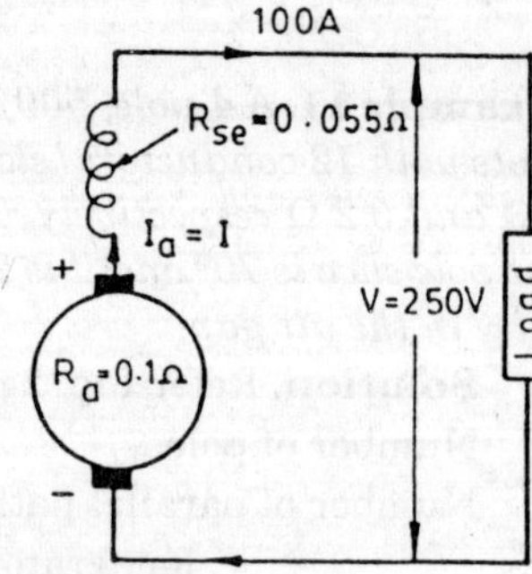

Fig. 25

Example 13. *A short shunt compound generator has armature, series field and shunt field resistances of 0.06 Ω, 0.03 Ω and 110 Ω respectively. It supplies 100 lamps rated at 250 V, 40 W. Find the generated e.m.f. Assume that contact drop / brush = 1 V.*

Solution. *See* Fig. 26.

Armature resistance, $R_a = 0.06\ \Omega$

Series field resistance, $R_{se} = 0.03\ \Omega$

Shunt field resistance, $R_{sh} = 110\ \Omega$

Terminal voltage, $V = 250$ Volts

Lamp load, $P = 100 \times 40 = 4000$ W

Contact drop/brush = 1 V

Generated e.m.f., E_g :

Load current, $I = \dfrac{P}{V} = \dfrac{4000}{250} = 16$ A

Voltage drop in series winding

$= IR_{se} = 16 \times 0.03 = 0.48$ V

Voltage across shunt field winding

$= V + IR_{se} = 250 + 0.48 = 250.48$ V

Shunt field current, $I_{sh} = \dfrac{250.48}{110} = 2.277$ A

Armature current, $I_a = I + I_{sh}$

$= 16 + 2.277 = 18.277$ A

Generated e.m.f., $E_g = V + IR_{se} + I_aR_a$ + brush drop

$= 250 + 0.48 + 18.277 \times 0.06 + 2 \times 1$

= **253.58 V. (Ans.)**

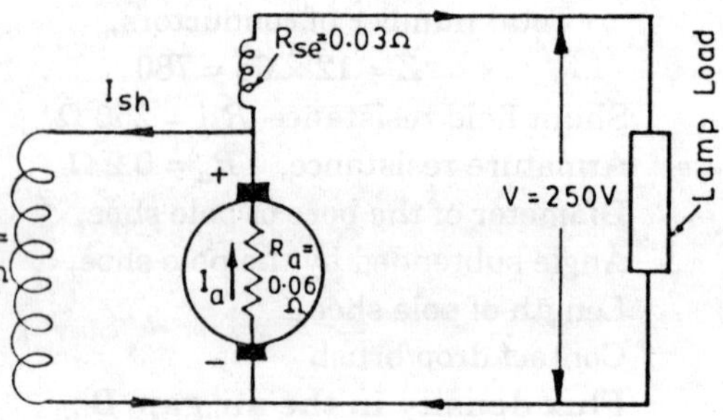

Fig. 26

Example 14. *A long shunt compound generator has an armature, series field and shunt field resistances of 0.04 Ω, 0.03 Ω and 200 Ω respectively. It supplies a load current of 180 A at 400 V. Calculate the generated e.m.f. Assume contact drop/brush = 1 V.*

Solution. Refer Fig. 27.

Armature resistance, $R_a = 0.04\ \Omega$

Series field resistance, $R_{se} = 0.03\ \Omega$

Shunt field resistance, $R_{sh} = 200\ \Omega$
Load current, $I = 180$ A
Terminal voltage, $V = 400$ Volts
Contact drop/brush $= 1$ V

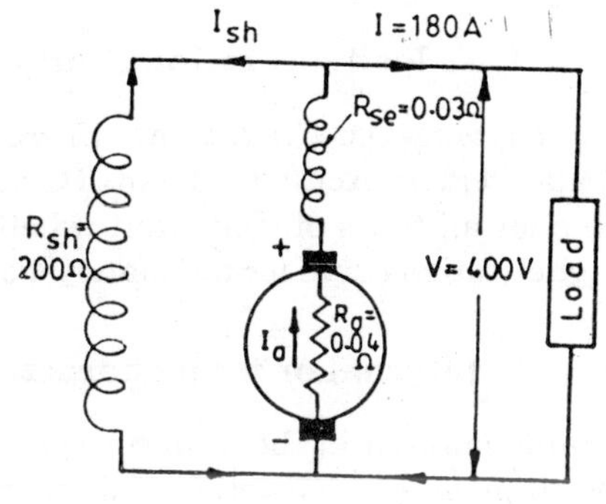

Fig. 27

Generated e.m.f., E_g :

Shunt current, $I_{sh} = \dfrac{V}{R_{sh}} = \dfrac{400}{200} = 2$ A

Armature current, $I_a = I + I_{sh}$
$= 180 + 2 = 182$ A

Generated e.m.f., $E_g = V + I_aR_a + I_aR_{se}$ + drop at brushes
$= 400 + 182 \times 0.04 + 182 \times 0.03 + 2 \times 1$
$= 400 + 7.28 + 5.46 + 2 = 414.74$ V

Hence, **generated e.m.f. = 414.74 V. (Ans.)**

4. POWER DIVISION IN A D.C. GENERATOR

Fig. 28 presents in graphical form power balances for D.C. generators with both shunt and series fields.

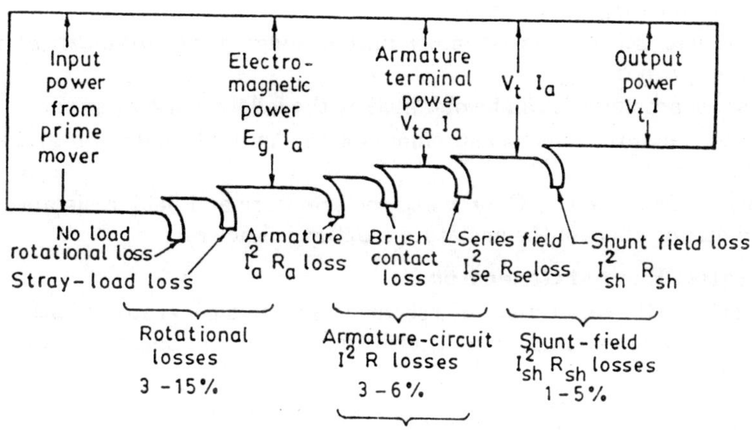

Fig. 28. Power division in a D.C. generator.

5. CHARACTERISTICS OF D.C. GENERATORS

The properties of generators are analysed with the aid of characteristics which give the relations between fundamental quantities determining the operation of a generator. These include the voltage across the generator terminals V, the field or exciting current I_f, the armature current I_a, and the speed of rotation N.

The three most important characteristics of D.C. generators are given below :

1. No load saturation characteristics $\left(\dfrac{E_0}{I_f}\right)$

2. Internal or total characteristics $\left(\dfrac{E}{I_a}\right)$

3. External characteristics $\left(\dfrac{V}{I}\right)$.

1. No load saturation characteristic $\left(\frac{E_0}{I_f}\right)$. It is also a known as *magnetic or open circuit characteristic* (O.C.C.). It shows the relationship between the no-load generated e.m.f. in armature, E_0 and field or exciting current I_f at a given fixed speed. The shape of the curve is practically the same for all types of generators whether they are separately excited or self-excited. It is just the magnetisation curve for the material of the electromagnets.

2. Internal or total characteristic $\left(\frac{E}{I_a}\right)$. It gives the relationship between the *e.m.f. E actually induced* in the armature after allowing for the demagnetising effect of armature reaction and the *armature current* I_a. This characteristic is of interest mainly to the designer.

3. External characteristic $\left(\frac{V}{I}\right)$

- This characteristic is also referred to as *performance characteristic* or sometimes *voltage-regulating curve.*
- It gives relation between the *terminal voltage V* and *load current I.*
- The curve lies below the internal characteristic because it takes into account the voltage drop over the armature circuit resistance. The *values of V are obtained by subtracting* I_aR_a *from corresponding values of E.*
- The characteristic is of great importance in judging the *suitability of a generator for a particular purpose.*

The external characteristic can be obtained by the following two ways :

(*i*) By making simultaneous measurements with a suitable voltmeter and an ammeter on a loaded generator.

(*ii*) Graphically from the O.C.C. provided the armature and field resistances are known and also if the demagnetising effect of the armature reaction is known.

5.1. Separately Excited Generator

- Fig. 29 shows the connections of a separately excited generator, a battery being indicated as the source of exciting current, although any other constant voltage source could be used.

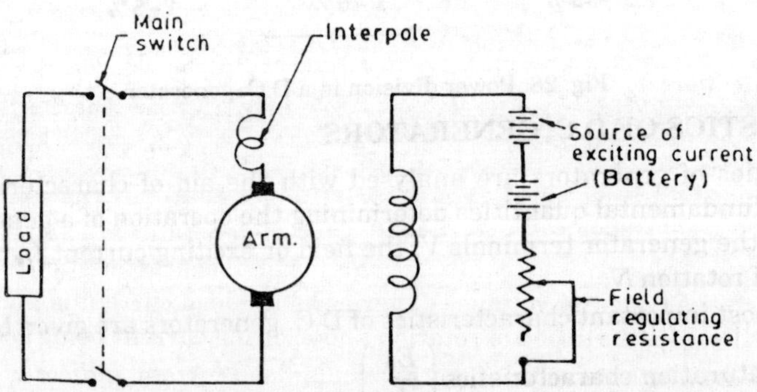

Fig. 29. Connection for a separately excited generator.

The field circuit is provided with a variable resistance and would normally contain a field switch and an ammeter, these being omitted from the diagram for simplicity. The armature is connected through 2-pole main switch to the bus bars, between which the load is connected.

5.1.1. No-load saturation characteristic (or O.C.C.) :

- If the generator is run at constant speed with the main switch open, and the terminal voltage is noted at various values of exciting or field current then the O.C.C. shown in Fig. 30 can be plotted. This is also referred to as the '*magnetisation curve*' since the same graph shows, to a suitably chosen scale, the amount of magnetic flux, there being a constant relationship (depending upon speed of rotation) between flux and induced voltage.
- It will be noticed that a small voltage is produced when the field current is zero, this being due to a small amount of permanent magnetism in the field poles. This is called *residual magnetism* and is usually sufficient to produce 2 or 3 per cent of normal terminal voltage, although in some special cases it is purposely increased to 10 per cent or more.
- The *first part of the curve is approximately straight* and *shows that the flux produced is proportional to the exciting current*; but *after a certain point, saturation of the iron becomes perceptible as the curve departs from straight line form.*

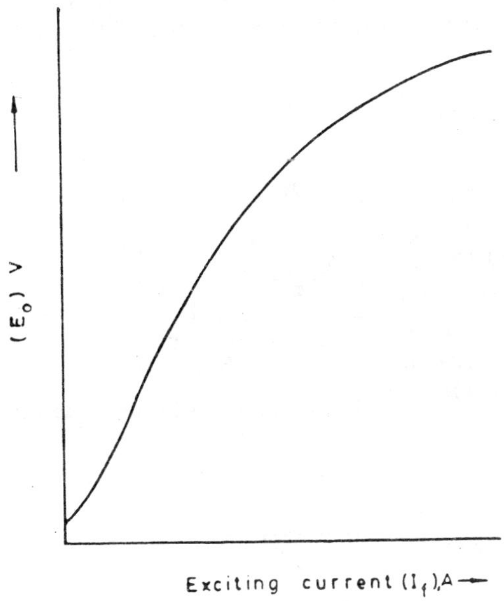

Fig. 30. Open-circuit characteristic of a separately excited generator.

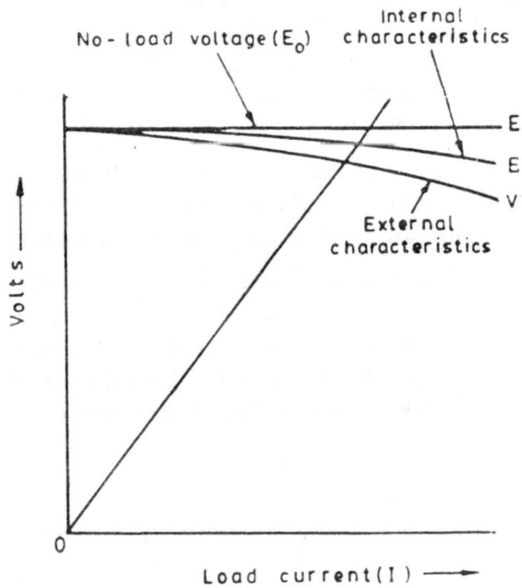

Fig. 31. Load characteristics of a separately excited generator.

5.1.2. Internal and external characteristics (or load characteristics) :

- Load characteristics for a separately excited generator are shown in Fig. 31. The most important is the '*external characteristic*' (or total characteristic), which indicates the way in which the terminal voltage (*V*) varies as the load current is increased from zero to its full load value, the speed of rotation and exciting current being constant.

The voltage drop (drop of volts) at any particular load current, *indicated by the vertical distance between the external characteristic and the no-load voltage* is brought about by two causes :

(*i*) *Armature reaction which has a demagnetising effect upon the field.*

(*ii*) *Resistance drop*, this being the product of the armature current and the total armature-circuit resistance, consisting of the armature resistance, interpole resistance and brush contact resistance.

- The '*internal characteristic*' is obtained by *calculating the resistance drop for a few values of current and adding this to the voltage shown by the external characteristic. The vertical distance between the internal characteristic and no load voltage then represents the effect of armature reaction alone.*
- When the resistance of load is R, then voltage across its terminals is $V = IR$, where I represents the current, so that if the values of V corresponding to various values of I are calculated, the values will all lie upon a straight line such as OL in Fig. 31. *The load current and terminal voltage corresponding to this resistance are given by the inter-section of the line OL with the external characteristics.*

Note. The great *advantage of separate excitation* over all other forms of excitation is that the *current is entirely independent of the load current in the armature.* It is however, rather inconvenient to have to depend upon a separate source of supply and, therefore, the *method is used only in special cases, where the generator has to operate over a wide range of terminal voltage.*

5.2. Building up the Voltage of Self-excited Shunt Generator

One of the simplest forms of 'self-excited' generator is the shunt-wound machine, the connection diagram (without load) of which is shown in Fig. 32. The manner in which a self-excited generator manages to excite its own field and build a D.C. voltage across its armature is described with reference to Fig. 33 in the following **steps** :

(*i*) Assume that the generator starts from rest, *i.e.*, prime-mover speed is zero. Despite a residual magnetism, the generated e.m.f. E, is zero.

(*ii*) As the prime-mover rotates the generator armature and the speed approaches rated speed, the voltage due to residual magnetism and speed increases.

(*iii*) At rated speed, the voltage across the armature due to residual magnetism is small, E_1, as shown in the figure. But this voltage is also across the field circuit whose resistance is R_f. Thus, the current which flows in the field circuit I_1, is also small.

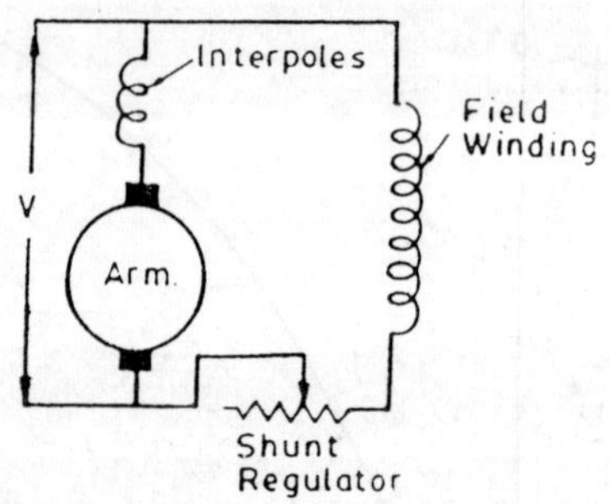

Fig. 32. Self-excited shunt generator.

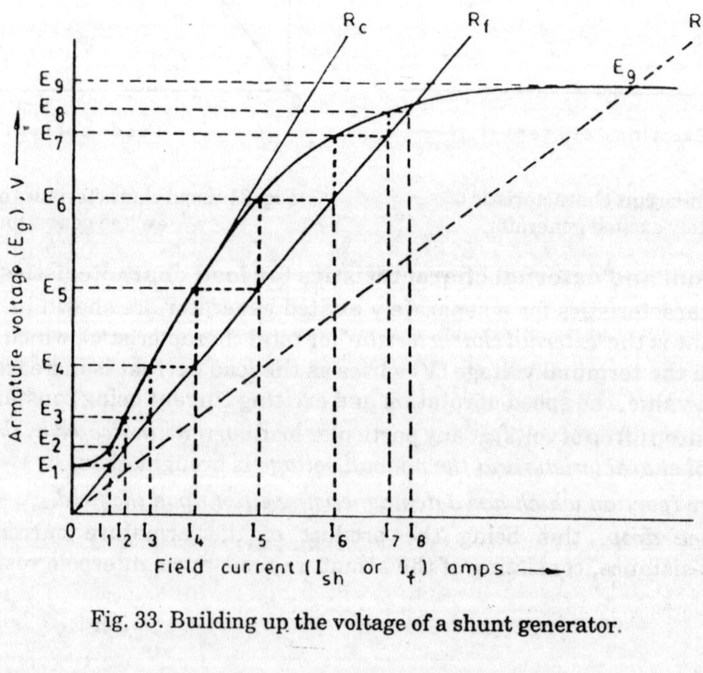

Fig. 33. Building up the voltage of a shunt generator.

(*iv*) When I_1 flows in the field circuit of the generator of Fig. 32, an increase in m.m.f. results (due to $I_f T_f$, T_f being field turns which aids the residual magnetism in increasing the induced voltage to E_2 as shown in Fig. 33.

(*v*) Voltage E_2 is now impressed across the field, causing a large current I_2 to flow in the field circuit. $I_2 T_f$ is an increased m.m.f., which produces generated voltage E_3.

(*vi*) E_3 yields I_3 in the field circuit, producing E_4. But E_4 causes I_4 to flow in the field producing E_5 ; and so on, up to E_8, the maximum value.

(*vii*) The *process continues until that point where the field resistance line crosses the magnetisation curve* in Fig. 33. Here the process stops. The induced voltage produced, when impressed across the field circuit, produces a current flow that in turn produces an induced voltage of the same magnitude, E_8, as shown in the figure.

Critical Resistance :

- In the above description a particular value of field resistance R_f was used for building up of self-excited shunt generator. If the field resistance were reduced by means of adjusting the field rheostat of Fig. 32 to a lower value say R_{f_1}. shown in Fig. 33, the build-up process would take place along field resistance line R_{f_1}, and build-up a somewhat higher value than E_8, *i.e.* the point where R_{f1} intersects the magnetisation curve, E_9. Since the curve is extremely saturated in the vicinity of E_9, reducing the field resistance (to its limiting field winding resistance) will not increase the voltage appreciably. Conversely, increasing the field rheostat resistance and the field circuit resistance (to a value having a higher slope than R_f in the figure) will cause a reduction of the maximum value to which build-up can possibly occur.
- The field resistance may be increased until the field circuit reaches a *critical field resistance. Field circuit resistance above the critical field resistance will fail to produce build-up.*

This critical field circuit resistance, R_c, is shown as tangent to the saturation curve passing through the origin, *O*, of the axes of the curve of Fig. 33. Thus a field circuit resistance higher than R_c will produce an armature voltage of E_1 approximately (and no more).

☞**Reasons for Failure of Self-excited Shunt Generator to Build-up Voltage.** The reasons why a self-excited generator may fail to build-up voltage are given below :

1. **No residual magnetism.** The start of the build-up process requires some residual magnetism in the magnetic circuit of the generator. If there is little or no residual magnetism, because of inactivity or jarring in shipment, no voltage will be generated that can produce field current. To overcome this difficulty, a separate source of direct current is applied to the field for a short period of time and then removed. The magnetic field should now be sufficient to allow the voltage to build-up. *The application of a separate source of direct current to the field is called 'flashing the field'.*

2. **Field connection reversed.** The voltage generated due to residual magnetism is applied to the field. Current should flow in the field coils in such a direction as to produce lines of flux in the same direction as the residual flux. If the field connections are reversed, the *lines of flux produced by the current flow will oppose the residual flux* so that the generated voltage will *decrease* rather than increase when the field circuit is closed. In this instance it is necessary to *reverse the field connections with respect to the armature.*

3. **Field circuit resistance too high.** A field circuit resistance greater than critical value will prevent an appreciable build-up. At no load, resistance greater than the critical may be caused by the following :

- **Open field circuit connection.** The effects of an open circuit are apparent. The field circuit resistance is *much greater than the critical value*; hence *generator will not build-up*.
- **Dirty commutator.** A dirty commutator does not permit good contact between the brushes and the commutator. This poor contact shows up as a high resistance to the flow of current in the field circuit and produces the same effect as a high field circuit resistance.

5.3. Shunt Generator Characteristics

In a shunt generator the field circuit is connected directly across the armature. Appliances, motors, light bulbs, and other electrical devices connected in *parallel* across the generator terminals represent a *load* on the generator. As more devices are connected in parallel, the load on the generator increases ; that is, the generator current increases. Because the generator current increases, the terminal voltage of the generator decreases. There are three factors that cause this decrease in voltage :

(*i*) Armature-circuit resistance (R_a),
(*ii*) Armature reaction, and
(*iii*) Reduction in field current.

(*i*) **Armature-circuit resistance.** The armature circuit of a generator, like every electrical circuit, contains resistance. This resistance includes the resistance of (*i*) *the copper conductors of the armature winding, (ii) the commutator, (iii) contact resistance between brushes and commutator,* and (*iv*) *the brushes themselves.* When no current flows through the armature, there is no *IR* drop in the armature and the voltage at the terminals is the same as the generated voltage. However, when there is current in the armature circuit, a voltage drop exists due to the armature resistance, and the terminal voltage is less than the generated voltage. The terminal voltage may be calculated from the following reaction :

$$V = E_g - I_a R_a$$

where V = voltage at terminals of generator,
E_g = generated or induced voltage,
I_a = total armature current, and
R_a = armature-circuit resistance.

(*ii*) **Armature reaction.** When current flows in the armature conductors a flux surrounds these conductors. The direction of this armature flux is such that it reduces the flux from the field poles, resulting in both a reduced generated voltage and terminal voltage.

(*iii*) **Reduction in the field current.** The field circuit is connected across the terminals of the generators. When the terminal voltage of the generator becomes smaller because of the armature-resistance volt drop and armature reaction, the voltage across the field circuit also becomes smaller and therefore field current will be less. A reduction in the magnitude of field current also reduces the flux from the field poles, which in turn reduces the generated voltage and also the terminal voltage.

External Characteristic

- See Figs. 34 and 35. The effect of the preceding three factors is shown in Fig. 35, which shows external (load-voltage) characteristic of a shunt generator.
- As shown in the circuit of Fig. 34, the readings of the voltage across the armature (and load), V are plotted

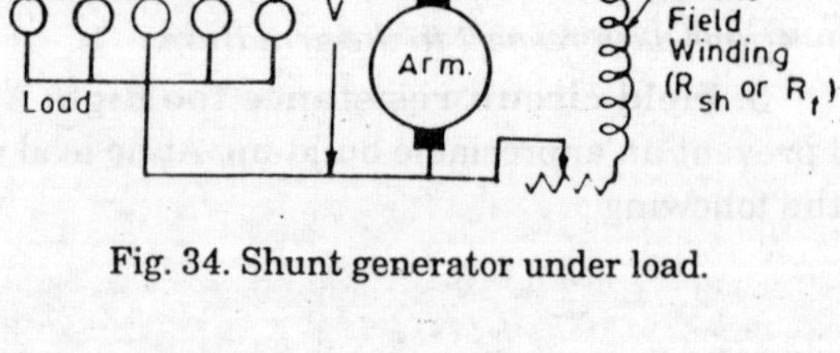

Fig. 34. Shunt generator under load.

as a function of load current, *I*. The voltage, *V*, is the same as E_g at no load (neglecting the I_aR_a and armature reaction drop produced by the field current). The effects of armature reaction, armature circuit voltage drop, and decrease in field current are all shown with progressive increase in load. Note that both the armature reaction and the I_aR_a drops are shown as dashed straight lines, representing theoretically linear voltage directly proportional to the increase in load current. The drop owing to decreased field current is a curved line, since it depends on the degree of saturation existing in the field at the value of load.

- Generally, the external load-voltage characteristic decreases with application of load only to *a small extent up to its rated load* (current) *value*. Thus, the shunt generator is considered as having a fairly constant output voltage with application of load, and in practice, is *rarely operated beyond the rated load current value continuously for any appreciable time.*

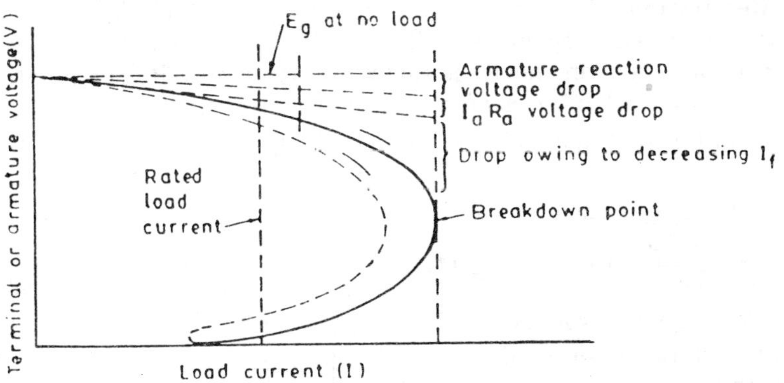

Fig. 35. Shunt generator load characteristics.

- As shown in Fig. 35 *further application of load causes the generator to reach a breakdown point beyond which further load causes it to 'unbuild' as it operates on the unsaturated portion of its magnetisation curve. This unbuilding process continues until the terminal voltage is zero, at which point the load current is of such magnitude that the internal armature circuit voltage drop equals the e.m.f. generated on the unsaturated or linear portion of its magnetisation curve.*
- It may be noted that if the *external load is decreased* (an increase of external load resistances), the *generator will tend to build up gradually along the dashed line* shown in Fig. 35. Note that for any value of load current, the terminal or armature voltage is less (as the voltage increases) compared to the solid lines which yield a higher voltage (as the voltage decreases). This *difference is due to hysteresis.*

Effect of Varying Excitation. Fig. 36 shows the effect of varying excitation upon the external characteristic of a shunt generator. With normal excitation the initial slope is small and heavy load current can be varied. With reduced excitation, the fall of voltage is more rapid and the maximum load current is reduced.

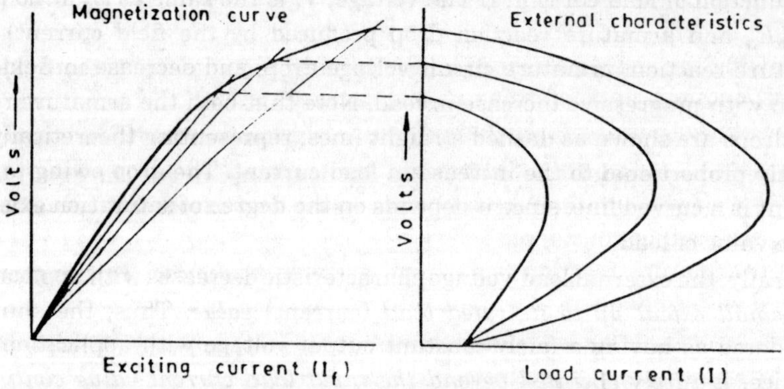

Fig. 36. Effect of varying excitation.

Voltage Regulation. The term *'voltage regulation'* is used to indicate the degree of change in armature voltage produced by application of load. If there is little change from no-load to full load, the generator or voltage-supplying device is said to possess good voltage regulation. If the *voltage changes appreciably with load, it is considered to have poor voltage regulation.*

'Voltage regulation' is defined as the *change in voltage from no-load to full load, expressed as a percentage of the rated terminal voltage* (armature voltage at full load).

i.e. Per cent voltage regulation

$$= \frac{V_{nl} - V_{fl}}{V_{fl}} \times 100 \qquad ...(1)$$

where V_{nl} = no load terminal voltage,
V_{fl} = full load (rated) terminal voltage.

Internal or Total Characteristic. To determine internal characteristic from external characteristic the following procedure is adopted [*see* Fig. 37].

Steps : 1. From the given data, draw the external characteristic (*I*).

2. Draw the shunt field resistance line *OL* and armature resistance line *OM*.

3. On the external characteristic take any point say *F*.

4. From point *F* draw vertical and horizontal lines intersecting *X* and *Y* axes respectively. Let these lines be *FC* and *FA* respectively.

5. Take point *D* on *X*-axis so that *CD* = *AB* representing the shunt field current, I_{sh}.

6. From point *D* draw vertical *DE* and produce it intersecting line *AF* produced at *H*.

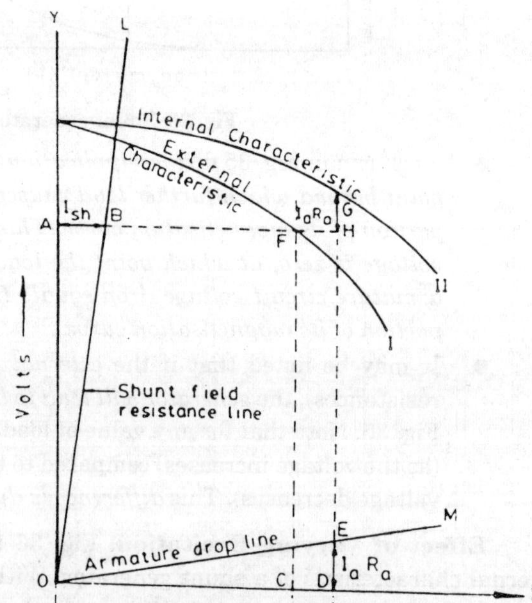

Fig. 37. Determination of internal characteristic from external characteristic.

7. Take point G on line DH produced so that $HG = DE$ $(= I_a R_a)$ representing the armature drop.
8. Following the above procedure take a number of points on external characteristic and find corresponding points lying on internal characteristic.
9. Draw a curve passing through these points which is the required internal characteristic (*II*).

External Characteristic and No-load Saturation Curve. The external characteristic of a shunt generator can be obtained directly from its no-load saturation curve as explained below. Following two cases will be considered :

(*A*) When armature reaction is so small as to be *negligible*. This is more or less true for generators fitted with compoles.

(*B*) When armature reaction is *not negligible.*

(A) Armature Reaction Negligible :

Steps 1. From given data draw O.C.C. [*see* Fig. 38].

2. Draw shunt field resistance line (say *OS*) meeting O.C.C. at any point (say *A*).
3. From point *A* draw horizontal line intersection *Y*-axis, say at point *B*.

   Hence, *OB* is the maximum no-load or open circuit voltage.
4. Take any point (say *L*) on the O.C.C. and draw an ordinate, say, *LMN* intersecting field resistance line at *M* and *X*-axis at *N*. Now *LN represents the generated e.m.f., MN represents the terminal voltage and LM represents voltage drop in armature.*
5. From points *L* and *M* draw horizontal lines cutting vertical axis say at point *D* and *E* respectively.
6. Draw armature resistance line *OC*.
7. From point *E*, draw line *EF* parallel to line *OC* cutting line *LD* produced at *F*. Hence, *F* is lying on the internal characteristic.

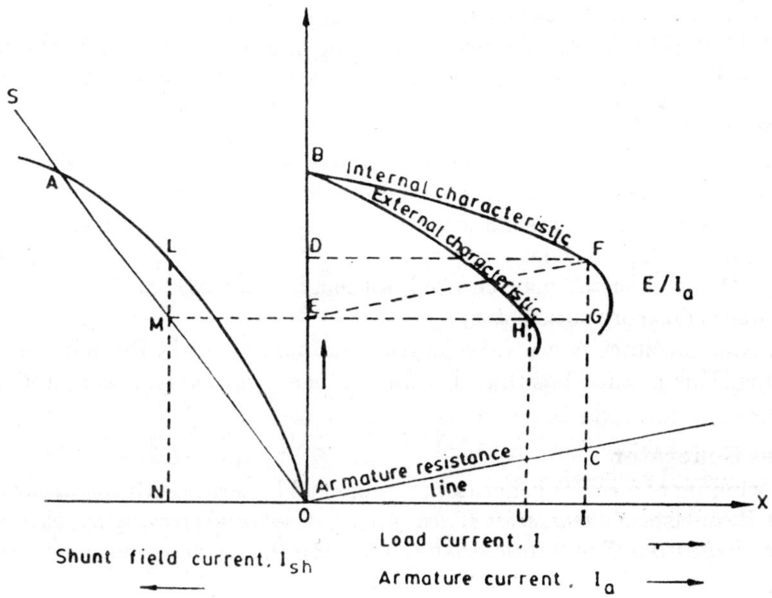

Fig. 38. Determination of external and internal characteristics from O.C.C.

Similarly other points can be obtained and internal characteristic may be drawn through these points :

8. From point F draw vertical line intersecting produced line ME at any point G and X-axis say, at point T. Since $FG = LM = CT$, hence point G lies on the *curve representing relation between armature current and terminal voltage.*

9. Take TU = shunt field current (I_{sh}) ON (scale being different). OU represents the load current corresponding to armature current represented by OT and terminal voltage OE.

10. From point U draw a vertical line intersecting line EG at H. Point *H lies on the external characteristic.*

Similarly other points may be obtained and curve may be drawn, which is the required *external characteristic.*

(B) Taking Armature Reaction into Account :

Here, in addition to considering the voltage drop in armature, voltage drop due to armature reaction is also taken into account.

Let I_aR_a = voltage drop in armature

I_{sh} = increase in shunt field current to counteract the demagnetising effect.

Now if a right-angled triangle say lmn is drawn as that ln = voltage drop in armature, and mn = shunt field current. The triangle lmn is called as the *drop reaction triangle.*

In order to draw external and internal characteristic repeat the process as in A with following modifications in steps 4 and 9 respectively.

4. Take any point L on the O.C.C. and draw line LM parallel to the line lm of triangle lmn and complete the triangle LMN. Now from the points L and M draw vertical lines cutting X-axis at points N' and M'. Now *LN' represents the generated e.m.f. MM'* represents the terminal voltage, *LN represents the voltage drop in armature due to armature resistance, ON' is the shunt field current to induce an e.m.f. represented by* LN' and $N'M'$ is the *increase in shunt field current to counteract the demagnetising effect.*

9. Take TU = shunt field current ON' (scale being different). OU represents the load current corresponding to the armature current represented by OT and terminal voltage MM'.

Voltage Control of Shunt Generators

- The terminal voltage of a shunt generator may be kept constant at all loads with the use of adjustable resistance, called a *field rheostat, connected in series with the shunt-field circuit.* By adjusting the resistance of the rheostat to suit the load on the machine, changes in terminal voltage with load may be prevented. When the load changes gradually, hand control of the rheostat may be used, although *automatic control* employing a *voltage regulator* is far more satisfactory.
- The terminal voltage may also be controlled *automatically* by the addition of a series-field winding. This method has the advantage of being *automatic cheap*, and *generally satisfactory.*

5.4. Series Generator

The field winding of a series generator is connected in series with the armature winding as shown in Fig. 39. It consists of a *few series of heavy wire, capable of carrying the output current of the machine without overheating.* The characteristic curves of a D.C. series generator are drawn as given below :

- The *saturation* curve (1) may be obtained in a manner exactly similar to that already described for the shunt machine. The series-connected generator is illustrated in Fig. 39.

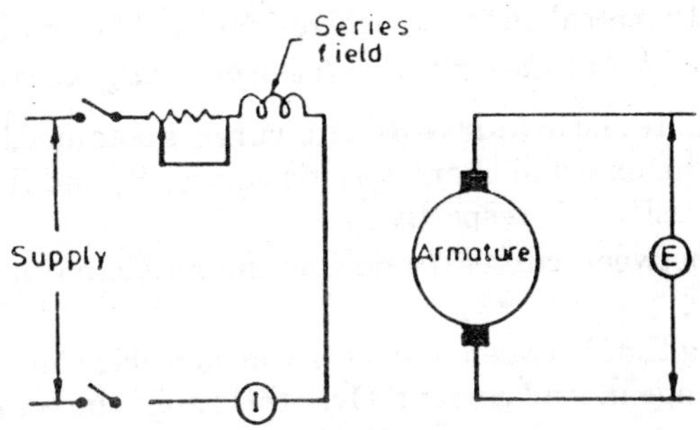

Fig. 39. Connection diagram for obtaining the saturation curve of a series generator.

The series connected generator illustrated in Fig. 40 must be capable of safely carrying the maximum current to be used, or about 125 per cent of rated load current. A plot of simultaneous readings of generated voltage and field current, taken at a rated speed, yields the magnetisation curve 1 of Fig. 41.

- **External Characteristic (curve 2).** To obtain the data for this curve, the machine is connected to the load as shown in Fig. 40, ammeter and voltmeter being inserted to read the load current I (= I_a) and the terminal voltage V respectively. The machine is run at constant (rated) speed, a series of simultaneous readings of voltage and current is taken while the load is varied from a minimum value to perhaps 125 per cent of rated load. When these readings are plotted, using V as co-ordinate and I_a (= I) as abscissa curve 2 (*see* Fig. 41) is obtained.

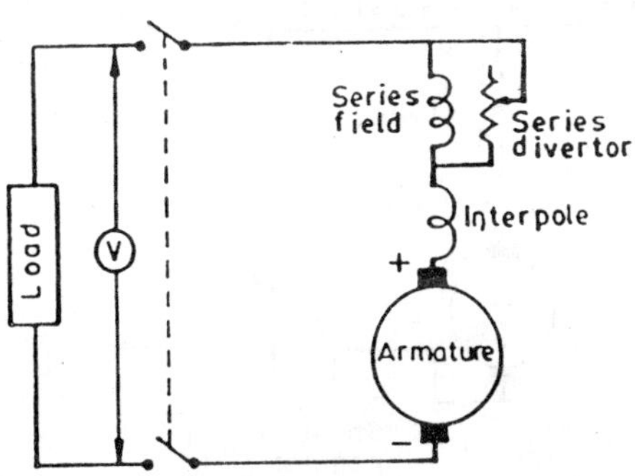

Fig. 40. Circuit for loading a series generator.

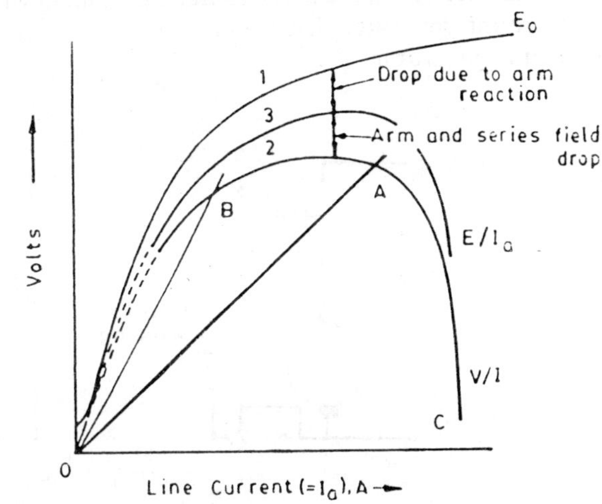

Fig. 41. Characteristic curves of a D.C. series generator.

It may be noted that the *readings cannot begin at zero load* as with the shunt generator, for if *the resistance of the circuit including armature, series field and load is increased beyond a certain critical value, the generator unbuilds and loses its load entirely.* Thus, if OA is the resistance line for the circuit the terminal voltage is the ordinate to the curve at A. When the *resistance of the circuit* is gradually *increased*, the *load falls off* along the curve, and A approaches B. When the resistance line finally becomes tangent to the curve, however, *operation becomes unstable*, and any slight further increase in the resistance causes the machine to unbuild its voltage and lose its load. The resistance that brings about this condition is called the *critical resistance*. Therefore, to *begin with, the resistance of the circuit must be reduced below the critical value before the generator delivers any load.*

- **Internal Characteristic (curve 3).** This curve is obtained by adding the resistance drop $I_a (R_a + R_{se})$ to the external characteristic curve ; R_a and R_{se} being armature resistance and series field resistance respectively.
  The difference between curves 1 and 3 is the *reduction in voltage caused by armature reaction.*
- It is worth noting that between *A* and *C* a considerable change in resistance brings about only a slight change in load current. Over this range the *voltage decreases rapidly,* owing to increasing armature reaction (particularly when the brushes are shifted forward), while the *current remains nearly constant*. Thus, between *A* and *C* the machine may be used to supply power to a *constant current variable-voltage circuit, such as series arc circuit.*
- Owing to initially *rising characteristic,* the series generator is often used as a *voltage booster* to give an increase of voltage practically proportional to the current.
- A series generator also finds application in electric traction where *'dynamic braking'* is employed. The connections of the series traction motors are changed by means of a controller so that they act as generators ; the power absorbed in braking the vehicle being dissipated in resistances, which are also used for starting purposes when the machines are reconnected as motors.

5.5. Compound Wound Generator

In case of a series generator the voltage regulation is very poor but the ability of the series field to produce additional useful magnetisation in response to increased load cannot be denied. This useful characteristic of the series field, combined with the relatively constant voltage characteristic of the shunt generator, led to the compound generator. Fig. 42 shows connection diagrams for a compound generator.

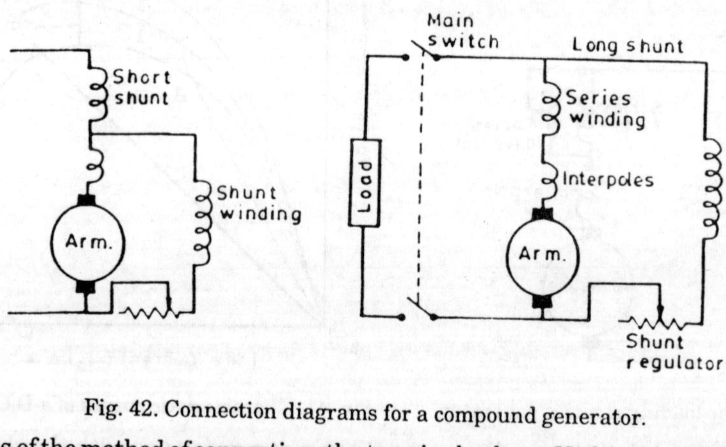

Fig. 42. Connection diagrams for a compound generator.

Regardless of the method of connection, the terminal voltage *V* of the short shunt or long shunt compound generator in same,

$$V = E_g - (I_{se} \cdot R_{se} + I_a R_a).$$

The generated voltage, E_g, of a compound generator is the result of the combination of m.m.f.s produced by the series ($I_{se}T_{se}$) and shunt ($I_{sh}T_{sh}$) ampere-turns due to currents which flow in their field windings. In a compound generator, the *shunt field predominates and is much stronger of the two. When the series field m.m.f. aids the shunt field m.m.f.,* the generator is said to be *'cumulatively compounded* ' [*see* Fig. 43 (*a*)]. When the *series field m.m.f. opposes the shunt field m.m.f.*, the generator is said to be '*differentially compounded*' [*see* Fig. 43 (*b*)].

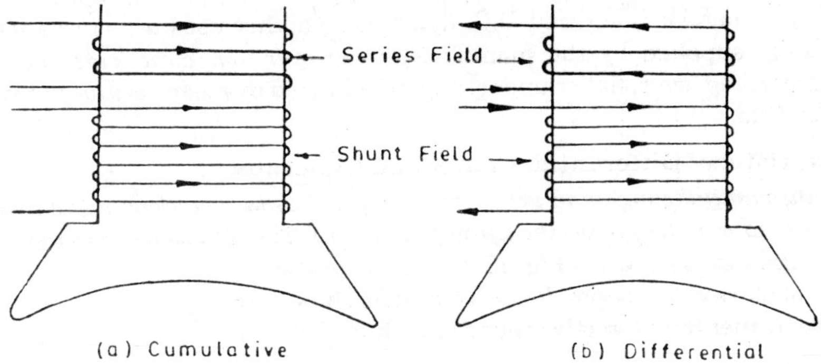

Fig. 43. Current directions in series and shunt-field coils of cumulative and differential-compound generators.

Characteristics of Cumulative Compound Generator

See Fig. 44. Depending on the relative additional aiding m.m.f., produced by the series field there are three types of load characteristics possible for the *cumulative* compound generator.

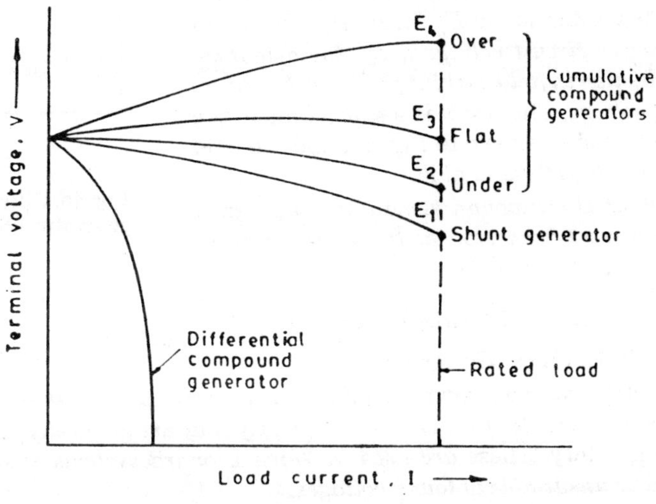

Fig. 44. External load voltage characteristics of cumulative and differential compound generators.

These types are called :

(*i*) Over-compound (*ii*) Flat-compound

(*iii*) Under-compound.

Over compound generator. An over-compound generator is one whose terminal voltage *rises with the application of load* so that its full-load voltage *exceeds* its no-load voltage (*negative regulation*).

Flat-compound generator. A flat compound generator has a load-voltage characteristic in which the no-load and full-load voltages are equal (*zero per cent regulation*).

Under-compound generator. An under-compound generator has a load characteristic in which the full load voltage is somewhat *less* than no-load voltage, but whose aiding series-field ampere-turns cause its characteristic to have *better regulation than an equivalent shunt generator*.

- Most commercial compound D.C. dynamos, whether used as generators or motors, are normally supplied by the manufacturer as *over-compound machines.* The *degree of compounding* (over, flat or under) may be adjusted by means of *divertor* which shunts the series field.

Characteristic of Differential Compound Generator

- The *differential compound generator is defined as that compounding produced when the series field m.m.f. opposes the shunt field m.m.f.* The difference in current direction of the two windings is shown in Fig. 43 (*b*), where for the sake of clarity, the series field winding is shown above (rather than directly around) the shunt field winding.
- Fig. 45 shows the load characteristic of differential compound generator.

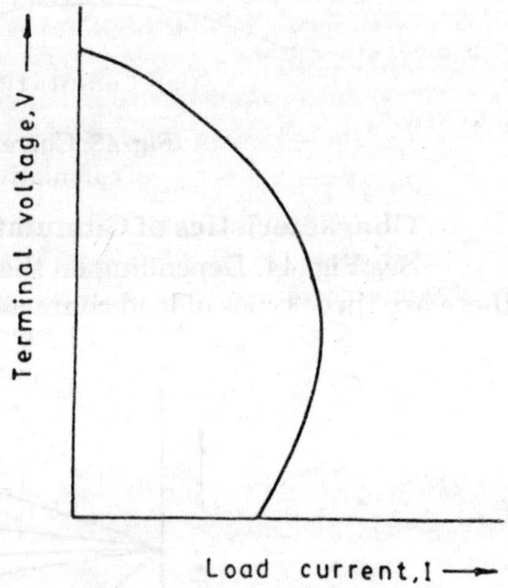

Fig. 45. Differential compound generator—load characteristic.

When the differential compound generator is *without load* it builds up and self-excites its shunt field in much the same manner as the shunt generator. However, *when a load is applied, the generated voltage E_g is now reduced by the* reduction in the main field flux created by the *opposing m.m.f. of the series field.* This reduction in E_g occurs *in addition to the armature and series circuit voltage drop, the armature reaction, and the reduction in field current produced by reduction of the armature voltage.* The result is a sharp drop in the terminal voltage with load as shown in Fig. 45, and the field is below saturation and rapidly unbuilds.

- *The differential compound generator is used as a constant-current generator for the same constant-current applications as the series generator.*

6. APPLICATIONS OF D.C. GENERATORS

6.1. Separately Excited Generators :

(*i*) The separately excited generators are usually *more expensive than self-excited generators* as they require *a separate source of supply*. Consequently they are generally used where self-excited are relatively unsatisfactory. These *are used in Ward Leonard systems of speed control, because self-excitation would be unsuitable at lower voltages.*

(*ii*) These generators are also used where quick and requisite response to control is important (since separate excitation gives a quicker and more precise response to the changes in the resistance of the field circuit).

6.2. Shunt Generators :

(*i*) These generators are used to advantage, in conjunction with automatic regulators, *as exciters for supplying the current required to excite the fields of A.C. generators.* The regulator controls the voltage of the exciter by cutting in and out some of the resistance of the shunt-field rheostat, thereby holding the voltage at whatever value is demanded by operating conditions. This is one of the most important applications of shunt generators.

(*ii*) Shunt generators are used *to charge batteries.* In this application the voltage should drop off slightly as the load increases, because the voltage of a lead battery is lower when battery is discharged than when battery is charged. When it is discharged, however, the battery can stand a large charging current than when it is charged. *Because of its drooping characteristic the shunt generator is admirably suited to battery charging service,* for, in a general way, the voltage curve of the generator has the same shape as the voltage curve of the battery itself. In both cases, as the load falls the voltage rises.

- Shunt generators can be operated in parallel without difficulty, and the wiring of parallel-operated shunt machines is quite a bit simpler than the corresponding wiring for compound machines. When a slight drop in voltage is not objectionable, as when a motor load is fed directly from the generator terminals, shunt machines may be used to advantage.

6.3. Series Generators. The field of application of series generator is limited. These are used for the following purposes :

(*i*) Series arc lighting.

(*ii*) Series incandescent lighting.

(*iii*) As a series booster for increasing the voltage across the feeder carrying current furnished by some other sources.

(*iv*) Special purposes such as supplying the field current for regenerative braking of D.C. locomotives.

6.4. Compound Generators. The compound generator is used for more than any other type.

(*i*) It may be built and adjusted automatically to supply an approximately *constant voltage at the point of use.,* throughout the entire range of load. This is *very great advantage.* It is possible to provide a constant supply voltage at the end of a long feeder by the simple expedient of *overcompounding* the generator, because the resistance drop in the line is compensated for by the rising characteristic of the generator.

When the point of utilisation is near the generator, a flat-compounded machine may be used.

(*ii*) *Differentially compounded generator* finds an useful application *as an arc welding generator* where the generator is practically short circuited every time the electrode touches the metal plates to be welded.

(*iii*) Compound generators are used to supply power to :

- Railway circuits,
- Motors of electrified steam rail-roads,
- Industrial motors in many fields of industry,
- Incandescent lamps, and
- Elevator motors etc.

Example 15. *Draw armature drop line if armature resistance is 0.2 Ω*

Solution. The *procedure* of drawing an armature drop line is as follows (*see* Fig. 46) :

(*i*) Take any value of armature current and find corresponding voltage in the armature resistance. For example, take armature current of 200 A, voltage drop in the armature for this value of armature current is $200 \times 0.2 = 40$ V.

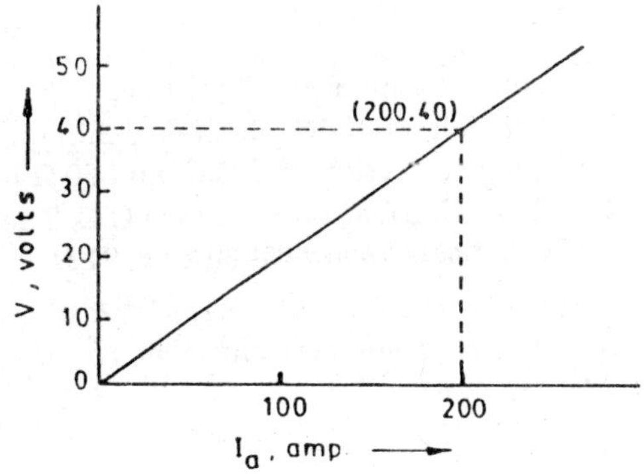

Fig. 46

So a point (200, 40) lying on this line is obtained.

(*ii*) Join the above point with origin O, the armature drop or ohmic drop line is obtained.

Example 16. *The open circuit characteristic (O.C.C.) of a separately excited generator, at 600 r.p.m., is as under:*

Field Current, A	:	*1.6*	*3.2*	*4.8*	*6.4*	*8.0*	*9.6*	*11.2*
E.m.f., V	:	*148*	*285*	*390*	*460*	*520*	*560*	*590*

Find :

(*i*) *The voltage to which the machine will excite as a shunt generator with a field circuit resistance of 60 ohm.*

(*ii*) *The critical resistance, at this speed.*

Solution.

- Plot O.C.C. (E_0/I_f) as shown in Fig. 47.

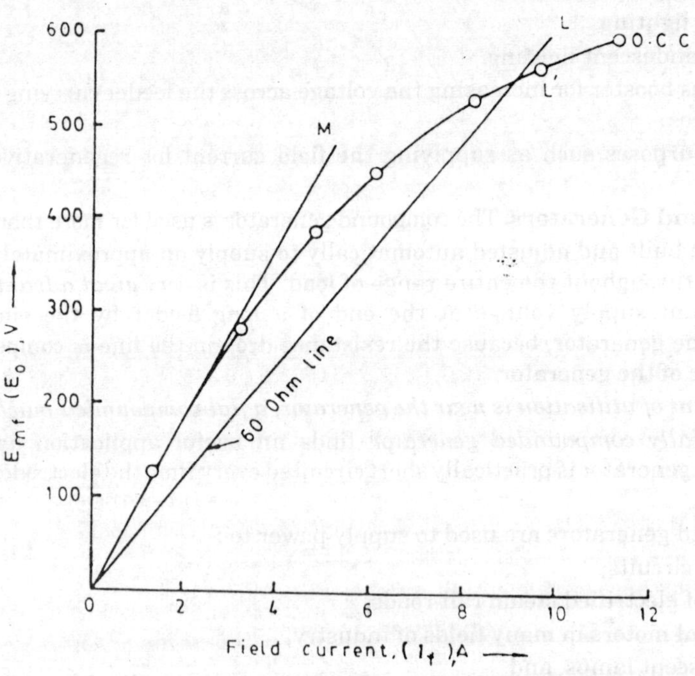

Fig. 47

- Line OL represents 60 Ω line.
- The voltage to which the machine will excite a shunt generator is given by point L' *i.e.* the intersection of O.C.C. and 60 Ω line. The machine will *excite* at 550 V.
- Draw line OM tangent to O.C.C. The *slope of this line represents critical resistance.* This value of critical resistance is 91 Ω.

Example 17. *The open circuit characteristic of a 4-pole, 250 V shunt generator having 610 lap-connected armature conductors running at 750 r.p.m. is as follows :*

Field current, A	:	*0*	*0.5*	*1.0*	*2.0*	*3.0*	*4.0*	*5.0*
E.m.f., V	:	*10*	*50*	*100*	*175*	*220*	*245*	*262*

Calculate :

(i) Field circuit critical resistance.

(ii) Critical speed for field circuit resistance of 80 Ω.

(iii) Residual flux per pole.

Solution. Number of poles of the generator, $p = 4$

Number of parallel paths, $a = p = 4$ [Generator being lap connected]

Number of armature conductors, $Z = 610$

- Plot O.C.C. from the given data as shown in Fig. 48.

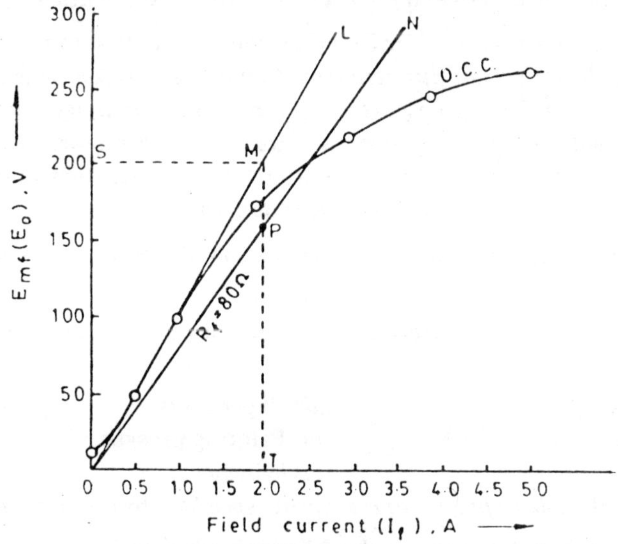

Fig. 48

- Draw line *OL* tangent to O.C.C. at the origin. To determine the slope of the line *OL*, take any point *M* on this line and from *M* draw horizontal and vertical lines cutting O.C. voltage axis and field current axis at points *S* and *T* respectively.

(*i*) Critical resistance of field circuit = slope of line *OL*

$$= \frac{OS \text{ in volts}}{OT \text{ in amperes}} = \frac{200}{2} = \textbf{100 ohms. (Ans.)}$$

(*ii*) Draw line *ON* representing resistance of 80 ohms. Let ordinate drawn from point *M* intersect line *ON* at *P*.

$$\textbf{Critical speed, } N_c = N \times \frac{TP}{TM} = 750 \times \frac{160}{200} = \textbf{600 r.p.m. (Ans.)}$$

(*iii*) **Residual flux per pole, ϕ :**

With no exciting current, the e.m.f. induced due to residual flux is 10 V at 750 r.p.m. (given).

Using the relation, $$E_g = \frac{p\phi ZN}{60a}$$

$$10 = \frac{4 \times \phi \times 610 \times 750}{60 \times 4}$$

$$\therefore \quad \phi = \frac{10 \times 60 \times 4}{4 \times 610 \times 750} = 0.00131 \text{ Wb.}$$

Hence, **residual flux/pole = 0.00131 Wb. (Ans.)**

7. DIRECT CURRENT MOTOR

7.1. General Aspects

The electric motor is a machine which converts electric energy into mechanical energy. It depends for its operation on the force which is known to exist on a conductor carrying current while situated in a magnetic field.

Construction. A D.C. motor is similar in construction to a D.C. generator. As a matter of fact any D.C. generator will run as a motor when its field and armature windings are connected to a source of direct current. The field winding produces the necessary magnetic field. *The flow of current through the armature conductors produces a force which rotates the armature.*

Though the essential construction of D.C. motor is identical to that of a generator, the *external appearance of a motor may be somewhat different from that of a generator.* This is mainly due to the fact that the *frame of a generator may be partially open* because it is located in relatively clean environment and only skilled operators are present in its vicinity. A motor, on the other hand, may be operating in a rather dusty environment and only unskilled operators may be working in its vicinity. Therefore, frames of motors are to a large extent closed.

The body of *D.C. mill motors* is made in two halves bolted together for easy access to the field windings and inter-poles.

Applications. Because of their inherent characteristics D.C. motors find extensive application in :

(*i*) Steel plants
(*ii*) Paper mills
(*iii*) Textile mills
(*iv*) Printing presses
(*v*) Cranes
(*vi*) Winches
(*vii*) Excavators etc. *where precise and accurate speed control over a wide range is required.*

Advantages. The D.C. motors possess the following *advantages* :

(*i*) High starting torque.
(*ii*) Speed control over a wide range, both below and above the normal speed.
(*iii*) Accurate stepless speed control with constant torque.
(*iv*) Quick starting, stopping, reversing and accelerating.

Disadvantages. The *disadvantages* of D.C. motors are :

(*i*) High initial cost.
(*ii*) Increased operating and maintenance costs because of the commutators and brushgear.

7.2. Principle of Operation of D.C. Motor

The principle of motor action can be stated as follows :

"Whenever a current carrying conductor is placed in a magnetic field, it experiences a force whose direction is given by Fleming's left hand rule".

- Fig. 49 illustrates this principle.
  Fig. 49 (*a*) shows the field set up by the poles.
  Fig. 49 (*b*) shows the conductor field due to flow of current in the conductor.

Fig. 49 (*c*) shows the resultant field produced when the current carrying conductor wire of Fig. 49 (*b*) is inserted in the air gap of Fig. 49 (*a*) with the *axis of the conductor at right angles to the direction of the flux.*

On the *upper side* of the conductor in Fig. 49 (*c*) the magnetizing forces of the field and of the current in the conductor are *additive* while on the *lower side* these are *subtractive*. This explains why the resultant field is strengthened above and weakened below the conductor (wire).

The above experiment shows that the wire in Fig. 49 (*c*) has a force on it which tends to move it downward. Thus the *force acts in the direction of the weaker field*. When the current in the wire is reversed, the direction of the force is also reversed, as in Fig. 49 (*d*).

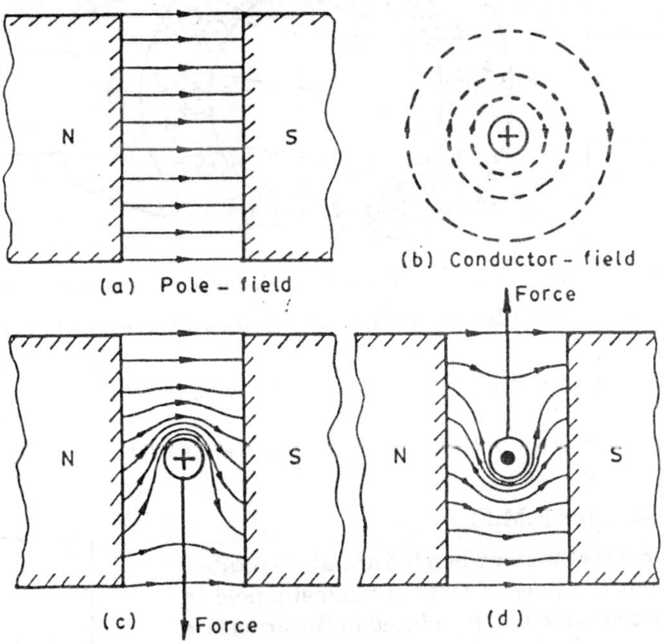

Fig. 49. The principle of motor action.

The force (F) developed in the conductor is given by the relation,

$$F = BIl \text{ newtons}$$

where B = flux density, T (Wb/m^2),

I = current in conductor, A, and

l = exposed length of conductor, m.

Now consider the magnetic field of a D.C. motor in which there is no current in the armature conductors ; the lines of force will be distributed as shown in Fig. 50.

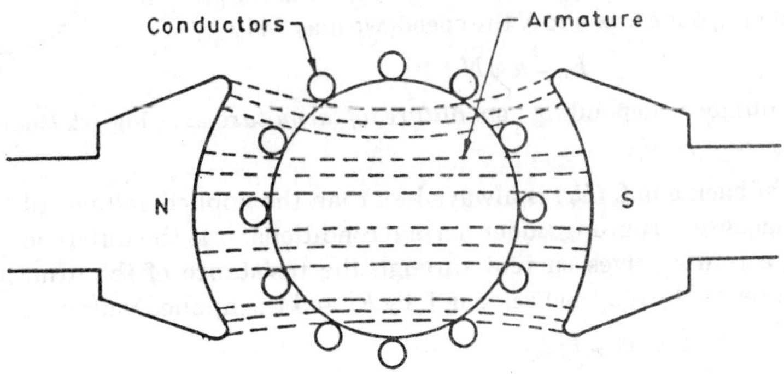

Fig. 50. Distribution of lines of force in a motor due to magnetic field only.

If, now, the armature carries current, each of its conductors will produce a magnetic field which, when super-imposed on the main field, causes a distribution of magnetic lines as shown in Fig. 51.

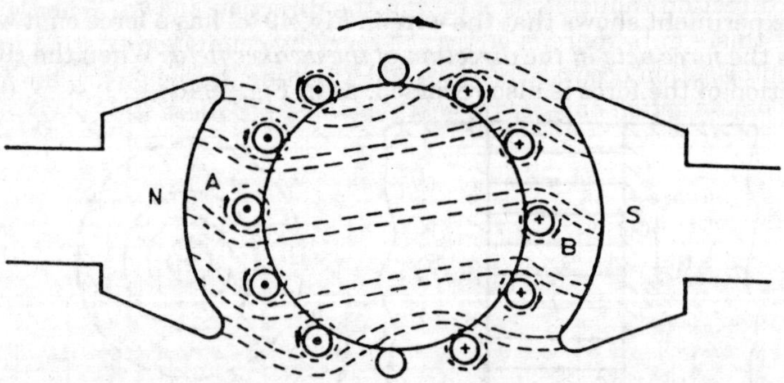

Fig. 51. Distribution of lines of force in a motor, on load, due to the armature and magnetic field.

The magnetic field is said to be distorted, since the lines of force no longer follow approximately straight paths.

These lines of force have the *property of tending to shorten* themselves, so that they may be regarded as *being in tension. Each conductor* in Fig. 51 *will experience a force like that exerted on a stone in a catapult*. Since these conductors are embedded in slots in the armature, the latter is caused to rotate in a clockwise direction.

7.3. Back or Counter E.M.F.

Refer Fig. 52. In a D.C. motor when the armature rotates, the conductors on it cut the lines of force of magnetic field in which they revolve, so that an e.m.f. is induced in the armature as in a generator. The *induced e.m.f. acts in opposition to the current in the machine* and, therefore, *to the applied voltage*, so that *it is customary to refer to this voltage as the 'back e.m.f.'* That this is so can be deduced by Lenz's law, which states that the *direction of an induced e.m.f. is such as to oppose the change causing it, which is, of course, the applied voltage.*

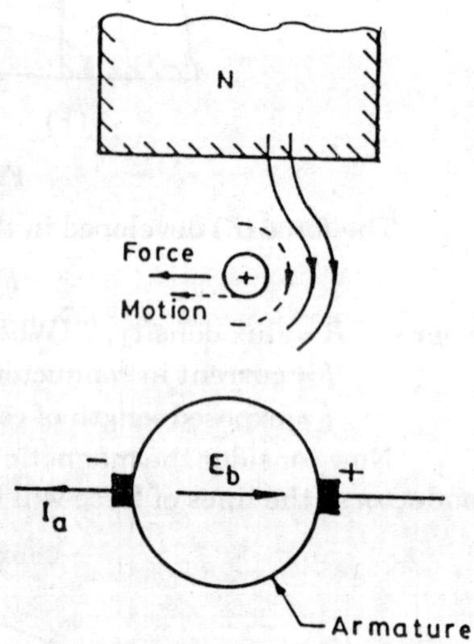

Fig. 52. Motoring operation.

The magnitude of the back or counter e.m.f. can be calculated by using formula for the induced e.m.f. in a generator, and it is important in the case of the motor, to appreciate that this is *proportional to the product of the flux and the speed.* Thus if E_b denotes the back e.m.f., ϕ the flux and N the speed, we may write,

$$E_b = k\,\phi\,N$$

where k is a number depending on *nature of armature winding.*

The value of back e.m.f. (E_b) is always less than the applied voltage, although difference is small when the machine is running under normal conditions. It is the difference between these two quantities which actually drives current through the resistance of the armature circuit. If this resistance is represented by R_a, the back e.m.f. by E_b and the applied voltage by V, then we have

$$V = E_b + I_aR_a$$

where I_a is the current in the armature circuit.

7.4. Comparison Between Motor and Generator Action

— Fig. 53 (*b*) shows a generator action, where a mechanical force moves a conductor in upward direction inducing an e.m.f. in the direction shown. When a current flows as a result of this e.m.f., there is a current-carrying conductor existing in a magnetic field ; hence motor action occurs. Shown as a dotted line in Fig. 53 (*b*), the *force developed as a result of motor action opposes the motion which produced it.*

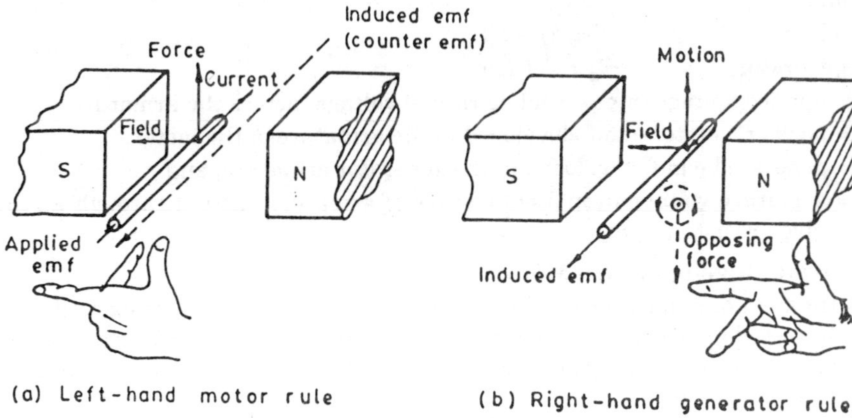

Fig. 53. Comparison of motor and generator action.

Thus it can be stated categorically that in rotating electric machines *generator action and motor action occur simultaneously*. Hence the same dynamo may be operated either as a motor or a generator or both (as in dynamo motor or synchronous converter).

- Fig. 54 presents a more graphic representation in terms of rotational elements, which compares the elementary motor and generator for the same direction of rotation and shows the electric circuits of each which is self-explanatory.

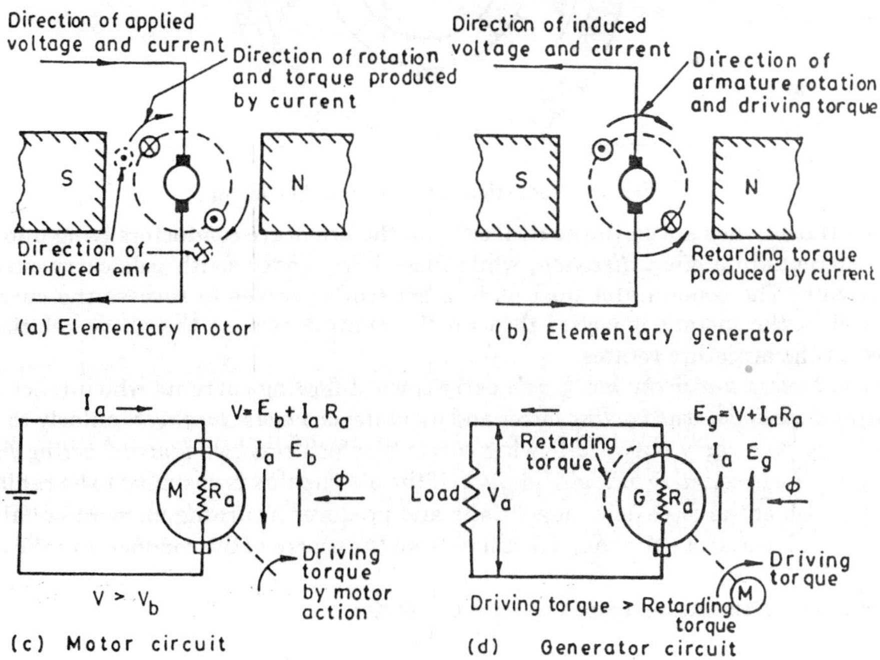

Fig. 54. Elementary motor action *versus* generator action.

It may be noted that when a dynamo is operating as a *motor*, the generated e.m.f. is *always less than the terminal voltage* (that produces motor action) and it *opposes the armature current.* On the other hand when a dynamo is operating as a *generator*, the *armature current is in the same direction as the generated e.m.f.*, and the *generated e.m.f.* E_g *exceeds the terminal voltage V applied across the load.* This distinction between generator and motor, in which the armature-generator voltage aids or opposes the armature current, respectively give rise to the following basic armature circuit equations :

For a motor, $V = E_b + I_aR_a$ ...(2)

For a generator, $E_g = V + I_aR_a$ ...(3)

where V = applied voltage (measurable terminal voltage) across the armature,

E_b = back or counter e.m.f. developed in the armature of the motor,

E_g = generated e.m.f. developed in the generator armature, and

I_aR_a = armature voltage drop due to a flow of armature current through an armature of a given resistance, R_a.

7.5. Torque Developed in a Motor

When the field of a machine (of the type described as generator) is excited and a potential difference is impressed upon the machine terminals, the current in the armature winding reacts with the air-gap flux to produce a turning moment or *torque* which tends to cause the armature to revolve. Fig. 55 illustrates production of torque in a motor.

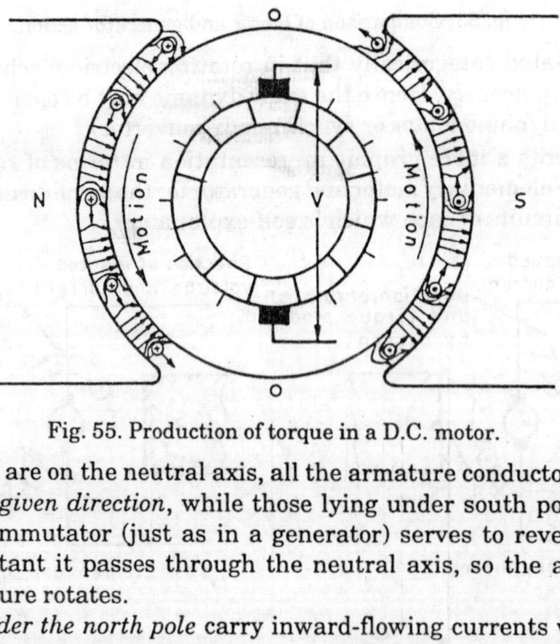

Fig. 55. Production of torque in a D.C. motor.

When the brushes are on the neutral axis, all the armature conductors lying under the north pole carry currents in a *given direction*, while those lying under south pole carry currents in the *reverse direction*. The commutator (just as in a generator) serves to reverse the current in each armature coil at the instant it passes through the neutral axis, so the above relation is always maintained as the armature rotates.

All *conductors under the north pole* carry inward-flowing currents which react with the air gap flux to *produce down-ward acting forces* and a *counter clockwise torque*. Similarly the *conductors under the south pole* carry outward-flowing currents which *produce upward-acting forces*. These *forces also give rise to counter clockwise torques*. If the air-gap flux is assumed to be radially directed at all points, each of the force acts tangentially and produces a turning moment equal to the force multiplied by its lever arm—the radial distance from the centre of the conductor to the centre of the shaft.

Magnitude of torque developed by each conductor

$$= BIlr \text{ Nm}$$

If the motor contains Z conductors, the total torque developed by the armature

$$T_a = BIlrZ \text{ Nm} \quad ...(4)$$

where B = gap density, T (Wb/m^2)
I = armature current in a conductor, A
l = active length of each conductor, m
r = average lever arm of a conductor or the average radius at which conductors are placed, m
Z = total number of armature conductors.

It is more convenient to express T_a in terms of armature current I_a, total flux per pole ϕ and number of poles p.

$$I = \frac{I_a}{a}$$

and

$$B = \frac{\phi}{A}$$

where a = number of parallel paths,
and A = the cross-sectional area of flux path at radius r.

$$A = \frac{2\pi rl}{p}$$

Then

$$T_a = \frac{Z\phi lI_a r}{2\pi rl} \times \frac{p}{a}$$

$$= \frac{Z\phi I_a p}{2\pi a} \text{ Nm}$$

i.e.,

$$T_a = 0.159 Z\phi p \times \frac{I_a}{a} \quad ...(5)$$

or

$$T_a = k\phi I_a \text{ Nm} \quad ...(6)$$

where $k = \frac{Zp}{2\pi a}$ is a *constant for any machine.*

Alternative proof :

The expression for the torque developed by the motor armature may also be deduced as follows :
Let T_a be the torque developed in Nm by the motor armature running at N r.p.m.
Power developed = work done per second

$$= T_a \times 2\pi N \text{ watts} \quad ...(i)$$

Electrical equivalent of mechanical power developed by the armature also

$$= E_b I_a \text{ watts} \quad ...(ii)$$

Equating (*i*) and (*ii*), we get

$$T_a \times \frac{2\pi N}{60} = E_b I_a$$

or

$$T_a = \frac{E_b I_a}{2\pi \left(\frac{N}{60}\right)}$$

Also since

$$E_b = \frac{p\phi ZN}{60a}$$

$$\therefore \quad T_a \times 2\pi \frac{N}{60} = \frac{p\phi ZN}{60a} \, . \, I_a \quad \text{or} \quad T_a = \frac{Z\phi p}{2\pi} \, . \, \frac{I_a}{a} \text{ Nm}$$

i.e.,

$$T_a = 0.159\, Z\phi p \, . \, \frac{I_a}{a} \, .$$

Note. From the above equation for torque, we find that

$$T \propto \phi I_a$$

Then

(*i*) In the case of shunt motors, ϕ is practically constant,

hence $$T \propto I_a$$

(*ii*) In the case of series motors, ϕ is proportional to I_a before saturation (because field windings carry full armature current)

$$\therefore \quad T \propto I_a^2.$$

Shaft torque (T_{sh}). The torque developed by the armature is the *gross torque*. Whole of this torque is not available at the pulley, since certain percentage of torque developed by the armature is lost to overcome the iron and friction losses. *The torque which is available for useful work is known as* **shaft torque** T_{sh}. It is so called because it is *available at the shaft*. The horse power obtained by using shaft torque is called brake horse power (B.H.P.).

$$\text{B.H.P. (metric)} = \frac{T_{sh} \times 2\pi N}{735.5}$$

$$\therefore \quad T_{sh} = \frac{\text{B.H.P. (metric)} \times 735.5}{\frac{2\pi N}{60}} \quad \text{...(7)}$$

where N = speed of armature in r.p.m.

The difference $T_a - T_{sh}$ is known as **lost torque** (*i.e.* torque lost in iron and friction losses)

$$= 0.159 \times \frac{\text{iron and friction losses}}{\frac{N}{60}} \text{ Nm.}$$

7.6. Mechanical Power Developed by Motor Armature

Refer Fig. 56. The voltage V applied across the motor armature has to (*i*) overcome back e.m.f. E_b and (*ii*) supply the armature ohmic drop I_aR_a.

$$\therefore \quad V = E_b + I_aR_a$$

This is known as *voltage equation of a motor.*

Multiplying both sides by I_a, we get

$$VI_a = E_bI_a + I_a^2R_a$$

Here VI_a = electrical input to the armature,

E_bI_a = electrical equivalent of mechanical power P_m developed in the armature, and

$I_a^2R_a$ = copper loss in the armature.

The power available at the pulley for doing useful work is *somewhat less than the mechanical power developed by the armature*.

This is evident, since there are certain *mechanical losses* (such as bearing and windage friction and iron losses) that must be supplied by the driving power of the motor.

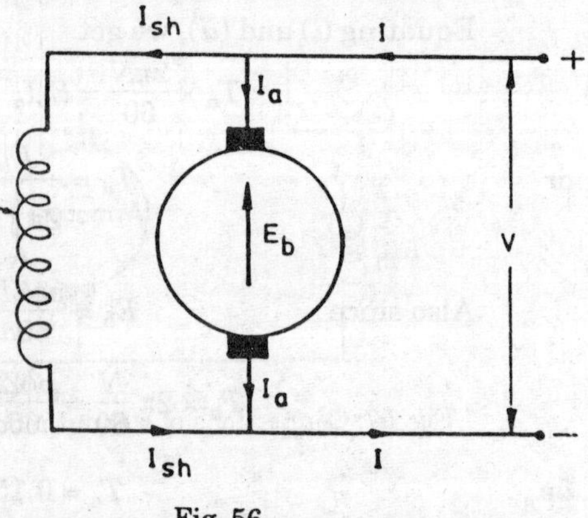

Fig. 56

Condition for maximum power. We know that, the mechanical power developed by the motor,

$$P_m = VI_a - I_a^2 R_a$$

Differentiating both sides with respect to I_a

$$\frac{d(P_m)}{dI_a} = V - 2I_aR_a = 0$$

$$\therefore \qquad I_aR_a = \frac{V}{2}$$

As $$V = E_b + I_aR_a$$

$$\therefore \qquad V = E_b + \frac{V}{2}$$

i.e., $$E_b = \frac{V}{2} \qquad ...(8)$$

Hence, *mechanical power developed by a motor is maximum when back e.m.f. is equal to half the applied voltage.* In practice, however, *this condition is not realized* because in that case current would be much beyond the normal current of the motor. Moreover, half the input would be wasted in the form of heat and taking other losses, such as mechanical and magnetic, into consideration, the efficiency of the motor will be below 50 per cent.

7.7. Types of D.C. Motors

There are three main types of motors characterised by the connection of the field winding in relation to the armature. These are :

1. **Shunt wound motor** or the shunt motor, in which the *field winding is connected in parallel with the armature.*

2. **Series motor,** in which the *armature and field windings are connected in series.*

3. **Compound motor,** which has two field windings, *one of which is connected in parallel with the armature and the other in series with it.*

Shunt wound motor. Fig. 57 shows the connections of a shunt motor. From these connections it may be observed at once that the *field current is constant,* since it is connected directly to the supply which is assumed to be at constant voltage. Hence the flux is approximately constant and, since also the *back e.m.f. is almost constant under normal conditions the speed is approximately constant.* This is not strictly true, but nevertheless, *it is usual for all practical purposes to regard the shunt motor as a constant speed machine.* It is, therefore, employed in practice for drives, the speeds of which are required to be *independent of the loads.* The speed can, of course, be varied when necessary and this is done by the inclusion of a variable resistor in series with the field winding, as shown in Fig. 57.

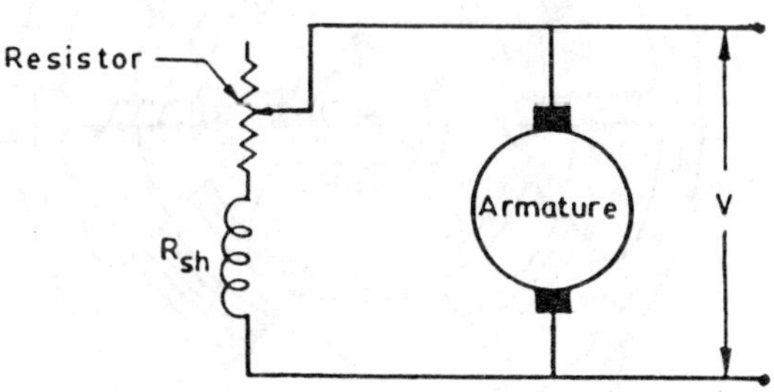

Fig. 57. Connections of a shunt motor.

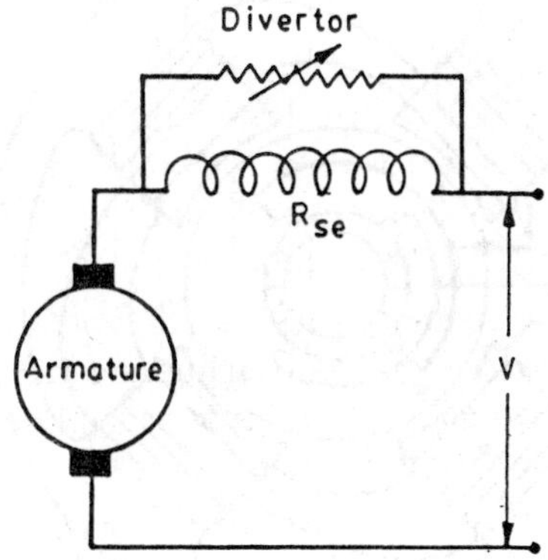

Fig. 58. Connections of a series motor.

Series motor. Fig. 58 shows the connections for the series motor. The current passing through the field winding is the same as that in the armature, since the armature current increases as the mechanical load on the shaft increases, so also does the field current. The resultant increase in magnetic flux causes a reduction in the speed, as can be observed from a consideration of the formula :

$$E_b = \frac{p\phi ZN}{60a} \quad \text{or} \quad E_b = k\phi N$$

where $k = \dfrac{pZ}{60a}$ being constant

or $$N = \frac{E_b}{k\phi}.$$

This is a useful property for many drives in which it is desirable that a heavy increase in the load should automatically bring about a compensating reduction in speed. As with the shunt motor, the speed may also be varied independently of the load by the inclusion of a variable resistor in the field circuit. In this case, however, it is connected in parallel with the series winding as shown in Fig. 58, and is called a *divertor compound motor*.

Refer Fig. 59. The compound motor has a shunt field winding in addition to the series winding so that the number of magnetic lines of force produced by each of its poles is the resultant of the flux produced by the shunt coil and that due to the series coil. The *flux so produced depends not only on the current and number of turns of each coil, but also on the winding direction of the shunt coil in relation to that of the series coil*. When the two fluxes *assist each other* the machine is a **cumulative compound motor**, while if they *oppose each other*, it is said to be a **differential compound motor.**

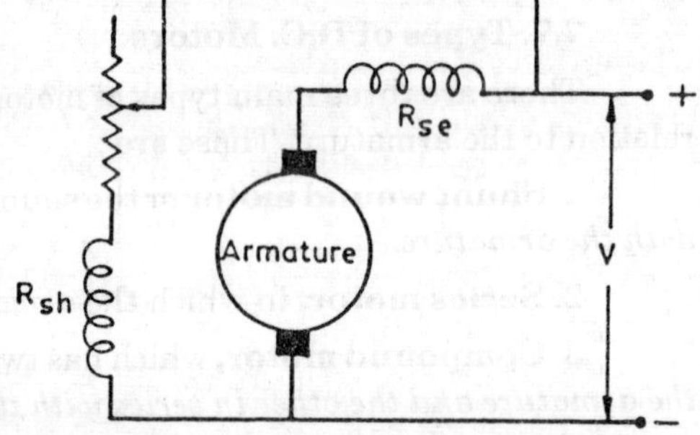

Fig. 59. Connections of a compound motor.

Fig. 60 shows the field windings and interpole connections of a *differential compound wound motor*. The shunt coil is made up of *many turns of fine wire,* whilst the series coil comprises relatively *few turns of thickwire.*

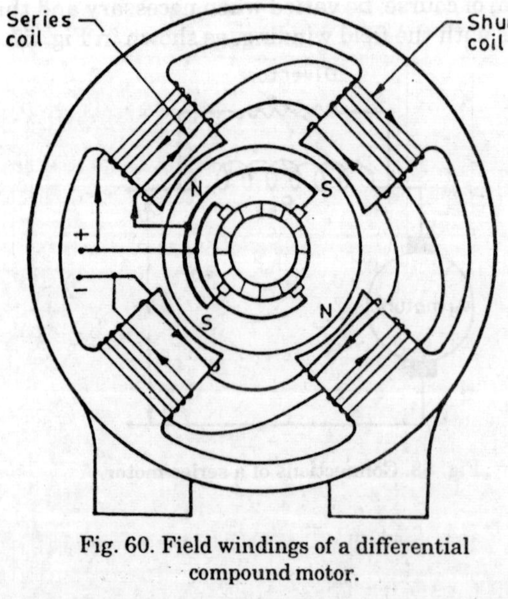

Fig. 60. Field windings of a differential compound motor.

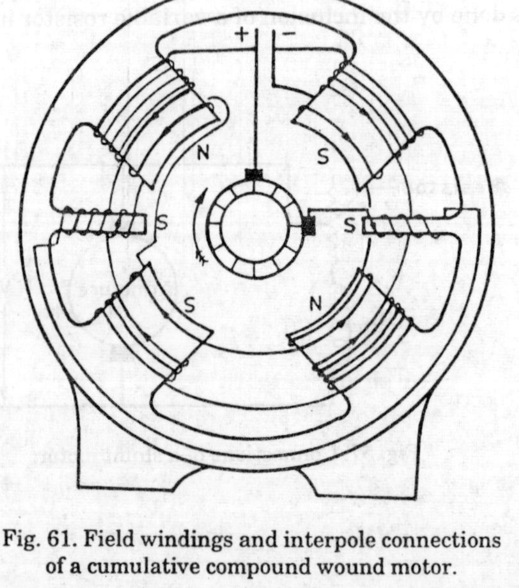

Fig. 61. Field windings and interpole connections of a cumulative compound wound motor.

Fig. 61 shows the field windings of a *cumulative compound motor*.

The flow of currents in the shunt and series coils is worth noting in Figs. 60 and 61.

7.8. Speed of a D.C. Motor

We know that the voltage equation of a motor is given by

$$V = E_b + I_a R_a$$

or
$$E_b = V - I_a R_a$$

or
$$\frac{p\phi ZN}{60a} = V - I_a R_a$$

$\therefore$
$$N = \frac{(V - I_a R_a)}{\phi} \times \frac{60a}{Zp}$$

But
$$V - I_a R_a = E_b$$

$\therefore$
$$N = \frac{E_b}{\phi} \times \frac{60a}{Zp}$$

or
$$N = k' \frac{E_b}{\phi} \quad ...(9)$$

where $k' = \frac{60a}{Zp}$ = constant.

The equation (9) shows that speed is *directly proportional to back e.m.f.* and *inversely to the flux*

or
$$N \propto \frac{E_b}{\phi}.$$

For series motor :

Let N_1 = speed in the first case,

I_{a1} = armature current in the first case, and

ϕ_1 = flux in the first case.

N_2, I_{a2} and ϕ_2 = corresponding quantities in the second case.

Using the above relation $\left(i.e.\ N \propto \frac{E_b}{\phi} \right)$, we get

$$N_1 \propto \frac{E_{b1}}{\phi_1}$$

and
$$N_2 \propto \frac{E_{b2}}{\phi_2}$$

where $E_{b1} = V - I_{a1} R_a$

and $E_{b2} = V - I_{a2} R_a$

$\therefore$
$$\frac{N_2}{N_1} = \frac{E_{b2}}{E_{b1}} \times \frac{\phi_1}{\phi_2} \quad ...(10)$$

Prior to saturation of poles :

$$\phi \propto I_a$$

$\therefore$
$$\frac{N_2}{N_1} = \frac{E_{b2}}{E_{b1}} \times \frac{I_{a1}}{I_{a2}} \quad ...[10(a)]$$

For shunt motor :

Applying the same equation in this case also, we get

$$\frac{N_2}{N_1} = \frac{E_{b2}}{E_{b1}} \times \frac{\phi_1}{\phi_2}$$

If
$$\phi_1 = \phi_2$$
$$\frac{N_2}{N_1} = \frac{E_{b2}}{E_{b1}} \quad \text{...(11)}$$

7.9. Speed Regulation

The speed regulation of a D.C. motor is defined as follows :

"The change in speed when the load on the motor is reduced from rated value to zero, expressed as percent of the rated load speed."

$\therefore$ Percent speed regulation $= \dfrac{\text{no load speed} - \text{full load speed}}{\text{full load speed}}$.

SHUNT MOTORS

Example 18. *Determine the torque developed when a current of a 30 A passes through the armature of a motor with the following particulars : lap winding, 310 conductors, 4-pole, pole-shoes 16.2 cm long subtending an angle of 60° at the centre, bore radius 16.2 cm, flux density in air gap 0.7 tesla.*

Solution. Number of poles, $p = 4$

Number of parallel paths, $a = p = 4$ [Motor being lap-wound]

Number of armature conductors, $Z = 310$

Length of pole shoe = 16.2 cm

Bore radius = 16.2 cm

Flux density in air gap, $B = 0.7$ tesla

Armature current, $I_a = 30$ A

Torque developed, T :

We know that,

Pole-shoe arc $= \pi D \times \dfrac{60}{360} = \pi(2 \times 16.2) \times \dfrac{60}{360} = 16.967$ cm

$\therefore$ Pole area = pole-shoe length × pole-shoe arc

$= (16.2 \times 10^{-2}) \times (16.967 \times 10^{-2})$

$= 16.2 \times 16.967 \times 10^{-4}\ \text{m}^2 = 0.02748\ \text{m}^2$

Also flux, $\phi = B \times$ pole area

$= 0.7 \times 0.02748 = 0.01924$ Wb

Using the relation, $T = 0.159 Z\phi p \left(\dfrac{I_a}{a}\right) = 0.159 \times 310 \times 0.01924 \times 4 \times \dfrac{30}{4}$

$= 28.45$ Nm.

Hence, **torque developed = 28.45 N-m. (Ans.)**

Example 19. *A 230 V D.C. shunt motor takes 32 A at full load. Find the back e.m.f. on full load if the resistances of motor armature and shunt field windings are 0.2 ohm and 115 ohms respectively.*

Solution. Supply voltage, $V = 230$ Volts

Full load current, $I = 32$ A

Armature resistance, $R_a = 0.2$ ohm

Shunt field windings' resistance, $R_{sh} = 115$ ohms

Back e.m.f., E_b :

Shunt field current, $I_{sh} = \dfrac{230}{115} = 2$ A

$\therefore$ Armature current, $I_a = I - I_{sh} = 32 - 2 = 30$ A

Back e.m.f. on full load, $E_b = V - I_aR_a$
$= 230 - 30 \times 0.2 = 224$ V

Hence, **back e.m.f. on full load = 224 V. (Ans.)**

Example 20. *The power input to a 230 volts D.C. shunt motor is 8.477 kW. The field resistance is 230 Ω and armature resistance is 0.28 Ω. Find the input current, armature current and back e.m.f.*

(P.T.U., May, 2001)

Solution. *Given* : $P_{in} = 8.477$ kW ; $R_{sh} = 230\ \Omega$; $R_a = 0.28\ \Omega$

I : ; I_a : E_b :

Input current, $I = \dfrac{P_{in}}{V} = \dfrac{8.477 \times 1000}{230}$

$= \mathbf{36.86\ A.}$ **(Ans.)**

Shunt field current, $I_{sh} = \dfrac{V}{R_{sh}} = \dfrac{230}{230} = 1$ A

$\therefore$ *Armature current,* $I_a = I - I_{sh}$
$= 36.86 - 1$
$= \mathbf{35.86\ A.}$ **(Ans.)**

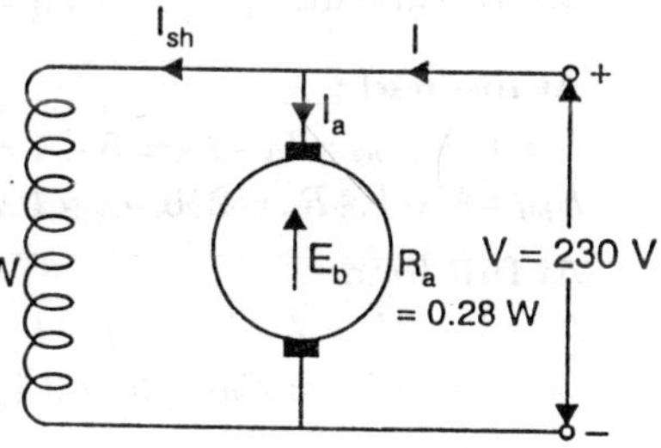

Fig. 62

Back e.m.f. $E_b = V - I_aR_a$
$= 230 - 35.86 \times 0.28 = \mathbf{219.96\ V.}$ **(Ans.)**

Example 21. *A six-pole lap-connected 230 V shunt motor has 410 armature conductors. It takes 41 A on full load. The flux per pole is 0.05 weber. The armature and field resistances are 0.1 ohm and 230 ohms respectively. Contact drop per brush = 1 V.*

Determine the speed of motor at full load.

Solution. Number of poles, $p = 6$

Number of parallel paths, $a = 6$ [Motor being lap-connected]

Number of armature conductors, $Z = 410$

Full load current, $I = 41$ A

Flux per pole, $\phi = 0.05$ weber

Armature resistance, $R_a = 0.1$ ohm

Shunt field resistance, $R_{sh} = 230$ ohms

Contact drop/brush $= 1$ V

Speed of motor on full load, N :

Shunt field current, $I_{sh} = \dfrac{V}{R_{sh}} = \dfrac{230}{230} = 1$ A

Armature current, $I_a = I - I_{sh} = 41 - 1 = 40$ A

Back e.m.f. on full load, $E_b = V - I_aR_a$ – drop at brushes
$= 230 - 40 \times 0.1 - 2 \times 1 = 224$ V

We know that, $E_b = \dfrac{p\phi ZN}{60a}$

$$224 = \frac{6 \times 0.05 \times 410 \times N}{60 \times 6}$$

$$\therefore \quad N = \frac{224 \times 60 \times 6}{6 \times 0.05 \times 410} = 655.6 \text{ r.p.m.}$$

Hence, **speed of motor on full load = 655.6 r.p.m. (Ans.)**

Example 22. *A 250 volt d.c. shunt motor, on no load, runs at 1000 rpm and takes 5 A. The field and armature resistances are 250 ohms and 0.25 ohm respectively. Calculate the speed when the motor is loaded such that it takes 41 A if the armature reaction weakens the field by 3%.* **[P.T.U., May, 2001]**

Solution. *Given* : $V = 250$ V ; $N_0 = 1000$ r.p.m. ; $I_0 = 5$ A ; $R_{sh} = 250\ \Omega$;

$$R_a = 0.25\ \Omega, I = 41\text{ A}, \phi = \left(1 - \frac{3}{100}\right)\phi_0 = 0.97\ \phi_0$$

Speed of the motor, N :

Shunt current, $I_{sh} = \dfrac{V}{R_{sh}} = \dfrac{250}{250} = 1$ A

At no-load :

$I_0 = 5$ A ; $I_{as} = I_0 - I_{sh} = 5 - 1 = 4$ A

$E_{b0} = V - I_{a0} R_a = 250 - 4 \times 0.25 = 249$ V

At full-load :

$I = 41$ A ; $I_a = I - I_{sh} = 41 - 1 = 40$ A

$E_b = V - I_a R_a = 250 - 40 \times 0.25 = 240$ V

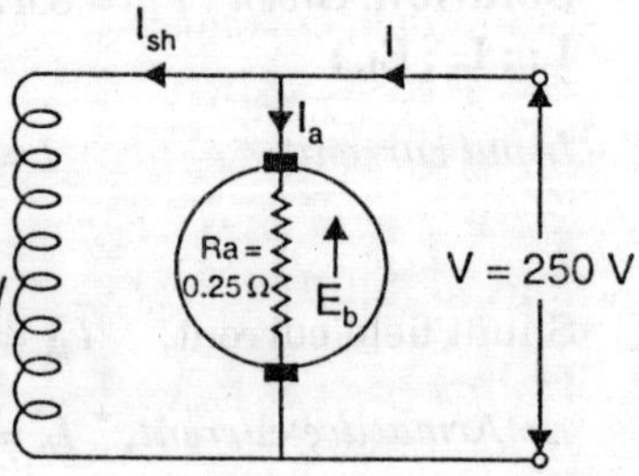

Fig. 63

Using the relation, $\dfrac{N}{N_0} = \dfrac{E_b}{E_{b0}} \times \dfrac{\phi_0}{\phi}$

or $$\frac{N}{1000} = \frac{240}{249} \times \frac{\phi_0}{0.97\phi_0}$$

∴ $$\mathbf{N} = 1000 \times \frac{240}{249} \times \frac{1}{0.97} = \mathbf{993.67\ r.p.m.}\ \textbf{(Ans.)}$$

Example 23. *A 120 volt d.c. shunt motor has an armature resistance of 0.2 ohms and a field resistance of 60 ohms. The full-load line current is 60 A and full-load speed is 1800 r.p.m. If the brush contact drop is 3 V, find the speed of the motor at half-load.* **(P.T.U., June, 2000)**

Solution. *Given* : $V = 120$ V ; $R_a = 0.2\ \Omega$; $R_{sh} = 60\ \Omega$

$I_1 = 60$ A, $N_1 = 1800$ r.p.m. ;

Brush contact drop = 3 V

Speed of the motor at half-load, N$_2$:

$$I_{sh} = \frac{V}{R_{sh}} = \frac{120}{60} = 2\text{ A}$$

Full load armature current,

$$I_{a1} = I_1 - I_{sh} = 60 - 2 = 58\text{ A}$$

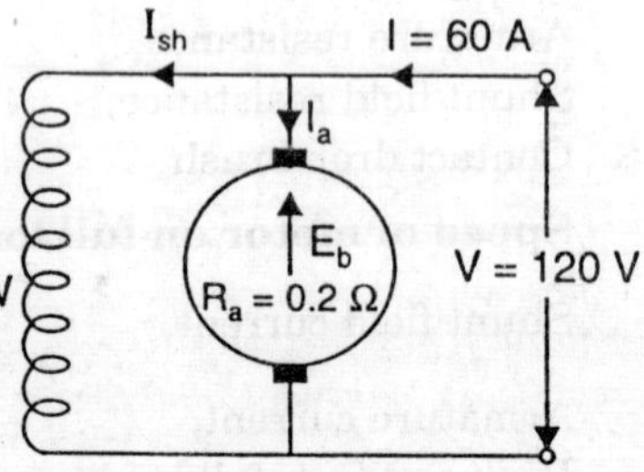

Fig. 64

Back e.m.f. $E_{b1} = V - I_{a1} R_a$ – brush contact drop

$= 120 - 58 \times 0.2 - 3 = 105.4$ V

Armature current at half-load, $I_{a2} = \dfrac{I_{a1}}{2} = \dfrac{58}{2} = 29$ A

Back e.m.f. at half-load, $E_{b2} = V - I_{a2} R_a$ – brush contact drop

$= 120 - 29 \times 0.2 - 3 = 111.2$ V

Now, $\dfrac{N_2}{N_1} = \dfrac{E_{b2}}{E_{b1}} \times \dfrac{\phi_1}{\phi_2} = \dfrac{E_{b2}}{E_{b1}}$ [∵ $\phi_1 = \phi_2$ ...being shunt motor]

∴ $$\frac{N_2}{1800} = \frac{111.2}{105.4}$$

or $$\mathbf{N_2} = \frac{1800 \times 111.2}{105.4} = \mathbf{1899\ r.p.m.}\ \textbf{(Ans.)}$$

☞**Example 24.** *A 4-pole 500 V shunt motor takes 7 A on no-load, the no-load speed being 750 r.p.m. It has a shunt field current of 2 A. Calculate the full-load speed of the motor if it takes 122 A at full-load. Armature resistance = 0.2 ohm. Contact drop/brush = 1 V. Armature reaction weakens the field by 4% on full-load.*

Solution. Supply voltage, $V = 500$ Volts

No-load current, $I_0 = 7$ A

No-load speed, $N_0 = 750$ r.p.m.

Shunt field current, $I_{sh} = 2$ A

Full-load current, $I = 122$ A

Armature resistance, $R_a = 0.2$ ohm

Contact drop/ brush $= 1$ V

Flux at full-load, $\phi = \left(\frac{100-4}{100}\right)\phi_0 = 0.96\phi_0$

Full-load speed, N :

At no-load :

$$I_{a0} = I_0 - I_{sh} = 7 - 2 = 5 \text{ A}$$

$$E_{b0} = V - I_{a0}R_a - \text{contact drop at brushes}$$

$$= 500 - 5 \times 0.2 - 2 \times 1 = 497 \text{ V.}$$

At full-load :

$$I_a = I - I_{sh} = 122 - 2 = 120 \text{ A}$$

$$E_b = V - I_aR_a - \text{contact drop at brushes}$$

$$= 500 - 120 \times 0.2 - 2 \times 1 = 500 - 24 - 2 = 474 \text{ V}$$

We know that, $E_{b0} \propto \phi_0 N_0$

and $E_b \propto \phi N$

$$\therefore \quad \frac{E_b}{E_{b0}} = \frac{\phi}{\phi_0} \times \frac{N}{N_0}$$

$$\frac{474}{497} = \frac{0.96\phi_0}{\phi_0} \times \frac{N}{750}$$

$$\therefore \quad N = \frac{474 \times 750}{497 \times 0.96} = 745 \text{ r.p.m. (app.)}$$

Hence, **full-load speed of the motor = 745 r.p.m. (app.) (Ans.)**

☞**Example 25.** *A 440 V, 4-pole, lap connected shunt motor has a no-load input current of 15 A and a shunt field current of 10 A. At full-load it takes a current of 150 A. If armature resistance = 0.1 ohm, flux per pole on no-load = 0.05 weber, number of armature conductors = 750 and contact drop per brush = 1 V, calculate :*

(i) No-load speed *(ii) Full-load speed*

(iii) Speed regulation.

Armature reaction weakens the field by 1.5% on full-load.

Solution. Supply voltage, $V = 440$ Volts

Number of poles, $p = 4$

Number of parallel paths, $a = p = 4$ [Motor being lap-connected]

Armature resistance, $R_a = 0.1$ ohm

Flux per pole on no-load, $\phi_0 = 0.05$ weber

Number of armature conductors, $Z = 750$

Contact drop per brush, $= 1$ V

No-load input current, $I_0 = 15$ A
Shunt field current, $I_{sh} = 10$ A
Full-load current, $I = 150$ A

No-load speed, N_0 :

Full-load speed, N :

Speed regulation :

(*i*) **No-load speed :**

Armature current, $I_{a0} = I_0 - I_{sh} = 15 - 10 = 5$ A

$$E_{b0} = V - I_{a0}R_a - \text{brush contact drop}$$
$$= 440 - 5 \times 0.1 - 2 \times 1 = 437.5 \text{ V}$$

Using the relation : $E_{b0} = \dfrac{p\phi_0 ZN_0}{60a}$

or $$437.5 = \frac{4 \times 0.05 \times 750 \times N_0}{60 \times 4}$$

$$\therefore \quad N_0 = \frac{437.5 \times 60 \times 4}{4 \times 0.05 \times 750} = 700 \text{ r.p.m.}$$

Hence, **no-load speed = 700 r.p.m. (Ans.)**

(*ii*) **Full-load speed :**

Armature current, $I_a = I - I_{sh} = 150 - 10 = 140$ A

$$E_b = V - I_aR_a - \text{brush contact drop}$$
$$= 440 - 140 \times 0.1 - 2 \times 1 = 424 \text{ V}$$

$\phi = (1 - 0.015)\ \phi_0$ [Since armature reaction weakens the field by 1.5 per cent]

$$= 0.985\ \phi_0$$

Using the relation : $\dfrac{N}{N_0} = \dfrac{E_b}{E_{b0}} \times \dfrac{\phi_0}{\phi}$

or $$\frac{N}{700} = \frac{424}{437.5} \times \frac{\phi_0}{0.985\phi_0}$$

$$\therefore \quad N = 700 \times \frac{424}{437.5} \times \frac{\phi_0}{0.985\phi_0} = 688.73 \text{ r.p.m.}$$

Hence, **full-load speed = 688.73 r.p.m. (Ans.)**

(*iii*) **Percentage speed regulation**

$$= \frac{\text{no-load speed} - \text{full-load speed}}{\text{full-load speed}} \times 100$$

$$= \frac{700 - 688.73}{688.73} \times 100 = 1.637$$

Hence, **percentage speed regulation = 1.637%. (Ans.)**

Example 26. *A 250 V shunt motor takes a line current of 60 A and runs at 800 r.p.m. Its armature and field resistances are 0.2 ohm and 125 ohms respectively. Contact drop/brush = 1 V. Calculate :*

(*i*) *No-load speed if the no-load current is 6 A.*

(*ii*) *The percentage reduction in the flux per pole in order that the speed may be 1000 r.p.m. when the armature current is 40 A.*

Neglect the effects of armature reaction.

Solution. Supply voltage, $V = 250$ Volts

Load current, $I_1 = 60$ A

Load speed, $N_1 = 800$ r.p.m.

Armature resistance, $R_a = 0.2$ ohm

Shunt field resistance, $R_{sh} = 125$ ohms

Contact drop $= 2$ V

Armature current, $I_{a2} = 40$ A

Load speed, $N_2 = 1000$ r.p.m.

No-load current, $I_0 = 6$ A

No-load speed, N_0 :

Percentage reduction in flux/pole :

Shunt field current, $I_{sh} = \frac{250}{125} = 2$ A

At no-load :

$$I_{a0} = I_0 - I_{sh} = 6 - 2 = 4 \text{ A}$$

$$E_{b0} = V - I_{a0}R_a - \text{brush drop}$$

$$= 250 - 4 \times 0.2 - 2 \times 1 = 247.2 \text{ V}$$

At load (I_1 = 60 A) :

$$I_{a1} = I_1 - I_{sh} = 60 - 2 = 58 \text{ A}$$

$$E_{b1} = V - I_{a1}R_a - \text{brush drop}$$

$$= 250 - 58 \times 0.2 - 2 \times 1 = 236.4 \text{ V}$$

Using the relation, $\frac{N_1}{N_0} = \frac{E_{b1}}{E_{b0}} \times \frac{\phi_0}{\phi_1}$

Since the motor is shunt wound and armature reaction effect is negligible so *flux may be assumed to be constant* and therefore speed is directly proportional to the back e.m.f. developed.

$$\therefore \quad \frac{N_1}{N_0} = \frac{E_{b1}}{E_{b0}}$$

$$\frac{800}{N_0} = \frac{236.4}{247.2}$$

$$N_0 = \frac{800 \times 247.2}{236.4} = 836.5 \text{ r.p.m.}$$

Hence, **no-load speed = 836.5 r.p.m. (Ans.)**

At armature current, I_{a2} = 40 A

Let the flux be reduced to ϕ_2

$$E_{b2} = V - I_{a2}R_a - \text{brush drop}$$

$$= 250 - 40 \times 0.2 - 2 \times 1 = 240 \text{ V}$$

Using the relation, $\frac{N_2}{N_1} = \frac{E_{b2}}{E_{b1}} \times \frac{\phi_1}{\phi_2}$

or

$$\frac{1000}{800} = \frac{240}{236.4} \times \frac{\phi_1}{\phi_2}$$

$$\therefore \quad \frac{\phi_2}{\phi_1} = \frac{240 \times 800}{1000 \times 236.4} = 0.812$$

∴ Percentage reduction in flux

$$= \frac{\phi_1 - \phi_2}{\phi_1} \times 100 = \left(1 - \frac{\phi_2}{\phi_1}\right) \times 100 = (1 - 0.812) \times 100 = 18.8\%$$

Hence, **%age reduction in flux = 18.8%. (Ans.)**

☞**Example 27.** *A 220 V shunt motor takes 60 A when running at 800 r.p.m. It has an armature resistance of 0.1 ohm. Determine the speed and armature current if the magnetic flux is weakened by 20%. Contact drop per brush = 1 V.*

Total torque developed remains constant.

Solution. Supply voltage $= 220$ V

Load current, $= I_1 = I_{a1} = 60$ A [Neglecting shunt field current]

Speed at 60 *A*, $N_1 = 800$ r.p.m.

Armature resistance, $R_a = 0.1$ ohm

Contact drop/brush $= 1$ V

$\mathbf{N_2 = ?}$, $\mathbf{I_{a2} = ?}$

$$\phi_2 = \left(1 - \frac{20}{100}\right)\phi_1 \quad \text{[Since magnetic flux is weakened by 20\%]}$$

$$= (1 - 0.20)\phi_1 = 0.8\phi_1$$

Since $T \propto \phi I_a$

and torque remains constant [Given]

$$\therefore \quad \phi_1 I_{a1} = \phi_2 I_{a2}$$

$$\phi_1 \times 60 = 0.8\phi_1 I_{a2}$$

$$\therefore \quad I_{a2} = \mathbf{75\ A.}$$

Back e.m.f., $E_{b1} = V - I_{a1}R_a -$ brush drop

$= 200 - 60 \times 0.1 - 2 \times 1 = 212$ V

Back e.m.f., $E_{b2} = V - I_{a2}R_a -$ brush drop

$= 220 - 75 \times 0.1 - 2 \times 1 = 210.5$ V

Using the relation, $$\frac{N_2}{N_1} = \frac{E_{b2}}{E_{b1}} \times \frac{\phi_1}{\phi_2}$$

$$\frac{N_2}{800} = \frac{210.5}{212} \times \frac{\phi_1}{0.8\phi_1}$$

$$\therefore \quad N_2 = \frac{800 \times 210.5}{212 \times 0.8} = \mathbf{992.9\ r.p.m.}$$

Hence, **Speed = 992.9 r.p.m. (Ans.)**

Armature current = 75 A. (Ans.)

☞**Example 28.** *A 440 V shunt motor takes 105 A (armature current) from the supply and runs at 1000 r.p.m. Its armature resistance is 0.15 ohm. If total torque developed is unchanged, calculate the speed and armature current if the magnetic field is reduced to 70% of the initial value.*

(Panjab University)

Solution. Supply voltage, $V = 440$ Volts

Armature current, $I_{a1} = 105$ A

Speed, $N_1 = 1000$ r.p.m.

Armature resistance, $R_a = 0.15$ ohm

$\phi_2 = 0.7\ \phi_1.$

(*i*) **Armature current, I_{a2} :**

As the torque developed is same

$$\therefore \quad \phi_1 I_{a1} = \phi_2 I_{a2}$$

$$I_{a2} = \frac{\phi_1}{\phi_2} \times I_{a1} = \frac{\phi_1}{0.7\phi_1} \times 105 = 150 \text{ A}$$

Hence, **armature current = 150 A. (Ans.)**

Now, $\quad E_{b1} = V - I_{a1}R_a = 440 - 105 \times 0.15 = 424.25$

and $\quad E_{b2} = V - I_{a2}R_a = 440 - 150 \times 0.15 = 417.5$ V.

(*ii*) **Speed, N_2 :**

Using the relation, $\quad \dfrac{N_2}{N_1} = \dfrac{E_{b2}}{E_{b1}} \times \dfrac{\phi_1}{\phi_2}$

$$\frac{N_2}{1000} = \frac{417.5}{424.25} \times \frac{\phi_1}{0.7\phi_1}$$

$$\therefore \quad N_2 = 1000 \times \frac{417.5}{424.25} \times \frac{1}{0.7} = 1405 \text{ r.p.m.}$$

Hence, **speed of the motor when magnetic field is reduced to 70% of initial value = 1405 r.p.m. (Ans.)**

SERIES MOTORS

Example 29. *A 230 V series motor develops torque of 310 N-m at 800 r.p.m. The torque lost due to iron and friction loss is 10 N-m. If the efficiency is 85% determine the current taken by the motor at 800 r.p.m.*

Solution. Supply voltage $= 230$ V

Torque developed, $T_a = 310$ N-m

Torque lost due to iron and friction, $T_{lost} = 10$ N-m

Efficiency $= 85\%$

Speed of the motor, $N = 800$ r.p.m.

Current taken by motor at 800 r.p.m. :

$$T_{useful} = T_a - T_{lost} = 310 - 10 = 300 \text{ N-m}$$

Watts corresponding to useful torque

$$= \frac{T_{useful} \times 2\pi N}{60} = \frac{300 \times 2\pi \times 800}{60} = 25136 \text{ W}$$

Power in watts taken from mains $= \dfrac{25136}{0.85} = 29571.7$ W

Current taken by motor $= \dfrac{29571.7}{230} = 128.57$ A.

Example 30. *A 8-pole 240 V lap-wound, series motor has armature and series field resistances of 0.2 ohm and 0.02 ohm respectively. There are 660 armature conductors. If the flux/pole is 0.3 Wb and the total torque developed in the armature is 320 N-m, find the current taken by the motor and its speed.*

Solution. Number of poles, $p = 4$

Number of parallel paths, $a = p = 4$ [Motor being lap-wound.]

Armature resistance, $R_a = 0.2$ ohm

Series field resistance, $R_{se} = 0.02$ ohm

Number of armature conductors, $Z = 660$
Flux/pole, $\phi = 0.03$ Wb
Total torque developed, $I_a = 320$ N-m

(*i*) **Current taken by motor, I_a :**

$$E_b = \frac{p\phi ZN}{60a}$$

$$\therefore \quad \frac{E_b}{N} = \frac{p\phi Z}{60a} = \frac{4 \times 0.03 \times 660}{60 \times 4} = 0.33$$

Also, $$T_a \times \left(\frac{2\pi N}{60}\right) = E_b I_a$$

or $$T_a = \frac{E_b}{N} \cdot \frac{60 I_a}{2\pi} = 0.33 \times \frac{60 \times I_a}{2\pi}$$

$$\therefore \quad 320 = \frac{0.33 \times 60 \times I_a}{2\pi}$$

or $$I_a = \frac{320 \times 2\pi}{0.33 \times 60} = 101.56 \text{ A}$$

Hence, current taken by motor $=$ **101.56 A. (Ans.)**

(*ii*) **Speed of motor, N :**

$$E_b = V - I_a(R_a + R_{se})$$
$$= 240 - 101.56(0.2 + 0.02) = 217.66 \text{ V}$$

But $$\frac{E_b}{N} = 0.33$$ [Already calculated above]

$$\therefore \quad N = \frac{E_b}{0.33} = \frac{217.66}{0.33} = 659.6 \text{ r.p.m.}$$

☞ **Example 31.** *A six-pole, lap-wound 400 V series motor has the following data : Number of armature conductors = 920, flux / pole = 0.045 Wb, total motor resistance = 0.6 ohm, iron and friction losses = 2 kW. If the current taken by the motor is 90 A, find :*

(*i*) *Total torque ;* (*ii*) *Useful torque at the shaft ;*
(*iii*) *Power output ;*
(*iv*) *Pull at the rim of a pulley of 40 cm diameter connected to the shaft.*

Solution. Number of poles, $p = 6$
Supply voltage, $V = 400$ Volts
Number of parallel paths, $a = p = 6$ [Motor being lap-wound]
Number of armature conductors, $Z = 920$
Flux/pole, $\phi = 0.045$ Wb
Motor resistance, $R_m = 0.6$ ohm
Iron and friction losses $= 2$ kW or 2000 W
Current taken by the motor, $I_a = 90$ A
Radius of the pulley, $= 40/2 = 20$ cm or 0.2 m
Using the relation, $E_b = V - I_a R_m = 400 - 90 \times 0.6$
$= 346$ V

Also, $$E_b = \frac{p\phi ZN}{60a}$$

$$346 = \frac{6 \times 0.045 \times 920 \times N}{60 \times 6}$$

$$\therefore \qquad N = \frac{346 \times 60 \times 6}{6 \times 0.045 \times 920} = 501 \text{ r.p.m.}$$

(*i*) **Total torque, T_a :**

We know that,
$$T_a = 0.159 \times Z\phi p \times \left(\frac{I_a}{a}\right)$$
$$= 0.159 \times 920 \times 0.045 \times 6 \times \frac{90}{6} = 592.4 \text{ Nm}$$

Hence, **Total torque = 592.04 N-m. (Ans.)**

(*ii*) **Useful torque, T_{useful} :**

$$T_{lost} \times \left(\frac{2\pi N}{60}\right) = \text{Iron and friction loss} = 2000 \text{ W}$$

$$\therefore \qquad T_{lost} = \frac{2000 \times 60}{2\pi N} = \frac{2000 \times 60}{2\pi \times 501} = 38.11 \text{ Nm}$$

$$\therefore \qquad T_{useful} = T_a - T_{lost} = 592.4 - 38.11$$
$$= 554.29 \text{ Nm}$$

Hence, **useful torque = 554.29 Nm. (Ans.)**

(*iii*) **Power output :**

Power output
$$= T_{useful}\left(\frac{2\pi N}{60}\right) = \frac{554.29 \times 2\pi \times 501}{60}$$
$$= 29084.4 \text{ or } 29.08 \text{ kW}$$

Hence, **power output = 29.08 kW. (Ans.)**

(*iv*) If F is the pull at the rim of the pulley and r is the radius,

Torque at the shaft, $T_{useful} = F \times r$

i.e., $F \times 0.2 = 554.29$

$\therefore \qquad F = 2771.45$ N

Hence, **pull at the rim of the pulley = 2771.45 N. (Ans.)**

Example 32. *A 240 V series motor takes 40 A when giving its rated output at 1500 r.p.m. Its resistance is 0.3 Ω. Find what resistance must be added to obtain rated torque* (*i*) *at starting and* (*ii*) *at 1000 r.p.m.*

Solution. *Given* : $V = 240$ volts ; $I\ (= I_a) = 40$ A, $N = 1500$ r.p.m., $R = 0.3\ \Omega$.

(*i*) **Resistance to be added to obtain rated torque at starting, R_{add} :**

Since the torque remains the same in both the cases, it is obvious that the current drawn by the motor remains constant at 40 A.

$$\therefore \qquad 40 = \frac{240}{R_{add} + 0.3}$$

or
$$R_{add} = \frac{240}{40} - 0.3 = 5.7\ \Omega.$$

(*ii*) **Resistance to be added to obtain rated torque at 1000 r.p.m., R_{add} :**

We know that,
$$\frac{N_2}{N_1} = \frac{E_{b2}}{E_{b1}}$$
$$E_{b1} = 240 - 40 \times 0.3 = 228 \text{ V}$$

$$\therefore \qquad \frac{1000}{1500} = \frac{E_{b2}}{228}$$

or $$E_{b2} = \frac{1000 \times 228}{1500} = 152 \text{ V}$$

Now $$E_{b2} = V - 40(R_{add} + 0.3) \quad \text{or} \quad 152 = 240 - 40(R_{add} + 0.3)$$

$\therefore$ $$\mathbf{R_{add}} = \frac{240 - 152}{40} - 0.3 = \mathbf{1.9\ \Omega}. \quad \textbf{(Ans.)}$$

Example 33. *A 200 V D.C. series motor runs at 700 r.p.m. when operating at its full-load current of 20 A. The motor resistance is 0.5 Ω and the magnetic circuit can be assumed unsaturated. What will be the speed if (i) the load torque is increased by 44% (ii) the motor current is 10 A.*

Solution. *Given* : $V = 200$ volts ; $N_1 = 700$ r.p.m., $I_1 = 20$ A, $R_m = (R_a + R_{sh}) = 0.5\ \Omega$

Speeds, (N_2, N_3) :

(*i*) **When the load torque is increased by 44% :**

$$T_2 = 1.44 T_1$$

$$\phi_2 I_2 = 1.44 \phi_1 I_1 \quad (\because T \propto \phi I_a)$$

$$I_2^2 = 1.44 I_1^2 \quad (\because \phi \propto I)$$

or $$I_2 = I_1\sqrt{1.44} = 20 \times 1.2 = 24 \text{ A}.$$

Back e.m.f. $$E_{b1} = V - I_1(R_a + R_{se}) = 200 - 20 \times 0.5 = 190 \text{ V}$$

Back e.m.f. $$E_{b2} = V - I_2(R_a + R_{se}) = 200 - 24 \times 0.5 = 188 \text{ V}$$

Also $$\frac{N_2}{N_1} = \frac{E_{b2}}{E_{b1}} \times \frac{\phi_1}{\phi_2}$$

or $$\frac{N_2}{700} = \frac{188}{190} \times \frac{20}{24}$$

or $$\mathbf{N_2 = 577.2 \text{ r.p.m.}} \quad \textbf{(Ans.)}$$

(*ii*) **When the motor current is 10 A :**

Back e.m.f., $$E_{b3} = V - I_3(R_a + R_{se})$$
$$= 200 - 10 \times 0.5 = 195 \text{ V}$$

Also, $$\frac{N_3}{N_1} = \frac{E_{b3}}{E_{b1}} \times \frac{\phi_1}{\phi_3}$$

$$\frac{N_3}{700} = \frac{195}{190} \times \frac{20}{10}$$

or $$\mathbf{N_3 \simeq 1437 \text{ r.p.m.}} \quad \textbf{(Ans.)}$$

Example 34. *A 220 V D.C. series motor draws full-load line current of 38 A at the rated speed of 600 r.p.m. The motor has armature resistance of 0.4 Ω and the series field resistance is 0.2 Ω. The brush voltage drop irrespective of load is 3.0 volts, find :*

(*i*) *The speed of the motor when the load current drops to 19 A.*

(*ii*) *The speed on removal of load when the motor takes only 1 A from supply.*

(*iii*) *The internal horse power developed in each of the above cases. Neglect the effect of armature reaction and saturation.*

Solution. *Given* : $V = 220$ volts ; $I_1 = 38$ A, $N_1 = 600$ r.p.m., $R_a = 0.4\ \Omega$, $R_{se} = 0.2\ \Omega$, Brush drop = 3 V ; $I_2 = 19$ A, $I_0 = 1$ A.

(*i*) **Speed of the motor N_2 (at 19 A)**

On full-load :

Back e.m.f., $$E_{b1} = V - I_1(R_a + R_{se}) - \text{brush voltage drop}$$
$$= 220 - 38(0.4 + 0.2) - 3 = 194.2 \text{ V}.$$

On-load current of 19 A

Back e.m.f., $E_{b2} = 220 - 19(0.4 + 0.2) - 3 = 205.6$ V

No-load, back e.m.f., $E_{b0} = 220 - 1(0.4 + 0.2) - 3 = 216.4$ V

We know that
$$\frac{N_2}{N_1} = \frac{E_{b2}}{E_{b1}} \times \frac{\phi_1}{\phi_2}$$

$$\frac{N_2}{600} = \frac{205.6}{194.2} \times \frac{I_1}{I_2} \quad \left[\text{Assuming unsaturated field } i.e. \ \frac{\phi_1}{\phi_2} = \frac{I_1}{I_2}\right]$$

$\therefore$
$$\mathbf{N_2} = 600 \times \frac{205.6}{194.2} \times \frac{38}{19} = \mathbf{1270 \text{ r.p.m.}} \quad \textbf{(Ans.)}$$

(*ii*) **$\mathbf{N_0}$ (at 1 A) :**

Now,
$$\frac{N_0}{N_1} = \frac{E_{b0}}{E_{b1}} \times \frac{\phi_1}{\phi_0}$$
$$= \frac{E_{b0}}{E_{b1}} \times \frac{I_1}{I_0}$$

$\therefore$
$$N_0 = N_1 \times \frac{E_{b0}}{E_{b1}} \times \frac{I_1}{I_0} = 600 \times \frac{216.4}{194.2} \times \frac{38}{1} = \mathbf{25406 \text{ r.p.m.}} \quad \textbf{(Ans.)}$$

(*iii*) **Internal H.P. developed (in the above cases) :**

Internal H.P. developed at a load current 38 A
$$= \frac{194.2 \times 38}{735.5} = \mathbf{10.} \quad \textbf{(Ans.)}$$

Internal H.P. developed at a load current of 19 A
$$= \frac{205.6 \times 19}{735.5} = \mathbf{5.31.} \quad \textbf{(Ans.)}$$

Internal H.P. developed at no load current of 1 A
$$= \frac{216.4 \times 1}{735.5} = \mathbf{0.294.} \quad \textbf{(Ans.)}$$

Example 35. *A D.C. series motor draws a line current of 100 A from the mains while running at 1000 r.p.m. Its armature resistance is 0.15 Ω and the field resistance is 0.1 Ω. Assuming that the flux corresponding to a current of 25 A is 40% of that corresponding to 100 A, find the speed of the motor when it is drawing 25 A from 230 V supply.* **(GATE, 1996)**

Solution. *Given* : $V = 230$ volts ; $I_1 = 100$ A ; $N_1 = 1000$ r.p.m. ; $R_a = 0.15\ \Omega$; $R_{se} = 0.1\ \Omega$, $I_2 = 25$ A ; $\phi_2 = 0.4\phi_1$

Speed, $\mathbf{N_2}$:

While drawing line current of 100 A,
$$E_{b1} = V - I_1(R_a + R_{se})$$
$$= 230 - 100(0.15 + 0.1) = 205 \text{ V}$$

While drawing line current of 25 A,
$$E_{b2} = V - I_2(R_a + R_{se})$$
$$= 230 - 25(0.15 + 0.1) = 223.75 \text{ V}$$

We know that,
$$\frac{N_2}{N_1} = \frac{E_{b2}}{E_{b1}} \times \frac{\phi_1}{\phi_2} \quad \text{or} \quad \frac{N_2}{1000} = \frac{223.75}{205} \times \frac{\phi_1}{0.4\phi_1}$$

$\therefore$
$$\mathbf{N_2} = 1000 \times \frac{223.75}{205} \times \frac{1}{0.4} = \mathbf{2729 \text{ r.p.m.}} \quad \textbf{(Ans.)}$$

7.10. Motor Characteristics

The properties of all motors and, in particular, D.C. motors are defined as a totality of the following characteristics :

(i) Starting ; *(ii) Operating and mechanical* ;
(iii) Braking ; and *(iv) Regulation.*

Starting characteristics. The starting characteristics *determine the operation of starting from the moment the motor begins running to the moment when steady-state operation is established* and *include* :

- The starting current I_{start} generally determined by the ratio $\frac{I_{start}}{I_{run}}$;
- The starting torque T_{start}, determined by the ratio $\frac{T_{start}}{T_{run}}$;
- The duration of starting t_{start} ;
- The economy of operation determined by the amount of energy consumed in starting ; and
- The cost and reliability of the starting equipment.

Operating characteristics. The operating characteristics are those that give the relation between *speed, torque* and *efficiency* as *functions* of the *useful power* or the *armature* current for V = constant and constant resistances in the armature and field circuit.

Mechanical characteristics. Of major importance for industrial drive mechanisms are the mechanical characteristics, which are the relation $N = f(T)$ (where N and T stand for speed and torque respectively) for conditions of constant voltage and resistances in the armature and field circuits. These also include the *braking characteristics.*

Regulation characteristics. These characteristics determine the *properties of motors when their speed is controlled.* These include :

- The regulation range determined by the ratio $\frac{N_{max}}{N_{min}}$;
- The efficiency of regulation from the point of view of the initial cost of the equipment and maintenance ;
- The nature of regulation—continuous or stepped ; and
- The simplicity of the control apparatus and methods.

The *D.C. motors possess versatile* and diverse regulation characteristics, and for this reason are indispensable in installations where wide-range control of speed is necessary.

The characteristic curves of a motor are those curves which show relation between the following quantities :

1. *Torque and armature current i.e.,* T_a/I_a characteristic. This is also known as **electrical characteristic.**

2. *Speed* and *armature current i.e.,* N/I_a characteristic.

3. *Speed* and *torque i.e.,* N/T_a characteristic. This is also known as **mechanical characteristic.** This can be obtained from (1) and (2) above.

Following relations are worth *keeping in mind* while discussing motor characteristics :

$$N \propto \frac{E_b}{\phi} \text{ and } T_a \propto \phi I_a.$$

7.10.1. Torque-current characteristics

Shunt motor :

- When running on *no-load,* a small armature current flows to supply the field and to *drive the machine against the friction and other losses in it.*
- As the load is applied to the motor, and is increased, the torque rises *almost proportionally to the increase in current.* This is not quite true, because the flux has been assumed to be constant, whereas it decreases slightly owing to armature reaction. The effect of this is to cause the top of the curve connecting torque and line current to *bend over* as shown in Fig. 65.
- The starting torque of a motor is determined by the starting resistance, which in turn, governs the initial current through the machine when the main switch is closed. At this moment the speed is zero, so that the back e.m.f. is zero and the starting current is given by $I = V/R$, where V is the supply voltage and R is the total resistance, which includes the armature and starting resistance.

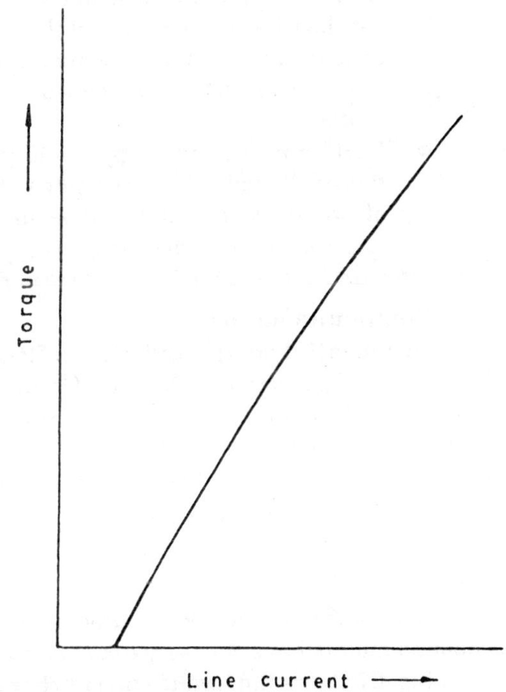

Fig. 65. Torque-current characteristic of a shunt motor.

If the starting current is *limited* by *heating considerations* to twice the full load current, then with normal supply voltage the starting torque of a shunt motor is twice the full load torque. If, however, the supply voltage is below normal, the flux is also less than twice full load torque. The importance of this will be appreciated when the starting torque of a series motor is compared with that of the shunt motor.

Series motor :

- In a series motor the torque ($T_a \propto \phi I_a$) increases much more than does the armature current. This is because the flux itself increases with the armature current, though, owing to the magnetic saturation, the two are not strictly proportional. Nevertheless, for all *but the heavy loads which tend to produce saturation of the field system,* it may be said that the *torque is approximately proportional to the square of the load current.*

Fig. 66 shows the relationship between torque and current. Here the current commences at the no-load value, rises *parabolically at first,* but *increases more slowly* as the *effects of armature reaction and magnetic saturation becomes appreciable.*

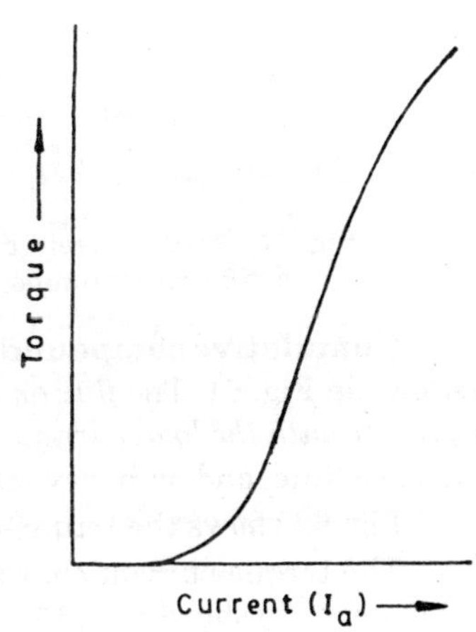

Fig. 66. Torque-current characteristic of a series motor.

- This property of a series motor, by virtue of which a *heavy current gives rise to a very high torque,* also influences its starting characteristics. In a case of a shunt motor, it has already been seen, that the current at the moment of starting may be as high as twice the full-load value ; if we allow for the armature of the magnetisation characteristic and for weakening effect of armature reaction and assume that the flux is increased to 1.5 times its full-load value, then it is obvious that the *starting torque of a series motor is three times the full-load torque.*

- Further more, if the supply voltage falls, the starting current may still be maintained at twice full-load value by cutting out some of the starting resistance, so that the high value of starting torque may still be maintained.

This type of motor (series motor) is *superior to shunt motor for drives in which machines have to be started and accelerated from rest when fully loaded,* as *is the case with traction equipment.*

Compound motors :

Differential compound motor. Refer Fig. 60. In this type of motor the two field windings (shunt and series) *oppose* each other. On *light loads,* such a machine *runs as a shunt motor*, since the series field winding, carrying only a small current, has relatively little effect.

On *heavy loads,* the series coils strengthen and since they are in opposition to the shunt winding, cause a *reduction in the flux* and a consequence *decrease in torque.*

On *heavy overloads* or *when starting up on load, the series winding may become as strong as shunt,* or *it may even predominate, in which the torque will be reduced to zero* or *may even be reversed.* In the latter case, the motor would tend to start-up in the *wrong direction.*

It is obvious that such characteristics may cause *dangerous results,* so that differential compound motors have only *very limited applications in practice.*

Fig. 67 shows the torque-current characteristic of a differential compound motor.

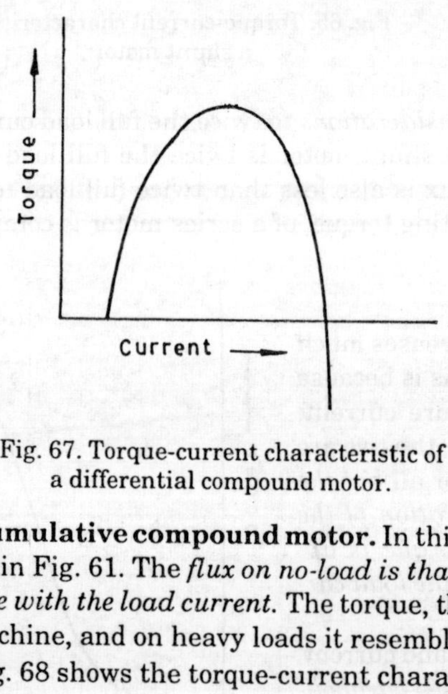

Fig. 67. Torque-current characteristic of a differential compound motor.

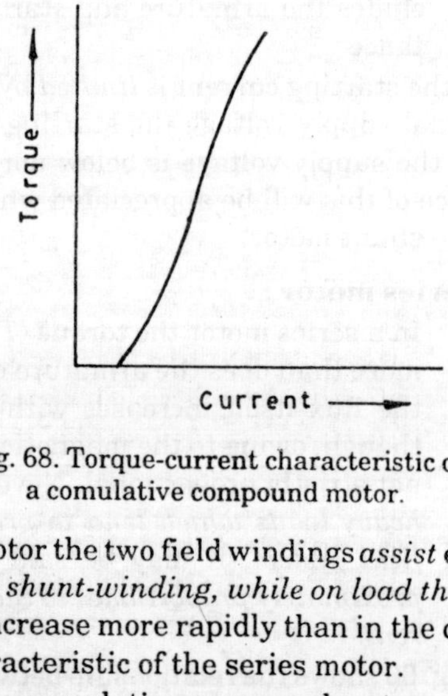

Fig. 68. Torque-current characteristic of a comulative compound motor.

Cumulative compound motor. In this type of motor the two field windings *assist* each other as shown in Fig. 61. The *flux on no-load is that due to the shunt-winding, while on load the flux and torque rise with the load current.* The torque, therefore, increase more rapidly than in the case of the shunt machine, and on heavy loads it resembles the characteristic of the series motor.

Fig. 68 shows the torque-current characteristic of a cumulative compound motor.

The torque-current characteristics of shunt, series and compound (differential and cumulative) motors are showing in Fig. 69 from which their properties, may be compared, as far as torque is concerned.

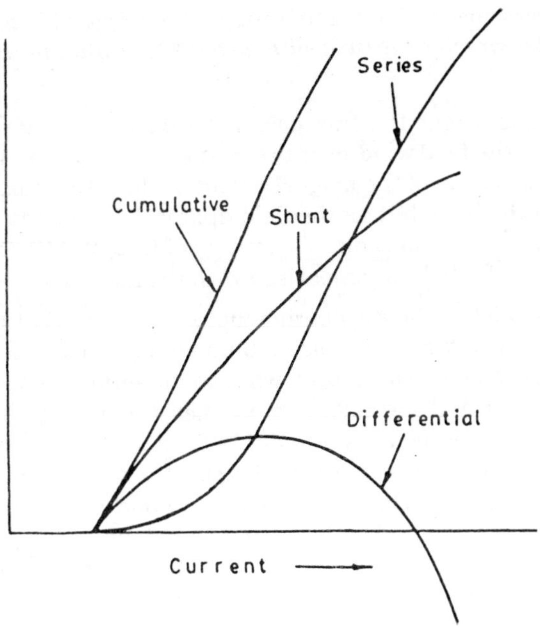

Fig. 69. Torque-current characteristics of shunt, series and compound motors.

7.10.2. Speed-current characteristics. The speed-current characteristics of various motors can be deduced from the following *motor equation,*

$$N \propto \frac{E_b}{\phi} \quad \text{or} \quad N \propto \frac{V - IR}{\phi}.$$

Shunt motor :

- In the shunt motor, the field circuit is connected to the supply terminals so that the exciting current remains constant as long the temperature of the machine is constant, and field regulator is not adjusted. Actually as the machine warms up, the field resistance increases and the exciting current decreases by about 4% for every 10°C rise in temperature. *Neglecting* this effect and also due to armature reaction, it is seen that the speed of a shunt machine falls slightly as the load increases. The *fall in speed is proportional to the volt drop IR in the armature circuit.* If, however, we *consider the effect of armature reaction, an increase of load causes a slight decrease in flux, unless the machine is fitted with compensating windings.* This weakening of the field *tends to raise the speed,* so that the actual fall in speed is less than that calculated by a consideration of the volt drop in the armature.

On the whole, the shunt motor may be regarded as one in which the *speed is approximately constant,* falling slightly as the load increases (*see* Fig. 70).

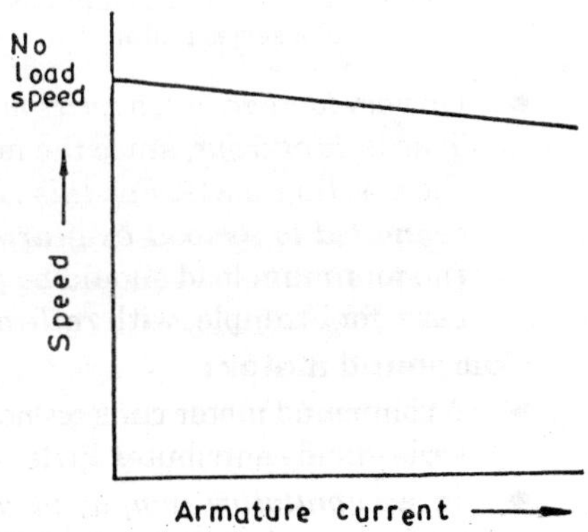

Fig. 70. Speed-current characteristic of a shunt motor.

- The *speed* of a shunt machine can be *increased by inserting resistance* in the field by means of a field regulator. This weakens the field and causes the motor to run faster in order to generate the necessary back e.m.f. of course, it is *impossible to reduce the speed by this method below that at which it runs with no field resistance in the circuit.*

Series motor :

- In case of a series motor the flux does not remain constant, or even approximately constant, because the field winding is in series with the load, so that as the load increases so also does the strength of the magnetic field. At first the flux increases approximately in proportion to the load, but as the field approaches saturation, owing to the heavier loads, the increase is not so rapid. The effects of temperature changes and of armature reaction may be neglected (in comparison with the above mentioned effect).
- It will be appreciated, while considering motor equation, that the back e.m.f. decreases as the armature current increases, as in shunt motor ; in the latter, however, the decrease is due to the volt drop in the armature, while in the series machine the loss in volts occurs in the field as well as in the armature, since they are in series. The back e.m.f. in a series motor, therefore, decreases more rapidly than in a corresponding shunt machine. *The speed,* however, is *proportional to the back e.m.f. divided by the flux ; the former decreases, while the latter increases with increasing load so that the speed decreases rapidly as the armature current increases.* This *property is a valuable feature in a drive of which the speed is required automatically to adjust to compensate for changes in load.*

The speed-current characteristic of a series motor are shown in Fig. 71 (*a*).

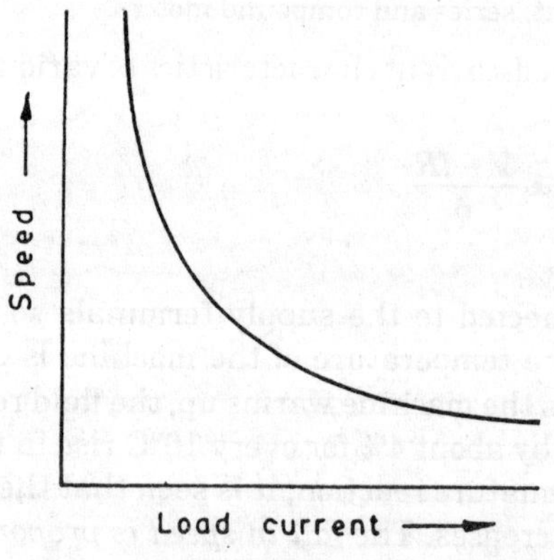

Fig. 71. (*a*) Speed-current characteristic of a series motor.

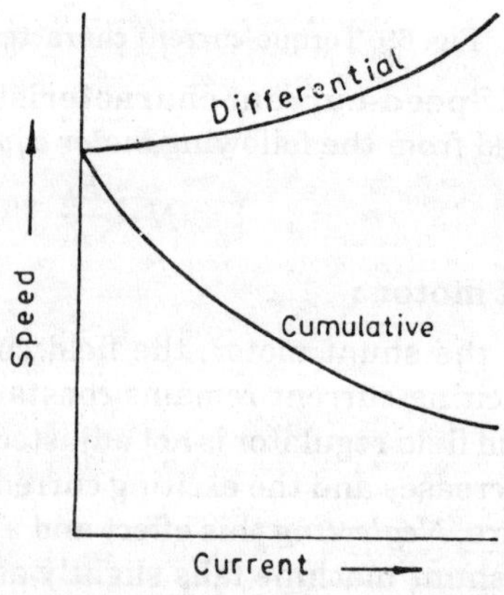

Fig. 71. (*b*) Speed-current characteristic of compound motors.

- On very *low current*, a *series motor runs at very high speeds,* or *tends to race*, as it is termed. This is *dangerous,* since the machine may be destroyed by the centrifugal forces set up in the rotating parts. *For this reason, when installing a series motor it must be positively connected to its load by gearing or by direct connection and never by belting.* Moreover, the minimum load should be great enough to keep the speed within safe limits, as is the case, for example, with *railway motors, hoists* and *rolling mills.*

Compound motor :

- A compound motor runs on *'no-load'* at a speed determined by its shunt winding since the series field contributes little to the total flux in this condition.
- In a *cumulative compound motor at 'no-load'*, the series field strengthens the shunt winding so that the speed falls as in a series machine. Since the flux at any load is equal to the shunt and series fluxes, the speed is less than it would be if running on either field

alone. The speed-current characteristic of such a machine is shown in Fig. 71 (*b*). Such a characteristic has two important **advantages.** These are :

(*i*) *The machine has the compensating action of reducing its speed on heavy loads, as is the case with series machine.*

(*ii*) The maximum speed on no-load is limited by the shunt winding, since this produces a maximum flux even on no-load.

This type of motor, therefore, is suitable for driving machines which operate *on a cycle consisting of a power or working stroke followed by a return or idle stroke.* The series winding produces a fall in speed on the working stroke, while the shunt winding permits the return stroke to be completed at a high, but safe, speed. *A fly-wheel is also provided to act as a load equilizer in such a drive.*

In case of a *differential compound motor,* since the series winding opposes the shunt, the resultant flux decreases as the load increases ; thus the machine runs at a higher speed than it would do as a shunt motor. If the series windings were relatively weak, this reduction in flux might be just sufficient for the fall in speed, brought about by the volt drop in the machine. *Such a motor would have a useful application in driving loads at a constant speed.* If the series field were strong, however, an increase in load would result in a decrease in the magnetic flux and a rise in speed would take place as shown in Fig. 49, the heavier the load, the faster would the motor tend to run. This is the property which may have *dangerous consequences,* since a heavy overload would result in such a high speed that the motor would destroy itself.

7.10.3. Speed-Torque (or Mechanical) characteristics. The speed-torque characteristics of the four types, *i.e.* shunt, series, cumulative and differential of motors drawn on the same diagram are shown in Fig. 72 for the purpose of comparison.

The main properties of individual motors, from this diagram, may be summarised as under :

1. **Shunt motor.** As the *load torque increases* the *speed falls somewhat,* but the machine *may be regarded as an approximately constant speed motor.*

The shunt motor is *used* :

- When the speed is required to remain approximately constant from no-load to full-load.
- When the load has to be driven at a number of speeds, any one of which is required to remain approximately constant.

2. **Series motor.** As the *load torque increases the speed falls rapidly.* At low torque the speed becomes very high and machine tends to race.

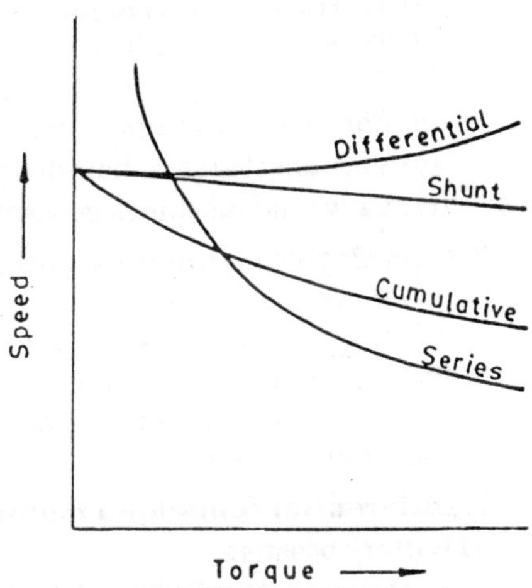

Fig. 72. Speed-torque characteristics of D.C. motors.

The series motors are used :

- When *large starting torque is required* (as in *traction motors*).
- When the load is subject to heavy fluctuations, and a reduced speed is desired to compensate for the high torque, provided that there is no possibility of the machine *'losing'* its load.

3. **Cumulative compound motor.** In this type of motor the *speed falls appreciably as the torque increases,* but on low torques the maximum speed is limited to a safe value. These motors are used :

- *When a large starting torque is required but when the load may fall so low that a series motor would race.*

- *When the load is of a fluctuating nature and a reduced speed is desirable on the heavy loads.*

In such a case a flywheel is usually fitted so that when speed is so reduced the kinetic energy stored in the flywheel at high speeds is given up to assist the motor in driving the heavy load.

- *When the supply voltage is subject to fluctuations* (as in traction systems).

4. **Differential compound motor.** The speed at low torque is limited by the shunt winding, as in the cumulative compound machine. At *high torques*, the speed may be arranged to remain constant or, with a stronger series field, the speed may rise with increasing load.

On *very heavy loads the machine may tend to race.*

Its use is usually restricted to applications *which require a constant speed.*

Industrial applications of D.C. Motors :

1. Shunt motors :

(*i*) Drills and milling machines
(*ii*) Line-shaft drives
(*iii*) Boring mills
(*iv*) Grinders and shapers
(*v*) Spinning and weaving machines
(*vi*) Wood working machines
(*vii*) Small printing presses
(*viii*) Light machine tools generally.

2. Series motors :

(*i*) Traction drives generally
(*ii*) Tram cars and railway cars
(*iii*) Cranes, derricks, hoists, elevators and winches
(*iv*) Fans and air compressors
(*v*) Vacuum cleaners, hair driers, sewing machines
(*vi*) Universal machines generally.

3. Cumulative compound motors :

(*i*) Punching, shearing and planing machines
(*ii*) Lifts, haulage gears and mine hoists
(*iii*) Pumps and power fans
(*iv*) Rolling mills, stamping presses and large printing presses
(*v*) Trolley buses.

4. Differential compound motors :

(*i*) Battery boosters
(*ii*) Experimental and research work.

7.11. Comparison of D.C. Motor Characteristics

Fig. 73 shows the comparison among shunt, series and compound operation for the *same motor* with its armature being acted upon by either or both fields. It is worth noting here that the outputs vary at the same armature current.

- Fig. 74 compares shunt, series and compound performance when the three are rated at the horse power (or kilowatts) and rpm (or radians/sec). In this type of plot all the characteristics cross at the design load point.
- Figs. 73 and 74 are drawn with the motors operating on their full rated voltage. *Different voltages will move the curves, but the relative shapes will remain similar.*

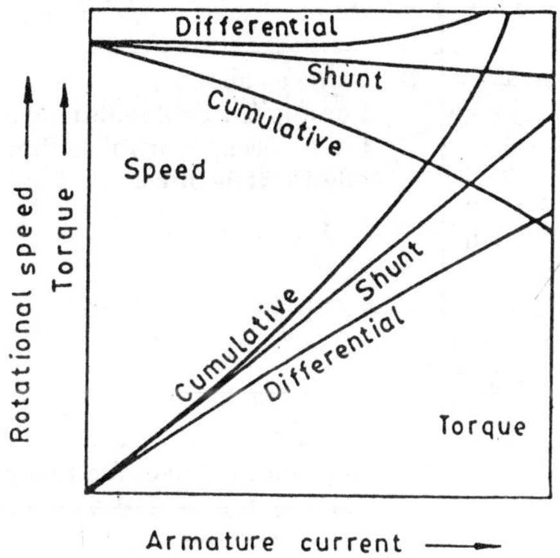

Fig. 73. Speed, torque relations for shunt and compound fields with the same armature current.

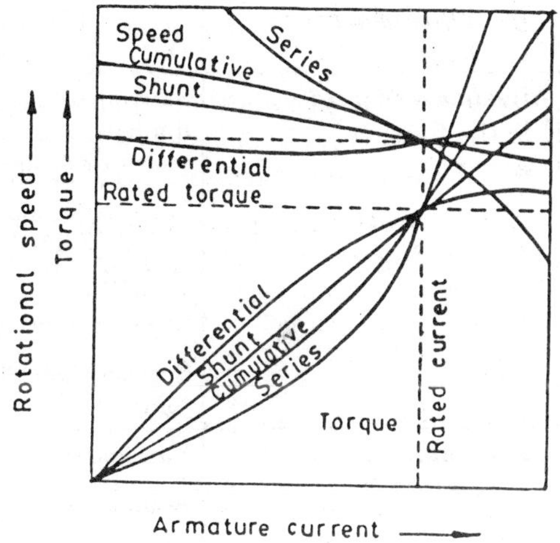

Fig. 74. Speed and torque relations for shunt, series and compound fields. Motors of same rated speed, torque and current.

The following may be deduced from Fig. 74.

(*i*) The curves diverge between no-load and full load. If the load torque were around mid-load, the shunt motor would draw the least current and the series the most.

(*ii*) At overloads the opposite is true. The series motor can develop an overload torque of around double the rated torque without drawing excessive current.

Fig. 73 shows that the speed characteristics for the same armature current are very much different. Therefore, the ability to take increasing torque loads that is the special contribution of a series field is accompanied by a much greater speed change.

7.12. Summary of Characteristics and Applications of D.C. Motors

The summary of characteristics and applications of D.C. motors is given in Table 1.

Table 1. Summary of Characteristics and Applications of D.C. motors

S No.	*Type of motor*	*Characteristics*	*Applications*
1.	**Separately excited D.C. motors :**	Possible to *obtain very accurate speeds.* Most suitable for applications requiring *speed variation from very low value to high value.*	● Paper machines. ● Steel rolling units. ● Diesel electric propultion of ships.
2.	**Shunt motors :** (*i*) *Constant speed :*	Starting torque-medium, usually limited to 250% by a starting resistor but may be increased. Maximum operating torque usually limited to about 200% by commutation. Speed control : — increase upto 200% speed by field control. — decrease by armature voltage control.	● Employed for constant-speed applications ; may be used for adjustable speed not greater than 2 : 1 range. Field of applications includes : — Lathes ; — Centrifugal pumps ; — Fans and blowers — Machine tools ; — Wood working machines ; — Reciprocating pumps ; — Spinning and weaving machines ; — Printing presses, etc.

S No.	*Type of motor*	*Characteristics*	*Applications*
	(ii) Adjustable speed :	Starting torque-medium, usually limited to 250% by a starting resistor but may be increased. Maximum operating torque-usually limited to 200% by commutation. Speed regulation – 10 to 15%. Speed control : 6 : 1 range by field control.	● Same as above, For applications requiring adjustable speeds control, either constant torque or constant output.
3.	**Series motors :**	Variable speed. Adjustable varying speed. *Starting torque* very high upto 500%. *Maximum momentary operating* torque upto 400%. *Speed regulation* : Widely variable, very high at no-lead. *Speed control* : By series resistance.	● Suitable for drives requiring high starting torque and where adjustable, varying speed is satisfactory. Fields of application include : — Cranes ; — Hoists ; — Trolley cars ; — Conveyors ; — Electric locomotives etc. **Loads must be positively connected, not belted.* *To prevent overspeed, lightest load should not be much less than 15 to 20% of full-load torque.
4.	**Compound motors :**		
	(i) Cumulative compound wound :	Variable speed. Adjustable varying speed. *Starting torque* high upto 450% depending upon the degree of compounding. *Maximum momentary operating torque* higher than shunt, upto 350%. *Speed regulation* : Varying, depending upon degree of compounding, upto 25-30%. *Speed control* : Usually not used but may be upto 125% by field control.	● Suitable for drives requiring high starting torque and only fairly constant speed, pulsuating loads with flywheel action. Fields of application includes : — Shears ; — Punches ; — Elevators ; — Conveyors ; — Rolling mills ; — Heavy planes etc.
	(ii) Differentially compound wound :	Torque and speed *almost constant.* Tendency towards speed instability with a possibility of motor running away and strong possibility of motor starting in wrong direction.	● Employed for experimental and research work.

7.13. Starting of D.C. Motors

7.13.1. Need for starters. A motor at *rest* has no back or counter e.m.f. At staring therefore, the *armature current is limited only by the resistance of the armature circuit.* The *armature resistance is very low*, however, and if full voltage were impressed upon the motor terminals at stand still, the resulting armature current would be *many times full-load* value—*usually sufficient to damage the machine.* For this reason, *additional resistance* is introduced into the armature circuit at starting. As the motor gains speed, its back e.m.f. builds up and the starting resistance is cut out.

Note. Very small D.C. motors, either shunt, series or compound wound, have sufficient armature resistance so that they may be started directly from the line without the use of a staring resistance and without injury to the motor.

Fig. 75 shows the connections of a starting resistance in three types of D.C. motors :
(*a*) A series motor ; (*b*) A shunt motor ; and
(*c*) A compound motor.

- In the case of *series motor* [Fig. 75 (*a*)], the armature, field and starting resistance are all in series.
- In the case of *shunt motor* [Fig. 75 (*b*)], it will be seen that the *top end of shunt field is connected to the first contact* on the starting resistance. This is to *ensure* that the *field winding receives the full supply at the moment of switching on.* If the fields were connected to the *last stud* of the starting resistance, then on starting, the field would receive only a *proportion of the supply voltage, the field current would be correspondingly weak and the torque might be too small to start the motor against the friction of the moving parts.*

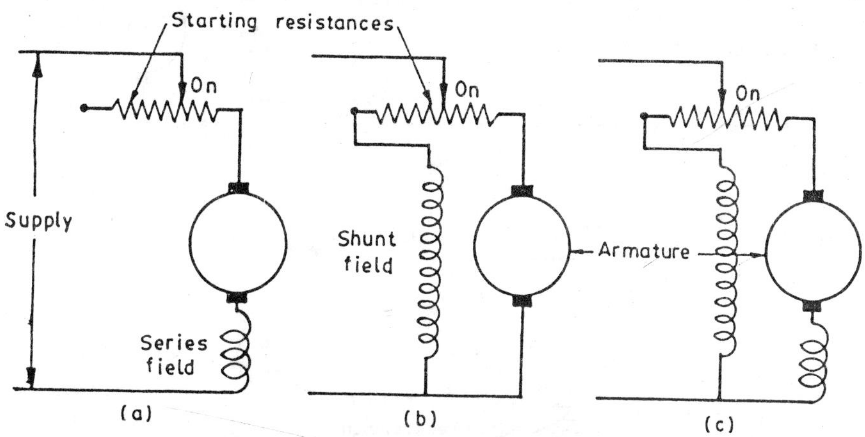

Fig. 75. Circuits incorporating starting resistances.

- The connections for the *compound motor* are seen from [Fig. 75 (*c*)] to be a combination of those of the series and the shunt connections.

7.13.2. Starters for shunt and compound motors

- The starters of D.C. motors are generally manufactured in convenient sizes and styles for use as auxiliaries with D.C. shunt and compound motors. Their *primary function is to limit the current in the armature circuit during the starting accelerating period.*
- The motor starters are always rated on the basis of output power and voltage of the motors with which they are to be used.

(*i*) Three-point type ; and (*ii*) Four-point type.

Three-point starters are not completely satisfactory when used with motors whose speeds must be controlled by inserting resistance in the shunt field circuit. However, when applications require *little or no speed control,* either may be employed.

Three-point starter. Refer Fig. 76. The starter has three terminals *L, F* and *A*. The line terminal *L* must be connected to either side, positive or negative of the D.C. source on the main switch ; the field terminal *F* is connected to one field terminal on the motor ; the armature terminal *A* must be connected to either one of the motor armature terminals. The final connection must then be made from the second line terminal on the main switch to a junction of the remaining two armature and field terminals of the motor. If it is desired that the speed of the motor is controlled, a field rheostat is added as shown in Fig. 76 (*a*).

When the motor is at rest, the starter arm [represented by an arrow in Fig. 76 (*a*)] is held in the OFF position by a strong spiral spring.

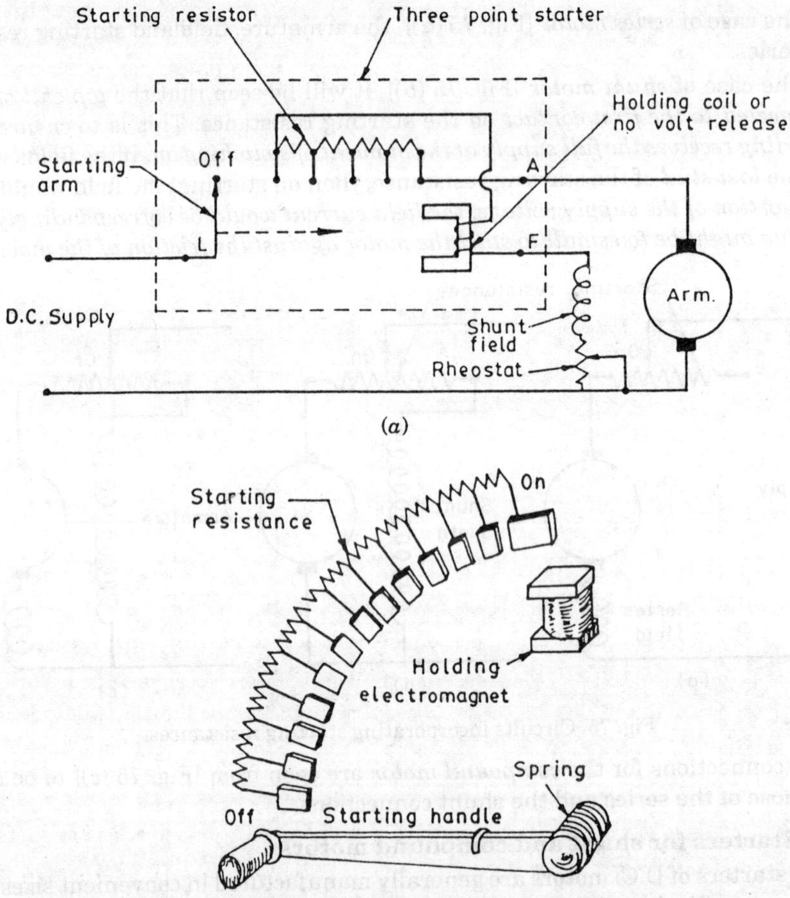

Fig. 76. Three-point starter connected to a shunt motor.

Starting of motor :

- In order to *start the motor*, one hand is held on the handle of the open main switch while the starter arm is moved to the first stud [Figs. 76 (*a*) and (*b*)] with the other hand ; then the main switch is closed. If all the wiring is correct and the armature is free to turn, the motor will start.

- After the armature has accelerated sufficiently on the first stud, the starter arm is slowly moved to studs 2, 3, 4 etc. until the arm rests firmly against the iron poles of the *holding-coil* electromagnet. The entire starting process should take from 5 to 10 seconds. In the final position, the electromagnetic pull exerted by the holding coil will be greater than the force exerted by the spiral spring. *Should there be a power failure or should the field circuit be opened accidentally, the starter arm will fall back to its OFF position.* This function of starter is particularly important because :

  (*i*) if the power fails and starter arm is not restored to the OFF position, the motor might be damaged should the power, come on again ; and

  (*ii*) if the shunt field circuit were opened accidentally and the starter arm did not return to the OFF position, the motor speed might become dangerously high.

- Often the motors are protected against *overloads* by thermal *overload relays* in which *bimetallic is heated by motor current at approximately the same rate at which the motor is itself heating up. Above a certain temperature, this relay trips and opens the line contractor thereby isolating the motor from the supply.*

Drawback of a three-point starter. The use of a three-point starter presents a problem. The speed of the motor is controlled by means of the field rheostat. To increase the speed of motor necessitates the setting of the field rheostat to a higher resistance value. The current through the shunt field is reduced, and so is the current through the coil of the holding electromagnet. The *reduced current through the coil weakens the strength of the magnet and makes susceptible to line-voltage variations.* In the weakened condition a slight reduction in line voltage would further weaken the holding magnet, releasing the arm of the starter and thus *disconnecting the motor from the line.* Unscheduled stoppages of the motor make the three-point starter quite unpopular.

Four-point starter :

- Fig. 77 shows a simplified diagram of a *four-point starter*.
- In this starter the drawback/disadvantage of the three-point starter is eliminated. In addition to the same three-points that were used with the three-point starter, the other

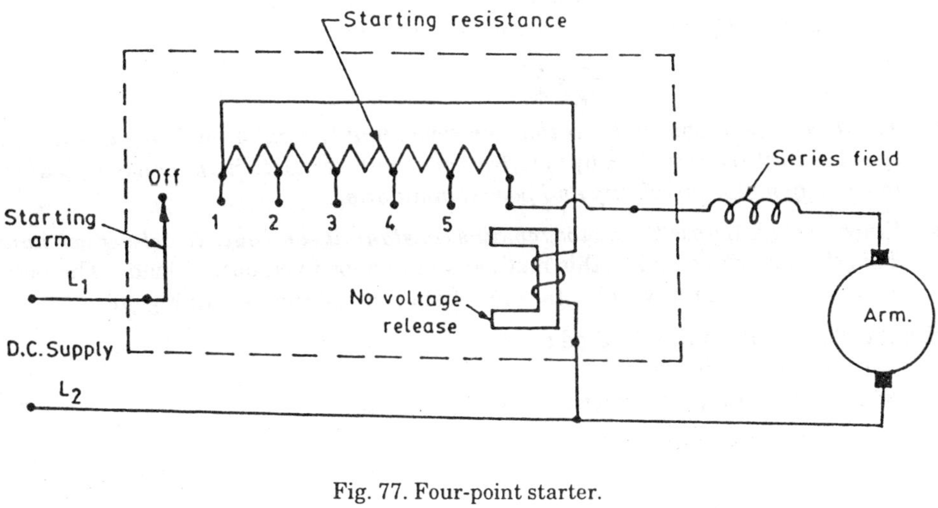

Fig. 77. Four-point starter.

side of the line, L_2 is the *fourth point* brought to the starter. The coil of the holding magnet is connected across the line when the arm is moved from the 'off' position. The holding magnet and starting resistors function as in the three-point starter. *The possibility of accidentally opening the field circuit is quite remote ; hence the greater acceptance of the four-point starter over the three-point starter.*

— The four-point starter *provides the motor with no voltage protection.* Should the power fail, the motor must be disconnected from the line. If not, full line voltage will be applied to the armature without the benefit of starting resistors when power is restored. The holding magnet, being connected across the line, releases the arm *when the voltage drops below a specific value, thus protecting the motor when the power is restored.*

8. SPEED CONTROL OF D.C. MOTORS

8.1. Factors Controlling the Speed

D.C. machines are generally much more adaptable to adjustable speed service. *The ready availability of D.C. motors to adjustment of their operating speed over wide ranges and by a variety of methods is one of the important reasons for the strong competitive position of D.C. machinery in modern industrial applications.*

The speed of a D.C. motor can be expressed by the following relationship.

$$N \propto \frac{V - I_a R_a}{\phi}.$$

Therefore, the speed of D.C. motor can be regulated by changing ϕ, R or V, or in other words, by,

1. Field control
2. Rheostatic control
3. Voltage control.

Note. Thyristor control of D.C. motor is out of scope of this book.

8.2. Field Control Method

- **Field control** is the most common method and forms one of the *outstanding advantages of shunt motors*. The method is, of course, also applicable to compound motors. Adjustment of field current and hence the flux and speed by adjustment of the shunt field circuit resistance or with a solid-state control when the field is *separately excited* is accomplished simply, *inexpensively*, and *without much change in motor losses*.

The *speed is inversely proportional to the field current*

i.e.
$$N \propto \frac{1}{I_f} \propto \frac{1}{\phi}.$$

- The *lowest speed obtainable is that corresponding to maximum field current* ; the *highest speed is limited electrically by the effects of armature reaction under weak-field conditions in causing motor instability and poor commutation.*
- *Since, voltage across the motor remains constant, it continues to deliver constant output.* This characteristic makes this method suitable for fixed output loads. The performance curve of a D.C. motor with voltage and field control is shown in Fig. 78.

Merits. The merits of this method are :
1. Good working efficiency.
2. Compact controlling equipment.

3. Capability of minute speed control.

4. The speed is not effected by load, and speed control can be performed effectively even at light loads.

5. Relatively inexpensive and simple to accomplish, both manually and automatically.

6. Within limits, field control does not affect speed regulation in the cases of shunt, compound, and series motors.

7. Provides relatively smooth and stepless control of speed.

Demerits. The demerits of field control as a method of speed control are :

1. Inability to obtain speeds below the basic speed.

2. Instability at high speeds because of armature reaction.

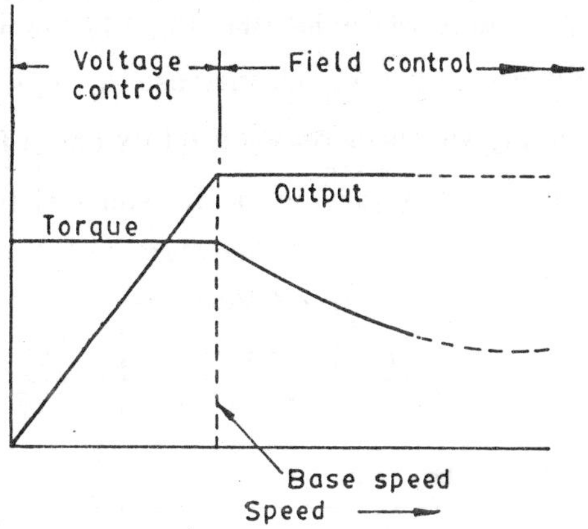

Fig. 78

3. Commutation difficulties and possible commutator damage at high speeds.

Shunt Motors :

- The flux of a D.C. shunt motor can be changed by changing shunt field current (I_{sh}) with the help of a shunt field rheostat as shown in Fig. 79. Since the field current is very small, the power wasted in the controlling resistance is very small.

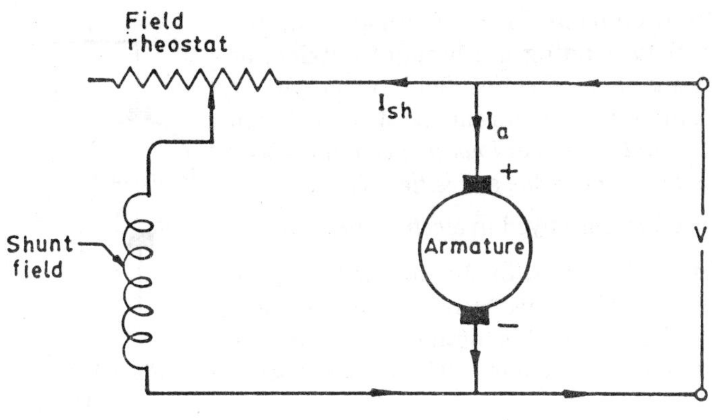

Fig. 79. Field rheostatic control for a D.C. shunt motor.

- In *non-interpolar machines* the speed can be increased by this method in the ratio **2 : 1**. In machines fitted *with interpoles* a ratio of maximum to minimum speeds of **6 : 1** is fairly common.

Series Motors. In a series motor, *variations of flux* can be brought about in any one of the following ways :

(*i*) Field divertors (*ii*) Armature divertor

(*iii*) Tapped field control (*iv*) Paralleling field coils.

(*i*) **Field divertors.** A variable resistance, known as field divertor (Fig. 80) *shunts* the series windings. Any desired amount of current can be passed through the divertor by adjusting its resistance. Hence, the *flux can be decreased and consequently the speed of the motor increased.*

(*ii*) **Armature divertor.** In order to *get speeds lower than the normal speed* a divertor across the armature can be used (Fig. 81). For a *given constant load torque*, if I_a is reduced due to armature divertor, then ϕ must increase ($\because\ T_a \propto I_a$). This results in an increase in current taken from the supply which increases the flux and a fall in speed $\left(\because\ N \propto \frac{1}{\phi}\right)$. The variations in speed can be controlled by varying the divertor resistance.

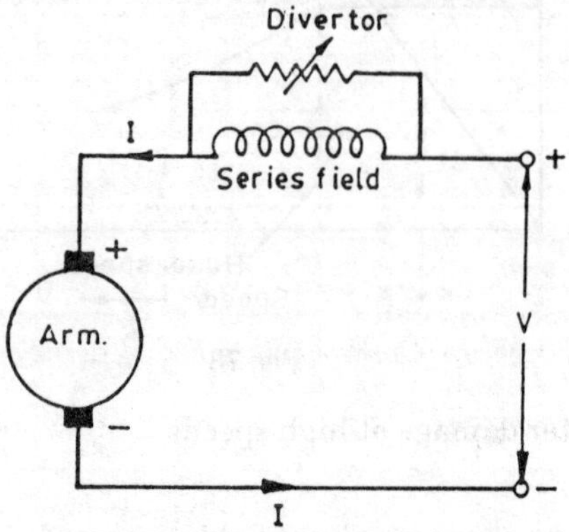

Fig. 80. Field divertor method of speed control for D.C. series motor.

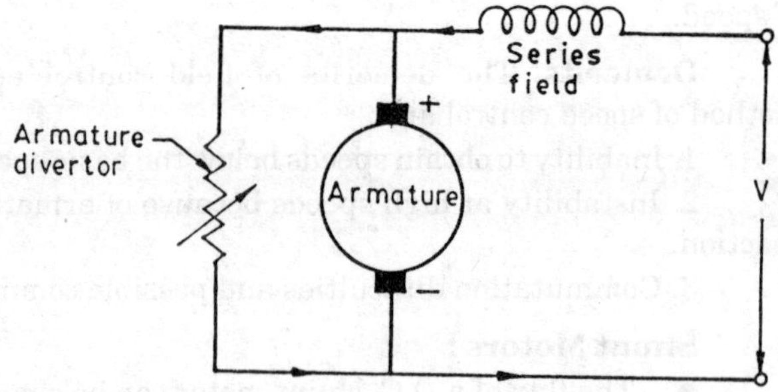

Fig. 81. Armature divertor method of speed control of D.C. series motor.

(*iii*) **Tapped field control.** In this method a number of tappings from the field winding are brought outside, as shown in Fig. 82. A number of series field turns can be short-circuited according to the requirement. *When all field turns are in circuit, the motor runs at lowest speed and speed increases with cutting out some of the series field turns.*

This method is often employed in *electric traction*.

(*iv*) **Paralleling field coils.** In this method of speed control several speeds can be obtained by regrouping the field coils as shown in Fig. 83 (*a*, *b*, *c*). This method is used for *fan motors*. It is seen that for a 4-pole motor, three fixed speeds can be obtained.

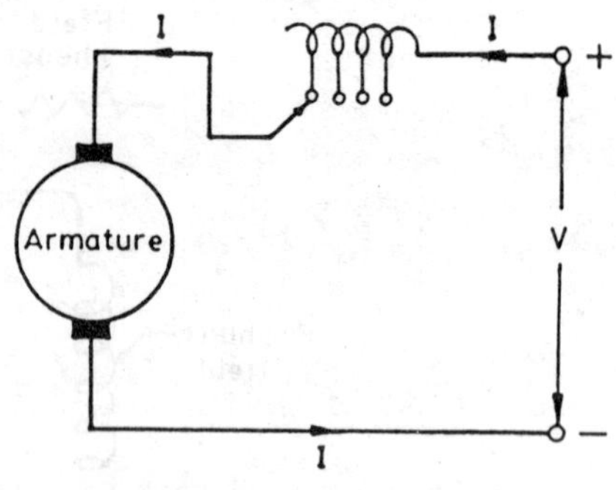

Fig. 82. Tapped field control for D.C. series motor.

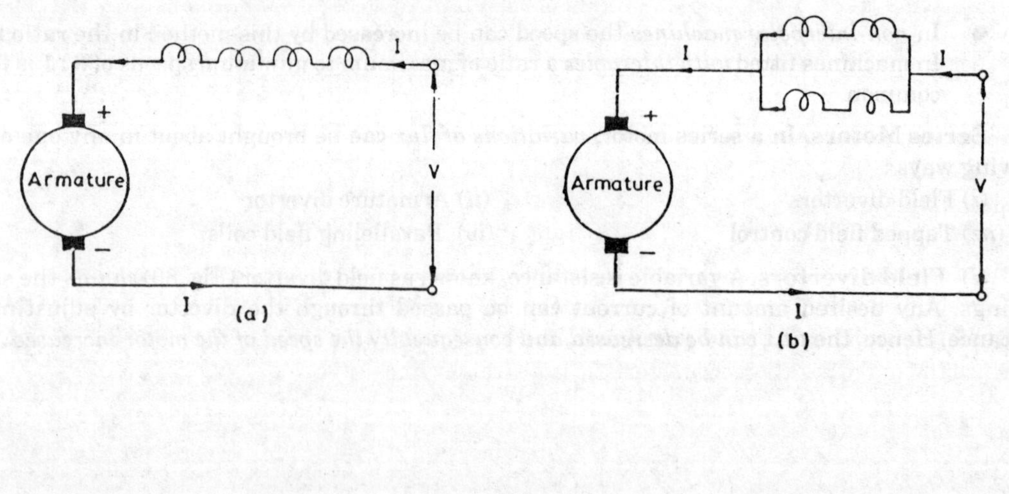

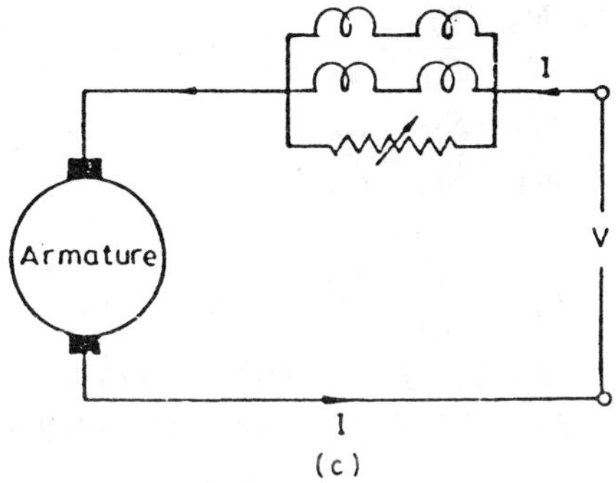

Fig. 83. Paralleling field coils method for speed control of D.C. series motor.

Example 36. *A 220 V D.C. shunt motor draws a no-load armature current of 2.5 A when running at 1400 r.p.m. Determine its speed when taking an armature current of 60 A, if armature reaction weakens the flux by 3 per cent.*

Take armature resistance = 0.2 Ω.

Solution. Supply voltage, $V = 220$ Volts

No-load current, $I_{a0} = 2.5$ A

No-load speed, $N_0 = 1400$ r.p.m.

Armature resistance, $R_a = 0.2\ \Omega$

Armature current, $I_a = 60$ A

Full-load flux, $\phi = 0.97\ \phi_0$

Load speed N :

Back e.m.f. at no-load, $E_{b0} = V - I_{a0}R_a = 200 - 2.5 \times 0.2 = 219.5$ V

Back e.m.f. on load, $E_b = V - I_aR_a = 220 - 60 \times 0.2 = 208$ V.

Now using the relation,

$$\frac{N}{N_0} = \frac{E_b}{E_{b0}} \times \frac{\phi_0}{\phi}$$

$$\frac{N}{1400} = \frac{208}{219.5} \times \frac{\phi_0}{0.97\phi_0}$$

$$\therefore \quad N = \frac{1400 \times 208}{219.5 \times 0.97} = 1367.7 \text{ r.p.m.}$$

Hence, **load speed = 1367.7 r.p.m. (Ans.)**

Example 37. *The armature and shunt field resistances of a 500 V shunt motor are 0.2 Ω and 100 Ω respectively. Find the resistance of the shunt field regulator to increase the speed from 800 r.p.m. to 1000 r.p.m., if the current taken by the motor is 450 A. The magnetisation characteristic may be assumed as a straight line.*

Solution. Supply voltage, $V = 500$ Volts

Armature resistance, $R_a = 0.2\ \Omega$

Shunt field resistance, $R_{sh} = 100\ \Omega$

Initial speed, $N_1 = 800$ r.p.m.

Final speed, $N_2 = 1000$ r.p.m.

Current drawn by the motor, $I = 450$ A

Resistance of shunt field regulator, R :

Since the magnetization characteristic is a straight line,

$$\phi \propto I_{sh}$$

Also, $$\frac{N_2}{N_1} = \frac{E_{b2}}{E_{b1}} \times \frac{\phi_1}{\phi_2}$$

$\therefore$ $$\frac{N_2}{N_1} = \frac{E_{b2}}{E_{b1}} \times \frac{I_{sh1}}{I_{sh2}}.$$

At 800 r.p.m.

$$I_{sh1} = \frac{V}{R_{sh}} = \frac{500}{100} = 5 \text{ A}$$

Armature current, $I_{a1} = I - I_{sh1} = 450 - 5 = 445$ A

Back e.m.f. $E_{b1} = V - I_{a1}R_a = 500 - 445 \times 0.2 = 411$ V

At 1000 r.p.m.

Armature current, $I_{a2} = I - I_{sh2} = 450 - I_{sh2}$

Back e.m.f., $E_{b2} = V - I_{a2}R_a = 500 - (450 - I_{sh2}) \times 0.2 = 410 + 0.2I_{sh2}$

Now $$\frac{1000}{800} = \frac{(410 + 0.2I_{sh2})}{411} \times \frac{5}{I_{sh2}}$$

$$410 + 0.2I_{sh2} = \frac{1000}{800} \times \frac{411}{5} I_{sh2} = 102.75I_{sh2}$$

or $$102.55I_{sh2} = 410$$

$\therefore$ $$I_{sh2} = 4 \text{ A app.}$$

Total resistance in the field circuit

$$= \frac{V}{I_{sh2}} = \frac{500}{4} = 125\ \Omega$$

Resistance of the shunt field $= 100\ \Omega$

Resistance of the shunt field regulator,

$$R = 125 - 100 = 25\ \Omega$$

Hence, **resistance of the shunt field regulator = 25 Ω. (Ans.)**

☞**Example 38.** *A 500 V series motor has an armature resistance of 0.4 ohm and series field resistance of 0.3 ohm. It takes a current of 100 A at a speed of 600 r.p.m. Find the speed of the motor if a divertor of resistance 0.6 ohm is connected across the field, the load torque being kept constant. Neglect armature reaction and assume that flux is proportional to the current.*

Solution. Supply voltage, $V = 500$ Volts

Armature resistance, $R_a = 0.4$ ohm

Series field resistance, $R_{se} = 0.3$ ohm

Divertor resistance, $R_{div.} = 0.6$ ohm

$I_{a1} = 100$ A, $N_1 = 600$ r.p.m.

Speed, N_2 :

Back e.m.f., $$E_{b1} = V - I_{a1}(R_a + R_{se})$$
$$= 500 - 100(0.4 + 0.3) = 430 \text{ V}.$$

Let I_{a2} be the current taken and ϕ_2 be the flux produced when a divertor is connected across the series field (Fig. 84)

Since the torque remains constant,

$\therefore$ $$\phi_1 I_{a1} = \phi_2 I_{a2} \quad ...(i)$$

But $\phi \propto$ current through series field

$\therefore$ $$\phi_1 \propto I_{a1}$$

Current through the series field when a divertor is connected

$$= I_{a2} \times \frac{R_{div.}}{R_{div.} + R_{se}}$$

$$= I_{a2} \times \frac{0.6}{0.6 + 0.3} = 0.667 I_{a2}$$

Flux in this case $\phi_2 \propto 0.667 I_{a2}$

Substituting in (*i*), we get

$$I_{a1}^2 = 0.667 I_{a2}^2$$

$$I_{a2}^2 = \frac{I_{a1}^2}{0.667}$$

$$\therefore \quad I_{a2} = \frac{I_{a1}}{\sqrt{0.667}} = \frac{100}{\sqrt{0.667}} = 122.44 \text{ A}$$

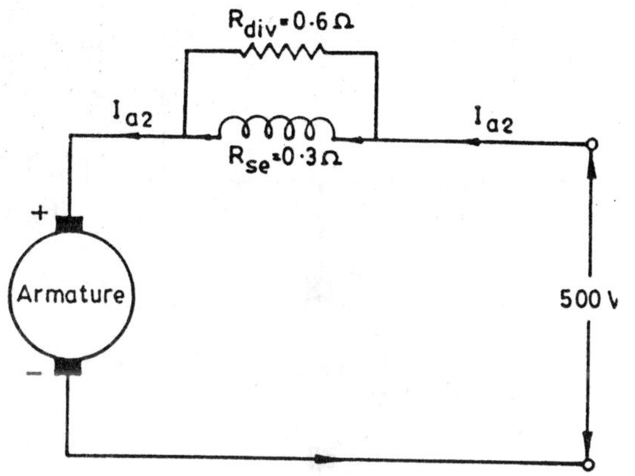

Fig. 84

Series field current, $I_{se} = 0.667 I_{a2}$

$= 0.667 \times 122.44 = 81.64$ A

Back e.m.f., $E_{b2} = V - I_{a2}R_a - I_{se}R_{se}$

$= 500 - 122.44 \times 0.4 - 81.64 \times 0.3$

$= 500 - 48.97 - 24.49 = 426.54$ V

Using the relation, $$\frac{N_2}{N_1} = \frac{E_{b2}}{E_{b1}} \times \frac{\phi_1}{\phi_2}$$

$$\frac{N_2}{600} = \frac{426.54}{430} \times \frac{100}{81.64}$$

$$\therefore \quad N_2 = 729 \text{ r.p.m.}$$

Hence, **speed of the motor = 729 r.p.m. (Ans.)**

☞**Example 39.** *A D.C. series motor drives a load the torque of which varies are square of the speed. The motor takes a current of 20 A when the speed is 800 r.p.m. Calculate the speed and current when the motor field winding is shunted by a divertor of the same resistance as that of the field winding. Neglect all motor losses and assume that the magnetic circuit is unsaturated.*

Solution. $I_{a1} = 20$ A, $N_1 = 800$ r.p.m.

$\mathbf{I_{a2} = ?,\ N_2 = ?}$

When the field winding is shunted by a divertor of equal resistance, then current through either is half the armature current. If I_{a2} is the new armature current, then $\frac{I_{a2}}{2}$ passes through the winding (Fig. 85).

$$\therefore \quad \phi_2 = \frac{I_{a2}}{2}$$

Now $T_1 \propto \phi_1 I_{a1} \propto N_1^2$ (Given)

and $T_2 \propto \phi_2 I_{a2} \propto N_2^2$ (Given)

From (*i*) and (*ii*), we get

$$\left(\frac{N_2}{N_1}\right)^2 = \frac{\phi_2 I_{a2}}{\phi_1 I_{a1}} \quad ...(i)$$

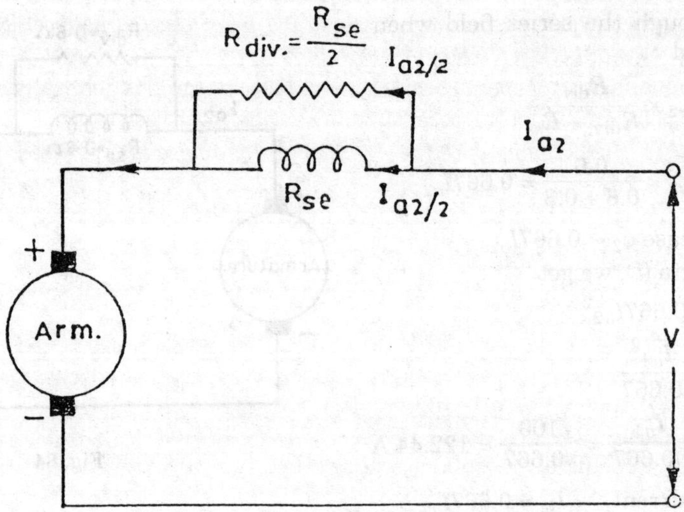

Fig. 85

Because all losses are negligible, hence the armature and series field resistances are negligible. This means that back e.m.f. in both cases is the same as the applied voltage.

$$\therefore \qquad \frac{N_2}{N_1} = \frac{E_{b2}}{E_{b1}} \times \frac{\phi_1}{\phi_2} \text{ becomes}$$

$$\frac{N_2}{N_1} = \frac{\phi_1}{\phi_2} \qquad \text{...}(ii)$$

$$[\because \; E_{b2} = E_{b1} = \text{applied voltage}]$$

Putting this value in (*i*) above, we get

$$\left(\frac{\phi_1}{\phi_2}\right)^2 = \frac{\phi_2 I_{a2}}{\phi_1 I_{a1}}$$

or

$$\frac{I_{a2}}{I_{a1}} = \left(\frac{\phi_1}{\phi_2}\right)^3$$

$$\frac{I_{a2}}{20} = \left(\frac{20}{I_{a2}/2}\right)^3 \qquad [\because \; \phi_2 \propto I_a/2]$$

$$\therefore \qquad I^4_{a2} = 20 \times 40^3$$

or

$$I_{a2} = 33.63 \text{ A}$$

Hence, **current = 33.63 A. (Ans.)**

From (*ii*), we get $\dfrac{N_2}{800} = \dfrac{20}{(33.63)/2}$

$$N_2 = 800 \times \frac{40}{33.63} = 951.52 \text{ r.p.m.}$$

Hence, **speed of the motor = 951.53 r.p.m. (Ans.)**

8.3. Rheostatic Control

- This method consists of obtaining reduced speeds by the insertion of external series resistance in the armature circuit. It can be used with series, shunt and compound motors ;

for the last two types, the *series resistor must be connected between the shunt field and the armature, not between line and the motor.*

- It is *common method of speed control for series motors* and is generally analogous in action to wound-rotor induction-motor control by series rotor resistance.
- *This method is used when speeds below the no-load speed is required.*

Advantages :

1. The ability to achieve speeds below the basic speed.
2. Simplicity and ease of connection.
3. The possibility of combining the functions of motor starting with speed control.

Disadvantages :

1. The relatively high cost of large, continuously rated, variable resistors capable of dissipating large amounts of power (particularly at higher power ratings).
2. Poor speed regulation for any given no-load speed setting.
3. Low efficiency resulting in high operating cost.
4. Difficulty in obtaining stepless control of speed in higher power ratings.

Shunt motors :

- In armature or rheostatic control method of speed the voltage across the armature (which in normally constant) is varied by inserting a variable rheostat or resistance, called *controller resistance,* in *series* with the armature circuit. As the controller resistance is increased, the potential difference across the armature is decreased thereby decreasing the armature speed. For a load of constant torque, speed is approximately proportional to the potential difference across the armature.

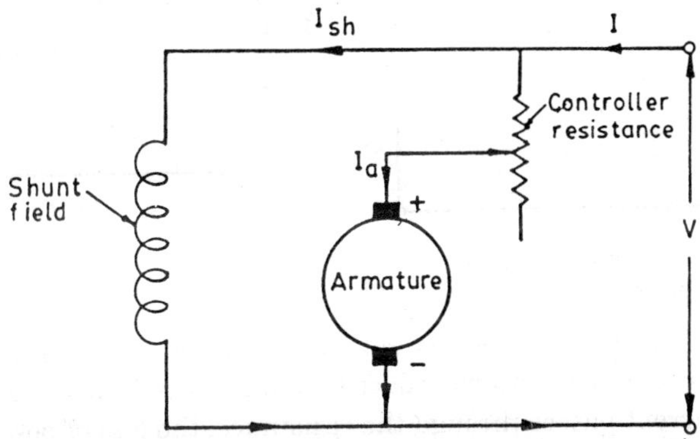

Fig. 86. Armature resistance control for D.C. shunt motor.

From the speed/armature current characteristic it is seen that *greater the resistance in armature, greater is the fall in speed.*

There is a particular load current for which the *speed would be zero. This is the maximum current and is known as 'stalling current'.*

- This method is *very wasteful, expensive and unsuitable for rapidly changing load,* because for a given value of R_t the speed will change with load. A more stable operation can be obtained by using a *divertor across the armature* (Fig. 88) in addition to armature control resistance. Now, the changes in armature current due to changes in the load torque *will not be so effective* in changing the potential difference across the armature and hence the speed of the armature.

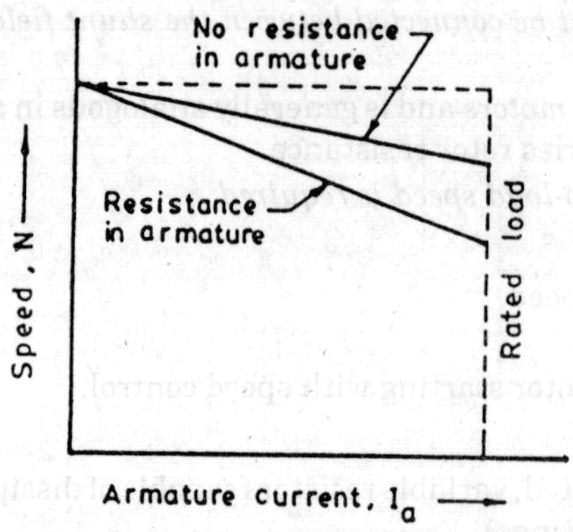

Fig. 87. Speed-current characteristic of D.C. shunt motor.

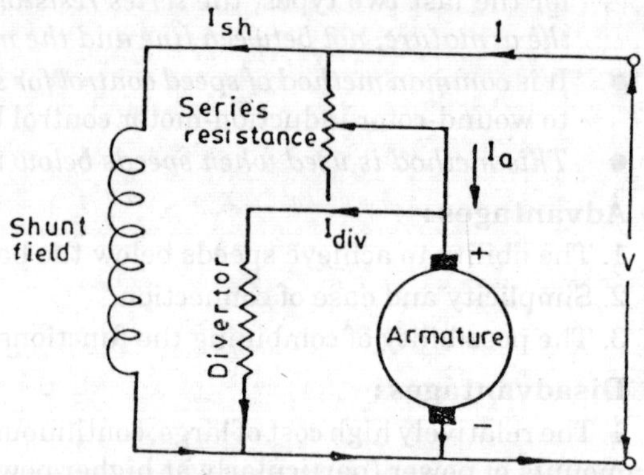

Fig. 88. Use of divertor across the armature for speed control of D.C. shunt motor.

Series motors. Armature resistance control is the most common method employed for D.C. series motors (Figs. 89 and 90).

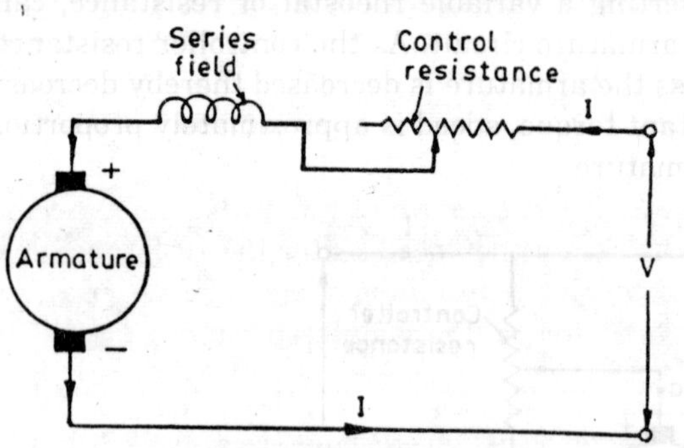

Fig. 89. Armature resistance control for D.C. series motor.

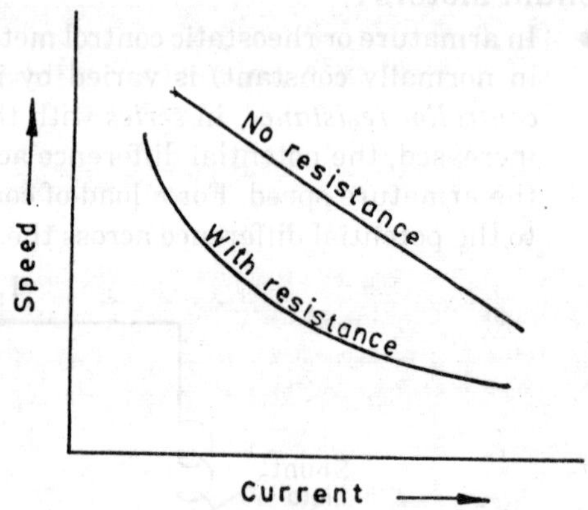

Fig. 90

By increasing the resistance in series with the armature the voltage applied across the armature terminals can be decreased. With the reduced voltage across the armature, the speed is reduced.

Since full motor current passes through the resistance, the loss of power is considerable.

Although terminal-voltage control by means of a variable voltage supply would effectively control the speed of a D.C. series motor, the high cost of the control equipment is seldom warranted.

Series-parallel control :

- This system is widely used in *electric traction.* Here two or more similar mechanically-coupled motors are employed.
- **At low speeds** the *motors are joined in series* as shown in Fig. 91 (*a*). The additional resistance is gradually cut out by controller as the motors attain the speed, and finally the resistance is totally removed, then *each motor has half of line voltage*. In this arrangement, for *any given value of armature current, each motor will run at half of its normal speed.* As there is no external resistance in the circuit, therefore *there is no waste of energy and so motors operate at an efficiency nearly equal to that obtainable with full line voltage across the terminals of each motor.*

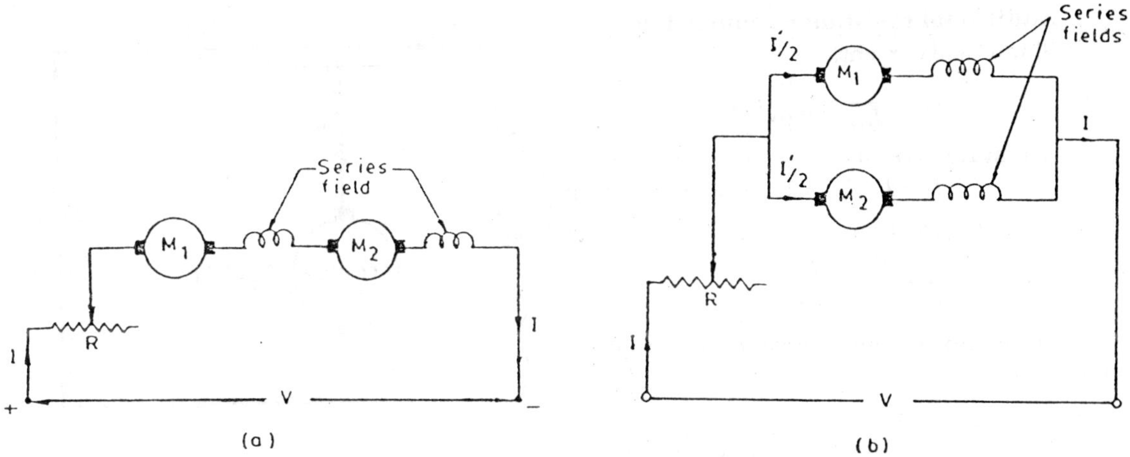

Fig. 91. Series-parallel control.

When the motors are connected in *series* and resistance R is *completely cut out* :

Voltage across each motor $= \frac{V}{2}$

Current through each motor $= I$

$$\text{Speed} \propto \frac{\text{voltage}}{\text{current}} \propto \frac{V/2}{I} \propto \frac{V}{2I}$$

$$\text{Torque} \propto \phi I \propto I^2 \qquad \text{[Since } \phi \propto I\text{, assuming unsaturated field]}$$

- **At high speeds the motors** *are joined in parallel* as shown in Fig. 91 (*b*). The variable resistance R is gradually cut out as motors attain the speed. After the resistance R is completely cut out each motor is connected across the full line voltage.

When the motors are connected in *parallel* and resistance R is completely cut out :

Voltage across each motor $= V$

Current through each motor $= \frac{I}{2}$ $\left[= \frac{I'}{2} \text{ when resistance } R \text{ is } \textit{not} \text{ completely cut out} \right]$

$$\text{Speed} \propto \frac{\text{voltage}}{\text{current}} \propto \frac{V}{I/2} \propto \frac{2V}{I}$$

Also, $\text{torque} \propto \phi I \propto I^2$ $[\because \phi \propto I]$

$$\therefore \qquad T \propto \left(\frac{I}{2}\right)^2 \propto \frac{I^2}{4}$$

The *torque is $\frac{1}{4}$ times that produced by motors when in series.*

Example 40. *The armature and shunt field resistances of a 230 V shunt motor are 0.1 ohm and 230 ohms respectively. It takes a current of 61 A at 1000 r.p.m. If the current taken remains unaltered find the resistance to be included in series with the armature circuit to reduce the speed to 750 r.p.m.*

Solution. Supply voltage, $V = 230$ Volts

Armature resistance, $R_a = 0.1$ ohm

Shunt field resistance, $R_{sh} = 230$ ohms

Load current, $I = 61$ A

Speed, $N_1 = 1000$ r.p.m.

Speed, $N_2 = 750$ r.p.m.

Additional resistance required, R :

Shunt field current,

$$I_{sh} = \frac{V}{R_{sh}} = \frac{230}{230} = 1 \text{ A}$$

Armature current,

$$I_1 = I - I_{sh} = 61 - 1 = 60 \text{ A}$$

Using the relation,

$$\frac{N_2}{N_1} = \frac{E_{b2}}{E_{b1}} \times \frac{\phi_1}{\phi_2} \quad ...(i)$$

Since flux remains constant in shunt motor

$$\phi_1 = \phi_2$$

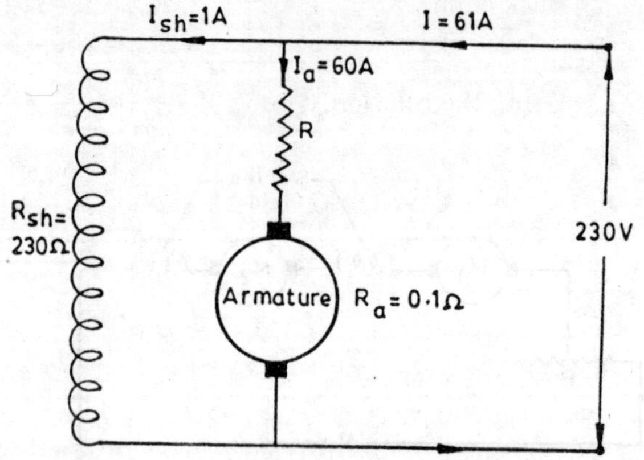

Fig. 92

Also it is given that the current taken by the motor remains constant

$\therefore$ $$I = I_1 = I_2 = 61 \text{ A}$$

and $$I_{a1} = I_{a2} = 60 \text{ A}$$

Thus eqn. (*i*) reduces to $$\frac{N_2}{N_1} = \frac{E_{b2}}{E_{b1}} \quad ...(ii)$$

$$E_{b1} = V - I_{a1}R_a = 230 - 60 \times 0.1 = 224 \text{ V}$$

$$E_{b2} = V - I_{a2}(R_a + R)$$

$$= 230 - 60(0.1 + R) = 230 - 6 - 60R$$

$$= 224 - 60R$$

Putting this value in (*ii*), we get

$$\frac{750}{1000} = \frac{224 - 60R}{224}$$

or $$60R = 224 - \frac{750}{1000} \times 224$$

$\therefore$ $$R = 0.933 \text{ ohm}$$

Hence, **additional resistance required = 0.933 ohm. (Ans.)**

Example 41. *The armature resistance of a 230 V D.C. shunt motor is 0.2 ohm. It takes 15 A at rated voltage and runs at 800 r.p.m. Calculate the value of additional resistance required in the armature circuit to reduce the speed to 600 r.p.m. when the load torque is independent of speed. Ignore the field current.*

Solution. Supply voltage, $V = 230$ Volts

Armature resistance, $R_a = 0.2$ ohm

Armature current, $I_1 = I_{a1} = 15$ A

Speed, $N_1 = 800$ r.p.m.

Speed, $N_2 = 600$ r.p.m.

Additional resistance required, R :

Back e.m.f., $$E_{b1} = V - I_{a1}R_a = 230 - 15 \times 0.2 = 227 \text{ V}$$

Since as per given data load torque is independent of speed and flux is constant

$$I_{a1} = I_{a2} = 15 \text{ A}$$

$$\phi_1 = \phi_2$$

Back e.m.f., $E_{b2} = V - I_{a2}(R_a + R) = 230 - 15(0.2 + R)$

$$= 230 - 3 - 15R = 227 - 15R$$

Using the relation, $\frac{N_2}{N_1} = \frac{E_{b2}}{E_{b1}} \times \frac{\phi_1}{\phi_2}$

$$\frac{N_2}{N_1} = \frac{E_{b2}}{E_{b1}} \quad [\because \phi_1 = \phi_2]$$

$$\therefore \quad \frac{600}{800} = \frac{227 - 15R}{227}$$

$$15R = 227 - \frac{600}{800} \times 227$$

$$\therefore \quad R = 3.783 \text{ ohms}$$

Hence, **additional resistance required = 3.783 ohms. (Ans.)**

8.4. Voltage Control

When the speed is controlled by regulating the motor terminal voltage while maintaining constant field current, it is called **voltage control.**

With voltage control, the change in speed is almost proportional to the change in voltage as shown in Fig. 78. The *output varies directly with speed and the torque remains constant.* Since the voltage has to be regulated without affecting the field, the application of voltage control is limited to *separately excited motors* (Fig. 93) *only.*

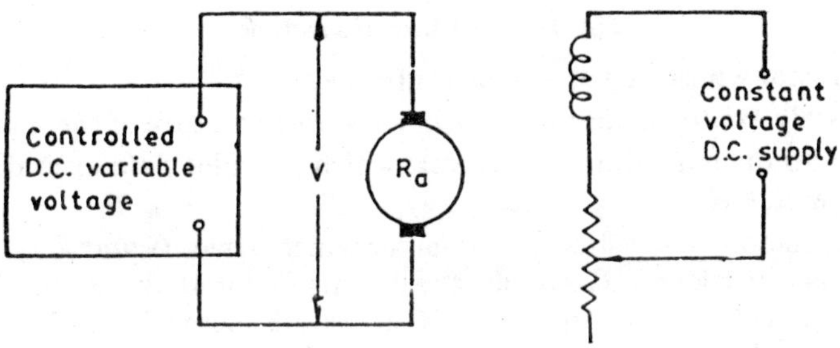

Fig. 93. Voltage control method.

- For D.C. motors of *fractional* and *relatively low power* rating, the variable D.C. voltage source may be a D.C. vacuum tube, a gas or thyraton tube, or a semi-conductor (silicon controlled rectifier) amplifier, operating from a three-phase or single phase A.C. supply.
- *Motors of moderate rating up to* 75 kW may be controlled by this method using Rototrol or Regulex or magnetic amplifiers as the adjustable D.C. voltage source.
- *Large D.C. motors* are controlled in this manner by means of rotary amplifiers such as the amplidyne or the Ward-Leonard control system.

Advantages :

1. Speed control over a wide range is possible.
2. This method eliminates the need for series armature starting resistance.
3. Uniform acceleration can be obtained.
4. Speed regulation is good.

Disadvantages :

1. Arrangement is costly as two extra machines are required.
2. The overall efficiency of the system is low, especially at light loads.

Applications :

Inspite of the high capital cost, this method finds wide applications in :

(*i*) Steel mills for reversing the rolling mills.

(*ii*) Seamless tube mills and shears.

(*iii*) High and medium speed elevators in tall buildings, mine hoists, paper machine drives and electric shovels.

Ward-Leonard System. This method of control not only gives a wide range of operating speeds, but reduces to the very minimum the wastage of energy that may take place at staring and stopping.

Fig. 94 shows the schematic arrangement of Ward-Leonard method.

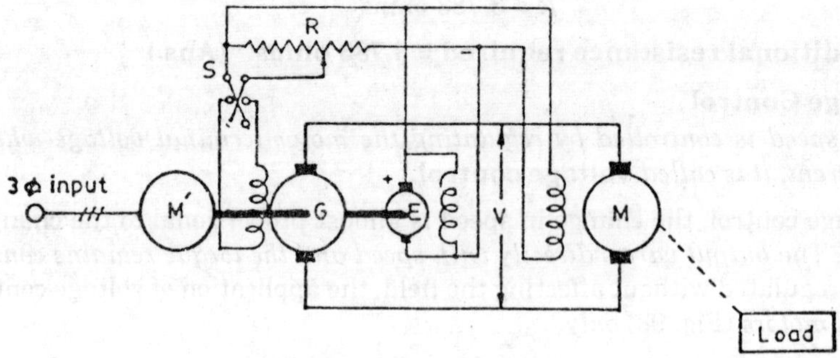

Fig. 94. Ward-Leonard method.

M = main motor whose speed is to be controlled

G = separately excited generator which feeds the armature of the motor M

E = an exciter (a small shunt generator) which provides field excitation to the generator G and motor M

M' = driving motor—a constant speed motor which drives G and E.

[If the system is to work on A.C. supply, the driving motor M' is a 3-phase induction motor. If the system is to work on D.C. supply, the motor M' is a shunt motor. In the latter case the exciter E is not necessary because the excitation for the generator G and motor M can be obtained from D.C. mains. A diesel engine can also be used in place of motor M'].

R = a potentiometer rheostat

S = a double throw switch.

The working of this system is as follows :

- The motor M' drives the generator G and excitor E at constant speed. The voltage fed to motor M can be controlled by varying the setting of R. A change in voltage applied to motor M changes its speed. The speed can be adjusted to any value from zero to maximum in either direction by means of a rheostat R and switch S.
- When the sliding contact of R is at extreme right, the motor is running at full speed in one direction. To decrease the speed the sliding contact is moved to the left. When the sliding contact is at the extreme left position, the speed of motor M is zero. In order to reverse the speed of the motor, the sliding contact is shifted to the extreme left, the switch S is reversed and the sliding contact shifted to right again.
- A modification of the Ward-Leonard system is known as *Ward-Leonard Ilgner system*, which uses a small motor-generator set with the *addition of a flywheel whose function is to reduce fluctuations in the power demand from the supply circuit*. When the main motor M becomes suddenly overloaded, the driving motor M' slows down, thus allowing the

inertia of the flywheel to supply a part of the overload. However, when the load is suddenly thrown off the main motor M, then M' speeds up thereby again storing energy in the flywheel. When Ilgner system is driven by means of an A.C. motor (whether induction or synchronous) another refinement in the form of a 'slip regulator' can be usefully employed thus giving an additional control.

- One important feature of the Ward-leonard system is its *regenerative action*. When a locomotive, fitted with this system, is descending a slope, it speeds up due to the action of gravity. The speed of motor M increases until its back e.m.f. exceeds the applied voltage. Motor M then runs as generator and feeds the machine G which now works as a generator and feeds electrical energy back into the trolley wire. This results in salvaging of considerable amount of energy and a superior and smooth braking action. Such an action is known as *regenerative braking*.

Advantages of Ward-Leonard system :

1. A wide range of speed from standstill to high speeds in either direction.
2. Rapid and instant reversal without excessively high armature currents.
3. Starting without the necessity of series armature resistances.
4. Stepless control from standstill to maximum speed in either direction.
5. Larger units employing generator field reversal eliminate the need for heavy armature conductors for reversing, and at the same time prevent motor runway since the motor field is always excited.
6. The method lends itself to adaptation of intermediate electronic, semi-conductor, and magnetic amplifiers to provide stages of amplification for an extremely large motor. Thus the power in the control circuit may be extremely small.
7. Extremely good speed regulation at any speed.

Disadvantages :

1. High initial cost.
2. Since the efficiency, neglecting the exciter efficiency, is essentially the product of the individual efficiencies of the two larger machines, the efficiency of this method is *not as high as rheostat speed control or the field control method.*

Example 42. *A series motor drives a fan for which the torque varies as square of the speed. Its resistance between terminals is 1.2 ohm. On 220 V, it runs at 350 r.p.m. and takes 30 A. The speed is to be raised to 450 r.p.m. by increasing the voltage. Find the voltage.*

Assume that flux varies directly as current.

Solution. Resistance between terminals = 1.2 ohm

$$I_{a1} = 30 \text{ A}, N_1 = 350 \text{ r.p.m.}, \ N_2 = 450 \text{ r.p.m.}$$

Since $\phi \propto I_a$

$\therefore$ $T \propto \phi I_a \propto I_a^2$ ...(*i*)

Also, $T \propto N^2$ ...[Given] ...(*ii*)

From (*i*) and (*ii*), we get $I_a^2 \propto N^2$ or $I_a \propto N$

or $$\frac{I_{a2}}{I_{a1}} = \frac{N_2}{N_1} = \frac{450}{350}$$

$\therefore$ $$I_{a2} = I_{a1} \times \frac{450}{350} = 30 \times \frac{450}{350} = 38.57 \text{ A}$$

Back e.m.f., $E_{b1} = 220 - 30 \times 1.2 = 184$ V

Back e.m.f., $E_{b2} = V - 38.57 \times 1.2 = V - 46.28$

$$\frac{\phi_1}{\phi_2} = \frac{I_{a1}}{I_{a2}} = \frac{30}{38.57}$$

Now using the relation, $\frac{N_2}{N_1} = \frac{E_{b2}}{E_{b1}} \times \frac{\phi_1}{\phi_2}$

$$\frac{450}{350} = \frac{V - 46.28}{184} \times \frac{30}{38.57}$$

$$\therefore \quad V - 46.28 = \frac{450}{350} \times 184 \times \frac{38.57}{30} = 304.15$$

$$\therefore \quad \mathbf{V = 304.15\ Volts.\ (Ans.)}$$

9. LOSSES AND EFFICIENCY OF D.C. MACHINES

9.1. Introduction

The output of a generator or motor is always less than the input because some of the energy supplied is lost as heat. These losses raise the temperatures of the machine parts above that of surrounding air until such temperatures are reached that the heat losses are radiated as fast as they are generated. Certain of the losses depend upon the load. The temperature rise therefore depends upon the load also, and the maximum allowable temperature rise determines the maximum permissible load that the machine may carry. The limit of output occurs at the load for which the temperature rise becomes high enough to endanger the insulation of the windings.

Thus the consideration of machine losses is important for the following three reasons :

1. Losses appreciably influence the operating cost of the machine.

2. Losses determine the heating of the machine and hence the rating or power output that can be obtained without undue deterioration of the insulation.

3. The voltage drops or current components associated with supplying the losses must be properly accounted for in a machine representation.

Machine efficiency is given by

$$\text{Efficiency} = \frac{\text{output}}{\text{input}} \qquad ...(12)$$

which can also be expressed as

$$\text{Efficiency} = \frac{\text{input} - \text{losses}}{\text{input}} = 1 - \frac{\text{losses}}{\text{input}} \qquad ...(13)$$

$$\text{Efficiency} = \frac{\text{output}}{\text{output} + \text{losses}} \qquad ...(14)$$

9.2. Losses

Losses in a D.C. machine can be classified as follows :

1. **Electrical losses** (*copper loss*) :

(*i*) Armature $= I_a^2 R_a$ (*ii*) Series field $= I_{se}^2 R_{se}$

(*iii*) Shunt field $= VI_{sh}$ (*iv*) Commutating field $= I_c^2 R_c$.

2. **Rotational losses** (*stray-power loss*) :

(*i*) Core loss :

(*a*) Hysteresis (*b*) Eddy currents.

(*ii*) Mechanical (or friction) loss :

(*a*) Bearings (*b*) Brushes

(*c*) Windage.

- The *core losses and friction losses* are supplied from the mechanical power developed by the machine. They are put into a single group called *mechanical losses* or more generally, *stray-power losses*. When a generator or motor runs at a fixed speed and generates a given voltage, the *stray-power loss is constant regardless* of the electrical output or input of the machine for *speed and flux density are the only factors that influence the stray-power loss*.
- The *electrical losses* are supplied from the electrical power generated by or delivered to the machine, as the case may be. Of these, the shunt field loss is somewhat, though not entirely independent of the load, while the remaining electrical losses are nearly proportional to the square of the load current. Fig. 95 shows the combined power-flow diagram for motor or generator action which is self-explanatory.

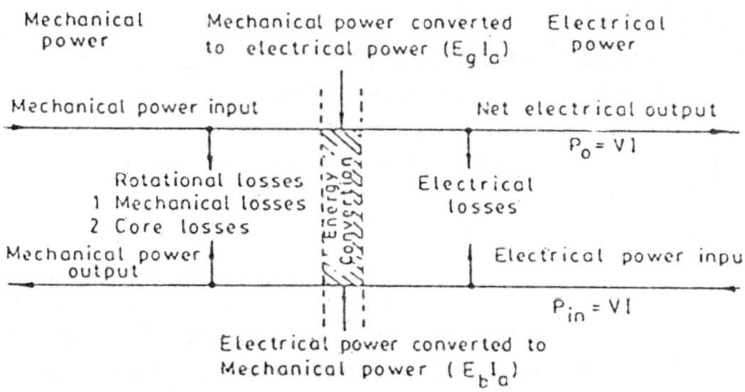

Fig. 95. Combined power-flow diagram for motor or generator action.

9.2.1. Copper (or electrical) losses. When an electric current of I ampere flows in a resistance of R ohms, heat energy is lost at the rate of I^2R joules/sec, and the power loss is I^2R watts. Generators and motors have one or more field circuits and an armature circuit in which such losses occur. All resistance losses of kind are classed as *copper loss*.

(*i*) **Armature copper loss** $= I_a^2 R_a$. This is about 30% to 40% of total full-load losses.

(*ii*) **Field copper loss.** This loss equals $I_{sh}^2 R_{sh}$ for the shunt field winding and $I_{se}^2 R_{se}$ for the series field winding. This loss constitutes about 30% of the total full-load losses.

(*iii*) **Brush contact loss.** This loss is due to the resistance of brush contacts. The voltage drop at the brush is almost independent of I_a. For carbon brushes the voltage drop is around 1 volt per brush. The power loss due to brush contact resistance is $2e_b I_a$ where e_b is the voltage drop at one brush.

(*iv*) **Loss in commutating pole winding.** This loss equals I_a^2 × resistance of commutating pole winding.

(*v*) **Loss in compensating winding.** This loss equals I_a^2 × resistance of compensating winding.

9.2.2. Iron losses. *Iron losses are a function of both flux and speed.*

Hysteresis loss. The hysteresis loss P_h is a *measure of the electric energy required to overcome the retentivity of the iron in the magnetic flux path,* using watts as unit,

$$P_h = K_h B^x f V \qquad \text{...(15)}$$

where V = volume of iron in dynamo subject to change of flux

K_h = constant for the grade of iron employed

B = flux density raised to the Steinmetz exponent. With modern values of dynamo x is no longer 1.6 but closer to 2.0. This is not to imply that for a given volume, V, of iron the loss has increased, because K_h has been reduced considerably

f = frequency (Hz) of reversal of flux.

Eddy current losses. These losses occur not only in the dynamo iron but in all conductive materials with the flux path of the rotating or varying magnetic field of the dynamo. The eddy current loss P_e, in watts is

$$P_s = K_e t^2 B^2 f^2 V \qquad \text{...(16)}$$

where K_e = an eddy current constant for the grade of iron employed

t = thickness of the laminations of the pole core and armature

B = flux density

f = frequency (Hz) of reversal of flux

V = volume of iron subject to change of flux.

For a D.C. dynamo the frequency, f, reversal of flux varies with speed. Thus the hysteresis loss varies directly with speed, whereas the eddy current loss varies as the square of speed. Both hysteresis loss and eddy current loss vary approximately as the square of the flux density. For this reason core losses are considered a function of both flux and speed.

9.2.3. Mechanical (or friction) losses. When a machine is running, there are various frictional forces to be overcome, each of which requires a continuous expenditure of energy and results in heating the rubbed parts. There is friction loss in the machine bearings, at the surface of the commutator due to the rubbing of the brushes, and in the armature core due to its fanning action. *These losses depend upon the speed* but are independent of the load on the machine. They are difficult to estimate by direct calculation but may be found by measurement.

9.3. Efficiency of D.C. Machines :

Generator :

1. *Mechanical efficiency* (η_m)

$$\eta_m = \frac{\text{Electrical power developed by armature}}{\text{Total mechanical power input}}$$

$$= \frac{E_g I_a}{\text{B.H.P. of prime mover} \times 735.5} \qquad \text{...(17)}$$

2. *Electrical efficiency* (η_e)

$$\eta_e = \frac{\text{Useful electrical power output}}{\text{Electrical power developed}} = \frac{VI}{E_g I_a} \qquad \text{...(18)}$$

3. *Overall or commercial efficiency* ($\eta_{og} = \eta_m \times \eta_e$)

$$\eta_{og} = \frac{\text{Useful electrical power output}}{\text{Total mechanical power input}}$$

$$= \frac{VI}{\text{B.H.P. of prime mover} \times 735.5} \qquad \text{...(19)}$$

The overall efficiency of generator can also be expressed as follows :

$$\eta_{og} = \frac{\text{Useful power output}}{\text{Useful power output + total losses}}$$

$$= \frac{VI}{VI + \text{total losses}} \qquad \text{...(20)}$$

where E_g = generated e.m.f. V = terminal voltage

I = load current, I_a = armature current.

For good generators the value of overall or commercial efficiency may be *as high as 95%*.

Motor :

1. *Electrical efficiency* (η_e)

$$\eta_e = \frac{\text{Mechanical power developed}}{\text{Total electrical power input}} = \frac{E_b I_a}{VI} \quad ...(21)$$

2. *Mechanical efficiency* (η_m)

$$\eta_m = \frac{\text{Useful mechanical power output}}{\text{Mechanical power developed}}$$

$$= \frac{\text{B.H.P. of motor} \times 735.5}{E_b I_a} \quad ...(22)$$

3. *Overall or commercial efficiency* ($\eta_{om} = \eta_m \times \eta_e$)

$$\eta_{om} = \frac{\text{Useful mechanical power output}}{\text{Total electrical power input}}$$

$$= \frac{\text{B.H.P. of motor} \times 735.5}{VI} \quad ...(23)$$

η_{om} can also be expressed as follows :

Overall efficiency of motor

$$\eta_{om} = \frac{\text{Useful power output}}{\text{Total power input}} = \frac{\text{Total power input} - \text{total losses}}{\text{Total power input}}$$

$$= \frac{VI - \text{total losses}}{VI} \quad ...(24)$$

where E_b = back e.m.f. V = supply voltage

I = load current I_a = armature current.

9.3.1. Condition for maximum efficiency. Condition for maximum efficiency for a D.C. generator or D.C. motor is same. For a D.C. generator the condition for maximum efficiency is derived as follows :

Generator power output $= VI$.

If flux and speed are constant all losses except armature copper loss are constant.

Losses = armature copper loss + constant loss

$$= (I + I_{sh})^2 R_a + P_c = I^2 R_a + P_c$$

[Neglecting I_{sh} in comparison with load current I]

where I_{sh} is the shunt field current and P_c denotes constant losses which include iron loss, field winding loss and mechanical loss.

$$\text{Efficiency, } \eta = \frac{\text{Output}}{\text{Input}} = \frac{\text{Output}}{\text{Output + losses}}$$

$$= \frac{VI}{VI + I^2 R_a + P_c} \quad ...(25)$$

η is maximum when

$$\frac{d\eta}{dI} = 0 = \frac{(VI + I^2R_a + P_c)\,V - VI(V + 2IR_a)}{(VI + I^2R_a + P_c)^2}$$

or

$$I^2 R_a = P_c \quad ...(26)$$

Hence, *efficiency will be maximum when variable losses are equal to constant losses*

Further (from eqn. 26) $$I = \sqrt{\frac{P_c}{R_a}} \quad ...(27)$$

Thus the efficiency increases with increase in load current, reaches a maximum value when load current equals the value given by eqn. (27) and then starts decreasing.

Efficiency curve. The efficiency of a machine is different at different values of power output. *As the output increases*, the *efficiency increases till it reaches a maximum value*. As the *output is further increased, the efficiency starts decreasing*. A graph of efficiency *vs*. output is called efficiency curve. A typical efficiency curve is shown in Fig. 96. *The machines are so designed as to give maximum efficiency at or near the rated output of the machine.* Since the generators operate at a constant terminal voltage V, the efficiency curve of a generator can be drawn between efficiency and load current *I*.

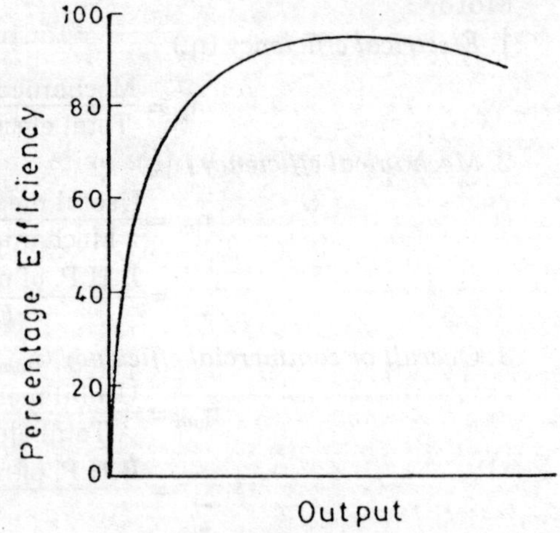

Fig. 96. Efficiency curve.

☞**Example 43.** *The armature and shunt field resistances of 110 V shunt generator running at 800 r.p.m. are 0.12 Ω, 55 Ω respectively. It supplies 220 lamps each rated at 60 W, 110 V. The friction, windage and core loss of the machine is equal to 410 W. Calculate :*

(*i*) *Armature copper loss on full-load and shunt field copper loss.*

(*ii*) *Torque in Nm exerted by the prime-mover.*

(*iii*) *Mechanical, electrical and overall efficiencies.*

Solution. Terminal voltage, $V = 110$ V

Speed of the generator, $N = 800$ r.p.m.

Armature resistance, $R_a = 0.12\ \Omega$

Shunt field resistance, $R_{sh} = 55\ \Omega$

Friction, windage and core loss = 410 W

Full-load current, $I = \dfrac{220 \times 60}{110} = 120$ A

Shunt field current, $I_{sh} = \dfrac{V}{R_{sh}} = \dfrac{110}{55} = 2$ A

$I_a = I + I_{sh} = 120 + 2 = 122$ A

(*i*) Armature copper loss $= I_a^2 R_a$

$= 122^2 \times 0.12 = 1786$ W

∴ **Armature copper loss = 1786 W. (Ans.)**

Shunt field copper loss $= VI_{sh} = 110 \times 2 =$ **220 W. (Ans.)**

[Shunt field copper loss $= I_{sh}^2 R_{sh} = 2^2 \times 55 = 220$ W]

(*ii*) Total copper loss = 1786 + 220 = 2006 W

Core, friction and windage loss = 410 W

∴ Total loss = 2006 + 410 = 2416 W

Output = 220 × 60 = 13200 W

Input = output + losses = 13200 + 2416 = 15616 W

If *T* is the torque in Nm exerted by the prime mover

$$\frac{2\pi NT}{60} = \text{input in watts}$$

$\therefore \quad \frac{2\pi NT}{60} = 15616$

i.e., $T = \frac{15616 \times 60}{2\pi N} = \frac{15616 \times 60}{2\pi \times 800} = 186.4$ Nm

$\therefore$ **Torque exerted by the prime mover**

= **186.4 Nm. (Ans.)**

(*iii*) Electrical power generated in armature

$= E_b I_a$

= mechanical input – stray losses

= 15616 – 410 = 15206 W

Mechanical power input = 15616 W.

$$\therefore \quad \eta_{\text{mechanical}} = \frac{\text{Power generated in armature}}{\text{Mechanical power input}} \times 100$$

$$= \frac{15206}{15616} \times 100 = \mathbf{97.4\%}. \text{ (Ans.)}$$

Electrical power generated = 220 × 60 = 13200 W

Electrical efficiency, $\eta_{\text{electrical}} = \frac{\text{Electrical power output}}{\text{Total power developed in armature}} \times 100$

$$= \frac{13200}{15206} \times 100 = \mathbf{86.8\%}. \text{ (Ans.)}$$

$$\eta_{\text{overall}} = \frac{\text{Electrical power output}}{\text{Mechanical power input}} \times 100$$

$$= \frac{13200}{15616} \times 100 = \mathbf{84.53\%}. \text{ (Ans.)}$$

Example 44. *Find the efficiency at full-load and at half-load for a 1200 kW, 600 V shunt generator using the following data :*

Armature resistance = 0.005 Ω, shunt field resistance = 60 Ω, brush contact drop = 1 V per brush, mechanical and iron losses at rated load = 20 kW, stray load loss = 1.2% of output.

Solution. Terminal voltage, $V = 600$ V

Full-load output = 1200 kW

Armature resistance, $R_a = 0.005\ \Omega$

Shunt field resistance, $R_{sh} = 60\ \Omega$

Brush contact drop = 1 V per brush

Mechanical losses at rated load = 20 kW

Stray load loss = 1.2% of output

Efficiency at full-load and at half-load :

(*i*) **At full-load**

Full load current, $I = \frac{\text{Full-load output}}{\text{Voltage}} = \frac{1200 \times 1000}{600} = 2000$ A

Shunt field current, $I_{sh} = \frac{V}{R_{sh}} = \frac{600}{60} = 10$ A

$\therefore$ Armature current, $I_a = I + I_{sh} = 2000 + 10 = 2010$ A

Armature copper loss $= I_a^2 R_a = 2010^2 \times 0.005 = 20200$ W

Shunt field loss $= VI_{sh} = 600 \times 10 = 6000$ W

Brush contact loss = 2 × 2010 = 4020 W

Mechanical and iron losses = 20 × 1000 = 20000 W

Stray load losses $= 1.2\%$ of full-load output

$$= \frac{1.2}{100} \times 1200 \times 1000 = 14400 \text{ W}$$

Efficiency $= \dfrac{\text{Output}}{\text{Output + Losses}}$

$$= \frac{1200 \times 1000}{1200 \times 1000 + 20200 + 6000 + 4020 + 20000 + 14400}$$

$$= \frac{12 \times 10^5}{1200000 + 64620} = \frac{12 \times 10^5}{1264620} = 0.949 \text{ or } 94.9\%$$

Hence, **efficiency at full-load** $=$ **94.9%. (Ans.)**

At half-load :

Output $= \dfrac{1200 \times 1000}{2} = 600000$ W

Load current $= \dfrac{2000}{2} = 1000$ A

Armature current, $I_a = 1000 + 10 = 1010$ A

Armature copper loss, $I_a^2 R_a = 1010^2 \times 0.005 = 5100$ W

Shunt field loss $= VI_{sh} = 600 \times 10 = 6000$ W

Brush contact loss $= 2 \times 1010 = 2020$ W

Mechanical and iron losses $= 20000$ W

Stray load losses $= 1.2\%$ of $600000 = 7200$ W

Efficiency $= \dfrac{\text{Output}}{\text{Output + Losses}}$

$$= \frac{600000}{600000 + 5100 + 6000 + 2020 + 20000 + 7200}$$

$$= \frac{600000}{640320} = 0.937 \text{ or } 93.7\%.$$

Hence, **efficiency at half-load** $=$ **93.7%. (Ans.)**

Example 45. *The following data pertains to a short-shunt compound generator :*

Armature resistance = 0.04 Ω, series field resistance = 0.025 Ω, shunt field resistance = 45 Ω, contact drop / brush = 1 V, magnetic losses = 2.5 kW, mechanical losses = 1 kW.

Find the efficiency of the generator when it delivers 400 A at 440 V.

Solution. Terminal voltage, $V = 440$ V

Load current, $I = 400$ A

Armature resistance, $R_a = 0.04\ \Omega$

Series field resistance, $R_{se} = 0.025\ \Omega$

Shunt field resistance, $R_{sh} = 45\ \Omega$

Copper drop/brush $= 1$ V

Magnetic losses $= 2.5$ kW $= 2500$ W

Mechanical losses $= 1$ kW $= 1000$ W

Refer Fig. 97

Output of the generator $= VI = 440 \times 400 = 176000$ W

Voltage across shunt field winding $= V + IR_{se} = 440 + 400 \times 0.025 = 450$ V

Shunt field current, $I_{sh} = \dfrac{V}{R_{sh}} = \dfrac{450}{45} = 10$ A

Armature current, $I_a = I + I_{sh} = 400 + 10 = 410$ A

Armature copper loss $= I_a^2 R_a = 410^2 \times 0.04 = 6724$ W

Brush contact loss $=$ voltage drop/pair of brush $\times I_a$

$= 2 \times 410 = 820$ W

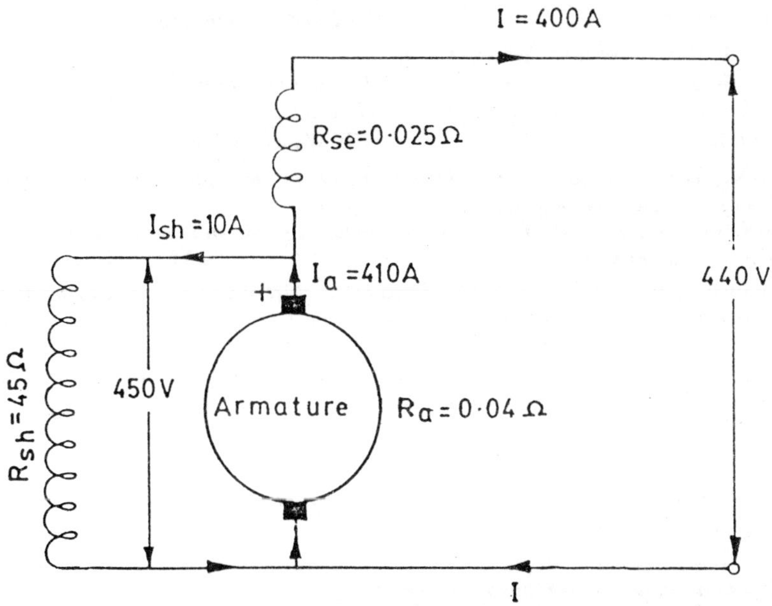

Fig. 97

Series field copper loss $= I^2 R_{se} = 400^2 \times 0.025 = 4000$ W

Shunt field copper loss $= VI_{sh} = 450 \times 10 = 4500$ W

Magnetic losses $= 2500$ W [Given]

Mechanical losses $= 1000$ W [Given]

Total losses $= 6724 + 820 + 4000 + 4500 + 2500 + 1000 = 19544$ W

Efficiency $= \dfrac{\text{output}}{\text{output + losses}} = \dfrac{176000}{176000 + 19544}$

$= \dfrac{176000}{195544} = 0.90$ or 90%

Hence, **efficiency of the generator = 90%. (Ans.)**

HIGHLIGHTS

1. **Basic type of D.C. machine** *is that of commutator type. This is actually an alternating current (A.C.) machine, but furnished with a special device, a commutator, which under certain conditions converts alternating current into direct current.*
2. *An electrical* **generator** *is a machine which converts mechanical energy (or power) into electrical energy (or power). It works on the following principle :*

   "Whenever a conductor cuts magnetic flux, dynamically induced e.m.f. is produced in it according to Faraday's Laws of Electromagnetic induction".
3. *A D.C. machine consists of two* **main** *parts :*
   — **Stationary Part :** *designed mainly for producing magnetic flux.*
   — **Rotating Part :** *called armature, where mechanical energy is converted into electrical (electric generator) or, conversely, electrical energy into mechanical (electric motor).*

The various parts of a D.C. machine are enumerated below :

Frame, Field poles, Commutating poles, Armature, Commutator, Brush gear, Armature shaft bearings, Armature windings.

4. *According to the degree of closure produced by winding, armature windings are of following two types :*
   1. Open coil winding *2. Closed coil winding*
   The closed armature windings are of two types :
   (i) Ring winding *(ii) Drum winding*
   In general there are two types of drum armature windings :
   (i) Lap winding *(ii) Wave winding.*
5. **'Lap winding'** *is suitable for comparatively low voltage but high current generators whereas* **'wave winding'** *is used for high voltage, low current machines.*
6. *In the* **'lap winding'** *the finish of each coil is connected to the start of the next coil so that winding or commulator pitch is unity.*
   In **'wave winding'** *the finish of coil is connected to the start of another coil well away from the first coil.*
7. *The* **rated output** *of an electrical machine is its duty in the conditions for which it has been designed by the manufacturer. The rated duty is characterised by values given on the machine's rating plate, and termed rated value, as, for instance : rated output, rated voltage, rated current, rated speed, etc.*
8. *E.m.f. equation of a generator is given as follows :*

$$E_g = \frac{p\phi ZN}{60a}$$

where E_g = *generated e.m.f. per parallel path in armature.*
p = *number of poles*
ϕ = *flux/pole, Wb*
Z = *total number of conductors*
= *number of slots* × *number of conductors/slot*
N = *rotational speed of armature, r.p.m.*
a = *number of parallel paths in armature*

[*For a* **wave wound** *generator :* $a = 2$
For a **lap wound** *generator :* $a = p$]

9. *According to* **method of excitation** *D.C. generators are classified as follows :*
   1. Separately excited generators *2. Self excited generators.*
   Self excited generators can be divided in accordance with how the field winding is connected into generators as follows :
   (i) Shunt wound generators *(ii) Series wound generators*
   (iii) Compound wound generators :
   (a) Short shunt
   (b) Long shunt.
10. *The electric motor is a machine which converts electrical energy into mechanical energy.*
11. *Voltage equation of a motor is* $V = E_b + I_aR_a$.
12. *Condition for maximum power developed in armature is* $E_b = \frac{V}{2}$.
13. *Armature torque of a motor is given by* $T_a = 0.159\ Z\phi p \times \frac{I_a}{a}$ *Nm.*
14. *Shaft or useful torque is given by* $T_{sh} = \frac{B.H.P.\ (metric) \times 735.5}{2\pi N/60}$ *Nm*
   Torque lost in iron and friction losses $= 0.159 \times \frac{\text{iron and friction losses}}{N/60}$ *Nm.*
15. *Speed of a D.C. motor is given by* $N = \frac{E_b}{\phi}\left(\frac{60a}{Zp}\right)$ *r.p.m.*

16. *Speed regulation of a motor is given by*

$$\text{Percentage speed regulation} = \frac{\text{no-load speed} - \text{full-load speed}}{\text{full-load speed}}.$$

17. *In a D.C. motor reversal is accomplished by changing the polarity of either the armature or the field, but not by changing both.*

OBJECTIVE TYPE QUESTIONS

D.C. GENERATOR

Choose the Correct Answer :

1. Laminations of core are generally made of
(*a*) cast iron (*b*) carbon
(*c*) silicon steel (*d*) stainless steel.
2. Which of the following could be approximately the thickness of laminations of a D.C. machine ?
(*a*) 0.005 mm (*b*) 0.05 mm
(*c*) 0.5 mm (*d*) 5 mm.
3. The armature of D.C. generator is laminated to
(*a*) reduce the bulk (*b*) provide passage for cooling air
(*c*) insulate the core (*d*) reduce eddy current loss.
4. The resistance of armature winding depends on
(*a*) length of conductor (*b*) cross-sectional area of the conductor
(*c*) number of conductors (*d*) all of the above.
5. The field coils of D.C. generator are usually made of
(*a*) mica (*b*) copper
(*c*) cast iron (*d*) carbon.
6. The commutator segments are connected to the armature conductors by means of
(*a*) copper lugs (*b*) resistance wires
(*c*) insulation pads (*d*) brazing.
7. In a commutator
(*a*) copper is harder than mica (*b*) mica and copper are equally hard
(*c*) mica is harder than copper (*d*) none of the above.
8. In D.C. generators the pole shoes are fastened to the pole core by
(*a*) rivets (*b*) counter sunk screws
(*c*) brazing (*d*) welding.
9. According to Fleming's right-hand rule for finding the direction of induced e.m.f., when middle finger points in the direction of induced e.m.f., forefinger will point in the direction of
(*a*) motion of conductor (*b*) lines of force
(*c*) either of the above (*d*) none of the above.
10. Fleming's right-hand rule regarding direction of induced e.m.f., correlates
(*a*) magnetic flux, direction of current flow and resultant force
(*b*) magnetic flux, direction of motion and the direction of e.m.f. induced
(*c*) magnetic field strength, induced voltage and current
(*d*) magnetic flux, direction of force and direction of motion of conductor.
11. While applying Fleming's right-hand rule to find the direction of induced e.m.f., the thumb points towards
(*a*) direction of induced e.m.f.
(*b*) direction of flux
(*c*) direction of motion of the conductor if forefinger points in the direction of generated e.m.f.
(*d*) direction of motion of conductor, if forefinger points along the lines of flux.

12. The bearings used to support the rotor shafts are generally
(*a*) ball bearings (*b*) bush bearings
(*c*) magnetic bearings (*d*) needle bearings.

13. In D.C. generators, the cause of rapid brush wear may be
(*a*) severe sparking (*b*) rough commutator surface
(*c*) imperfect contact (*d*) any of the above.

14. In lap winding, the number of brushes is always
(*a*) double the number of poles (*b*) same as the number of poles
(*c*) half the number of poles (*d*) two.

15. For a D.C. generator when the number of poles and the number of armature conductors are fixed, then which winding will give the higher e.m.f. ?
(*a*) Lap winding (*b*) Wave winding
(*c*) Either of (*a*) and (*b*) above (*d*) Depends on other features of design.

16. A 4 pole D.C. generator rotates at 1500 r.p.m. How many times in a second will the e.m.f. induced in each armature conductor change its direction ?
(*a*) 100 (*b*) 50
(*c*) 400 (*d*) 200.

17. If B is the flux density, l the length of conductor and v the velocity of conductor, then induced e.m.f. is given by
(*a*) Blv (*b*) Blv^2
(*c*) Bl^2v (*d*) Bl^2v^2.

18. In case of a wave wound generator, the average pitch
(*a*) must be even (*b*) must be odd
(*c*) may be odd or even (*d*) is generally fractional.

19. In case of a 4 pole D.C. generator provided with a two layer lap winding with sixteen coils, the pole pitch will be
(*a*) 4 (*b*) 8
(*c*) 16 (*d*) 32.

20. The material for commutator brushes is generally
(*a*) mica (*b*) copper
(*c*) cast iron (*d*) carbon.

21. The insulating material used between the commutator segments is normally
(*a*) graphite (*b*) paper
(*c*) mica (*d*) insulating varnish.

22. In a D.C. generator, if p be the number of poles and N be the r.p.m. of rotor, then the frequency of magnetic reversals will be
(*a*) $\frac{Np}{2}$ (*b*) $\frac{Np}{60}$
(*c*) $\frac{Np}{120}$ (*d*) $\frac{Np}{3000}$

23. For generating large currents on D.C. generators which winding is generally preferred ?
(*a*) Progressive wave winding (*b*) Lap winding
(*c*) Retrogressive wave winding (*d*) Current depends on design.

24. The purpose of providing dummy coils in a generator is
(*a*) to enhance flux density (*b*) to amplify voltage
(*c*) to provide mechanical balance for the rotor (*d*) to reduce eddy currents.

D.C. MOTOR

25. No-load speed of which of the following motor will be highest ?
(*a*) Shunt motor (*b*) Series motor
(*c*) Cumulative compound motor (*d*) Differentially compound motor.

26. The direction of rotation of a D.C. series motor can be changed by
(*a*) interchanging supply terminals (*b*) interchanging field terminals
(*c*) either of (*a*) and (*b*) above (*d*) none of the above.

27. Which of the following applications requires high starting torque ?
(*a*) Lathe machine (*b*) Centrifugal pump
(*c*) Locomotive (*d*) Air blower.

28. In case a D.C. motor is connected across A.C. supplies
(*a*) the motor will burn (*b*) the motor will run at rated speed
(*c*) the motor will run at corresponding synchronous speed
(*d*) the motor will run at slow speed.

29. If a D.C. motor is to be selected for conveyors, which motor would be preferred ?
(*a*) Series motor (*b*) Shunt motor
(*c*) Differentially compound motor (*d*) Cumulative compound motor.

30. Which D.C. motor will be preferred for machine tools ?
(*a*) Series motor (*b*) Shunt motor
(*c*) Cumulative compound motor (*d*) Differentially compound motor.

31. Which D.C. motor will be preferred for constant speed line shafting ?
(*a*) Cumulatively compound motor (*b*) Differentially compound motor
(*c*) Shunt motor (*d*) Series motor.

32. Differentially compound D.C. motors can find applications requiring
(*a*) high starting torque (*b*) low starting torque
(*c*) variable speed (*d*) frequent on-off cycles.

33. Which D.C. motor is preferred for elevators ?
(*a*) Shunt motor (*b*) Series motor
(*c*) Differentially compound motor (*d*) Cumulatively compound motor.

34. According to Fleming's left-hand rule, when the forefinger points in the direction of the field or flux, the middle finger will point in the direction of
(*a*) current in the conductor (*b*) movement of conductor
(*c*) resultant force on conductor (*d*) none of the above.

35. If the field of a D.C. shunt motor gets opened while motor is running
(*a*) the speed of motor will be reduced (*b*) the armature current will reduce
(*c*) the motor will attain dangerously high speed
(*d*) the motor will continue to run at constant speed.

36. Starters are used with D.C. motors because
(*a*) these motors have high starting torque (*b*) these motors are not self-starting
(*c*) back e.m.f. of these motors is zero initially
(*d*) to restrict armature current as there is no back e.m.f. while staring.

37. Which of the following D.C. motors has self-load properties ?
(*a*) Series motor (*b*) Shunt motor
(*c*) Cumulatively compound motor (*d*) Differentially compound motor.

38. In D.C. shunt motors as load is reduced
(*a*) the speed will increase abruptly
(*b*) the speed will increase in proportion to reduction in load
(*c*) the speed will remain almost constant
(*d*) the speed will reduce.

39. What will happen if the back e.m.f. of a D.C. motor vanishes
(*a*) the motor will stop (*b*) the motor will continue to run
(*c*) the armature may burn (*d*) the motor will run noisy.

40. In case of D.C. shunt motors the speed is dependent on back e.m.f. only because
(*a*) back e.m.f. is equal to armature drop (*b*) armature drop is negligible
(*c*) flux is proportional to armature current (*d*) flux is practically constant in D.C. shunt motors.

41. In a D.C. shunt motor, under the conditions of maximum power, the current in the armature will be
(*a*) almost negligible (*b*) rated full-load current
(*c*) less than full-load current (*d*) more than full-load current.

42. Which D.C. motor will have least percentage increase of input current, for the same percentage increase in the torque ?
(*a*) Shunt motor (*b*) Series motor
(*c*) Cumulatively compound motor (*d*) Separately excited motor.

43. These days D.C. motors are widely used in
(*a*) pumping sets (*b*) air compressors
(*c*) electric traction (*d*) machine shops.

44. By looking at which part of the motor, it can be easily confirmed that a particular motor is D.C. motor ?
(*a*) frame (*b*) shaft
(*c*) commutator (*d*) stator.

45. In which of the following applications D.C. series motor is invariably used ?
(*a*) Starter for a car (*b*) Drive for a water pump
(*c*) Fan motor (*d*) Motor operation in A.C. or D.C.

46. In D.C. machines fractional pitch winding is used
(*a*) to improve cooling (*b*) to reduce copper losses
(*c*) to increase the generated e.m.f. (*d*) to reduce the sparking.

47. A three point starter is considered as suitable for
(*a*) shunt motors (*b*) shunt as well as compound motors
(*c*) shunt, compound and series motors (*d*) all D.C. motors.

48. Small D.C. motors up to 5 H.P. usually have
(*a*) 2 poles (*b*) 4 poles
(*c*) 6 poles (*d*) 8 poles.

49. In case the conditions for maximum power for a D.C. motor are established, the efficiency of the motor will be
(*a*) 100% (*b*) around 90%
(*c*) anywhere between 75% and 90% (*d*) less than 50%.

50. A shearing machine has cyclic load consisting of intermittent light and heavy loads. Which of the following D.C. motors will be suitable for this purpose ?
(*a*) Series motor (*b*) Shunt motor
(*c*) Cumulative compound motor (*d*) Differentially compound motor.

51. The ratio of starting torque to full-load torque is least in case of
(*a*) series motor (*b*) shunt motor
(*c*) compound motors (*d*) none of the above.

52. In D.C. motor which of the following can sustain the maximum temperature rise ?
(*a*) Slip rings (*b*) Commutator
(*c*) Field winding (*d*) Armature winding.

53. Which of the following law/rule can be used to determine the direction of rotation of D.C. motor ?
(*a*) Lenz's law (*b*) Faraday's law
(*c*) Coloumb's law (*d*) Fleming's left-hand rule.

54. Which of the following loads normally needs staring torque more than the rated torque ?
(*a*) Blowers (*b*) Conveyors
(*c*) Air compressors (*d*) Centrifugal pumps.

55. The starting resistance of a D.C. motor is generally

(*a*) low (*b*) around 500 Ω
(*c*) 1000 Ω (*d*) infinitely large.

56. In case of conductively compensated D.C. series motors, the compensating winding is provided

(*a*) as separately wound unit (*b*) in parallel with armature winding
(*c*) in series with armature winding (*d*) in parallel with field winding.

57. Sparking at the commutator of a D.C. motor may result in

(*a*) damage to commutator segments (*b*) damage to commutator insulation
(*c*) increased power consumption (*d*) all of the above.

58. Which of the following motors is preferred for operation in highly explosive atmosphere ?

(*a*) Series motor (*b*) Shunt motor
(*c*) Air motor (*d*) Battery operated motor.

59. If the supply voltage for a D.C. motor is increased, which of the following will decrease ?

(*a*) Staring torque (*b*) Operating speed
(*c*) Full-load current (*d*) All of the above.

60. When the speed of a D.C. motor increases

(*a*) back e.m.f. increases and current drawn decreases
(*b*) back e.m.f. as well as current drawn both increase
(*c*) back e.m.f. as well as current drawn both decrease
(*d*) back e.m.f. decreases and current drawn increases.

61. As compared to an induction motor, the air gap in a D.C. motor is

(*a*) less than 50% (*b*) between 50% and 90%
(*c*) same (*d*) more.

62. Field winding of a D.C. series motor is usually provided with thick wire

(*a*) to provide large flux (*b*) to reduce the use of insulating materials
(*c*) as it carries large load current (*d*) in order to reduce eddy current.

63. Which one of the following is not the function of pole shoes in a D.C. machine ?

(*a*) To reduce eddy current loss (*b*) To support the field coils
(*c*) To spread out flux for better uniformity (*d*) To reduce the reluctance of the magnetic path.

64. The mechanical power developed by a shunt motor will be maximum when the ratio of back e.m.f. to applied voltage is

(*a*) 4.0 (*b*) 2.0
(*c*) 1.0 (*d*) 0.5.

65. The condition for maximum power in case of D.C. motor is

(*a*) back e.m.f. = 2 × supply voltage (*b*) back e.m.f. = $\frac{1}{2}$ × supply voltage
(*c*) supply voltage = $\frac{1}{2}$ × back e.m.f. (*d*) supply voltage = back e.m.f.

66. For which of the following applications a D.C. motor is preferred over an A.C. motor ?

(*a*) Low speed operation (*b*) High speed operation
(*c*) Variable speed operation (*d*) Fixed speed operation.

67. What will happen in case 220 V D.C. series motor is connected to 220 V A.C. supply ?

(*a*) The armature winding of motor will burn
(*b*) The motor will vibrate violently
(*c*) The motor will run with less efficiency and more sparking
(*d*) The motor will not run.

68. In D.C. machines the residual magnetism is of the order of

(*a*) 2 to 3 per cent (*b*) 10 to 15 per cent
(*c*) 20 to 25 per cent (*d*) 50 to 75 per cent.

69. If T_a be the torque and I_a the armature current for a D.C. motor, then which of the following relation is valid before saturation ?

(a) $T_a \propto I_a$
(b) $T_a \propto \frac{I}{I_a}$
(c) $T_a \propto I_a^2$
(d) $T_a \propto \frac{I}{(I_a)^2}$.

70. Which D.C. motor is generally preferred for cranes and hoists ?

(a) Series motor
(b) Shunt motor
(c) Cumulatively compound motor
(d) Differentially compound motor.

ANSWERS

1. (c)	2. (c)	3. (d)	4. (d)	5. (b)	6. (a)	7. (c)
8. (b)	9. (b)	10. (b)	11. (d)	12. (a)	13. (d)	14. (b)
15. (b)	16. (a)	17. (a)	18. (c)	19. (b)	20. (d)	21. (c)
22. (c)	23. (b)	24. (c)	25. (b)	26. (b)	27. (c)	28. (a)
29. (a)	30. (b)	31. (c)	32. (b)	33. (d)	34. (a)	35. (c)
36. (d)	37. (d)	38. (c)	39. (c)	40. (d)	41. (d)	42. (b)
43. (c)	44. (c)	45. (a)	46. (d)	47. (b)	48. (a)	49. (d)
50. (c)	51. (b)	52. (c)	53. (d)	54. (b)	55. (a)	56. (c)
57. (d)	58. (c)	59. (c)	60. (a)	61. (d)	62. (c)	63. (a)
64. (d)	65. (b)	66. (c)	67. (c)	68. (a)	69. (c)	70. (a).

THEORETICAL QUESTIONS

D.C. GENERATOR

1. What is a basic type of D.C. machine ?
2. Explain the principle on which a generator works.
3. Explain the construction and working of an elementary generator.
4. Describe briefly various parts of a D.C. machine.
5. Why is a commutator and brush arrangement necessary for operation of a D.C. machine ?
6. Name the main parts of a D.C. machine and state the materials of which each part is made.
7. Enumerate all the parts of a D.C. machine and indicate their functions.
8. Give the materials and functions of the following parts of a D.C. machine :
   (i) Field poles
   (ii) Yoke
   (iii) Commutator
   (iv) Commutating poles
   (v) Armature.
9. Give the constructional features and working principle of a D.C. generator. Draw a cross-sectional view of a 4 pole D.C. generator and label there on all parts. Also sketch the shapes of :
   (i) Commutator segment with riser, and
   (ii) Armature stamping.
10. Give the advantages and uses of lap and wave windings.
11. Explain the following terms as applied to a D.C. armature winding.
    (i) Front pitch
    (ii) Back pitch
    (iii) Pole pitch
    (iv) Commutator pitch.
12. Draw up the winding table for a 4 pole, wave connected armature having 30 coil sides and give a developed diagram of the winding showing the polarity and position of the brushes, the main poles and the direction of motion of the armature for a D.C. motor.
13. Draw the developed winding diagram of lap winding for 6 poles, 18 slots with two coil sides/slot, double layer showing therein position of poles, direction of motion, direction of generated e.m.f. and position of brushes.

14. Derive the e.m.f. equation of a D.C. generator.
15. Deduce an expression for the voltage induced in a D.C. generator.
16. Write a short note on equiliser rings.
17. How are ratings of electrical machines expressed ?
18. Draw the magnetic circuit of a 4 pole D.C. machine.
19. Explain different methods of excitation of D.C. generators with suitable diagrams.
20. How are D.C. generators classified ?
21. What is the difference between a separately excited and a self-excited generator ?
22. Sketch the following types of D.C. generators :
    (*i*) Shunt (*ii*) Series
    (*iii*) Compound.
    State with reason(s) where each is used ?
23. What is the difference between the short-shunt and long shunt compound generators ?
24. With the help of a neat diagram show power division in a D.C. generator.
25. Answer the following questions briefly :
    (*i*) State the principle on which generators operate.
    (*ii*) Name the materials of which the following parts of a D.C. machine are made and why ?
    (*a*) Frame (*b*) Armature
    (*c*) Bushes.
    (*iii*) Why are field coils provided in a D.C. generator ?
    (*iv*) What is the function of an armature in a D.C. generator ?
    (*v*) Of what material are the laminations of an armature made up of ?
    (*vi*) What is the function of a commutator ?
    (*vii*) What purpose is served by brushes in a D.C. machine ?

D.C. MOTOR

26. What is the working principle of a D.C. motor ?
27. "In every D.C. generator motor action occurs and in every D.C. motor a generator action occurs". Explain.
28. What is torque ? What is the source of the torque force in a D.C. motor ?
29. What is back e.m.f. or counter voltage ?
30. What is the effective voltage across a D.C. motor armature ?
31. What is the meaning of armature power ?
32. Is electrical power related to mechanical power ?
33. Explain the function of commutator in a D.C. motor. In what respects is the commutation process in D.C. motors different from that in generators ?
34. What is mean by speed regulation ?
35. What is meant by the term 'shunt motor' ?
36. What is meant by the term 'series motor' ?
37. What is the dominant speed characteristic of a shunt motor ?
38. Explain speed-current and torque-current characteristics of a series motor ?
39. What is meant by a compound motor ?
40. What conditions require the use of a compound motor ?
41. How is D.C. motor reversed ?
42. Why is a starter necessary for a motor ? Give the diagram and explain the working of a three-point starter for a shunt motor.
43. What are the functions of 'no-volt release' and 'overload release' in a starter ? Discuss the operation of these two features in a shunt motor starter.
44. Give the reasons for the following :
    (*i*) A series motor should not be connected to a load through a belt.
    (*ii*) A series motor develops a high starting torque.
    (*iii*) A differential compound motor is very rarely used.
    (*iv*) A shunt motor runs at almost constant speed irrespective of load current.

45. Derive an expression for the torque of a D.C. motor.
46. Differentiate between the generator action and motor action of a D.C. machine.
47. How may the direction of rotation of a D.C. shunt motor be reversed ? What is the effect of reversing the line terminals ?
48. Develop the general expression for the speed of a motor in terms of supply voltage, armature resistance and flux per pole.
49. Explain the 'armature reaction', in D.C. shunt motors, indicating also few remedies to its adverse effects.
50. Explain in detail how the shunt motor behaves as a constant speed motor and the series motor as a variable speed motor.
51. Draw the characteristic curves of D.C. shunt and series motors. Use these curves to explain the applications for which these motors are used.
52. Discuss the applications of shunt, series and compound motors.
53. State the types of loads for which shunt, series and compound D.C. motors are suitable. Give reasons in support of your answer.
54. Why is a starter necessary for D.C. motor ? Explain the working of a three-point starter with the help of a neat diagram.
55. What is the difference between three-point and four-point starters ? What are the additional features incorporated in a four-point starter.
56. Explain what happens, giving reasons thereof, in the following cases :
    (*i*) In a D.C. shunt motor armature terminals are reversed.
    (*ii*) In a D.C. shunt motor field terminals are reversed.
    (*iii*) The line terminals to a D.C. motor interchanged.
    (*iv*) A D.C. shunt motor is directly connected to the supply mains.
    (*v*) A D.C. motor is stopped by throwing back the starting arm.
57. Explain with neat diagrams construction and working of a series motor starter.
58. How does D.C. motor automatically adjust input to match the mechanical load on the motor ?

UNSOLVED EXAMPLES

D.C. GENERATOR

1. An eight-pole lap connected armature runs at 750 r.p.m. It has 800 conductors and flux/pole of 60 mWb. Calculate the e.m.f. generated. **[Ans.** 600 volts]
2. An 8-pole lap connected generator has a useful flux/pole of 0.05 Wb. If the no load voltage at 400 r.p.m. is 300 V find the conductors on the armature periphery. **[Ans.** 900]
3. An eight-pole wave connected armature has 300 conductors and runs at 400 r.p.m. Determine the useful flux/pole if the electromotive force generated on open circuit is 500 V. **[Ans.** 6.25×10^{-2} Wb]
4. A 6-pole wave connected D.C. generator has 52 slots and each slot has 20 conductors. Find the speed of the generator. The generated e.m.f. is 240 V and the flux per pole is 5 mWb. **[Ans.** 923 r.p.m.]
5. An 8-pole wave connected D.C. generator has 900 armature conductors and a flux/pole of 0.04 Wb. At what speed it must be driven to generate 500 V ? **[Ans.** 208.3 r.p.m.]
6. The lap wound armature of a 4-pole generator has 51 slots. Each slot contains 20 conductors. What will be the e.m.f. generated in machine when driven at 1500 r.p.m. if useful flux per pole is 0.01 Wb ? **[Ans.** 255 V]
7. An 8-pole lap wound generator armature has 960 conductors, a flux of 40 mWb and a speed of 400 r.p.m. Calculate the e.m.f. generated on open circuit. If the same armature is wave wound, at what speed must it be driven to generate 400 volts ? **[Ans.** 256 V, 156.25 r.p.m.]
8. An 8-pole D.C. shunt generator with 788 wave connected conductors and running at 500 r.p.m. supplies a load of 12.5 Ω resistance at a terminal voltage of 250 V. The armature resistance is 0.24 Ω and the field resistance is 250 Ω. Find :
    (*i*) Armature current (*ii*) Induced e.m.f.
    (*iii*) Flux/pole. **[Ans.** 21 A, 255.04 V, 9.83 mWb]

9. A 4-pole lap wound D.C. shunt generator has a useful flux/pole 0.07 Wb. The armature winding consists of 220 turns, each turn having a resistance of 0.004 Ω. Calculate the terminal voltage when running at 900 r.p.m. if the armature current is 50 A. [**Ans.** 459.25 V]

10. A shunt generator has a generated voltage on open circuit of 127 V. When the machine is on load, terminal voltage is 120 V. Find the load current if the field resistance be 15 Ω and the armature resistance 0.02 Ω. Neglect armature reaction. [**Ans.** 342 A]

11. A short shunt compound generator has armature, series field and shunt field resistances of 0.08 Ω, 0.04 Ω and 100 Ω respectively. It supplies 125 lamps each rated at 250 V, 40 W. Find the generated e.m.f. and the armature current. Drop under each brush may be taken as 1 V. [**Ans.** 22.508 A, 254.6 V]

D.C. MOTOR

12. A 250 V D.C. shunt motor takes 41 A at full-load. Find the back e.m.f. on full-load if the resistances of motor armature and shunt field windings are 0.1 ohm and 250 ohms respectively. [**Ans.** 246 V]

13. A 230 V motor has an armature circuit resistance of 0.6 ohm. If the full-load armature current is 30 A and the no-load armature current is 4 A, find the change in back e.m.f. from no-load to full-load. [**Ans.** 15.6 V]

14. A six-pole lap-connected 250 V shunt motor has 396 armature conductors. It takes 30 A on full-load. The flux per pole is 0.04 weber. The armature and field resistances are 0.1 ohm and 200 ohms respectively. Contact drop per brush = 1 V.
Determine the speed on full-load. [**Ans.** 928.5 r.p.m.]

15. A four-pole, 440 V shunt motor takes 8 A on no-load, the no-load speed being 750 r.p.m. It has a shunt field current of 1.5 A. Calculate the full-load speed of the motor if it takes 100 A at full-load. Armature resistance = 0.25 ohm. Contact drop per brush = 1 V. Armature reaction weakens the field by 5 per cent on full-load. [**Ans.** 748 r.p.m.]

16. A four-pole D.C. shunt motor working on 250 V takes a current of 2 A when running at 1,000 r.p.m. What will be its back e.m.f., speed and the percentage speed drop if the motor takes 51 A at a certain load ? Armature resistance = 0.2 ohm and shunt field resistance = 250 ohms. [**Ans.** 240 V, 963 r.p.m., 3.84%]

17. A four-pole lap wound 500 V shunt motor takes a full-load armature current of 50 A, the full-load speed being 800 r.p.m. There are 440 armature conductors and armature resistance is 0.3 ohm, contact drop per brush = 1 V.
Find the useful flux per pole. [**Ans.** 0.08233 weber]

18. A 500 V, 6-pole lap connected shunt motor has a no-load input current of 16 A and a shunt field current of 10 A. At full-load it takes a current of 160 A. If armature resistance = 0.1 ohm, flux/pole at no-load = 0.06 weber, number of armature conductors = 744 and contact drop/brush = 1 V, calculate :

(*i*) No-load speed (*ii*) Full-load speed
(*iii*) Speed regulation.

Armature reaction weakens the field by 2 per cent. [**Ans.** 669 r.p.m., 662.7 r.p.m., 0.9504%]

19. 250 V shunt motor takes a line current of 50 A and runs at 600 r.p.m. Its armature and field resistances are 0.4 ohm and 125 ohms respectively. Neglecting the effects of armature reaction and allowing 2 V brush drop, calculate :

(*i*) No-load speed if the no-load line current is 5 A
(*ii*) Percentage reduction in the flux per pole in order that the speed may be 800 r.p.m. when the armature current is 40 A. [**Ans.** 647 r.p.m., 23.7%]

20. A 250 V shunt motor on no-load runs at 1000 r.p.m. and takes 5 A. The total armature and shunt field resistances are respectively 0.2 ohm and 250 ohms. Calculate the speed when loaded and taking a current of 50 A, if armature reaction weakens the field by 3%. [**Ans.** 994 r.p.m.]

21. A D.C. series motor of resistance 1 ohm between terminals runs at 800 r.p.m. at 200 V with a current of 15 A. Find the speed at which it will run when connected in series with a 5 ohms resistance and taking the same current. [**Ans.** 476 r.p.m.]

22. A 250 V shunt motor takes 70 A when running at 1000 r.p.m. It has an armature resistance of 0.1 ohm. Determine the speed and armature current if the magnetic flux is weakened by 30 per cent. Contact drop per brush = 1 V.
Total torque developed remains constant. [**Ans.** 1411 r.p.m., 100 A]

23. A 500 V shunt motor takes a no-load current of 10 A at 500 V. It takes 100 A at 500 V on full-load. Armature and shunt field resistances are 0.2 ohm and 200 ohms respectively. Contact drop/brush = 1 V. If the no-load speed is 800 r.p.m. find the full-load speed.
Armature reaction weakens the field by 3 per cent on full-load. **[Ans.** 795 r.p.m.]

24. A 250 V shunt motor on no-load runs at 1000 r.p.m. and takes 5 A. The total armature and shunt field resistances are respectively 0.2 ohm and 250 ohms. Calculate the speed when loaded and taking a current of 50 A, if armature reaction weakens the field by 3 per cent. **[Ans.** 994 r.p.m.]

25. A 250 V shunt motor takes a current of 52 A and runs at 1000 r.p.m. at full-load. Armature and shunt field resistances are 0.3 ohm and 125 ohms respectively. Determine :
(*i*) Speed at full-load torque (*ii*) Speed at double full-load torque
(*iii*) Stalling torque in terms of the full-load torque.
Assume that the flux remains constant. **[Ans.** 681 r.p.m., 298 r.p.m., 2.78 times]

26. A 4-pole, 250 V lap connected shunt motor has armature and field resistances of 0.05 ohm and 100 ohms respectively. There are 492 armature conductors and the useful flux per pole is 0.03 weber. It takes a current of 7.5 A on no-load and a current of 120 A on full-load. Determine :
(*i*) No-load speed (*ii*) Full-load speed
(*iii*) Speed regulation.
Neglect drop/brush = 1 V. Armature reaction may be neglected. **[Ans.** 1007 r.p.m., 984 r.p.m., 2.337%]

27. A 500 V shunt motor has an armature resistance of 0.2 ohm. It takes an armature current of 60 A at 750 r.p.m. Find the speed and armature current if the flux is increased to 150% of its initial value. Assume the torque to be constant throughout. Contact drop/brush = 1 V. **[Ans.** 504.2 r.p.m., 40 A]

28. Determine the value of torque in Nm established by the armature of a 4-pole motor having 774 conductors, two paths in parallel , 24 m Wb flux per pole when the total armature current is 50 amperes.
[Ans. 295.3 Nm]

29. Calculate torque on an armature carrying a total of 25000 ampere-conductors, the diameter of the core being 0.9144 m, the length 30.5 cm, the polar arc 70% and the flux density in the air gap 0.6 tesla.
[Ans. 1464 Nm]

30. A 220 V series motor in which the total armature and field resistance is 0.1 ohm is working with unsaturated field, taking 100 A and running at 800 r.p.m. Calculate at what speed the motor will run when developing half the torque ? **[Ans.** 1147 r.p.m.]

31. A 4-pole wave wound shunt motor, running at 1000 r.p.m, takes 100 A at 250 V. There are 252 armature conductors. Useful flux per pole = 0.04 Wb. The shunt field current of motor is 5 A. Determine :
(*i*) Total torque developed
(*ii*) Actual torque if the friction and iron loss is 2 kW. **[Ans.** 304.7 Nm, 285.61 Nm]

32. A 4-pole 250 V D.C. shunt motor has lap connected 960 conductors. The flux per pole is 20 mWb. Determine the torque developed by the armature and the useful torque in Nm when current drawn by the motor is 32 A. The armature resistance is 0.1 ohm and shunt field resistance is 125 ohms. The rotational losses of the machine amount to 825 watts. **[Ans.** 91.6 Nm ; 81.4 Nm]

33. A shunt motor operating on 230 V takes an armature current of 6 A at no-load and runs at 1200 r.p.m. The armature resistance is 0.25 ohm. Determine the speed and electromagnetic torque when the armature takes 36 A with the same flux. **[Ans.** 1161 r.p.m, 65.44 Nm]

34. A 44.76 kW, 250 V, 4-pole lap-connected D.C. shunt motor has 30 slots with 12 conductors/slot. The armature and shunt field resistances are 0.06 ohm and 100 ohms respectively. The flux/pole is 0.03 Wb. If the full-load efficiency is 88%, find at full-load :
(*i*) The speed ;
(*ii*) Useful torque at shaft. **[Ans.** 1323 r.p.m., 323.3 Nm]

35. A 4-pole, lap-wound, 250 V shunt motor has the following particulars : No. of armature conductors = 720 ; armature resistance = 0.311 ohm, useful flux/pole = 0.04 Wb. If the total torque developed by the motor is 147.15 Nm, find the armature current taken and the speed. **[Ans.** 32.12 A ; 500 r.p.m.]

36. The input to a 220 V D.C. shunt motor is 11 kW. The other particulars of the motor are : no-load current = 5 A ; no-load speed = 1150 r.p.m. ; armature resistance = 0.5 ohm ; shunt field resistance = 110 ohms. Calculate :
(*i*) The torque developed (*ii*) The efficiency
(*iii*) The speed at this load. **[Ans.** 87.1 Nm ; 79.6% ; 1031 r.p.m.]

6

Transformers

1. GENERAL ASPECTS

Although the transformer is not classified as an electric machine, the principles of its operation are fundamental for the induction motor and synchronous machines. Since A.C. electric machines are normally built for low frequencies only the low-frequency power transformer will be considered in this text.

Function. *The function of a transformer, as the name implies, is to transform alternating current energy from one voltage into another voltage.* The transformer has *no rotating parts*, hence it is often called a *static transformer.*

When energy is transformed into a higher voltage the transformer is called a *step-up transformer* but when the case is otherwise it is called a *step-down transformer*. Most power transformers operate at constant voltage, *i.e.* if the power varies the current varies while the voltage remains fairly constant.

Applications. A transformer performs many important functions in prominent areas of electrical engineering.

- In *electrical power engineering* the transformer makes it possible to convert electric power from a generated voltage of about 11 kV (as determined by generator design limitations) to higher values of 132 kV, 220 kV, 400 kV, 500 kV and 765 kV thus permitting transmission of huge amounts of power along long distances to appropriate distribution points at tremendous savings in the cost of transmission lines as well as in power losses.
- At *distribution points* transformers are used to reduce these high voltages to a safe level of 400/230 volts for use in homes, offices etc.
- In *electric communication circuits* transformers are used for a variety of purposes *e.g.*, as an impedance transformation device to allow maximum transfer of power from the input circuit to the output device.

- In *radio and television circuits* input transformers, interstage transformers and output transformers are widely used.
- Transformers are also used in *telephone circuits, instrumentation circuits* and *control circuits*.

2. BASIC DEFINITIONS

- A *transformer* is a *static electromagnetic device designed for the transformation of the (primary) alternating current system into another (secondary) one of the same frequency with other characteristics, in particulars, other voltage and current*.
- As a rule a transformer consists of a core assembled of sheet transformer steel and two or several windings coupled *electromagnetically*, and in the case of *autotransformer, also electrically*.
- A transformer with two windings is called *double-wound transformer* ; a transformer with three or more windings is termed a *triple wound* or *multi-winding* one.
- According to the kind of current, transformers are distinguished as single-phase, three-phase and poly-phase ones. A *poly-phase transformer winding is a group of all phase windings of the same voltage*, connected to each other in a definite way.
- **Primary and secondary windings.** The transformer winding to which the energy of the alternating current is delivered is called the *primary winding* ; the other winding from which energy is received is called the *secondary winding*.
- In accordance with the names of the windings, all quantities pertaining to the primary winding as, for example, power, current, resistance etc., are also primary, and those pertaining to the secondary winding secondary.
- **h.v. and l.v. windings.** The winding connected to the circuit with the higher voltage is called the *high-voltage winding* (h.v.), the winding connected to the circuit with the lower voltage is called the *low-voltage winding*. (l.v.). If the secondary voltage is *less* than the primary one, the transformer is called a *step-down transformer* and if *more-a step-up transformer*.
- *A tapped transformer* is one whose windings are fitted with special taps for changing its voltage or current ratio.
- **Oil and dry transformers.** To avoid the deterimental effect of the air on the winding insulation and improve the cooling conditions of the transformer its core together with the windings assembled on it is immersed in a tank filled with transformer oil. Such transformers are called **oil transformers**. Transformers not immersed in oil are called **dry transformers.**

3. WORKING PRINCIPLE OF A TRANSFORMER

A transformer operates on the principle of *mutual inductance,* between two (and sometimes more) inductively coupled coils. It consists of two windings in close proximity as shown in Fig. 1. *The two windings are coupled by magnetic induction.* (There is no conductive connection between the windings). One of the windings called *primary* is energised by a sinusoidal voltage. The second winding, called *secondary* feeds the load. The alternating current in the primary winding sets up an alternating flux (ϕ) in the core. The secondary winding is linked by most of this flux and e.m.fs are induced in the two windings. The e.m.f. induced in the secondary winding drives a current through the load connected to the winding. Energy is transferred from the primary circuit to the secondary circuit through the medium of the magnetic field.

In brief, a transformer is a *device* that :

(*i*) *transfers electric power from one circuit to another* ;

(*ii*) *it does so without change of frequency ; and*

(*iii*) *it accomplishes this by electromagnetic induction (or mutual inductance).*

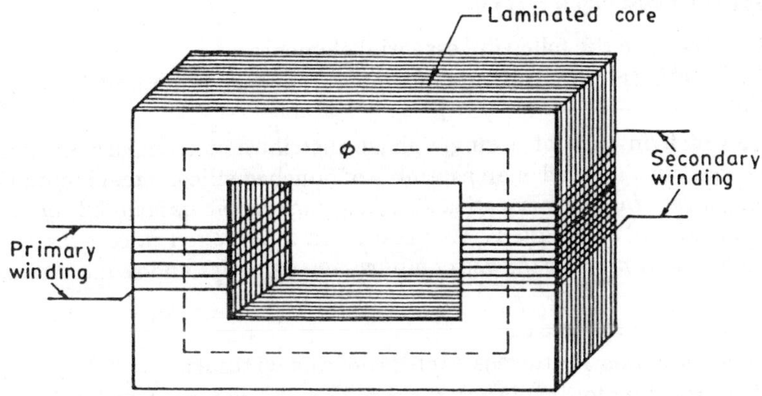

Fig. 1. Two winding transformer.

4. TRANSFORMER RATINGS

The rated quantities of a transformer, its power, voltage, frequency, etc. are given in Manufacturer's name plate, which should always be arranged so as to be accessible. But the term 'rated' can also be applied to quantities not indicated on the name plate, but relating to the rated duty, as for example, the rated efficiency, rated temperature conditions of the cooling medium, etc. :

- *The rated duty* of a transformer is determined by the quantities given in the name plate.
- The *rated power* of the transformer is the power at the secondary terminals, indicated in the name plate and expressed in kVA.
- *The rated primary voltage* is the voltage indicated in the transformer name plate ; if the primary is provided with taps, the rated tapped voltage is specially noted.
- *The rated secondary voltage* is the voltage across the transformer *secondary terminals at no-load* and with the rated voltage across the primary terminals ; if the secondary winding has taps, then their rated voltage is specially indicated.
- *The rated currents of the transformer*, primary and secondary, are the currents indicated in the name plate of the transformer and calculated by using the corresponding rated values of power and voltage.

5. KINDS OF TRANSFORMERS

The following kinds of transformers are the most important ones :

1. **Power transformers.** *For the transmission and distribution of electric power.*

2. **Auto-transformers.** *For converting voltages within relatively small limits to connect power systems of different voltages, to start A.C. motors etc.*

3. **Transformer for feed installations with static convertors.** (Mercury arc rectifiers, ignitrons, semi-conductor valves, etc.) *When converting A.C. into D.C.* (rectifying) *and convering D.C. into A.C. (inverting).*

4. **Testing transformers.** *For conducting tests at high and ultra-high voltages.*

5. **Power transformers for special applications.** Furnace, welding etc.

6. **Radio-transformers.** *It is used in radio engineering etc.*

Note. *Distribution transformers* should be designed to have maximum efficiency at a load much *lower than full-load (about 50 per cent).*

Power transformers should be designed to have maximum efficiency *at or near full-load.*

6. TRANSFORMER CONSTRUCTION

All transformers have the following essential elements :

1. Two or more **electrical windings** insulated from each other and from the core (except in auto-transformers).

2. A **core,** which in case of a single-phase distribution transformers usually comprises *cold-rolled silicon-steel strip* instead of an assembly of punched silicon-steel laminations such as are used in the larger power-transformer cores. The *flux path in the assembled core is parallel to the directions of steel's grain or 'orientation'.* This results in a *reduction in core losses* for a given flux density and frequency, or *it permits the use of higher core densities* and *reduced size of transformers for given core losses.*

Other necessary parts are :

- A *suitable container* for the assembled core and windings.
- A *suitable medium* for insulating the core and its windings from each other and from the container.
- Suitable *bushings* for insulating and bringing the terminals of the windings out of the case.

The two basic types of transformer construction are :

1. *The core type.*
2. *The shell type.*

The above two types differ in their relative arrangements of copper conductors and the iron cores. In the *'core type'*, the *copper virtually surrounds the iron core*, while *in the 'shell type', the iron surrounds the copper winding.*

6.1. Core Type Transformer. The completed magnetic circuit of the core-type transformer is in the shape of the hollow rectangle, exactly as shown in Fig. 2 in which I_0 is the no-load current and ϕ is the flux produced by it. N_1 and N_2 are the number of turns on the primary and secondary sides respectively.

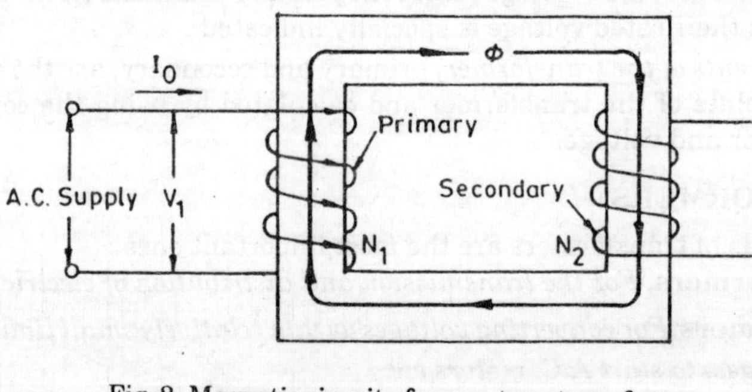

Fig. 2. Magnetic circuit of a core-type transformer.

The core is made up of *silicon-steel laminations* which are, either rectangular or L-shaped. With the coils wound on two legs the appearance is that of Fig. 3. If the two coils shown were the respective high and low-side coils as in Fig. 3, the *leakage reactance would be much too great.* In order to provide maximum *linkage* between windings, the group on *each leg is made up of both high-tension and low-tension coils.* This may be seen in Fig. 4, where a cross-sectional cut is taken across the legs of the core. By placing the high-voltage winding around the low-voltage winding, only one layer of high-voltage insulation is required, that between the two coils. *If the high-voltage coils were adjacent to the core, an additional high-voltage insulation layer would be necessary between the coils and the iron core.*

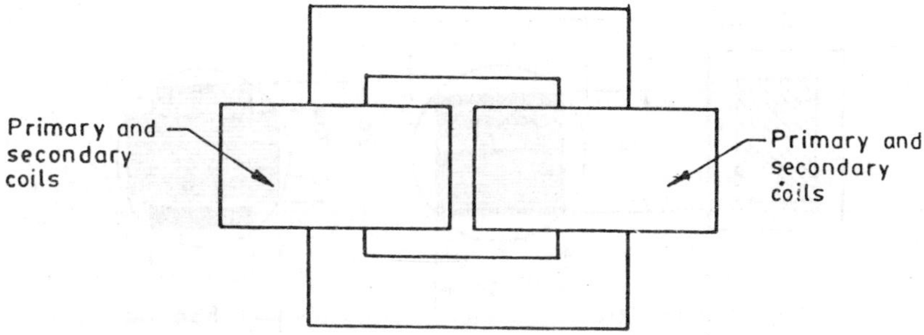

Fig. 3. Core-type transformer.

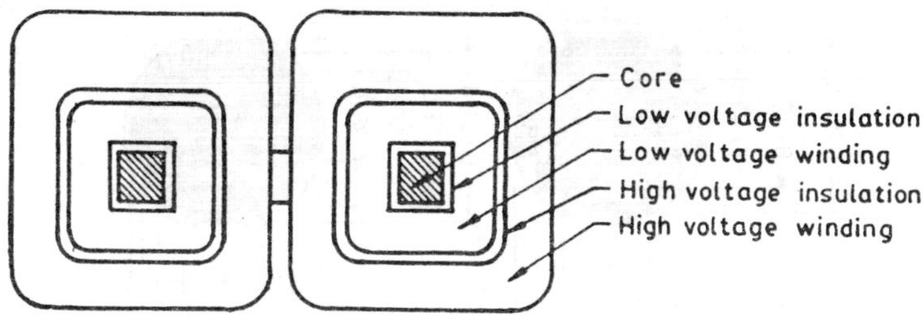

Fig. 4. Cross-section of a core-type transformer.

Fig. 5 shows the coils and laminations of a core-type transformer with a cruciform core and circular coils.

- Fig. 6 shows *the different types of cores used in core type transformers.*

Rectangular cores [Fig. 6 (*a*)] with rectangular cylindrical coils can be used for small size core-type transformers. For large size transformers it becomes wasteful to use rectangular cylindrical coils and so circular cylindrical coils are preferred. For such purposes, '*square cores*' may be used as shown in Fig. 6 (*b*) where circles represent the tubular former carrying the coils. Evidently a considerable amount of useful space is still wasted. A common improvement on the square core is to employ a '*cruciform core*' [Fig. 6 (*c*)] which demands, atleast, two sizes of core strips. For *very large transformers*, further core stepping is done as in Fig. 6 (*d*) where atleast three sizes of core plates are necessary. *Core stepping not only gives high space factor but also results in reduced length of the mean turn and the consequent I^2R loss. Three stepped core is the most commonly* used although more steps may be used for very large transformers as shown in Fig. [6 (*e*)].

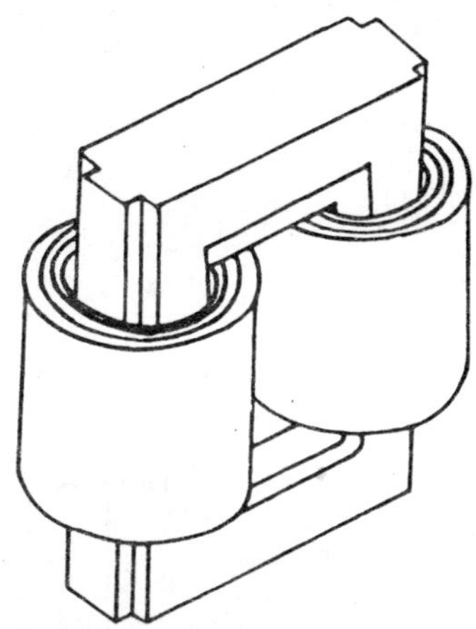

Fig. 5. Coils and laminations of a core-type transformer.

6.2. Shell Type Transformer. In the shell-type construction the iron almost entirely surrounds the copper (Fig. 7). The core is made up of E-shaped or F-shaped laminations which are stacked to give a rectangular figure eight. All the windings are placed on the centre leg, and in order to reduce leakage, each high-side coil is adjacent to a low-side coil. The coils actually occupy the entire space of both windows, are flat or pencake in shape, and are usually constructed of strip copper. Again, *to reduce the amount the high-voltage insulation required, the low-voltage coils are placed adjacent to the iron core.*

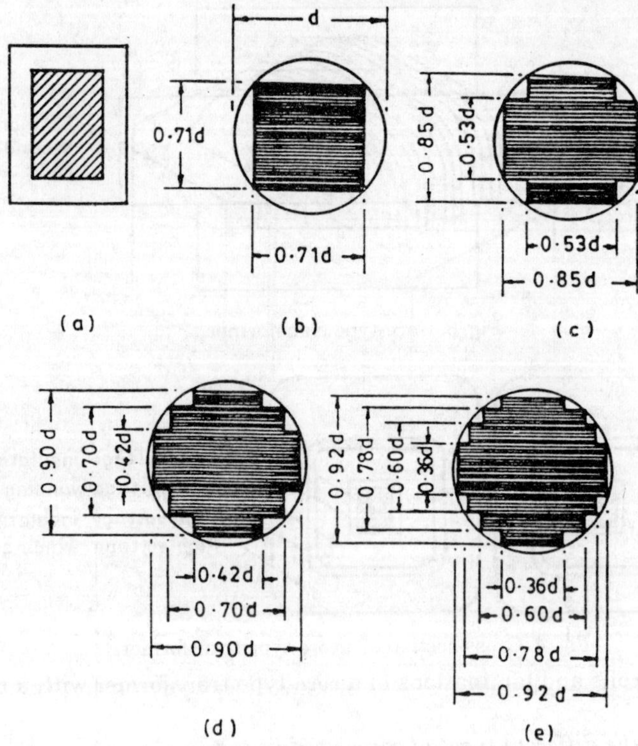

Fig. 6. Various types of cores.

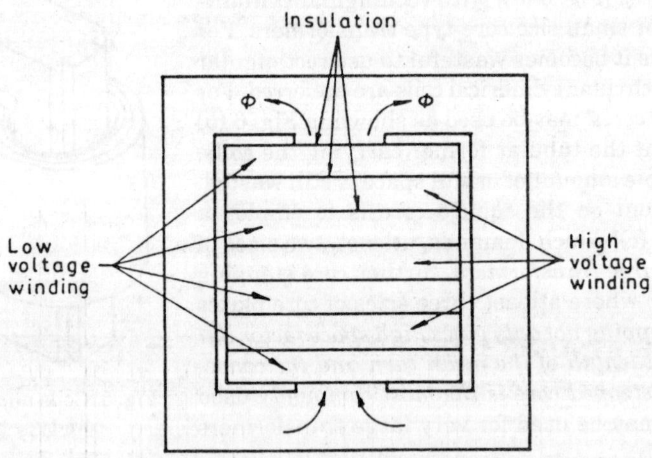

Fig. 7. Shell-type transformer.

Fig. 8 shows the coils and laminations of a typical shell-type transformer.

Choice of Core- or Shell Type Construction. In general, the core-type has a longer mean length of core and a shorter mean length of coil turn. The core type also has a smaller cross-section of iron and so will need a greater number of turns of wire, since, in general, not as high a flux may be reached in the core. However, *core type is better adopted for some high-voltage service since there is more room for insulation*. The *shell type has better provision for mechanically supporting and*

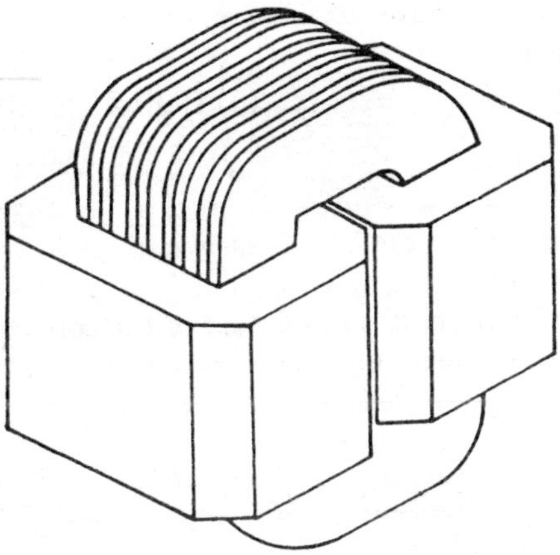

Fig. 8. Coils and laminations of a shell-type transformer.

bracing the coils. This allows better resistance to the very high mechanical forces that develop during a high-current short circuit.

The choice of core- or shell type construction is usually one of *cost*, for similar characteristics can be obtained with both types.

Both core and shell forms are used, and selection is based upon many factors such as voltage rating, kVA rating, weight, insulation stress, mechanical stress, and heat distribution.

6.3. Spiral Core Transformer. The typical spiral core is shown in Fig. 9. The core is assembled either of a *continuous* strip of transformer steel wound in the form of a circular or elliptical cylinder or of a group of short strips assembled to produce the same elliptical-shaped core. By using this construction the core flux always follows along the grain of the iron. Cold-rolled steel of high silicon content enables the designer to use higher operating flux densities with lower loss per kg. *The higher flux density reduces the weight per kVA.*

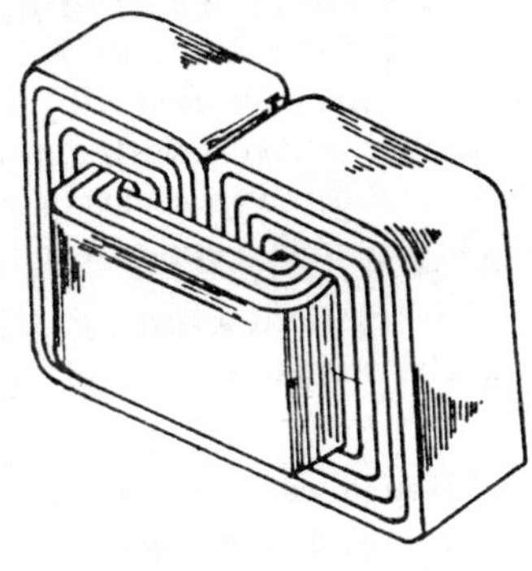

Fig. 9. Spiral-core transformer.

7. TRANSFORMER WINDINGS, TERMINALS, TAPPINGS AND BUSHINGS

7.1. Transformer Windings. The most important requirements of transformer windings are :

1. The winding should be *economical both as regards initial cost*, with a view to the market availability of copper, and the efficiency of the transformer in service.

2. The heating conditions of the windings should *meet standard requirements*, since departure from these requirements towards allowing higher temperature will drastically shorten the service life of the transformer.

3. The winding should be *mechanically stable* in respect to the forces appearing when sudden short circuit of the transformer occur.

4. The winding should have the *necessary electrical strength in respect to over voltages.*

The different *types of windings* are *classified* and briefly discussed below :

1. Concentric windings :

(*i*) Cross-over (*ii*) Helical

(*iii*) Disc.

2. Sandwich windings

1. Concentric windings. Refer Fig. 10. These windings are *used for core type transformers.* Each limb is wound with a group of coils consisting of both primary and secondary turns which may be concentric cylinders. The l.v. winding is placed next to the core and h.v. winding on the outside. But the two windings can be sub-divided, *and interlaced with high tension and low tension section alternately to reduce leakage reactance.* These windings can be further divided as follows :

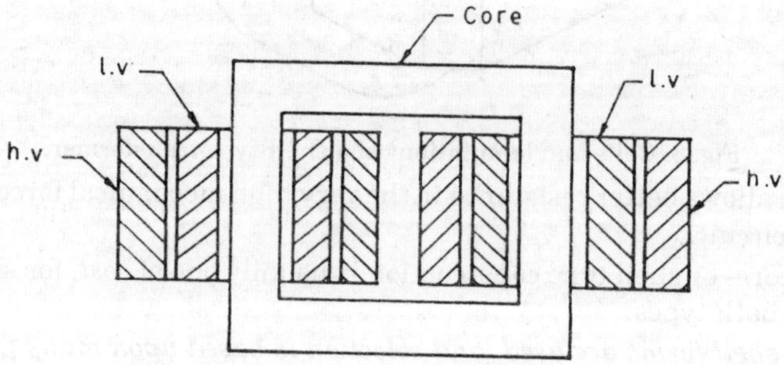

Fig. 10. Concentric coils.

(*i*) **Cross-over windings.** Cross-over windings are used for currents up to 20 A and so they are suitable for h.v. winding of small transformers. The conductors are either cotton covered round wires or strips insulated with paper. Cross-over coils are wound over formers and each coil consists of a number of layers with a number of turns per layer. The complete winding consists of a number of coils connected in series. Two ends of each coil are brought out, one from inside and one from outside. The inside end of a coil is connected to the outside end of the adjacent coil.

(*ii*) **Helical winding.** A helical winding consists of rectangular strips wound in the form of a helix. The strips are wound in parallel radially and each turn occupies the total radial depth of winding.

Helical coils are well suited for *l.v. windings of large transformers.* They can also be used for h.v. windings by putting extra insulation between layers in addition to insulation of conductors.

(*iii*) **Continuous disc winding.** This type of winding consists of a number of flat strips wound spirally from inside (radially) outwards. The conductor is used in such lengths as are sufficient for complete winding or section of winding between tappings. The conductor can either be a single strip or a number of strips in parallel, wound on the flat. This gives a robust construction for each disc. The discs are wound on insulating cylinders spaced from it by strips along the length of cylinder. The discs are separated from each other with press board sectors attached to the vertical strips. The vertical and horizontal spacers provide ducts for free circulation of oil which is in contact with every turn.

2. Sandwich coils. Sandwich coils (Fig. 11) are employed in transformers of *shell type.* Both high and low voltage windings are split into a number of sections. Each high voltage section lies between the low voltage sections.

The *advantage of sandwich coils is that their leakage can be easily controlled and so any desired value of leakage reactance can be had by the division of windings.*

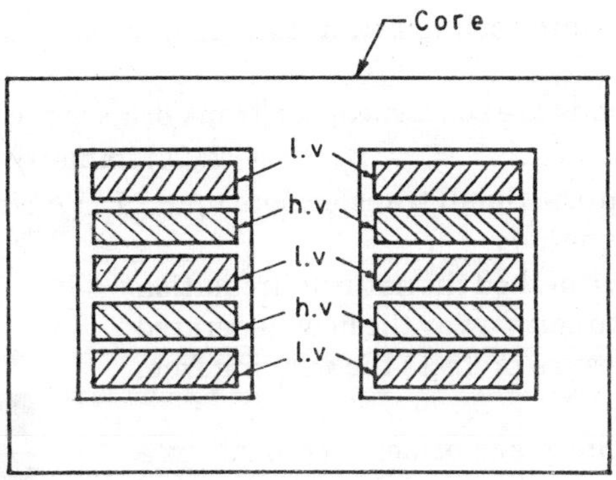

Fig. 11. Sandwich coils.

7.2. Terminals and Leads. The connection to the windings are of insulated copper rods or bars. The shape and size of leads is important in high voltage transformers owing to dielectric stress and corona which are caused at bends and corners. Connections from windings are directly taken to the busbars in the case of air-cooled transformers while they are taken to insulated bushings in the case of oil-cooled transformers.

7.3. Tappings. In a supply network the voltage can be controlled by changing the transformation ratio. This can be done by tapping the winding in order to alter the number of turns. The change in number of turns may be effected when the transformer is out of circuit (known as off load tap changing) or when on load (known as on load tap changing). *The tappings are provided on the high voltage winding because a fine voltage variation is obtained owing to large number of turns. It is difficult to obtain voltage variation within close percentage limits in low voltage winding as there are few turns and voltage per turn is a large percentage of the total voltage.*

In transformers, the tappings can be provided at :

(*i*) phase ends ; and

(*ii*) neutral point or in the middle of the windings.

- The advantage of providing tappings at *phase ends* is that the number of bushing insulators is reduced, this is important where the cover space is limited. Some transformers have reinforced insulation at the phase ends. It is essential that in such cases either the tapping should not be provided at end turns or the reinforcement should be carried beyond the lower tap.
- When the tappings are made at the *neutral point* the insulation between various parts is small. *This arrangement is economical especially in the case of high voltage transformers.*

7.4. Bushings. The bushings are *employed for insulating and bringing out terminals of the winding from the container to the external circuit.* For low-voltage transformers this is accomplished by employing *bushings of porcelain around the conductor at the point of entry.* For high voltages it is necessary to employ bushings of larger sizes. In modern transformers the problem is met by using *large porcelain or composition bushings for voltages as high as 33 kV, above that oil filled* or *condenser type bushings are used.*

8. TRANSFORMER COOLING

8.1. Cooling Methods. The transformers get heated due to iron and copper losses occurring in them. It is necessary to dissipate this heat so that the temperature of the windings is kept below the value at which the insulation begins to deteriorate. The *cooling of transformers is more difficult than that of rotating machines because the rotating machines create a turbulent air flow which assists in removing the heat generated due to losses.* Luckily the losses in transformers are comparatively small. Nevertheless the elaborate cooling arrangements have been devised to deal with the whole range of sizes.

As far as cooling methods are concerned, the transformers are of following two types :

1. Dry type.
2. Oil immersed type.

Dry Type Transformers. Small transformers upto *25 kVA size* are of the *dry type* and have the following cooling arrangements :

(*i*) **Air natural.** In this method the natural circulation of surrounding air is utilized to carry away the heat generated by losses. A sheet metal enclosure protects the winding from mechanical injury.

(*ii*) **Air blast.** Here the transformer is cooled by a continuous blast of cool air forced through the core and windings (Fig. 12). The blast is produced by a fan. The air supply must be filtered to prevent accumulation of dust in ventilating ducts.

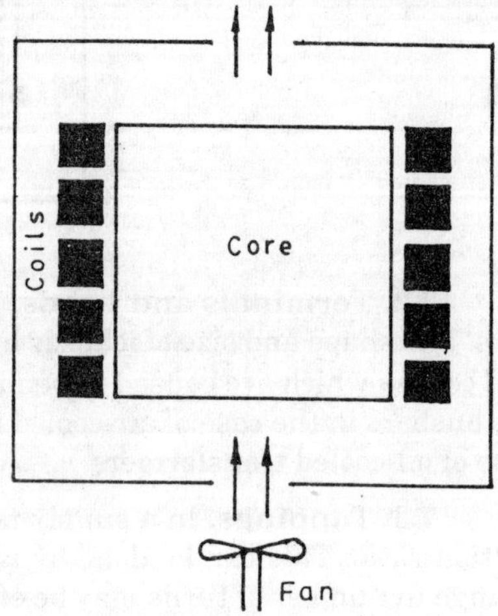

Fig. 12

Oil Immersed Transformers. In general most transformers are of oil immersed types. The *oil provides better insulation than air and it is a better conductor of heat than air*. Mineral oil is used for this purpose.

Oil immersed transformers are classified as follows :

(*i*) **Oil immersed self-cooled transformers.** The transformer is immersed in oil and heat generated in cores and windings is passed to oil by conduction. Oil in contact with heated parts rises and its place is taken by cool oil from the bottom. The natural oil transfers its heat to the tank walls from where heat is taken away by the ambient air. The oil gets cooler and falls to the bottom from where it is dissiptated into the surroundings. The tank surface is the best dissipator of heat but a plain tank will have to be excessively large, if used without any auxiliary means for high rating transformers. As both space and oil are costly, these auxiliary means should not increase the cubic capacity of the tank. The heat dissipating capacity can be increased by providing (*i*) corrugations, (*ii*) fins, (*iii*) tubes (Fig. 13) and (*iv*) radiator tanks.

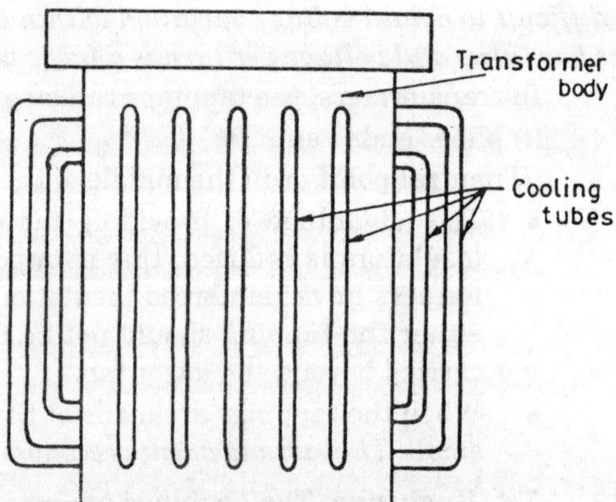

Fig. 13. Transformer with cooling tubes.

The *advantages of 'oil natural' cooling is that it does not clog the ducts and the windings are free from effects of moisture.*

(*ii*) **Oil immersed forced air-cooled transformers.** In this type of cooling, air is directed over the outer surfaces of the tank of the transformer immersed in oil.

(*iii*) **Oil immersed water-cooled transformers.** Heat is extracted from the oil by means of a stream of water pumped through a metallic coil immersed in the oil just below the top of the tank. The heated water is in turn cooled in a spray pond or a cooling tower.

(*iv*) **Oil immersed forced oil cooled transformers.** In such transformers heat is extracted from the oil by pumping the oil itself upward through the winding and then back by way of external radiators which may themselves be cooled by fans. The *extra cost of oil pumping equipment must of* *course be economically justified but it has incidentally the advantage of reducing the temperature difference between the top and bottom of enclosing tank.*

Fig. 14 shows the cooling of transformers having capacities from 10000 kVA and higher. In such cases air blast cooling of radiator is used.

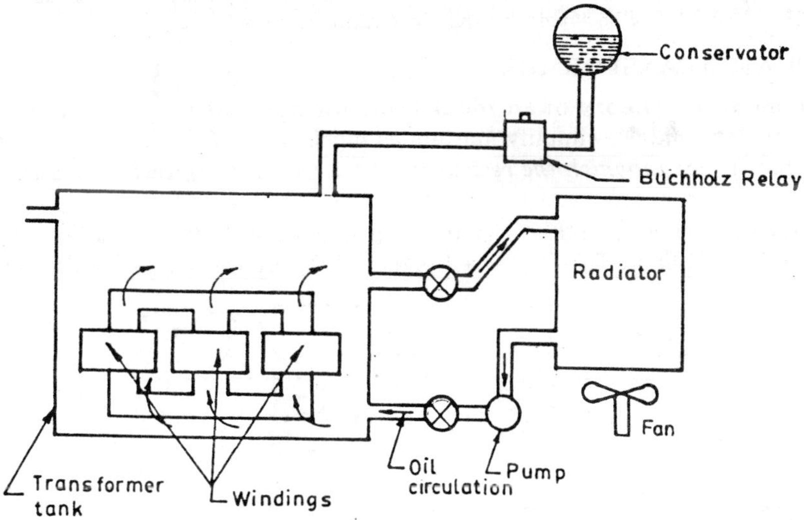

Fig. 14. Air blast cooling of radiator.

8.2. Transformer Oil. It is a *mineral oil obtained by refining crude petroleum*. It serves the following *purposes* :

(*i*) Provides additional insulation.

(*ii*) Carries away the heat generated in the core and coils.

(*iii*) Protects the paper from dirt and moisture.

The transformer oil should possess the following *properties* :

1. High dielectric strength.
2. Low viscosity to provide good heat transfer.
3. Good resistance to emulsion.
4. Free from inorganic acid, alkali and corrosive sulphur.
5. Free from sludging under normal operating conditions.
6. High flash/fire point.

8.3. Conservator and Breather

Conservator. The oil should not be allowed to come in contact with atmospheric air as it may take up moisture which may spoil its insulating properties. Also air may cause acidity and sludging of oil. To prevent this, many transformers are provided with conservators. The *function of a conservator (Fig. 14) is to take up contraction and expansion of oil without allowing it to come in contact with outside air*. The conservator consists of an air tight metal-drum fixed above the level of the top of the tank and connected with it by a pipe. The main tank is completely filled with oil when cold. The *conservator is partially filled with oil*. So the oil surface in contact with air is greatly reduced. The sludge thus formed remains in the conservator itself and does not go to the main tank.

Breather. When the temperature changes, the oil expands or contracts and there is a displacement of air. When the transformer cools, the oil level goes down, and air is drawn in. This is known as *breathing*. The air, coming in, is passed through an apparatus called *breather* for the purpose of extracting moisture. The *breather consists of a small vessel which contains a drying agent like silica get crystal impregnated with cobalt crystal*.

Note. *Sludging* means the slow formation of solid hydrocarbons due to heating and oxidation. The sludge deposit itself on the windings and cooling ducts producing overheating. This makes transformer still hotter producing more sludge. This process may continue till the transformer becomes unusable due to overheating. So the contact of oil with air should be avoided as the air contains oxygen.

9. SINGLE PHASE TRANSFORMER

9.1. Elementary Theory of an Ideal Transformer. The basic theory of a transformer is not difficult to understand. To simplify matters as much as possible, let us first consider *an ideal transformer*, that is, *one in which the resistance of the windings is negligible and the core has no losses*.

Let the secondary be open (Fig. 15), and let a sine wave of potential difference v_1 (Fig. 16) be impressed upon the primary. The impressed potential difference causes an alternating current to

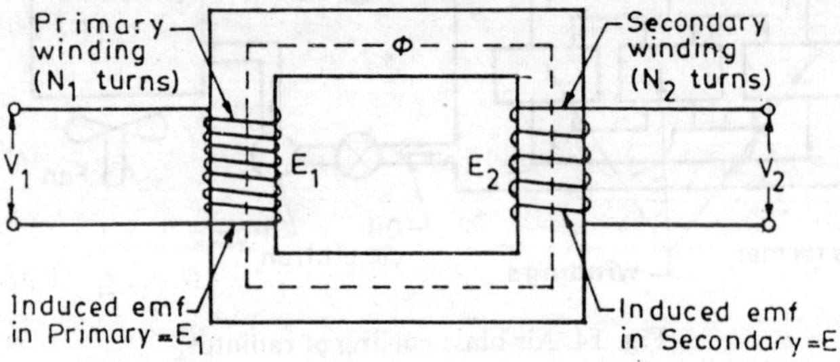

Fig. 15. Elementary diagram of an ideal transformer with an open secondary winding.

flow in the primary winding. Since the primary resistance is negligible and there are no losses in the core, the effective resistance is zero and the circuit is purely reactive. Hence the current wave i_m lags the impressed voltage wave v_1 by 90 time degrees, as shown in Fig. 17. The reactance of circuit is very high and the magnetizing current is very small. This current in the N_1 turns of the primary magnetizes the core and produces a flux ϕ that is at all times proportional to the current (if the permeability of the circuit is assumed to be constant), and therefore in time phase with the current. The flux, by its rate of change, induces in the primary winding E_1 which at every instant of time is equal in value and opposite in direction to V_1. It is called *counter e.m.f.* of the primary. The value which the primary current attains must be such that the flux which it produces in the core is of sufficient value to induce in the primary the required counter e.m.f.

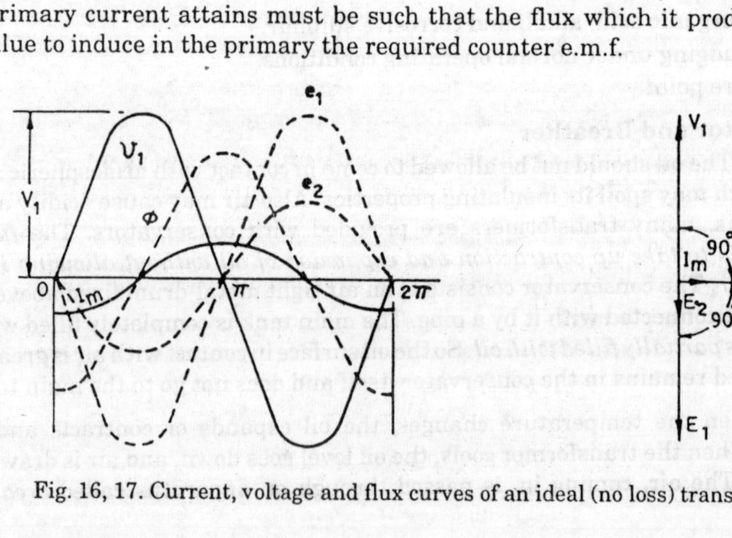

Fig. 16, 17. Current, voltage and flux curves of an ideal (no loss) transformer.

Since the flux also threads (or links) the secondary winding a voltage e_2 is induced in the secondary. This voltage is likewise proportional to the rate of change of flux and so is in time phase with e_1, but it may have any value depending upon the number of turns N_2 in the secondary.

9.2. E.M.F. Equation of a Transformer

Let N_1 = number of turns in primary,

N_2 = number of turns in secondary,

ϕ_m = maximum flux in the core, Wb.

$= B_m \times A$, [where B_m is the maximum flux density in the core and A is the core area], and

f = frequency of a.c. input, Hz.

Refer Fig. 18. Since the flux increases from its zero value to maximum value ϕ_m in one quarter of the cycle *i.e.* in $\frac{T}{4}$ or $\frac{1}{4f}$ second (T being time-period of the cycle),

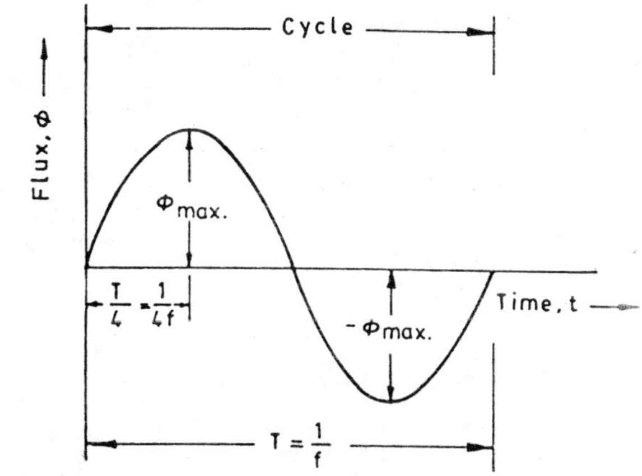

Fig. 18

∴ Average rate of change of flux

$$= \frac{\phi_m}{\frac{1}{4f}} = 4f\phi_{max} \text{ Wb/s or volt}$$

If flux ϕ varies *sinusoidally,* then r.m.s. (root mean square) value of inducted e.m.f. is obtained by multiplying the average value with form factor.

But, form factor $= \frac{\text{r.m.s. value}}{\text{average value}} = 1.11$

∴ r.m.s. value of e.m.f./turn

$$= 1.11 \times 4f\phi_{max} = 4.44f\phi_{max} \text{ volt}$$

Now, r.m.s. value of induced e.m.f. in the whole of primary winding,

$$E_1 = 4.44f\phi_{max} N_1 \quad ...(1)$$

Similarly r.m.s. value of induced e.m.f. in secondary is,

$$E_2 = 4.44f\phi_{max} N_2 \quad ...(2)$$

In an ideal transformer on no-load $V_1 = E_1$ and $V_2 = E_2$ (Fig. 15).

9.3. Voltage Transformation Ratio (K). The *transformation ratio is defined as the ratio of the secondary voltage to primary voltage*. It is denoted by the letter K.

From eqns. (1) and (2), $$\frac{E_2}{E_1} = \frac{N_2}{N_1} = K \quad ...(3)$$

- If $N_2 > N_1$ *i.e.* $K > 1$, then transformer is called *step-up transformer.*
- If $N_2 < N_1$ *i.e.* $K < 1$, then transformer is called *step-down transformer.*

Again for an *ideal transformer*

$$\text{Input } (VA) = \text{Output } (VA)$$

$$V_1I_1 = V_2I_2 \quad \text{or} \quad E_1I_1 = E_2I_2$$

or

$$\frac{I_2}{I_1} = \frac{E_1}{E_2} = \frac{N_1}{N_2} = \frac{1}{K} \quad ...(4)$$

i.e., Primary and secondary currents are inversely proportional to their respective turns.

Example 1. *A 40 kVA, single phase transformer has 400 turns on the primary and 100 turns on the secondary. The primary is connected to 2000 V, 50 Hz supply. Determine :*

(i) The secondary voltage on open circuit.

(ii) The current flowing through the two windings on full-load.

(iii) The maximum value of flux.

Solution. Rating $= 40$ kVA

Primary turns, $N_1 = 400$

Secondary turns, $N_2 = 100$

Primary induced voltage, $E_1 = V_1 = 2000$ V

(*i*) **Secondary voltage on open, circuit V_2 :**

Using the relation, $\dfrac{E_2}{E_1} = \dfrac{N_2}{N_1}$

$$E_2 = E_1 \times \frac{N_2}{N_1}$$

$$\therefore \quad E_2 = V_2 = 2000 \times \frac{100}{400} = 500 \text{ V}$$

Hence, $\mathbf{V_2 = 500}$ **V. (Ans.)**

(*ii*) **Primary current, I_1 :**

Secondary current, I_2 :

Primary full-load current, $I_1 = \dfrac{\text{kVA} \times 100}{V_1} = \dfrac{40 \times 1000}{2000} = $ **20 A. (Ans.)**

Secondary full-current, $I_2 = \dfrac{\text{kVA} \times 1000}{V_2} = \dfrac{40 \times 1000}{500} = $ **80 A (Ans.)**

(*iii*) **Maximum value of flux, ϕ_{max} :**

Using e.m.f. equation, $E_1 = 4.44 f \phi_{max} N_1$

$$\therefore \quad 2000 = 4.44 \times 50 \times \phi_{max} \times 400$$

$$\therefore \quad \phi_{max} = \frac{2000}{4.44 \times 50 \times 400} = 0.0225 \text{ Wb.}$$

Hence, $\phi_{max} = $ **0.0225 Wb. (Ans.)**

Example 2. *The no-load ratio required in a single-phase 50 Hz transformer is 6600/600 V. If the maximum value of flux in the core is to be about 0.08 Wb, find the number of turns in each winding.*

Solution. Primary, $E_1 = V_1 = 6600$ V

Secondary, $E_2 = V_2 = 600$ V

Maximum value of flux $\phi_{max} = 0.08$ Wb.

Primary turns, N_1 :

Secondary turns, N_2 :

Using the relation, $E_1 = 4.44 f \phi_{max} N_1$

$$6600 = 4.44 \times 50 \times 0.08 \times N_1$$

$$\therefore \quad N_1 = \frac{6600}{4.44 \times 50 \times 0.08} \simeq 372$$

Hence, $\mathbf{N_1 = 372.}$ **(Ans.)**

Also $$\frac{E_2}{E_1} = \frac{N_2}{N_1}$$

$\therefore$ $$N_2 = \frac{E_2 N_1}{E_1} = \frac{600 \times 372}{6600} \simeq 34$$

Hence, $\mathbf{N_2 = 34.}$ **(Ans.)**

Example 3. *A single-phase transformer is connected to a 230 V, 50 Hz supply. The net cross-sectional area of the core is 60 cm*^{2}*. The number of turns in the primary is 500 and in the secondary 100. Determine :*

(*i*) *Transformation ratio.* (*ii*) *E.m.f. induced in secondary winding.*

(*iii*) *Maximum value of flux density in the core.*

Solution. Primary turns, $N_1 = 500$

Secondary turns, $N_2 = 100$

Primary, $E_1 = V_1 = 230$ V

Core area, $a = 60 \text{ cm}^2 = 60 \times 10^{-4} \text{ m}^2$

(*i*) **Transformation ratio, K :**

$$K = \frac{N_2}{N_1} = \frac{100}{500} = 0.2$$

Hence, $\mathbf{K = 0.2.}$ **(Ans.)**

(*ii*) **Maximum value of flux density, $\mathbf{B_{max}}$:**

Using the e.m.f. equation, $E_1 = 4.44 f \phi_{max} N_1$

$\therefore$ $$230 = 4.44 \times 50 \times \phi_{max} \times 500$$

or $$\phi_{max} = \frac{230}{4.44 \times 50 \times 500} = 0.00207 \text{ Wb}$$

Now, $$B_{max} = \frac{\phi_{max}}{A} = \frac{0.00207}{60 \times 10^{-4}} = 0.345 \text{ T}$$

[where T stands for tesla (Wb/m^2)]

Hence, $\mathbf{B_{max} = 0.345\ T.}$ **(Ans.)**

(*iii*) **E.m.f. induced in the secondary winding, $\mathbf{E_2}$:**

Using the relation, $$\frac{E_2}{E_1} = \frac{N_2}{N_1}$$

$$\frac{E_2}{230} = \frac{100}{500}$$

$\therefore$ $\mathbf{E_2 = 46\ V.}$ **(Ans.)**

Example 4. *3300/300 V single-phase 300 kVA transformer has 1100 primary turns. Find :*

(*i*) *Transformation ratio.* (*ii*) *Secondary turns.*

(*iii*) *Voltage/turn.*

(*iv*) *Secondary current when it supplies a load of 200 kW at 0.8 power factor lagging.*

Solution. Primary, $E_1 = 3300$ V

$N_1 = 1100$

Secondary, $E_2 = 300$ V

Rating of the transformer $= 300$ kVA

Output $= 200$ kW

377

(*i*) **Transformation ratio, K = ?**

$$K = \frac{N_2}{N_1} = \frac{E_2}{E_1} = \frac{300}{3300} = \frac{1}{11}$$

Hence, $\mathbf{K = \frac{1}{11}}$. **(Ans.)**

(*ii*) **Secondary turns, N_2 = ?**

Using the relation, $\frac{E_2}{E_1} = \frac{N_2}{N_1}$

$\therefore$ $N_2 = \frac{E_2 N_1}{E_1} = \frac{300 \times 1100}{3300} = 100$

Hence, $\mathbf{N_2 = 100.}$ **(Ans.)**

(*iii*) **Voltage/turn = ?**

Voltage/turn $\frac{E_1}{N_1} = \frac{3300}{1100} = \mathbf{3\ V.}$ **(Ans.)**

(*iv*) **Secondary current,** $I_2 = \frac{\text{Output}}{V_2 \cos \phi} = \frac{200 \times 1000}{300 \times 0.8} = 833.33$ A

Hence, $\mathbf{I_2 = 833.33\ A.}$ **(Ans.)**

Example 5. *The voltage per turn of a single-phase transformer is 1.1 V. When the primary winding is connected to a 220 V, 50 Hz A.C. supply, the secondary voltage is found to be 550 V. Find :*

(*i*) *Primary and secondary turns.*

(*ii*) *Core area if the maximum flux density is 1.1 T.*

Solution. Voltage per turn = 1.1 V

Primary, $E_1 = 220$ V

Secondary, $E_2 = 550$ V

Max. flux density, $B_{max} = 1.1$ T

(*i*) **Primary turns,** $\mathbf{N_1} = \frac{E_1}{1.1} = \frac{220}{1.1} = \mathbf{200.}$ **(Ans.)**

Secondary turns, $\mathbf{N_2} = \frac{E_2}{1.1} = \frac{550}{1.1} = \mathbf{500.}$ **(Ans.)**

(*ii*) **Core area A :**

Using the relation, $E_1 = 4.44 f \phi_{max} N_1$

$$220 = 4.44 \times 50 \times \phi_{max} \times 200$$

$\therefore$ $\phi_{max} = \frac{220}{4.44 \times 50 \times 200} = 0.004955$ Wb

Core area, $A = \frac{\phi_{max}}{B_{max}} = \frac{0.004955}{1.1} = 0.004504\ \text{m}^2 = \mathbf{45.04\ cm^2}.$ **(Ans.)**

Example 6. *The core of 1000 kVA, 11000/550 V, 50 Hz, single-phase transformer has a cross-section of 20 cm × 20 cm. If the maximum core density is not to exceed 1.3 tesla, calculate :*

(*i*) *The number of h.v. and l.v. turns per phase.*

(*ii*) *The e.m.f. per turns.*

Assume a stacking factor of 0.9.

Solution. *Given* : $A = 20\ \text{cm} \times 20\ \text{cm} = 400 \times 10^{-4}\ \text{m}^2$; $B_{max} = 1.3\ \text{Wb/m}^2$, $f = 50$ Hz ; $E_1 = 11000$ V ; $E_2 = 550$ V ; Stacking factor = 0.9

(*i*) **Number of turns N_1 (h.v.) ; N_2 (l.v.) :**

We know that, Flux = flux density × area of cross-section × stacking factor

$$\therefore \quad \phi_{max} = 1.3 \times 400 \times 10^{-4} \times 0.9 = 0.0468 \text{ Wb}$$

$$E_1 = 4.44 f \phi_{max} N_1$$

or
$$11000 = 4.44 \times 50 \times 0.0468 \times N_1$$

$$\therefore \quad N_1 = \frac{11000}{4.44 \times 50 \times 0.0468} = 1058.75 \simeq \mathbf{1059.} \quad \textbf{(Ans.)}$$

and
$$N_2 = \frac{E_2}{E_1} \times N_1 = \frac{550}{11000} \times 1059 = 52.95 \simeq \mathbf{53.} \quad \textbf{(Ans.)}$$

(*ii*) **The e.m.f. per turn :**

$$\text{E.M.F. per turn} = \frac{E_1}{N_1} = \frac{11000}{1059} = \mathbf{10.387\ V.} \quad \textbf{(Ans.)}$$

9.4. Transformer with Losses but no Magnetic Leakage. We shall consider the following two cases :

1. When transformer is on no-load.
2. When transformer is on load.

9.4.1. Transformer on no-load. A transformer is said to be on *no-load* if its secondary side is open and primary is connected to a sinusoidal alternating voltage V_1. The alternating applied voltage will cause *flow of alternating current in the primary winding which will create alternating flux*. This primary input current (I_0) under no-load conditions supply :

(*i*) *Iron losses in the core* (*i.e.* hysteresis loss and eddy current loss).

(*ii*) A very *small amount of copper losses in primary* (there being no copper loss in the secondary as it is open).

Thus I_0 is not at 90° behind V_1, but lags it by an angle $\phi_0 < 90°$.

No-load power input, $\quad P_0 = V_1 I_0 \cos \phi_0$ ...(5)

where $\cos \phi_0$ = primary power factor under no-load conditions.

As is evident from Fig. 19, primary current I_0 has the following two components :

(*i*) *Active* or *working or iron loss component* I_w. This component is in *phase with* V_1 and mainly supplies the iron loss plus small quantity of primary copper loss.

$$I_w = I_0 \cos \phi_0 \quad ...(i)$$

(*ii*) *Magnetising component* I_m. This component is *in quadrature with* V_1 and its function is to *sustain the alternating flux in the core*. It is *wattless*.

$$I_m = I_0 \sin \phi_0 \quad ...(ii)$$

Also
$$I_0 = \sqrt{I_w^2 + I_m^2} \quad ...(iii)$$

The following points are *worthnoting* :

- The no-load primary current I_w is *very small as compared to the full-load primary current.*
- As I_0 is very small, the no-load primary copper loss is negligibly small which means that *no-load primary input is practically equal to the iron loss in the transformer.*

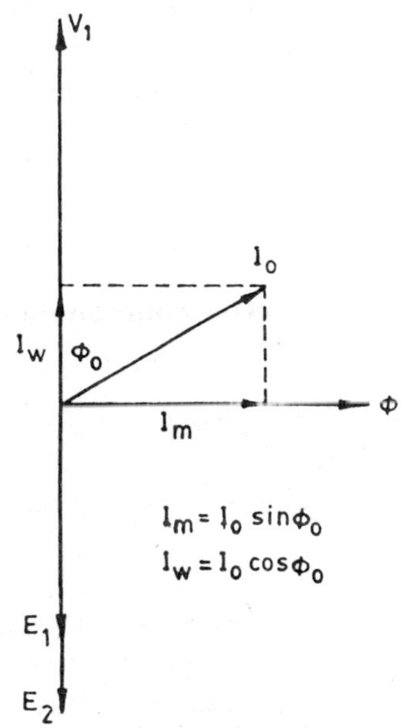

Fig. 19. No-load vector diagram.

- Since, it is primarily the core loss which is responsible for shift in the current vector, angle ϕ_0 is known as **hysteresis angle of advance.**

☞**Example 7.** *A 3300/300 V single-phase transformer gives 0.6 A and 60 W as ammeter and wattmeter readings when supply is given to the low voltage winding and high voltage winding is kept open, find :*

(i) Power factor of no-load current. *(ii) Magnetising component.*
(iii) Iron loss component.

Solution. No-load losses, $P_0 = 60$ W

No-load current, $I_0 = 0.6$ A

(i) $P_0 = VI_0 \cos \phi_0$

$\therefore$ $60 = 300 \times 0.6 \times \cos \phi_0$

or $\cos \phi_0 = \dfrac{60}{300 \times 0.6} = 0.33$ (lagging)

Hence, **power factor of no-load current = 0.33. (Ans.)**

(ii) Magnetising component, $I_m = I_0 \sin \phi_0 = 0.6 \sqrt{1 - \cos^2 \phi_0} = 0.6 \sqrt{1 - (0.33)^2} = 0.566$ A

$\therefore$ $I_m =$ **0.566 A. (Ans.)**

(iii) Iron loss component, $I_w = I_0 \cos \phi_0 = 0.6 \times 0.33 =$ **0.198 A. (Ans.)**

[or $I_w = \sqrt{I_0^2 - I_m^2} = \sqrt{(0.6)^2 - (0.566)^2} = 0.198$ A].

☞**Example 8.** *Find (i) active and reactive components of no-load current ; and (ii) no-load current of a 440/220 V single-phase transformer if the power input on no- load to the high voltage winding is 80 W and power factor of no- load current is 0.3 lagging.*

Solution. Primary, $E_1 = 440$ V

Secondary, $E_2 = 220$ V

Power factor, $\cos \phi_0 = 0.3$ (lagging)

No-load losses, $P_0 = 80$ W

(i) **Active component** (or *wattful component*),

$$I_w = (I_0 \cos \phi_0) = \frac{P_0}{V_1} = \frac{80}{440} = \mathbf{0.182\ A.\ (Ans.)}$$

$$\cos \phi_0 = 0.3 \ ; \ \phi_0 = \cos^{-1}(0.3) = 72.54°$$

$\therefore$ $\tan \phi_0 = 3.18$

Reactive component (or *magnetising components*)

$$I_m = I_w \tan \phi_0 = 0.182 \times 3.18 = \mathbf{0.578\ A.\ (Ans.)}$$

(ii) $I_w = I_0 \cos \phi_0$

$\therefore$ $I_0 = \dfrac{I_w}{\cos \phi_0} = \dfrac{0.182}{0.3} =$ **0.606. (Ans.)**

or $[I_0 = \sqrt{I_w^2 + I_m^2} = \sqrt{(0.182)^2 + (0.578)^2} = 0.606$ A].

Example 9. *A 3300/220 V, 30 kVA, single-phase transformer takes a no-load current of 1.5 A when the low voltage winding is kept open. The iron loss component is equal to 0.4 A find :*

(i) No-load input power.

(ii) Magnetising component and power factor of no-load current.

Solution. Rating of transformer $= 30$ kVA

Primary, $E_1 = 3300$ V

Secondary, $E_2 = 220$ V

No-load current, $I_0 = 1.5$ A

Iron loss component, $I_m = 0.4$ A

(*i*) **No-load input power,**

$$P_0 = V_1 I_0 \cos \phi_0 = V_1 I_w \qquad (\because\ I_w = I_0 \cos \phi_0)$$

$$= 3300 \times 0.4 = \mathbf{1320\ W.}\ \ \textbf{(Ans.)}$$

(*ii*) **Magnetising component,**

$$I_m = \sqrt{I_0^2 - I_w^2} = \sqrt{1.5^2 - 0.4^2} = \mathbf{1.44\ A.}\ \ \textbf{(Ans.)}$$

No-load power factor,

$$\cos \phi_0 = \frac{I_w}{I_0} = \frac{0.4}{1.5} = \mathbf{0.267.}\ \ \textbf{(Ans.)}$$

9.4.2. Transformer on load. *The transformer is said to be loaded when the secondary circuit of a transformer is completed through an impedance or load.* The magnitude and phase of secondary current I_2 with respect to secondary terminal voltage will depend upon the characteristic of load, *i.e.* current I_2 will be in phase, lag behind and lead the terminal voltage V_2 respectively when the load is purely resistive, inductive and capacitive.

The secondary current I_2 sets up its own ampere-turns ($= N_2 I_2$) and creates its own flux ϕ_2 opposing the main flux ϕ_0 created by no-load current I_0. The opposing secondary flux ϕ_2 weakens the primary flux ϕ_0 momentarily hence primary counter or back e.m.f. E_1 tends to be reduced. V_1 gains the upper hand over E_1 momentarily and hence causes more current to flow in primary. Let this additional primary current be I_2'. It is known as *load component of primary current*. The additional primary m.m.f. $N_1 I_2'$ sets up its own flux ϕ_2' which is in opposition to ϕ_2 (but is in the same direction as ϕ_0) and is equal to it in magnitude. Hence they *cancel* each other. Thus we find that the magnetic effects of secondary current I_2 are immediately neutralised by the additional primary current I_2' which is brought into existence exactly at the same instant as I_2.

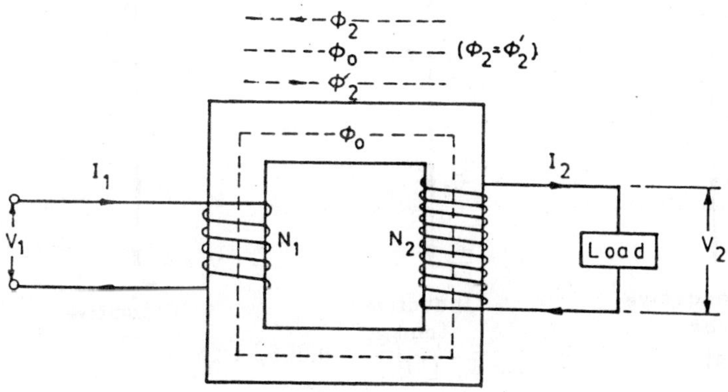

Fig. 20. An ideal transformer on load.

From above discussion it can be concluded that :

- Whatever be the load conditions, the *net flux passing through the core is approximately the same as at no-load.*
- Since the core flux remains constant at all loads, the core *loss almost remains constant under different loading conditions.*

Since $\qquad \phi_2 = \phi_2'$

$\therefore \qquad N_2I_2 = N_1I_2'$

i.e., $\qquad I_2' = \dfrac{N_2}{N_1} \times I_2 = KI_2 \qquad \left(\because \dfrac{N_2}{N_1} = K\right)$

The total primary current is the vector sum of I_0 and I_2' ; the current I_2' is in *antiphase* with I_2 and K times in magnitude.

The components of primary current can be shown as below :

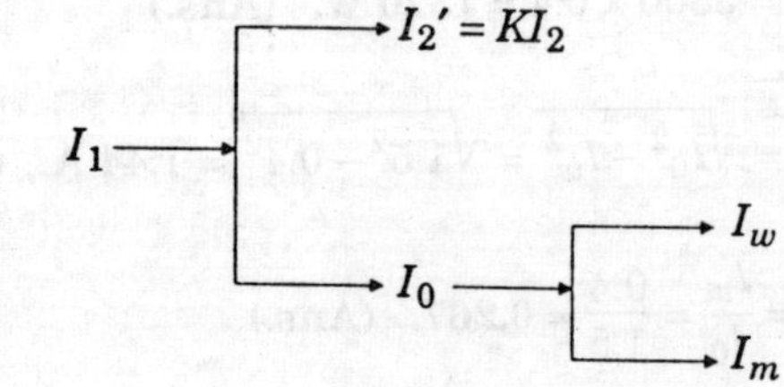

The vector diagrams for transformer on non-inductive, inductive and capacitive loads are shown in Fig. 21 (*a, b, c*).

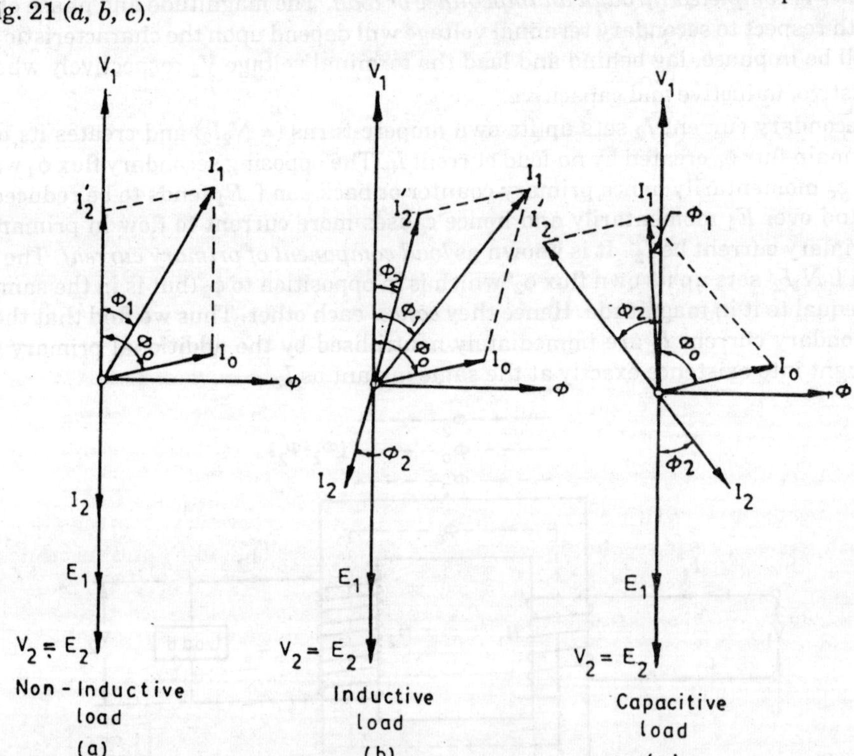

Fig. 21

9.5. Resistance and Magnetic Leakage. In an ideal transformer it is assumed that windings have got no resistance and there is no leakage flux. But in actual practice it is not possible to have an ideal transformer.

Resistance. In an actual transformer the primary as well secondary winding possess some resistance due to which some voltage drop takes place in them.

Magnetic leakage. Magnetic flux cannot be confined into a desired path. The greater portion of the flux (*i.e.*, the mutual flux) remains confined to the core and links both the windings but a small portion, called the *leakage flux,* completes its path through the air surrounding the coils. As shown in Fig. 22 each of the winding is associated with a leakage flux. Since the path of the leakage flux is

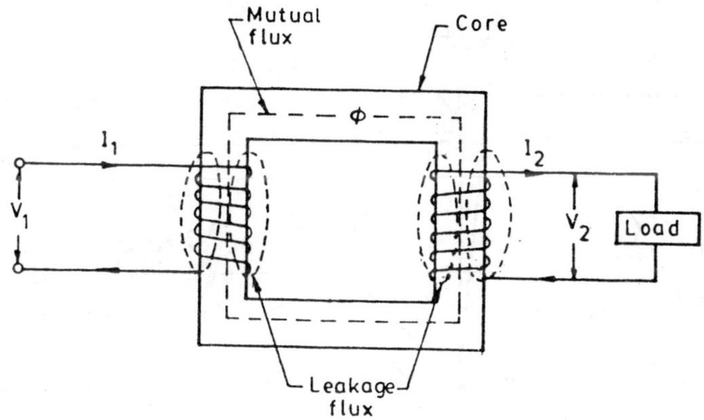

Fig. 22. Magnetic flux in a transformer.

largely in air the leakage flux and the voltage induced by it vary *linearly with the current. The primary leakage flux varies linearly with the primary current and the secondary leakage flux varies linearly with the secondary current.* The effect of primary leakage may be simulated by assigning to the primary a leakage inductance (equal to primary leakage flux per ampere of primary current). The reactance corresponding to this primary leakage inductance (*i.e.* $2\pi f \times$ primary leakage inductance) is known as primary leakage reactance X_1. Similarly the effect of secondary leakage flux may be simulated by a secondary leakage inductance and the corresponding leakage reactance X_2.

The terminal voltage V_1 applied to the primary must have a component $jI_1 X_1$ to balance the primary leakage e.m.f. Similarly V_2 *i.e.* the output voltage from secondary will be less than the induced e.m.f. E_2 by a component $jI_2 X_2$ to account for the secondary leakage flux.

A transformer with winding resistance and magnetic leakage is equivalent to an ideal transformer (having no resistance and leakage resistance) having resistive and inductive coils connected in series with each winding as shown in Fig. 23.

9.6. Transformer with Resistance and Leakage Reactance. Let us consider transformer (Fig. 23) having primary and secondary windings of resistances R_1 and R_2 and reactances X_1 and X_2 respectively.

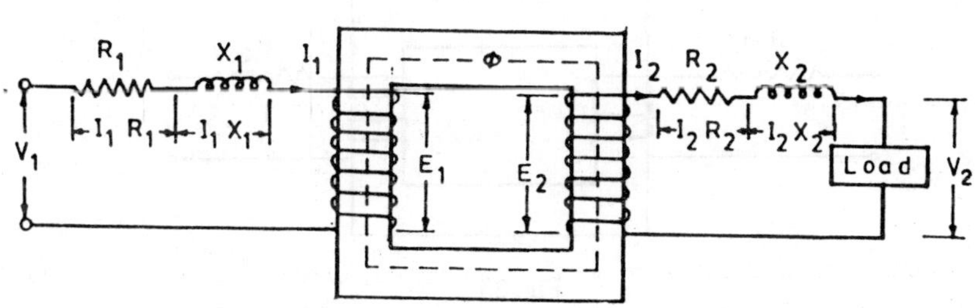

Fig. 23. An equivalent diagram of an actual transformer.

Impedance of primary winding, $Z_1 = R_1 + jX_1$

Impedance of secondary winding, $Z_2 = R_2 + jX_2$

The *vector diagrams* for different kinds of loads for such a transformer are shown in Fig. 24 (*a, b, c*). *In these diagrams, vectors for resistive drops are drawn parallel to current vectors* whereas *reactive drops are drawn perpendicular to the current vectors*. The angle ϕ_1 between V_1 and I_1 gives the power factor angle of the transformer.

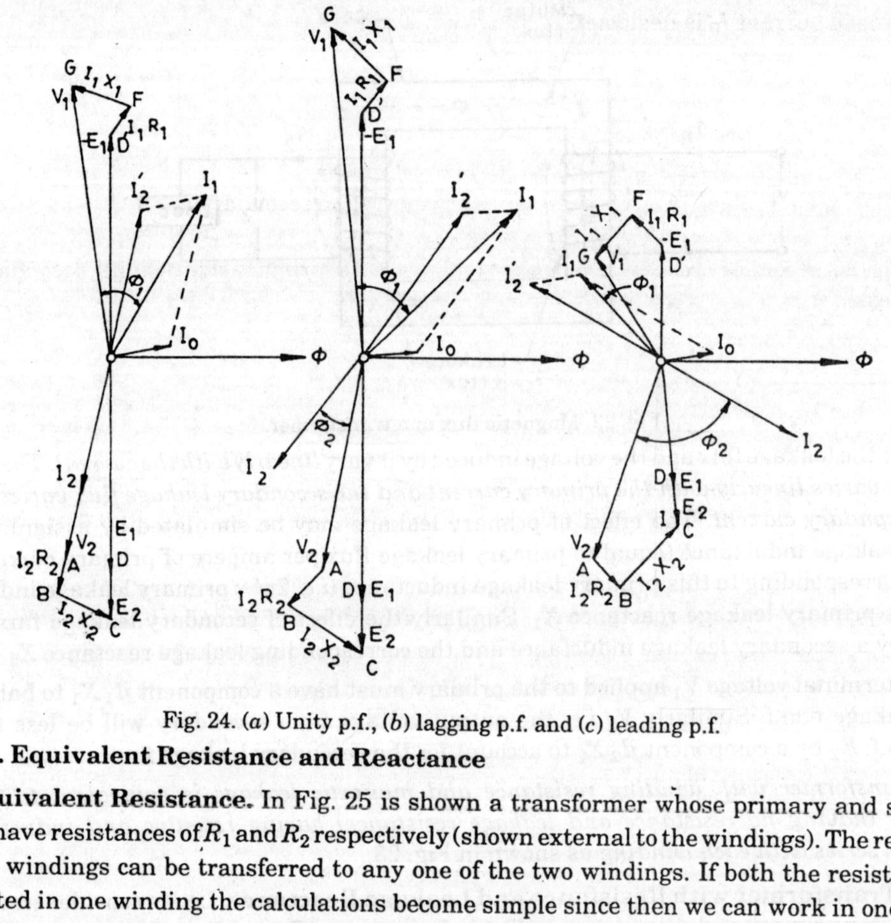

Fig. 24. (*a*) Unity p.f., (*b*) lagging p.f. and (*c*) leading p.f.

9.7. Equivalent Resistance and Reactance

Equivalent Resistance. In Fig. 25 is shown a transformer whose primary and secondary windings have resistances of R_1 and R_2 respectively (shown external to the windings). The resistances of the two windings can be transferred to any one of the two windings. If both the resistances are concentrated in one winding the calculations become simple since then we can work in one winding only.

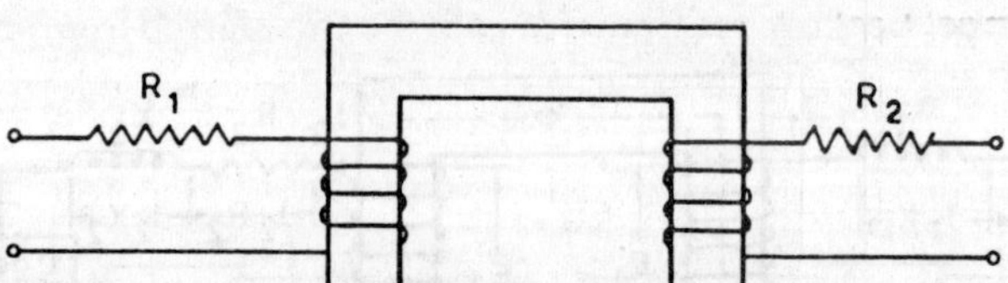

Fig. 25

We shall now prove that a resistance of R_2 in secondary is equivalent to $\dfrac{R_2}{K^2}$ in primary and its value will be denoted by R_2', — *the equivalent secondary resistance referred to primary.*

In the secondary, copper loss is $I_2^2R_2$. This loss is supplied by primary which takes a current of I_1. Hence, if R_2' is the equivalent resistance referred to primary which would have caused the same loss as R_2 in secondary, then

$$I_1^2R_2' = I_2^2R_2 \quad \text{or} \quad R_2' = \left(\frac{I_2}{I_1}\right)^2 R_2$$

If no-load current I_0 is neglected,

then, $$\frac{I_2}{I_1} = \frac{1}{K}$$

Hence, $$R_2' = \frac{R_2}{K^2}$$

Similarly, equivalent primary resistance as referred to secondary is

$$R_1' = K^2R_1$$

In Fig. 26, secondary resistance has been transferred to primary side (leaving secondary circuit resistanceless).

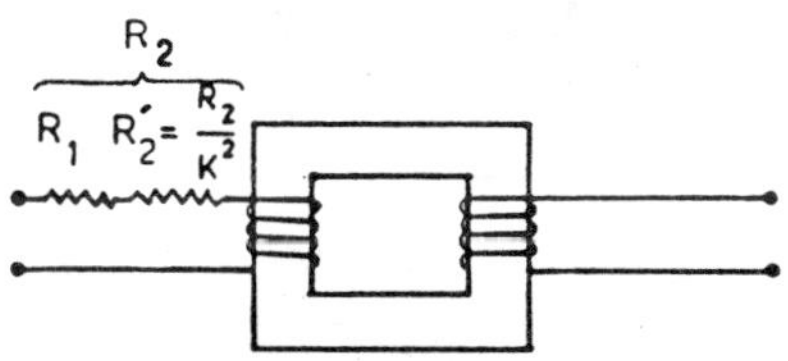

Fig. 26

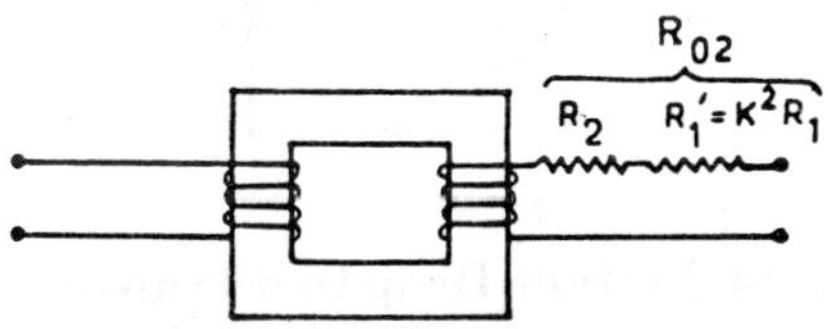

Fig. 27

The resistance $R_1 + R_2' = R_1 + \frac{R_2}{K^2}$ is known as the *equivalent or effective resistance of the transformer as referred to the primary* and may be designated as R_{01}.

Thus $$R_{01} = R_1 + R_2' = R_1 + \frac{R_2}{K^2}. \qquad ...(6)$$

Similarly, the equivalent resistance of the transformer as referred to the secondary (Fig. 27) is

$$R_{02} = R_2 + R_1' = R_2 + K^2R_1 \qquad ...(7)$$

The following points are worth remembering :

(*i*) When shifting resistance *to the secondary, multiply* it by K^2.

(*ii*) When shifting resistance *to the primary, divide* by K^2.

Leakage reactance. Leakage reactance can also be transferred from one winding to the other in the same way as resistance (Fig. 28 and Fig. 29).

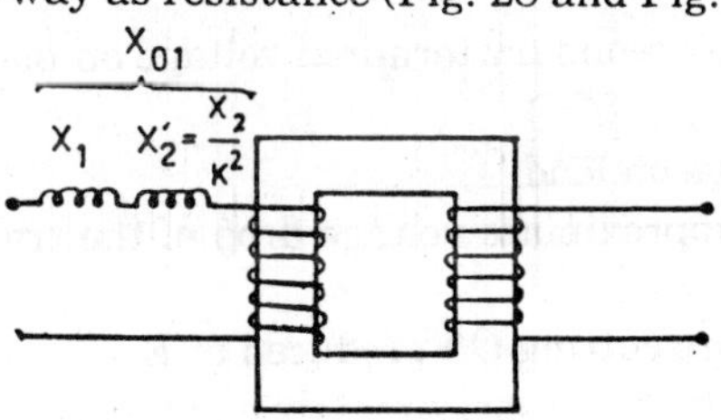

Fig. 28

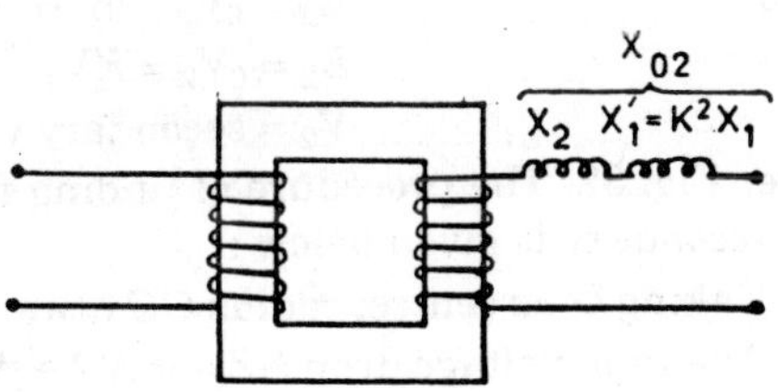

Fig. 29

$$X_2' = \frac{X_2}{K^2} \quad \text{and} \quad X_1' = K^2 X_1$$

and

$$X_{01} = X_1 + X_2' = X_1 + \frac{X_2}{K^2} \qquad \text{...(8)}$$

$$X_{02} = X_2 + X_1' = X_2 + K^2 X_1 \qquad \text{...(9)}$$

Total impedance. It is obvious that total impedance of the transformer as referred to primary is given by

$$Z_{01} = \sqrt{R_{01}^2 + X_{01}^2} \qquad \text{(Fig. 30) ...(10)}$$

and

$$Z_{02} = \sqrt{R_0^2 + X_{02}^2} \qquad \text{(Fig. 31) ...(11)}$$

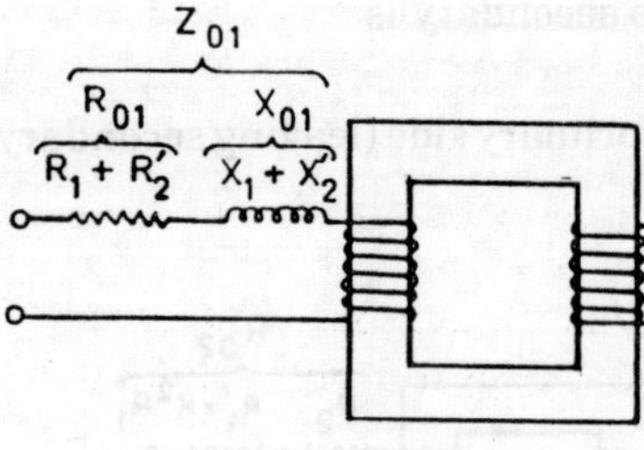

Fig. 30

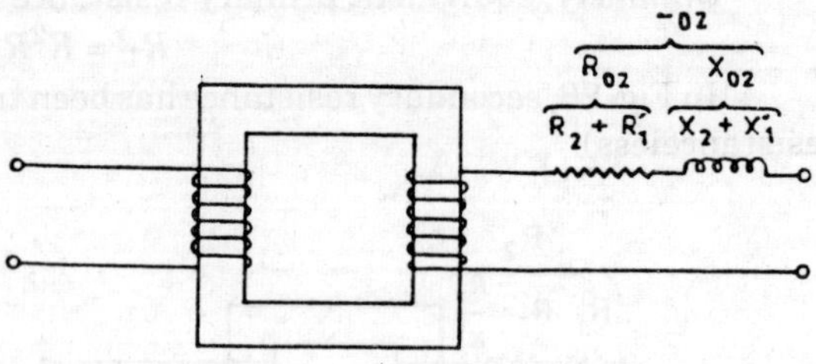

Fig. 31

9.8. Total Voltage Drop in a Transformer

(i) **Approximate voltage drop.** When there is no-load on the transformer, then,

$$V_1 = E_1 \text{ (approximately).}$$

and

$$E_2 = KE_1 = KV_1$$

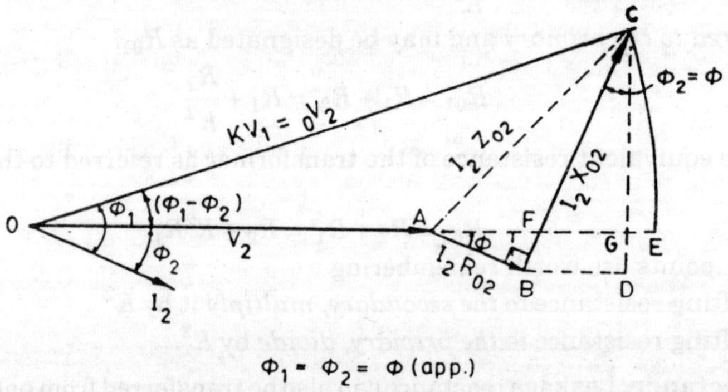

Fig. 32. Lagging power factor.

Also $E_2 = {}_0V_2$, where ${}_0V_2$ is secondary terminal voltage on no-load

or $E_2 = {}_0V_2 = KV_1$

V_2 = secondary voltage on load.

Refer Fig. 32. The procedure of finding the approximate voltage drop of the transformer as *referred to secondary* is given below :

- Taking O as centre, radius OC draw an arc cutting OA produced at E. The total voltage drop $I_1 Z_{02} = AC = AE$ which is approximately equal to AG.
- From B draw BF perpendicular on OA produced. Draw CG perpendicular to OE and draw BD parallel to OE.

Approximate voltage drop $= AG = AF + FG = AF + BD$ $[\because\ FG = BD]$

$= I_2R_{02} \cos \phi + I_0X_{02} \sin \phi$

This is the value of approximate voltage drop for a *lagging power factor*.

Figs. 33 and 34 refer to *unity* and *leading power factor* respectively.

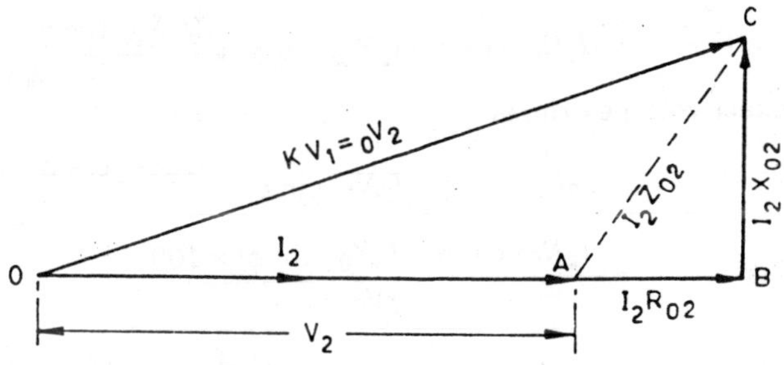

Fig. 33. Unity power factor.

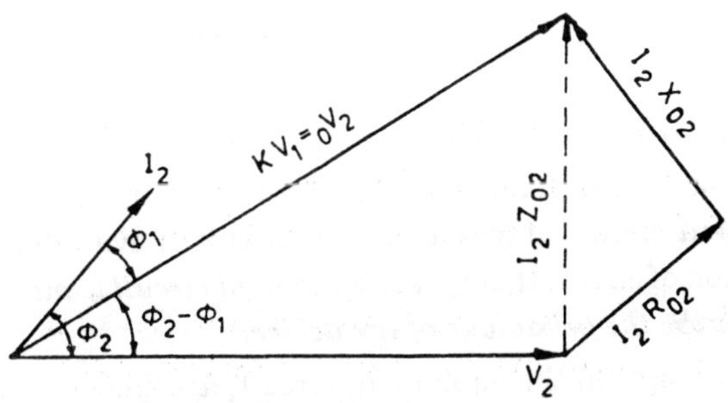

Fig. 34. Leading power factor.

- The approximate voltage drop for a *leading power factor* becomes :

$$(I_2R_{02} \cos \phi - I_2X_{02} \sin \phi)$$

- The approximate voltage drop for a transformer *in general* is given by :

$$(I_2R_{02} \cos \phi \pm I_1X_{02} \sin \phi) \quad ...(12)$$

- The voltage drop as *referred to primary* is given by :

$$(I_1R_{01} \cos \phi \pm I_1X_{02} \sin \phi)$$

- *Percentage* voltage drop in *secondary*

$$= \left(\frac{I_2R_{02} \cos \phi \pm I_2X_{02} \sin \phi}{{}_0V_2} \right) \times 100 \quad ...(13)$$

(*ii*) **Exact voltage drop.** When we refer to Fig. 32 we find that exact drop is AE and not AG. If quantity GE is added to AG the exact value of voltage drop will be obtained.

From right angled triangle OCG, we get

$$CG^2 = OC^2 - OG^2 = (OC + OG)(OC - OG)$$

$$= (OC + OG)(OE - OG) = GE \times 2OC$$

$$\therefore \quad GE = \frac{CG^2}{2OC}$$

Now $\quad CG = I_2X_{02} \cos \phi - I_2R_{02} \sin \phi$

$$\therefore \quad GE = \frac{(I_2X_{02} \cos \phi - I_2R_{02} \sin \phi)^2}{2\,{}_0V_2}$$

∴ The exact voltage drop for a *lagging* power factor is

$$= AG + GE$$

$$= (I_2R_{02} \cos \phi + I_2X_{02} \sin \phi) + \frac{(I_2X_{02} \cos \phi - I_2R_{02} \sin \phi)^2}{2_0V_2} \quad ...(14)$$

For a *leading* power, the expression becomes

$$= (I_2R_{02} \cos \phi - I_2X_{02} \sin \phi) + \frac{(I_2X_{02} \cos \phi + I_2R_{02} \sin \phi)^2}{2_0V_2} \quad ...(15)$$

In *general*, the exact voltage drop is

$$= (I_2R_{02} \cos \phi \pm I_2X_{02} \sin \phi) + \frac{(I_2X_{02} \cos \phi \mp I_2R_{02} \sin \phi)^2}{2_0V_2}$$

Percentage drop is

$$= \frac{(I_2R_{02} \cos \phi \pm I_2X_{02} \sin \phi) \times 100}{_0V_2} + \frac{(I_2X_{02} \cos \phi \mp I_2R_{02} \sin \phi)^2 \times 100}{2_0V_2{}^2} \quad ...(16)$$

[Upper signs ... for *lagging* power factor
Lower signs ... for *leading* power factor]

9.9. Equivalent Circuit. In transformers, the problems concerning voltages and currents can be solved by the use of phasor diagrams. However, it is more convenient to represent the transformer by an equivalent circuit. If an equivalent circuit is available the computations can be done by the direct application of circuit theory. An *equivalent circuit is merely a circuit interpretation of the equations which describe the behaviour of the device.*

The transformer windings, in the equivalent circuit, are shown as ideal. The resistance and leakage reactance of the primary and secondary are shown separately in the primary and secondary circuits. The effect of magnetising current is represented by X_0 connected in parallel across the winding. The effect of core loss is represented by a non-inductive resistance R_0 as shown in Fig. 35.

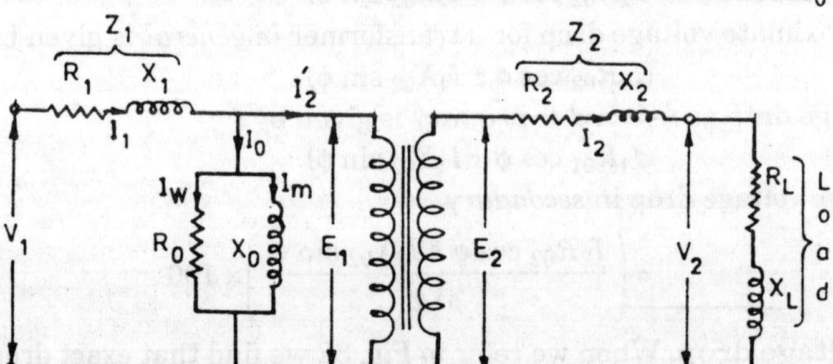

Fig. 35. Equivalent circuit of a transformer.

The equivalent circuit can be simplified by transferring the secondary resistances and reactances to the primary side in such a way that the ratio of E_2 to E_1 is not affected in magnitude or phase. Let R_2' be the resistance which must be placed in the primary circuit to produce the same drop as produced by R_2 in the secondary. Then R_2' causes a voltage drop in primary equal to $I_2'R_2'$.

The ratio of $I_2'R_2'$ and I_2R_2 must be the same as the turn ratio $\frac{N_1}{N_2}$.

Thus $$\frac{I_2' R_2'}{I_2 R_2} = \frac{N_1}{N_2} = \frac{1}{K} \quad \text{or} \quad R_2' = R_2 \times \frac{I_2}{I_2'} \times \frac{1}{K}$$

But $$\frac{I_2}{I_2'} = \frac{N_1}{N_2} = \frac{1}{K}$$

$\therefore$ $$R_2' = \frac{R_2}{K^2}$$

Similarly $$X_2' = \frac{X_2}{K^2}$$

The load resistance and reactance can be transferred to the primary side in the same way. When all the secondary impedances have been transferred to the primary side, the winding need not be shown in the equivalent circuit. The exact equivalent circuit of the transformer with all impedances transferred to the primary side is shown in Fig. 36.

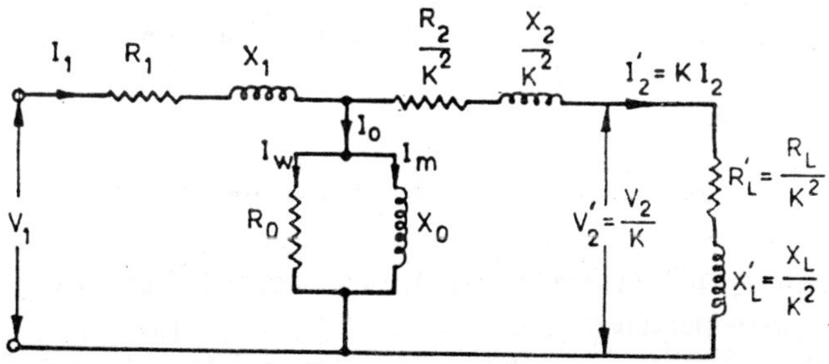

Fig. 36. Equivalent circuit with secondary impedances transferred to primary.

Just as the secondary impedances have been transferred to the primary it is possible to transfer the primary winding resistance and reactance to the secondary side. The primary resistance R_1 and leakage reactance X_1 when transferred to secondary are denoted by R_1' and X_1' and are given by

$$R_1' = K^2 R_1$$
$$X_1' = K^2 X_1.$$

Approximate equivalent circuit. It is seen that E_1 differs from V_1 by a very small amount. Moreover, the no-load current I_0 is only a small fraction of full-load primary current so that I_2' is practically equal to I_1. Consequently the equivalent circuit can be simplified by transferring the parallel branch consisting of R_0 and X_0 to the extreme left position of the circuit as shown in Fig. 37. This circuit is known as *approximate equivalent circuit*. Analysis with the approximate equivalent circuit gives almost the same results as the analysis with the exact equivalent circuit. *However, the analysis*

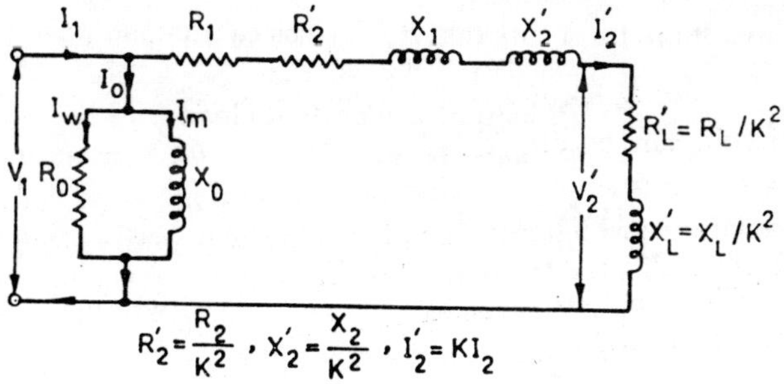

Fig. 37. Approximate equivalent circuit of transformer.

with the approximate equivalent circuit is simple because the resistances R_1 and R_2' and leakage reactances X_1 and X_2' can be combined.

9.10. Transformer Tests. The performance of a transformer can be calculated on the basis of its equivalent circuit which contains the following four main parameters :

(*i*) Equivalent resistance R_{01} as referred to primary (or secondary R_{02}).

(*ii*) Equivalent leakage resistance X_{01} as referred to primary (or secondary X_{02}).

(*iii*) Core loss conductance G_0 (or resistance R_0).

(*iv*) Magnetising susceptance B_0 (or reactance X_0).

These parameters or constants can be determined by the following *two tests* :

1. Open-circuit or no-load test.
2. Short-circuit or impedance test.

The above two tests are *convenient to perform and very economical because they furnish the required information without actually loading the transformer.*

9.10.1. Open-circuit or no-load test (O.C. Test). An open-circuit or no-load test is conducted to find :

(*i*) No-load loss or core loss.

(*ii*) No-load current I_0 which is helpful in finding R_0 and X_0.

The connections for this test are made as shown in Fig. 38. One winding of the transformer (*usually high voltage winding*) is *left open* and the other is connected to its supply of *normal voltage and frequency*. Ammeter *A* and wattmeter *W* are connected to measure *no-load current* (I_0) *and no-load input power* (P_0) respectively.

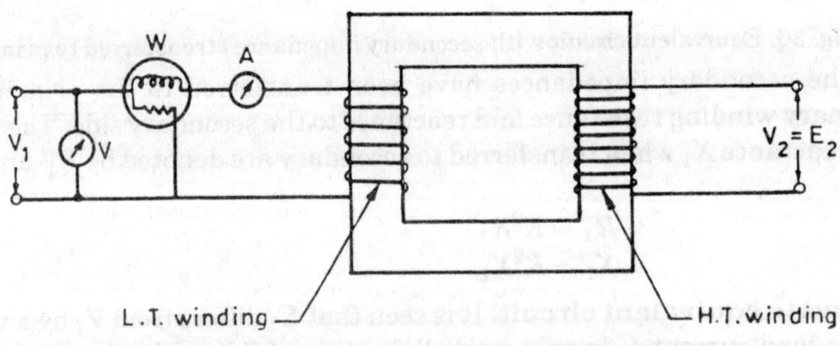

Fig. 38. Circuit diagram for open-circuit test.

As the primary no-load current I_0 (as measured by ammeter) is small (usually 3 to 10% of rated load current) copper loss is negligibly small in primary (L.T. winding) and nil in secondary winding (it being open). Hence the *wattmeter reading represents practically the core-loss under no-load conditions* (and *this loss is same for all loads*).

From the data available from this test R_0, X_0, $\cos \phi_0$ (no-load power factor), I_w and I_m can be calculated as follows :

Now, Iron loss $= P_i =$ input power on no-load $= P_0$ watts (say)

No-load current $= I_0$

Applied primary voltage $= V_1$

Also $P_0 = V_1 I_0 \cos \phi_0$ (where $\cos \phi_0$ = no-load power factor)

$$\cos \phi_0 = \frac{P_0}{V_1 I_0} \quad \text{...(17)}$$

or $$\phi_0 = \cos^{-1} \frac{P_0}{V_1 I_0}$$

No-load current *wattful component,*

$$I_w = I_0 \cos \phi_0 = \frac{P_0}{V_1} \quad ...(18)$$

No-load current *magnetising component,*

$$I_m = \sqrt{I_0^2 - I_w^2} \quad ...(19)$$

The no-load resistance, $R_0 = \frac{V_1}{I_w} = \frac{V_1^2}{P_0}$ ...(20)

The no-load reactance, $X_0 = \frac{V_1}{I_m} = \frac{V_1}{\sqrt{I_0^2 - I_w^2}}$ ...(21)

The no-load vector diagram is shown in Fig. 19.

Since the current is practically all-exciting current when a transformer is on no-load (*i.e.* $I_0 = I_m$) and the voltage drop in primary leakage impedance is small, hence the exciting admittance Y_0 of the transformer is given by

$$I_0 = V_1 Y_0 \quad \text{or} \quad Y_0 = \frac{V_1}{I_0} \quad ...(22)$$

The exciting conductance $G_0 = \frac{P_0}{V_1^2}$ ...(23)

The exciting susceptance $B_0 = \sqrt{Y_0^2 - G_0^2}$ ...(24)

Separation of core losses. The core loss is made up of the following two parts :

(*i*) Eddy current loss.

(*ii*) Hysteresis loss.

Eddy current loss, $P_e = AB^2_{max} f^2$, where A is constant.

Hysteresis loss, $P_h = BB^{1.6}_{max} f$, where B is constant.

Total loss $= P_e + P_h = AB^2_{max} f^2 + BB^{1.6}_{max} f$.

The values of constants A and B can be found out by conducting two experiments using two different frequencies but the same maximum flux density ; thereafter eddy current and hysteresis loss can be found separately.

9.10.2. Short-circuit or impedance test (S.C. Test). This test is conducted to determine the following :

(*i*) Full-load copper loss.

(*ii*) Equivalent resistance and reactance *referred to metering side.*

In this test (Fig. 39) the terminals of the secondary winding (*usually low voltage winding*) are short-circuited by a thick conductor or through an ammeter which may serve the additional purpose of indicating rated load current. A low voltage, usually 5 to 10% of normal primary voltage, at correct frequency is applied to the primary and *is continuously increased till full-load currents flow in the primary as well as secondary windings* (as indicated by the respective ammeters).

Since applied voltage is very low so flux linking with the core is very small and therefore, iron losses are so small that these can be neglected, the reading of the wattmeter gives total copper losses at full-load.

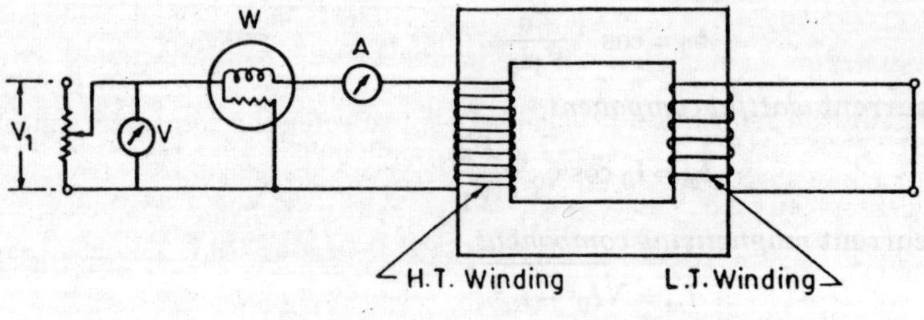

Fig. 39. Short-circuit test.

The equivalent circuit of the transformer under short-circuit condition is shown in Fig. 40.

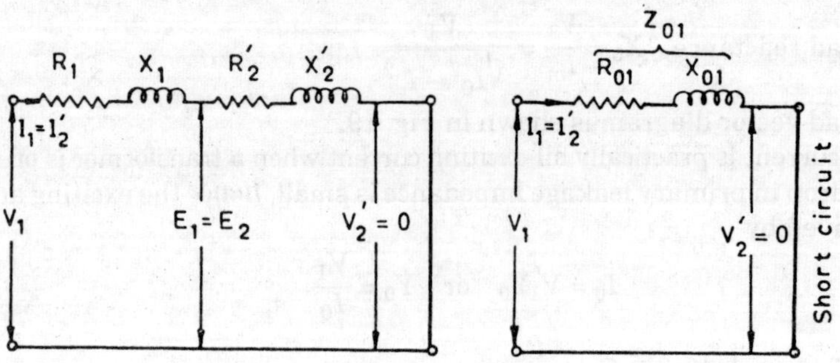

Fig. 40. Equivalent circuit of transformer under short-circuit condition.

Let V_{SC} = voltage required to circulate rated load currents

I_1 = reading of the ammeter on the primary side

Z_{01} = total impedance as referred to primary side

R_{01} = total resistance as referred to primary side

X_{01} = total reactance as referred to primary side.

Then, equivalent impedance as referred to primary side,

$$Z_{01} = \frac{V_{SC}}{I_1} \quad \text{...(25)}$$

Also $P = I_1^2 R_{01}$

$$\therefore \quad R_{01} = \frac{P}{I_1^2} \quad \text{...(26)}$$

and $$X_{01} = \sqrt{Z_{01}^2 - R_{01}^2} \quad \text{...(27)}$$

9.11. Regulation of a Transformer

- Due to the resistances of the windings and leakage reactances voltage drop takes place in a transformer. Accordingly the output voltage under load conditions is different from the output voltage under no-load conditions. **Voltage regulation** is defined as :

"The change in secondary voltage when rated load at a specified power is removed".

It is specified as a percentage of the rated secondary voltage.

Thus, if $_0V_2$ = secondary terminal voltage at *no-load*

$= E_2 = KE_1 = KV_1$ because at no-load the impedance drop is negligible

V_2 = secondary terminal voltage on *full-load*

Then, % regulation (down) $$= \frac{_0V_2 - V_2}{_0V_2} \times 100 \quad \text{...(28)}$$

% regulation (up) $= \dfrac{{}_0V_2 - V_2}{V_2} \times 100$ ...(29)

For calculating the regulation it is convenient to refer the total resistance and reactance to the secondary side.

- Refer Fig. 32.

% regulation $= \left(\dfrac{I_2R_{02} \cos \phi \pm I_2X_{02} \sin \phi}{{}_0V_2} \right) \times 100$ (app.) ...(30)

$\begin{bmatrix} \text{+ sign for } \textit{lagging} \text{ power factor} \\ \text{– sign for } \textit{leading} \text{ power factor} \end{bmatrix}$

The less this value, the better the transformer, because a good transformer should keep its secondary terminal voltage as constant as possible under all conditions of load.

The regulation may also be expressed in terms of primary values as follows :

% regulation $= \dfrac{V_1 - V_2'}{V_1} \times 100$ (neglecting angle between V_1 and V_2')

$$= \left(\frac{I_1R_{01} \cos \phi \pm I_1X_0 \sin \phi}{V_1} \right) \times 100 \qquad \text{...(31)}$$

- In the above definitions of regulation, *primary voltage was supposed to be kept constant* and changes in secondary terminal were considered. But as the transformer is loaded, the secondary terminal voltage falls (for lagging power factor) ; hence to keep the circuit voltage constant, the primary voltage must be increased. The rise *in primary voltage required to maintain rated output voltage from no-load to full-load at a given power factor expressed as percentage of rated primary voltage gives the regulation of the transformer.*

If the primary voltage has to be raised from its rated value V_1 to V_1', then

% regulation $= \dfrac{V_1' - V}{V_1} \times 100$ ...[31 (*a*)]

9.12. Percentage Resistance and Reactance

- Invariably the equivalent resistance and reactance of a transformer are expressed in percent.

Percentage resistance is the *resistance drop in volts at rated current and frequency as a percentage of the rated voltage i.e.,*

$$\text{Percent } R = \frac{IR}{V} \times 100 \qquad \text{...}(i)$$

Percentage reactance is the *reactance drop in volts at rated current and frequency expressed as a percentage of the rated voltage i.e.,*

$$\text{Percent } X = \frac{IX}{V} \times 100 \qquad \text{...}(ii)$$

and

$$\text{Percent } Z = \frac{IZ}{V} \times 100 \qquad \text{...}(iii)$$

The important **advantage** of expressing resistance and reactance of a transformer in percentage is that the *percentage resistance and percentage reactance have same values whether determined referred to primary or secondary whereas when expressed in ohms they have different values when referred to primary and secondary.*

Further, *percent reactances and impedances are convenient for computing the current that will flow in a transformer when the secondary is accidentally shorted.* When the secondary is shorted,

$I_1 = \frac{E_1}{Z_0}$ or $E_1 = I_1 Z_0$ *i.e.*, when secondary is shorted, $I_1 Z_0$ is 100 percent of E_1. Therefore, the short-circuit current is given by the equation,

$$\text{Short circuit current}\ ;\ I_{SC} = \frac{100}{\text{Percent impedance}} \times \text{full load rated current} \quad \text{...(iv)}$$

Unless the transformer is quite small, equivalent resistance is negligible and percent reactance may be substituted for percent impedance in the above equation (*iv*).

- **Another method of designating the resistances and reactances** is by the **per unit values.** The per unit values are equal to the percentage values divided by 100.

$$\therefore \quad \text{Per unit } R = \frac{IR}{V} \quad \text{...(v)}$$

and

$$\text{Per unit } X = \frac{IX}{V} \quad \text{...(vi)}$$

9.13. Transformer Losses. The losses in a transformer are classified as follows :

1. Iron losses (or core losses).
2. Copper losses.

1. **Iron or core losses.** It includes *hysteresis loss* and *eddy current loss.*

(*i*) **Hysteresis loss.** Since the flux in a transformer core is alternating, power is required for the continuous reversals of the elementary magnets of which the iron is composed. This loss is known as *hysteresis loss.*

$$\text{Hysteresis loss} = K_h f B^{1.6}_{max} \quad \text{...(32)}$$

where f is the frequency in Hz, B_{max} is the maximum flux density in core and K_h is a constant.

(*ii*) **Eddy current loss.** This is due to the flow of eddy currents in the core. Thin laminations, insulated from each other, reduce the eddy current loss to small proportion.

$$\text{Eddy current loss} = K_e f^2 B^2_{max} \quad \text{...(33)}$$

where K_e is a constant.

Iron or core loss is found from open circuit test. The input of the transformer when on no-load measures the core loss.

2. **Copper losses.** These losses are due to the ohmic resistance of the transformer windings.

$$\text{Total copper loss} = I_1^2 R_1 + I_2^2 R_2 = I_1^2 R_{01} = I_2^2 R_{02}$$

These losses, as is evident, are proportional to square of the current (or kVA)^2.

The value of copper losses is found from the *short-circuit test.*

9.14. Transformer Efficiency. The efficiency of a transformer at a particular load and power factor is defined as *the ratio of power output to power input.*

$$\therefore \quad \text{Efficiency} = \frac{\text{output}}{\text{input}}$$

$$= \frac{\text{output}}{\text{output + losses}} = \frac{\text{output}}{\text{output + cu loss + iron loss}} \quad \text{...(34)}$$

or

$$\text{Efficiency} = \frac{\text{input} - \text{losses}}{\text{input}} = 1 - \frac{\text{losses}}{\text{input}} \quad \text{...(35)}$$

It may be noted that efficiency is based on power output in watts and not in volt-amperes, although losses are proportional to volt-amperes. Hence at any volt-ampere load, the efficiency depends on power factor, being *maximum at unity power factor.*

Efficiency can be calculated by determining core losses from open-circuit test and copper losses from short-circuit test.

Condition for maximum efficiency :

Iron losses, P_i = hysteresis loss + eddy current loss

$= P_h + P_e$

Copper losses, $P_c = I_1^2 R_{01}$ or $I_2^2 R_{02}$

Considering primary side :

Input to primary $= V_1 I_1 \cos \phi_1$

Efficiency,
$$\eta = \frac{V_1 I_1 \cos \phi_1 - \text{losses}}{V_1 I_1 \cos \phi_1} = \frac{V_1 I_1 \cos \phi_1 - I_1^2 R_{01} - P_i}{V_1 I_1 \cos \phi_1}$$

$$= 1 - \frac{I_1 R_{01}}{V_1 \cos \phi_1} - \frac{P_i}{V_1 I_1 \cos \phi_1}$$

Differentiating both sides w.r.t. I_1, we get

$$\frac{d\eta}{dI_1} = 0 - \frac{R_{01}}{V_1 \cos \phi_1} + \frac{P_i}{V_1 I_1^2 \cos \phi_1}$$

For η to be maximum,

$\dfrac{d\eta}{dI_1} = 0$. Hence the above equation reduces to

$$\frac{R_{01}}{V_1 \cos \phi_1} = \frac{P_i}{V_1 I_1^2 \cos \phi_1}$$

or
$$P_i = I_1^2 R_{01} \quad \text{or} \quad I_2^2 R_{02} \qquad ...(36)$$

or **Copper losses = Iron losses**

The output current corresponding to maximum efficiency is

$$I_2 = \sqrt{\frac{P_i}{R_{02}}} \qquad ...(37)$$

By proper design it is possible to make the maximum efficiency occur at any desired load.

Variation of efficiency with power factor. We know that transformers efficiency,

$$\eta = \frac{\text{output}}{\text{input}} = \frac{\text{input} - \text{losses}}{\text{input}}$$

$$= 1 - \frac{\text{losses}}{\text{input}} = 1 - \frac{\text{losses}}{(V_2 I_2 \cos \phi + \text{losses})}$$

Let,
$$\frac{\text{losses}}{V_2 I_2} = \beta$$

$\therefore$
$$\eta = 1 - \left(\frac{\text{losses}/V_2 I_2}{\cos \phi + \text{losses}/V_2 I_2} \right) \quad \text{or} \quad = 1 - \frac{\beta}{(\cos \phi + \beta)}$$

or
$$\eta = 1 - \frac{(\beta / \cos \phi)}{1 + (\beta / \cos \phi)} \qquad ...(38)$$

The variations of efficiency with power factor at different loadings on a typical transformer are shown in Fig. 41.

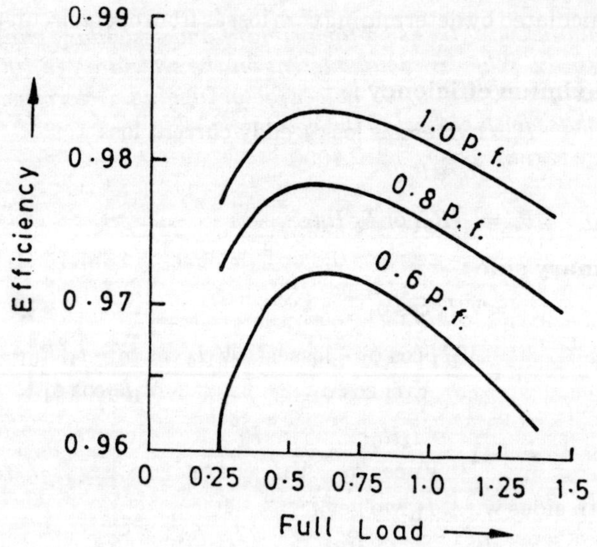

Fig. 41. Variations of efficiency with power factor at different loadings.

Transformer on Load

Example 10. *A 230 V/115 V single-phase transformer takes a no-load current of 2 A at a power factor of 0.2 lagging with low voltage winding kept open. If the low voltage winding is now loaded to take a current of 15 A at 0.8 power factor lagging find the current taken by high voltage winding.*

Solution. Primary, $E_1 = V_1 = 230$ V

Secondary, $E_2 = V_2 = 115$ V

No-load current $I_0 = 2$ A

No-load power factor, $\cos \phi_0 = 0.2$ or $\phi_0 = 78.46°$ or $78° 29'$

Load power factor, $\cos \phi_2 = 0.8$ or $\phi_2 = 36.9°$ or $36° 54'$

Current taken by h.v. winding, I_1 :

Now, transformation ratio,

$$K = \frac{V_2}{V_1} = \frac{E_2}{E_1} = \frac{115}{230} = \frac{1}{2}$$

$\therefore$ Secondary current referred to primary,

$$I_2' = KI_2 = \frac{1}{2} \times 15 = 7.5 \text{ A}$$

Phase angle, between I_0 and I_2'

$$= 78.46° - 36.9° = 41.56° \text{ or } 41° 34'$$

Using parallelogram law of forces (Fig. 42), we get

$$I_1 = \sqrt{2^2 + 7.5^2 + 2 \times 2 \times 7.5 \times \cos 41.56°}$$

$$= 9.09 \text{ A}$$

Hence, **current taken by h.v. winding,**

I_1 = 9.09 A. (Ans.)

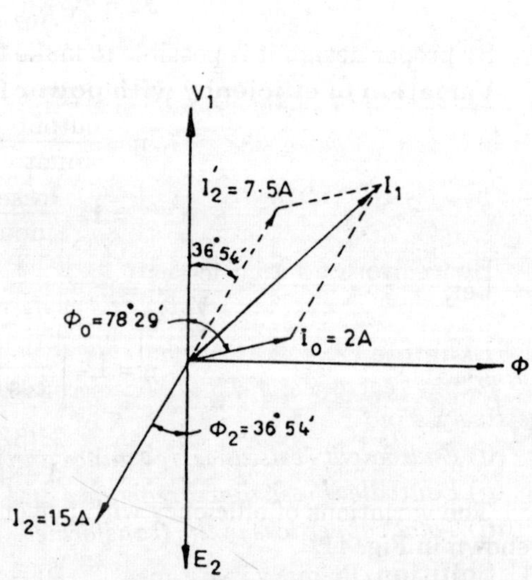

Fig. 42

Example 11. *The number of turns on the primary and secondary windings of a transformer are 1000 and 200 respectively. When the load current on the secondary is 100 A at 0.8 power factor lagging, the primary current is 30 A at 0.707 power factor lagging. Determine the no-load current of the transformer and its phase with respect to the voltage.*

Solution. Primary turns, $N_1 = 1000$

Secondary turns, $N_2 = 200$

Secondary current, $I_2 = 100$ A

Power factor, $\cos \phi_2 = 0.8$ or $\phi_2 = 36.9°$ or $36° \, 54'$

Primary current, $I_1 = 30$ A

Power factor, $\cos \phi_1 = 0.707$ or $\phi_1 = 45°$

No-load current, I_0, ϕ_0 :

Transformation ratio, $K = \frac{200}{1000} = \frac{1}{5}$

Secondary current referred to primary,

$$I_2' = KI_2 = \frac{1}{5} \times 100 = 20 \text{ A}$$

Refer Fig. 43. I_1 is the vector sum of I_0 and I_2'. Let I_0 lag behind V_1 by an angle ϕ_0.

Resolving currents into their X-and Y-components, we get

$$I_0 \cos \phi_0 + 20 \cos 36.9° = 30 \cos 45°$$

$$\therefore \quad I_0 \cos \phi_0 = 30 \cos 45° - 20 \cos 36.9°$$

$$= 21.21 - 16 = 5.21 \text{ A} \quad \quad ...(i)$$

$$I_0 \sin \phi_0 + 20 \sin 36.9° = 30 \sin 45°$$

$$\therefore \quad I_0 \sin \phi_0 = 30 \sin 45° - 20 \sin 36.9°$$

$$= 21.21 - 12 = 9.21 \text{ A} \quad \quad ...(ii)$$

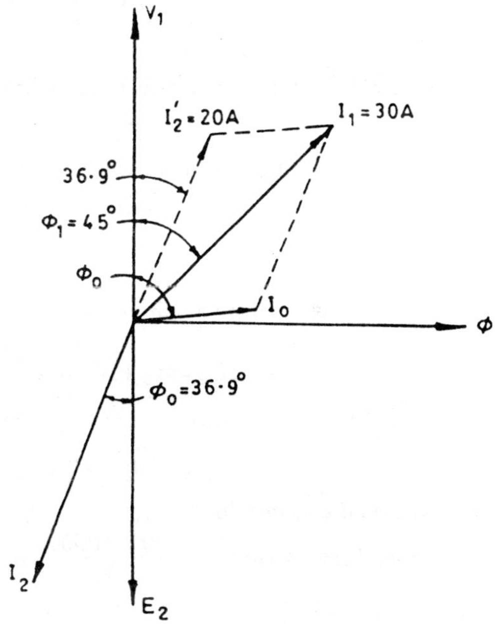

Fig. 43

From (*i*) and (*ii*), we get

$$\tan \phi_0 = \frac{9.21}{5.21} = 1.767$$

$$\therefore \quad \phi_0 = 60.5°$$

Putting the value of ϕ_0 in (*i*), we get

$$I_0 \cos 60.5° = 5.21$$

$$\therefore \quad I_0 = \frac{5.21}{\cos 60.5°} = 10.58 \text{ A}$$

Hence, **no-load current = 10.58 A**

and $\phi_0 = 60.5°$. **(Ans.)**

Example 12. *A 30 kVA, 2000/200 V, single-phase, 50 Hz transformer has a primary resistance of 3.5 Ω and reactance of 4.5 Ω. The secondary resistance and reactance are 0.015 Ω and 0.02 Ω respectively. Find :*

(*i*) *Equivalent resistance, reactance and impedance referred to primary.*

(*ii*) *Equivalent resistance, reactance and impedance referred to secondary.*

(*iii*) *Total copper loss of the transformer.*

Solution. Primary resistance, $R_1 = 3.5 \, \Omega$

Primary reactance, $X_1 = 4.5 \, \Omega$

Secondary resistance, $R_2 = 0.015\ \Omega$
Secondary reactance, $X_2 = 0.02\ \Omega$

Transformation ratio, $K = \dfrac{200}{2000} = \dfrac{1}{10}$

(*i*) Equivalent resistance *referred to primary*,

$$R_{01} = R_1 + R_2' = R_1 + \frac{R_2}{K^2} = 3.5 + \frac{0.015}{\left(\frac{1}{10}\right)^2} = \mathbf{5.0\ \Omega.}\quad \textbf{(Ans.)}$$

Equivalent reactance *referred to primary*,

$$X_{01} = X_1 + X_2' = X_1 + \frac{X_2}{K^2} = 4.5 + \frac{0.02}{\left(\frac{1}{10}\right)^2} = \mathbf{6.5\ \Omega.}\quad \textbf{(Ans.)}$$

Equivalent impedance *referred to primary*,

$$Z_{01} = \sqrt{(R_{01})^2 + (X_{01})^2} = \sqrt{(5)^2 + (6.5)^2} = \mathbf{8.2\ \Omega.}\quad \textbf{(Ans.)}$$

(*ii*) Equivalent resistance *referred to secondary*,

$$R_{02} = R_2 + R_1' = R_2 + K^2R_1$$
$$= 0.015 + \left(\frac{1}{10}\right)^2 \times 3.5 = \mathbf{0.05\ \Omega.}\quad \textbf{(Ans.)}$$

Equivalent reactance *referred to secondary*,

$$X_{02} = X_2 + X_1' = X_2 + K^2 X_1 = 0.02 + \left(\frac{1}{10}\right)^2 \times 4.5 = \mathbf{0.065\ \Omega.}\quad \textbf{(Ans.)}$$

Equivalent impedance *referred to secondary*,

$$Z_{02} = \sqrt{(R_{02})^2 + (X_{02})^2} = \sqrt{(0.05)^2 + (0.065)^2} = \mathbf{0.082\ \Omega.}\quad \textbf{(Ans.)}$$

(*iii*) **Total copper loss :**

Secondary current $= \dfrac{30 \times 1000}{200} = 150$ A

Total copper loss $= I_2^2 R_{02} = 150^2 \times 0.05 = \mathbf{1125\ W.}$ **(Ans.)**

☞**Example 13.** *A single-phase transformer has the following data :*

Turn ratio 20 : 1 ; $R_1 = 20\ \Omega$, $X_1 = 80\ \Omega$; $R_2 = 0.04\ \Omega$; $X_2 = 0.2\ \Omega$. No-load current = 1.2 A leading the flux by 30°.

The secondary delivers 180 A at a terminal voltage of 400 V and at a power factor of 0.8 lagging. Determine by the aid of a vector diagram :

(*i*) *The primary applied voltage.* (*ii*) *The primary power factor.*
(*iii*) *The efficiency.*

Solution. Refer Fig. 44 :

(*i*) **Primary applied voltage, $\mathbf{V_1}$:**

Taking V_2 as the reference vector

$$\therefore \quad V_2 = 400 \angle 0° = 400 + j0$$
$$I_2 = 180(0.8 - j0.6) = 144 - j108$$
$$Z_2 = (0.04 + j0.2)$$
$$E_2 = V_2 + I_2Z_2 = (400 + j0) + (144 - j108) \times (0.04 + j0.2)$$
$$= 400 + (5.76 + j28.8 - j4.32 + 21.6) = (427.36 + j24.48) = 428.1 \angle 3.28°.$$

Obviously,

$$\beta = 3.28°$$

$$E_1 = \frac{E_2}{K} = 20E_2 = 20(427.36 + j24.48) = 8547 + j490$$

$\therefore \quad -E_1 = -8547 - j490 = 8561 \angle 183.28°$

Secondary current referred to primary,

$$I_2' = -KI_2 = \frac{(-144 + j\,108)}{20} = -7.2 + j5.4$$

As seen from Fig. 44, I_0 leads V_2 by an angle $3.28° + 90° + 30° = 123.28°$

$$\therefore \quad I_0 = 1.2 \angle 123.28° = 1.2(\cos 123.28° + j \sin 123.28°)$$
$$= 1.2(-0.548 + j0.836) = -0.657 + j1.003.$$

Primary current,

$$I_1 = -I_2' + I_0 = (-7.2 + j5.4) + (-0.657 + j1.003)$$
$$= -7.857 + j6.403 = 10.14 \angle 140.8°$$
$$V_1 = -E_1 + I_1Z_1 = (-8547 - j490) + (-7.857 + j6.403)(20 + j80)$$
$$= -8547 - j490 + (-157.14 - j628.56 + j128.06 - 512.24)$$
$$= -9216.38 - j990.5 = \mathbf{9269 \angle 186.13°}. \quad \textbf{(Ans.)}$$

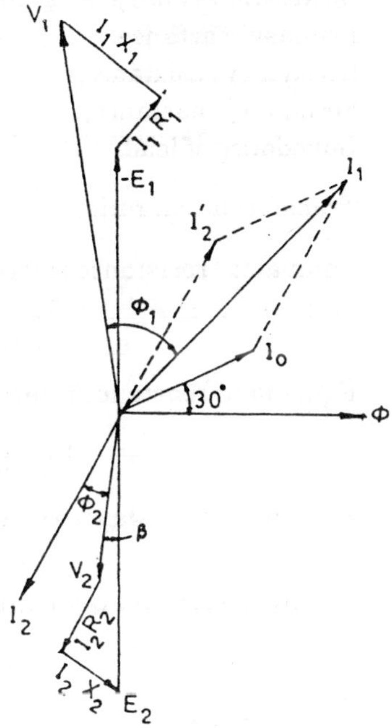

Fig. 44

(*ii*) **Primary power factor, cos ϕ_1 :**

Phase angle between V_1 and I_1,

$$\phi_1 = 186.13° - 140.8° = 45.33°$$

$\therefore$ Primary power factor $= \cos 45.33° = \mathbf{0.703}$ **(lag). (Ans.)**

(*iii*) **Efficiency :**

No-load primary input power

$$= V_1I_1 \cos \phi_0 = 9269 \times 1.2 \times \cos 60° = 5561.4 \text{ W}$$

$$R_{02} = R_2 + K^2R_1 = 0.04 + \left(\frac{1}{20}\right)^2 \times 20 = 0.09 \ \Omega$$

Total copper losses as referred to secondary

$$= I_2^2R_{02} = (180)^2 \times 0.09 = 2916 \text{ W}$$

Output $= V_2I_2 \cos \phi_2 = 400 \times 180 \times 0.8 = 57600$ W

Total losses $= 5561.4 + 2916 = 8477.4$ W

Input $=$ Output + Losses $= 57600 + 8477.4 = 66077.4$ W

$\therefore$ Efficiency, $\eta = \dfrac{\text{output}}{\text{input}} = \dfrac{57600}{66077.4} = 0.872$ or **87.2%. (Ans.)**

Example 14. *The high voltage and low voltage windings of a 2200/220 V single-phase 50 Hz transformer has resistances of 4.8 Ω and 0.04 Ω and reactances 2 Ω and 0.018 Ω respectively. The low voltage winding is connected to a load having an impedance of (6 + j4) Ω. Determine :*

(*i*) *Current in l.v. winding,* (*ii*) *Current in h.v. winding,*

(*iii*) *Load voltage, and* (*iv*) *Power consumed by the load.*

Solution. Primary resistance, $R_1 = 4.8\ \Omega$

Primary reactance, $X_1 = 2\ \Omega$

Secondary resistance, $R_2 = 0.04\ \Omega$.

Secondary reactance, $X_2 = 0.018\ \Omega$

Impedance of load, $Z_L = (6 + j4)$

Transformation ratio, $K = \frac{E_2}{E_1} = \frac{220}{2200} = \frac{1}{10}$

Equivalent resistance referred to l.v. side,

$$R_{02} = R_2 + K^2 R_1 = 0.04 + \left(\frac{1}{10}\right)^2 \times 4.8 = 0.088\ \Omega$$

Equivalent reactance referred to l.v. side,

$$X_{02} = X_2 + K^2 X_1 = 0.018 + \left(\frac{1}{10}\right)^2 \times 2 = 0.038\ \Omega$$

Equivalent impedance referred to l.v. side,

$$Z_{02} = R_{02} + jX_{02} = (0.088 + j0.038)\ \Omega$$

(*i*) **Current in l.v. winding,**

$$I_2 = \frac{220}{Z_{02} + Z_L} = \frac{220}{(0.088 + j0.038) + (6 + j4)}$$

$$= \frac{220}{6.088 + j4.038} = \frac{220}{\sqrt{(6.088)^2 + (4.038)^2}} = \mathbf{30.11\ A.}\ \textbf{(Ans.)}$$

(*ii*) **Current in h.v. winding,**

$$I_1 = I_2' = KI_2 = \frac{1}{10} \times 30.11 = \mathbf{3.011\ A.}\ \textbf{(Ans.)}$$

(*iii*) **Load voltage,** $= I_2 Z_L = 30.11(6 + j4) = 30.11 \times \sqrt{(6)^2 + (4)^2} = \mathbf{217.12\ V.}\ \textbf{(Ans.)}$

(*iv*) **Power consumed by the load,**

$$= I^2 R_L = 30.11^2 \times 6 = \mathbf{5439.6\ W.}\ \textbf{(Ans.)}$$

☞**Example 15.** *A single-phase transformer has $Z_1 = 1.4 + j5.2\ \Omega$ and $Z_2 = 0.0117 + j0.0465\ \Omega$. The input voltage is 6600 V and the turn ratio is 10.6 : 1. The secondary feeds a load which draws 300 A at 0.8 power factor lagging. Find the secondary terminal voltage and the kW output. Neglect no-load current I_0.*

Solution. Impedance $Z_1 = 1.4 + j5.2\ \Omega$

Impedance $Z_2 = 0.0117 + j0.0465\ \Omega$

Input voltage $V_1 = 6600$ V

Turn ratio, $K = 10.6 : 1$

Secondary load current, $I_2 = 300$ A

Power factor, $\cos \phi_2 = 0.8$

Secondary terminal voltage, V_2 :

kW output :

Let the secondary terminal voltage $= V_2 \angle 0°$

$$I_2 = 300 \angle -36.87°\ \text{A}$$

$$E_2 = V_2 \angle 0° + I_2 Z_2$$

$$= V_2 \angle 0° + (300 \angle -36.87°)(0.0117 + j0.0465) = V_2 \angle 0° + 14.4 \angle 39°$$

$$E_1 = 10.6E_2 = 10.6V_2 \angle 0° + 152.64 \angle 39°$$

$$I_1 = \frac{300}{10.6} \angle -36.87° = 28.3 \angle -36.87°$$

$$\begin{aligned} V_1 &= E_1 + I_1 Z_1 \\ &= 10.6V_2 \angle 0° + 152.64 \angle 39° + (28.3 \angle -36.87°)(1.4 + j5.2) \\ &= 10.6V_2 + 152.64 \angle 39° + 152.4 \angle 38° \\ &= 10.6V_2 + 118.62 + j96.06 + 120 + j93.83 \\ &= 10.6V_2 + 238.62 + j189.89 \end{aligned}$$

Since $|V_1| = 6600$ V

$\therefore \quad (10.6V_2 + 238.62)^2 + (189.89)^2 = 6600^2$

or $\quad V_2 \simeq 600$ **V**

Hence, **secondary terminal voltage = 600 V. (Ans.)**

Output :

Output $= V_2 I_2 \cos \phi_2 = 600 \times 300 \times 0.8 = 144000$ W or **144 kW. (Ans.)**

Example 16. *The full-load copper loss on h.v. side of 100 kVA, 11000/317 V 1-phase transformer is 0.62 kW and on the l.v. side is 0.48 kW.*

(i) Calculate R_1, R_2 and R_2' in ohms :

(ii) The total reactance is 4 percent, find X_1, X_2 and X_2' in ohms if the reactance is divided in the same proportion as resistance.

Solution. *Given* : Rated kVA = 100 ; $V_1 = 11000$ V ; $V_2 = 317$ V ; loss on *h.v.* side = 0.62 kW ; loss of *l.v.* side = 0.48 kW ; total reactance = 4%.

(i) $\mathbf{R_1}$**,** $\mathbf{R_2}$ **and** $\mathbf{R_2'}$ **:**

h.v. side full-load current, $\quad I_1 = \dfrac{\text{rated kVA} \times 1000}{V_1} = \dfrac{100 \times 1000}{11000} = 9.09$ A

l.v. side full-load current, $\quad I_2 = \dfrac{\text{rated kVA} \times 1000}{V_2} = \dfrac{100 \times 1000}{317} = 315.46$ A

Primary winding resistance, $\quad \mathbf{R_1} = \dfrac{\text{full-load copper loss in primary winding}}{I_1^2}$

$$= \frac{0.62 \times 1000}{(9.09)^2} = \mathbf{7.5\ \Omega.\ (Ans.)}$$

Secondary winding resistance, $\quad \mathbf{R_2} = \dfrac{\text{full-load copper loss in secondary winding}}{I_2^2}$

$$= \frac{0.48 \times 1000}{(315.46)^2} = \mathbf{0.00482\ \Omega.\ (Ans.)}$$

Transformation ratio, $\quad K = \dfrac{V_2}{V_1} = \dfrac{317}{11000} = 0.02882$

Secondary resistance referred to primary,

$$R_2' = \frac{R_2}{K^2} = \frac{0.00482}{(0.02882)^2} = \mathbf{5.8\ \Omega.\ (Ans.)}$$

(ii) $\mathbf{X_1}$**,** $\mathbf{X_2}$ **and** $\mathbf{X_2'}$ **:**

Percentage reactance referred to *h.v.* side $= \dfrac{I_1 X_{01}}{V_1} \times 100 = 4$

or transformer total reactance referred to primary,

$$X_{01} = \frac{4 \times V_1}{100 I_1} = \frac{4 \times 11000}{100 \times 9.09} = 48.4\ \Omega.$$

Now, $$\frac{R_1}{R_2'} = \frac{X_1}{X_2'}$$ ...(Given)

or $$\frac{R_1 + R_2'}{R_2'} = \frac{X_1 + X_2'}{X_2'}$$

or $$\frac{7.5 + 5.8}{5.8} = \frac{48.4}{X_2'}$$ $(\because\ X_1 + X_2' = X_{01})$

or $$\mathbf{X_2'} = \frac{48.4 \times 5.8}{(7.5 + 5.8)} = \mathbf{21.1\ \Omega.\ (Ans.)}$$

Primary winding reactance, $\mathbf{X_1} = X_{01} - X_2' = 48.4 - 21.1 = \mathbf{27.3\ \Omega.}$ **(Ans.)**

Secondary winding reactance, $\mathbf{X_2} = K^2 X_2' = \left(\frac{377}{11000}\right)^2 \times 21.1 = \mathbf{0.0175\ \Omega.}$ **(Ans.)**

Hysteresis and Eddy Current Losses

Example 17. *A 230 V, 2.5 kVA single-phase transformer has an iron loss of 100 W at 40-Hz and 70 W at 30-Hz. Find the hysteresis and eddy current losses at 50-Hz.*

Solution. Iron loss at 40-Hz = 110 W

Iron loss at 30-Hz = 75 W

Hysteresis and eddy current loss at 50-Hz :

We know that hysteresis loss,

$$P_h \propto f = Af$$

Eddy current loss, $$P_e \propto f^2 = Bf^2$$

Iron loss, $$P_i = P_h + P_e = Af + Bf^2$$

At 40-Hz :

$$100 = 40A + (40)^2 B$$

i.e., $$40A + 1600B = 100$$

or $$A + 40B = 2.5 \qquad ...(i)$$

At 30-Hz :

$$70 = 30A + (30)^2 B$$

i.e., $$30A + 900B = 70$$

or $$3A + 90B = 7 \qquad ...(ii)$$

Solving (*i*) and (*ii*), we get $A = \frac{11}{6}, B = \frac{1}{60}.$

Hysteresis loss at 50-Hz

$$= Af = \frac{11}{6} \times 50 = \mathbf{91.67\ W.\ (Ans.)}$$

Eddy current loss at 50-Hz

$$= Bf^2 = \frac{1}{60} \times 50^2 = \mathbf{41.67\ W.\ (Ans.)}$$

Example 18. *When a transformer is supplied at 400 V, 50-Hz the hysteresis loss is found to be 310 W and eddy current loss is found to be 260 W. Determine the hysteresis loss and eddy current loss when the transformer is supplied at 800 V, 100-Hz.*

Solution. Hysteresis loss at 400 V, 50 Hz = 310 W

Eddy current loss at 400 V, 50-Hz = 260 W

We know that,

Induced e.m.f., $E = 4.44 f \phi_{max} N = 4.44 (B_{max} A) f N$

i.e. $E \propto B_{max} f$

or $B_{max} \propto \frac{E}{f} = a \frac{E}{f}$ [a = constant]

In the **Ist case,** $B_{max_1} = a \cdot \frac{E_1}{f_1} = \frac{400}{50} a = 8a.$

In the **IInd case,** $B_{max_2} = a \frac{E_2}{f_2} = \frac{800}{100} a = 8a$

Hence, $B_{max_1} = B_{max_2}$

[∵ Core area A, and number of turns N remain the same]

Hence, $P_h = Af$ and $P_e = Bf^2$

$P_{h1} = 310$ W at $f_1 = 50$-Hz

∴ $A = \frac{310}{50} = 6.2$

Hysteresis loss at 100-Hz

$= Af_2 = 6.2 \times 100 =$ **620 W. (Ans.)**

$P_{e1} = 260$ W at $f_1 = 50$-Hz

∴ $B = \frac{260}{50^2} = 0.104$

Eddy current loss at 100-Hz

$= Bf_2^2 = 0.104 \times 100^2 =$ **1040 W. (Ans.)**

Transformer Tests (O.C. and S.C.) and Equivalent Circuit

☞**Example 19.** *A 50-Hz, single-phase transformer has a turn ratio of 5. The resistances are 0.8 Ω, 0.02 Ω and reactances are 4 Ω and 0.12 Ω for high-voltage and low-voltage windings respectively. Find :*

(i) The voltage to be applied to the h.v. side to obtain full-load current of 180 A in the l.v. winding on short-circuit.

(ii) The power factor on short-circuit.

Draw the equivalent circuit and vector diagram.

Solution. Turn ratio, $\frac{N_1}{N_2} = 5, K = \frac{1}{5}$

Primary resistance, $R_1 = 0.8\ \Omega$

Primary reactance, $X_1 = 4\ \Omega$

Secondary resistance, $R_2 = 0.02\ \Omega$

Secondary reactance, $X_2 = 0.12\ \Omega$

Total resistance referred to primary,

$$R_{01} = R_1 + R_2' = R_1 + \frac{R_2}{K^2} = 0.8 + 0.02 \times (5)^2 = 0.8 + 0.5 = 1.3\ \Omega$$

Total reactance referred to secondary,

$$X_{01} = X_1 + X_2' = X_1 + \frac{X_2}{K^2} = 4 + 0.12(5)^2 = 4 + 3 = 7\ \Omega$$

Total impedance referred to primary,

$$Z_{01} = \sqrt{R_{01}^2 + X_{01}^2} = \sqrt{(1.3)^2 + (7)^2} = 7.12 \text{ ohm}$$

Also

$$I_1 = KI_2 = \frac{1}{5} \times 180 = 36 \text{ A}$$

(*i*) **Voltage to be applied** on h.v. side to obtain full-load current I_1,

$$V_{SC} = I_1 Z_{01} = 36 \times 7.12 = \mathbf{256.3\ V.}\ \textbf{(Ans.)}$$

(*ii*) The **power factor** on short-circuit,

$$\cos \phi_0 = \frac{R_{01}}{Z_{01}} = \frac{1.3}{7.12} = \mathbf{0.1825.}\ \textbf{(Ans.)}$$

Equivalent circuit and vector diagram are shown in Figs. 45, 46 and 47.

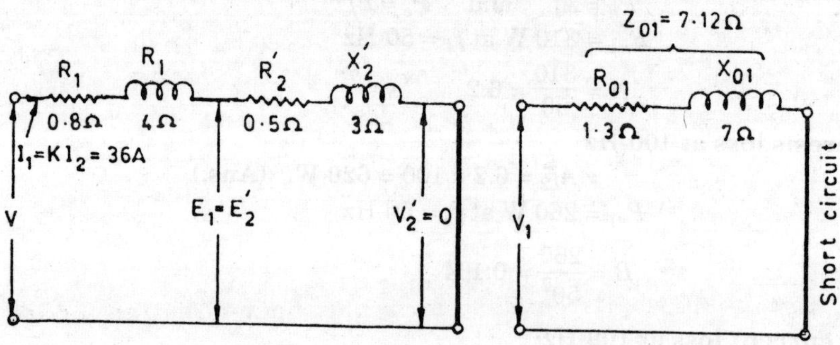

Fig. 45

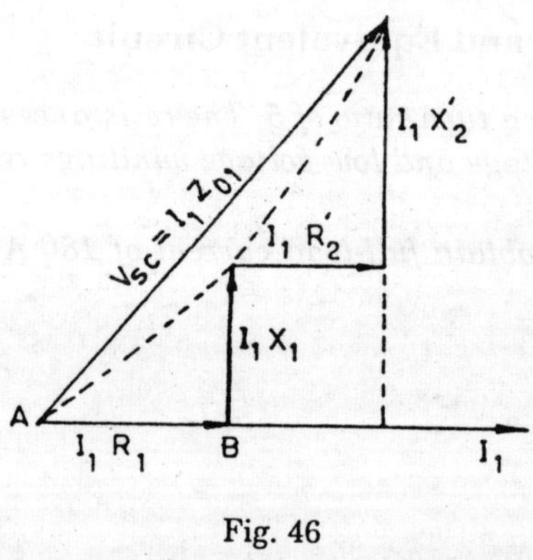

Fig. 46

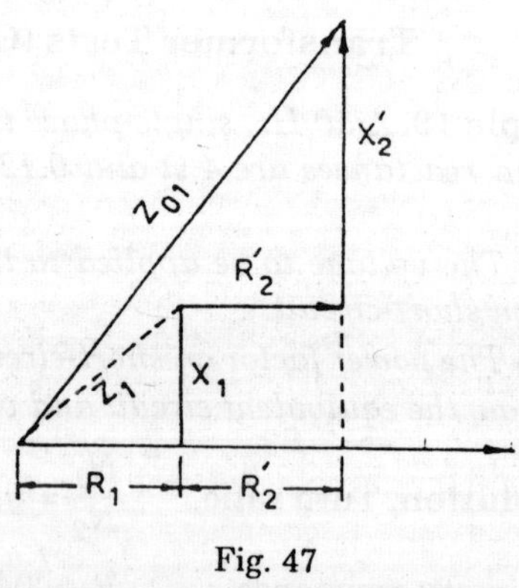

Fig. 47

Example 20. *A 12 kVA 4000/400 V transformer has primary and secondary winding resistances of 13 Ω and 0.15 Ω and leakage reactances of 20 Ω and 0.25 Ω respectively. The magnetising reactance is 6000 Ω and the resistance representing core loss is 12000 Ω. Determine :*

(*i*) *Equivalent resistance and reactance as referred to primary.*

(*ii*) *Input current with secondary terminals open circuited.*

(*iii*) *Input current when the secondary load current is 30 A at 0.8 power factor lagging.*

Solution. Transformation ratio, $K = \frac{400}{4000} = \frac{1}{10}$

Primary resistance, $R_1 = 13\ \Omega$

Primary reactance, $X_1 = 20\ \Omega$

Secondary resistance, $R_2 = 0.15\ \Omega$

Secondary reactance, $X_2 = 0.25\ \Omega$

Magnetising reactance, $X_0 = 6000\ \Omega$

Resistance representing the core loss, $R_0 = 12000\ \Omega$

(*i*) **Equivalent resistance referred to primary,**

$$R_{01} = R_1 + R_2' = R_1 + \frac{R_2}{K^2}$$

$$= 13 + \frac{0.15}{\left(\frac{1}{10}\right)^2} = 13 + 15 = \mathbf{28\ \Omega.}\ \ \textbf{(Ans.)}$$

Equivalent reactance referred to primary,

$$X_{01} = X_1 + X_2' = X_1 + \frac{X_2}{K^2}$$

$$= 20 + \frac{0.25}{\left(\frac{1}{10}\right)^2} = 20 + 25 = \mathbf{45\ \Omega.}\ \ \textbf{(Ans.)}$$

(*ii*) **Input current with secondary terminals open-circuited (or input current at no-load) :**

Magnetising component, $I_m = \frac{4000}{6000} = 0.667$ A

Wattful component, $I_w = \frac{4000}{12000} = 0.333$ A

Hence input current at no-load,

$$I_0 = 0.333 - j0.667 = \mathbf{0.745 \angle -63.5° \ A.}\ \ \textbf{(Ans.)}$$

(*iii*) **Input current $\mathbf{I_1}$:**

Secondary output current, $I_2 = 30$ A (*given*)

Power factor, $\cos \phi_2 = 0.8$ (lagging)

i.e., $I_2 = 30 \angle -36.9°$

Secondary current referred to primary,

$$I_2' = KI_2 = \frac{30}{10} \angle -36.9° = 3 \angle -36.9° = 2.4 - j1.8$$

$\therefore$ Input current, $I_1 = I_2' + I_0 = (2.4 - j1.8) + (0.333 - j0.667)$

$$= 2.733 - j2.467 = \mathbf{3.68 \angle -42.07° \ A.}\ \ \textbf{(Ans.)}$$

☞**Example 21.** *Obtain the approximate equivalent circuit of a given 200/2000 V single-phase 30 kVA transformer having the following test results :*

O.C. test : 200 V, 6.2 A, 360 W on l.v. side

S.C. test : 75 V, 18 A, 600 W on h.v. side.

Solution. O.C. test (l.v. side)

Primary voltage, $V_1 = 200$ V

No-load current, $I_0 = 6.2$ A

No-load loss, $P_0 = 360$ W

Now, $P_0 = V_1 I_0 \cos \phi_0$

$$360 = 200 \times 6.2 \times \cos \phi_0$$

$$\therefore \quad \cos \phi_0 = \frac{360}{200 \times 6.2} = 0.29$$

$$\sin \phi_0 = 0.957$$

Wattful component of no-load current,

$$I_w = I_0 \cos \phi_0 = 6.2 \times 0.29 = 1.8 \text{ A.}$$

Resistance representing the core loss,

$$R_0 = \frac{V_1}{I_w} = \frac{200}{1.8} = \mathbf{111.11\ \Omega.\ (Ans.)}$$

Magnetising component of no-load current,

$$I_m = I_0 \sin \phi_0 = 6.2 \times 0.95 = 5.93 \text{ A}$$

Magnetising reactance, $X_0 = \dfrac{V_1}{I_m} = \dfrac{200}{5.93} = \mathbf{33.7\ \Omega.\ (Ans.)}$

S.C. test (h.v. side) :

Short-circuit voltage, $V_{SC} = 75$ V

Short-circuit current, $I_{SC} = 18$ A

Losses, $P_{SC} = 600$ W

Impedance of transformer referred to h.v. side,

$$Z_{02} = \frac{V_{SC}}{I_{SC}} = \frac{75}{18} = 4.167\ \Omega$$

$$P_{SC} = I_{SC}^2 \times R_{02}$$

$$600 = 18^2 \times R_{02}$$

$$\therefore \quad R_{02} = \frac{600}{18^2} = 1.85\ \Omega$$

Transformation ratio, $K = \dfrac{2000}{200} = 10$

Referred to 200 V side :

$$Z_{01} = \frac{Z_{02}}{K^2} = \frac{4.167}{(10)^2} = 0.04167\ \Omega$$

$$R_{01} = \frac{R_{02}}{K^2} = \frac{1.85}{(10)^2} = 0.0185\ \Omega$$

$$\therefore \quad X_{01} = \sqrt{Z_{01}^2 - R_{01}^2} = \sqrt{(0.04167)^2 - (0.0185)^2} = \mathbf{0.0373\ \Omega.\ (Ans.)}$$

Approximate equivalent circuit is shown in Fig. 48.

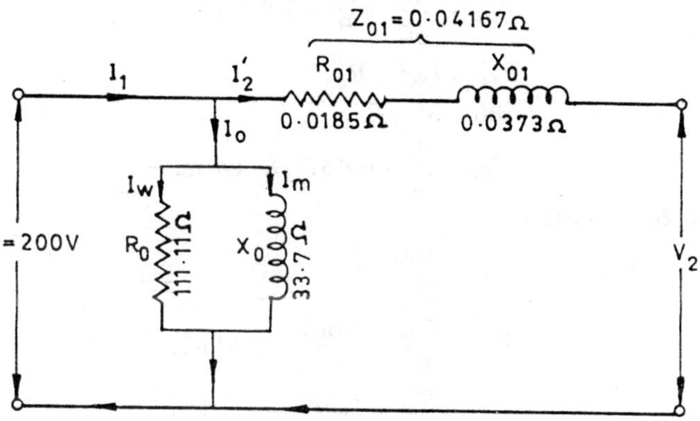

Fig. 48

Example 22. *The following readings were obtained on O.C. and S.C. tests on a 200/400 V, 50 Hz, single-phase transformer.*

O.C. test (l.v. side) : 200 V, 0.6 A, 60 W

S.C. test (h.v. side) : 15 V, 9 A, 80 W

Calculate the secondary voltage when delivering 4 kW at 0.8 power factor lagging, the primary voltage being 200 V.

Solution. O.C. test—l.v. side (*Instruments in the primary side* and secondary open)

Primary voltage, $V_1 = 200$ V

No-load current, $I_0 = 0.6$ A

No-load loss, $P_0 = 60$ W

Now, $P_0 = V_1 I_0 \cos \phi_0$

$$60 = 200 \times 0.6 \times \cos \phi_0$$

$$\therefore \quad \cos \phi_0 = \frac{60}{200 \times 0.6} = 0.5$$

$$\sin \phi_0 = 0.866$$

Wattful component of no-load current

$$I_w = I_0 \cos \phi_0 = 0.6 \times 0.5 = 0.3 \text{ A}$$

Magnetising component of no-load current,

$$I_m = I_0 \sin \phi_0 = 0.6 \times 0.866 = 0.52 \text{ A}$$

∴ Resistance representing the core loss,

$$R_0 = \frac{V_1}{I_w} = \frac{200}{0.3} = \mathbf{666.67\ \Omega.} \quad \textbf{(Ans.)}$$

Magnetising reactance, $X_0 = \frac{V_1}{I_m} = \frac{200}{0.52} = \mathbf{384.6\ \Omega.}$ **(Ans.)**

S.C. test—h.v. side (*Instruments in the secondary side* and primary short-circuited)

Short circuit voltage, $V_{SC} = 15$ V

Short circuit current, $I_{SC} = 9$ A

Losses, $P_{SC} = 80$ W

Impedance of transformer referred to secondary,

$$Z_{02} = \frac{V_{SC}}{I_{SC}} = \frac{15}{9} = 1.667\ \Omega$$

Also, $$P_{SC} = I_{SC}^2 \times R_{02}$$

$$80 = 9^2 \times R_{02}$$

$\therefore$ $$R_{02} = \frac{80}{81} = \mathbf{0.987\ \Omega.\ (Ans.)}$$

Referred to 200 V side :

Transformation ratio, $$K = \frac{400}{200} = 2$$

$$Z_{01} = \frac{Z_{02}}{K^2} = \frac{1.667}{2^2} = 0.417\ \Omega$$

$$R_{01} = \frac{R_{02}}{2^2} = \frac{0.987}{4} = 0.247\ \Omega$$

$\therefore$ $$X_{01} = \sqrt{Z_{01}^2 - R_{01}^2} = \sqrt{(0.417)^2 - (0.247)^2} = 0.336\ \Omega$$

Output, $$\text{kVA} = \frac{4}{0.8} = 5\ \text{kVA}$$

Output current, $$I_2 = \frac{5 \times 1000}{400} = \mathbf{12.5\ A.\ (Ans.)}$$

This value of I_2 is *approximate* because V_2 (which is to be calculated as yet) has been taken equal to 400 V (which in fact is equal to E_2 or V_2)

Now, $$Z_{02} = 1.667\ \Omega$$

$$R_{02} = 0.987\ \Omega$$

$\therefore$ $$X_{02} = \sqrt{Z_{02}^2 - R_{02}^2} = \sqrt{1.667^2 - 0.987^2} = 1.343\ \Omega$$

Total drop as referred to secondary

$$= I_2 (R_{02} \cos \phi_2 + X_{02} \sin \phi_2)$$

$$= 12.5(0.987 \times 0.8 + 1.343 \times 0.6) = 19.94\ \text{V}$$

$\therefore$ $$V_2 = 400 - 19.94 \simeq 380\ \text{V}$$

Hence, **secondary voltage = 380 V. (Ans.)**

Approximate equivalent circuit is shown in Fig. 49.

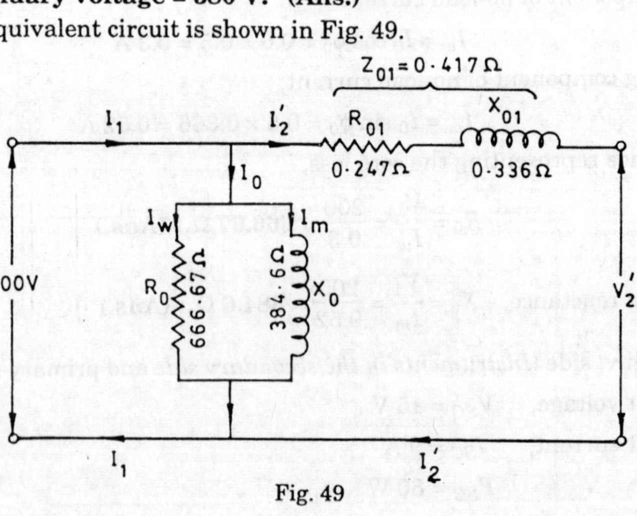

Fig. 49

Example 23. *A 4 kVA, 400/200 V, 50-Hz single-phase transformer has the following test data :*

O.C. test (l.v. side) 200 V, 1 A, 64 W
S.C. test (h.v. side) 15 V, 10 A, 80 W
Determine :
(i) Equivalent circuit referred to l.v. side, and
(ii) Secondary load voltage on full-load at 0.8 power factor lagging.

Solution. (*i*) **O.C. test—l.v. side :**

Voltage, $V_0 = 200$ V
No-load current, $I_0 = 1$ A
No-load loss, $P_0 = 64$ W
Now, $P_0 = V_0 I_0 \cos \phi_0$

$64 = 200 \times 1 \times \cos \phi_0$

$\therefore \quad \cos \phi_0 = 0.32$

$\sin \phi_0 = 0.9474$

Wattful component of no-load current,

$$I_w = I_0 \cos \phi_0 = 1 \times 0.32 = 0.32 \text{ A}$$

Magnetising component of no-load current,

$$I_m = I_0 \sin \phi_0 = 1 \times 0.9474 = 0.9474 \text{ A}$$

$\therefore$ Resistance representing the core loss,

$$R_0 = \frac{V_0}{I_w} = \frac{200}{0.32} = \mathbf{625\ \Omega.\ (Ans.)}$$

Magnetising reactance, $X_0 = \dfrac{V_0}{I_m} = \dfrac{200}{0.9474} = \mathbf{211.1\ \Omega.\ (Ans.)}$

S.C. test—h.v. side :
Short-circuit voltage, $V_{SC} = 15$ V
Short-circuit current, $I_{SC} = 10$ A
Losses, $P_{SC} = 80$ W
Impedance of the circuit referred to h.v. side,

$$Z_{01} = \frac{V_{SC}}{I_{SC}} = \frac{15}{10} = \mathbf{1.5\ \Omega.\ (Ans.)}$$

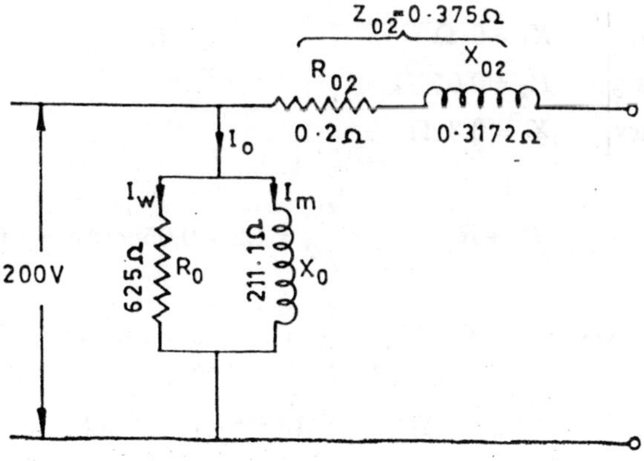

Fig. 50

Also $P_{SC} = I_{SC}^2 \times R_{01}$

$\therefore$ $80 = 10^2 \times R_{01}$

$\therefore$ $R_{01} = \dfrac{80}{100} = \mathbf{0.8\ \Omega}$. **(Ans.)**

Referred to l.v. side :

Transformation ratio, $K = \dfrac{200}{400} = \dfrac{1}{2}$

$$Z_{02} = K^2 Z_{01} = (\tfrac{1}{2})^2 \times 1.5 = \mathbf{0.375\ \Omega}. \text{ (Ans.)}$$

$$R_{02} = K^2 R_{01} = (\tfrac{1}{2})^2 \times 0.8 = \mathbf{0.2\ \Omega}. \text{ (Ans.)}$$

$$\therefore \quad X_{02} = \sqrt{Z_{02}^2 - R_{02}^2} = \sqrt{(0.375)^2 - (0.2)^2} = \mathbf{0.3172\ \Omega}. \text{ (Ans.)}$$

The approximate equivalent circuit is shown in Fig. 52.

(*ii*) **Secondary load voltage, V_2 :**

Secondary full-load current, $I_2 = \dfrac{4 \times 1000}{200} = 20$ A

$$\cos \phi_2 = 0.8$$

$$\sin \phi_2 = 0.6$$

$$\therefore \quad I_2 = 20(0.8 - j0.6) = (16 - j12)$$

$$Z_{02} = R_{02} + jX_{02} = 0.2 + j0.3172$$

Secondary load voltage, $V_2 = 200 - I_2 Z_{02} = 200 - (16 - j12)(0.2 + j0.3712)$

$$= 200 - (3.2 + j5.075 - j2.4 + 3.806)$$

$$= 200 - (7 + j2.675) = 193 - j2.675 \simeq \mathbf{193\ V}. \text{ (Ans.)}$$

Example 24. *A single-phase step-down transformer has a turn ratio of 3. The resistance and reactance of the primary winding are 1.2 Ω and 6 Ω and those of the secondary winding are 0.05 Ω and 0.03 Ω respectively. If the h.v. winding is supplied at 230 V, 50-Hz with l.v. winding short circuited, find :*

(*i*) *Current in the l.v. winding,* (*ii*) *Copper loss in the transformer, and*

(*iii*) *Power factor.*

Solution. Turn ratio $= 3$, *i.e.*, $K = \frac{1}{3}$

Resistance of primary, $R_1 = 1.2\ \Omega$

Reactance of primary, $X_1 = 6\ \Omega$

Resistance of secondary, $R_2 = 0.05\ \Omega$

Reactance of secondary, $X_2 = 0.3\ \Omega$

Referred to *h.v. side*

$$R_{01} = R_1 + R_2' = R_1 + \frac{R_2}{K^2} = 1.2 + 0.05 \times (3)^2 = 1.65\ \Omega$$

$$X_{01} = X_1 + X_2' = X_1 + \frac{X_2}{K^2} = 6 + 0.3 \times (3)^2 = 8.7\ \Omega$$

$$Z_{01} = \sqrt{R_{01}^2 + X_{01}^2} = \sqrt{(1.65)^2 + (8.7)^2} = 8.855\ \Omega$$

Current in h.v. winding when l.v. winding is short-circuited,

$$I_{SC} = \frac{V_{SC}}{Z_{01}} = \frac{230}{8.855} = 25.97 \text{ A}$$

Neglecting I_0, $I_1 = I_2' = 25.97$ A

(*i*) **Current in the l.v. winding,**

$$I_2 = \frac{I_2'}{K} = \frac{25.97}{\left(\frac{1}{3}\right)} = \mathbf{77.91\ A.} \quad \textbf{(Ans.)}$$

(*ii*) **Total copper loss** $= I_1^2 R_{01} = (25.97)^2 \times 1.65 =$ **1112.8 W. (Ans.)**

(*iii*) **Primary power factor :**

Power input on short-circuit,

$$P_{SC} = V_{SC} I_{SC} \cdot \cos \phi_{SC}$$

$$1112.8 = 230 \times 25.97 \times \cos \phi_{SC}$$

$$\therefore \quad \cos \phi_{SC} = \frac{1112.8}{230 \times 25.97} = \mathbf{0.1863.} \quad \textbf{(Ans.)}$$

☞**Example 25.** *A single-phase, 3 kVA, 230/115 V, 50-Hz transformer has the following constants :*

Resistance : Primary 0.3 Ω, secondary 0.09 Ω.

Reactance : Primary 0.4 Ω, secondary 0.1 Ω.

Resistance of equivalent exciting circuit referred to primary, $R_0 = 600\ \Omega$.

Reactance of equivalent exciting circuit-referred to primary, $X_0 = 200\ \Omega$.

What would be the readings of the instruments when the transformer is connected for (i) O.C. test, (ii) S.C. test. In both tests supply is given to h.v. side.

Solution. Primary resistance, $R_1 = 0.3\ \Omega$

Primary reactance, $X_1 = 0.4\ \Omega$

Secondary resistance, $R_2 = 0.09\ \Omega$

Secondary reactance, $X_2 = 0.1\ \Omega$

$R_0 = 600\ \Omega$

$X_0 = 200\ \Omega$

(*i*) **O.C. test :**

Wattful component of no-load current,

$$I_w = \frac{V_1}{R_0} = \frac{230}{600} = 0.383 \text{ A}$$

Magnetising component of no-load current,

$$I_w = \frac{V_1}{X_0} = \frac{230}{200} = 1.15 \text{ A}$$

$$\therefore \text{ No-load current,} \quad I_0 = \sqrt{I_w^2 + I_m^2} = \sqrt{(0.383)^2 + (1.15)^2} = 1.212 \text{ A}$$

Input on no-load to h.v. winding

$$= V_1 I_w = 230 \times 0.383 = 88.09 \text{ W}$$

Hence, the readings of the instruments are :

Voltmeter reading = 230 V. (Ans.)

Ammeter reading = 1.212 A. (Ans.)

Wattmeter reading = 88.09 W. (Ans.)

(*ii*) **S.C. test :**

Transformation ratio, $K = \frac{115}{230} = \frac{1}{2}$

Referred to *h.v. side*, $R_{01} = R_1 + R_2' = R_1 + \frac{R_2}{K^2} = 0.3 + \frac{0.09}{(\frac{1}{2})^2} = 0.66\ \Omega$

$$X_{01} = X_1 + X_2' = X_1 + \frac{X_2}{K^2} = 0.4 + \frac{0.1}{(\frac{1}{2})^2} = 0.8\Omega$$

$$\therefore \quad Z_{01} = \sqrt{R_{01}^2 + X_{01}^2} = \sqrt{(0.66)^2 + (0.8)^2} = 1.037\ \Omega$$

Full-load current in the h.v. winding

$$I_1 = \frac{3 \times 1000}{230} = 13.04 \text{ A}$$

Voltage to be applied to h.v. winding on short-circuiting l.v. winding to pass full-load current of 13.04 A through h.v. winding.

$$V_{SC} = I_1 Z_{01} = 13.04 \times 1.037 = 13.52 \text{ V}$$

Power supplied to h.v. winding on short-circuit

$$= I_1^2 \cdot R_{01} = (13.04)^2 \times 0.66 = 112.2 \text{ W}$$

Hence, the readings of the instruments are :

Voltmeter reading = 13.52 V. (Ans.)

Ammeter reading = 13.04 A. (Ans.)

Wattmeter reading = 112.2 W. (Ans.)

Regulation and Efficiency of a Transformer

Example 26. *A 4kVA 220/440 V, 50 Hz, single-phase transformer gave the following test figures. No load test performed on 220 V side keeping 440 V side open : 220 V, 0.7 A, 60 W. Short circuit test performed short circuiting the 440 V side through an ammeter : 9V, 6A, 21.6W calculate.*

(*a*) *The magnetising current and the component corresponding to iron loss at normal voltage and frequency.*

(*b*) *The efficiency at full-load at unity power factor and the corresponding secondary terminal voltage.*

(*c*) *Draw the phasor diagram corresponding to the full load operation.* **(PTU, May, 2000)**

Solution. *Open circuit test* on 220 V (l.v.) side : $V_1 = 220$ V ; $P_0 = 60$ W ; $I_0 = 0.7$ A

Short circuit test on 440 V (h.v.) side : $V_{SC} = 9$V ; $I_{SC} = 6$A ; $P_{SC} = 21.6$ W

(*a*) $P_0 = V_1 I_0 \cos \phi_0$ or $60 = 220 \times 0.7 \times \cos \phi_0$

or $\cos \phi_0 = 0.3896$, and $\sin \phi_0 = 0.921$

$\therefore$ Iron loss component of current, $I_w = I_0 \cos \phi_0 = 0.7 \times 0.3896 = $ **0.273 A. (Ans.)**

Magnetising component of current, $I_m = I_0 \sin \phi_0 = 0.7 \times 0.921 = $ **0.6447 A. (Ans.)**

(*b*) Full-load secondary current, $I_2 = \frac{4 \times 1000}{440} = 9.1$ A

Full-load primary current, $I_1 = \frac{4 \times 1000}{220} = 18.2$ A

Full-load copper losses, $P_c = \left(\frac{I_1}{I_{sc}}\right)^2 \times P_{SC} = \left(\frac{18.2}{6}\right)^2 \times 21.6 = 198.7$ W

∴ Full-load efficiency at unity power factor,

$$\eta = \frac{4 \times 1000 \times 1}{4 \times 1000 \times 1 + 60 + 198.7} \times 100 = \mathbf{93.92\%.} \quad \textbf{(Ans.)}$$

From short circuit test, we have

$$Z_{01} = \frac{V_{SC}}{I_{sc}} = \frac{9}{6} = 1.5\ \Omega$$

$$R_{01} = \frac{P_{SC}}{I_{sc}^2} = \frac{21.6}{6^2} = 0.6\ \Omega$$

$$\therefore \quad X_{01} = \sqrt{Z_{01}^2 - R_{01}^2} = \sqrt{1.5^2 - 0.6^2} = 1.375\ \Omega$$

Now, $$R_{02} = K^2 R_{01} = (2)^2 \times 0.6 = 2.4\ \Omega \qquad \left(\because\ K = \frac{440}{220} = 2\right)$$

$$X_{02} = K^2 X_{01} = 2^2 \times 1.375 = 5.5\ \Omega$$

Since $\cos \phi_2 = 1 \quad \therefore \quad \sin \phi_2 = \sin^{-1}(\cos^{-1} 1) = 0$

$${}_0V_2^2 = (V_2 + I_2R_{02})^2 + (I_2X_{02})^2$$

$$(440)^2 = (V_2 + 9.1 \times 2.4)^2 + (9.1 \times 5.5)^2$$

or $$(V_2 + 21.84)^2 = (440)^2 - (9.1 \times 5.5)^2 = 191095$$

or $$\mathbf{V_2} = (\sqrt{191095} - 21.84) = \mathbf{415.3\ V.} \quad \textbf{(Ans.)}$$

(c) The phase diagram of transformer corresponding to full-load operation is shown in Fig. 51.

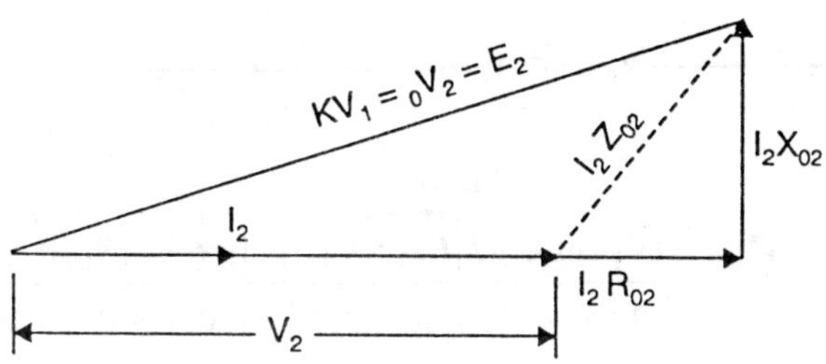

Fig. 51

Example 27. *The parameters of the equivalent circuit of a 100 kVA, 2000/200 volt single-phase transformer are as follows :*

Primary resistance	*= 0.2 ohms ;*
Secondary resistance	*= 2 milli ohms.*
Primary leakage reactance	*= 0.45 ohms ;*
Secondary leakage reactance	*= 4.5 milli ohms.*
Core loss resistance	*= 10 kilo ohms ;*
Magnetizing reactance	*= 1.55 kilo ohms.*

Using the circuit referred to primary, determine the :

(i) *Voltage regulation.*

(ii) *Efficiency of the transformer operating at rated load with 0.8 lagging power factor.*

(PTU, June and Dec. 2000)

Solution. *Given* : Rating = 100 kVA ; $R_1 = 0.2\ \Omega$; $R_2 = 0.002\ \Omega$;

$X_1 = 0.45\ \Omega$; $X_2 = 0.0045\ \Omega$; $R_0 = 10000\ \Omega$; $X_0 = 1550\ \Omega$.

(*i*) **Voltage regulation :**

Transformation ratio, $K = \dfrac{200}{2000} = 0.1$

$$R_{01} = R_1 + R'_2 = R_1 + \frac{R_2}{K^2} = 0.2 + \frac{0.002}{0.1^2} = 0.4\ \Omega$$

$$X_{01} = X_1 + X'_2 = X_1 + \frac{X_2}{K^2} = 0.45 + \frac{0.0045}{0.1^2} = 0.9\ \Omega$$

Primary full-load current, $I_1 = \dfrac{100 \times 1000}{2000} = 50$ A,

$$\cos\phi = 0.8,\ \sin\phi = \sin[\cos^{-1}(0.8)] = 0.6$$

Refer Fig. 52.

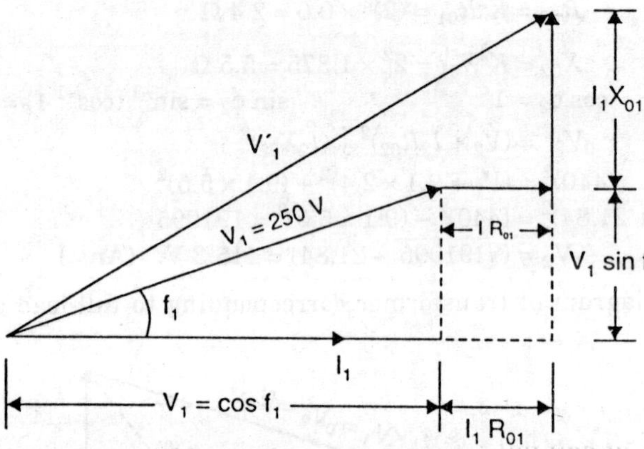

Fig. 52

Applied voltage, $$V'_1 = \sqrt{(V_1\cos\phi + I_1R_{01})^2 + (V_1\sin\phi + I_1X_{01})^2}$$

$$= \sqrt{(2000 \times 0.8 + 50 \times 0.4)^2 + (2000 \times 0.6 + 50 \times 0.9)^2} = 2043 \text{ V}$$

$$\therefore \ \% \text{ voltage regulation} = \frac{V'_1 - V_1}{V_1} \times 100 = \frac{2043 - 2000}{2000} \times 100 = \mathbf{2.15\%}. \ \textbf{(Ans.)}$$

(*ii*) **Efficiency of the transformer, η :**

Working component of no-load current, $I_w = \dfrac{V}{R_0} = \dfrac{2000}{10000} = 0.2$ A

Iron loss = $V_1I_0\cos\phi_0 = V_1I_w = 2000 \times 0.2 = 400$ W

Full-load copper losses, $P_c = I_1^2R_{01} = (50)^2 \times 0.4 = 1000$ W

Efficiency of the transformer at 0.8 p.f.,

$$\%\ \eta = \frac{\text{kVA} \times 1000 \times \cos\phi}{(\text{kVA} \times 1000 \times \cos\phi) + P_i + P_c} \times 100$$

$$= \frac{(100 \times 1000 \times 0.8)}{(100 \times 1000 \times 0.8 + 400 + 1000)} \times 100$$

$$= \mathbf{98.28\%}. \ \textbf{(Ans.)}$$

Example 28. *In a 25 kVA, 2000/200 V transformer, the constant and variable losses are 350 W and 400 W respectively. Calculate the efficiency on u.p.f. at*

(*i*) *Full load, and* (*ii*) *Half full load.* **(PTU, 1999)**

Solution. Rating of transformer $= 25$ kVA

Constant or iron losses, $P_i = 350$ W or 0.35 kW

Variable or copper losses, $P_c = 400$ W or 0.4 kW

Power factor, $\cos\phi = 1$

Efficiency, η :

(*i*) **At full load :**

$$\eta = \frac{\text{output}}{\text{output} + \text{losses}} = \frac{kVA \cos\phi}{kVA \cos\phi + P_i + P_c}$$

$$= \frac{25 \times 1}{25 \times 1 + 0.35 + 0.4} = 0.9708 \text{ or } \mathbf{97.08\%}. \text{ (Ans.)}$$

(*ii*) **At half full load :**

$$\eta = \frac{(0.5 \times 25) \times 1}{(0.5 \times 25 \times 1) + 0.35 + \left[\left(\frac{1}{2}\right)^2 \times 0.4\right]}$$

$$= 0.9653 \text{ or } \mathbf{96.53\%}. \text{ (Ans.)}$$

☞**Example 29.** *The following readings were obtained from O.C. and S.C. tests on 8 kVA 400/120 V, 50-Hz transformer.*

O.C. test (l.v. side) : 120 V ; 4 A ; 75 W.

S.C. test (h.v. side) : 9.5 V ; 20 A ; 110 W.

Calculate :

(*i*) *The equivalent circuit (approximate) constants,*

(*ii*) *Voltage regulation and efficiency for 0.8 lagging power factor load, and*

(*iii*) *The efficiency at half full-load and 0.8 power factor load.*

Solution. (*i*) Transformation ratio, $K = \frac{120}{400} = \frac{3}{10}$

O.C. test :

It is seen from the O.C. test, that *primary open*, the secondary draws a no-load current of 4 A. Since $K = \frac{3}{10}$, the corresponding no-load current $I_0 = 4 \times \frac{3}{10} = 1.2$ A.

Also $P_0 = V_1 I_0 \cos\phi_0$

$\therefore$ $75 = 400 \times 1.2 \times \cos\phi_0$

i.e., $\cos\phi_0 = \frac{75}{400 \times 1.2} = 0.156$

and $\sin\phi_0 = 0.987$

Now, wattful component of no-load current, I_0,

$$I_w = I_0 \cos\phi_0 = 1.2 \times 0.156 = 0.187 \text{ A}$$

and magnetising component of no-load current I_0,

$$I_m = I_0 \sin\phi_0 = 1.2 \times 0.987 = 1.184 \text{ A}$$

$\therefore$ Resistance representing the core loss,

$$R_0 = \frac{V_1}{I_w} = \frac{400}{0.187} = \mathbf{2139\ \Omega}. \text{ (Ans.)}$$

Magnetising reactance, $X_0 = \frac{V_1}{I_m} = \frac{400}{1.184} = \mathbf{337.8\ \Omega}.$ **(Ans.)**

S.C. test :

During S.C. test the *instruments have been placed in primary.*

Here, $V_{SC} = 9.5$ V

$I_{SC} = 20$ A

$P_{SC} = 110$ W

Now, $Z_{01} = \frac{V_{SC}}{I_{SC}} = \frac{9.5}{20} = \mathbf{0.475\ \Omega}$. **(Ans.)**

$R_{01} = \frac{P_{SC}}{I_{SC}^2} = \frac{110}{(20)^2} = \mathbf{0.25\ \Omega}$. **(Ans.)**

$\therefore$ $X_{01} = \sqrt{Z_{01}^2 - R_{01}^2} = \sqrt{(0.475)^2 - (0.25)^2} = \mathbf{0.404\ \Omega}$. **(Ans.)**

The equivalent circuit is shown in Fig. 53.

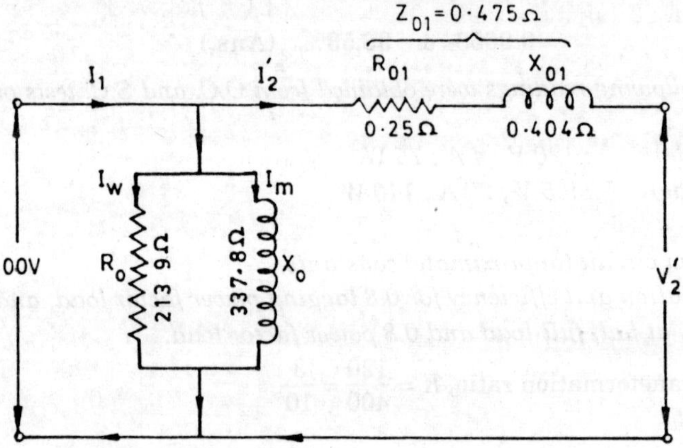

Fig. 53

(*ii*) Total *approximate voltage drop* as referred to primary is

$= I_1 (R_{01} \cos \phi + X_{01} \sin \phi)$

Now, full-load, $I_1 = \frac{8 \times 1000}{400} = 20$ A

$\therefore$ Voltage drop $= 20 (0.25 \times 0.8 + 0.404 \times 0.6) = 8.85$ V

% Voltage regulation $= \frac{8.85}{400} \times 100 = \mathbf{2.21\%}$. **(Ans.)**

Iron loss, $P_i = 75$ W

Copper loss, $P_c = 110$ W

Full-load losses $= P_i + P_c = 75 + 110 = 185$ W

$\therefore$ Full-load output $= 8 \times 1000 \times 0.8 = 6400$ W

$\therefore$ **Full-load efficiency,** $\eta = \frac{\text{output}}{\text{output + losses}} = \frac{6400}{6400 + 185} = 0.972$ or **97.2%. (Ans.)**

(*iii*) **Half full-load :**

Iron loss, $P_i = 75$ W

Copper loss, $P_c = (\frac{1}{2})^2 \times 110 = 27.5$ W

Total losses $= P_i + P_c = 75 + 27.5 = 102.5$ W

Output at half full-load $= \frac{1}{2} \times 8 \times 1000 \times 0.8 = 3200$ W

∴ **Efficiency** $= \dfrac{\text{output}}{\text{output + losses}} = \dfrac{3200}{3200 + 102.5} = 0.969$ or **96.9%. (Ans.)**

Example 30. *Consider a 20 kVA, 2200/220 V, 50 Hz transformer. The OC/SC test results are as follows :*

O.C. Test : 220 V, 4.2 A, 148 W (l.v. side)

S.C. Test : 86 V, 10.5 A, 360 W (h.v. side)

(i) Determine the regulation at 0.8 p.f. lagging at full load.

(ii) What is the p.f. on short-circuit. **(Nagpur University, 1998)**

Solution. *Given :*

S.C. Test :

$$V_{SC} = 86 \text{ V}, I_{SC} = 10.5 \text{ A}, P_C = 360 \text{ W}$$

(i) **Regulation :** Equivalent impedance referred to primary, $Z_{01} = \dfrac{V_{SC}}{I_{SC}} = \dfrac{86}{10.5} = 8.19\ \Omega$

(∵ Short-circuit test has been conducted on the h.v. (primary) side)

Equivalent resistance referred to primary, $R_{01} = \dfrac{P_{SC}}{I_{SC}^2} = \dfrac{360}{(10.5)^2} = 3.27\ \Omega$

Equivalent reactance referred to primary, $X_{01} = \sqrt{Z_{01}^2 - R_{01}^2} = \sqrt{8.19^2 - 3.27^2} = 7.51\ \Omega$

Full load primary current, $I_1 = \dfrac{20 \times 1000}{2200} = 9.09$ A

Percentage regulation $= \dfrac{I_1 (R_{01} \cos \phi + X_{01} \sin \phi)}{V_1} \times 100$

$= \dfrac{9.09(3.27 \times 0.8 + 7.51 \times 0.6)}{2200} \times 100 =$ **2.94%. (Ans.)**

(ii) **P.f. on short-circuit, cos ϕ_{SC} :**

$$\cos \phi_{SC} = \frac{R_{01}}{Z_{01}} = \frac{3.27}{8.19} = \mathbf{0.399\ (lag).\ \ (Ans.)}$$

Example 31. *High voltage side short circuit test data for 20 kVA, 2300/230 V transformer are :*

Power = 250 watts ; Current = 8.7 A ; Voltage = 50 V.

Calculate equivalent impedance, resistance, reactance referred to h.v. side. Find the transformer regulation at 0.7 lagging power factor.

Solution. *Given :* **h.v. side :** $P_{SC} = 250$ W ; $I_{SC} = 8.7$ A ; $V_{SC} = 50$ V ; $\cos \phi = 0.7$.

From S.C. test with measurements performed on h.v. side the various parameters referred to *h.v.* side are :

$$R_{01} = \frac{P_{SC}}{I_{SC}^2} = \frac{250}{8.7^2} = 3.303\ \Omega$$

$$Z_{01} = \frac{V_{SC}}{I_{SC}} = \frac{50}{8.7} = 5.747\ \Omega$$

$$X_{01} = \sqrt{Z_{01}^2 - R_{01}^2} = \sqrt{5.747^2 - 3.303^2} = 4.703\ \Omega$$

We know that, % Regulation $= \dfrac{I_1R_{01} \cos \phi + I_1X_{01} \sin \phi}{V_1} \times 100$

where $I_1 = \dfrac{20 \times 10^3}{2300} = 8.696$ A (rated current) ; and

$\cos \phi = 0.7$ (given) and hence $\sin \phi = \sin (\cos^{-1} 0.7) = 0.714$.

$\therefore$ **% Regulation** $= \dfrac{8.696 \times 3.303 \times 0.7 + 8.696 \times 4.703 \times 0.714}{2300} \times 100 =$ **2.14%. (Ans.)**

Example 32. *A 10 kVA, 2300/230 V, 50 Hz, distribution transformer has h.v. winding resistance 3.96 Ω and leakage resistance of 15.8 Ω. The l.v. winding has corresponding value of 0.0396 Ω and 0.158 Ω respectively. The transformer has a core loss of 58 W under normal operating conditions. Find :*

(i) Load terminal voltage when transformer delivers rated current at 0.8 p.f. lagging to a load on l.v. side, with h.v. side voltage held at rated value and compute efficiency at this load.

(ii) The h.v. side voltage necessary to maintain rated voltage at load terminals when the transformer is loaded as above. Is efficiency same as in (i) ? **(AMIE Summer, 1999)**

Solution. *Given :* $R_1 = 3.96\ \Omega$; $X_1 = 15.8\ \Omega$; $R_2 = 0.0396\ \Omega$; $X_2 = 0.158\ \Omega$;

$P_i = 58$ W ; *p.f.* = 0.8 lagging

(*i*) **Load terminal voltage :**

Rated secondary current, $I_2 = \dfrac{10 \times 10^3}{230} = 43.48$ A

$$R_{02} = R_2 + K^2R_1 = 0.0396 + \left(\frac{230}{2300}\right)^2 \times 3.96 = 0.0792\ \Omega$$

$$X_{02} = X_2 + K^2X_1 = 0.158 + \left(\frac{230}{2300}\right)^2 \times 15.8 = 0.316\ \Omega$$

$\therefore$ Voltage drop on load side $= I_2 (R_{02} \cos \phi + X_{02} \sin \phi)$

$= 43.48 (0.0792 \times 0.8 + 0.316 \times 0.6) = 11$ V

$\therefore$ **Load terminal voltage** = 230 – 11 = **219 V. (Ans.)**

Copper loss $(P_C) = I_2^2R_{02} = 43.48^2 \times 0.0792 = 149.73$ W

$$\therefore \quad \eta = \frac{\text{Output}}{\text{Output + Core loss } (P_i) + \text{Copper loss } (P_C)}$$

$$= \frac{10 \times 10^3 \times 0.8}{(10 \times 10^3 \times 0.8) + 58 + 149.73} = 0.9747 \text{ or } \mathbf{97.47\%.\ (Ans.)}$$

(*ii*) To maintain rated voltage at load terminals of 230 V, the O.C. voltage required at secondary terminals would be 230 + 11 = 241 V. Hence primary induced voltage required

$= 241 \times \dfrac{2300}{230} =$ **2410 V. (Ans.)**

Example 33. *The following test results were obtained for a 1000/100 V, 100 kVA single-phase transformer :*

O.C. test :

Primary volts = 1000, secondary volts = 100, watts in primary = 950.

S.C. test :

Primary volts for full-load current = 20, watts in primary = 1000.

Determine the regulation and efficiency of the transformer at full-load and at 0.8 power factor lagging.

Solution. Transformation ratio, $K = \frac{100}{1000} = \frac{1}{10}$

O.C. test :

Primary volts, $V_1 = 1000$ V

Iron loss, $P_i = 950$ W $= 0.95$ kW

S.C. test :

Full-load copper loss, $P_c = 1000$ W $= 1$ kW

In S.C. test 20 V in primary are equivalent to $20 \times \frac{1}{10} = 2$ V in secondary.

Now, full-load secondary current,

$$I_2 = \frac{100 \times 1000}{100} = 1000 \text{ A}$$

Also $$Z_{02} = \frac{2}{1000} = 0.002\ \Omega$$

and $$R_{02} = \frac{1000}{(1000)^2} = 0.001\ \Omega$$

$$\therefore \quad X_{02} = \sqrt{Z_{02}^2 - R_{02}^2} = \sqrt{0.002^2 - 0.001^2} = 0.00173\ \Omega$$

Approximate voltage drop at 0.8 power factor lagging

$$= I_2(R_{02} \cos \phi + X_{02} \sin \phi)$$
$$= 1000(0.001 \times 0.8 + 0.00173 \times 0.6) = 1.838 \text{ V}$$

$\therefore$ % **regulation down** $= \frac{1.838}{100} \times 100 =$ **1.838%. (Ans.)**

Full-load output $= 100 \times 0.8 = 80$ kW

Total losses $= P_i + P_c = 0.95 + 1 = 1.95$ kW

$\therefore$ **Efficiency,** $\eta = \frac{\text{output}}{\text{output + losses}} = \frac{80}{80 + 1.95} = 0.976$ or **97.6%. (Ans.)**

Example 34. *The primary and secondary winding resistances of a 30 kVA, 6600/250 V single-phase transformer are 8 Ω and 0.015 Ω respectively. The equivalent leakage reactance as referred to the primary winding is 30 Ω. Find the full-load regulation for load power factors of :*

(i) unity *(ii) 0.8 lagging* *(iii) 0.8 leading.*

Solution. Transformer ratio, $K = \frac{250}{6600} = \frac{1}{26.4}$

Primary resistance, $R_1 = 8\ \Omega$

Secondary resistance, $R_2 = 0.015\ \Omega$

Equivalent leakage reactance as referred to primary,

$$X_{01} = 30\ \Omega$$

Now, $$R_{02} = R_2 + K^2 R_1 = 0.015 + \left(\frac{1}{26.4}\right)^2 \times 8 = 0.02648\ \Omega$$

$$X_{02} = K^2 X_{01} = \left(\frac{1}{26.4}\right)^2 \times 30 = 0.04304\ \Omega$$

Secondary full-load current, $I_2 = \frac{30 \times 1000}{250} = 120$ A.

(*i*) **Unity power factor,** $\cos \phi_2 = 1$:

Voltage drop $= I_2R_{02} \cos \phi_2 + I_2X_{02} \sin \phi_2 = 120 \times 0.02648 \times 1 + 0 = 3.177$ V

% **regulation** $= \frac{3.177}{250} \times 100 =$ **1.27%. (Ans.)**

(*ii*) **0.8 power factor (lagging) :**

Voltage drop $= I_2R_{02} \cos \phi_2 + I_2X_{02} \sin \phi_2$

$= 120 \times 0.02648 \times 0.8 + 120 \times 0.04304 \times 0.6 = 5.64$ V

% **regulation** $= \frac{564}{250} \times 100 =$ **2.256%. (Ans.)**

(*iii*) **0.8 power factor (leading) :**

Voltage drop $= I_2R_{02} \cos \phi_2 - I_2X_{02} \sin \phi_2$

$= 120 \times 0.02648 \times 0.8 - 120 \times 0.04304 \times 0.6 = -0.557$ V

% **regulation** $= -\frac{0.557}{250} \times 100 =$ **– 0.223%. (Ans.)**

Example 35. *Short-circuit test is conducted on a 5 kVA, 400 V/100 V single phase transformer with 100 V winding shorted. The input voltage at full load current is 40 V. The wattmeter, on the input reads 250 W. Find the power factor for which regulation at full load is zero.* **(GATE, 1994)**

Solution. *Given* : Rating of transformer : 5 kVA, 400 V/100 V

Input power, on short-circuit, $P_{SC} = 250$ W

The input voltage at full load current, $V_{SC} = 40$ V

Power factor for which regulation at full load is zero :

Input current I_{SC} = full load current, $I_1 = \frac{5 \times 1000}{400} = 12.5$ A

Equivalent resistance as referred to primary, $R_{01} = \frac{P_{SC}}{I_{SC}^2} = \frac{250}{(12.5)^2} = 1.6\ \Omega$

Equivalent impedance referred to primary, $Z_{01} = \frac{V_{SC}}{I_{SC}} = \frac{40}{12.5} = 3.2\ \Omega$

Equivalent reactance referred to primary, $X_{01} = \sqrt{Z_{01}^2 - R_{01}^2} = \sqrt{3.2^2 - 1.6^2} = 2.77\ \Omega$

Regulation will be zero when phase angle,

$$\phi = \tan^{-1} \frac{R_{02}}{X_{02}} = \tan^{-1} \frac{R_{01}}{X_{01}} = \tan^{-1} \left(\frac{1.6}{2.77} \right) = 30°$$

or **power factor** $= \cos \phi = \cos (30°) =$ **0.866. (Ans.)**

Example 36. *A 50 MVA, 76.2 V/33 kV, 1-phase, 50 Hz, two-winding transformer with tap changer has percentage impedance of 0.5 + j7.0. What tapping must be used to maintain rated voltage at the secondary on*

(*i*) *full load at 0.8 lagging power factor, and*

(*ii*) *40 MVA load at 0.6 lagging power factor.*

Assume that the tap changer is provided on the h.v. side.

Solution. *Given* : Rated MVA = 50 ; $f = 50$ Hz ; $E_1 = 76.2$ V ; $E_2 = 33$ kV ;

percentage impedance = $0.5 + j7.0$.

(i) Percentage regulation $= \dfrac{L_2R_{02} \cos \phi + I_2X_{02} \sin \phi}{E_2} \times 100$

$= \dfrac{I_2R_{02}}{E_2} \times 100 \cos \phi + \dfrac{I_2 X_0}{E_2} \times 100 \sin \phi$

$= \% \, R \cos \phi + \% \, X \sin \phi$

$= 0.5 \times 0.8 + 7 \times 0.6 = 4.6\%$

i.e., voltage to be raised for maintaining rated voltage = 4.6%

So tap setting required on h.v. side = – 4.6% or **4.6% down. (Ans.)**

(ii) At load of 40 MVA *i.e.* 80% of full load or 0.8 times full load,

voltage regulation $= 0.8 \, (\% \, R \cos \phi + \% \, X \sin \phi)$

$= 0.8(0.5 \times 0.6 + 7 \times 0.8) = 4.72\%$

Tap setting required on h.v. side = **4.72% down. (Ans.)**

Note. *If the tap changer is on the primary side, tap setting will be down and if it is to be provided on secondary side, tap setting will be up for raising the secondary voltage on load.*

Example 37. *The percentage resistance and reactance of a transformer are 2% and 4% respectively. Find the approximate regulation on full-load at :*

(i) *Unity power factor,* (ii) *0.8 power factor lagging, and*

(iii) *0.8 power factor leading.*

Solution. Percentage resistance = 2%

Percentage reactance = 4%

Approximate % regulation = % resistance × cos ϕ ± % reactance × sin ϕ

(i) **Unity power factor :**

Approximate % regulation = % resistance cos ϕ ($\because$ sin ϕ = 0)

= 2 × 1 = **2%. (Ans.)**

(ii) **0.8 power factor lagging :**

Approximate % regulation = % resistance × cos ϕ + % reactance × sin ϕ

= 2 × 0.8 + 4 × 0.6 = 1.6 + 2.4 = **4%. (Ans.)**

(iii) **0.8 power factor leading :**

Approximate % regulation = % resistance × cos ϕ – % reactance × sin ϕ

= 2 × 0.8 – 4 × 0.6 = 1.6 – 2.4 = **– 0.8%. (Ans.)**

Example 38. *A single-phase 80 kVA, 2000/200 V, 50-Hz transformer has impedance drop of 8% and resistance drop of 4% :*

(i) *Find the regulation at full-load 0.8 power factor lagging.*

(ii) *At what power factor is the regulation zero.*

Solution. Impedance drop = 8%

Resistance drop = 4%

i.e., $$\frac{I_2Z_{02}}{V_2} = \frac{8}{100} \quad \text{or} \quad I_2Z_{02} = \frac{200 \times 8}{100} = 16 \text{ V}$$

and $$\frac{I_2R_{02}}{V_2} = \frac{4}{100} \quad \text{or} \quad I_2R_{02} = \frac{200 \times 4}{100} = 8 \text{ V}$$

$\therefore$ $$I_2X_{02} = \sqrt{(I_2Z_{02})^2 - (I_2R_{02})^2} = \sqrt{16^2 - 8^2} = 13.86 \text{ V.}$$

(i) % regulation $= \dfrac{I_2R_{02}\cos\phi_2 + I_2X_{02}\sin\phi_2}{V_2}$

$$= \frac{8\times 0.8 + 13.86\times 0.6}{200} = \frac{6.4\times 8.316}{200} = \textbf{7.36\%.}\ \textbf{(Ans.)}$$

(ii) *Regulation can be zero only when the power factor is leading.* Then

$$\frac{I_2R_{02}\cos\phi_2 - I_2X_{02}\sin\phi_2}{V_2} = 0$$

or $$I_2R_{02}\cos\phi_2 - I_2X_{02}\sin\phi_2 = 0$$

or $$I_2R_{02}\cos\phi_2 = I_2X_{02}\sin\phi_2$$

or $$\tan\phi_2 = \frac{I_2R_{02}}{I_2X_{02}} = \frac{8}{13.86} = 0.577 \quad \text{or} \quad \phi_2 = 30°$$

$\therefore$ **Power factor = cos ϕ_2 = 0.866 leading. (Ans.)**

Example 39. *The high voltage of a single-phase 200 kVA 4400/220 V transformer takes a current of 35 A and power of 1250 W at 80 V when the low voltage winding is short-circuited. Determine :*

(i) The voltage to be applied to the high voltage winding on full-load at 0.8 power factor lagging if the full-load secondary terminal voltage is to be kept at 220 V.

(ii) % regulation.

Solution. (i) Transformation ratio, $K = \dfrac{220}{4400} = \dfrac{1}{20}$

Voltage applied to h.v. winding in S.C. test

$$V_{SC} = 80\ \text{V}$$

and $$I_{SC} = 35\ \text{A}$$

$$P_{SC} = 1250\ \text{W}$$

Referred to *h.v. side,*

$$Z_{01} = \frac{V_{SC}}{I_{SC}} = \frac{80}{35} = 2.28\ \Omega$$

$$R_{01} = \frac{P_{SC}}{I_{SC}^2} = \frac{1250}{35^2} = 1.02\ \Omega$$

$$X_{01} = \sqrt{Z_{01}^2 - R_{01}^2} = \sqrt{2.28^2 - 1.02^2} = 2.04\ \Omega$$

Full-load current in h.v. winding,

$$I_1 = \frac{200\times 1000}{4400} = 45.45\ \text{A}$$

Voltage drop in resistance $= I_1R_{01} = 45.45\times 1.02 = 46.36$ V

Voltage drop in reactance $= I_1X_{01} = 45.45\times 2.04 = 92.72$ V

Full-load secondary voltage referred to primary,

$$V_2' = \frac{V_2}{K} = \frac{220}{\frac{1}{20}} = 4400\ \text{V}$$

To get full-load secondary voltage equal to 220 V, supply voltage V_1 must be equal to the vector sum of 4400 V, I_1R_{01} and IX_{01}.

From Fig. 54,

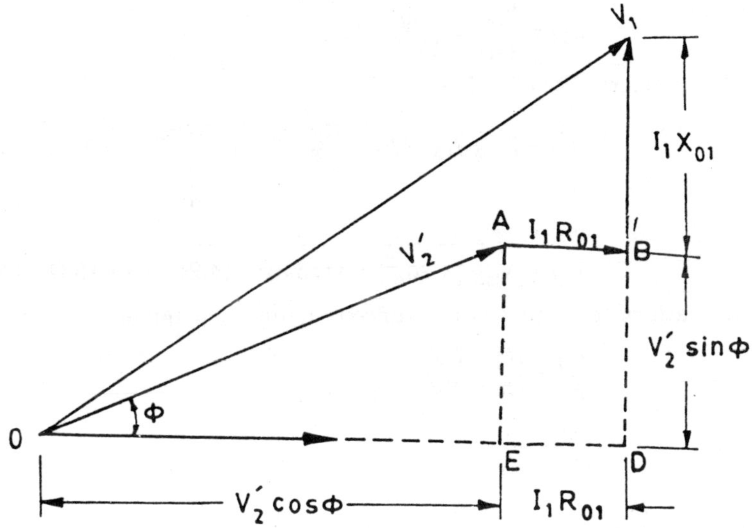

Fig. 54

$$OC^2 = OD^2 + CD^2 = (OE + ED)^2 + (BD + BC)^2$$

$$V_1{}^2 = (V_2' \cos \phi + I_1R_{01})^2 + (V_2' \sin \phi + I_1X_{01})^2$$

$$= (4400 \times 0.8 + 46.36)^2 + (4400 \times 0.6 + 92.72)^2$$

or $$V_1 = 4493 \text{ V}$$

Voltage to be applied to h.v. winding to get a full-load secondary voltage of 220 V = 4493 V. (Ans.)

(*ii*) % **regulation** $= \dfrac{V_1 - V_2'}{V_1} \times 100 = \dfrac{4493 - 4400}{4493} \times 100 =$ **2.07%. (Ans.)**

Example 40. *The high voltage and low voltage windings of a 50 kVA, 4400/220 V, 50-Hz transformer have resistances of 2.2 Ω and 0.005 Ω respectively. The full-load current is obtained when 160 V at 50-Hz is applied to h.v. winding with l.v. winding short-circuited. Find :*

(*i*) *The equivalent resistance and reactance of the transformer referred to h.v. side, and*

(*ii*) *Reactance of each winding.*

Assume that the ratio of resistance to reactance is the same for each winding and the full-load efficiency of transformer is 0.98.

Solution. (*i*) Transformation ratio, $K = \dfrac{220}{4400} = \dfrac{1}{20}$

Primary resistance, $R_1 = 22\ \Omega$

Secondary resistance, $R_2 = 0.005\ \Omega$

Full-load output $= 50 \text{ kVA} = 50 \times 10^3 \text{ VA}$

Full-load input $= \dfrac{\text{output}}{\text{efficiency}} = \dfrac{50 \times 10^3}{0.98} \text{ VA}$

Full-load input current, $I_1 = \dfrac{\text{input in VA}}{\text{supply voltage}} = \dfrac{50 \times 10^3}{0.98 \times 4400} = 11.6 \text{ A}$

Impedance referred to h.v. side

$$Z_{01} = \frac{V_{SC}}{I_{SC}} = \frac{160}{11.6} = 13.8\ \Omega$$

Total resistance referred to h.v. side,

$$R_{01} = R_1 + R_2' = R_1 + \frac{R_2}{K^2} = 2.2 + \frac{0.005}{\left(\frac{1}{20}\right)^2} = 2.2 + 2 = \mathbf{4.2\ \Omega.}\ \ \textbf{(Ans.)}$$

$$X_{01} = \sqrt{Z_{01}^2 - R_{01}^2} = \sqrt{(13.8)^2 - (4.2)^2} = \mathbf{13.15\ \Omega.}\ \ \textbf{(Ans.)}$$

(*ii*) Since the ratio of resistance to reactance remains same (given),

$$\therefore \quad \frac{R_1}{X_1} = \frac{R_2}{X_2} = \frac{R_2'}{X_2'}$$

i.e.,

$$\frac{R_1}{R_2'} = \frac{X_1}{X_2'}$$

or

$$\frac{R_1 + R_2'}{R_2'} = \frac{X_1 + X_2'}{X_2'} \quad \text{or} \quad \frac{R_{01}}{R_2'} = \frac{X_{01}}{X_2'}$$

or

$$X_2' = \frac{R_2' X_{01}}{R_{01}} = \frac{2 \times 13.15}{4.2} = 6.26\ \Omega$$

Reactance of primary winding,

$$X_1 = X_{02} - X_2' = 13.15 - 6.26 = \mathbf{6.85\ \Omega.}\ \ \textbf{(Ans.)}$$

Reactance of secondary winding,

$$X_2 = K^2 X_2' = \left(\frac{1}{20}\right)^2 \times 6.26 = \mathbf{0.01565\ \Omega.}\ \ \textbf{(Ans.)}$$

☞**Example 41.** *The following test results were obtained in a 250/500 V transformer :*

O.C. test (l.v. side) : 250 V, 1 A, 80 W.

S.C. test (l.v. winding short-circuited) : 20 V, 12 A, 100 W.

Determine :

(i) The circuit constants.

(ii) The applied voltage and efficiency when the output is 10 A at 500 V and 0.8 power factor lagging.

Solution. Transformation ratio, $K = \frac{500}{250} = 2$

(*i*) **Circuit constants :**

O.C. test *(l.v. side)*

Voltage, $V_1 = 250$ V

No-load current, $I_0 = 1$ A

No-load loss, $P_0 = 80$ W

Also, $P_0 = V_1 I_0 \cos \phi_0$

$$80 = 250 \times 1 \times \cos \phi_0$$

$$\cos \phi_0 = \frac{80}{250} = 0.32$$

Wattful component of no-load current I_0,

$$I_w = I_0 \cos \phi_0 = 1 \times 0.32 = 0.32 \text{ A}$$

Magnetising component of no-load current, I_0,

$$I_m = \sqrt{I_0^2 - I_w^2} = \sqrt{1^2 - 0.32^2} = 0.95 \text{ A}$$

Now, resistance representing the core loss,

$$R_0 = \frac{V_1}{I_w} = \frac{250}{0.32} = \mathbf{781.25\ \Omega.\ (Ans.)}$$

Magnetising reactance, $X_0 = \frac{V_1}{I_m} = \frac{250}{0.95} = \mathbf{263.16\ \Omega.\ (Ans.)}$

S.C. test *(l.v. winding short-circuited)*

Short-circuit voltage, $V_{SC} = 20$ V

Short-circuit current, $I_{SC} = 12$ A

Losses, $P_{SC} = 100$ W

As the primary is short-circuited all *values refer to secondary winding.*

$$\therefore \quad R_{02} = \frac{P_{SC}}{I_{SC}^2} = \frac{100}{(12)^2} = 0.694\ \Omega$$

$$Z_{02} = \frac{V_{SC}}{I_{SC}} = \frac{20}{12} = 1.677\ \Omega$$

and

$$X_{02} = \sqrt{Z_{02}^2 - R_{02}^2} = \sqrt{(1.667)^2 - (0.694)^2} = 1.516\ \Omega$$

As R_0 and X_0 refer to primary, let us transfer these values to primary as follows :

$$R_{01} = \frac{R_{02}}{K^2} = \frac{0.694}{(2)^2} = \mathbf{0.174\ \Omega}$$

$$X_{01} = \frac{X_{02}}{K^2} = \frac{1.516}{(2)^2} = \mathbf{0.38\ \Omega}$$

$$Z_{01} = \frac{Z_{02}}{K^2} = \frac{1.669}{(2)^2} = \mathbf{0.417\ \Omega}$$

The equivalent circuit is shown in Fig. 55.

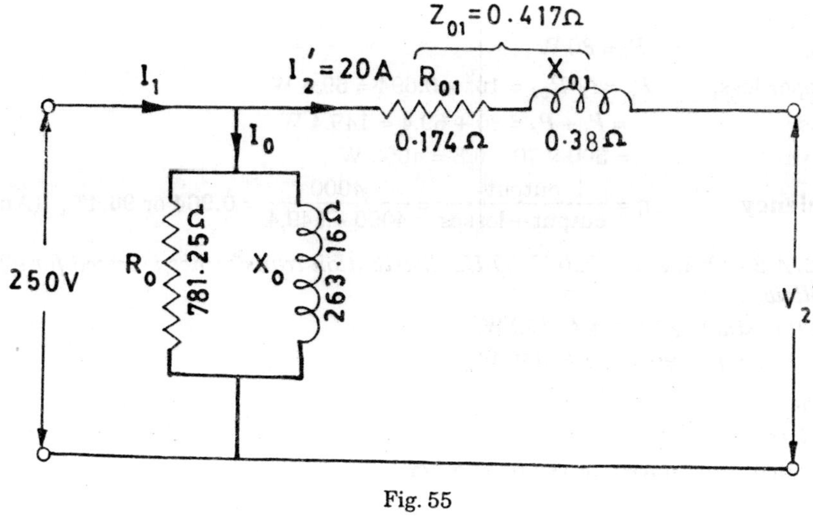

Fig. 55

(*ii*) **Applied voltage, V_1' :**

The applied voltage V_1' is the vector sum of V_1 and I_1Z_{01} (Fig. 56).

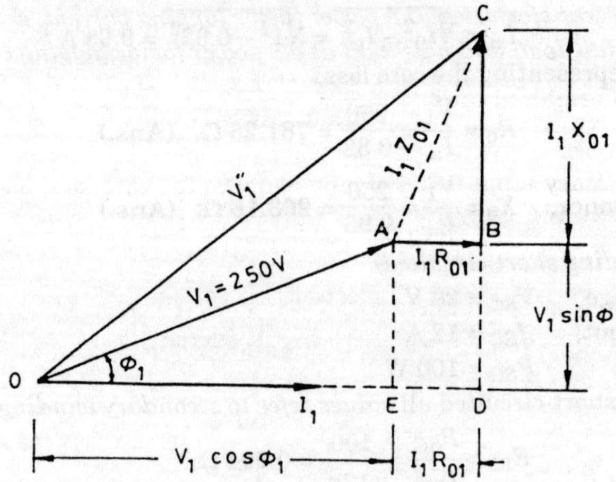

Fig. 56

Output current, $I_2 = 10$ A

$\therefore$ $I_1 = KI_2 = 2 \times 10 = 20$ A

Now, $I_1R_{01} = 20 \times 0.174 = 3.48$ V

$I_1X_{01} = 20 \times 0.38 = 7.6$ V

Refer Fig. 56. Neglecting the angle between V_1 and V_1', we have

$$OC^2 = OD^2 + DC^2$$

$$OC = V_1' = \sqrt{(OE + ED)^2 + (DB + BC)^2}$$

$$= \sqrt{(V_1 \cos \phi_1 + I_1R_{01})^2 + (V_1 \sin \phi_1 + I_1X_{01})^2}$$

$$= \sqrt{(250 \times 0.8 + 3.48)^2 + (250 \times 0.6 + 7.6)^2} = 257.4 \text{ V}$$

Hence, **applied voltage = 257.4 V. (Ans.)**

Efficiency :

Iron loss, $P_i = 80$ W

Total copper loss, $P_c = I_2^2R_{02} = 10^2 \times 0.694 = 69.4$ W

Total loss, $= P_i + P_c = 80 + 69.4 = 149.4$ W

Output $= 500 \times 10 \times 0.8 = 4000$ W

$\therefore$ **Efficiency,** $\eta = \dfrac{\text{output}}{\text{output + losses}} = \dfrac{4000}{4000 + 149.4} =$ **0.964** or **96.4%. (Ans.)**

☞**Example 42.** *A 25-kVA, 2200/220 V, 50-Hz distribution transformer is tested for efficiency and regulation as follows :*

O.C. test (l.v. side) : 220 V, 4 A, 150 W.

S.C. test (h.v. side) : 90 V, 10 A, 350 W.

Determine :

(*i*) *Core loss,*

(*ii*) *Equivalent resistance referred to primary,*

(*iii*) *Equivalent reactance referred to secondary,*
(*iv*) *Equivalent reactance referred to primary,*
(*v*) *Equivalent reactance referred to secondary,*
(*vi*) *Regulation of transformer at 0.8 power factor lagging current, and*
(*vii*) *Efficiency at full-load and half-load at 0.8 power factor lagging current.*

Solution. Transformation ratio, $K = \frac{220}{2200} = \frac{1}{10}$

(*i*) **Core loss :**

Since no-load primary input is practically equal to the core loss, hence, **core loss as found from no-load test, is 150 W. (Ans.)**

(*ii*) **From S.C. test :**

$$V_{SC} = 90 \text{ V (short-circuit voltage)}$$
$$I_{SC} = 10 \text{ A (short-circuit current)}$$
$$P_{SC} = 350 \text{ W (copper loss)}$$

∴ **Equivalent resistance referred to primary,**

$$R_{01} = \frac{P_{SC}}{I_{SC}^2} = \frac{350}{(10)^2} = \mathbf{3.5\ \Omega.\ (Ans.)}$$

(*iii*) **Equivalent resistance referred to secondary,**

$$R_{02} = K^2R_{01} = \left(\frac{1}{10}\right)^2 \times 3.5 = \mathbf{0.035\ \Omega.\ (Ans.)}$$

(*iv*) Also $Z_{01} = \frac{V_{SC}}{I_{SC}} = \frac{90}{10} = 9\ \Omega$

∴ **Equivalent reactance referred to primary,**

$$X_{01} = \sqrt{Z_{01}^2 - R_{01}^2} = \sqrt{9^2 - 3.5^2} = \mathbf{8.29\ \Omega.\ (Ans.)}$$

(*v*) **Equivalent reactance referred to secondary,**

$$X_{02} = K^2X_{01} = \left(\frac{1}{10}\right)^2 \times 8.29 = \mathbf{0.0829\ \Omega.\ (Ans.)}$$

(*vi*) **% regulation :**

Let us find the rise in voltage necessary to maintain the output terminal voltage constant from no-load to full-load.

Rated primary current $= \frac{25 \times 1000}{2200} = 11.36$ A

Now, using the relation,

$$V_1' = \sqrt{(V_1 \cos\phi + I_1R_0)^2 + (V_1 \sin\phi + I_1X_{01})^2}$$
$$= \sqrt{(2200 \times 0.8 + 11.36 \times 3.5)^2 + (2200 \times 0.6 + 11.36 \times 8.29)^2}$$
$$= 2289 \text{ V (app.)}$$

∴ **% regulation** $= \frac{V_1' - V_1}{V_1} \times 100$ (Art. 9.11)

$$= \frac{2289 - 2200}{2200} \times 100 = \mathbf{4.045\%.\ (Ans.)}$$

We can get the same result by working in the secondary :

Rated secondary current,

$$I_2 = \frac{I_1}{K} = \frac{11.36}{1/10} = 113.6 \text{ A}$$

$$_0V_2 = \sqrt{(V_2 \cos\phi + I_2R_{02})^2 + (V_2 \sin\phi + I_2X_{02})^2}$$

$$= \sqrt{(220 \times 0.8 + 113.6 \times 0.035)^2 + (220 \times 0.6 + 113.6 \times 0.0829)^2}$$

$$= 228.9 \text{ V}$$

$\therefore$ % regulation $= \dfrac{_0V_2 - V_2}{V_2} \times 100 = \dfrac{228.9 - 220}{220} \times 100 =$ **4.045%. (Ans.)**

(*vii*) **Efficiency :**

Core loss, $P_i = 150$ W

Copper loss at full-load,

$$P_c = I_1^2 R_{01} = 11.36^2 \times 3.5 = 451.7 \text{ W}$$

Copper loss at half-load,

$$P_c = \left(\frac{11.36}{2}\right)^2 \times 3.5 = 112.9 \text{ W}$$

$\therefore$ Efficiency at *full-load*

$$= \frac{\text{output}}{\text{output} + P_i + P_c} = \frac{25 \times 1000 \times 0.8}{(25 \times 1000 \times 0.8) + 150 + 451.7}$$

= **0.9707** or **97.07%. (Ans.)**

Efficiency at half *full-load*

$$= \frac{(25 \times 1000/2) \times 0.8}{(25 \times 1000/2) \times 0.8 + 150 + 112.9}$$ = **0.9744** or **97.44%. (Ans.)**

☞**Example 43.** *Two similar 100 kVA, single-phase transformers gave the following test readings when tested by Sumpner's test.*

Supply power = 2.4 kW.

Power supplied to secondary circuit in passing full-load current through it = 3.2 kW.

Find the efficiency and regulation of each transformer at unity power factor.

Solution. Core loss (or iron loss) of the transformers

= 2.4 kW

Iron loss of each transformer,

$$P_i = \frac{2.4}{2} = 1.2 \text{ kW}$$

Full-load copper loss of both transformers

= 3.2 kW

Full-load copper loss of each transformer

$$P_c = \frac{3.2}{2} = 1.6 \text{ kW}$$

Total full-load losses of each transformer

$= P_i + P_c = 1.2 + 1.6 = 2.8$ kW

Output at full-load at unity power factor

$= 100 \times 1 = 100$ kW

$\therefore$ **Full-load efficiency,** $\eta = \frac{\text{output}}{\text{output + losses}} = \frac{100}{100 + 2.8} = \mathbf{0.9727}$ or **97.27%. (Ans.)**

Example 44. *A 10 kVA, 2500/250 V, single phase transformer gave the following test results :*

Open circuit test : 250 V, 0.8 A, 50 W

Short circuit test : 60 V, 3 A, 45 W

(i) Calculate the efficiency of half full-load at 0.8 p.f.

(ii) Calculate the load kVA at which maximum efficiency occurs and also the maximum efficiency at 0.8 p.f.

(iii) Compute the voltage regulation at 0.8 p.f. leading. **(AMIE Winter, 1998)**

Solution. Full rated current $= \frac{10 \times 1000}{2500} = 4$ A

Hence reading of wattmeter corresponding to full-load current of 4 A

$$= 45 \times \left(\frac{4}{3}\right)^2 = 80 \text{ W}$$

$\therefore$ Full-load copper losses $= 80$ W, and

Full-load iron losses $= 50$ W

(*i*) **Efficiency at half-load at 0.8 p.f. :**

$$\eta = \frac{\text{half-load output}}{\text{half-load output + losses}}$$

$$= \frac{(10 \times 1000 \times 0.5) \times 0.8}{[(10 \times 1000 \times 0.5) \times 0.8] + 50 + 80 \times (\frac{1}{2})^2}$$

$$= \frac{4000}{4000 + 50 + 20} = 0.983 \text{ or } \mathbf{98.3\%.\ (Ans.)}$$

(*ii*) **Load kVA for maximum efficiency, and its value :**

For maximum efficiency :

Copper loss = Iron loss = 50 W

$\therefore$ Current at which maximum efficiency occurs $= \frac{50 \times 4}{80} = 2.5$ A

$\therefore$ **Load kVA** $= 10 \times \frac{2.5}{4} = \mathbf{6.25\ kVA.\ (Ans.)}$

η_{max} (at 0.8 p.f.) $= \frac{(6.25 \times 1000 \times 0.8)}{(6.25 \times 1000 \times 0.8) + 50 + 50} = 0.98$ or **98%. (Ans.)**

(*iii*) **Voltage regulation at 0.8 p.f. leading :**

From short circuit test :

$$V_{SC} = 60 \text{ V} ; I_{SC} = 3 \text{ A} ; P_{SC} = 45 \text{ W}$$

Now,
$$R_{01} = \frac{P_{SC}}{I_{SC}^2} = \frac{45}{3^2} = 5\ \Omega$$

$$Z_{01} = \frac{V_{SC}}{I_{SC}} = \frac{60}{3} = 20\ \Omega$$

$\therefore$
$$X_{01} = \sqrt{Z_{01}^2 - R_{01}^2} = \sqrt{20^2 - 5^2} = 19.36\ \Omega$$

∴ Voltage drop at leading p.f. of 0.8

$$= I_1R_{01}\cos\phi - I_1X_{01}\sin\phi$$

$$= 4 \times 5 \times 0.8 - 4 \times 19.36 \times 0.6 = -30.46 \text{ V}$$

Hence % **voltage regulation** $= \frac{-30.46}{2500} \times 100 =$ **– 1.218%. (Ans.)**

Example 45. *A 50 kVA single-phase transformer has a full-load primary current of 250 A and total resistance referred to primary is 0.006 ohm. If the iron loss amounts to 200 W, find the efficiency on full-load and half-load at (i) unity power factor and (ii) 0.8 power factor.*

Solution. Full-load primary current,

$$I_1 = 250 \text{ A}$$

Total resistance referred to primary,

$$R_{01} = 0.006\ \Omega$$

Iron loss, $P_i = 200$ W

Full-load copper loss, $P_c = I_1^2R_{01} = (250)^2 \times 0.006 = 375$ W

Total full-load loss $= P_i + P_c = 200 + 375 = 575 \text{ W} = 0.575$ kW

Half full-load copper loss $= (\frac{1}{2})^2 P_c = (\frac{1}{2})^2 \times 375 = 93.75$ W

Total loss on half-load $= 200 + 93.75 = 293.75 \text{ W} = 0.294$ kW (say)

(*i*) **At unity power factor :**

Full-load output, $= 50 \times 1 = 50$ kW

Efficiency at full-load $= \frac{\text{output}}{\text{output + losses}} = \frac{50}{50 + 0.575} =$ **0.9886** or **98.86%. (Ans.)**

Half-full-load output $= 50 \times \frac{1}{2} \times 1 = 25$ kW

Efficiency at half full-load at unity power factor

$$= \frac{25}{25 + 0.294} = \textbf{0.9984 or 99.84\%.} \quad \textbf{(Ans.)}$$

(*ii*) **At 0.8 power factor :**

Full-load output $= 50 \times 0.8 = 40$ kW

Efficiency on full-load at 0.8 power factor

$$= \frac{40}{40 + 0.575} = \textbf{0.9858 or 98.58\%.} \quad \textbf{(Ans.)}$$

Half full-load output $= 50 \times \frac{1}{2} \times 0.8 = 20$ kW

Efficiency on half full-load at 0.8 power factor

$$= \frac{20}{20 + 0.294} = \textbf{0.9855 or 98.55\%.} \quad \textbf{(Ans.)}$$

☞**Example 46.** *A 40 kVA, single-phase transformer has an iron loss of 300 W and full-load copper loss of 600 W,*

(*i*) *Find the load at which maximum efficiency occurs and the value of maximum efficiency at unity power factor.*

(*ii*) *If the maximum efficiency occurs at 80% of full-load, find the new core loss and full-load copper loss assuming that total full-load loss is a constant.*

Solution. Rating of transformer = 40 kVA

Iron loss, $P_i = 300$ W

Full-load copper loss, $P_c = 600$ W

(*i*) Let the maximum efficiency occurs at x times full-load, then

$$x^2P_c = P_i$$

or
$$x = \sqrt{\frac{P_i}{P_c}} = \sqrt{\frac{300}{600}} = 0.707 = 70.7\%$$

Hence, **efficiency occurs at 70.7% of full-load. (Ans.)**

Maximum efficiency :

Output at unity power factor

$$= 0.707 \times 40 \times 1 = 28.28 \text{ kW}$$

Total losses $= P_i + P_c = 2P_i$ $\quad$ [Because when efficiency is maximum, $P_i = P_c$]

$$= 2 \times 300 = 600 \text{ W}$$
$$= 0.6 \text{ kW}$$

$\therefore$ **Maximum efficiency**

$$= \frac{\text{output}}{\text{output + losses}} = \frac{28.28}{28.28 + 0.6} = \mathbf{0.9792} \text{ or } \mathbf{97.92\%}. \quad \textbf{(Ans.)}$$

(*ii*) **New core loss, P_i' :**

New copper loss, P_c' :

Maximum efficiency occurs at 80% of full-load

Now, $\quad P_i' + P_c' = P_i + P_c$ $\quad$ (*given*)

$$= 300 + 600 = 900 \text{ W}$$

Also
$$0.8 = \sqrt{\frac{P_i'}{P_c'}} \quad \text{or} \quad \frac{P_i'}{P_c'} = [0.8]^2 = 0.64$$

or
$$\frac{P_i'}{P_i' + P_c'} = \frac{0.64}{1 + 0.64} = 0.3902 \quad \text{or} \quad \frac{P_i'}{900} = 0.3902$$

$\therefore$
$$P_i' = 351.2 \text{ W}$$

and
$$P_c' = 900 - P_i' = 900 - 351.2 = 548.8 \text{ W}$$

Hence, **new iron loss = 351.2 W. (Ans.)**

new copper loss = 548.8 W. (Ans.)

Example 47. *The primary and secondary resistances of a 1100/220 V transformer are 0.3 Ω and 0.02 Ω respectively. If iron loss amounts to 260 W determine the secondary current at which maximum efficiency occurs and find the maximum efficiency at 0.8 power factor.*

Solution. Transformation ratio, $K = \frac{220}{1100} = \frac{1}{5}$

Primary resistance, $\quad R_1 = 0.3\ \Omega$

Secondary resistance, $\quad R_2 = 0.02\ \Omega$

Iron loss, $\quad P_i = 260$ W

Total resistance referred to secondary,

$$R_{02} = R_2 + K^2R_1 = 0.02 + \left(\frac{1}{5}\right)^2 \times 0.3 = \mathbf{0.032\ \Omega} \quad \textbf{(Ans).}$$

Let I_2 be the secondary current at maximum efficiency.

At maximum efficiency,

$$\text{Copper loss} = \text{Iron loss}$$

$$I_2^2R_{02} = 260$$

$\therefore$
$$I_2^2 = \frac{260}{R_{02}} = \frac{260}{0.032} = 8125 \quad \text{or} \quad I_2 = 90.14 \text{ A}$$

Hence, **secondary current at maximum efficiency**

$$= 90.14 \text{ A. (Ans.)}$$

Maximum efficiency at 0.8 power factor :

Output at maximum efficiency at 0.8 power factor

$$= V_2 I_2 \cos\phi_2 = 220 \times 90.14 \times 0.8 = 15865 \text{ W}$$

Losses $= P_i + P_c = 2P_i$ $(\because P_i = P_c)$

$$= 2 \times 260 = 520 \text{ W}$$

$$\therefore \textbf{ Maximum efficiency} = \frac{\text{output}}{\text{output + losses}} = \frac{15865}{15865 + 520} = \textbf{0.9683 or 96.83\%. (Ans.)}$$

9.15. All-day Efficiency. *All-day efficiency* is the *ratio of energy (kWh) delivered in a 24 hour period divided by the energy (kWh) input in the same length of time.*

$$\therefore \quad \eta_{\text{all-day}} = \frac{\text{output in kWh}}{\text{input in kWh}} \text{ (for 24 hours)} \quad ...(39)$$

Transformers used on residence-lighting circuits (and distribution circuits generally) are either idle or only lightly loaded during much of 24-hour period. However, they must at all times be connected to the line and ready to serve, so that the core losses are being supplied continually. *It is therefore very important that such transformers be designed for minimum core loss.* The copper losses are relatively less important, since they depend on the load. Because they are lightly loaded much of the time, *distribution transformers are designed for relatively large full-load copper loss and have their maximum power efficiencies at light loads.* This design results in improved all-day efficiency for these transformers. *Power transformers,* on the other hand are loaded more or less continuously and *are designed for full-load copper losses equal to about twice the no-load losses.*

To calculate all-day efficiency, it is necessary to know how the load on the transformer varies from hour to hour. The quotient obtained by dividing the energy output by the energy output plus energy losses over a 24-hour period yields the efficiency expressed as a decimal fraction.

The use of a *load factor* facilitates practical calculations.

Example 48. *A 15 kVA, 2000/200 V transformer has an iron loss of 250 W and full-load copper loss 350 W. During the day it is loaded as follows :*

No. of hours	*Load*	*Power factor*
9	$\frac{1}{4}$ *load*	*0.6*
7	*full-load*	*0.8*
6	$\frac{3}{4}$ *load*	*1.0*
2	*no-load*	—

Calculate the all-day efficiency.

Solution. Rating of transformer = 15 kVA

Iron loss, $P_i = 250 \text{ W} = 0.25 \text{ kW}$

Full-load copper loss, $P_c = 350 \text{ W} = 0.35 \text{ kW}$

Iron loss/day $= 0.25 \times 24 = 6 \text{ kWh}$

Copper loss at $\frac{1}{4}$ *load* $= \left(\frac{1}{4}\right)^2 \times P_c = \frac{1}{16} \times 0.35 = 0.0218 \text{ kW}$

Copper loss for 9 hours at $\frac{1}{4}$ load

$$= 9 \times 0.0218 = 0.196 \text{ kWh}$$

Copper loss at *full-load* $= P_c = 0.35 \text{ kW}$

Copper loss for 7 hours on full-load

$= 7 \times 0.35 = 2.45$ kWh

Copper loss at $\frac{3}{4}$ *load* $= \left(\frac{3}{4}\right)^2 \times P_c = \frac{9}{16} \times 0.35 = 0.197$ kW

Copper loss for 6 hours at $\frac{3}{4}$ load $= 0.197 \times 6 = 1.18$ kWh

Copper loss/day $= 0.196 + 2.45 + 1.18 = 3.826$ kWh

Total loss/day = iron loss/day + copper loss/day $= 6 + 3.826 = 9.826$ kWh

Total output/day $= \frac{1}{4} \times 15 \times 0.6 \times 9 + 15 \times 0.8 \times 7 + \frac{3}{4} \times 15 \times 1.0 \times 6$

$= 20.25 + 84 + 67.5 = 171.75$ kWh

All-day efficiency $= \frac{\text{output}}{\text{output + losses}} = \frac{171.75}{171.75 + 9.826} =$ **0.9459** or **94.59%. (Ans.)**

9.16. Transformer Noise. The "**hum**" caused by energized power transformer, under *no-load conditions*, originates in the *core where the laminations tend to vibrate by magnetic forces*. The noise is transmitted through the oil to the tank side and thence to the surroundings.

The following are the *main factors which produce noise in transformers* :

1. *Magnetostriction* (occurrence of dimensional changes both parallel to, and perpendicular to the direction of magnetisation).
2. The mechanical vibrations caused by the laminations, depending upon the tightness of clamping, size, gauge, associated structural parts, etc.
3. The mechanical vibration of trank walls.
4. The damping.

The noise emission may be reduced by the following methods/means :

1. Prevention of vibration of core-plate by the use of a lower flux density and giving attention to constructional feature (such as clamping bolts, proportions and dimensions of the '*steps*' in plate width, tightness of clamping and uniformity of plates).
2. Using cushions, padding, or oil barriers to sound insulate the transformer from tank.
3. Designing suitably the tank and stiffeners to check tank wall vibration.
4. Sound insulating the tank from the ground or surrounding air.

However, the *noise problem cannot be solved completely.*

9.17. Auto-transformer. *A transformer in which part of the winding is common to both the primary and secondary circuits is known as an* **auto-transformer.** *The primary is electrically connected to the secondary, as well as magnetically coupled to it.*

Refer Fig. 57. *LM* is primary winding having N_1 turns and *MS* is secondary winding having N_2 turns. If no-load current and iron losses are neglected.

$$\frac{V_2}{V_1} = \frac{N_2}{N_1} = K.$$

The current in the section *MS* is vector difference of I_2 and I_1 ; but since the two currents are practically in phase opposition, the resultant is $(I_2 - I_1)$ where $I_2 > I_1$.

Saving of copper (in comparison to conventional two winding transformer) :

The volume and hence weight of copper is proportional to the length and area of cross-section of the conductors. But the *length of conductor is proportional to the number of turns and cross-section depends on current*. Hence the weight of copper is proportional to the *product of number of turns and current to be carried.*

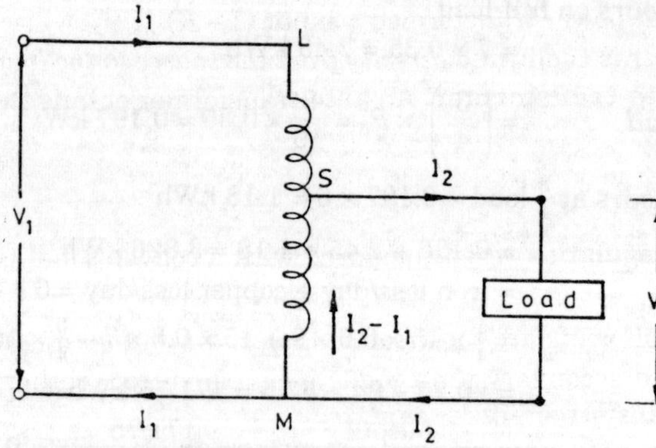

Fig. 56A. Auto-transformer.

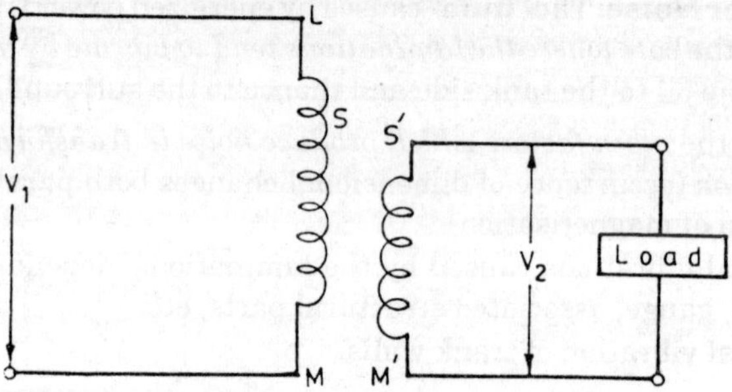

Fig. 57. Conventional two winding transformer.

Weight of copper in conventional two winding transformer

$$\propto (N_1I_1 + N_2I_2).$$

Weight of copper in auto-transformer

= weight of copper in section LS + weight of copper is section MS.

But weight of copper in section $LS \propto (N_1 - N_2)I_1$

and weight of copper in section $MS \propto N_2 (I_2 - I_1)$

∴ Weight of copper in auto-transformer

$$\propto (N_1 - N_2) I_1 + N_2 (I_2 - I_1)$$

$$\therefore \frac{\text{Weight of copper in auto-transformer } (W_a)}{\text{Weight of copper in ordinary transformer } (W_0)} = \frac{(N_1 - N_2) I_1 + N_2 (I_2 - I_1)}{N_1I_1 + N_2I_2} = \frac{(N_1 - 2N_2) I_1 + N_2I_2}{N_1I_1 + N_2I_2}$$

$$= \frac{\frac{N_1}{N_2} - 2 + \frac{I_2}{I_1}}{\frac{N_1}{N_2} + \frac{I_2}{I_1}} = \frac{\frac{1}{K} - 2 + \frac{1}{K}}{\frac{1}{K} + \frac{1}{K}}$$

$$= 1 - K \qquad \ldots(40) \quad \left[\because \frac{N_2}{N_1} = K, \frac{I_2}{I_1} = \frac{1}{K}\right]$$

∴ Saving in copper $= W_0 - W_a = W_0 - (1 - K) W_0 = KW_0$

∴ Saving in copper $= K \times$ weight of copper in ordinary transformer

It can be proved that power transformed = input $(1 - K)$

The rest of the power is conducted *directly from the source to the load.*

Advantages of auto-transformer. An auto-transformer entails the following *advantages* :

(*i*) Higher efficiency

(*ii*) Small size

(*iii*) Lower cost

(*iv*) Better voltage regulation when compared with a conventional two-winding transformer of the same rating.

Disadvantages. Following are the *disadvantages/limitations* of auto-transformers :

(*i*) The primary and secondary are conductively connected, rather than isolated as in the conventional (ordinary) transformer. Because of this, *both sides are subject to any stresses set up by disturbances on either side.* The low-voltage side is subject to high-voltage stress and should be insulated for the higher voltage. In the case of a step down transformer, the high voltage may still be impressed upon equipment connected to the low-voltage side. This is shown in Fig. 58 where the low-voltage coil has accidentally opened. The voltage on the load is nearly 2300 V, being less than that by the impedance volt drop in the primary coil.

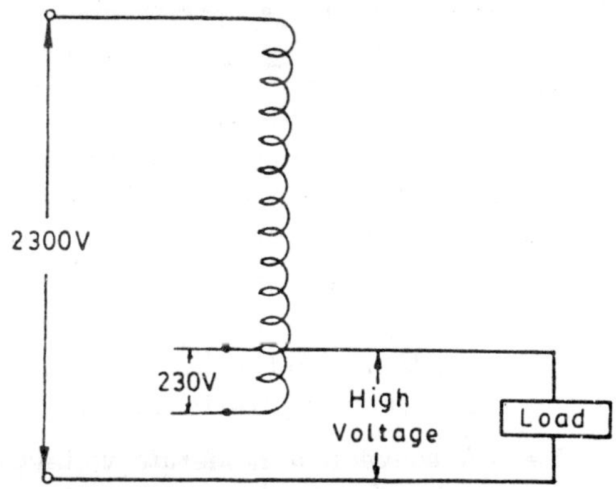

Fig. 58. Open circuit on auto-transformer.

(*ii*) As the voltage ratio of an auto-tranformer increases, the common coil is much smaller compared with the entire winding. This means that the economy gained is only a small part of the transformer and therefore this advantage is minimized.

Thus, because of the above disadvantages, lack of isolation and decreased economy, *auto-transformers are rarely used in ratios greater than 4 : 1, except for low-power devices on low- voltage systems.*

Uses. The auto-transformers find the following *applications* :

1. To tie together transmission or distribution circuits of slightly different voltages (*e.g.* 11000 V system with a 13200 V system).

2. To obtain partial line voltages for starting induction and synchronous motors with squirrel-cage windings.

3. To give a small boost to a distribution cable to correct for the voltage drop.

4. As furnace transformers for getting a convenient supply to suit the furnace winding from a 230 V supply.

5. As regulating transformers.

6. To obtain a neutral in a 3-wire A.C distribution system in the same way as a balancer set is used in a 3-wire D.C. distribution system.

7. A continuously variable auto-transformer finds useful applications in electrical testing laboratory.

9.18. Polarity of Transformers. In a transformer each of the primary terminals becomes alternately positive and negative with respect to the other and the same is true about the secondary terminals. If the transformer is to be used alone, the polarity is not important but if the transformer is to be used in parallel with another transformer the *instantaneous polarity is important* because the terminals having identical instantaneous polarity have to be connected together.

Refer Figs. 59 and 60. The terminals of the high voltage (h.v.) winding (*i.e.* H_1 and H_2) and the terminals of the low voltage (l.v.) winding (*i.e.* X_1 and X_2) are so marked that when the instantaneous voltage is directed from H_1 to H_2 in the high voltage winding, it is directed from X_1 to X_2 in the low voltage winding. In other words if the high voltage terminal H_1 is positive with respect to H_2 at any instant when the low voltage terminal X_1 will be positive with respect to X_2 at the same instant. It follows that when the terminals are arranged as in Fig. 59 the polarity is subtractive whereas the additive polarity is represented as in Fig. 60.

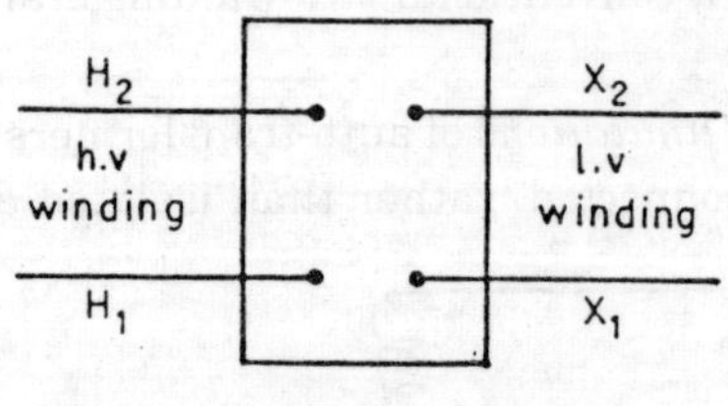

Fig. 59. Subtractive polarity.

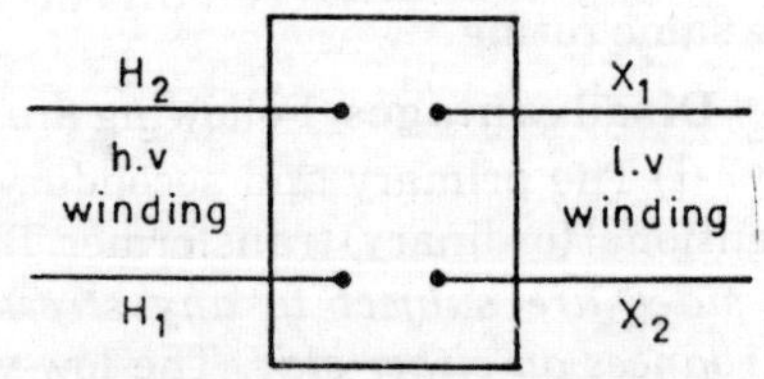

Fig. 60. Additive polarity.

Polarity Test. A polarity test is carried out to *find out the terminals having the same instantaneous polarity* assuming that the terminals are not marked. The connections are shown in Fig. 61.

- One h.t. and one l.t. terminals are joined together. A voltmeter is placed between the remaining two terminals.
- A convenient moderate voltage is impressed on the h.t. winding.

If the voltage V', is 'greater' than the applied voltage V, then the transformer has 'additive' polarity.

If V' is 'less' than V, the transformer has 'subtractive' polarity.

The terminals are then marked accordingly.

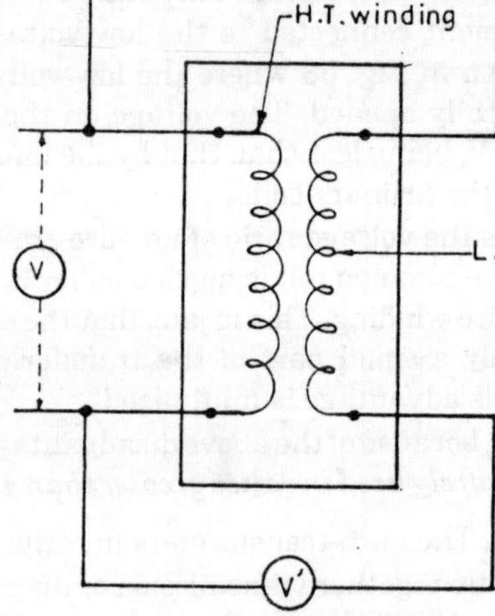

Fig. 61. Polarity test.

Example 49. *A load of 6 kW is supplied by an anto-transformer at 120 V and at unity power factor. If the primary voltage is 240 V, determine :*

(*i*) *Transformation ratio,* (*ii*) *Secondary current,*
(*iii*) *Primary current,*
(*iv*) *Number of turns across secondary if the total number of turns is 280.*
(*v*) *Power transformed, and*
(*vi*) *Power conducted directly from supply mains to load.*

Solution. Load suppied = 6 kW
Output voltage, $V_2 = 120$ V
Primary voltage, $V_1 = 240$ V
Total number of turns, $N_1 = 280$

(*i*) **Transformation ratio,** $K = \frac{V_2}{V_1} = \frac{120}{240} = \frac{1}{2}$. **(Ans.)**

(*ii*) **Secondary current,** $I_2 = \frac{6 \times 1000}{V_2 \cos \phi} = \frac{6000}{120 \times 1} = 50$ **A. (Ans.)**

(*iii*) **Primary current,** $I_1 = KI_2 = \frac{1}{2} \times 50 =$ **25 A. (Ans.)**

(*iv*) **Turns across secondary,** $N_2 = kN_1 = \frac{1}{2} \times 280 =$ **140. (Ans.)**

(*v*) **Power transformed** $= \text{Load} \times (1 - K) = 6 \times (1 - \frac{1}{2}) =$ **3 kW. (Ans.)**

(*vi*) **Power conducted directly from supply mains**

$= 6 - 3 =$ **3 kW. (Ans.)**

Example 50. *The primary and secondary voltages of an auto-transformer are 600 V and 500 V respectively. Show with the aid of a diagram the current distribution in the windings when the secondary current is 210 A. Calculate the economy in copper.*

Solution. Primary voltage, $V_1 = 600$ V

Secondary voltage, $V_2 = 500$ V

Secondary current, $I_2 = 210$ A

Economy in copper :

Transformation ratio, $\frac{V_2}{V_1} = \frac{500}{600} = \frac{5}{6}$

Primary current, $I_1 = KI_2 = \frac{5}{6} \times 210 = 175$ A

The current distribution is shown in Fig. 62.

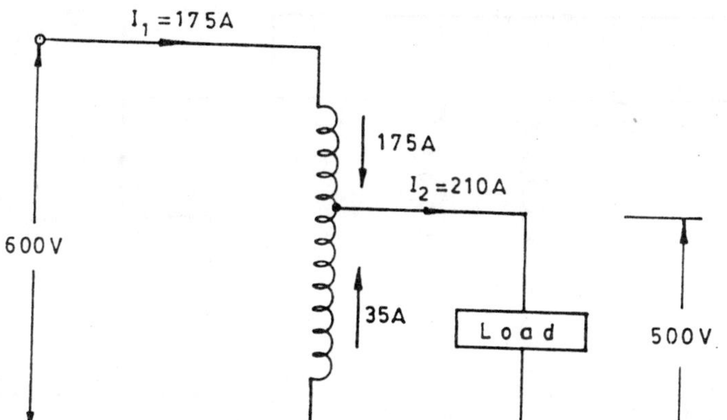

Fig. 62

We know that,

$$\frac{\text{Weight of copper on auto-transformer}}{\text{Weight of copper on ordinary transformer}} = 1 - K$$

Economy in copper

$$= \frac{\text{wt. of Cu on ordinary transformer} - \text{wt. of Cu on auto-transformer}}{\text{wt. of Cu on ordinary transformer}}$$

$$= 1 - \frac{\text{wt. of Cu on auto-transformer}}{\text{wt. of Cu on ordinary transformer}}$$

$= 1 - (1 - K) = K = \frac{5}{6}$ or **83.33%. (Ans.)**

Example 51. *A 4-kVA single-phase 50-Hz transformer has a full-load efficiency of 95.5% and iron loss of 45 W. The transformer is now connected as an auto-transformer to 220 V supply. If it delivers 4 kW load at unity power factor to a 110 V circuit, calculate the efficiency of the operation and the current drawn by the high-voltage side.*

Solution. Fig. 63 shows the connections for a 2-winding transformer. In Fig. 64 the same unit has been connected as an auto-transformer to a 220 V supply. Since the two-windings are connected in series, hence voltage across each is 110 V.

In both connections the iron loss would remain the same. Since the auto-transformer windings will each carry but *half the current* as the conventional two-winding transformer, the copper losses will be $\frac{1}{4}$*th of previous value.*

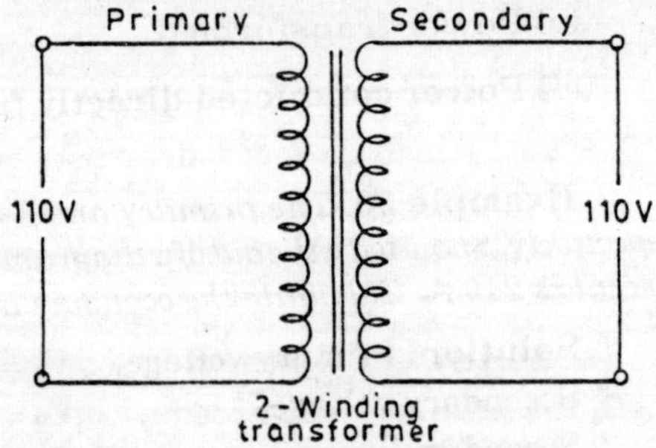

Fig. 63. Two-winding transformer.

Two-winding transformer :

Efficiency = 95.5% or 0.955

$$\therefore \quad 0.955 = \frac{\text{output}}{\text{output + losses}} = \frac{4000}{4000 + 45 + \text{copper loss}}$$

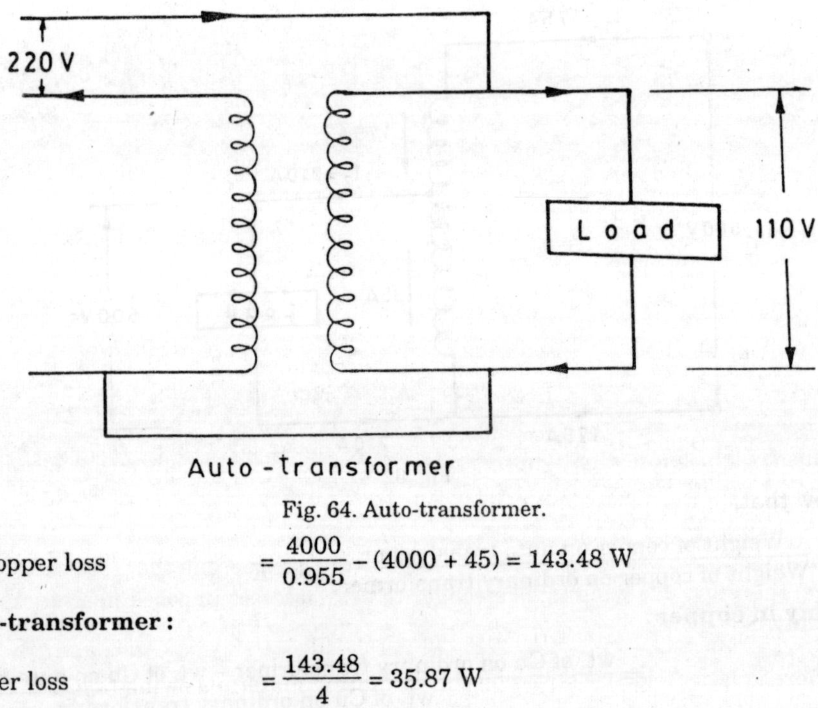

Fig. 64. Auto-transformer.

$$\therefore \text{ Copper loss} = \frac{4000}{0.955} - (4000 + 45) = 143.48 \text{ W}$$

Auto-transformer :

$$\text{Copper loss} = \frac{143.48}{4} = 35.87 \text{ W}$$

Iron loss = 45 W.

$$\therefore \textbf{ Efficiency} = \frac{4000}{4000 + 35.87 + 45} = 0.9802 \text{ or } \mathbf{98.02\%}. \quad \textbf{(Ans.)}$$

10. THREE PHASE TRANSFORMER

10.1. Three-phase Transformer Connections. Virtually all power distribution is by poly-phase system of voltages. Three-phase transformations may be made with the use of properly connected single-phase transformers. These connections are in extensive commercial use. The most frequently used connections are the following :

(*i*) Primary *Y*—secondary *Y*. (*ii*) Primary Δ—secondary Δ.

(*iii*) Primary Δ—secondary *Y*, *or vice versa*. (*iv*) Primary and secondary open Δ.

(*v*) Primary *T* secondary *T* (Scott connection).

Thus the most common connections are *Y-Y*, Δ-Δ, *Y*-Δ, Δ-*Y*, open delta or *V-V* and Scott connection or *T-T* connection.

10.1.1. The Y-Y connection. Fig. 65 shows a bank of three transformers connected in *Y* on both the primary and secondary sides. If the ratio of transformation of each transformer is *K*, the same ratio will exist between the *line* voltages on the primary and secondary sides. This *connection will give satisfactory service only if the three-phase load is balanced* ; when the load is unbalanced, the electrical neutral will shift from its exact centre to a point that will make the line-to-neutral voltages unequal.

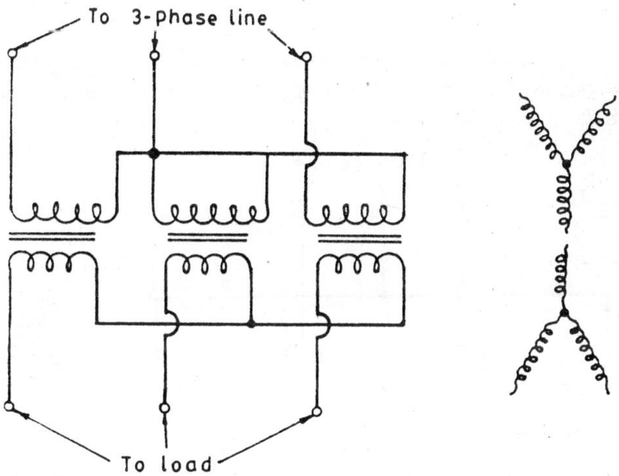

Fig. 65. The *Y-Y* connection of transformers.

Advantages :

1. This type of connection requires fewer turns per winding since the voltage across each is $\frac{1}{\sqrt{3}}$ times the line voltage ; hence it is *cheaper*.

2. The phase or winding current being equal to the line current, the cross-section of the winding wire is large, therefore the winding is stronger to bear stresses imposed upon it during heavy load or short circuit.

3. There is less dielectric stress on the insulating materials owing to lesser voltage *i.e.* $\frac{1}{\sqrt{3}}$ of line voltage.

The above advantages are enumerated with the understanding that, other things being equal, its rival is the delta-delta connection.

Disadvantages :

1. In case the load on the secondary side is unbalanced, as in the case of distributing network, the potential of the star-point will assume any value if the star-point is not earthed. This may impose full-line voltage on secondary windings. The shifting of the neutral point must be prevented by connecting the primary star-point to the star-point of the alternator winding.

2. In spite of grounding the star-point, if there is a third harmonic in the form of the alternator voltage, the third harmonic will appear in the voltages of the secondary side. This will cause triple frequency currents in the three-phase circuits. These currents when they flow in the neutral wire are additive and do not cancel out. Hence they will *cause interference to telephone lines located along the same route.*

3. The magnetising current in a transformer has third harmonic components. These currents will find a return path via the connection between the primary star-point of transformer and the neutral point of the alternator. However, if this connection is missing, these components will distort the flux wave which will produce a voltage having a third harmonic in each of the transformer, both on the primary and secondary sides. And, as before, if the star-point on the secondary is earthed, or grounded, triple harmonic currents will appear in the secondary circuit, and they will flow through the neutral wire causing interference to telephone lines in the vicinity.

4. If the star-points of both the primary and the secondary sides are not earthed, the regulation of the phases will be very poor if the load happen to be unbalanced as the case of distribution network.

10.1.2. The Δ-Δ connection. Fig. 66 shows a bank of transformers connected in Δ on both the primary and secondary sides. *This arrangement is generally used in systems in which the voltages are not very high and especially when continuity of service must be maintained even though one of the transformers should fail.*

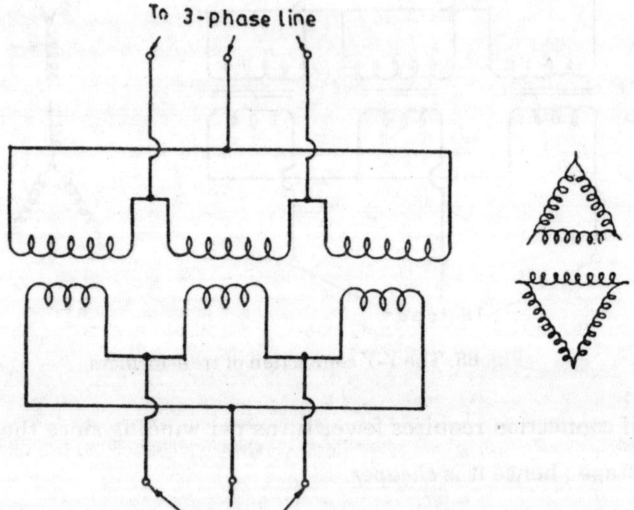

Fig. 66. The Δ-Δ connection of transformers.

Advantages :

1. The system voltages are more stable in relation to an unbalanced load.

2. If one transformer fails it may be switched out of the line and operation continued at a reduced power level. This is known as open-delta or *V-V* operation.

3. There is no distortion of flux, because the third harmonic component of magnetising current can flow in the delta connected primary windings without flowing in the line wires.

4. No difficulty is experienced due to unbalancing of loads on secondary side.

Disadvantages :

1. In comparison to *Y-Y* connections it requires more insulation.

2. The absence of star-point may be disadvantageous. If one line gets earthed due to fault, maximum voltage between windings and core will be full line voltage.

10.1.3. The Y-Δ connection. The *Y*-Δ connection is shown in Fig. 67.

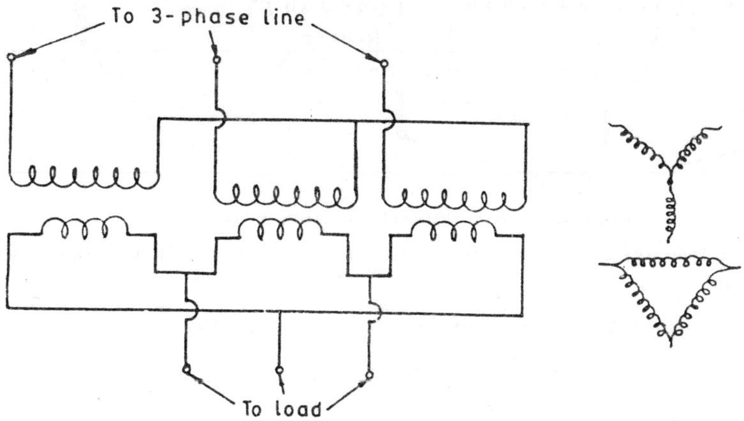

Fig. 67. The *Y*-Δ connection.

It is principally used where the voltage is to be stepped down, as for example, at the end of a transmission line. It is also employed in *moderately low-voltage distribution circuits for stepping down* from transmission voltages of 4000 or 8000 V to 230 (and 115 V).

- The *Y* connection takes advantage of the fact that one that leg of a *Y*, or the line-to-neutral voltage, is less than the line-to-line voltage by a $\sqrt{3}$ factor. This is especially important when the primary voltage is a few hundred thousand volts.
- The *Y*-Δ does have a phase shift between the primary and secondary voltages. This 30° phase shift means that a *Y*-Δ transformer bank *cannot* be paralleled with either a *Y-Y* or a Δ-Δ. The phasor voltage differences between the two systems would be around sin 30° = 0.5 times the secondary voltages. This would cause an excessive circulating current between transformer banks.

10.1.4. The Δ-Y connection

- The three-phase Δ-*Y* connections are shown in Fig. 68.

This type of connection is employed where it is necessary to *step up the voltage*, as for example, at the beginning of a high-tension transmission system.

- The ratio of secondary to primary voltage is $\sqrt{3}$ times the transformation ratio of each transformer.
- The neutral of the secondary is grounded for providing 3-phase 4-wire service. This connection is popular since it can be used to serve both the 3-phase power equipment and single-phase lighting circuit.
- This connection is not open to the objection of a floating neutral and voltage distortion because the existence of a Δ-connection allows a path for the third-harmonic currents. It would be observed that the primary and secondary line voltages and line currents are out of phase with each other by 30°. Because of this 30° shift, it is impossible to parallel such

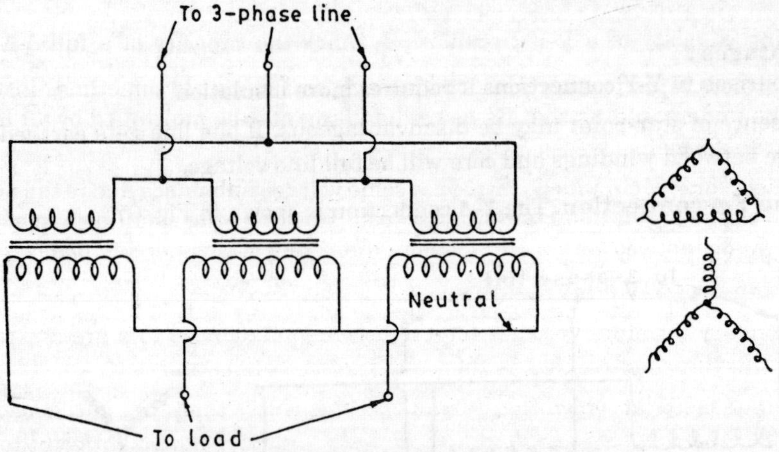

Fig. 68. The Δ-*Y* connection of transformers.

a bank with a Δ-Δ or *Y-Y* bank of transformers even though the voltage ratio are correctly adjusted.

10.1.5. The V-V (open-Δ) connections. Fig. 69 shows the *V-V* connection.

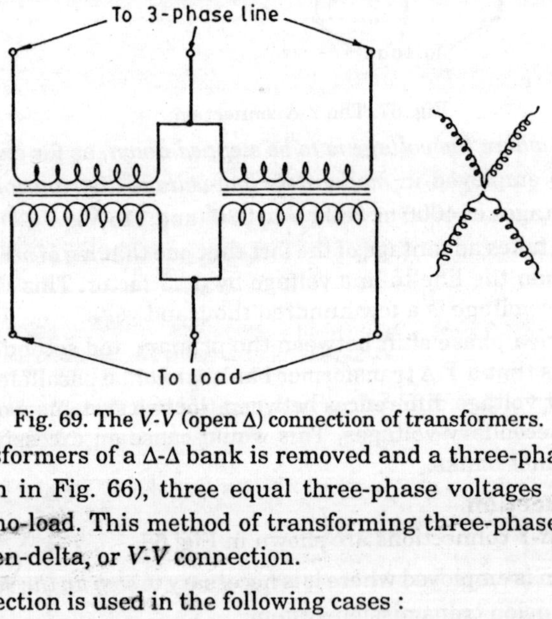

Fig. 69. The *V-V* (open Δ) connection of transformers.

If one of the transformers of a Δ-Δ bank is removed and a three-phase source is connected to the primaries (as shown in Fig. 66), three equal three-phase voltages will be measured at the secondary terminals at no-load. This method of transforming three-phase power, using two transformers, is called the open-delta, or *V-V* connection.

This type of connection is used in the following cases :

(*i*) When the three-phase load is comparatively small so that the installing does not warrant a three transformer bank.

(*ii*) When one of the transformers in a Δ-Δ bank fails, so that the service may be continued until the faulty transformer is repaired or a good one is substituted.

(*iii*) When it is anticipated that the future load will increase to warrant the chosing of the open Δ at some time later.

V-V connection has a number of *features* that are *advantageous* :

- Upon failure of the primary or secondary of one transformer of a complete Δ-Δ transformer circuit, the system reverts to a *V-V* circuit, so this is an *automatic standby*. The power-handling capacity of a *V-V* circuit is $\frac{1}{\sqrt{3}}$ times the capacity of a full Δ-Δ of the same transformers. This feature works both ways, so a circuit is sometimes installed as *V-V* with the understanding that its power handling may be multiplied by $\sqrt{3}$ by adding one more transformer.

- Open delta or *V-V* circuits do introduce some voltage unbalance due to the non-symmetry of the voltage regulation effects under load. However, the small degree of unbalance is not normally noticed by a motor load or other types of commercial load.

Disadvantages of V-V connection :

1. The secondary terminal voltages tend to become unbalanced to a great extent when the load is increased, this happens even when the load is perfectly balanced.

2. The average factor at which the *V*-bank operates is less than that of the load. This power factor is actually *86.6 per cent of the balanced load factor*. Another important point to note is that, except, for a balanced unity power factor load, the two transformers in the *V-V* bank operate at different power factors.

Uses of V-V connection :

(*i*) The *V-V* circuit is frequently used for two *auto-transformers.* Here advantage is taken of power handling of auto-transformers and their superior voltage regulation and efficiency.

(*ii*) Another major use of *V-V* transformer banks is in *A.C. motor starting.*

10.1.6. Scott or T-T connection. The connection of one polyphase system into another polyphase system is possible by suitably connecting the windings of transformers. One of the early types that was used is Scott or *T-T* connection, by which a 2-phase system is available from a 3-phase system or *vice-versa.*

In Fig. 70 two single-phase transformers *M* and *T*, the primaries of which are connected to a 3-phase supply. The secondary of *M* forms one phase and the secondary of *T* the other phase of a true 2-phase system. *M* is called the *main transformer* and *T* is called the *teaser*. One end of the teaser primary is connected to the mid-point of the main primary. The two ends of the main primary are connected to two lines wires of a 3-phase, 3-wire system, and the third line wire is connected to a tapping *X* on the teaser primary.

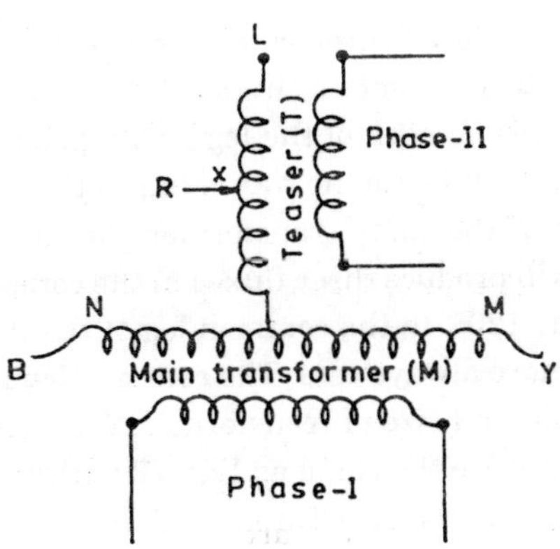

Fig. 70. Scott or *T-T* connection.

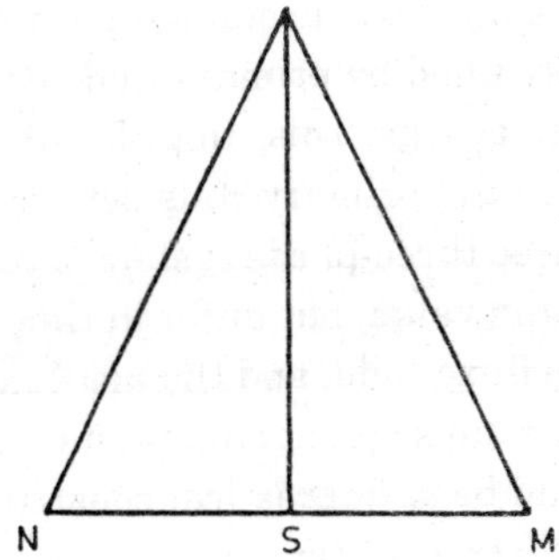

Fig. 71

If the supply voltages are assumed symmetrical, the triangle of voltages is *equilateral* as shown in Fig. 71. The vertical line $LS = \frac{\sqrt{3}}{2} NM$, so that it gives the relationship between the number of turns of two primaries, as

$$\frac{\text{Number of turns for the main transformer } M}{\text{Number of turns for the teaser } T} = \frac{100}{86.6}$$

Hence, if two identical single-phase transformers are to be used for Scott-connection, the primary one must have a tapping point brought out from its *mid-point of the primary,* and the secondary transformer must have a tapping X brought out from *86.6% of its primary turns.*

As the two-phase side is asymmetrical, there cannot be perfect balanced conditions on the three-phase side. However, neglecting the impedance of the windings, it can be shown that if load on the 2-phase side is balanced the 3-phase side is also balanced.

10.2. Three-phase Transformer Construction. The windings of three single-phase transformers can be wound on a common core. The *advantages* are :

1. One 3-phase transformer is *cheaper* than three single-phase transformers.
2. It has *slightly better efficiency and regulation.*
3. A 3-phase transformer takes *less floor space.*

On the other hand, from the point of view of standby, or same capacity, it is economical to have 3 single-phase transformers *plus* one spare rather than two 3-phase transformers one of which is a spare. However, in large central stations 3-phase transformers are often advantageous.

Disadvantages :

1. Three-phase transformers are much more difficult and costly to repair than are single-phase units.

2. When failure does occur and it becomes necessary to substitute a replacement unit to maintain service, the cost of spare is much greater than it would be were a single-phase transformer to be used as a replacement in a three-transformer bank.

3. There is a difficulty in transporting a heavier three-phase transformer compared with the moving of each of the three single-phase transformers.

Two general kinds of three-phase transformers are recognized, similar to single-phase transformers,depending upon the relative arrangements of windings and cores. These are the *core type* and the *shell type.*

Three-phase core type transformer. Fig. 72 shows three core-type transformers placed together so that they have a common path for the return magnetic circuit. Although the windows should be entirely filled by primary and secondary coils on each of the legs, only primary coils are shown on the outside legs. This simplifies the diagram, while it in no way changes the actual theory that follows, since the primary coils set up the flux. If the three transformers are identical in all respects, a balanced three-phase system of voltages will produce three fluxes in the cores which have the same maximum value, but differ in time phase by 120°. In the common leg of the three cores of Fig. 73, the three fluxes add, and the net flux is therefore always zero. The common leg may then be eliminated, with a subsequent saving in core material and size of transformer. A single polyphase transformer would be of impractical construction if it were the same as Fig. 72 with the centre leg omitted. Instead, the core-type polyphase transformer is manufactured so that it looks like that shown in Fig. 73. Actually, what we have done is, that axes of three coils have been moved into one plane. This causes the magnetic reluctance of coil M to differ somewhat from that of L and N. This

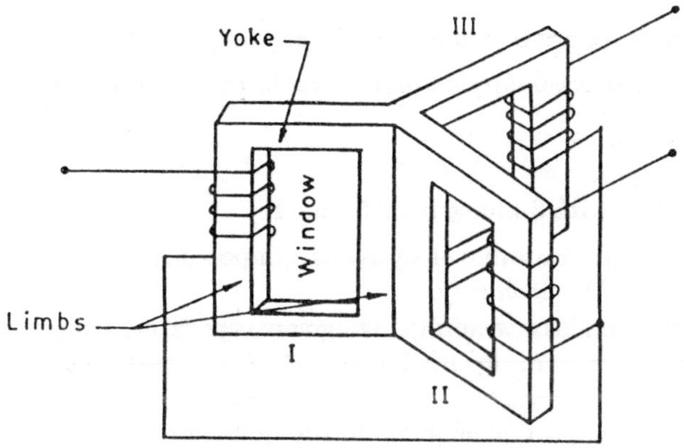

Fig. 72. Core-type transformers for polyphase transformation.

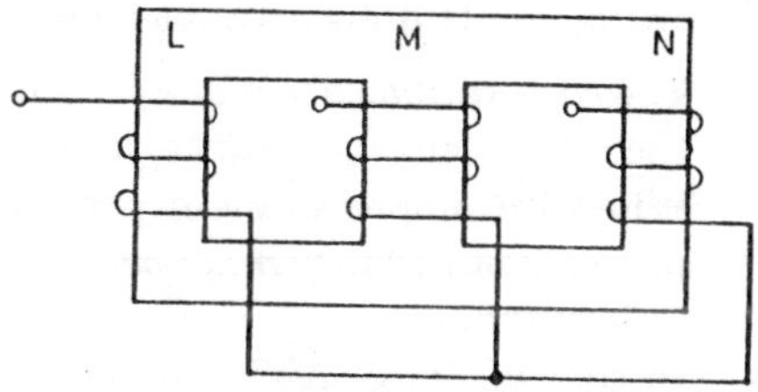

Fig. 73. Core-type three-phase transformer.

produces a slight unbalance in the three magnetizing currents, but the effect is not serious, especially under load.

Three-phase shell type transformer :

- In Fig. 74 are shown three shell-type transformers stacked one above the other, with only the primary coils shown.

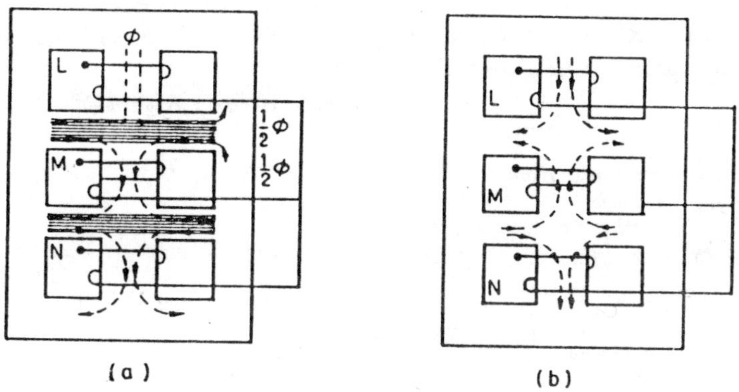

Fig. 74. Shell-type three-phase transformer (*a*) coil *M* wound in the same direction, (*b*) coil *M* wound in opposite direction.

- In Fig. 74 (*a*) the three coils are wound in the same direction. The flux in the common core area between adjacent phases such as the shaded portions is thus equal to the *difference* of two of the phase fluxes. Since the fluxes are 120° apart in time, this mutual flux is equal to $\sqrt{3} \times (\frac{1}{2}\phi)$, or 0.866 of the flux in the centre leg. If the same flux density is maintained throughout, less iron is required in the common legs. Now, however if the centre coil, phase *M* is reversed, the flux in the common core is equal to the *sum* of the fluxes of two adjacent phases. This is shown in Fig. 74 (*b*). As in any three-phase system, the sum of two fluxes is equal to either flux alone, which in this case is 0.5 ϕ. This represents a further saving in iron, and for this reason shell-type three-phase transformers are *usually wound with centre coil opposing the two outside ones.*

10.3. Parallel Operation of 3-phase Transformers. The conditions for paralleling 3-phase transformers are same as that required for parallel operation of single-phase transformers with the following *additions* :

(i) The voltage ratio must refer to the terminal voltage of primary and secondary.

(ii) The phase displacement between primary and secondary voltages must be the same for all transformers which are to be paralleled.

(iii) The phase sequence must be the same.

Following points are worthnoting while dealing with 3-phase transformers :

- *The calculations are made for one phase only.* The value of equivalent impedance used is the equivalent impedance per phase *referred to secondary.*
- When the impedances of primary and secondary windings are given separately then primary impedance *must be transferred to secondary by multiplying it with (transformation ratio)^2.*
- In case of Y/Δ or Δ/Y transformers the voltage ratios as given in the questions, *refer to terminal voltages and are quite difference from turn ratio.*

HIGHLIGHTS

1. *The function of a transformer is to transform alternating current energy from one voltage into another voltage. It operates on the principle of mutual inductance (between two or more inductively coupled coils).*
2. **Distribution transformers** *should be designed to have maximum efficiency at a load much lower than full-load (about 50%).*

   Power transformers *should be designed to have maximum efficiency at or near full-load.*
3. *The transformation ratio (K) is defined as the ratio of the secondary voltage to primary voltage.*
4. **Approximate** *voltage drop* $= I_2R_{02} \cos \phi \pm I_2 X_{02} \sin \phi$

   Exact *voltage drop* $= (I_2R_{02} \cos \phi \pm I_2X_{02}) + \dfrac{(I_2X_{02} \cos \phi \mp I_2R_{02} \sin \phi)^2}{2\ {}_0V_2}$
5. *Transfer of resistance or reactance from*

   Primary to secondary ...... $\times K^2$

   Secondary to primary ...... $\times \dfrac{1}{K^2}$
6. *The change in secondary voltage when rated load at a specified power is removed.*

$$\% \text{ regulation} = \frac{I_2R_{02} \cos \phi \pm I_2X_{02} \sin \phi}{{}_0V_2} \times 100$$

$$\eta_{all\text{-}day} = \frac{\text{output in kWh}}{\text{input in kWh}} \text{ (for 24 hours).}$$

7. *A transformer in which part of the winding is common to both the primary and secondary circuit is known as an* **auto-transformer.**

OBJECTIVE TYPE QUESTIONS

(A) Choose the Correct Answer :

1. Which of the following does not change in a transformer ?
   (*a*) Current (*b*) Voltage
   (*c*) Frequency (*d*) All of the above.
2. In a transformer the energy is conveyed from primary to secondary
   (*a*) through cooling coil (*b*) through air
   (*c*) by the flux (*d*) none of the above.

3. A transformer core is laminated to
(*a*) reduce hysteresis loss (*b*) reduce eddy current losses
(*c*) reduce copper losses (*d*) reduce all above losses.

4. Which loss is not common between a transformer and rotating machines ?
(*a*) Eddy current loss (*b*) Copper loss
(*c*) Windage loss (*d*) Hysteresis loss.

5. The degree of mechanical vibrations produced by the laminations of a transformer depends on
(*a*) tightness of clamping (*b*) gauge of laminations
(*c*) size of laminations (*d*) all of the above.

6. The no-load current drawn by transformer is usually what percent of the full-load current ?
(*a*) 0.2 to 0.5 per cent (*b*) 2 to 5 per cent
(*c*) 12 to 15 per cent (*d*) 20 to 30 per cent.

7. In case there are burrs on the edges of the laminations of the transformer, it is likely to result in
(*a*) vibrations (*b*) noise
(*c*) higher eddy current loss (*d*) higher hysteresis loss.

8. The path of a magnetic flux in a transformer should have
(*a*) high resistance (*b*) high reluctance
(*c*) low resistance (*d*) low reluctance.

9. No-load test on a transformer is carried out to determine
(*a*) copper loss (*b*) magnetising current
(*c*) magnetising current and loss (*d*) efficiency of the transformer.

10. The dielectric strength of transformer oil is expected to be
(*a*) 1 kV (*b*) 33 kV
(*c*) 100 kV (*d*) 330 kV.

11. Sumpner's test is conducted on transformers to determine
(*a*) temperature (*b*) stray losses
(*c*) all-day efficiency (*d*) none of the above.

12. The permissible flux density in case of cold rolled grain oriented steel is around
(*a*) 1.7 Wb/m^2 (*b*) 2.7 Wb/m^2
(*c*) 3.7 Wb/m^2 (*d*) 4.7 Wb/m^2.

13. During the short-circuit test on a small transformer the frequency is increased from 50 Hz to 200 Hz. The copper losses will increase by a factor of
(*a*) 16 (*b*) 4
(*c*) 1 (*d*) $\frac{1}{4}$.

14. The efficiency of a transformer will be maximum when
(*a*) copper losses = hysteresis losses (*b*) hysteresis losses = eddy current losses
(*c*) eddy current losses = copper losses (*d*) copper losses = iron losses.

15. No-load current in a transformer
(*a*) lags behind the voltage by about 75° (*b*) leads the voltage by about 75°
(*c*) lags behind the voltage by about 15° (*d*) leads the voltage by about 15°.

16. The purpose of providing an iron core in a transformer is to
(*a*) provide support to windings (*b*) reduce hysteresis loss
(*c*) decrease the reluctance of the magnetic path
(*d*) reduce eddy current losses.

17. Which of the following is not a part of transformer installation ?
(*a*) Conservator (*b*) Breather
(*c*) Buchholz relay (*d*) Exciter.

18. While conducting short-circuit test on a transformer the following side is short-circuited
(*a*) high voltage side (*b*) low voltage side
(*c*) primary side (*d*) secondary side.

19. In a transformer the toppings are generally provided on
(*a*) primary side (*b*) secondary side
(*c*) low voltage side (*d*) high voltage side.

20. The use of higher flux density in the transformer design
(*a*) reduces weight per kVA (*b*) reduces iron losses
(*c*) reduces copper losses (*d*) increases part load efficiency.

21. The chemical used in breather for transformer should have the quality of
(*a*) ionizing air (*b*) absorbing moisture
(*c*) cleansing the transformer oil (*d*) cooling the transformer oil.

22. The chemical used in breather is
(*a*) asbestos fibre (*b*) silica sand
(*c*) sodium chloride (*d*) silica gel.

23. If a pump motor is run on 2/3rd of its supply voltage, it will
(*a*) continue to deliver same power (*b*) burn
(*c*) stall (*d*) continue to run at lower speed.

24. An ideal transformer has infinite values of primary and secondary inductances. The statement is
(*a*) true (*b*) false.

25. The transformer ratings are usually expressed in terms of
(*a*) volts (*b*) amperes
(*c*) kW (*d*) kVA.

26. The noise resulting from vibrations of laminations set by magnetic force, is termed as
(*a*) magnetostriction (*b*) boo
(*c*) hum (*d*) zoom.

27. Hysteresis loss in a transformer varies as (B_{max} = maximum flux density)
(*a*) B_{max} (*b*) $B_{max}^{1.6}$
(*c*) B_{max}^{2} (*d*) $B_{max}^{2.4}$.

28. Material used for construction of transformer core is usually
(*a*) wood (*b*) copper
(*c*) aluminium (*d*) silicon steel.

29. The thickness of laminations used in a transformer is usually
(*a*) 0.4 mm to 0.5 mm (*b*) 4 mm to 5 mm
(*c*) 14 mm to 15 mm (*d*) 25 mm to 40 mm.

30. The function of conservator in a transformer is
(*a*) to protect against internal fault (*b*) to reduce copper as well as core losses
(*c*) to cool the transformer oil
(*d*) to take care of the expansion and contraction of transformer oil due to variation of temperature of surroundings.

31. The highest voltage for transmitting electrical power in India is
(*a*) 33 kV (*b*) 66 kV
(*c*) 132 kV (*d*) 400 kV.

32. In a transformer the resistance between its primary and secondary is
(*a*) zero (*b*) 1 Ω
(*c*) 1000 Ω (*d*) infinite.

33. A transformer oil must be free from
(*a*) sludge (*b*) odour
(*c*) gases (*d*) moisture.

34. A Buchholz relay can be installed on
(*a*) auto-transformers (*b*) air-cooled transformers
(*c*) welding transformers (*d*) oil cooled transformers.

35. Gas is usually not liberated due to dissociation of transformer oil unless the oil temperature exceeds
(*a*) 50°C (*b*) 80°C
(*c*) 100°C (*d*) 150°C.

36. The main reason for generation of harmonics in a transformer could be
(*a*) fluctuating load (*b*) poor insulation
(*c*) mechanical vibrations (*d*) saturation of core.

37. Distribution transformers are generally designed for maximum efficiency around
(*a*) 90% load (*b*) zero load
(*c*) 25% load (*d*) 50% load.

38. Which of the following properties is not necessarily desirable in the material for transformer core ?
(*a*) Mechanical strength (*b*) Low hysteresis loss
(*c*) High thermal conductivity (*d*) High permeability.

39. Helical coils can be used on
(*a*) low voltage side of high kVA transformers (*b*) high frequency transformers
(*c*) high voltage side of small capacity transformers
(*d*) high voltage side of high kVA rating transformers.

40. High frequency transformers sometimes make use of ferrite cores because it has
(*a*) high specific gravity (*b*) high resistance
(*c*) high hysteresis (*d*) low permeability.

41. Cross over windings are used on
(*a*) low voltage side of high kVA rating transformers
(*b*) current transformers
(*c*) high voltage side of high kVA rating transformers
(*d*) high voltage side of low kVA rating transformers.

42. While conducting short-circuit test on a transformer the following side is short-circuited.
(*a*) h.v. side (*b*) l.v. side
(*c*) primary side (*d*) secondary side.

43. A substance is placed in strong magnetic field. Which of the following will have maximum influence on the capacity of the substance to get magnetised ?
(*a*) Permeability (*b*) Susceptibility
(*c*) Permittivity (*d*) Resistivity.

44. Harmonics in transformer result in
(*a*) increased core losses (*b*) increased I^2R losses
(*c*) magnetic interference with communication circuits
(*d*) all of the above.

45. Tertiary winding is provided in transformers having
(*a*) mesh/star winding (*b*) mesh/mesh winding
(*c*) star/star winding (*d*) any of the above.

46. The core used in high frequency transformer is usually
(*a*) copper core (*b*) cost iron core
(*c*) air core (*d*) mild steel core.

47. The full-load copper loss of a transformer is 1600 W. At half-load the copper loss will be
(*a*) 6400 W (*b*) 1600 W
(*c*) 800 W (*d*) 400 W.

48. The value of flux involved in the e.m.f. equation of a transformer is
(*a*) average value (*b*) r.m.s. value
(*c*) maximum value (*d*) instantaneous value.

49. Silicon steel used in laminations mainly reduces

(*a*) hysteresis loss (*b*) eddy current losses
(*c*) copper losses (*d*) all of the above.

50. Which winding of the transformer has less cross-sectional area ?

(*a*) Primary winding (*b*) Secondary winding
(*c*) Low voltage winding (*d*) High voltage winding.

51. Power transformers are generally designed to have maximum efficiency around

(*a*) no-load (*b*) half-load
(*c*) near full-load (*d*) 10% overload.

52. Which of the following is the main advantage of an auto-transformer over a two winding transformer ?

(*a*) Hysteresis losses are reduced (*b*) Saving in winding material
(*c*) Copper losses are negligible (*d*) Eddy losses are totally eliminated.

53. During short-circuit test iron losses are negligible because

(*a*) the current on secondary side is negligible (*b*) the voltage on secondary side does not vary
(*c*) the voltage applied on primary side is low (*d*) full-load current is not supplied to the transformer.

54. Two transformers are connected in parallel. These transformers do not have equal percentage impedance. This is likely to result in

(*a*) short-circuiting of the secondaries
(*b*) power factor of one of the transformers is leading while that of the other lagging
(*c*) transformers having higher copper losses will have negligible core losses
(*d*) loading of the transformers not in proportion to their kVA ratings.

55. The changes in volume of transformer cooling oil due to variation of atmospheric temperature during day and night is taken care of by which part of transformer ?

(*a*) Conservator (*b*) Breather
(*c*) Bushings (*d*) Buchholz relay.

56. The transformer laminations are insulated from each other by

(*a*) mica strip (*b*) thin coat of varnish
(*c*) paper (*d*) any of the above.

57. Which type of winding is used in 3-phase shell-type transformer ?

(*a*) Circular type (*b*) Sandwich type
(*c*) Cylindrical type (*d*) Rectangular type.

58. During open circuit test of a transformer

(*a*) primary is supplied rated voltage (*b*) primary is supplied full-load current
(*c*) primary is supplied current at reduced voltage
(*d*) primary is supplied rated kVA.

59. Open circuit test on transformers is conducted to determine

(*a*) hysteresis losses (*b*) copper losses
(*c*) core losses (*d*) eddy current losses.

60. Short circuit test on transformers is conducted to determine

(*a*) hysteresis losses (*b*) copper losses
(*c*) core losses (*d*) eddy current losses.

61. For the parallel operation of single-phase transformers it is necessary that they should have

(*a*) same efficiency (*b*) same polarity
(*c*) same kVA rating (*d*) same number of turns on the secondary side.

62. The transformer oil should have ...... voltatility and ...... viscosity

(*a*) low ...... low (*b*) high ...... high
(*c*) low ...... high (*d*) high ...... low.

63. The function of breather in a transformer is
(*a*) to provide oxygen inside the tank (*b*) to cool the coils during reduced load
(*c*) to filter the transformer cooling oil
(*d*) to arrest flow of moisture when outside air enters the transformer.

64. The secondary winding of which of the following transformers is always kept closed ?
(*a*) Step-up transformer (*b*) Step-down transformer
(*c*) Potential transformer (*d*) Current transformer.

65. The size of a transformer core will depend on
(*a*) frequency (*b*) area of the core
(*c*) flux density of the core material (*d*) (*a*) and (*b*) both.

66. Natural air cooling is generally restricted for transformers up to
(*a*) 1.5 MVA (*b*) 5 MVA
(*c*) 15 MVA (*d*) 50 MVA.

67. A shell-type transformer has
(*a*) high eddy current losses (*b*) reduced magnetic leakage
(*c*) negligible hysteresis losses (*d*) none of the above.

68. A transformer can have regulation closer to zero
(*a*) on full-load (*b*) on overload
(*c*) on leading power factor (*d*) on zero power factor.

69. A transformer transforms
(*a*) voltage (*b*) current
(*c*) current and voltage (*d*) power.

70. Which of the following is not the standard voltage for power supply in India
(*a*) 11 kV (*b*) 33 kV
(*c*) 66 kV (*d*) 122 kV.

ANSWERS

1. (*c*)	**2.** (*c*)	**3.** (*b*)	**4.** (*a*)	**5.** (*d*)	**6.** (*b*)	**7.** (*c*)
8. (*d*)	**9.** (*c*)	**10.** (*b*)	**11.** (*a*)	**12.** (*a*)	**13.** (*a*)	**14.** (*d*)
15. (*a*)	**16.** (*c*)	**17.** (*d*)	**18.** (*b*)	**19.** (*c*)	**20.** (*a*)	**21.** (*b*)
22. (*d*)	**23.** (*c*)	**24.** (*b*)	**25.** (*d*)	**26.** (*c*)	**27** (*b*)	**28.** (*d*)
29. (*a*)	**30.** (*d*)	**31.** (*d*)	**32.** (*d*)	**33.** (*d*)	**34.** (*d*)	**35.** (*d*)
36. (*d*)	**37.** (*d*)	**38.** (*c*)	**39.** (*a*)	**40.** (*b*)	**41.** (*d*)	**42.** (*b*)
43. (*b*)	**44.** (*d*)	**45.** (*c*)	**46.** (*c*)	**47.** (*d*)	**48.** (*c*)	**49.** (*a*)
50. (*d*)	**51.** (*c*)	**52.** (*b*)	**53.** (*c*)	**54.** (*d*)	**55.** (*a*)	**56.** (*b*)
57. (*b*)	**58.** (*a*)	**59.** (*c*)	**60.** (*b*)	**61.** (*b*)	**62.** (*a*)	**63.** (*d*)
64. (*d*)	**65.** (*d*)	**66.** (*a*)	**67.** (*b*)	**68.** (*c*)	**69.** (*d*)	**70.** (*d*)

(B) Say 'Yes' or 'No'

1. When a transformer raises the voltage it is called the step-up transformer.
2. A transformer must not be connected to D.C. source.
3. The ratio of primary voltage to secondary voltage is known as transformation ratio.
4. An ideal transformer is one in which the resistance of the windings is negligible and the core has no losses.
5. Primary and secondary currents are directly proportional to their respective turns.
6. The function of the magnetising component of no-load current is to sustain the alternating flux in the core.
7. The no-load primary input is practically equal to the iron loss in the transformer.
8. A transformer is said to be loaded when the secondary circuit of a transformer is completed through an impedance load.

9. Magnetic flux can be confined into a designed path.
10. When shifting resistance to the secondary, divide it by K^2
11. An open-circuit test is conducted to find no-load or core loss.
12. Short-circuit test is conducted to find full-load copper loss.
13. The change in voltage when rated load at a specified power is removed is termed as voltage regulation.
14. Iron or core losses include copper loss and eddy current loss.
15. Iron or core losses are found from short-circuit test.
16. The efficiency of a transformer at a particular load and power factor is defined as the ratio of power output to power input.
17. Copper losses = iron losses is the condition for maximum efficiency of a transformer.
18. $\eta_{\text{all-day}} = \dfrac{\text{output in kWh}}{\text{input in kWh}}$ (for 24 hours)
19. A transformer in which part of the winding is common to both the primary and secondary circuits is known as auto-transformer.
20. The Δ-Δ connection is generally used in systems in which the voltages are not very high and especially when continuity of service must be maintained even though one of the transformers should fail.
21. The *Y*-Δ connection is principally used where the voltage is to be stepped up.
22. The Δ-*Y* connection is employed where it is necessary to step-up the voltage.
23. The *V*-*V* circuit is frequently used for two auto-transformers.
24. It is practicable to connect instruments and meters directly to the lines in high voltage circuits.
25. A potential transformer is a step down transformer used along with a low range voltmeter for measuring a high voltage.
26. The current transformer ratio is not equal to the ratio of secondary to primary turns, mainly because of the effect of the magnetising current.
27. A constant-current transformer is used to supply power to street lights which are connected in series.
28. In case of an induction regulator primary winding is stationary.

ANSWERS

1. Yes	2. Yes	3. No	4. Yes	5. No	6. Yes	7. Yes
8. Yes	9. No	10. No	11. Yes	12. Yes	13. Yes	14. No
15. No	16. Yes	17. Yes	18. Yes	19. Yes	20. Yes	21. No
22. Yes	23. Yes	24. No	25. Yes	26. Yes	27. Yes	28. No.

THEORETICAL QUESTIONS

1. What is a transformer ? How does it transfer electric energy from one circuit to another ?
2. Explain the principle of operation of a transformer.
3. What is meant by *transformer action* ? Under what conditions will it take place ?
4. If an alternating current is impressed on one coil, what will be the frequency of the induced voltage in another coil with which it is coupled ?
5. Enumerate the various kinds of transformers.
6. Explain the need for stepping up and stepping down voltages in a power system. How does a transformer accomplish ?
7. Why are transformer windings divided into several coils ?
8. What properties should a good transformer oil possess ?
9. What purposes are served by placing transformers in oil-filled tanks ?
10. Why are the tanks of some large transformers corrugated ?
11. Write a short note on 'transformer cooling'.
12. Enumerate and explain briefly different types of windings.

13. Derive an expression for induced e.m.f. in a transformer in terms of frequency, the maximum value of flux and the number of turns on the windings.
14. Derive an expression for the e.m.f. of an ideal transformer winding.
15. Explain the function of the active and reactive components of the no-load current of a static transformer.
16. Why are some transformers constructed with primary and secondary windings divided into two halves ?
17. Distinguish between *power* and *distribution* transformers.
18. Draw and explain the no-load phasor diagram for a single-phase transformer.
19. Draw the vector diagram of a power transformer under full-load condition.
20. What is meant by *equivalent resistance* of a transformer ? How may it be calculated in primary terms and secondary terms ?
21. What is meant by the *equivalent reactance* of a transformer ? How may it be calculated in primary terms and secondary terms ?
22. How can the equivalent impedance of a transformer be determined ?
23. Outline carefully the procedure for performing the short-circuit test.
24. What useful information is obtained from the short-circuit test ?
25. Outline carefully the procedure for performing the open-circuit test.
26. What useful information is obtained from open-circuit test ?
27. What are the two components of the core loss ?
28. How is the hysteresis loss affected by a change in flux density ?
29. Develop the equivalent circuit of a single phase transformer.
30. Draw the equivalent circuit of a transformer and show how the constants of the primary and secondary windings may be combined to give a simplified equivalent circuit with the values of constants given in terms of the secondary winding.
31. Describe the method of calculating the regulation and efficiency of a single-phase transformer by open circuit and short circuit tests.
32. Distinguish between the 'efficiency' and the 'regulation' of a transformer. Show how power factor affects both of them.
33. Explain with circuit diagrams, the open-circuit and short-circuit tests to be carried out in the laboratory for the determination of the parameters of a single-phase transformer. Derive the condition for maximum efficiency.
34. What is Sumpner's test ? Draw a circuit diagram to conduct this test and explain its principle.
35. Define all-day efficiency.
36. What is an auto-transformer ? What advantages are possessed by auto-transformers ?
37. Derive an expression for saving of copper when an auto-transformer is used.
38. State the applications of auto-transformers. Why is this transformer not used as a distribution transformer ? Prove that for the same capacity and voltage ratio, an auto-transformer requires less copper than an ordinary transformer.
39. What are the sources of heat in a transformer ? Describe briefly the various methods used for cooling of transformers.
40. To what does the '*polarity*' of a transformer refer ?

UNSOLVED EXAMPLES

E.m.f. Equation-Turn Ratio

1. The no-load ratio required in a single-phase 50-Hz transformer is 6600/300 V. If the maximum value of flux in the core is to be about 0.09 weber, find the number of turns in each winding.

[Ans. $N_1 = 330$; $N_2 = 15$]

2. A 20 kVA, single-phase transformer has 200 turns on the primary and 40 turns on the secondary. The primary is connected to 1000 V, 50-Hz supply. Determine :
   (*i*) the secondary voltage on open circuit ;
   (*ii*) the currents flowing through the two windings on full-load ; and
   (*iii*) the maximum value of flux. **[Ans.** (*i*) 200 V ; (*ii*) 20 A, 100 A ; (*iii*) 0.0225 Wb]
3. A single-phase transformer is connected to a 230 V, 50-Hz supply. The net cross-sectional area of the core is 50 cm^2. The number of turns in the primary is 460 and in the secondary 80. Determine :
   (*i*) Transformation ratio ; (*ii*) Peak value of flux density in the core ; and
   (*iii*) E.m.f. in the secondary winding. **[Ans.** 0.1739 ; 0.4504 T ; 40 V]
4. A 6600/440 V single-phase 600 kVA transformer has 1200 primary turns. Find :
   (*i*) Transformation ratio ; (*ii*) Secondary turns ;
   (*iii*) Voltage/turn ; and
   (*iv*) Secondary current when it supplies a load of 400 kW at 0.8 power factor lagging.
   $\left[\textbf{Ans.}\ \frac{1}{15}\ ;\ 80\ ;\ 5.5\ \text{V}\ ;\ 1136\ \text{A}\right]$
5. Find the primary and secondary turns of a 3300/300 V, single-phase, 50-Hz, 30 kVA transformer if the flux in the core is to be about 0.06 Wb. Also determine the primary and secondary currents if the losses are to be neglected. **[Ans.** 264 ; 24 ; 100 A ; 9.09 A]
6. The voltage/turn of a single-phase transformer is 1.1 V, when the primary winding is connected to a 220 V, 50-Hz A.C. supply, the secondary voltage is found to be 660 V. Find :
   (*i*) Primary and secondary turns ; and
   (*ii*) Core area if the maximum flux density is 1.2 T. **[Ans.** 200 ; 600 ; 41.29 cm^2]

Transformer on No-load

7. A 2000/200 V single-phase transformer gives 0.5 A and 40 W as ammeter and wattmeter readings when supply is given to the low voltage winding and high voltage winding is kept open. Find :
   (*i*) The magnetising component, (*ii*) The iron loss component, and
   (*iii*) The power factor of no-load current. **[Ans.** 0.4582 A ; 0.2 A ; 0.4 lagging]
8. Find (*i*) active component and reactive components of no-load current and ; (*ii*) no-load current of a 230 V/115 V single-phase transformer if the power input on no-load to the high voltage winding is 70 W and power factor of no-load current is 0.25 lagging. **[Ans.** 0.3044 A ; 1.179 A ; 1.2176 A]
9. A single-phase transformer has 500 turns on the primary and 40 turns on the secondary winding. The mean length of the magnetic path in the core is 150 cm and the joints are equivalent to an air-gap of 0.1 mm. When a potential difference of 3000 V is applied to the primary, maximum flux density is 1.2 T. Calculate :
   (*i*) The cross-sectional area of the core, (*ii*) No-load secondary voltage,
   (*iii*) The no-load current drawn by the primary, and
   (*iv*) Power factor on no-load.
   Given that AT/cm for a flux density of 1.2 T in the iron to be 5, the corresponding iron loss to be 2 W/kg at 50 Hz and density of iron as 7.8 g/cm^3. **[Ans.** 225 cm^2 ; 240 V ; 1.208 A ; 0.1457]

Transformer on Load

10. A 230 V/115 V single phase transformer takes a no-load current of 1.7 A at a power factor of 0.18 lagging with low voltage winding kept open. If the low voltage winding is now loaded to take a current of 13 A at 0.8 power factor lagging find the current taken by high voltage winding. **[Ans.** 7.834 A]
11. A transformer has a primary winding of 800 turns and a secondary winding of 200 turns. When the load current on the secondary is 80 A at 0.8 power factor lagging, the primary current is 25 A at 0.707 power factor lagging. Determine graphically or otherwise the no-load current of the transformer and its phase with respect to the voltage. **[Ans.** I_0 = 5.93 A ; ϕ_0 = 73.3°]

12. A 8 : 1 step down, single-phase transformer takes a no-load current of 0.6 A at a power factor of 0.8 lagging with 1.v. winding kept open. If the secondary is connected to a load taking a current of 80 A at (*i*) 0.8 power factor lagging ; and (*ii*) 0.8 power factor leading find the primary current and power factor.
[**Ans.** 10.49 A, 0.7996 lagging ; 9.817 A, 0.8333 leading]

13. A 20 kVA, 2000/200 V, single-phase, 50-Hz transformer has a primary resistance of 2.5 Ω and reactance of 4.8 Ω. The secondary resistance and reactance are 0.01 Ω and 0.018 Ω respectively. Find :
(*i*) Equivalent resistance referred to primary,
(*ii*) Equivalent impedance referred to primary,
(*iii*) Equivalent resistance, reactance and impedance referred to secondary, and
(*iv*) Total copper loss of the transformer.
[**Ans.** (*i*) 3.5 Ω ; (*ii*) 7.47 Ω ; (*iii*) 0.035, 0.066 Ω, 0.0747 Ω ; (*iv*) 350 W]

14. A 50 kVA, 4400/220 V transformer has R_1 = 3.45 Ω, R_2 = 0.009 Ω. The values of reactances are X_1 = 5.2 Ω and X_2 = 0.015 Ω. Calculate for the transformer :
(*i*) Equivalent resistance as referred-to primary,
(*ii*) Equivalent resistance as referred to secondary,
(*iii*) Equivalent reactance as referred to both primary and secondary,
(*iv*) Equivalent impedance as referred to both primary and secondary, and
(*v*) Total copper loss, first using individual resistances of the two windings and secondly, using equivalent resistances as referred to each side. [**Ans.** (*i*) 7.05 Ω ; (*ii*) 0.0176 Ω ; (*iii*) 11.2 Ω ; (*iv*) 13.23 Ω ; 0.0331 Ω ; (*iv*) 909 W]

15. A single-phase transformer has the following data :
Turn ratio = 19.5 : 1, R_1 = 25 Ω , X_1 = 100 Ω, R_2 = 0.06 Ω, X_2 = 0.25 Ω. No-load current = 1.25 A leading the flux by 30°.
The secondary delivers 200 A at a terminal voltage of 500 V and a power factor of 0.8 lagging. Determine by the aid of a vector diagram the primary applied voltage, the primary power factor and the efficiency.
[**Ans.** 12540 ∠186.7°, 0.698 (lag), 86.74%]

16. The high voltage and low voltage windings of a 1100/110 V single-phase 50 Hz transformer has resistances of 2.4 Ω and 0.02 Ω and reactances 1 Ω and 0.009 Ω respectively. The low voltage winding is connected to a load having an impedance of (3 + *j*2) Ω. Determine :
(*i*) Current in 1.v. winding,
(*ii*) Current in h.v. winding,
(*iii*) The load voltage, and
(*iv*) Power consumed by the load.
[**Ans.** (*i*) 30.11 A ; (*ii*) 3.011 A ; 108.5 V (*iv*) 2720.7 W]

Equivalent Circuit and O.C. and S.C. Tests

17. The parameters of a 2300/230 V, 50 Hz transformer are given below :

$$R_1 = 0.286\ \Omega \qquad R_2' = 0.319\ \Omega$$
$$X_1 = 0.73\ \Omega \qquad X_2' = 0.73\ \Omega$$
$$R_0 = 250\ \Omega \qquad X_0 = 1250\ \Omega$$

The secondary load impedance $Z_L = 0.387 + j0.29$. Solve the exact equivalent circuit with normal voltage across the primary. [**Ans.** η = 78.8%, % Regulation = 2.7%]

18. A 230 V, 3 kVA single-phase transformer has an iron loss of 100 W at 40-Hz and 70 W at 30-Hz. Find the hysteresis and eddy current losses at 50-Hz. [**Ans.** 91.67 W, 41.67 W]

19. When a transformer is supplied at 400 V, 50-Hz the hysteresis loss is found to be 300 W and eddy current loss is found to be 250 W. Determine the hysteresis loss and eddy current loss when the transformer is supplied at 800 V, 100 Hz. [**Ans.** 600 W ; 1000 W]

20. When the primary of a transformer rated at 2200 V, 50-Hz is supplied at 2200 V, 50-Hz the wattmeter gives a reading of 1200 W on no-load. When it is supplied at 1100 V, 25-Hz, the wattmeter gives a reading of 400 W on no-load. If the wattmeter is connected in the input circuit find the hysteresis loss and eddy current loss at normal voltage and frequency. [**Ans.** 400 V ; 800 W]

21. A 4400 V, 50-Hz transformer has a hysteresis loss of 1000 W, eddy current loss of 1500 W and full-load copper loss of 3500 W. If the transformer is supplied at 6600 V, 75-Hz, what will be the losses ? Assume that the full-load current remains the same. **[Ans.** P_h = 1500 W, P_e = 3375 W]

22. A 50 Hz, single-phase transformer has a turn ratio of 6. The resistances are 0.9 Ω, 0.03 Ω and reactances are 5 Ω and 0.13 Ω for high-voltage and low voltage windings respectively. Find :
    (*i*) The voltage to be applied to the h.v. side to obtain full-load current of 200 A in the l.v. winding on short-circuit.
    (*ii*) The power factor on short-circuit.
    Draw the equivalent circuit and show therein all the values. **[Ans.** 329.3 V, 0.2]

23. A 200/2000 V transformer is fed from a 200 V supply. The total winding resistance and leakage reactance as referred to low voltage side are 0.16 Ω and 7.0 Ω respectively. The representing core loss is 400 Ω and the magnetising reactance is 231 Ω. A load of impedance 596 + j444 Ω is connected across the secondary terminals. Calculate :
    (*i*) Input current ; (*ii*) The secondary terminal voltage ;
    (*iii*) Primary power factor. **[Ans.** (*i*) 25.96 $\angle -40.78°$ A ; (*ii*) 1859 V ; (*iii*) 0.757 lagging]

24. Determine the approximate equivalent circuit of a given 200/2000 V single-phase 40 kVA transformer having the following test results :
    O.C. test : 200 V, 6.4 A, 384 W .... on l.v. side.
    S.C. test : 78 V, 20 A, 620 W ...... on h.v. side. **[Ans.** R_{01} = 0.0155 Ω, X_{01} = 0.0358 Ω, R_0 = 104.2 Ω, X_0 = 32.75 Ω]

25. Determine the equivalent circuit of a 200/400 V, 50-Hz, single-phase transformer from the following test data :
    O.C. test (l.v. side) : 200 V, 0.7 A, 70 W
    S.C. test (h.v. side) : 15 V, 10 A, 85 W
    Calculate the secondary voltage when delivering 5 kW at 0.8 power factor, lagging, the primary voltage being 200 V. **[Ans.** R_0 = 571.4 Ω, X_0 = 330 Ω, R_{01} = 0.21 Ω, X_{01} = 0.31 Ω, V_2 = 377.8 V]

26. A single-phase, 10 kVA, 500/250 V, 50-Hz transformer has the following constants :
    Resistance : Primary 0.2 Ω, Secondary 0.5 Ω
    Reactance : Primary 0.4 Ω, Secondary 0.1 Ω
    Resistance or equivalent exciting circuit referred to primary, R_0 = 1500 Ω, reactance of the equivalent exciting circuit referred to primary, X_0 = 750 Ω.
    What would be the readings of the instruments when the transformer is connected for the open-circuit and short-circuit tests ? **[Ans.** O.C. test : 500 V, 0.745 A, 167 W ; S.C. test : 46.8 V, 20 A, 880 W]

27. Find the approximate equivalent circuit of a single-phase 400/4400 V transformer having the following test readings :
    O.C. test (l.v. side) : 400 V, 5.2 A, 600 W
    S.C. test (h.v. side) : 155 V. 50 A, 1850 W **[Ans.** R_0 = 266.7 Ω, X_0 = 80.34 Ω, Z_{01} = 0.02563 Ω, R_{01} = 0.006115 Ω, X_{01} = 0.02488 Ω]

28. The following readings were taken in open-circuit and short-circuit tests on a single-phase 20 kVA, 2000/200 V transformer :
    O.C. test (l.v. side) : 200 V, 2 A, 120 W
    S.C. test (h.v. side) : 30 V, 10 A, 140 W
    Determine :
    (*i*) Equivalent circuit referred to l.v. side
    (*ii*) Secondary terminal voltage on full-load at 0.8 power factor leading.
    [Ans. (*i*) R_0 = 333.3 Ω, X_0 = 104.8 Ω, R_{02} = 0.014 Ω, X_{02} = 0.02654 Ω ; (*ii*) 200.47 V]

Regulation and Efficiency of a Transformer

29. The corrected instrument readings obtained from open and short-circuit tests on 10 kVA, 450/120 V, 50-Hz transformer are :

O.C. Test :

$V_1 = 120$ V, $I_1 = 4.2$ A, $W_1 = 80$ W ; V_1, W_1 and I_1 were read on the low-voltage side.

S.C. Test :

$V_1 = 9.65$ V ; $I_1 = 22.2$ A ; $W_1 = 120$ W with low-voltage winding short-circuited.

Calculate :

(*i*) The equivalent circuit (approximate) constants.

(*ii*) Efficiency and voltage regulation for 0.8 lagging power.

(*iii*) The efficiency at half full-load and 0.8 lagging power factor load.

[**Ans.** (*i*) $R_0 = 25.30\ \Omega$, $X_0 = 409\ \Omega$; (*ii*) 97.57%, 2.04% ; (*iii*) 97.34%]

30. The primary and secondary winding resistance of a 40 kVA, 6600/250 V single-phase transformer are 10 Ω and 0.02 Ω respectively. The equivalent leakage reactance as referred to the primary winding is 35 Ω. Find the full-load regulation for load power factors of (*i*) unity ; (*ii*) 0.8 lagging ; and (*iii*) 0.8 leading.

[**Ans.** (*i*) 2.202% ; (*ii*) 3.69% ; (*iii*) ; – 0.166 %]

31. The percentage resistance and reactance of a transformer are 2.5% and 4% respectively. Find the approximate regulation on full-load at

(*i*) unity power factor, (*ii*) 0.8 power factor lagging, and

(*iii*) 0.8 power factor leading. [**Ans.** (*i*) 2.5% ; (*ii*) 4.4% ; (*iii*) – 0.4%]

32. A single-phase 100 kVA, 2000/200 V, 50 Hz transformer has impedance drop of 10% and resistance drop of 5%.

(*i*) Find the regulation at full-load, 0.8 power factor lagging.

(*ii*) At what power factor is the regulation zero ? [**Ans.** 9.196% ; 0.866 leading]

33. The high voltage of a single-phase 200 kVA 4400/220 V transformer takes a current of 30 A and power of 1200 W at 75 V when the low voltage winding is short-circuited. Determine :

(*i*) The voltage to be applied to the high voltage winding on full-load 0.8 power factor lagging if the secondary terminal voltage is to be kept at 220 V, and

(*ii*) % regulation. [**Ans.** (*i*) 4506 V ; (*ii*) 2.355%]

34. A 20 kVA, 2200/220 V 50-Hz distribution transformer is tested for efficiency and regulation as follows ;

O.C. test (l.v. side) : 220 V, 4.2 A, 148 W.

S.C. test (h.v. side) : 86 V, 10.5 A, 360 W.

Determine :

(*i*) Core loss

(*ii*) Equivalent resistance referred to primary.

(*iii*) Equivalent resistance referred to secondary.

(*iv*) Equivalent reactance referred to primary.

(*v*) Equivalent reactance referred to secondary.

(*vi*) Regulation of transformer at 0.8 power factor lagging current.

(*vii*) Efficiency at full-load and half full-load at 0.8 power factor lagging current.

[**Ans.** 148 W, 3.26 Ω, 0.0326 Ω, 7.51 Ω, 0.0751 Ω, 2.95%, 97.4%, 97.3%]

35. A 100 kVA single-phase transformer has a full-load primary current of 400 A and total resistance referred to primary is 0.006 Ω. If the iron loss amounts to 500 W, find the efficiency on full-load and half-load at

(*i*) unity power factor, and (*ii*) 0.8 power factor.

[**Ans.** 98.58%, 98.56%, 98.22%, 98.19%]

Maximum Efficiency

36. A 50 kVA, single phase transformer has an iron loss of 400 W and full-load copper loss of 800 W.

(*i*) Find the load at which maximum efficiency occurs and the value of maximum efficiency at unity power factor.

(*ii*) If the maximum efficiency occurs at 80% of full-load, find the new core loss and full-load copper loss assuming that total full-load loss is constant.

[**Ans.** (*i*) 70.7% of full-load, 97.79% ; (*ii*) P_i = 468.36 W, P_c = 731.64 W]

37. A 1100/220 V transformer has a primary resistance of 0.25 Ω and secondary resistance of 0.03 Ω. If iron loss amounts to 250 W, determine the secondary current at which maximum efficiency occurs and find the maximum efficiency at 0.8 power factor. [**Ans.** 111.8 A ; 97.52%]

38. A single-phase 250 kVA transformer has an efficiency of 96% on full-load at 0.8 power factor and on half full-load at 0.8 power factor. Find :

(*i*) Iron loss, and (*ii*) full-load copper loss.

[**Ans.** P_i = 2.767 kW, P_c = 5.533 kW]

39. The efficiency of a 300 kVA, single-phase transformer is 97.8% when delivering full-load at 0.8 power factor lagging and 98.4% when delivering half full-load at unity power factor. Determine the efficiency at 80% of full-load at 0.8 power factor lagging. [**Ans.** 97.95%]

40. A 12 kVA, 400/200 V single-phase 50-Hz transformer has maximum efficiency of 95% at 85% of full-load at unity power factor. Determine the efficiency at full-load at 0.8 power factor lagging. [**Ans.** 93.76%]

41. A 50 kVA, 2000/250 V single-phase transformer has resistances of 1.1 Ω and 0.015 Ω and reactances of 4.4 Ω and 0.06 Ω for the high voltage and low voltage windings respectively. It has a maximum efficiency of 96.05% at 80% of full-load at unity p.f. The magnetising current for the h.v. side at 2000 V is 1.25 A. Find the readings of suitable instruments for open circuit test and short-circuit tests, supply being given to h.v. side in both cases.

[**Ans.** O.C. test : 2000 V, 1.316 A, 824 W
S.C. test : 212.3 V, 25 A, 1287 W]

42. A 100 kVA, 3300/300 V single-phase transformer has an efficiency of 97.5% both on full-load at unity p.f. and on half full-load at unity power factor. The power factor of the no-load current is 0.3 lagging and the regulation on full-load at 0.8 p.f. lagging is 3 per cent. Draw the equivalent circuit referred to the l.v. side.

[**Ans.** R_{01} = 105.3 Ω, X_0 = 33.11 Ω,
R_{02} = 0.01538 Ω, X_{02} = 0.02451 Ω]

43. The maximum efficiency of a single-phase 240 kVA, 2000/250 V transformer occurs at 70% of full-load and is equal to 98% at 0.8 p.f. lagging. Determine the efficiency and regulation on full-load at 0.8 p.f. lagging if the impedance of the transformer is 8 per cent. [**Ans.** 97.81%, 5.699%]

44. A 5 kVA, 230/115 V transformer takes 1.2 A and 75 W when 230 V is applied to h.v. winding and l.v. winding is kept open. It takes 21.75 A and 150 W when the l.v. winding is short-circuited and 17.4 V is applied to the h.v. winding. Find :

(*i*) The no-load current as a fraction of full-load input current at 0.8 p.f. lagging ;

(*ii*) Percentage regulation on full-load at 0.8 p.f. lagging ; and

(*iii*) The load at which maximum efficiency occurs and the maximum efficiency at unity p.f.

[**Ans.** 5.162% of full-load input current ; 6.566% ; 70.7% of full-load, 95.94%]

45. The efficiency of a 20 kVA, 2000/200 V transformer is 96.8% at full-load at unity p.f. and 96% at 60% of a load at 0.8 p.f. Find the regulation at full-load at :

(*i*) 0.8 p.f. lagging ; and (*ii*) 0.8 p.f. leading if the impedance is 7 per cent.

[**Ans.** (*i*) 5.469%, (*ii*) – 2.385%]

46. A 100 kVA, 2000/200 V single-phase transformer takes a current of 50 A and 2400 W at 100 V when the low voltage winding is short circuited. Determine the load-voltage and percentage regulation when delivering full-load current at 0.8 p.f. leading, the supply voltage being 2000 V.

[**Ans.** 200.142 V ; – 0.071%]

47. A single-phase, 25 kVA transformer has an iron loss of 240 W and full-load copper loss of 600 W.

(*i*) Find the load at which maximum efficiency occurs and maximum efficiency at 0.8 p.f.

(*ii*) If the maximum efficiency occurs at 70% of full-load, find the core loss and copper loss assuming the total loss to be the same as in the previous case.

[**Ans.** (*i*) 63.25% of full-load, 96.36% ;
(*ii*) 276.3 W, 563.7 W]

All-Day Efficiency

48. A 20 kVA, 2000/200 V transformer has an iron loss of 300 W and full-load copper loss of 400 W. During the day it is loaded as follows :

No. of hours	*Load*	*Power factor*
8	$\frac{1}{4}$ load	0.5
6	Full-load	0.8
6	$\frac{3}{4}$ load	unity
4	No-load	—

Find the all-day efficiency. [**Ans.** 94.84%]

49. A 30 kVA transformer has got a maximum efficiency of 97% at 80% of load at unity p.f. During the day it is loaded as follows :

No. of hours	*Load*	*Power factor*
10	4 kW	0.6 lag
9	20 kW	0.8 lag
5	24 kW	0.85 lag

Find the all-day efficiency. [**Ans.** 95.68%]

50. The all-day efficiency of 200 kVA transformer is 96% when it is loaded as follows :

No. of hours	*Load*	*Power factor*
12	120 kW	0.8
8	150 kW	unity
4	No-load	—

If maximum efficiency of this transformer occurs at 80% of full-load, find iron loss and full-load copper loss. [**Ans.** P_i = 2.645 kW, P_c = 4.133 kW]

Auto Transformer

51. An auto transformer supplies a load of 5 kW at 125 V and at unity power factor. If the primary voltage is 250 V, determine :

(*i*) Transformer ratio, (*ii*) Secondary current,

(*iii*) Primary current,

(*iv*) Number of turns across secondary if the total number of turns is 250,

(*v*) Power transformed, and

(*vi*) Power conducted directly from the supply mains to load. [**Ans.** $\frac{1}{2}$, 40 A, 20 A, 125, 9.5 kW, 2.5 kW]

52. The primary and secondary voltages of an auto-transformer are 500 V and 400 V respectively. Show with the aid of diagram the current distribution in the windings when the secondary current is 100 A and calculate the economy in copper. [**Ans.** 80 per cent]

53. A 200/250 V auto-transformer draws power from a 200 V line and supplies a 5 kW load with a power factor of 0.8 lagging. A second load of 2 kW is supplied at unity power factor from 100 V winding. Neglecting losses, calculate the current drawn by the transformer from the 200 V line and its power factor. [**Ans.** 42.64 A, 0.898]

54. A 5 kVA, single-phase 50-Hz transformer has full-load efficiency of 95 per cent and an iron loss of 50 W. The transformer is now connected as an auto-transformer to a 220 V supply. If it delivers a 5 kW load at unity power factor to a 110 V circuit, calculate the efficiency of the operation and the current drawn by the high-voltage side. [**Ans.** 79.27%, 23.2 A]

55. A 480/120 V, 5 kVA two winding transformer is to be used as an auto-transformer to supply power at 480 V from 600 V source. Draw the connection diagram and determine the kVA capacity as an auto-transformer. [**Ans.** 25 kVA]

UNIT-IV : A.C. MACHINES

7

Synchronous Machines

1. Synchronous generator or alternator—Introduction—Classification and operating principle—Constructional details—frequency—Armature windings—Chording of windings—Pitch factor—Distribution or breadth or winding factor—E.M.F. equation—Alternator on-load—Voltage regulation. 2. Synchronous Motor—Introduction—Characteristic features : advantanges and disadvantages—Applications —Construction—Principle of operation—Synchronous motors—Starting—Effect of load on a synchronous motor—Torque developed by the motor—Effect of excitation on armature current and power factor (V-curves)—Highlights—Objective type questions—Theoretical questions—Unsolved examples.

1. SYNCHRONOUS GENERATOR OR ALTERNATOR

1.1. Introduction

- *A machine for generating alternating currents is referred to as an alternator.*
- The term 'A.C. generator' is also frequently used, in place of alternator and this is often contracted to just 'generator' when it is obvious that an A.C. machine is meant. In older literature, the term 'alternating current dynamo' will also be found, but the present tendency is to reserve the use of the word dynamo for D.C. generators.
- *High-speed* alternators driven by steam turbines differ considerably in their construction from the *slow speed* types and are distinguished by the use of the terms *'turbo-alternator'* or *'turbo-generator'* whilst the slow engine-driven machines are often described as being of the *'flywheel-type'*.

1.2. Classification and Operating Principle

- In *D.C. generators,* the field poles are stationary and the armature conductors rotate. The *alternating* voltage induced in armature conductors is converted to a *direct* voltage at the brushes by means of the commutator.
- *A.C. generators* commonly called alternators, have *no commutators* as they are required to supply electrical energy with an *alternating* voltage. Therefore, it is not necessary that armature be the rotating member.

Alternators, according to their construction, are divided into the following two classifications :

1. Revolving-armature type.
2. Revolving-field type.

1. **Revolving-armature type alternator**

- It has *stationary field poles* and *revolving armature.*
- It is usually of relatively *small* kVA *capacity* and *low-voltage rating.* It resembles a D.C. generator in general appearance except that it *has slip-rings instead of a commutator.* The field excitation must be direct current and therefore, must be supplied from an external direct current source.

2. **Revolving-field type alternator :**

- It has a *stationary* armature or *stator,* inside of which the *field poles rotate.*
- Most alternators are of the revolving-field type, in which the *'revolving-field structure'* or *'rotor'* has slip rings and brushes to *supply the excitation current from an outside D.C. source.* The armature coils are placed in slots in a laminated core, called the *'stator'*, which is made up of thin steel punchings or laminations securely clamped and held in place in the steel frame of the generator. Usually the field voltage is between 100 and 250 volts and the amount of *power delivered to the field circuit is relatively small.*

The following are the principal **advantages** of the revolving-field type alternators :

1. The armature windings are more easily braced to prevent deformation under the mechanical stresses due to short-circuit currents and centrifugal forces.

2. The armature (stator) winding must be insulated for a high voltage, while the voltage of field circuit is low (100 to 250 volts). *It is much easier to insulate the high-voltage winding when it is mounted on the stationary structure.*

3. Only a small amount of power at low voltage is handled by the slip ring contacts.

4. It is easier to build and properly balance high-speed rotors when they carry the field structure.

5. The armature winding is cooled more readily because the stator core can be made large enough and with many air passages or cooling ducts for forced air circulation.

Operating principle (*Revolving-field type*). When the rotor rotates, the stator conductors (being stationary) are cut by the magnetic flux, hence they have induced e.m.f. produced in them. Because the magnetic poles are alternately N and S, they induce an e.m.f. and hence current in armature conductors, which first flows in one direction and then in the other. Hence, an alternating e.m.f. is produced in the stator conductors whose frequency depends on the number of N and S poles moving past a conductor in one second and whose direction is given by Fleming's right-hand rule.

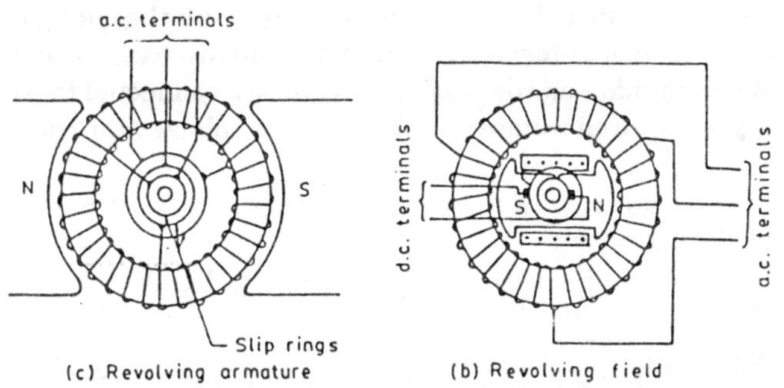

Fig. 1. Operating principle of a three-phase alternator.

Fig. 1 shows the operating principle of a three-phase alternator.

Note. All synchronous A.C. generators and motors require direct current for excitation. Excitation is supplied by a D.C. generator called an *exciter*. The capacity of the exciter is only a *small percentage* of the rated capacity of the alternator. The exciter may be directly connected to the shaft of the alternator, or it may be driven by a separate electric motor, water wheel, or small turbine. Large power stations usually have several exciters employing different methods of drive as insurance against the failure of excitation.

1.3. Constructional Details

Refer Fig. 2.

1.3.1. Stator. The *stator* of an alternator consists essentially of a cast iron or a welded-steel frame supporting a slotted ring made of soft laminated sheet-steel punchings (Fig. 3) in the slots of which the armature coils are assembled.

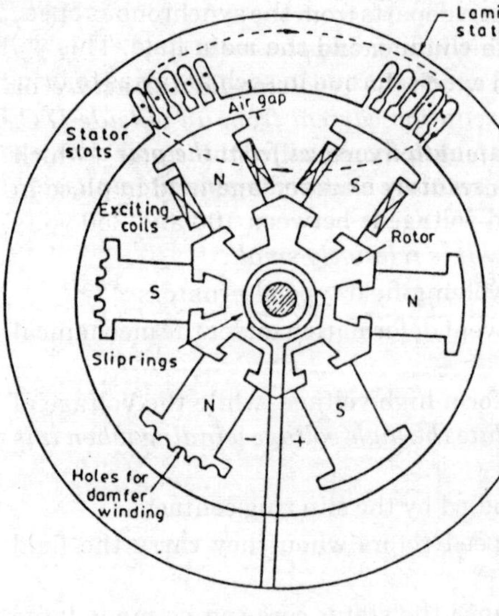

Fig. 2. Alternator.

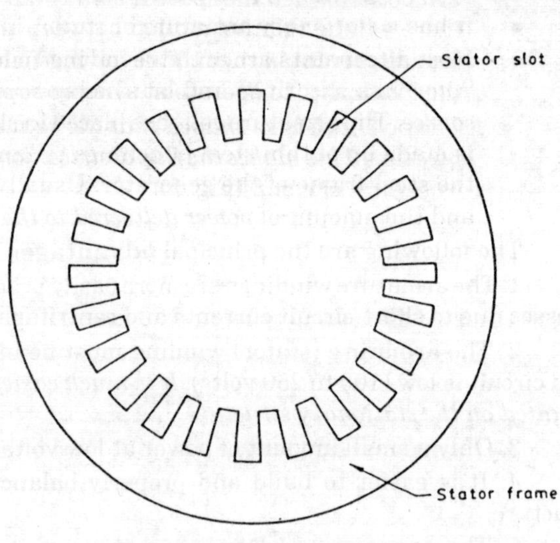

Fig. 3. Alternator stator.

- The *laminations* are annealed and are insulated from each other by a thin coating of oxide and an enamel (as in D.C. machines, transformers etc.)
- *Open slots* are used, permitting easy installation of stator coils and easy removal in case of repair. Suitable spacing blocks are inserted at intervals between laminations to leave radial *air ducts,* open at both ends, through which cooling air may circulate.
- The *coils* are shaped much like the coils of a D.C. generator, the two sides of the coil being approximately a pole pitch apart. All coils are alike, and therefore, interchangeable. They are insulated before being inserted in the slots and are further protected by a horn-fibre slot lining. When in place on the stator, the coils are connected together in groups to form a winding of the required number of phases, three phase star-connected windings being common.
- A fractional rather an *integral number of slots per pole is often used in order to eliminate harmonics in the waveform.*

1.3.2. Rotor. The revolving field structure is usually called the rotor. There are two types of rotors :

1. Salient pole type rotor.
2. Smooth cylindrical type rotor.

Salient pole type rotor :

This type of rotor is used for *slow-speed machines which have large diameters and small axial lengths.*

- The poles are made of thick steel laminations rivetted together and attached to a rotor by a devetail joint as shown in Fig. 4. The overhang of the pole gives mechanical support to the field coil.
- In most of the alternators, where the oscillation or the hunting effect is very high, the *damper winding* in the pole faces is provided. The copper bars short circuited at both ends are placed in the specially provided holes. The relative velocity of the damping winding with respect to main field

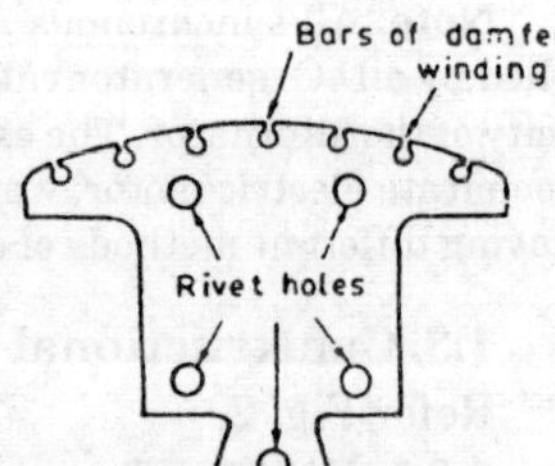

Fig. 4. Typical lamination of a salient pole rotor.

will be zero when the speed is steady but as soon as it departs from the synchronous speed, there will be relative motion between the damper winding and the main field. This will induce current in them. This induced current will exert a torque in such a way as to bring the alternator to operate at synchronous speed.

- *The pole face is so shaped that the radial air gap length increases from the pole centre to pole tips. This makes the flux distribution over the armature uniform to generate sinusoidal waveform of e.m.f.* (Fig. 5).

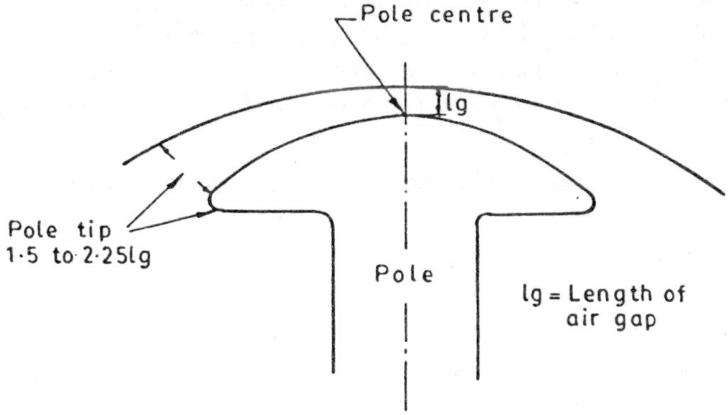

Fig. 5

The salient pole field structure has the following **special features :**

(*i*) They have *large diameter* and *short axial length*.

(*ii*) The pole shoe covers about $\frac{2}{3}$ rd of pole pitch.

(*iii*) Poles are laminated to reduce eddy current losses.

(*iv*) These are employed with hydraulic turbines or diesel engines. The speed is 100 to 375 r.p.m.

Smooth cylindrical rotor :

- This type of rotor is used for alternators which are coupled to *steam turbines* which run at very high speeds. To reduce the peripheral speed of the alternator the *diameter of the*

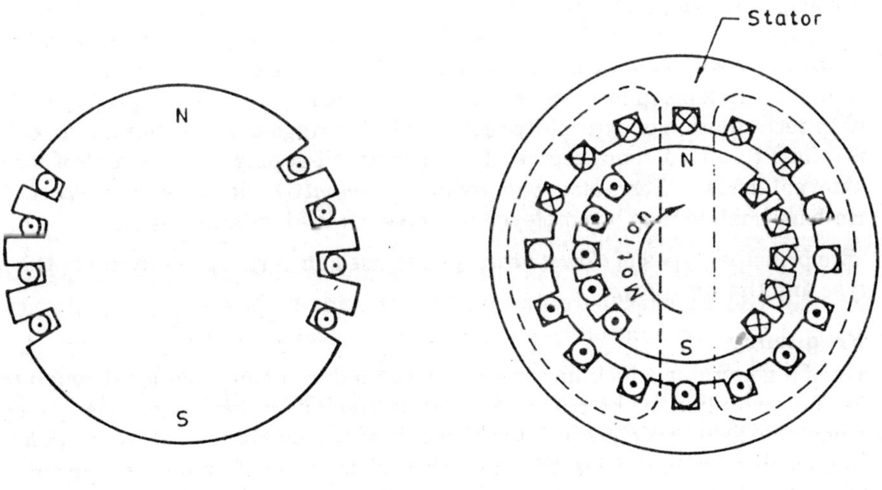

(*a*) Cylindrical rotor (two pole). (*b*) Cylindrical rotor alternator (two pole).

Fig. 6

rotor is reduced and axial length is increased. The number of poles of the rotor are *two* or *four*.

Figs. 5 and 6 show a cylindrical rotor and cylindrical rotor alternator respectively.

- *These rotors are made from solid forgings of alloy steel.* The outer periphery of rotor has slots in which the field winding is placed. About 2/3rd of rotor pole pitch is slotted, leaving the 1/3rd unslotted for the pole centre. Heavy wedges of non-magnetic steel are forced into the grooves in the teeth outside the field coils to keep the field coils in position.
- *Cylindrical rotor machines have always horizontal configuration.*
- Since these rotors have large lengths of core forced ventilation is necessary for proper cooling. *Forced air cooling is used up to about 50 MVA sizes and for bigger sizes hydrogen cooling is invariably employed because the conductivity of hydrogen is about 7 times that of air.*

The non-salient field structure has the following **special features** :

(*i*) They are of small diameter and of very long axial length.
(*ii*) Robust construction.
(*iii*) High operating speed (3000 r.p.m.).
(*iv*) Noiseless operation.
(*v*) Dynamic balancing is better.
(*vi*) Less windage (air-resistance) loss.
(*vii*) No need to provide damper windings, except in special cases to assist in synchronising.
(*viii*) Better e.m.f. waveform.

Note. It may be worth mentioning that *cylindrical rotors* will most likely be located on alternators where *steam power* is readily available. *Salient-pole rotors* will be found where *water power* is the prime-mover source of energy. Diesel engine, gas engine, and gas turbine prime-movers are considered medium-speed machines, and their alternators will also have salient poles. Where alternators are driven by other electrical machines, either A.C. or D.C. motors, there are no such restrictions on the rotor construction. A design is developed that is compatible with the space limitations, speed considerations, and heat dissipation for both electrical machines. If may be noted here that the terms high-speed and low-speed rotors are sometimes used synonymously with cylindrical and salient-pole rotors, respectively.

1.3.3. Bearings

- Although antifriction bearings are occasionally used on alternators of the smaller ratings, the great majority are furnished with oil-lubricated babbitted bearings. For *horizontal shafts* these will be self-contained ring-oiled bearings wherever design conditions permit. At higher shaft peripheral speeds and higher bearing loadings, ring oiling is supplemented with recirculation of externally cooled coil. The rings may be eliminated, or they may be retained to afford some degree of emergency oil supply in the event of a failure of the external system. *Lead-base babbits are commonly used for journal bearings,* although tin-base babbits may be employed for some heavy-duty application.
- Two principal types of *thrust bearings* are used on vertical alternators : the *pivoted shoe type* and the *spring type*.

1.4. Frequency

In case of a generator which has two poles, the induced e.m.f. passes through one complete revolution in one revolution of the machine. In a multipolar machine one cycle of e.m.f. would be generated when the field structure rotates through an angle subtended by a double pole pitch. Therefore, in a machine with p-poles, the number of cycles of e.m.f. in one revolution will be $p/2$. If a machine has a speed of N_s revolutions per minute, the frequency will be

$$\left(\frac{p}{2}\right)\times\frac{N_s}{60} \text{ per second.}$$

Thus, $$f=\frac{p}{2}\times\frac{N_s}{60}=\frac{N_s p}{120} \text{ Hz} \quad ...(1)$$

In order to keep the frequency constant, the speed N_s must remain unchanged. Therefore, a synchronous generator (*i.e.*, alternator) runs at a constant speed known as synchronous speed.

The following table gives the number of poles and speeds for three frequencies : 60-Hz, 50-Hz and 25-Hz.

Higher frequencies are today being used for specific purposes, especially when it is necessary to use high-speed induction motors, but these frequencies are not for general distribution. A great deal of aircraft equipment operates with voltage having a frequency of 400 Hz.

Table 1

Number of poles	*Speed, r.p.m.*		
	60 Hz	*50 Hz*	*25 Hz*
2	3600	3000	1500
4	1800	1500	750
6	1200	1000	500
8	900	750	375
10	720	600	300
12	600	500	250

Note. *The standard frequency in India is 50 Hz.*

1.5. Armature Windings

1.5.1. Winding types. A wide variety of winding types are possible to produce a desired voltage in the proper number of phases and with a suitable waveshape. Practical considerations, *mainly economic,* limit the usual alternator winding to a *double-layer 3-phase lap winding, arranged in 60° phase belts in open slots*. The number of coils, the number of turns per coil, the coil pitch, the number of circuits, and the connection of the phases are selected to give the desired voltage and waveform.

Double-layer windings in open slots permit the use of form-wound coils which are all alike in a given machine. These coils have the characteristic diamond shape in the end area. Three-phase windings may be either Δ-or *Y*-connected ; Y-connected machines are much more common, particularly in the larger sizes. The winding may be arranged to be connected either *Y* or Δ, with leads brought out from both ends of each phase to make this possible.

- **Double-layer windings** in open-slots have the following *advantages* over single layer windings in semi-enclosed slots :
  (*i*) Ease in manufacture of coils and lower cost of winding.
  (*ii*) Less number of coils are required as spare in the case of winding repairs.
  (*iii*) Fractional slot windings can be employed.
  (*iv*) Fractional pitch coils can be used.
- **Single-layer windings** have the following *advantages* :
  (*i*) Higher efficiency and quieter operation because of narrow slot openings.
  (*ii*) Space factor for slots is higher owing to absence of inter layer separator.

Modern practice all over the world favours use of double layer windings. Single layer windings are popular only in the continent.

1.5.2. Number of coils. For integral slot windings, where the number of slots per phase per pole is a whole number, all coil groups are identical, and the phase voltages are balanced with respect to magnitude and angle. Fractional slot windings, where the number of slots per phase per pole is not an integer, have unequal coil groups. These can be arranged to produce balanced voltages if the number of phases is a factor of the number of slots per phase at least as many times as it is a factor of the number of the poles. For example, a balanced 3-phase winding for a 36-pole alternator will require a number of slots per phase which is a multiple of 3-times 3, since 3 appears twice as a factor of 36. Similarly, a balanced 2-phase winding for a 16-pole alternator will require a number of slots per phase which is a multiple of 16 (2 times, 2 times, 2 times).

Other numbers of slots than those which satisfy the above criterion may be used by making coils which do not have the same number of turns. Special groupings of the coils may also permit a balanced wiring when the number of slots does not conform to this criterion. These expedients may be used because of limitations in available punching dies.

1.5.3. Conductor design. *Conductors are stranded* except in the smallest ratings to *enable easier shaping of the coil and to limit eddy-current losses* which result from the flux that crosses the field. The effect of this flux is to produce a voltage within the strand which results in circulating currents. The thinner the strand, the less the voltage and the resulting eddy currents. Since the strands ordinarily are all connected together at the joints between coils, there will also be eddy currents circulating between strands because of the difference in flux linked by the various strands. This source of loss can often be reduced sufficiently by the *use of more and thinner conductors per coil, with a corresponding increase in the number of parallel circuits.* As an alternative, some kind of *transposition* is to arrange the strands in one part of the coil with respect to another so that the induced voltages cancel each other.

1.5.4. Skewing. Sometimes *slots are skewed to minimize the effects of voltages resulting from the ripple in the air-gap flux produced by the stator slots. Skewing a slot, essentially a full slot pitch in length of the core, will eliminate voltages due to slot ripple.* The effect of a number of slots which produces undesirable ripple voltages can be limited by some degree of skew.

1.5.5. Dead coils. Coils can be left unconnected in the armature circuit during manufacture *to obtain operating characteristics not readily obtainable from adjustment of number of coils and turns.* They may also be cut out of machines in service in *order to permit operation after one coil or more has been damaged.* Where the winding has parallel circuits, the number of coils cut out must be the same in each circuit in order to avoid damaging circulating currents. If more than a few per cent of the coils are cut out, it may also be necessary to cut out a like number of coils in the other phases of the machine. The controlling consideration is the voltage balance and the circulating ground currents. This situation would be most critical in a Δ-connected winding.

1.5.6. Coil pitch. Most economical use of the flux in the magnetic circuit results when the stator coil is full pitch, or as near to full pitch as a fractional slot winding will permit. In a fractional slot winding, the maximum pitch ordinarily used is full pitch for the integer, *i.e.*, for $7\frac{5}{7}$ slots per pole winding, a coil pitch of 1 to 8 would be the maximum ordinarily used. *Less than full pitch coils are used to obtain adjustments in the voltage generated or to limit harmonics.*

1.5.7. Single phase windings. These are usually 3-phase windings with 1-phase not used. Sometimes the coils for the third phase are omitted, but most often they are wound in the machine and might be considered as spares. In small sizes a special concentric winding may be used. A *low-resistance damper winding* is generally necessary on single phase machines to *reduce the flux pulsations* that are set up by the single-phase armature reaction and to reduce the effective armature reactance. Single-phase machines have an inherent torque pulsation at twice rated frequency. The resulting noise and vibration are noticeable on small machines and may require special construction on large machines.

1.5.8. Two-phase windings. These windings differ from 3-phase windings only in the grouping of the stator coils. Ninety-degree phase belts are ordinarily used.

1.5.9. Double windings. These windings are sometimes *used to reduce short-circuit currents and to simplify switch gear and bus structure problems*. The windings have standard coil design with special end connections. The electrical designs must allow for the effects of unbalanced armature reaction if the two windings are not equally loaded and for the mutual reactance between the windings.

1.5.10. Multispeed windings. For some applications a winding is required which will permit operation at more than one speed, but at the same frequency. An example is a hydraulic pump-turbine unit, which may generate at one speed and pump at another. Such machines may have two windings in the same slot or in adjacent slots or a single armature winding may be reconnected by means of suitable switches to serve this purpose. Special rotor construction and field-winding reconnections are also necessary.

1.6. Chording of Windings

Chording of the pole windings is one of the design factors. If, on the 36-slot, four pole machine, an individual coil enters slot 1 and comes back in slot 10, it will have spanned 90 mechanical degrees of the stator circular structure. Since there are four poles, 90 mechanical degrees is 180 electrical degrees (because *total electrical degrees in one revolution = 180 p,* where *p*, is the number of poles).

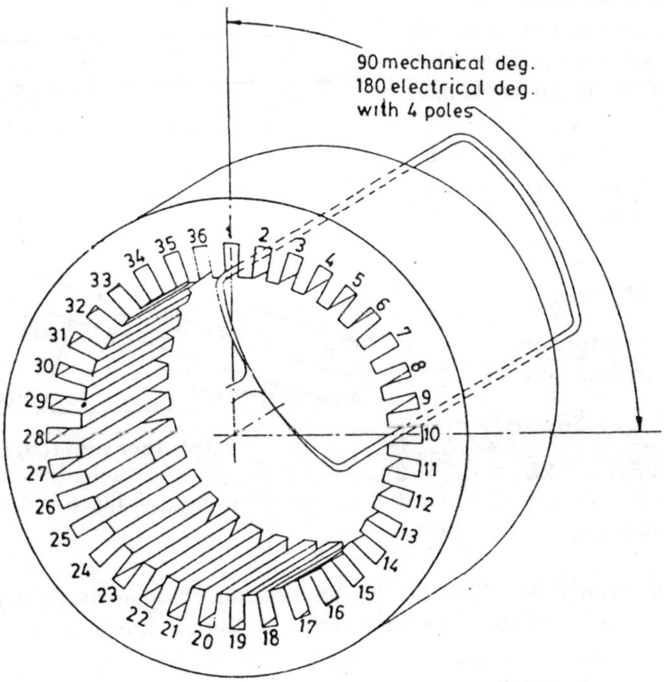

Fig. 7. Four-pole A.C. stator with full pitch coils.

Thus, the two sides of the coil are in the same relative position on the adjacent north and south pole positions. This is a full-pitch coil construction (Fig. 7).

The more usual A.C. machine coil will cover less of the periphery of the machine and is then said to be *fractional pitch.* A typical situation might have a coil enter slot 1 and leave slot 7. This then covers six out of a possible nine slot pitches, and is a 6/9 or 66.7% pitch. *The majority of A.C. machine coils are of fractional-pitch type.*

Advantages of fractional pitch winding :

(*i*) The ends of the coils are shorter, which means *less copper loss due to less total length.*

(*ii*) The end coils can be formed more compactly. The end belts will need less winding space resulting in a shorter unit.

(*iii*) Improved waveform of the induced e.m.f.

(*iv*) Fractional number of slots per pole can be used which in turn reduces tooth ripples.

(*v*) Mechanical strength of the coils is increased.

(*vi*) There is distinct *reduction in machine harmonics* due to cancellation of higher harmonics. Since all A.C. equipment is designed to operate on a pure sine wave, the generation of harmonics is to be avoided. This is especially so when the factor that achieves it is otherwise desirable.

1.7. Pitch Factor

In a full pitch coil, the e.m.fs. in the two coil sides are in phase and therefore the coil e.m.f. is twice the e.m.f. of each coil side. In a *short pitch coil* the e.m.fs. of the two coil sides are not in phase and must be *added vectorially* to give the coil e.m.f. The *factor by which the e.m.f. per coil is reduced, because of the pitch being less, is known as pitch factor* (or coil span factor) k_p. Thus,

$$k_p = \frac{\text{vector or phasor sum of induced e.m.fs. per coil}}{\text{arithmetic sum of the induced e.m.fs per coil}} \quad ...(2)$$

It is *always less than unity*.

Let E_s be induced e.m.f. in each side of the coil. If the coil were full-pitched, then the total induced e.m.f. in the coil would have been $2E_s$ [Fig. 8 (*a*)].

If it is short-pitched by angle α° (electrical) then their resultant [Fig. 8 (*b*)] is E which is vector sum of two voltages, α° (electrical) apart.

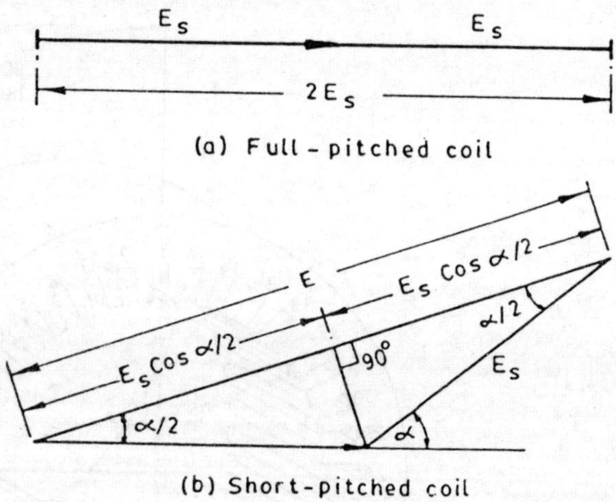

Fig. 8

$$\therefore \quad E = 2E_s \cos\frac{\alpha}{2}$$

$$\therefore \quad k_p = \frac{\text{vector sum}}{\text{arithmetic sum}}$$

$$= \frac{E}{2E_s} = \frac{2E_s \cos\frac{\alpha}{2}}{2E_s} = \cos\frac{\alpha}{2}$$

Hence, $\quad k_p = \cos\frac{\alpha}{2}$ ...(3)

The pitch factor given by eqn. (3) is for the *fundamental component of e.m.f.* If the flux density distribution contains space harmonics, the pitch factor for the *n*th harmonic is given by,

$$k_{pn} = \cos\frac{n\alpha}{2} \quad ...(4)$$

The nth harmonic e.m.f. is reduced to zero if the angle α is such that

$$\cos \frac{n\alpha}{2} = 0 \qquad \text{or} \qquad \frac{n\alpha}{2} = 90° \qquad ...(5)$$

This enables the windings to be designed such that specified harmonics will not be generated *e.g.* if $\alpha = 60°$, there can be no third harmonic generation. *Thus fractional pitch windings result in a voltage wave form which resembles a sinusoidal to a better degree than would a full-pitch winding.*

Note. If the value of α is not given in question, then assume $k_p = 1$.

1.8. Distribution or Breadth or Winding Factor

When the coils comprising a phase of the winding are distributed in two or more slots per pole, the e.m.fs. in the adjacent coils will be out of phase with respect to one another and their resultant will be less than their algebraic sum. *The ratio of the vector sum of the e.m.fs. induced in all the coils distributed in a number of slots under one pole to the arithmetic sum of the e.m.fs. induced (or to the resultant of the e.m.fs. induced in all the coils concentrated in one slot under one pole) is known as distributed factor k_d*

or
$$k_d = \frac{\text{e.m.f. induced in a distributed winding}}{\text{e.m.f. induced if the winding would have been concentrated}}$$

$$= \frac{\text{vector sum}}{\text{arithmetic sum}}.$$

The distribution factor is always less than unity.

Let
n = number of slots/pole
q = number of slots/pole/phase
E_s = induced e.m.f. in each coil side
β = angular displacement between the slots
$= \frac{180°}{n}$

The e.m.fs. induced in different coils of one phase under one pole are represented by side AB, BC, CD, DE ..... which are equal in magnitude (say each equal to E_s) and differ in phase (say $\beta°$) from each other. If bisectors are drawn on AB, BC, CD, DE ...... they would meet at common point (say at O). The point O would be the circumcentre of the circle having AB, BC, CD, DE, ... as the chords and representing the e.m.fs. induced in the coils in different slots.

E.m.f. induced in each coil side,

$$E_s = AB = 2OA \sin \frac{\beta}{2}$$

Arithmetic sum
$$= q \times 2 \times OA \sin \frac{\beta}{2}$$

The resultant e.m.f. induced in one polar group of one phase would be the vector sum as represented by the vector AF as shown in Fig. 9.

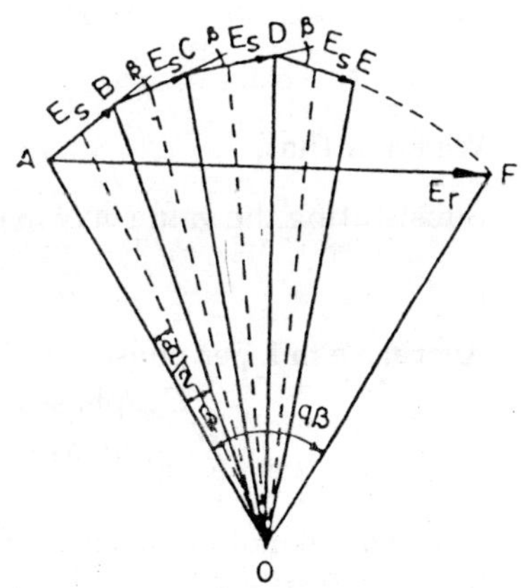

Fig. 9. Calculation of distribution factor.

$\therefore$ The resultant e.m.f.,

$$E_r = AF$$

$$= 2 \times OA \sin \frac{AOF}{2} = 2 \times OA \sin \left(\frac{q\beta}{2}\right)$$

Distribution factor,

$$k_d = \frac{\text{vector sum}}{\text{arithmetic sum}}$$

$$= \frac{2 \times OA \sin\left(\frac{q\beta}{2}\right)}{q \times 2 \times OA \sin \frac{\beta}{2}} = \frac{\sin\left(\frac{q\beta}{2}\right)}{q \sin \frac{\beta}{2}}$$

Hence,
$$k_d = \frac{\sin\left(\frac{q\beta}{2}\right)}{q \sin \frac{\beta}{2}} \quad ...(6)$$

$q\beta$ is also known as the *phase spread* and is expressed in electrical radians.

Value of the distribution factor for *nth harmonic* is,

$$k_d = \frac{\sin\left(\frac{qn\beta}{2}\right)}{q \sin\left(\frac{n\beta}{2}\right)}.$$

1.9. E.M.F. Equation

Let Z_{ph} = number of conductors or coil sides in series *per phase*
$= 2T_{ph}$, where T_{ph} is the number of coils or turns per phase (one turn coil has two sides),
p = number of poles,
ϕ = useful flux per pole, webers,
N = rotational speed of rotor, r.p.m.,
f = frequency, Hz,
k_d = distribution factor, and
k_p = pitch factor.

Consider a conductor on the stator of the alternator. Let the alternator rotor move through one revolution in $t \left(= \frac{60}{N}\right)$ seconds, the flux cut by the conductor

$$= p\phi \text{ webers}$$

The average e.m.f. induced in the conductor,

$$E_{av} = \frac{d}{dt} \text{(flux) volts} = p \cdot \frac{\phi}{t} = \frac{p\phi}{\frac{60}{N}} = \frac{p\phi N}{60} \quad ...(7)$$

We know that, $f = \frac{Np}{120}$ or $N = \frac{120f}{p}$

Substituting the value of N in eqn. (7), we get

$$E_{av} = \frac{p\phi \times 120f}{60 \times p} = 2f\phi \text{ volts/conductor}$$

Average e.m.f. per phase,

$$E_{av}\text{/phase} = 2f\phi \times 2T_{ph} = 4f\phi T_{ph}$$

$$E_{\text{r.m.s.}}\text{/phase} = E_{av} \times \text{form factor} = 4f\phi T_{ph} \times 1.11$$

$$= 4.44 f\phi T_{ph} \text{ volts}$$

The above equation of e.m.f. induced per phase is true only, *if the winding is concentrated in one slot but practically it is not true*, as the winding for each phase under each pole is *distributed* and for such cases k_p and k_d must be considered.

Thus, e.m.f. induced (for sinusoidal wave) per phase will be

$$E_{r.m.s.}/\text{phase} = 4.44 f\phi T_{ph} k_p k_d \text{ volts} \quad ...(8)$$

For full-pitched and concentrated windings $k_p = 1$, $k_d = 1$.

If the alternator is star-connected (as is usually the case) then the line voltage,

$$E_L = \sqrt{3}\, E_{r.m.s.} \text{ phase} = \sqrt{3} \times 4.44 f\phi T_{ph} k_p k_d \text{ volts.}$$

Example 1. *Determine the pitch (or coil span) factors for the following windings :*
(i) 36 stator slots, 4 poles, coil span 1 to 8. (ii) 96 stator slots, 6 poles, coil span 1 to 12. Sketch the two coil spans.

Solution. (*i*) **36 stator slots, 4 poles, coil span 1 to 8** : [Fig. 10 (*a*)]

In this case coil span falls short by,

$$\left(\frac{2}{9}\right) \times 180° = 40° \quad i.e., \quad \alpha = 40°$$

$\therefore$ Pitch factor, $\quad k_p = \cos \frac{40°}{2} = \cos 20° = 0.94.$

(*ii*) **96 stator slots, 6 poles, coil span 1 to 12 :** [Fig. 10 (*b*)]

Here, coil span falls short by,

$$\left(\frac{5}{16}\right) \times 180° = 56° \quad i.e., \quad \alpha = 56°$$

$\therefore$ **Pitch factor,** $\quad k_p = \cos \frac{\alpha}{2} = \cos \frac{56°}{2} = \cos 28° = \mathbf{0.883}$. **(Ans.)**

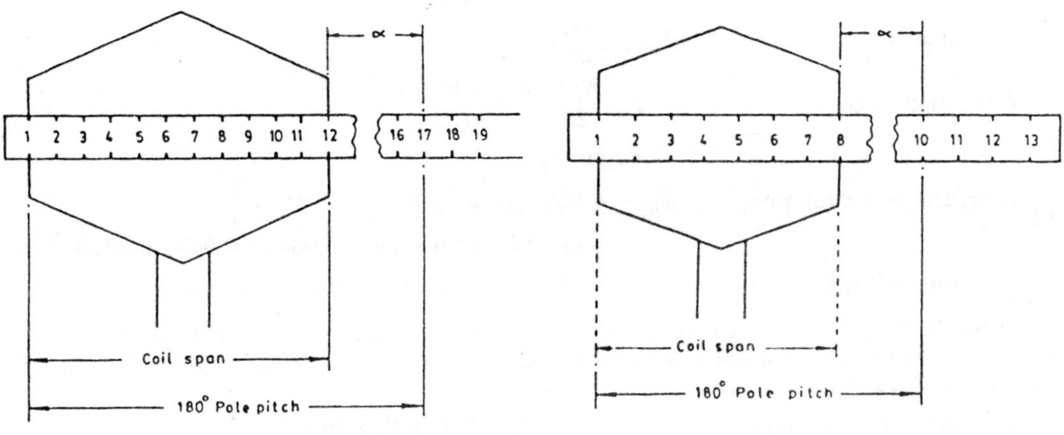

(*a*) 36 stator slots, 4 poles, coil span 1 to 8. (*b*) 96 stator slots, 6 poles, coil span 1 to 12.

Fig. 10

☞**Example 2.** *A 3-phase, 16-pole alternator has the following data :*

Number of slots = 192 ; Conductors/slot = 8 (conductors of each phase are connected in series) ; coil span = 160 electrical degrees ; speed of the alternator = 375 r.p.m. ; flux/pole = 55 m Wb. Calculate the phase and line voltages.

Solution. Number of poles, $\quad p = 16$

Number of slots $\quad = 192$

Conductors/slot $\quad = 8$

Coil span $\quad = 160°$ (electrical)

Speed of the alternator $\quad = 375$ r.p.m.

Flux/pole, $\phi = 55$ mWb $= 0.055$ Wb

Here, $\alpha = 180° - 160° = 20°$

$\therefore$ Pitch factor, $k_p = \cos\frac{\alpha}{2} = \cos\frac{20°}{2} = \cos 10° = 0.9848$

Number of slots/pole, $n = \frac{192}{16} = 12$

$\therefore$ $\beta = \frac{180°}{n} = \frac{180°}{12} = 15°$

q = number of slots/pole/phase $= \frac{192}{16 \times 3} = 4$

$\therefore$ Distribution factor, $k_d = \frac{\sin\frac{q\beta}{2}}{q\sin\frac{\beta}{2}} = \frac{\sin\left(\frac{4 \times 15°}{2}\right)}{4 \times \sin\left(\frac{15°}{2}\right)} = \frac{\sin 30°}{4\sin 7.5°} = 0.9577$

Now, total number of slots/phase $= \frac{192}{3} = 64$

Number of conductors/slot $= 8$ *(given)*

$\therefore$ Total number of conductors/phase

$= 64 \times 8 = 512$

$\therefore$ Turns/phase, $T_{ph} = \frac{512}{2} = 256$

Also, frequency, $f = \frac{Np}{120} = \frac{375 \times 16}{120}$

$f = 50$-Hz

Now, Using the relation, $E_{ph} = 4.44 f \phi T_{ph} k_p k_d$ volts

$= 4.44 \times 50 \times 0.55 \times 256 \times 0.9848 \times 0.9577 = 2948$ V

$\therefore$ **Line voltage,** $E_L = \sqrt{3}\, E_{ph} = \sqrt{3} \times 2948 =$ **5106 V. (Ans.)**

☞**Example 3.** *A 3-phase, star-connected alternator has the following data : voltage required to be generated on open-circuit = 4000 V (at 50-Hz) ; speed = 500 r.p.m. ; stator slots/pole/phase = 3 ; conductors/slot = 12. Calculate :*

(i) Number of poles, and *(ii) Useful flux pole.*

Assume all conductors per phase to be connected in series and coil to be full pitch.

Solution. Line voltage, $E_L = 4000$ V

$\therefore$ $E_{r.m.s.}$/phase, $E_{ph} = \frac{4000}{\sqrt{3}} = 2309$ V

Speed of the alternator, $N = 500$ r.p.m.

Number of slots/pole/phase,

$q = 3$

Number of conductors/slot $= 12$

Frequency, $f = 50$-Hz.

(i) **Number of poles, p :**

Using the relation, $f = \frac{Np}{120}$

or $$p = \frac{120f}{N} = \frac{120 \times 50}{500} = \textbf{12 poles. (Ans.)}$$

(*ii*) **Useful flux/pole, ϕ :**

Pitch factor, $k_p = 1$ (*given*) [$\because$ Coils are full pitched]

Number of slots/pole, $n = 3 \times 3 = 9$

Angular displacement between slots,

$$\beta = \frac{180}{n} = \frac{180}{9} = 20°$$

$\therefore$ Distribution factor, $$k_d = \frac{\sin \frac{q\beta}{2}}{q \sin \frac{\beta}{2}} = \frac{\sin\left(\frac{3 \times 20°}{2}\right)}{3 \times \sin\left(\frac{20°}{2}\right)} = \frac{\sin 30°}{3 \sin 10°} = 0.96$$

Number of turns/phase, $$T_{ph} = \frac{Z_{ph}}{2} = \frac{12 \times 9 \times 12}{2 \times 3} = 216$$

Now, using the relation,

$$E_{r.m.s.}/\text{phase} = 4.44 f \phi T_{ph} k_p k_d \text{ volts}$$

$$2309 = 4.44 \times 50 \times \phi \times 216 \times 1 \times 0.96$$

$\therefore$ $$\phi = \frac{2309}{4.44 \times 50 \times 216 \times 0.96} = 0.05 \text{ Wb}$$

Hence, **useful flux/pole = 0.05 Wb. (Ans.)**

Example 4. *A 3-phase, 10-pole alternator has 2 slots per pole per phase on its stator with 10 conductors per slot. The air gap flux is sinusoidally distributed and equals to 0.05 Wb. The stator has double layer winding with a coil span of 150° electrical degrees. If the alternator is running at 600 r.p.m., calculate the e.m.f. generated per phase at no-load.*

Solution. *Given* : $p = 10$; q number of slots/pole/phase = 2, conductors/slot = 10, $\phi = 0.05$ Wb, coil span = 150° (electrical), $N = 600$ r.p.m.

E.m.f. generated per phase, E_{ph} :

Number of slots/pole, $n = 2 \times 3 = 6$

Angular displacement between the slots, $\beta = \frac{180°}{n} = \frac{180°}{6} = 30°$

Angle of chording $= 180° - 150° = 30°$

$\therefore$ Pitch factor, $$k_p = \cos\left(\frac{30°}{2}\right) = 0.9659$$

Distribution factor, $$k_d = \frac{\sin\left(\frac{q\beta}{2}\right)}{q \sin\left(\frac{\beta}{2}\right)} = \frac{\sin\left(\frac{2 \times 30°}{2}\right)}{2 \times \sin\left(\frac{30°}{2}\right)} = 0.9659$$

Number of conductors in series/phase, $Z_{ph} = 10 \times \text{slots/phase} = 10 \times (2 \times 10) = 200$

$\therefore$ Number of turns/phase, $$T_{ph} = \frac{Z_{ph}}{2} = \frac{200}{2} = 100$$

$\therefore$ $$E_{ph} = 4.44 f \phi T_{ph} k_p k_d \text{ volts}$$
$$= 4.44 \times 50 \times 0.05 \times 100 \times 0.9659 \times 0.9659$$
$$= \textbf{1035.6 V. (Ans.)}$$

1.10. Alternator on Load

When load on an alternator varies, its terminal voltage also varies. This variation in terminal voltage is due to the following reasons :

(*i*) Voltage drop due to armature resistance,
(*ii*) Voltage drop due to armature leakage reactance X_L, and
(*iii*) Voltage drop due to armature reaction.

(*i*) **Armature Resistance.** The voltage drop caused by armature resistance per phase R_a is IR_a and is in phase with current I. This drop, however, is *practically negligible*.

Effective Resistance. The *effective resistance* of the armature winding is greater than the conductor resistance as measured by direct current. This is because additional energy over purely I^2R value is expended inside and sometimes outside of the conductor, owing to the alternating current.

The *chief sources of this additional energy loss are* :
(*i*) eddy currents in the surrounding material ;
(*ii*) magnetic hysteresis in the surrounding material ; and
(*iii*) eddy currents or unequal currents distribution in the conductor itself.

In many cases it is sufficiently accurate to measure the resistance of an armature by direct current and increase it to a fictitious value, called the *effective resistance*, which is large enough to take care of these extra losses. The exact value can vary widely from 1.25. to 1.75, or more × D.C. resistance, depending upon design. *Extreme accuracy is not necessary in this factor,* since the voltage change due to resistance is small in comparison to the changes produced by leakage reactance and armature reaction.

(*ii*) **Armature Leakage Reactance.** The load current, flowing through the armature winding, builds up local flux which on cutting the winding *generates a counter* (or *reactance*) *e.m.f.* This effect gives the armature a *reactance* which is numerically equal to $2\pi fL$. L, in henries, is the leakage inductance of the winding. This armature reactance is called the *leakage reactance* X_L (also known as Potier reactance X_P) since the flux *which causes it is around the armature turns only and does not effect the field flux directly. This leakage flux* (Fig. 11) is proportional to the armature current, since the magnetic path it covers is not normally saturated. X_L varies somewhat with the position of the armature and field poles. It is usually assumed to be constant.

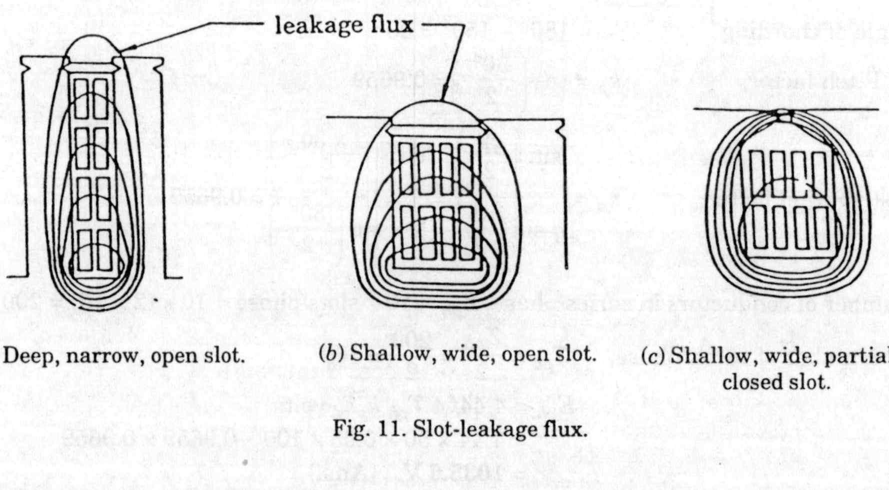

(*a*) Deep, narrow, open slot. (*b*) Shallow, wide, open slot. (*c*) Shallow, wide, partially closed slot.

Fig. 11. Slot-leakage flux.

The voltage drop due to X_L is IX_L. A part of generated e.m.f. is used up in overcoming this voltage drop (reactance e.m.f.).

$$\therefore \quad E = V + I(R_a + jX_L)$$

The fact is illustrated in Fig. 12.

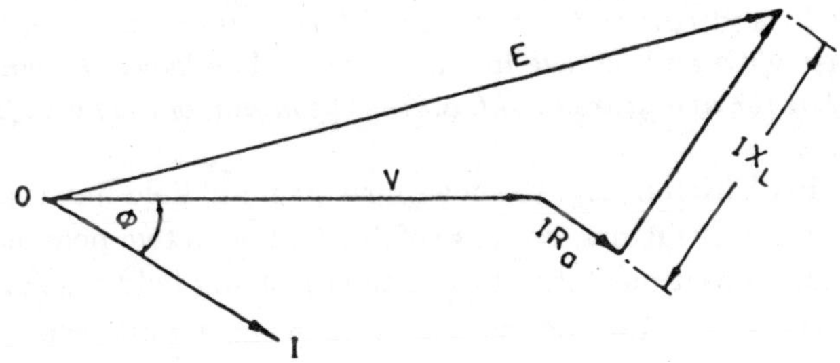

Fig. 12

(*iii*) **Armature Reaction.** Armature reaction is the magneto-motive force produced by the armature currents in the armature conductors.

Unity Power Factor. At unity power factor, armature reaction consists of distortion of the main field flux.

Lagging Power Factor. When an alternator operates with a lagging power factor a m.m.f. is set up by the current in the armature conductors which opposes the m.m.f of the main field and

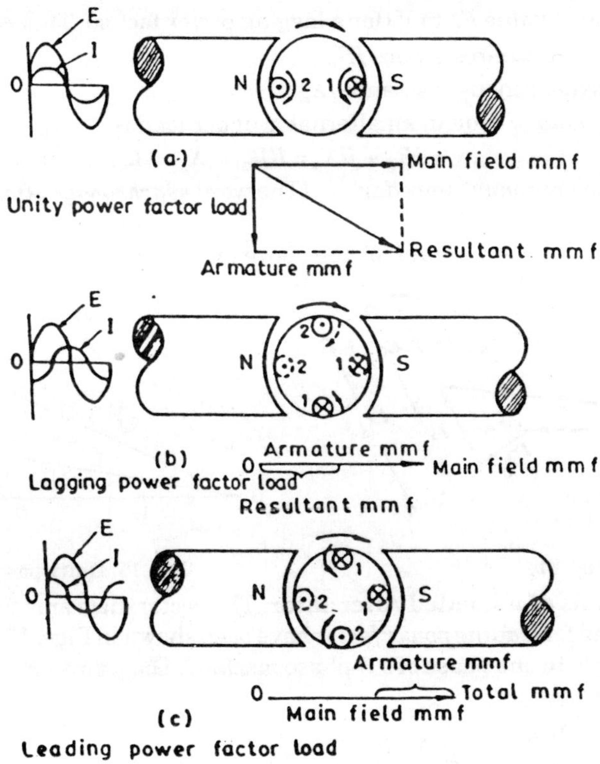

Fig. 13. Effects of armature reaction.

causes a *decrease in the main flux*. This, in turn, results in decrease in the induced voltage. The *lower the value of lagging power factor*, the *greater will be the armature m.m.f. which will oppose and weaken the field*.

Leading Power Factor. In those limited applications where an alternator supplies a load with a leading power factor, the armature current sets up m.m.f. in the armature which aids the m.m.f. of the main field and *causes the main field flux to increase*. Therefore, the voltage of the alternator will increase with an increase in load current. The *lower the value of the leading power factor, the greater will be the armature m.m.f. aiding the m.m.f. of the main field to cause an increase in the main field flux*.

The following illustration in Fig. 13 shows how the main field flux is affected by the armature m.m.f. for different load power factors. For ease of illustration, a two-pole rotating type alternator is used. However, the same conditions take place in the revolving field type alternator.

A study of illustrations in Fig. 13 and vector diagrams for different power factor loads, show that the *voltage drop due to inductive reactance and armature reaction are the same in their effect on the terminal voltage*. Both of these effects are proportional to the armature current.

Synchronous Reactance. The leakage reactance X_L (or X_p) and the armature reactance X_a may be combined to give '*synchronous reactance*' X_s.

Hence, $X_s = X_L + X_a$

[The ohmic value of X_a varies with the power factor of the load because armature reaction depends on load power factor].

From the above discussion, it is clear that for the same field excitation, terminal voltage is decreased from its no-load value E_0 to V (for a lagging power factor). This is because of :

(*i*) Drop due to armature resistance IR_a.

(*ii*) Drop due to synchronous reactance IX_s.

Therefore, *total voltage drop* in an alternator under load is

$$= IR_a + jIX_s = I(R_a + jX_s) = IZ_s$$

where Z_s is known as '*synchronous impedance*'. (The word *synchronous simply refers to the working conditions*)

Fig. 14 shows the vector diagram for the above fact.

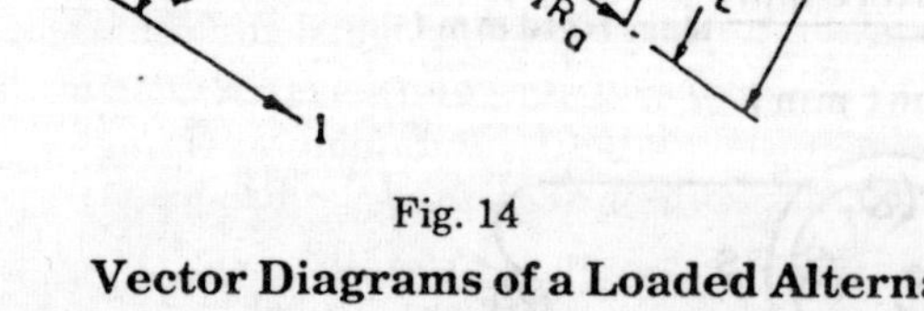

Fig. 14

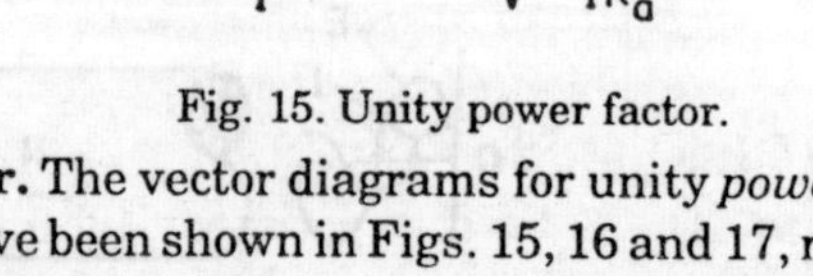

Fig. 15. Unity power factor.

Vector Diagrams of a Loaded Alternator. The vector diagrams for unity *power factor, for lagging power factor* and for *leading power factor* have been shown in Figs. 15, 16 and 17, respectively. All these diagrams apply to one phase of a 3-phase machine. Diagrams for the other phases can also be drawn in the same way.

In all these diagrams,

E_0 = no-load e.m.f. This being the voltage induced in armature in the absence of IR_a, IX_L and IX_a. Hence it represents the maximum value of induced e.m.f.

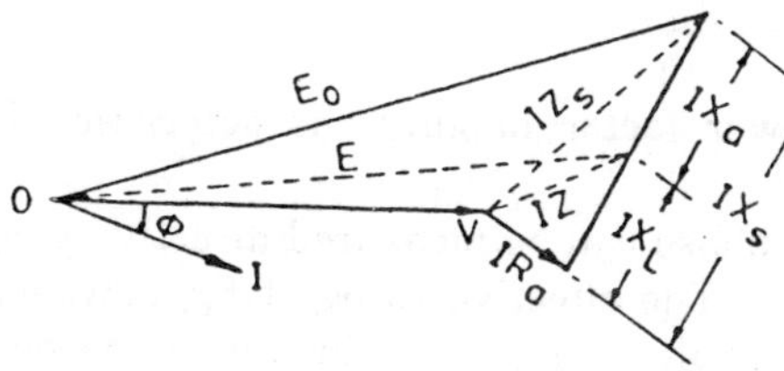

Fig. 16. Lagging power factor.

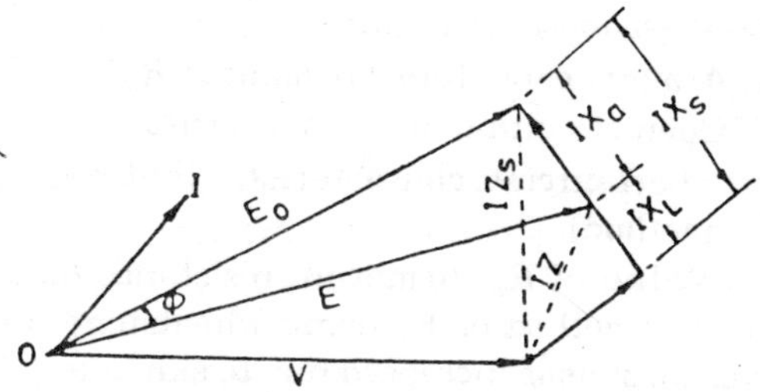

Fig. 17. Leading power factor.

E = it is the induced e.m.f. after allowing for armature reaction. E is vectorially less than E_0 by IX_a

V = terminal voltage. It is vectorially less than E by IZ where $Z = \sqrt{R^2 + X_L^2}$

I = armature current/phase and

ϕ = load power factor angle.

1.11. Voltage Regulation

- When an alternator is subjected to a varying load, the voltage at the armature terminals varies to a certain extent, the amount of this variation determines the *regulation* of the machine. The numerical value of *regulation* is defined as *the percentage rise in voltage when full-load at the specified power-factor is switched off, the excitation being adjusted initially to give normal voltage.* Thus,

$$\% \text{ regulation 'up'} = \frac{E_0 - V}{V} \times 100 \quad ...(9)$$

This gives a lower figure than the percentage drop in voltage when full-load is *switched on*, because in the latter case the excitation is only that required for normal voltage on open circuit. As actually defined, the excitation is greater, since it is that required to give normal voltage on full-load, and owing to the greater saturation of magnetic circuit in these circumstances, the rise in voltage when the load is switched off is less than the drop in voltage when the load is switched on.

- A normal alternator has a regulation of about 8 to 10 percent at unity power factor, but the *voltage rise is considerably increased at lagging power-factors.* At 0.8 lagging power-factor the value lies between 25 and 35 per cent, or even more.
- *Close regulation is not desired, since such an alternator would deliver an excessive current if accidentally short-circuited. Coarse regulation adds to the protection of the machine,* and it is usual to design an alternator with a considerable amount of internal reactance, since this limits the short-circuit current, a most important point where alternators of high power are concerned. Indeed, large alternators are now designed to withstand a dead short-circuit with impunity. *The disadvantage of coarse regulation is obviated by the usual practice of operating an alternator in conjunction with an automatic voltage regulator, which maintains an approximately constant voltage at all loads.*

Determination of Voltage Regulation. It is not usually possible or desirable to measure the regulation by direct experiment, indirect methods being adopted which do not necessitate the loading of the alternator. For this purpose, open circuit and short-circuit characteristics are required.

The following *indirect methods* are used to determine voltage regulation :

1. Synchronous impedance or E.M.F. method.
2. The Ampere-turn or M.M.F. method.
3. Zero power-factor or Potier method.

All these methods require :

(*i*) Armature (or stator) resistance R_a.

(*ii*) Open-circuit/no-load characteristic.

(*iii*) Short-circuit characteristic (but zero power factor lagging characteristic for Potier method).

(*i*) **Value of R_a.** Armature resistance R_a per phase can be measured directly by voltmeter and ammeter method or by using wheatstone bridge. The effective value of R_a, however, under working conditions is increased due to skin effect. The value of R_a so obtained is increased by sixty per cent or so to allow this effect. *Generally, a value 1.6 times D.C. value is taken.*

(*ii*) **Open-circuit Characteristic.** This type of characteristic (as in D.C. machines) is plotted by running the machine on no-load and noting the values of induced voltage and field excitation current.

(*iii*) **Short-circuit Characteristic.** Short-circuit characteristic is obtained by short-circuiting the armature (*i.e.*, stator) windings through a low resistance ammeter. The excitation is so adjusted as to give 1.5 to 2 times the value of full-load current. *The speed* (not necessarily synchronous) *is kept constant during the test.*

1.11.1. Synchronous impedance method. This method involves the following **steps** :

(*i*) Plot the open-circuit characteristic (O.C.C.) from a given data as shown in Fig. 18.

(*ii*) Plot short-circuit characteristic (S.C.C.) from the data given by short-circuit test.

Both these curves are drawn on a *common field-current base.*

Consider a field current I_f. Corresponding to this field current the O.C. voltage is E_1. When winding is short-circuited, the terminal voltage is zero. Hence, it may be assumed that the whole of this voltage E_1 is being used to circulate the armature short-circuit current I_1 against the synchronous impedance Z_s.

$$\therefore \quad E_1 = I_1 Z_s$$

or
$$Z_s = \frac{E_1 \text{ (open-circuit)}}{I_1 \text{ (short-circuit)}}$$

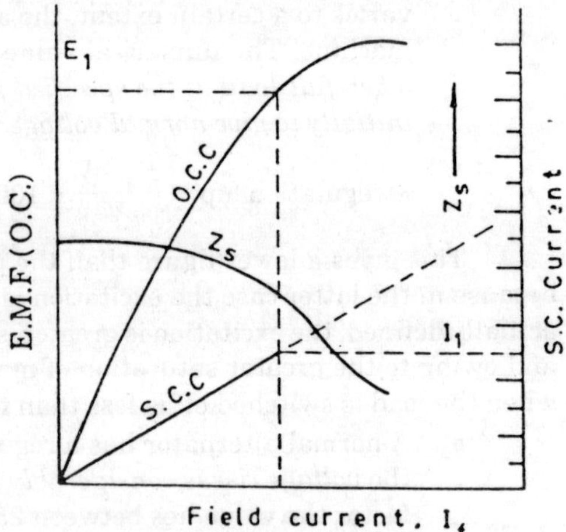

Fig. 18. O.C. and S.C. test curves of an alternator.

(*iii*) Find synchronous reactance X_s as follows :

$$X_s = \sqrt{{Z_s}^2 - {R_a}^2}$$

(R_a can be found as discussed earlier)

(*iv*) After finding R_a and X_s vector diagrams for any load any power factor may be drawn. Three cases are considered in Fig. 19, (*a*) *unity power factor*, (*b*) *lagging power factor*, and (*c*) *leading power factor*.

- **Unity power factor.** Refer Fig. 19 (*a*).

$$OC^2 = (OA + AB)^2 + BC^2$$

$$E_0^2 = (V + IR_a)^2 + (IX_s)^2$$

$$\therefore \quad E_0 = \sqrt{(V + IR_a)^2 + (IX_s)^2} \quad \text{...(10)}$$

- **Lagging power factor.** Refer Fig. 19 (*b*).

$$OC^2 = (OA + AB)^2 + (BD + DC)^2$$

$$E_0^2 = (V \cos \phi + IR_a)^2 + (V \sin \phi + IX_s)^2$$

$$\therefore \quad E_0 = \sqrt{(V\cos\phi + IR_a)^2 + (V\sin\phi + IX_s)^2} \quad ...(11)$$

- **Leading power factor.** Refer Fig. 19 (*c*).

$$OC^2 = (OF + FD)^2 + (BD - BC)^2$$

$$E_0^{\,2} = (V\cos\phi + IR_a)^2 + (V\sin\phi - IX_s)^2$$

$$\therefore \quad E_0 = \sqrt{(V\cos\phi + IR_a)^2 + (V\sin\phi - IX_s)^2} \quad ...(12)$$

The procedure is similar to that adopted in the case of transformers. The *approximate method* of calculating the total drop as was used for transformers *must not be used for alternators* where the *magnitudes involved are much greater in percentages.*

The results obtained by this method are *too high,* owing to the fact that the synchronous impedance determined at short-circuit condition is too *large due to a very low degree of saturation.* Hence this method is called the **'pessimistic'** method.

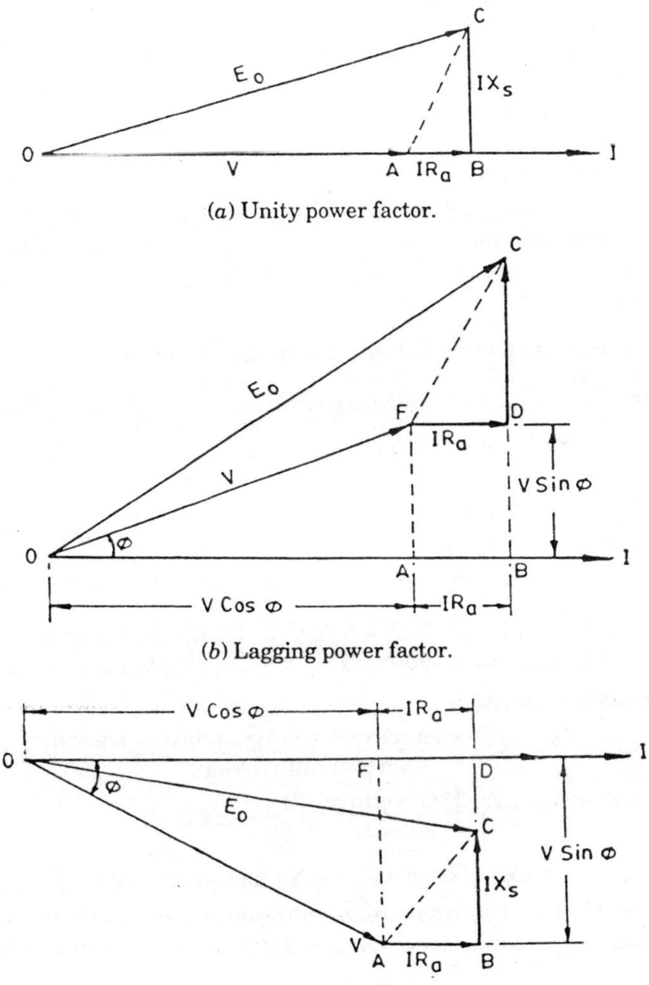

(*a*) Unity power factor.

(*b*) Lagging power factor.

(*c*) Leading power factor.

Fig. 19. Regulation : (*a*) Unity p.f., (*b*) Lagging p.f., (*c*) Leading p.f.

Assumptions inherent in the synchronous impedance method. The assumptions are as follows :

(*i*) *The effect of the armature-reaction flux can be replaced by a voltage drop proportional to the armature current.* The substitution of voltage for flux is the reason that the synchronous-impedance method is also called *e.m.f. method.*

(*ii*) *Since the voltage caused by the main-field flux is added vectorially to that caused by the armature flux, it is also assumed that both fluxes have sinusoidal distribution.* Little error is introduced because of this with non-salient-pole machines, but the error is much greater in the case of the salient-pole ones which have concentrated field windings.

(*iii*) *The magnetic reluctance to the armature flux is constant regardless of the power factor.* This is substantially true for a non-salient, or round-rotor machine, whose air gap is almost constant but introduces considerable error with salient poles, since the position of the armature flux relative to the field poles is determined by the power factor.

(*iv*) There is no saturation effect.

(*v*) The flux under test conditions is the same as that under load conditions.

Example 5. (*a*) *Compare the constructional features of hydel and turbo-generators.*

(*b*) *A 200 kVA, 480 V, 50 Hz Y-connected synchronous generator with a rated field current of 5 A was tested, and the following data were obtained :*

— *Open circuit terminal voltage 540 V (line-to-line)*

— *Short circuit current 300 A.*

When a d.c. voltage of 10 V was applied to two of its terminals, a current of 25 A was measured. Find the value of the synchronous reactance.

(Bombay University)

Solution. (*a*) A comparison of constructional features of hydel and turbo-generators in a tabular form is given below :

Constructional features of hydel and turbo-generators :

S. No.	Aspects	Hydel generators	Turbo-generators
1.	*Numbers of poles*	A large numbers of poles	Only two poles
2.	*Speed*	Low	High (normally 3000 r.p.m.)
3.	*Pole construction*	Salient	Smooth rotor construction
4.	*Length, diameter*	Less, more	Large, small
5.	*Cooling*	Indirect	Direct
6.	*Rating*	Less, lower rated comparatively	11 kV, upto 550 MW

(*b*) *Given* : Rating : 200 kVA, 480 V, $I_f = 5$ A, O.C. terminal voltage = 540 V (line-to-line) ; S.C. current = 300 A ; D.C. voltage = 10 V, current = 25 A.

Synchronous reactance, X_s :

For an alternator, $$Z_s = \frac{\text{open-circuit voltage/phase}}{\text{short-circuit current}} = \frac{540/\sqrt{3}}{300} = 1.04\ \Omega$$

Resistance, $$R_a = \frac{\text{D.C. voltage}}{\text{current}} = \frac{10}{25} = 0.4\ \Omega$$

$$\therefore \quad \mathbf{X_s} = \sqrt{Z_s^2 - R_a^2} = \sqrt{1.04^2 - 0.4^2} = \mathbf{0.96\ \Omega.} \quad \textbf{(Ans.)}$$

Example 6. *A 600 V, 60 kVA, single-phase alternator has an effective resistance of 0.2 Ω. A field current of 10 A produces an armature current of 210 A on short-circuit and an e.m.f. of 480 V on open-circuit. Calculate :*

(*i*) *Synchronous impedance and reactance.*

(*ii*) *Full-load regulation with 0.8 p.f. lagging.*

Solution. Terminal voltage, $V = 600$ V
Alternate output in kVA $= 60$ kVA
Armature effective resistance, $R_a = 0.2\ \Omega$
On short-circuit : Field current $= 10$ A
Armature current $= 210$ A
Open-circuit voltage $= 480$ V.

(*i*) **Synchronous impedance and reactance :**
Synchronous impedance,

$$Z_s = \frac{\text{open-circuit voltage on short-circuit}}{\text{short-circuit current}}$$

$$= \frac{480}{210} = \mathbf{2.28\ \Omega.}\ \textbf{(Ans.)}$$

∴ **Synchronous reactance,**

$$X_s = \sqrt{Z_s{}^2 - R_a{}^2} = \sqrt{(2.28)^2 - (0.2)^2} = \mathbf{2.27\ \Omega.}\ \textbf{(Ans.)}$$

(*ii*) **Full-load regulation with 0.8 p.f. lagging :**
Refer Fig. 20.

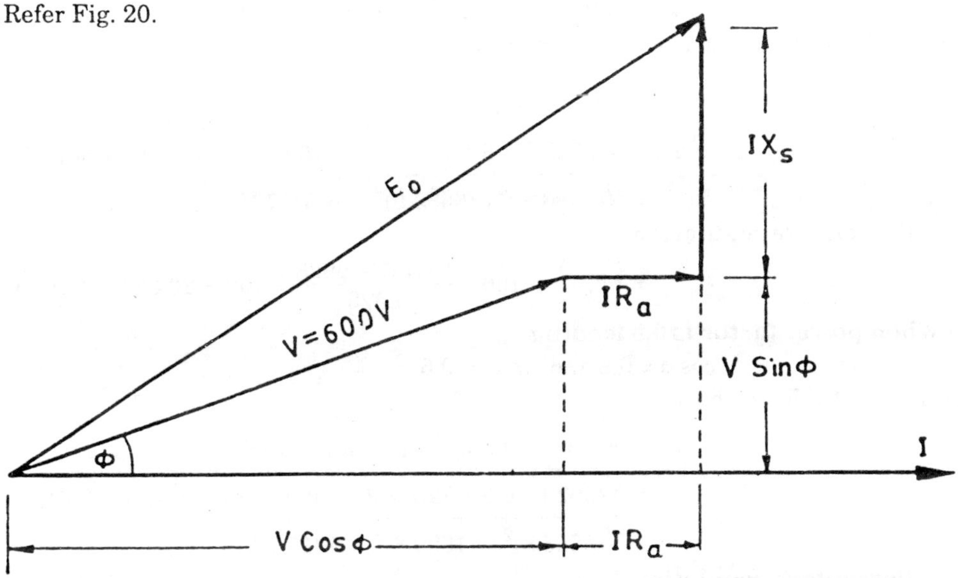

Fig. 20

$$E_0 = \sqrt{(V\cos\phi + IR_a)^2 + (V\sin\phi + IX_s)^2}$$

Now $\cos\phi = 0.8$ and $\sin\phi = 0.6$

Full-load current $= \dfrac{60 \times 1000}{600} = 100$ A

∴ $$E_0 = \sqrt{(600 \times 0.8 \times + 100 \times 0.2)^2 + (600 \times 0.6 + 100 \times 2.27)^2}$$

$$= \sqrt{(480 + 20)^2 + (360 + 227)^2} = 771 \text{ V}$$

∴ Percentage regulation $= \dfrac{E_0 - V}{V} = \dfrac{771 - 600}{600} \times 100 = \mathbf{28.5\%.}$ **(Ans.)**

☞**Example 7.** *A 3-phase star-connected alternator is rated at 1500 kVA, 12000 V. The armature effective resistance and synchronous reactance are 2 Ω and 35 Ω respectively per phase. Calculate the percentage regulation for a load of 1200 kW at power factors of :*

(i) 0.8 lagging *(ii) 0.8 leading.*

Solution. Terminal voltage, $E_L = 12000$ V

Alternator rating $= 1500$ kVA

Load $= 1200 \text{ kW} = \frac{1200}{0.8} = 1500$ kVA

Armature effective resistance/phase, $R_a = 2\ \Omega$

Synchronous reactance/phase, $X_s = 35\ \Omega$

Phase voltage, $E_{ph} (= V) = \frac{12000}{\sqrt{3}} = 6928$ V.

(*i*) **When power factor is 0.8 lagging :**

$$\cos\phi = 0.8 \text{ and } \sin\phi = 0.6$$

∴ Open-circuit voltage/phase,

$$E_0 = \sqrt{(V\cos\phi + IR_a)^2 + (V\sin\phi + IX_s)^2}$$

Now load current at a load of 1200 kW, 0.8 p.f.

$$= \frac{1200 \times 1000}{0.8 \times \sqrt{3} \times 12000} = 72.2 \text{ A}$$

$$\therefore \quad E_0 = \sqrt{(6928 \times 0.8 + 72.2 \times 2)^2 + (6928 \times 0.6 + 72.2 \times 35)^2}$$

$$= \sqrt{(5686.8)^2 + (6683.8)^2} = 8775.7 \text{ V}$$

∴ **Percentage regulation**

$$= \frac{E_0 - V}{V} \times 100 = \frac{8775.7 - 6928}{6928} \times 100 = \mathbf{26.67\%}. \quad \textbf{(Ans.)}$$

(*ii*) **When power factor is 0.8 leading :**

$$\cos\phi = 0.8 \text{ and } \sin\phi = 0.6$$

Open-circuit voltage/phase,

$$E_0 = \sqrt{(V\cos\phi + IR_a)^2 + (V\sin\phi - IX_s)^2}$$

$$= \sqrt{(6928 \times 0.8 + 72.2 \times 2)^2 + [6928 \times 0.6 - 72.2 \times 35]^2}$$

$$= \sqrt{(5686.8)^2 + (1629.8)^2} = 5915.7 \text{ V}$$

∴ **Percentage regulation**

$$= \frac{E_0 - V}{V} \times 100 = \frac{5915.7 - 6928}{5915.7} \times 100 = \mathbf{-\,17.11\%}. \quad \textbf{(Ans.)}$$

☞**Example 8.** *A 3300 V, 3-phase star-connected alternator has a full load current of 100 A. On short-circuit a field current of 5 amperes was necessary to produce full-load current. The e.m.f. on open-circuit for the same excitation was 900 volts. The armature resistance was 0.8 Ω/phase. Determine the full-load voltage regulation for :*

(i) 0.8 p.f. lagging *(ii) 0.8 p.f. leading.*

Solution. Rated voltage (line) = 3300 V

Rated voltage (phase) $= \frac{3300}{\sqrt{3}} = 1905$ V

E.m.f. on open-circuit corresponding to a 5 A excitation = 900 V

Short-circuit current for the same excitation = 100 A

∴ Synchronous impedance/phase,

$$Z_s = \frac{900}{100} = 9\ \Omega$$

Resistance/phase, $R_a = 0.8\ \Omega$

Full-load current, $I = 100$ A

∴ Synchronous reactance, $X_s = \sqrt{Z_s^2 - R_a^2} = \sqrt{9^2 - (0.8)^2} = 8.96\ \Omega$

(*i*) **At 0.8 p.f. lagging.** Refer Fig. 21.

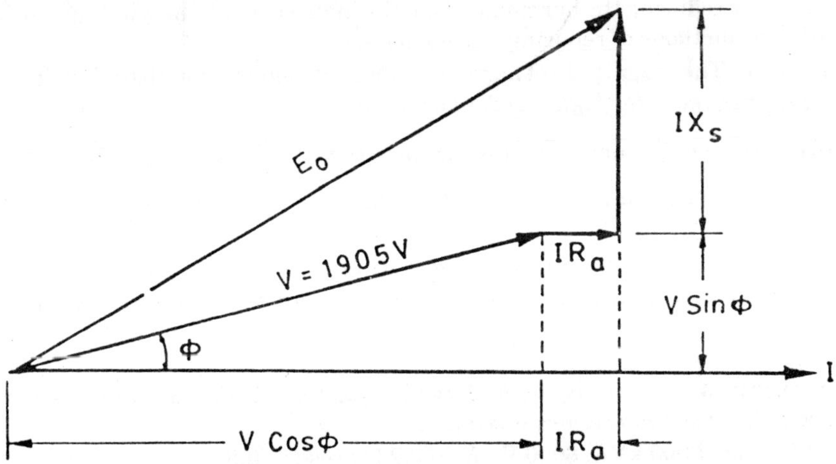

Fig. 21

$$E_0 = \sqrt{(V\cos\phi + IR_a)^2 + (V\sin\phi + IX_s)^2}$$

$$= \sqrt{(1905 \times 0.8 + 100 \times 0.8)^2 + (1905 \times 0.6 + 100 \times 8.96)^2}$$

$$= \sqrt{(1604)^2 + (2039)^2} = 2594.3 \text{ V.}$$

∴ **Percentage regulation** $= \frac{E_0 - V}{V} = \frac{2594.3 - 1905}{1905} \times 100 =$ **36.18%. (Ans.)**

(*ii*) **At 0.8 p.f. leading.** Refer Fig. 22.

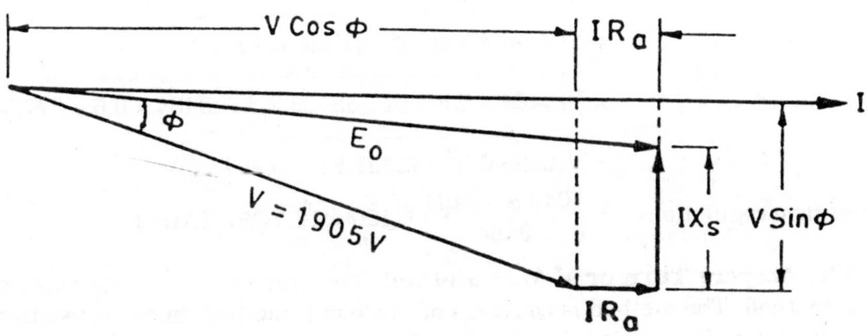

Fig. 22

$$E_0 = \sqrt{(V\cos\phi + IR_a)^2 + (V\sin\phi - IX_s)^2}$$

$$= \sqrt{(1905 \times 0.8 + 100 \times 0.8)^2 + (1905 \times 0.6 - 100 \times 8.96)^2}$$

$$= \sqrt{1604)^2 + (247)^2} = 1622.9 \text{ V}$$

$\therefore$ **% Regulation** $= \dfrac{E_0 - V}{V} \times 100 = \dfrac{1622.9 - 1905}{1905} \times 100 = -$ **14.8%. (Ans.)**

Example 9. *(a) Develop the terms generally used for reducing harmonic voltages from the induced e.m.f.*

(b) A 3-phase, 50 Hz, 6000 kVA, 6000 V star-connected alternator has an effective resistance of 0.2 Ω. A field current of 10 A produces 480 V on open circuit and a field current of 5 A gives armature current of 105 A.

Calculate the voltage regulation of this alternator at 0.8 power for lagging.

Solution. (*a*) For reducing the harmonics from the induced e.m.f. the winding is to be designed properly. The different methods of reducing harmonics are :

1. **Distribution.** The magnitude of harmonic e.m.fs. depends upon their distribution factors. *The distribution factor is small for higher order harmonics.*

2. **Chording.** The e.m.f. generated is proportional to $\cos\left(\dfrac{n\alpha}{2}\right)$ where α is the angle of chording and *n* is the order of harmonic. *The harmonic e.m.f. can therefore be considerably reduced by choosing a proper value of α.*

3. **Skewing.** The *slot harmonics can be eliminated by skewing the pole face.*

4. **Fractional slot winding.** The *harmonic e.m.fs. can be considerably reduced by using fractional slot windings on account of the fact that these windings give a much smaller distribution factor.*

5. **Large length of air gap.** By *using large air gap length, the reluctance is increased and therefore the magnitude of slot harmonics is reduced.*

(*b*) *Given* : Rating : 6000 kVA, 6000 V ; $R_a = 0.2\ \Omega$; $\cos\phi = 0.8$

Full-load current, $I = \dfrac{6000 \times 1000}{\sqrt{3} \times 6000} = 577.35$ A

As the O.C. voltage of 480 V appears at a field current of 10 A, hence the field current of 10 A under S.C. test will give an armature current of 105 × 2 = 210 A.

Phase voltage $= \dfrac{6000}{\sqrt{3}} = 3464$ V

Synchronous impedance, $Z_s = \dfrac{\text{O.C. voltage/phase}}{\text{S.C. current/phase}} = \dfrac{(480/\sqrt{3})}{210} = 1.32\ \Omega$

$$\therefore \quad X_s = \sqrt{Z_s^2 - R_a^2} = \sqrt{1.32^2 - 0.2^2} = 1.305\ \Omega$$

Now,
$$E_o = \sqrt{(V\cos\phi + IR_a)^2 + (V\sin\phi + IX_s)^2}$$
$$= \sqrt{(3464 \times 0.8 + 577.35 \times 0.2)^2 + (3464 \times 0.6 + 577.35 \times 1.305)^2}$$
$$= \sqrt{(2886.67)^2 + (2831.84)^2} = 4043.8\ \text{V}$$

$\therefore$ Percentage regulation $= \dfrac{4043.8 - 3464}{3464} \times 100 =$ **16.74%. (Ans.)**

1.11.2. The Ampere Turn or M.M.F. Method. The graphs of open and short circuit tests are used in this method. The method is converse of the 'e.m.f. method' in the sense that *armature leakage reactance is treated as an additional armature reaction.*

- From O.C. test field current I_f' is determined to give rated voltage *V* on no-load, neglecting armature resistance drop, and I_f'' is determined to cause short-circuit current, equal to full-load current, on short-circuit.

- The field excitation I_f'', on short-circuit, *balances the impedance drop in addition to armature reaction on full-load.* But since R_a is usually very small and X_L is also small for low voltage on short circuit, so impedance drop can be neglected. Hence *power factor on short-circuit is almost zero lagging and field ampere turns are used entirely to overcome the armature reaction. Therefore, I_f'' gives demagnetising ampere-turns on full-load.*

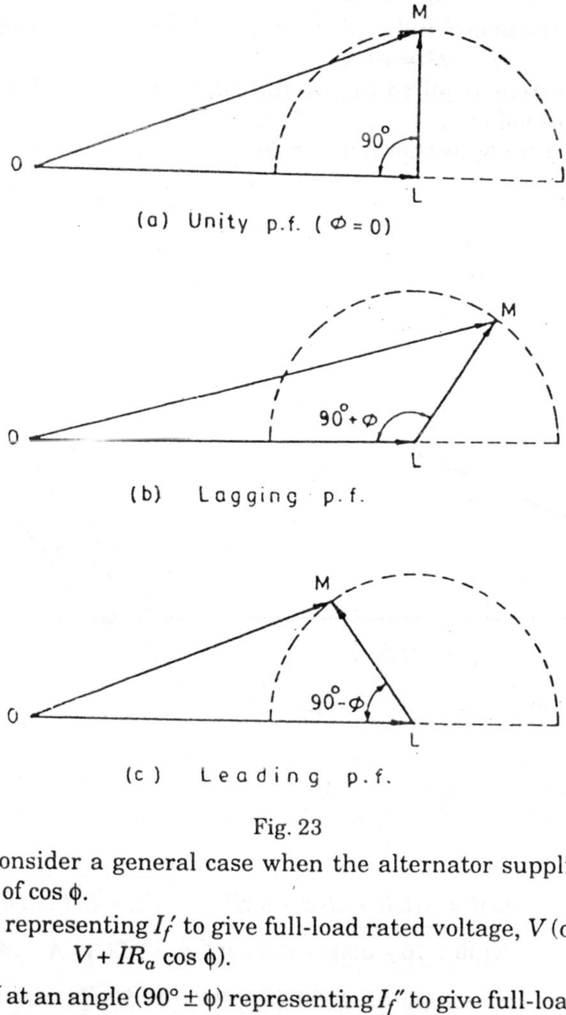

Fig. 23

- Now let us consider a general case when the alternator supplies full-load current at a power factor of cos ϕ.
  (*i*) Draw *OL* representing I_f' to give full-load rated voltage, *V* (or more exactly $V + IR_a \cos \phi$).
  (*ii*) Draw *LM* at an angle $(90° \pm \phi)$ representing I_f'' to give full-load current on short circuit (+ ve sign for lagging p.f. and – ve sign for leading p.f.).
  (*iii*) Find field current I_f, measuring *OM*, which will give open circuit e.m.f. E_0 which can be determined from open circuit characteristic (O.C.C.).

The percentage regulation then, can be obtained from the following relation :

$$\% \text{ Regulation} = \frac{E_0 - V}{V} \times 100.$$

- *Regulation given by this method is much lower than that given by the synchronous impedance method, but it is nearer the correct value. This method is called the* **'Optimistic'** *method.*

Example 10. *The no-load excitation of an alternator required to give rated voltage is 90 A. In a short-circuit test with full-load current flowing in the armature., the field excitation was 70 A Determine the approximate excitation that will be required to give full-load current at 0.8 p.f. lag and at the rated terminal voltage.*

Solution. Field current required to give full-load rated voltage on open circuit,

$$I_f' = 90 \text{ A}$$

Field current required to give full-load current on short circuit in armature,

$$I_f'' = 70 \text{ A}$$

To find the field current required to give full-load current at 0.8 p.f. lag and at the rated terminal voltage proceed as follows :

(*i*) Draw *OL* equal to 90 A (as shown in Fig. 24)

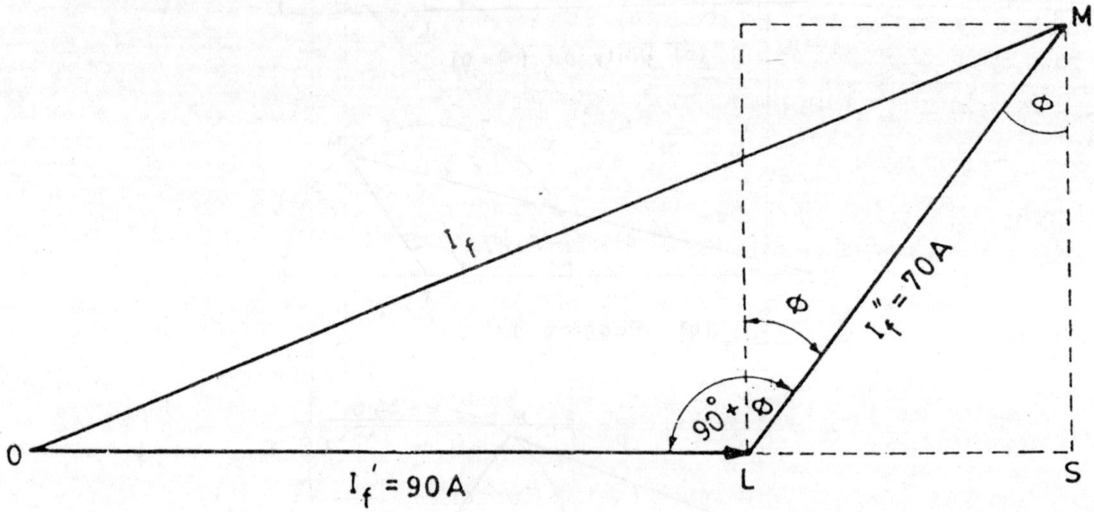

Fig. 24

(*ii*) Draw *LM* at an angle (90 + ϕ) with *OL*, where $\phi = \cos^{-1} 0.8$.

(*iii*) The required field current will be *OM*, the vector sum of *OL* and *LM*.

∴ The required field current I_f to give full-load current at 0.8 p.f. lag and at the rated terminal voltage

$$= OM = \sqrt{(OL + LS)^2 + MS^2} = \sqrt{(OL + LM \sin \phi)^2 + (LM \cos \phi)^2}$$

$$= \sqrt{(90 + 70 \times 0.6)^2 + (70 \times 0.8)^2} = \mathbf{143.39 \text{ A. (Ans.)}}$$

☞**Example 11.** *A 3-phase star-connected, 1000 kVA, 2000 V, 50-Hz, alternator gave the following open-circuit and short-circuit test readings :*

Field current, A :	*10*	*20*	*25*	*30*	*40*	*50*
Open-circuit voltage, V :	*800*	*1500*	*1760*	*2000*	*2350*	*2600*
Short-circuit armature current, A :	—	*200*	*250*	*300*	—	—

Draw the characteristic curves and estimate the full-load percentage regulation at :

(i) 0.8 power factor lagging ; and *(ii) 0.8 power factor leading.*

The armature effective resistance/phase may be taken as 0.2 Ω.

Solution. The phase voltages are :

462 V, 866 V, 1016 V, 1154 V, 1357 V, 1502 V

Full-load phase voltage $= 1154$ V $\left[i.e., \frac{2000}{\sqrt{3}} \right]$

Plot the O.C.C and S.C.C. as shown in Fig. 25.

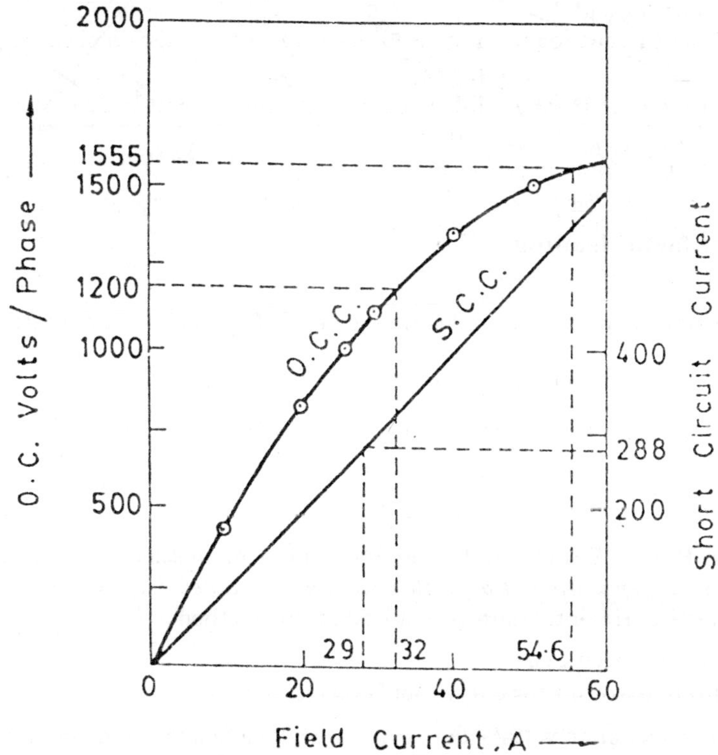

Fig. 25

Full-load current $= \frac{1000 \times 1000}{\sqrt{3} \times 2000} = 288.7$ A

Voltage/phase at full-load at 0.8 power factor

$$= V + IR_a \cos \phi = 1154 + (288.7 \times 0.2 \times 0.8) = 1200 \text{ V}.$$

— From O.C.C., the field current required to produce this voltage (*i.e.*, 1200 V) = 32 A.

— From S.C.C., the field current required to produce full-load current of 288.7 A = 29 A

(*i*) **0.8 power factor lagging :**

$$\cos \phi = 0.8 \quad \text{or} \quad \phi = \cos^{-1} 0.8 = 36° \, 52' \text{ (lag)}.$$

Refer Fig. 26 (Graphical method).

Draw to some scale ;

$$OL = 32 \text{ A}, LM = 29 \text{ A at an angle of } (90 + \phi)$$

or $(90 + 36° \, 52') = 126° \, 52'$

Then OM = total field current at full-load 0.8 p.f. lagging = 54.6 A.

From O.C.C. the open circuit volts corresponding to a field current of 54.6 A = 1555 V

$\therefore$ % **Regulation** $= \frac{E_0 - V}{V} \times 100 = \frac{1555 - 1154}{1154} \times 100 =$ **34.75%. (Ans.)**

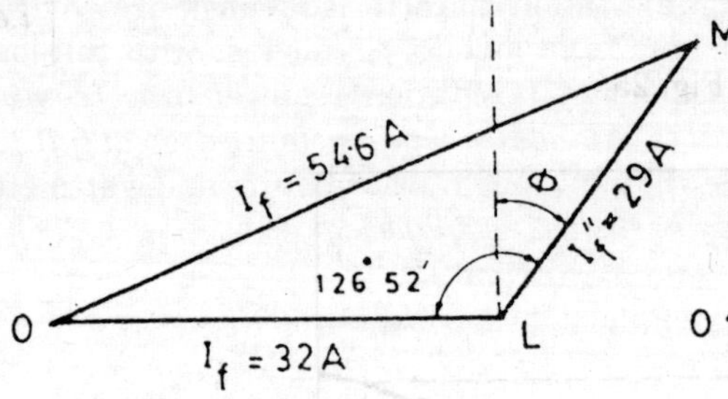

Fig. 26

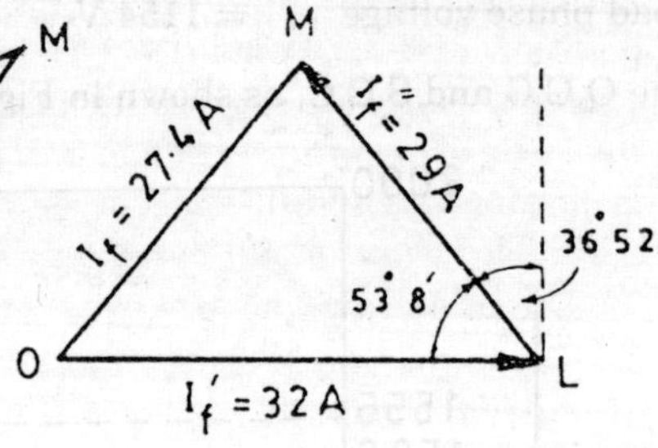

Fig. 27

(*ii*) **0.8 power factor leading :**

Refer Fig. 27.

In this case since power factor is leading, draw LM with OL at an angle of (90° – 36° 52′) = 53° 8′.

$$OM = 27.4 \text{ A}$$

From O.C.C. the open circuit volts corresponding to a field current of 27.4 A = 1080 V

$$\therefore \quad \% \textbf{ Regulation} \quad = \frac{1080 - 1154}{1154} \times 100 = -\mathbf{6.41\%}. \quad \textbf{(Ans.)}$$

1.11.3. Zero Power Factor or Potier Method. This method *gives more accurate results* since it is based on the *separation of armature leakage reactance drop and the armature reaction effects*. The following experimental data is required in this method :

(*i*) No-load or open circuit curve.
(*ii*) Full-load zero power factor curve (not S.C.C.).

From (*ii*) the reduction in voltage due to armature reaction is found out and voltage drop due to armature leakage reactance (also called Potier reactance) X_L is found from both (*i*) and (*ii*). By combining the two, E_0 can be calculated.

The above two curves are similar and displaced horizontally by the m.m.f. due to armature reaction in terms of the field current.

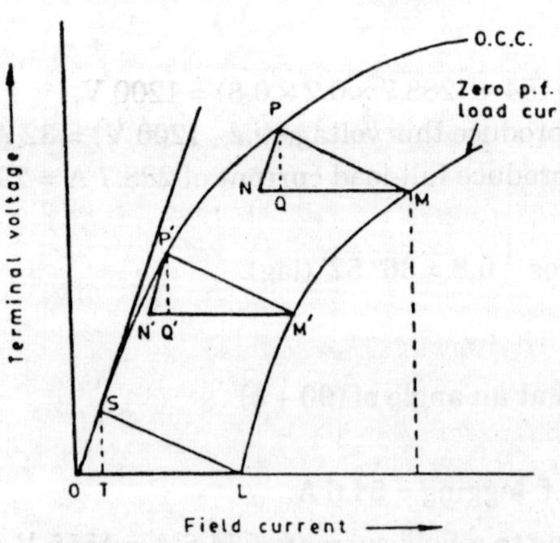

Fig. 28

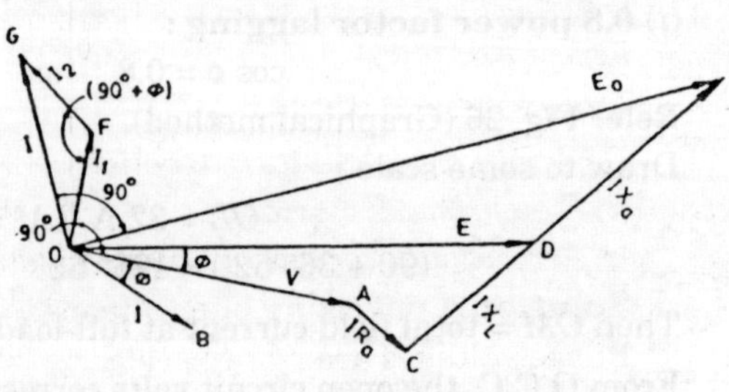

Fig. 29

- Zero power factor, full-load voltage excitation characteristic can be drawn by knowing two points L and M. Point L is obtained from a short circuit test with full-load armature current. Hence OL represents field current (excitation) required to *overcome demagnetising effect of armature reaction and to balance leakage reactance drop at full-load.* Point M is obtained when full-load current flows through the armature and wattmeter reading is zero. *Zero power factor curve may be drawn as follows :*

  (*i*) From M draw line MN equal and parallel to OL.

  (*ii*) Through point N draw a line parallel to initial straight part of O.C.C. (parallel to OS), cutting the O.C.C. at P.

  (*iii*) Join PN and drop a perpendicular PQ on MN.

  (*iv*) Impose the triangle MPQ at various-points of O.C.C. to obtain corresponding points on the zero power factor curve.

  In triangle MPQ :

  Length PQ represents *leakage reactance drop* (IX_L)

  Length MQ represents *armature reaction excitation.*

Potier Regulation Diagram. Following is the procedure to draw Potier regulation diagram :

(*i*) Draw OA horizontally to represent terminal voltage V on full load and OB to represent full load current (I) at a given power factor.

(*ii*) Draw AC ($= IR_a$), voltage drop due to resistance R_a (if resistance is given) parallel to OB.

(*iii*) Draw CD perpendicular to AC and equal to reactance drop IX_L.

Now OD represents generated e.m.f. E.

(*iv*) From O.C.C., find the field current I_1 corresponding to this generated e.m.f. E and draw OF (equal to I_1) perpendicular to OD. Draw FG parallel to load current OB (*i.e.*, I) to represent excitation (field current) *equivalent to full load armature reaction.*

OG gives *total field current* required.

(*v*) If the load is thrown off, then terminal voltage will be equal to generated e.m.f. corresponding to field excitation OG. Hence e.m.f. E_0 may be obtained from O.C.C. corresponding to field excitation OG. Vector OJ will lag behind vector OG by 90°. DJ represents *voltage drop due to armature reaction.*

Now regulation may be obtained from the following relation :

$$\% \text{ Regulation} = \frac{E_0 - V}{V} \times 100.$$

☞**Example 12.** *The following data pertains to a 15000 kVA, 11 kV, 3-phase, 50 Hz star-connected turbo-alternator :*

O.C. line kV	:	*4.9*	*8.4*	*10.1*	*11.5*	*12.8*	*13.3*	*13.65*
Field AT in 10^3	:	*10*	*18*	*24*	*30*	*40*	*45*	*50*
Zero p.f. full-load line kV	:	—	*0*	—	—	—	*10.2*	—

Determine : (i) Armature reaction *(ii) Armature reactance*

(iii) Synchronous reactance

(iv) Percentage regulation for full-load at 0.8 p.f. lagging.

Solution. Converting line voltages to phase voltages.

Phase voltages are :

2829 ; 4850 ; 5831 ; 6640 ; 7390 ; 7679 ; 7880

Plot the O.C.C. as shown in Fig. 30.

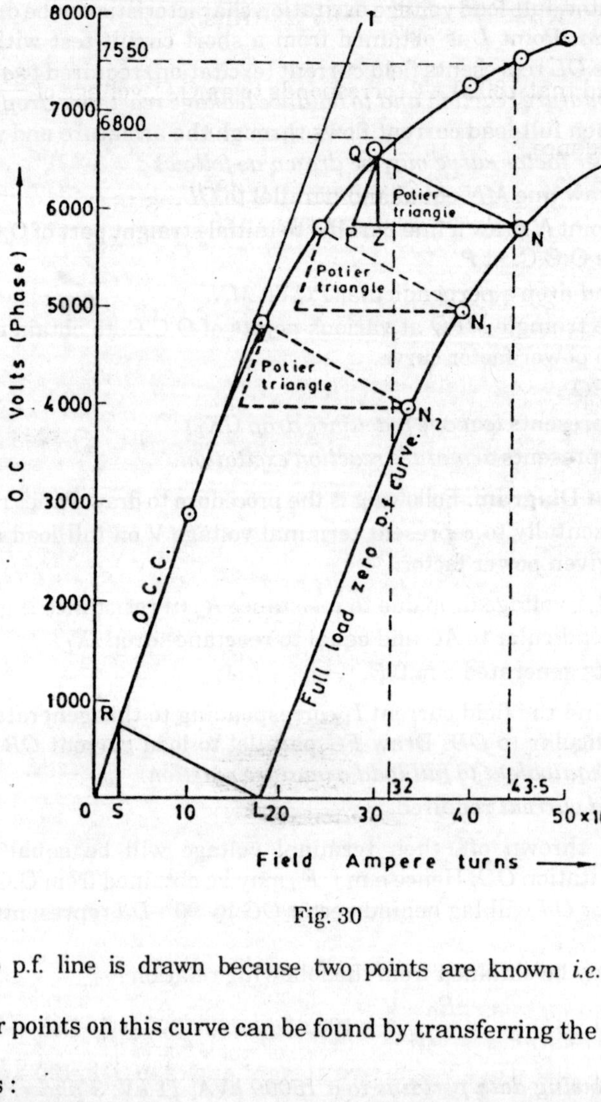

Fig. 30

Full-load zero p.f. line is drawn because two points are known *i.e.*, (18, 0) and $\left[45, \frac{10200}{\sqrt{3}}\right.$ *i.e.*, 5890 $\left.\right]$ and other points on this curve can be found by transferring the Potier triangle (as shown in Fig. 30) as follows :

- At N, draw $NP \parallel$ to and equal to OL and from P draw $PQ \parallel$ to OR. Join QN. Hence NPQ is *Potier triangle*.
- Draw line $QZ \perp$ to PN

(i) NZ = field ampere-turns for balancing armature reaction only

$$= \mathbf{15500.} \quad \textbf{(Ans.)}$$

$$QZ = RS = 670 \text{ V}$$

$$= \text{leakage reactance drop/phase}$$

Short-circuit AT required $= OL = 18000$

Full-load current, $I = \dfrac{15000 \times 1000}{\sqrt{3} \times 11000} = 787 \text{ A}$

$\therefore \quad IX_L = 670$

(*ii*) ∴ $$X_L = \frac{670}{I} = \frac{670}{787} = \mathbf{0.85\ \Omega.}\ \ \textbf{(Ans.)}$$

(*iii*) From O.C.C. we find that 18000 AT corresponds to an O.C. voltage of $\frac{8400}{\sqrt{3}} = 4850$ V

∴ Synchronous impedance,

$$Z_s = \frac{\text{O.C. volts}}{\text{short-circuit current}} = \frac{4850}{787} = 6.16\ \Omega$$

Since R is negligible, hence

$$Z_s = X_s = 6.16\ \Omega$$

Hence, **synchronous reactance, X_s = 6.16 Ω. (Ans.)**

(*iv*) **Regulation at 0.8 p.f. lag :** Refer Fig. 31.

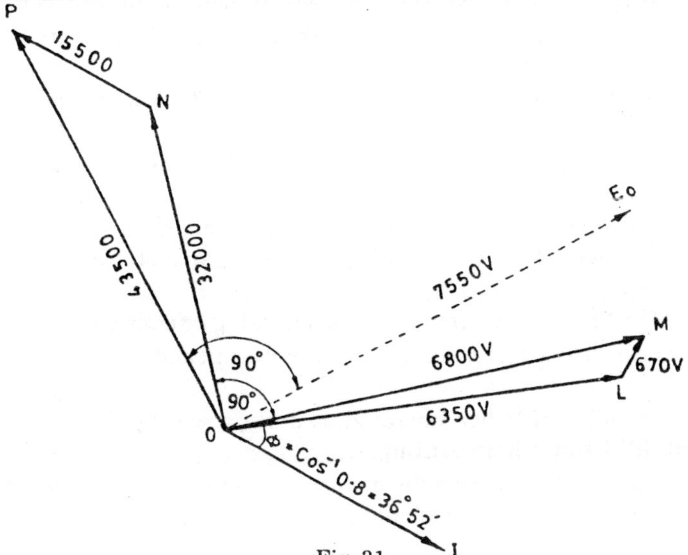

Fig. 31

$$OL = \text{phase voltage} = \frac{11000}{\sqrt{3}} = 6350 \text{ V}$$

LM = 670 V is drawn at right angles to OI or $(90 + \phi)$ to OL. Resultant is OM = 6800 V

From O.C.C., field ampere-turns corresponding to 6800 V is

ON = 32000 and is drawn ⊥ OM

NP = armature reaction at full-load = 15500 and is drawn ∥ to OI at $(90 + \phi)$ to ON

Hence, OP = 43500

From O.C.C. e.m.f. corresponding to 43500 AT of rotor

= 7550 V

∴ **% Regulation** $= \frac{7550 - 6350}{6350} \times 100 =$ **18.89 %. (Ans.)**

2. SYNCHRONOUS MOTOR

2.1. Introduction

- The synchronous motor is the one type of 3-phase A.C. motors *which operates at a constant speed from no-load to full-load.* It is similar in construction to 3-phase A.C. generator in that *it has a revolving field which must be separately excited from a D.C. source. By changing the D.C. field excitation current, the power factor of this type of motor can be varied over a wide range of lagging and leading values.*
- This motor is used in many individual applications because of its *fixed speed from no-load to full-load, its high efficiency* and *low initial cost.* It is also used *to improve the power factor of 3-phase A.C. industrial circuits.*

2.2. Characteristic Features, Advantages and Disadvantages

Characteristic Features :

The following **characteristic features** of a synchronous motor are worth noting :

1. *It runs either at synchronous speed or not at all.* The speed can be changed by *changing the frequency only* (since $N_s = 120f/p$).

2. *It is not inherently self-starting.* It has to be *run up to synchronous or near synchronous speed* by some means before it can be synchronized to the supply.

3. *It can operate under a wide range of power factors both lagging and leading.*

4. On no-load the motor draws very little current from the supply to meet the internal losses. With fixed excitation the input current increases with the increase in load. After the input current reaches maximum no further increase in load is possible. If the motor is further loaded, the motor will stop.

Advantages. Synchronous motors entail the following *advantages* :

1. These motors can be used for *power factor correction* in addition to supplying torque to drive loads.

2. They are more efficient (when operated at unity power factor) than induction motors of corresponding output (kW) and voltages rating.

3. The field pole rotors of synchronous motors *can permit the use of wider air-gaps* than the squirrel-cage designs used on induction motors, requiring *less bearing tolerance and permitting greater bearing wear.*

4. They may be less costly for the same output, speed, and voltage ratings as compared to induction motors.

5. They give constant speed from no-load to full-load.

6. Electro-magnetic power varies linearly with the voltage.

Disadvantages. The *disadvantages* of synchronous motors are :

1. They require D.C. excitation which must be supplied from external source.
2. They have a tendency to *hunt.*
3. They cannot be used for variable speed jobs as speed adjustment cannot be done.
4. They require collector rings and brushes.
5. They cannot be started under load. Their starting torque is zero.
6. They may fall out of synchronism and stop when overloaded.

2.3. Applications

The synchronous motors have the following *fields of application* :

1. **Power houses and sub-stations.** Used in power houses and sub-stations in parallel to the bus-bars to improve the power factor.

2. **Factories.** Used in factories having large number of induction motors or other power apparatus, operating at *lagging power factor, to improve the power factor.*

3. **Mills-industries etc.** Used in textile mills, rubber mills, mining and other big industries, cement factories for power applications.

4. **Constant speed equipments.** Used to drive continuously operating and constant speed equipment such as :

- Fans.
- Blowers.
- Centrifugal pumps.
- Motor generator sets.
- Ammonia and air compressors etc.

2.4. Construction

A three-phase synchronous motor consists of the following *essential parts* :

1. **Laminated stator core** with three-phase armature winding.
2. **Revolving field complete with amortisseur winding and slip rings.**
3. **Brushes and brush holders.**
4. **Two end shields** to house the bearings that support the shaft.

- The stator core and windings of a synchronous motor are similar to those of a 3-phase squirrel-cage induction motor or a wound-rotor induction motor. The leads for the stator winding, marked T_1, T_2 and T_3, terminate in a terminal box usually mounted on the side of the motor frame.
- The rotor is generally a salient pole rotor. *The number of rotor field poles must equal the number of stator field poles.* In order *to eliminate hunting* and *to develop the necessary starting torque* when A.C. voltage is applied to the stator, the rotor poles contain pole-face conductors which are short-circuited at their ends as shown in Fig. 32. This amortisseur or damper winding consists of *solid copper bars* embedded at the surface of the pole face and short-circuited at each end by means of a shorting strip as shown in Fig. 33.
- The field circuit leads are brought out to two *slip rings* mounted on the rotor shaft. Carbon brushes mounted in brush holders make contact with the two slip rings. The terminals of the field circuit are brought out from the brush holders to a second terminal box mounted on the motor frame. The two leads for the field circuit are marked F_1 and F_2.

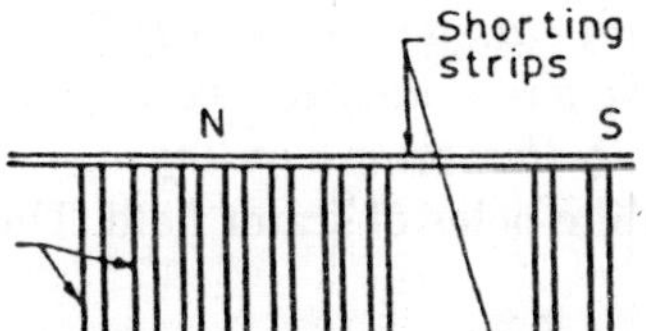

Fig. 32. Pole of an A.C. synchronous motor.

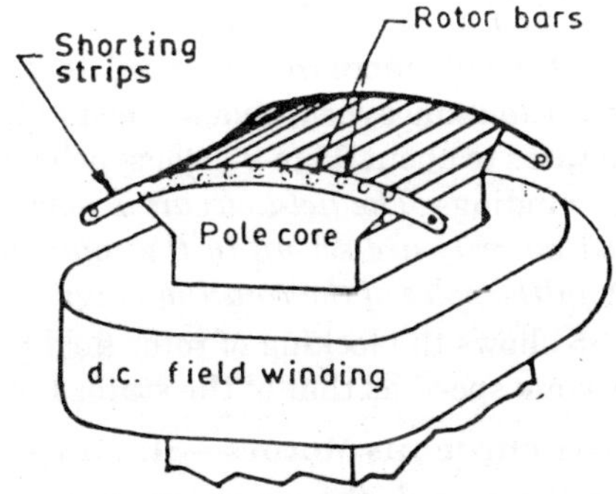

Fig. 33. Damper winding.

Fig. 34 shows a typical arrangement of synchronous motor, with overhung exciter for providing the D.C. supply to the **rotor of the machine.**

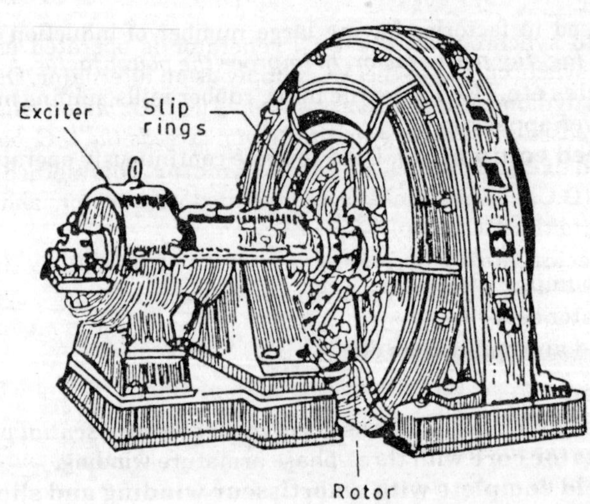

Fig. 34. Synchronous motor.

2.5. Principle of Operation

When the stator windings of a 3-phase synchronous motor are supplied with rated 3-phase voltage, a rotating field travelling at synchronous speed is set up. The synchronous speed is found from the relation $N_s = \frac{120f}{p}$ where N_s (in r.p.m.), f and p are synchronous speed, frequency and number of poles respectively. This rotating magnetic field cuts across the amortisseur or squirrel-cage winding of the rotor and induces voltages and currents in the bars of this winding. *The resultant magnetic field of squirrel-cage winding embedded in the rotor field poles reacts with the stator field in such a manner as to cause the rotation of the motor.* The rotor will increase its speed to a point *slightly below* the synchronous speed of the stator field. The rotor of the typical synchronous motor accelerates to about 95 to 97% of synchronous speed when started as an induction motor with amortisseur windings. *The field circuit is now excited from an outside source of D.C. and magnetic poles of fixed polarity are set up in the rotor field cores. The fixed magnetic poles of the rotor are attracted to unlike poles of the rotating magnetic field set up by the stator windings.*

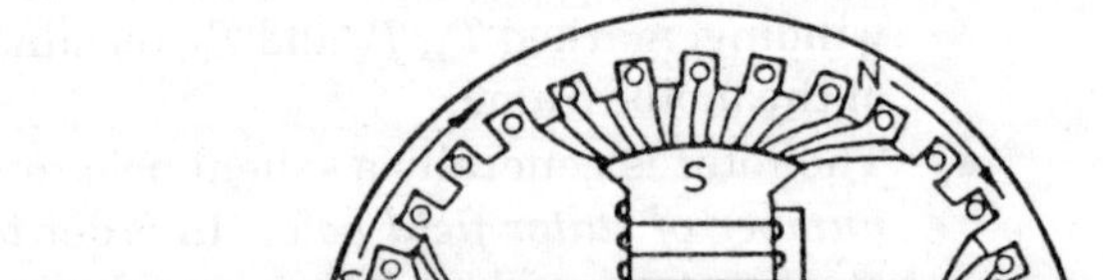

Fig. 35. Operating principle of a synchronous motor.

Fig. 35 shows the locking of rotor field poles with unlike poles of stator field. The rotor then runs at the same speed as that of the stator field *i.e.*, N_s.

2.6. Synchronous Motors—Starting

As earlier stated, the synchronous motor must be brought to a speed sufficiently *close to synchronous speed* in order to lock into synchronism with the rotating field. The means by which it is brought up to speed are :

1. A D.C. motor coupled to the synchronous motor shaft. This method is sometimes used in laboratories with synchronous motors not equipped with damper windings. Generally, the synchronous motor is intended as the constant speed primemover for the D.C. generator. But in order to bring the motor upto synchronism, the D.C. generator is operated as motor, and the A.C. synchronous dynamo is synchronized to the A.C. supply as an alternator. Once in parallel with the supply, the synchronous dynamo is operated as a motor. The D.C. *'motor' will now act as a generator if its field current is increased so that its generated e.m.f. exceeds the D.C. bus.*

2. **Using the field excited generator as a D.C. motor.** This method is the same as the first, except that the *exciter* (D.C. shunt generator) *is operated as a motor,* and the A.C. synchronous dynamo is synchronized to the A.C. supply.

3. **A small induction motor of at least one pair of poles less than the synchronous motor.** This method involves the same synchronizing procedure for A.C. synchronous motor as an alternator. At least one pair of poles fewer is required on the induction motor to compensate for the loss in induction motor speed due to slip.

In the above three methods the following *conditions* should be met with :

- There should be *little or no-load* on the synchronous motor.
- The *capacity of the starting motor* (D.C. or A.C.) should be between *5 and 10% of the rating of the synchronous motor* coupled to it.

4. **Using the damper windings as a squirrel-cage induction motor.**

Note. It is practically impossible to start a synchronous motor with its D.C. field energized. Even when left de-energized, the rapidly rotating magnetic field of the stator will induce extremely high voltages in the many turns of the field winding. It is customary, therefore, *to short-circuit the D.C. field winding during the starting period* ; whatever voltage and current are induced in it may then aid in producing induction motor action. In very large synchronous motors, field sectionalising or field-splitting switches are used which short-circuit individual field windings to prevent cumulative addition of induced voltages from pole to pole.

2.7. Effect of Load on a Synchronous Motor

When mechanical load on a D.C. motor or an A.C. motor is increased, the speed decreases. This, is turn, decreases the back or counter e.m.f. (E_b) so that the source is able to supply more current to meet the increased load demands. However, this action *cannot take place in the synchronous motor for the rotor must turn at synchronous speed at all loads.*

Fig. 36 (*a*) shows the relative position of a stator and rotor pole at *no-load*, poles centres are directly *in line* with each other.

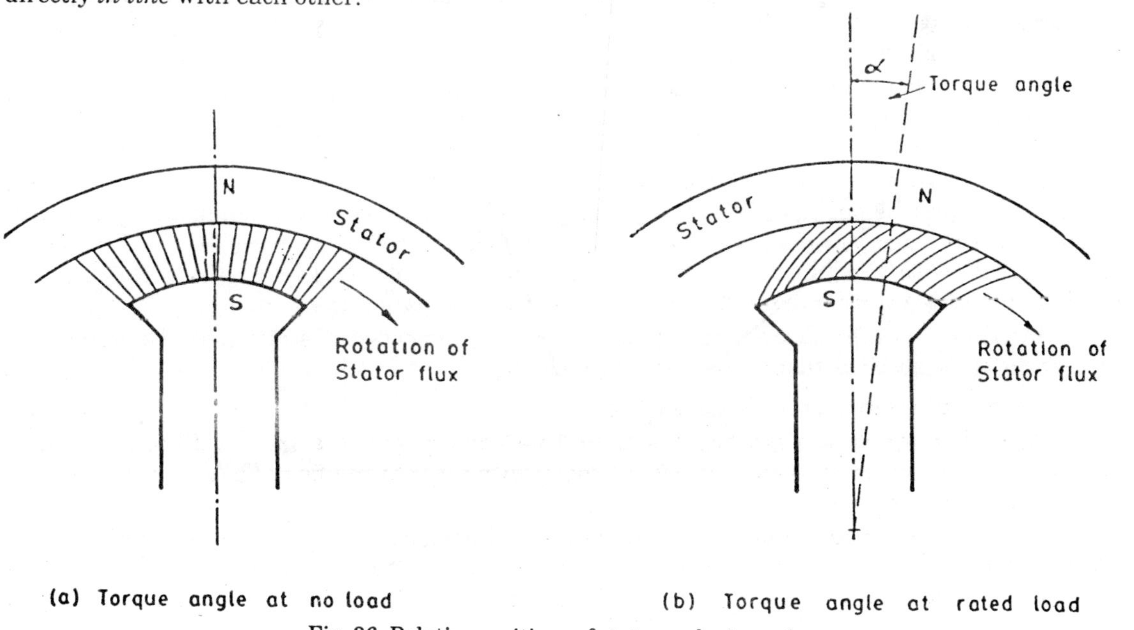

Fig. 36. Relative positions of stator and rotor poles.

Fig. 36 (*b*) represents the relative position of the stator and the rotor poles after mechanical load has been added to the motor. Now there has been a *shift* of the rotor pole in a direction opposite to that of the stator field flux and the direction of the rotor. It may be kept in mind that there has been no change in speed as the rotor will continue to rotate at synchronous speed. There is *only an angular displacement between the centres of the stator and rotor field poles*. The angular displacement shown in Fig. 36 (*b*) is called the **'torque angle'.**

No-load condition-vector diagrams. Fig. 37 shows the conditions when the motor (properly synchronised to the supply) is running on *no-load* and is having no losses. It is seen that $V = E_b$, hence their vector difference is zero and so is the armature current. Motor intake is zero, as there is neither load nor losses to be met by it. In other words, the motor just *floats*.

Fig. 37. No-load (no losses).

Fig. 38 shows the vector diagram when the motor is on no-load but has losses. The vector for E_b falls back by a certain angle α_0, so that a resultant voltage E_r and hence current I_0 is brought into existence which supplies losses.

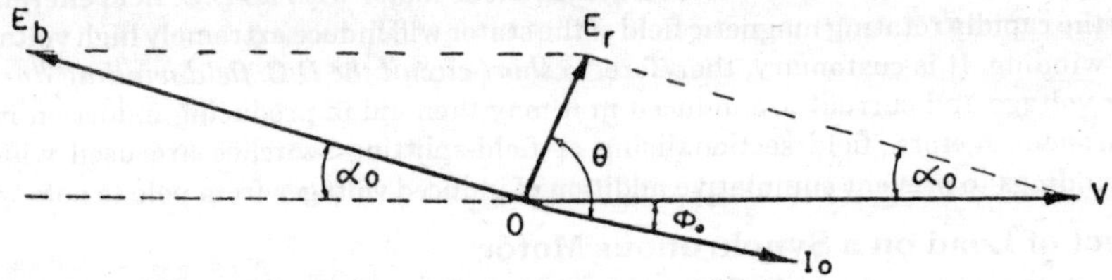

Fig. 38. No-load (with losses).

Load condition-vector diagram. When the motor is loaded, it slows down momentarily to adjust itself to the change in load condition, so the rotor pole falls back a little more relative to the stator pole, as shown in Fig. 39. Hence the torque angle increases with the increase in load. Due to increase in load or torque angle α, the resultant voltage E_r across the armature (or stator) circuit increases, and, therefore, current drawn from the supply mains increases. Thus a motor is able to supply increased mechanical load, not by reduction in speed, but by shift in relative positions of the rotor and rotating magnetic field (or stator flux). From Fig. 39 it is obvious that *for increasing load with a constant value of back e.m.f.* E_b *the phase angle* ϕ *Increases in lagging direction*.

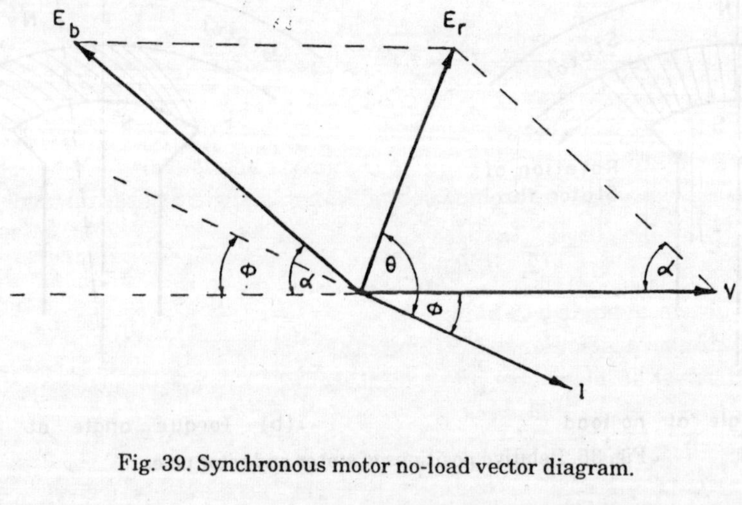

Fig. 39. Synchronous motor no-load vector diagram.

If *the angle between stator and rotor pole centres become too great, due to a serious over-load then the rotor will pull out of synchronism and operate as an induction motor with the aid of the amortisseur winding.* The maximum value of torque which a synchronous motor can develop without dropping out of synchronism is called the **'pull-out torque'.** *In most synchronous motors this is 150 to 200 per cent of rated torque output.*

2.8. Torque Developed by the Motor

Refer Fig. 40.

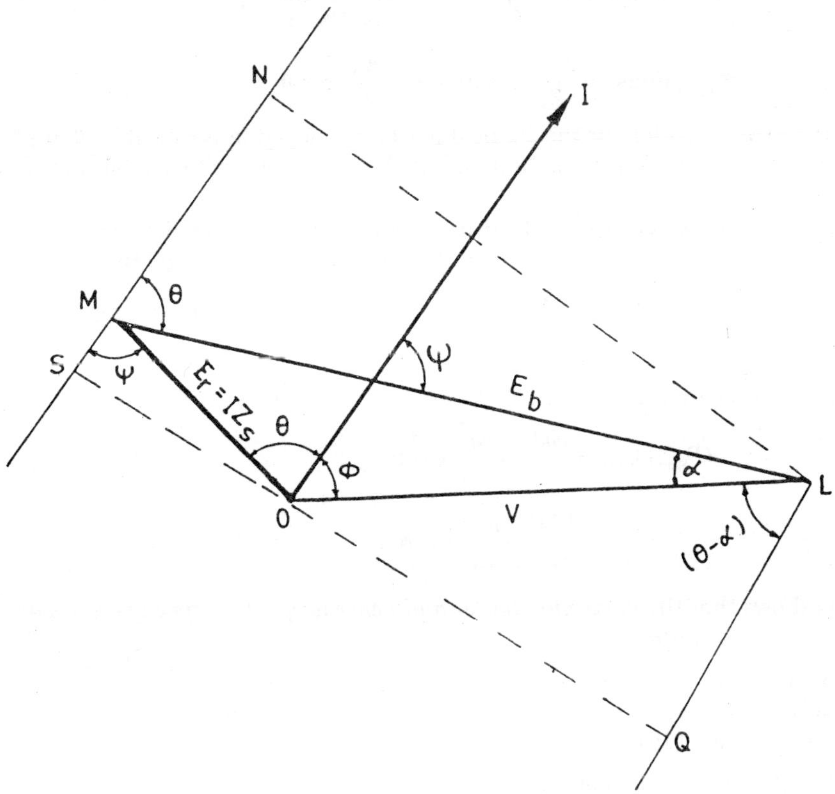

Fig. 40

OL = supply voltage/phase

I = armature current

LM = back e.m.f. at a load angle of α

OM = resultant voltage, $E_r = IZ_s$ (or IX_s if R_a is negligible)

- I lags/leads V by an angle ϕ and lags behind E_r by an angle θ (internal angle)

$$= \tan^{-1}\left(\frac{X_s}{R_a}\right).$$

- Line NS is drawn at angle θ to LM.
- LN and QS are perpendicular to NS (hence to LQ also).

Mechanical power developed *per phase* in the rotor,

$$P_{mech} = E_b I \cos \psi \quad \text{...(13)}$$

In ΔOMS, $MS = IZ_s \cos \psi$

Now, $MS = NS - NM = LQ - NM$

$\therefore \quad IZ_s \cos \psi = V \cos (\theta - \alpha) - E_b \cos \theta$

$$\therefore \quad I \cos \psi = \frac{V}{Z_s} \cos (\theta - \alpha) - \frac{E_b}{Z_s} \cos \theta$$

Putting this value in (13), we get

$$P_{mech}/\text{phase} = E_b \left[\frac{V}{Z_s} \cos (\theta - \alpha) - \frac{E_b}{Z_s} \cos \theta \right]$$

or
$$P_{mech}/\text{phase} = \frac{E_b V}{Z_s} \cos (\theta - \alpha) - \frac{E_b^2}{Z_s} \cos \theta \quad ...(14)$$

This is the expression for the mechanical power developed in terms of load angle (α) and the internal angle (θ) of the motor for a constant voltage V and E_b (or excitation because E_b depends on excitation only).

Maximum power developed. Condition for maximum power developed can be found by differentiating the above expression (eqn. 14) with respect to load angle and then equating it to zero.

$$\therefore \quad \frac{dP_{mech}}{d\alpha} = \frac{E_b V}{Z_s} \sin (\theta - \alpha) = 0 \quad \text{or} \quad \sin (\theta - \alpha) = 0$$

$$\therefore \quad \theta = \alpha$$

$\therefore$ Value of maximum power,

$$(P_{mech})_{max} = \frac{E_b V}{Z_s} - \frac{E_b^2}{Z_s} \cos \alpha$$

or
$$\frac{E_b V}{Z_s} - \frac{E_b^2}{Z_s} \cos \theta \quad ...(15)$$

- This shows that the *maximum power and hence torque* ($\because$ speed is constant) *depends on V and E_b i.e., excitation.*
- Maximum value of θ and hence α is 90°. For *all values of V and E_b, this limiting value of α is the same but maximum torque will be proportional to the maximum power developed* as given in eqn. (15).
- In Fig. 41 eqn. (14) is plotted.
- If R_a is neglected, then

$$Z_s \simeq X_s \text{ and } \theta = 90°$$

$$\therefore \quad \cos \theta = 0$$

$$P_{mech} = \frac{E_b V}{X_s} \cos (90° - \alpha) \quad \text{[from eq. (14)]}$$

i.e.,
$$P_{mech} = \frac{E_b V}{X_s} \sin \alpha \quad ...(16)$$

This gives the value of mechanical power developed in terms of α—the basic variable of a synchronous machine.

$$\therefore \quad (P_{mech})_{max} = \frac{E_b V}{X_s} \quad \text{[from eqn. (15)]}$$

This corresponds to the *'pull-out'* torque.

The above value can be obtained by putting $\alpha = 90°$ in eqn. (16).

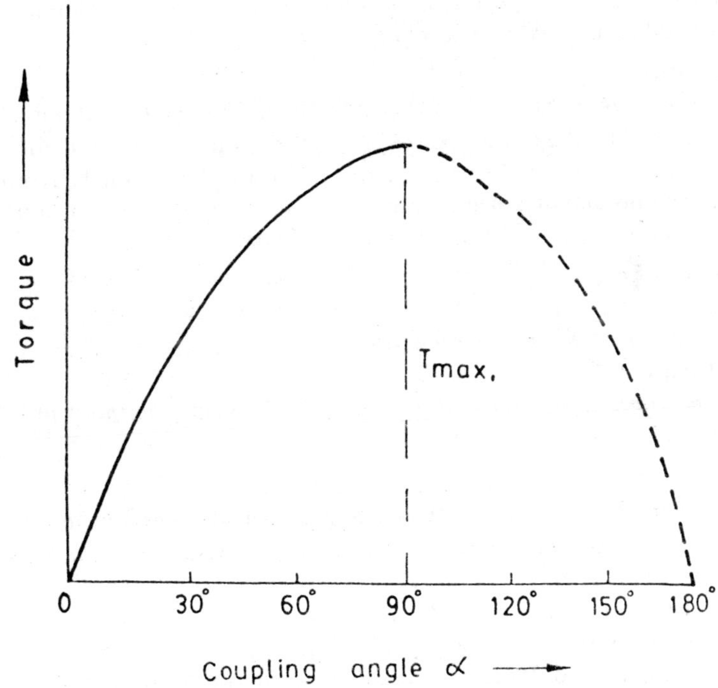

Fig. 41

- To determine the value of excitation or induced e.m.f. E_b to give maximum power developed possible, differentiate eqn. (15) with respect to E_b and equate to zero

$$\therefore \qquad \frac{d(P_{\text{mech}})_{\max}}{dE_b} = \frac{V}{Z_s} - \frac{2E_b}{Z_s}\cos\theta = 0 \quad \text{or} \quad E_b = \frac{V}{2\cos\theta}$$

Substituting $E_b = \dfrac{V}{2\cos\theta}$ in eqn. (15), we get

$$(P_{\text{mech}})_{\max} = \frac{V^2}{2Z_s\cos\theta} - \frac{V^2}{4Z_s\cos\theta}$$

$$= \frac{V^2}{4Z_s\cos\theta} = \frac{V^2}{4R_a} \qquad [\because\ Z\cos\theta = R_a]$$

where R_a = effective resistance of the motor.

Hence
$$(P_{\text{mech}})_{\max} = \frac{V^2}{4R_a} \qquad \text{...(17)}$$

Torques of a synchronous motor :

The various torques associated with a synchronous motor are described below :

1. Starting torque :

- It indicates the ability of the motor to accelerate the load. It is also sometimes called "*break away torque*".
- It may be as low as 10 percent in case of centrifugal pumps, and as high as 200 or 250 percent of full-load torque, as in case of loaded reciprocating two-cylinder compressors.
- Although the synchronous motor possesses no self-starting torque yet in modern

synchronous motors, by making proper changes in the design of damper windings, almost any reasonable torque can be developed.

2. Running torque :

- Running torque is the torque developed by the motor *under running conditions.*
- It is determined by the output power and speed of the driven machine.
- Peak output power determines the maximum torque that would be required by the driven machine. The breakdown or maximum running torque of a motor must be greater than this value in order to avoid stalling of the machine.

3. Pull-in torque :

- It pertains to the *ability of the motor to pull-into synchronism when changing from induction to synchronous motor operation.*

4. Pull-out torque :

- It *is the maximum torque which the motor will develop without pulling out of step or synchronism.*
- Its value varies from 1.25 to 3.5 times the full-load torque.

2.9. Effect of Excitation on Armature Current and Power Factor (V-Curves)

Consider a synchronous motor in which the *mechanical load is constant* and hence output is also constant if losses are neglected.

Case 1. 100% Excitation :

The case for 100% excitation *i.e.,* when $E_b = V$ is shown in Fig. 42 (*i*). Here the armature current I lags behind V by a small angle ϕ. Its angle with E_R is fixed by stator constants *i.e.,*

$$\tan\theta = \frac{X_s}{R_a}.$$

Case 2. Excitation less than 100% :

Fig. 42 (*ii*) represents the condition for *under-excited* motor *i.e.,* $E_b < V$. Here E_r is advanced clockwise and so is the armature current (because it lags behind E_r by a fixed angle θ). We find that :

The magnitude of I is increased, but *its power factor is decreased* ($\because$ ϕ has increased) :

— Since input as well as V are constant, hence the power component of I *i.e.,* $I\cos\phi$ remains the same as before, but wattless component $I\sin\phi$ is increased. Hence, as excitation is decreased, I will increase, but power factor will decrease so that power component *i.e.,* $I\cos\phi = OL$ will remain constant. *The locus of the extremity of current vector would be a horizontal straight line.*

Case 3. Excitation greater than 100% :

In Fig. 42 (*iii*) excitation is greater than 100% *i.e.,* $E_b > V$ (*i.e.,* motor is over-excited). Here the resultant voltage vector E_r is pulled-anticlockwise and so is I. It may be noted that now motor is drawing a *leading current*. It may also happen for some value of excitation, that I may be in phase with V *i.e.,* power factor *is unity* [Fig. 42 (*iv*)]. At that time the *current drawn by motor would be minimum.*

V-curves of a synchronous motor :

- It has been stated above that, when the field current (*i.e.,* excitation) of a synchronous motor is reduced, a lagging armature current [Fig. 42 (*ii*)] I is produced which exceeds the minimum current at unity power or at normal excitation.

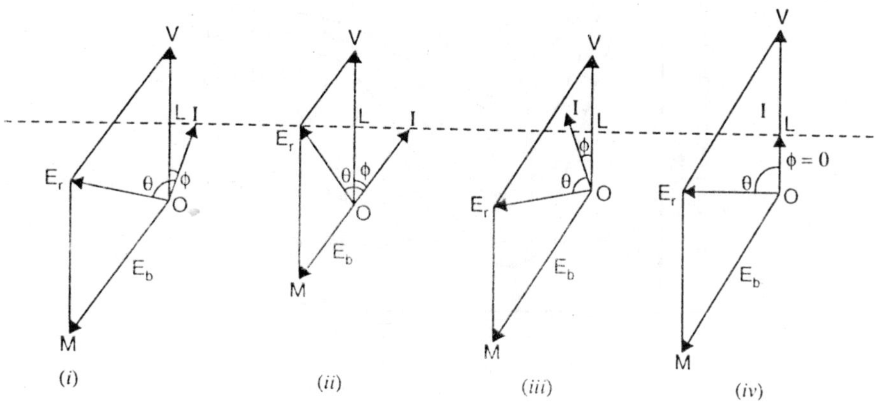

Fig. 42. Effect of excitation on armature current and power factor.

- Similarly, when motor is *over-excited,* the armature current also rises [Fig. 42 (*iii*)] and exceeds the current required at normal excitation to develop the necessary torque, at any given load.
- By applying a given *constant load* to the shaft of a synchronous motor and varying the field current from under-excitation to *over-excitation,* recording the armature current at each step, the curves of Fig. 43 (*a*) are obtained. The A.C. armature current is plotted against the D.C. field current for *no-load, half-load,* and *full-load* values, respectively.
- The wattmeter connections shown in Fig. 44 will also yield the power factor for each value of armature and field current at any given load condition. (The method employed in Fig. 44 is the two-wattmeters method. Since the synchronous motor is a balanced three-phase load, the one-wattmeter method, three-wattmeters method, industrial analyzer, or polyphase wattmeter would do as well in providing the power factor of the load).

Thus, as shown in Fig. 43 (*b*), the power factor as determined from the wattmeter readings, is plotted against the field current for the various given loads. It is worth noting that both curves show that a *slightly increased field current is required to produce normal excitation as the load is increased* (points 1, 2 and 3, respectively).

— At *no-load, the armature current at unity power factor* (normal excitation) *is not zero* but *some small value of A.C. armature current per phase, necessary to produce torque to counter balance rotational losses.* As load is applied (neglecting armature reaction) not only does the armature current rise but it is also necessary to increase the excitation to bring the armature current back in phase with the bus phase voltage. Each of the curves in the family, therefore, will have a shift to the right as the load is increased, as shown in Figs. 43 (*a*) and (*b*), to provide the excitation required to obtain the same power factor (0.8 lagging, unity, or 0.8 leading) at an increased load. Thus, the *V-curves represent the phasor diagrams,* and *vice versa* for various conditions of load and power factor.

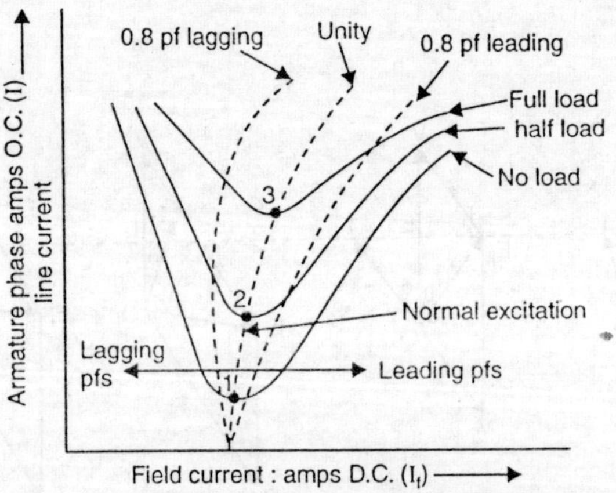

(a) Relation between armature current and field current for various loads

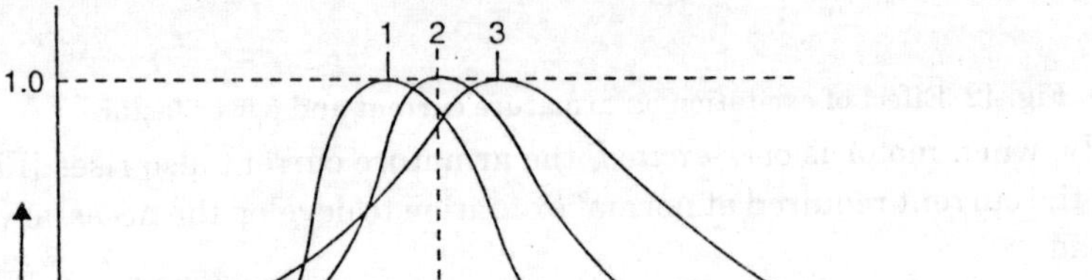

(b) Relation between power factor and field current for various loads

Fig. 43. Families of *V*-curves for a synchronous motor.

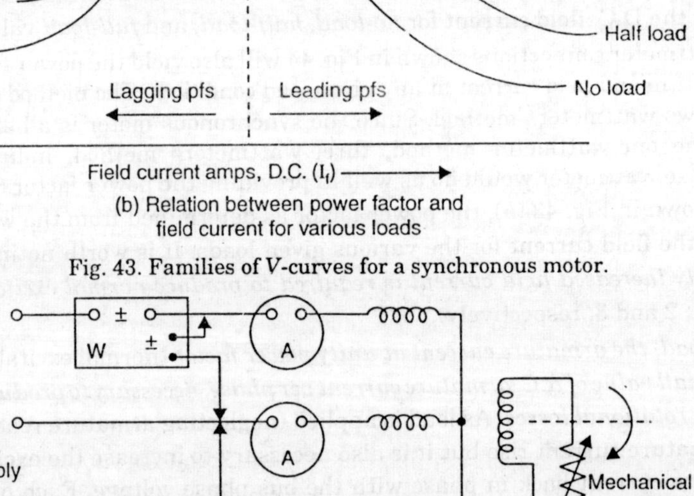

Fig. 44. Laboratory connections for obtaining *V*-curves.

Constant power lines :

Refer Fig. 45.

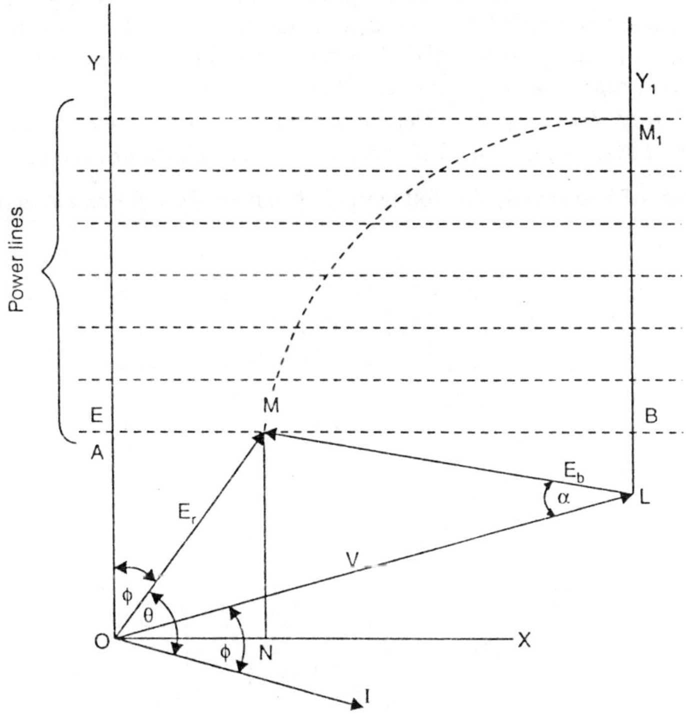

Fig. 45

OL = applied voltage/phase of the motor, V

LM = back e.m.f./phase, E_b

OM = resultant voltage, E_r.

OI = armature current lagging behind E_r by an angle

$$\theta = \tan^{-1} \frac{X_s}{R_a}, \text{ value of } I = \frac{E_r}{Z_s}$$

- OY is drawn at an angle ϕ with OM (or at an angle θ with OL).
- MN is drawn perpendicular to OX which is at right angles to OY. Vector OM when referred to OY also represents, on a different scale, the current both in magnitude and phase.
- The power input/phase of the motor

$$= VI \cos \phi = V \times MN$$

As V is constant, power input is dependent on MN. If the motor is working with a constant intake then locus of M is a straight line || to OX and ⊥ OY *i.e.*, line AY for which MN is constant. Hence AB represents a *constant power input line for a given voltage but varying excitation.* Similarly, a series of such parallel lines can be drawn each representing a definite power intake of the motor. Regarding these power lines the following points are worth noting :

(*i*) For equal increase in intake, the power lines are parallel and equally spaced.

(*ii*) The line OX represents zero power line.

(*iii*) The perpendicular distance from working point M to OX represents the motor intake.

(*iv*) If the excitation is fixed *i.e.*, LM is constant in length, then, as the load on the motor is increased, α increases. In other words, locus of M is a *circle* with radius = LM and centre at L. With increasing load, M goes on to the lines of higher power till point M_1 is reached. Any further increase in load on the motor will bring point M down to a lower power line. It means that as load increases beyond the value corresponding to point M_1, the motor intake decreases which is impossible. The area to the right of LY_1 *represents unstable condition.*

For a given voltage and excitation the maximum power the motor can develop, is determined by the location of point M_1, beyond which the motor pulls out of synchronism.

Construction of V-curves. The following is the *procedure for construction of V-curves* : Refer Fig. 46.

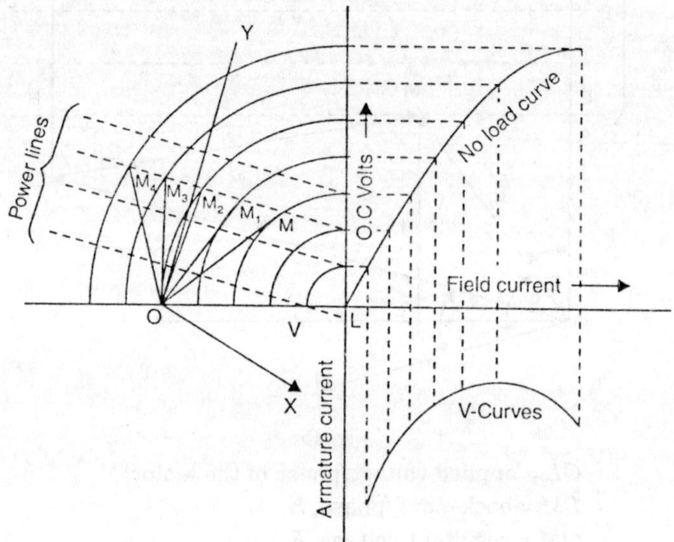

Fig. 46

- Draw constant-power lines as discussed above.
- With L as centre, draw concentric circles of different radii LM, LM_1, LM_2 etc. where LM, LM_1, LM_2 etc. are the back e.m.fs. corresponding to different excitations. The intersection of these circles with lines of constant power give *positions of the working points for specific loads and excitations.*

The vectors OM, OM_1, OM_2 etc. represent different values of E_r for different excitations. (Back e.m.fs. vectors LM, LM_1 etc. have not been drawn to avoid confusion).

- Project the different values of back e.m.f. on the magnification curve and read the corresponding values of the field amperes from it.
- Plot the field amperes against the corresponding armature currents to get V-curves.

2.10. Synchronous Condensers

- Several synchronous motors are manufactured without any shaft extensions whatever, ostensibly designed *solely for power-factor correction* and to be operated without any mechanical load. *Any over-excited synchronous motor operated without any mechanical load,* however, may be classed as a *synchronous condenser or synchronous capacitor.*

- Although there is no mechanical load to contribute to armature current, the *V*-curves of Fig. 46 will show that, even without load, the armature current is high. This is not a disadvantage, however, because the synchronous capacitor is over-excited to that point where its armature current is equal to or exceeds its rated current.

Alternating-current synchronous dynamos can be constructed *less expensively* in extremely high ratings and voltages than commercial fixed capacitors with the result that synchronous capacitors driving no mechanical load, whatever, are permitted to *'float'* on the line for the purposes of power factor improvement.

- *As compared with a synchronous motor with equal armature voltage and current ratings, a 'synchronous condenser' requires more copper in the field winding to carry the increased field current.* The synchronous condenser does not require so large shaft and bearings as the synchronous motor because no shaft torque is required.
- Synchronous condensers are sometimes operated at power factors ranging from lagging through unity to leading for voltage control. When applied in this manner, a synchronous condenser is called a *synchronous reactor.*
- It is customary not to attempt correction of the power factor of the system all the way to a unity power factor. There is an *economic reason* for this, despite the fact that large, high capacity synchronous capacitors are available and for the same kVA rating are usually *less expensive* than synchronous motors. The capacitors are less expensive because (*i*) they are started and run without load, requiring no special heavy cage windings, and (*ii*) they require smaller shaft diameters and lighter bearings, although their field windings are some-what heavier.

HIGHLIGHTS

1. *An alternator consists of a* **stator** *and* **rotor.** *The stator provides the armature windings whereas rotor provides the rotating magneting field.*
2. *There are two types of rotors :*

   (*i*) **Salient pole type.** *Used for low and medium-speed (engine-driven) alternators.*

   (*ii*) **Smooth cylindrical type.** *Used for turbo-alternators i.e. for turbine-driven alternators.*
3. *The frequency of the alternating current produced is,* $f = \frac{N_s P}{120}$ *Hz.*
4. *The e.m.f. induced (for sinusoidal wave) per phase will be,* $E_{r.m.s.}/phase = 4.44 f \phi T_{ph} k_p k_d$ *volts where* k_p = *pitch factor,* k_d = *distribution or breadth factor*

$$= \frac{\sin\left(\frac{q\beta}{2}\right)}{q \sin \frac{\beta}{2}}$$

Value of k_d *for nth harmonic is,* $k_d = \frac{\sin\left(\frac{qn\beta}{2}\right)}{q \sin\left(\frac{n\beta}{2}\right)}$.

5. *A synchronous motor is electrically identical with an alternator or A.C. generator.*
6. *Some characteristic features of a synchronous motor are :*

   — *It is not inherently self-starting.*

— *It runs at synchronous speed,* $N_s = \frac{120f}{p}$.

— *It can be operated under a wide range of power factors both lagging and leading.*

7. *Mechanical power developed/phase,*

$$(P_{mech})/phase = \frac{E_b V}{Z_s} \cos(\theta - \alpha) - \frac{E_b^2}{Z_s} \cos\theta$$

where α = *load angle,* θ = *internal angle.*

OBJECTIVE TYPE QUESTIONS

(A) Choose the Correct Answer :

1. The speed of a 4-pole 60 Hz synchronous machine will be
(*a*) 1800 r.p.m. (*b*) 2400 r.p.m.
(*c*) 3000 r.p.m. (*d*) 3600 r.p.m.
2. The speed of a *p*-pole synchronous machine in r.p.m. is given by
(*a*) $\frac{120f}{p}$ (*b*) $\frac{120p}{f}$
(*c*) $\sqrt{120fp}$ (*d*) $120fp$.
3. What is the largest size of alternator being manufactured in India ?
(*a*) 500 MW (*b*) 250 MW
(*c*) 210 MW (*d*) 110 MW.
4. Which of the following organisations is engaged in the manufacture of large size alternators for power plants in India ?
(*a*) Department of Science and Technology (*b*) Electricity Authority of India
(*c*) National Thermal Power Corporation Ltd. (*d*) Bharat Heavy Electricals Ltd.
5. An exciter is a nothing but a
(*a*) D.C. series motor (*b*) D.C. shunt motor
(*c*) D.C. shunt generator (*d*) D.C. series generator.
6. Hydrogen is used in large alternators mainly to
(*a*) reduce distortion of waveform (*b*) cool the machine
(*c*) strengthen the magnetic field (*d*) reduce eddy current losses.
7. An alternator coupled to which primemover will usually have the highest rotating speed ?
(*a*) Steam engine (*b*) Reciprocating diesel engine
(*c*) Francis turbine (*d*) Steam turbine.
8. In an alternator the voltage generated per phase is proportional to
(*a*) number of turns in coil (*b*) flux per pole
(*c*) frequency of waveform (*d*) all of the above.
9. Salient pole type alternators are generally used on
(*a*) low voltage alternators (*b*) hydrogen cooled primemovers
(*c*) high speed primemovers (*d*) low and medium speed primemovers.
10. Turbo-alternators are generally used to run at
(*a*) 1500 r.p.m. (*b*) 3000 r.p.m.
(*c*) 5000 r.p.m. (*d*) 15000 r.p.m.
11. The rotor preferred for alternators applied to hydraulic turbines are
(*a*) salient pole type (*b*) cylindrical rotor type
(*c*) solid rotor type (*d*) any of the above.

12. The frequency of voltage generated in large alternators is
(*a*) 50 Hz (*b*) 60 Hz (*c*) in kilo cycles (*d*) in mega cycles.
13. Which of the following primemovers is least efficient ?
(*a*) Gas turbine (*b*) Petrol engine (*c*) Diesel engine (*d*) Steam engine.
14. In which coil the harmonic component of the generated e.m.f. will be more ?
(*a*) Full pitch coil (*b*) Short pitch coil (*c*) Long pitch coil (*d*) Same in all coils.
15. In case of turbo-alternators the rotor is usually made of
(*a*) cast iron (*b*) forged steel
(*c*) laminated stainless steel (*d*) manganese steel.
16. In case a three-phase alternator supplies resistive load
(*a*) the armature reaction flux will be almost zero
(*b*) the armature reaction flux will be along the axis of the field
(*c*) the armature reaction flux will be at 90° with the field axis
(*d*) the armature reaction flux will be at 180° with the field axis.
17. A 12-pole alternator will pass through how many electrical degrees in one complete revolution
(*a*) 60° (*b*) 360° (*c*) 1080° (*d*) 2160°.
18. The number of poles in turbo-alternators is usually
(*a*) 2 (*b*) 4 (*c*) 12 (*d*) 50.
19. Which harmonic will be totally eliminated in an alternator by using a fractional pitch of 4/5 ?
(*a*) Third (*b*) Fifth (*c*) Seventh (*d*) Ninth.
20. Dirt accumulation in generators can cause all of the following *except*
(*a*) low power factor (*b*) poor voltage regulation
(*c*) flashovers (*d*) overheating.
21. Three-phase alternators are usually star-connected
(*a*) to save on copper (*b*) to reduce conductor size
(*c*) to obtain higher terminal voltage (*d*) to reduced windage losses.
22. Synchronous motors are generally not self-starting because
(*a*) the direction of rotation is not fixed.
(*b*) the direction of instantaneous torque reverses after half cycle
(*c*) starters cannot be used on these machines
(*d*) starting winding is not provided on these machines.
23. Zero power factor test is generally conducted on synchronous machines of
(*a*) large size (*b*) less than 10 kW
(*c*) miniature size (*d*) multiple speed type.
24. Zero power factor test can be used to ascertain
(*a*) pull in torque (*b*) the temperature rise
(*c*) the synchronous reactance (*d*) all of the above.
25. In case one phase of a three-phase synchronous motor is short-circuited, the motor will
(*a*) not start (*b*) run at $\frac{2}{3}$ of synchronous speed
(*c*) run with excessive vibrations (*d*) take less than the rated load.
26. A pony motor is basically a
(*a*) small induction motor (*b*) D.C. series motor
(*c*) D.C. shunt motor (*d*) double winding A.C./D.C. motor.
27. A synchronous motor can develop synchronous torque
(*a*) when under loaded (*b*) while over-excited
(*c*) only at synchronous speed (*d*) below or above synchronous speed.
28. A synchronous motor can be started by
(*a*) pony motor (*b*) D.C. compound motor
(*c*) providing damper winding (*d*) any of the above.

29. As compared to an induction motor of the same size the air-gap in a synchronous motor is
(*a*) three to five times (*b*) same
(*c*) half to three half (*d*) less than half.
30. A three-phase synchronous motor will have
(*a*) no slip-rings (*b*) one slip-ring
(*c*) two slip-rings (*d*) three slip-rings.
31. Under which of the following conditions hunting of synchronous motor is likely to occur ?
(*a*) Periodic variation of load (*b*) Over-excitation
(*c*) Over-loading for long periods (*d*) Small and constant load.
32. When the excitation of an unloaded salient pole synchronous motor suddenly gets disconnected
(*a*) the motor stops (*b*) it runs as a reluctance motor at the same speed
(*c*) it runs as a reluctance motor at a lower speed
(*d*) none of the above.
33. When *V* is the applied voltage, then the breakdown torque of a synchronous motor varies as
(*a*) V (*b*) $V^{3/2}$
(*c*) V^2 (*d*) $1/V$.
34. The power developed by a synchronous motor will be maximum when the load angle is
(*a*) zero (*b*) 45°
(*c*) 90° (*d*) 120°.
35. A synchronous motor can be used as a synchronous capacitor when it is
(*a*) under-loaded (*b*) over-loaded
(*c*) under-excited (*d*) over-excited.
36. A synchronous motor is running on a load with normal excitation. Now if the load on the motor is increased
(*a*) power factor as well as armature current will decrease
(*b*) power factor as well as armature current will increase
(*c*) power factor will increase but armature current will decrease
(*d*) power factor will decrease and armature current will increase.

ANSWERS

1. (*a*)	2. (*a*)	3. (*a*)	4. (*d*)	5. (*c*)	6. (*b*)	7. (*d*)
8. (*d*)	9. (*d*)	10. (*b*)	11. (*a*)	12. (*a*)	13. (*d*)	14. (*a*)
15. (*b*)	16. (*c*)	17. (*d*)	18. (*a*)	19. (*b*)	20. (*a*)	21. (*a*)
22. (*b*)	23. (*a*)	24. (*b*)	25. (*a*)	26. (*a*)	27. (*c*)	28. (*d*)
29. (*a*)	30. (*b*)	31. (*a*)	32. (*a*)	33. (*a*)	34. (*c*)	35. (*d*)
36. (*d*).						

THEORETICAL QUESTIONS

1. What is an alternator ? What is its operating principle ?
2. How are alternators classified ?
3. Give reasons :
(*i*) The speed of a synchronous generator should be kept constant at synchronous speed.
(*ii*) Generally cylindrical rotor construction is used for high speed machines and salient pole construction for low speed machines.
(*iii*) Cylindrical rotor alternators have small diameters and large lengths of core.
(*iv*) Salient pole alternators have large diameters and small core lengths.
(*v*) Both axial and radial ventilating ducts are necessary in cylindrical rotor machines where only radial ducts are usually sufficient in salient pole machines.
4. Describe the main constructional features of cylindrical rotor and salient pole alternators.

5. What is distribution factor ? What is its effect ? Derive an expression for distribution factor of a winding having q slots per pole per phase and a slot angle of β.
6. What is the meaning of 'electrical angle' ? How is it different from mechanical angle ? Can they be equal ?
7. What is a fractional pitch winding ?
8. Derive an expression for the voltage induced in an alternator. Discuss the role of different factors which appear in the expression.
9. What is the working principle of a synchronous motor ?
10. Is the synchronous motor self-starting ? If not, what are the methods adopted for starting it ?
11. What is the difference between a synchronous motor and an induction motor ?
12. State the characteristic features of a synchronous motor.
13. What are the advantages and disadvantages of a synchronous motor ?
14. What are the applications of a synchronous motor ?
15. Does the speed of the synchronous motor change when loaded ? If not explain how the motor takes load ?
16. What are V-curves ? How are they determined experimentally ?

UNSOLVED EXAMPLES

Synchronous generator

1. For a 3-phase winding with 4 slots/pole/phase and with the coil span of 10 slots pitch, calculate the values of the pitch factor and distribution factor. **[Ans.** 0.966, 0.9576]
2. A 2200 volt, 3-phase alternator is running at 300 r.p.m. and has 24 poles. Find the number of conductors in the stator winding, if the magnetic flux is 5×10^{-2} Wb/pole.
   Assume distribution factor as 0.96. **[Ans.** 600]
3. A 4-pole alternator has an armature with 25 slots and 8 conductors/slot and rotates at 1500 r.p.m. and the flux/pole is 0.05 Wb. Calculate the e.m.f. generated, if winding factor is 0.96 and all conductors are in series. **[Ans.** 1065.6 V]
4. The stator of a 3-phase, 20-pole alternator has 120 slots, and there are 4 conductors per slot accommodated in two layers. If the speed of the alternator is 300 r.p.m. calculate the e.m..f. induced per phase. Resultant flux in the air-gap is 0.055 Wb/ pole.
   Assume the coil span as 150° electrical. **[Ans.** 905 V]
5. A 16-pole, 3-phase alternator has a star-connected winding with 144 slots and 10 conductors per slot. The flux per pole is 0.03 Wb distributed sinusoidally and the speed is 375 r.p.m. Find the line voltage. **[Ans.** 2658 V]
6. Calculate the e.m.f. of a 4-pole, 3-phase star-connected alternator running at 1500 r.p.m., from the following data :

Flux per pole	= 0.1 Wb
Number of slots	= 48
Conductors/slot (in two layers)	= 4
Coil span	= 150°.

   [Ans. 1138 V]
7. A 3-phase, 16-pole, star-connected alternator has 192 stator slots with eight conductors/slot and the conductors of each phase are connected in series. The coil span is 150 electrical degrees. Determine the phase and line voltage if the machine runs at 375 r.p.m. and the flux per pole is 64 m Wb distributed sinusoidally over pole. **[Ans.** 3367 V, 5830 V]
8. A 3-phase, 4-pole star-connected alternator has 60 slots with 2 conductors/slot. The pitch of the coil is 3 slot less than the pole pitch. The flux/pole is 12.5×10^{-2} Wb sinusoidally distributed. Calculate the no-load terminal voltage for a frequency of 50-Hz.
9. A 3-phase, star-connected alternator on open-circuit is required to generate a line voltage of 3600 V, 50-Hz, when driven at 500 r.p.m. The stator has 3 slots per pole, and 10 conductors per slot. Calculate :
   (*i*) number of poles (*ii*) useful flux/pole.
   Assume all conductors per phase to be connected in series and coils to be full pitch.
   [Ans. (*i*) 12, (*ii*) 0.054 Wb]

Synchronous motor

10. A 500 V, single-phase synchronous motor gives a net output mechanical power of 7.46 kW and operates at 0.9 p.f. lagging. Its effective resistance is 0.8 Ω. If the iron and mechanical losses are 500 W and excitation losses are 800 W, estimate the armature current. Calculate the commercial efficiency. [**Ans.** 18.3 A, 82.56%]
11. A 2300 V, 3-phase, star-connected synchronous motor has a resistance of 0.2 Ω/phase and a synchronous reactance of 2.2 Ω/phase. The motor is operating at 0.5 p.f. leading with a line current of 200 A. Determine the value of the generated e.m.f./phase. [**Ans.** 1708 V/phase]
12. A 6600 V, 3-phase, star-connected synchronous motor draws a full-load current of 70 A at 0.8 p.f. leading. Find the e.m.f. induced, output power and efficiency of the machine if the stray losses total to 30 kW ; and the armature resistance is 2 Ω and reactance 20 Ω/phase. [**Ans.** 4696 V/phase, 580.6 kW, 90.72%]
13. The input to a 11000 V, 3-phase, star-connected synchronous motor is 50 A. The effective resistance and synchronous reactance per phase are 0.95 Ω and 29 Ω respectively. Calculate the power supplied to the motor and the induced e.m.f. for a power factor of :
(*i*) 0.8 lagging. (*ii*) 0.8 leading.
[**Ans.** (*i*) 9635 V, 762.1 kW, (*ii*) 12610 V 762.1 kW]
14. A 200 V, 3-phase star-connected synchronous motor has an effective resistance and synchronous reactance of 0.2 Ω and 2.2 Ω respectively. The input is 800 kW at normal voltage and the induced line e.m.f. is 2500 V. Calculate the line current and power factor. [**Ans.** 254.15 A, 0.909 (lead)]
15. A 3-phase star-connected synchronous motor has a synchronous reactance of 4 Ω per phase and is working on 1100 volts. Calculate the power factor of the machine when taking 90 kW from the mains, the excitation being adjusted to a value corresponding to an induced e.m.f. of 1200 volts.
Neglect armature resistance. [**Ans.** 0.986 A]
16. A synchronous motor takes 25 kW from 400 V supply mains. The synchronous reactance of the motor is 4 Ω. Find the power factor at which the motor would operate when the exciting current is so adjusted that the generated e.m.f. is 500 volts. [**Ans.** 0.9914 leading]
17. A 2500 V, 3-phase, 4-pole star-connected synchronous motor runs at 1500 r.p.m. The excitation is constant and corresponding to an open-circuit voltage of 2500 V. The resistance is negligible in comparison with synchronous reactance of 3 Ω/phase. Determine the power input, power factor and torque developed for an armature current of 250 A. [**Ans.** 1045.3 kW, 0.9656 (lag), 6655 N-m]
18. A 7.355 kW, 400 V, 3-phase star-connected synchronous motor has synchronous impedance per phase of $(0.33 + j2.8)$ Ω. Find the angle of retard and the voltage to which motor must be excited to give a full-load output at 0.866 leading power factor.
Assume an efficiency of 88%. [**Ans.** 431 V (line), 8.36°]
19. A 400 V, 6-pole, 3-phase, 50 Hz, star-connected synchronous motor has a resistance and synchronous impedance of 0.5 Ω and 4 Ω per phase respectively. It takes a current of 15 A at unity p.f. when operating with a certain field current. If the load torque is increased until the line current is increased to 60 A, the field current remaining unchanged, calculate the gross torque developed and the new power factor.
[**Ans.** 327 N-m ; 0.954 (lead)]
20. A 6600 V, 3-phase, star-connected synchronous motor draws a full-load current of 70 A at 0.8 p.f. leading. Find the e.m.f. induced and efficiency of the machine if the stray losses total 30 kW and the armature resistance is 2 Ω and the reactance is 20 Ω per phase. [**Ans.** 8130 V (line), 90.7%]
21. An 8-pole, 3300 V, 50-Hz, 3-phase, start-connected synchronous motor has armature resistance of 0.5 Ω per phase and synchronous reactance of 5 Ω/phase. While running on no-load, the excitation has been adjusted so as to make the e.m.f. numerically equal to and antiphase with the terminal voltage. With a certain load torque applied if the rotor gets retarded by 3 mechanical degrees calculate the armature current and the power factor of the motor. [**Ans.** 79.6 A, 0.9992 (lag)]
22. A 2300 V, 3-phase, 50 Hz, star-connected synchronous motor has a synchronous reactance of 11 Ω per phase. When it delivers 147.1 kW the efficiency is found to be 90% and the power angle is 15 electrical degrees. Neglecting ohmic resistance determine :
(*i*) Inducted excitation voltage/phase, and (*ii*) Line current.
[**Ans.** (*i*) 3018 V (line value), 52.18 A]

8

Polyphase Induction Motors

1. General aspects. 2. Classification of A.C. motors. 3. Constructional details. 4. Production of rotating magnetic field. 5. Theory of operation of an induction motor. 6. Slip. 7. Frequency of rotor current. 8. Rotor e.m.f. and rotor current. 9. Torque and power. 10. Effect of change in supply voltage on starting torque. 11. Effect of change in supply voltage on torque and slip. 12. Full-load torque and maximum torque. 13. Starting torque and maximum torque. 14. Torque-slip and torque-speed curves. 15. Operating characteristics of a 3-phase squirrel-cage induction motor. 16. Operating characteristics of a wound-rotor (slip ring) induction motor. 17. Power stages in an induction motor. 18. Induction motor as transformer. 19. Equivalent circuit of an induction motor. 20. Starting of induction motors. 21. Factors governing performance of induction motors. 22. Effects of operating conditions. 23. Ratings of 3-phase induction motors. 24. Squirrel-cage motors—Advantages, Disadvantages and applications. 25. Wound rotor (or Slip ring) induction motors—Advantages, disadvantages and applications. 26. Comparison of a squirrel-cage and a slip ring (or phase wound) induction motors. 27. Comparison between induction and synchronous motors—Highlights—Objective Type Questions—Theoretical Questions—Unsolved Examples.

1. GENERAL ASPECTS

Introduction. An induction motor is simply an *electric transformer* whose magnetic circuit is separated by an air gap into two relatively movable portions, one carrying the primary and the other the secondary winding. Alternating current supplied to the primary winding from an electric power system induces an opposing current in the secondary winding, when *latter is short-circuited* or *closed through an external impedance*. Relative motion between the primary and secondary structures is produced by the *electromagnetic forces corresponding to the power thus transferred across the air gap by induction*.

The *essential feature which distinguishes the induction machine from other types of electric motors is that the secondary currents are created solely by induction, as in a transformer instead of being supplied by a D.C. exciter or other external power source, as in synchronous and D.C. machines*.

Advantages. Three-phase induction motor is the *most commonly used motor* in industrial applications because of the *advantages* listed below :

1. Simple design
2. Rugged construction
3. Reliable operation
4. Low initial cost
5. Easy operation and simple maintenance
6. High efficiency
7. Simple control gear for starting and speed control.

Applications :

Induction motors are available with torque characteristics suitable for a *wide variety of loads* :

(*i*) The standard motor has a starting torque of about 120 to 150 per cent of full-load torque. Such motors are suitable for most applications.

(*ii*) For starting loads such as small refrigerating machines or plunger pumps operating against full pressure or belt conveyors, *high torque motors with a starting torque of twice normal full-load torque*, or more, are used.

(*iii*) For driving machines that use large flywheels to carry peak loads, such as punch presses and shears, a high-torque motor with a slip at full-load up to 10 per cent is available. *The high slip permits enough change in speed to make possible the proper functioning of the flywheel.*

(*iv*) By the use of a wound-rotor with suitable controller and external resistances connected in series with the rotor winding, it is possible to obtain any value of starting torque up to the maximum breakdown torque. *Such motors are well adapted as constant-speed drives for loads that have large friction loads to overcome at starting.*

2. CLASSIFICATION OF A.C. MOTORS

Different A.C. motors may be classified as follows :

1. **According to the 'type of current' :**
   - (*i*) Single-phase
   - (*ii*) Three-phase.
2. **According to 'speed' :**
   - (*i*) Constant speed
   - (*ii*) Variable speed
   - (*iii*) Adjustable speed.
3. **According to 'principle of operation' :**
   - (A) **Synchronous motors**
     - (*i*) Plain
     - (*ii*) Super.
   - (B) **Asynchronous motors**
     - (*a*) Induction motors :
       - (*i*) Squirrel cage
         - —single
         - —double
       - (*ii*) Slip-ring (external resistance)
     - (*b*) Commutator motors :
       - (*i*) Series
         - —single-phase
         - —universal
       - (*ii*) Compensated
         - —conductively
         - —inductively.
4. **According to 'structural features' :**
   - (*i*) Open
   - (*ii*) Enclosed
   - (*iii*) Semi-enclosed
   - (*iv*) Ventilated
   - (*v*) Pipe-ventilated
   - (*vi*) Riveted frame eye
   - (*vii*) Splash proof
   - (*viii*) Totally enclosed fan-cooled
   - (*ix*) Explosion proof
   - (*x*) Water proof.

3. CONSTRUCTIONAL DETAILS

3.1. The Stator

- The stator frame consists of a symmetrical and substantial casting, having feet cast integral with it. The stator core, consisting of high grade, low loss electrical sheet-steel stampings, is assembled in the frame under hydraulic pressure. The thickness of stampings/laminations is usually from 0.35 to 0.6 mm. The stator laminations are punched in one piece for small induction motor (Fig. 1). In induction machines of large size the stator core is assembled from a large number of *segmental laminations.*

The slots are *sometimes of the 'open type'* (*i.e.,* having parallel walls) for the accommodation of former wound coils. But the usual practice is to have practically *'enclosed slots' in order to reduce the effective* length of air-gap.

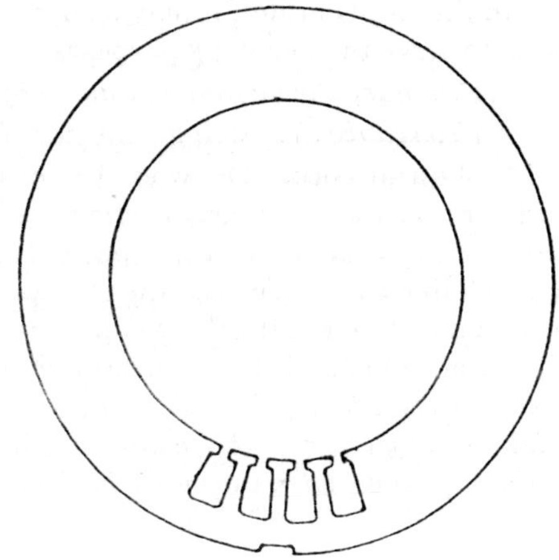

Fig. 1. Stator stamping.

- The *stator windings* are given the utmost care to make them mechanically and electrically sound, so as to ensure long life and high efficiency. After the winding is in position it is thoroughly dried out whilst still hot and is completely immersed in a high grade synthetic resin varnish. It is then *acid, alkali, moisture* and *oil proof.*

  For *small motors* working at ordinary voltages, *single layer mush winding* is used. For *medium size machines double layer lap winding with diamond shaped coils* is used. *Single layer concentric windings are used for large motors working at high voltages.*
- *Frames* of electrical machines house the stator core. Frames of small and medium sizes of induction motors have hollow cylindrical form and that of larger motors have the shape of a circular box. *In small induction motors,* having a frame diameter of up to about 150 cm, *the frame also supports the end shields.* The frame should be *strong and rigid* as rigidity is very important in the case of induction motors of large dimensions. *This is because the length of the air gap is very small and if the frame is not rigid, it would create an irregular air gap around the machine resulting in production of unbalanced magnetic pull.* Frames for small machines are made as a *single unit* and are usually cast. The frames of *medium and large sized* machines are *fabricated from rolled steel plates.*

3.2. The Rotor. The rotors are of two types :

1. Squirrel-cage ;
2. Wound rotor.

1. **Squirrel-cage.** The squirrel-cage rotor is made up of stampings (Fig. 2) which are keyed directly to the shaft. The slots are partially closed and the winding consists of embedded copper bars to which the short-circuited rings are brazed. The squirrel cage rotor is so robust that it is almost indestructible.

The great majority of present day induction motors are manufactured with squirrel-cage rotors, a *common practice being to employ winding of cast aluminium.* In this construction the assembled rotor laminations are placed in a mould after which molten aluminium is forced in, under pressure, to form bars, end rings and cooling fans as extension of end rings. This is known as *die cast rotor* and has become very popular as there are no joints and thus there is no possibility of high contact resistance.

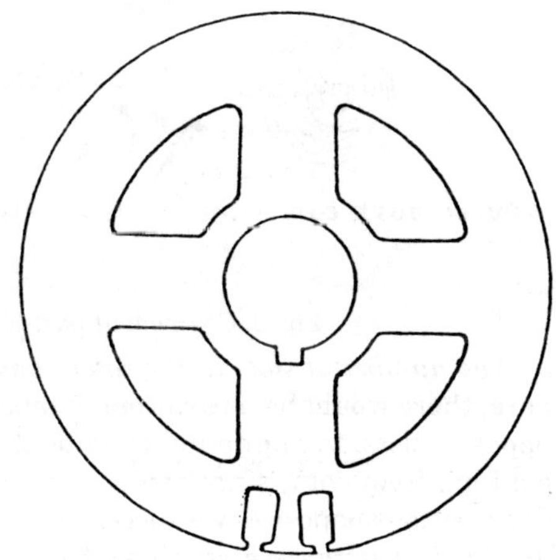

Fig. 2. Rotor stamping.

In this type of rotor, it may be noted that *slots are not made parallel to the shaft but they are 'skewed'* to serve the following purposes :

(*i*) *To make the motor run quietly by reducing the magnetic hum.*

(*ii*) *To reduce the locking tendency of the rotor.*

2. **Wound rotor.** The wound rotor has also slotted stampings and the *windings are former wound.* The *wound rotor construction is employed for induction motors requiring speed control or extremely high values of starting torque.* The wound rotor has completely insulated copper windings very much like the stator windings. The windings can be connected in star or delta and the three ends are brought out at the three slip rings. The current is collected from these slip rings with carbon brushes from which it is led to the *resistances for starting purposes.* When the motor is *running, the slip rings are short-circuited by means of a collar, which is pushed along the shaft and connects all the slip rings together on the inside.* Usually the *brushes* are *provided with a device for lifting them from the slip rings when the motor has started up, thus reducing the wear and the frictional losses.*

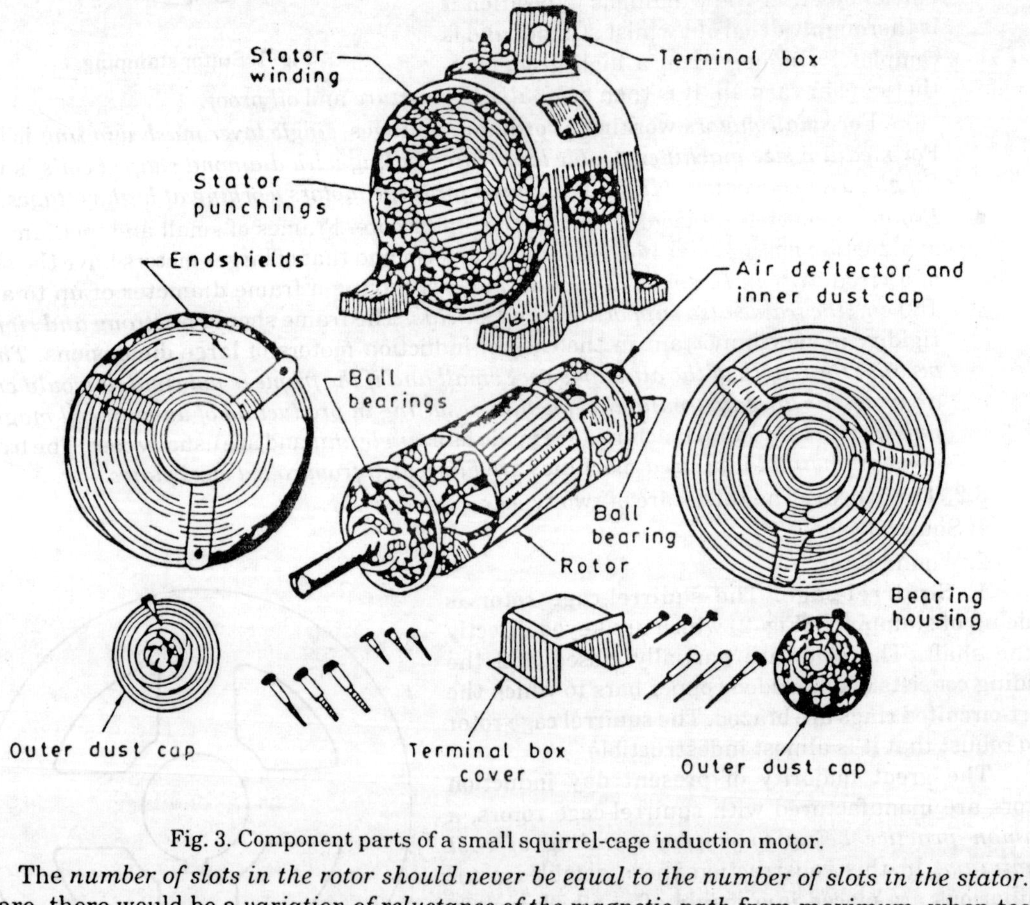

Fig. 3. Component parts of a small squirrel-cage induction motor.

The *number of slots in the rotor should never be equal to the number of slots in the stator.* If they are, there would be a *variation of reluctance of the magnetic path* from maximum, when teeth are opposite slots, to minimum when teeth are opposite teeth. The resulting flux pulsations would have a high frequency, since the periodic time would be the interval period for a tooth to occupy similar positions opposite two successive teeth. This will *not only cause extra iron loss but the rotor will tend to lock with the stator* if at the time of starting teeth are opposite teeth. The best plan is to make the number of the stator and the rotor teeth *prime to each other.*

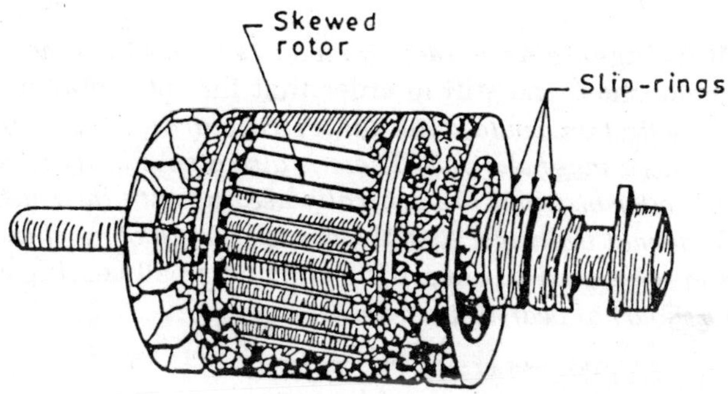

Fig. 4. Induction motor with phase-wound rotor, showing the three slip rings on the rotor shaft.

Figs. 5 and 6 show squirrel-cage and phase-wound induction motors respectively.

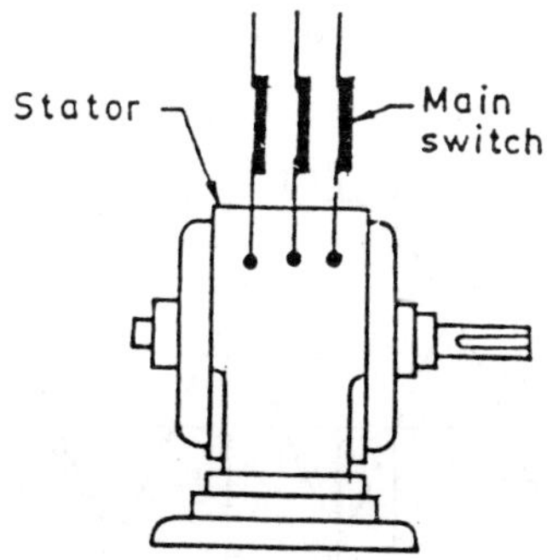

Fig. 5. Squirrel-cage motor.

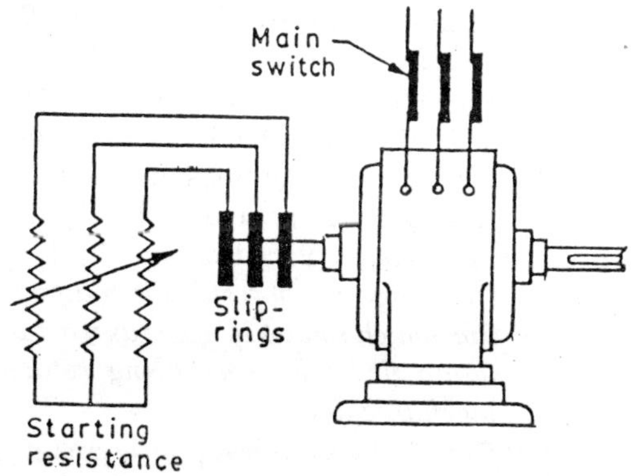

Fig. 6. Phase-wound motor connected to a three-phase star-connected starting resistance.

Advantages of a squirrel-cage motor over a phase-wound induction motor. As compared with a wound rotor a squirrel-cage induction motor entails the following *advantages* :

1. Slightly higher efficiency.
2. Cheaper and rugged in construction.
3. No slip rings, brush gear, short-circuiting devices, rotor terminals for starting rheostats are required. The star-delta starter is sufficient for starting.
4. It has better space factor for rotor slots, a shorter overhang and consequently a small copper loss.
5. It has a smaller rotor overhang leakage which gives a better power factor and a greater pull out torque and overload capacity.
6. It has bare end rings, a large space for fans and thus the cooling conditions are better.

The major *'disadvantage' of squirrel-cage motor* is that *it is not possible to insert resistance in the rotor circuit for the purpose of increasing the starting torque. The cage rotor has a smaller starting torque and large starting current as compared with wound rotor.*

3.3. Slip Rings. The slip rings for wound rotor machines are made of *either brass or phosphor bronze.* They are shrunk on to a cast iron sleeve with moulded silica insulation. This assembly is passed on to the rotor shaft. The slip rings are rotated either between the core and the bearing or on the shaft extension. In the latter case the shaft is made hollow to allow the three connections from rotor to slip rings to pass through bearings.

3.4. Shaft and Bearings. *In an induction motor the air gap is made as small as possible.* Therefore the shaft is made short and stiff in order that the rotor may not have any significant deflection, *as even a small deflection would create large irregularities in the air gap which would lead to production of an unbalanced magnetic pull. There is also a possibility of rotor and stator fouling with each other. Ball and roller bearings are generally used as with their use, accurate centering is much simpler than with journal bearings.* Also the *overall length of machine is reduced. For small motors,* a roller bearing may be used at the driving end and a ball bearing at the non-driving end. *For large and heavy rotors journal bearings are used.*

4. PRODUCTION OF ROTATING MAGNETIC FIELD

- *When a three-phase voltage is applied to the three-phase stator winding of the induction motor a rotating magnetic field is produced, which by transformer action induces a 'working' e.m.f. in the rotor winding. The rotor-induced e.m.f. is called a working e.m.f. because it causes a current to flow through the armature winding conductors. This combines with the revolving flux-density wave to produce torque. Thus revolving field is a key to the operation of the induction motor.*

  It will now be shown that when three-phase windings displaced in space by 120°, are fed by three-phase currents displaced in time by 120°, they *produce a resultant magnetic flux which rotates in space as if actual magnetic poles were being rotated mechanically.*

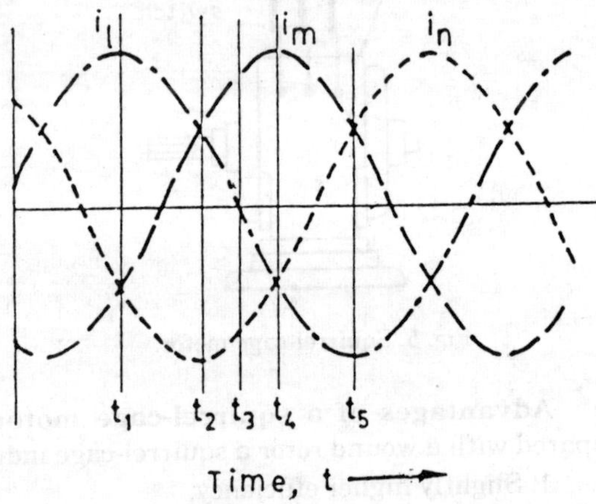

Fig. 7. Balanced three-phase alternating currents.

- Fig. 7 shows three-phase currents which are assumed to be flowing in phases '*l*', '*m*' and '*n*' respectively. These currents are time-displaced by the 120 electrical degrees.
- Fig. 8 shows the stator structure and the three-phase winding. Each phase (normally distributed over 60 electrical degrees) for convenience is represented by a single coil. Thus coil *l*-*l*′ represents the entire phase '*l*' winding having its flux axis directed along the vertical. This means that whenever phase '*l*' carries current it produces a flux field directed along the vertical-up or down. The right-hand flux rule readily verifies this statement. Similarly, the flux axis of phase '*m*' is 120 electrical degrees displaced from phase '*l*' ; and that of phase '*n*' is 120 electrical degrees displaced from phase '*m*'. The unprimed letters refer to the beginning terminals of each phase.
- **Magnitude and direction of the resultant flux corresponding to time instant t_1.** (Fig. 7). At this instant the current into phase '*l*' is at its *positive maximum* value while currents in phases '*m*' and '*n*' are at *one-half their maximum values.* In Fig. 8 it is arbitrarily assumed that when current in a given phase is positive, it flows into the paper with respect to the unprimed conductors. Thus since, at time t_1, i_l is positive, a cross is used for conductor '*l*' [Fig. 8 (*i*)]. Of course a dot is used for *l*′ because it refers to the return connection. Then by the right-hand rule it follows that phase '*l*' produces a flux contribution directed downward along the vertical. Moreover, the magnitude of this contribution is the maximum value because the current is at maximum. Hence, $\phi_l = \phi_{max}$, where ϕ_{max} is the maximum flux per pole of phase '*l*'. It is important to understand that phase '*l*' really

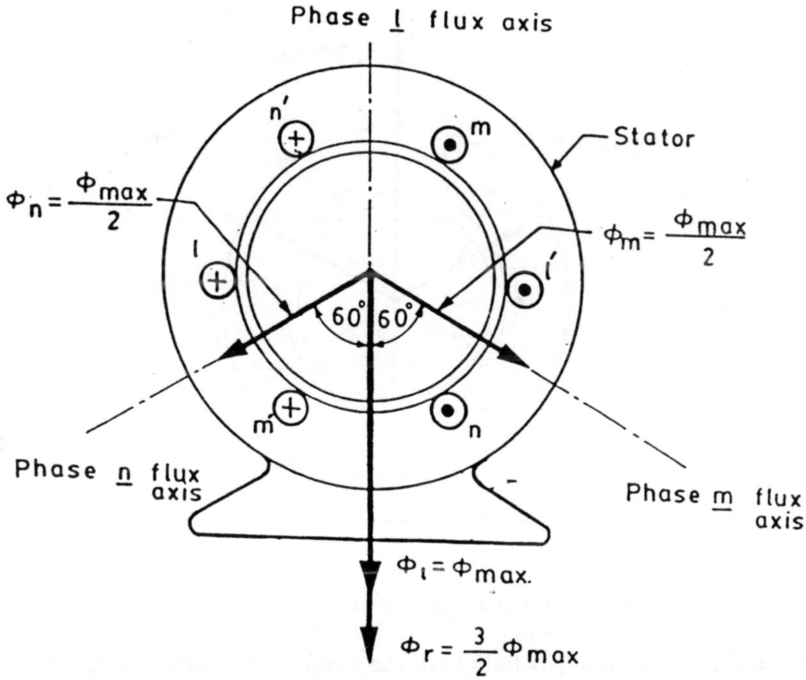
Phase l flux axis
n'
m
Stator
$\Phi_n = \frac{\Phi_{max}}{2}$
$\Phi_m = \frac{\Phi_{max}}{2}$
l
l'
60°
60°
m'
n
Phase n flux axis
Phase m flux axis
$\Phi_l = \Phi_{max.}$
$\Phi_r = \frac{3}{2}\Phi_{max}$

(i)

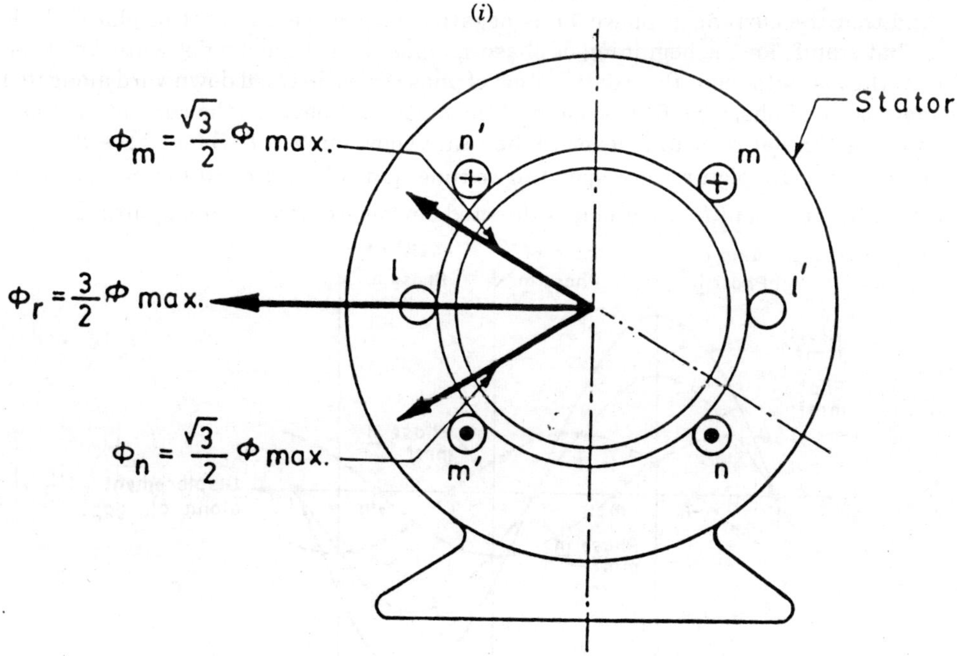
Stator
$\Phi_m = \frac{\sqrt{3}}{2}\Phi$ max.
n'
m
l
l'
$\Phi_r = \frac{3}{2}\Phi$ max.
$\Phi_n = \frac{\sqrt{3}}{2}\Phi$ max.
m'
n

(ii)

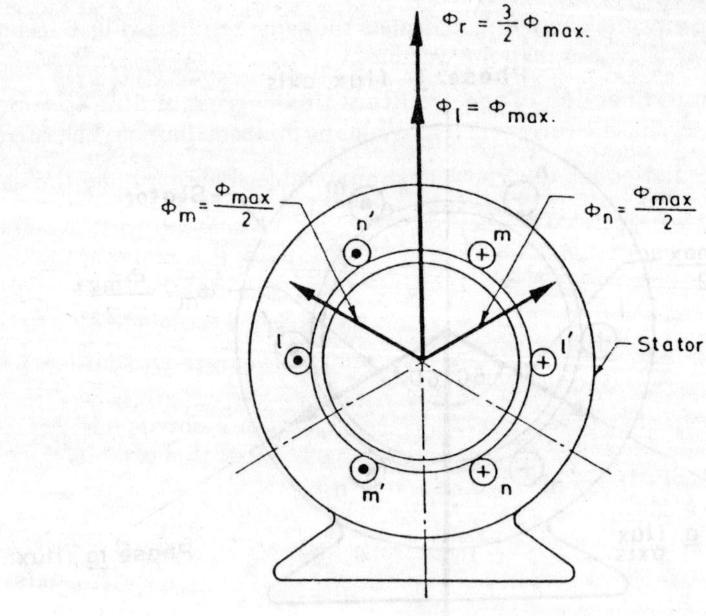

(iii)

Fig. 8. Representing the rotating magnetic field at three instants of time : (i) time t_1 in Fig. 7 ; (ii) time t_3 ; (iii) t_5.

produces a sinusoidal flux field with the amplitude located along the axis of phase '*l*' as shown in Fig. 9. However, in Fig. 8 (*i*) this sinusoidal distribution is conveniently represented by the vector ϕ_l.

For determining the magnitude and direction of the field contribution of phase '*m*' at time t_1, we first find that the current in phase '*m*' is negative with respect to that in phase '*l*'. Hence the conductor that stands for the beginning of phase '*m*' must be assigned a dot while '*m*' is assigned a cross. Hence the instantaneous flux contribution of phase '*m*' is directed downward along its flux axis and the magnitude of phase '*m*' flux is one-half the maximum because the current is at one-half its maximum value. Similar reasoning leads to the result shown in Fig. 8 (*i*) for phase '*n*'.

A glance at space picture corresponding to time t_1, as illustrated in Fig. 8 (*i*), should make it apparent that the resultant flux per pole is directed downward and has a magnitude of $\frac{1}{2}$ times the

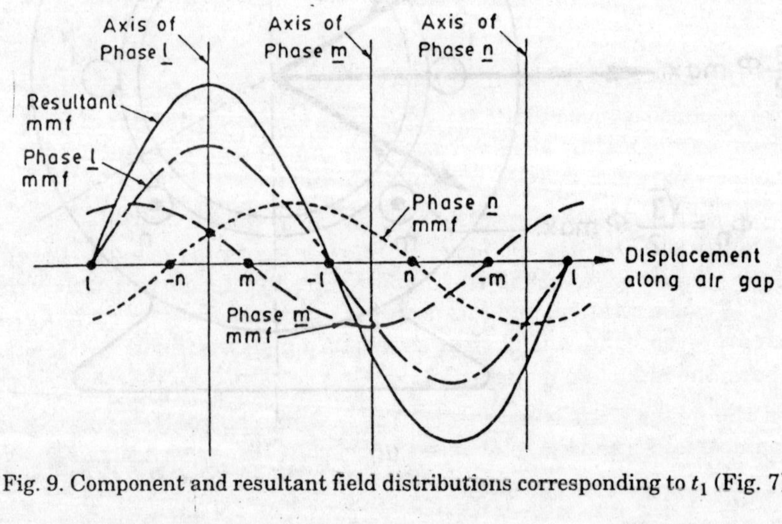

Fig. 9. Component and resultant field distributions corresponding to t_1 (Fig. 7).

maximum flux per pole of any one phase. Fig. 9 depicts the same results as Fig. 8 (*i*) but does so in terms of sinusoidal flux waves rather than flux vectors.

- **Magnitude and direction of the resultant flux corresponding to time instant t_3** [Fig. 8 (*ii*)]. Here phase '*l*' current is zero, yields no flux contribution. The current in phase '*m*' is positive and equal to $\frac{\sqrt{3}}{2}$ its maximum value. Phase '*n*' has the same current magnitude but is negative. Together phases '*m*' and '*n*' *combine to produce a resultant flux having the same magnitude as at time t_1*. See Fig. 8 (*ii*). It is important to note, too, that an *elapse of 90 electrical degrees in time results in rotation of the magnetic flux field of 90 electrical degrees.*
- **At time instant t_5.** A further elapse of time equivalent to an additional 90 electrical degrees leads to the situation depicted in Fig. 8 (*iii*).

From the above discussion it is apparent that the application of three-phase currents through a balanced three-phase winding gives rise to a rotating magnetic field that exhibits the following *two characteristics.*

1. It is of *constant amplitude.*

2. It is of *constant speed.* It follows from the fact that the resultant flux traverses through 2π electrical radians in space for every 2π electrical radians of variation in time for the phases currents. Hence for a 2-pole machines where electrical and mechanical degrees are identical, each cycle of variation of current produces one complete revolution of the flux field. Hence *this is a fixed relationship which is dependent upon the frequency of the currents* and number of poles for which the three-phase winding is designated. In the case where the winding is designed for *four* poles it requires *two cycles* of variation of the current to produce one revolution of the flux field. Therefore it follows that for a *p*-pole machine the relationship is :

$$N_s = \frac{120f}{p} \quad ...(1)$$

or

$$f = \frac{pN_s}{120} \quad ...[1\,(a)]$$

N_s (r.p.m.) is called the *synchronous speed* and all synchronous machines run at their respective synchronous speeds.

It may be noted that the speed of rotation of the field as described by eq. (1) is always given relative to the phase windings carrying the time-varying currents. Accordingly, if a situation arises where the *winding is itself revolving*, then the speed of rotation of the field relative to inertial force is different from it with respect to the winding.

5. THEORY OF OPERATION OF AN INDUCTION MOTOR

When a three-phase is given to the stator winding a rotating field is set-up. This field sweeps past the rotor (conductors) and by *virtue of relative motion,* an e.m.f. is induced in the conductors which form the rotor winding. Since this winding is in the form of a closed circuit, a current flows, the direction of which is, by Lenz's law, such as to oppose the change causing it.

Now, *the change is the relative motion of the rotating field and the rotor, so that, to oppose this, the rotor runs in the same direction as the field and* attempts to catch up with it. It is clear that torque must be produced to cause rotation, and *this torque is due to the fact that currents flow in the rotor conductors* which are situated in, and at right angles to, a magnetic field.

Fig. 10 shows the induction motor action.

- When the motor shaft is *not loaded,* the machine has only to rotate itself against the mechanical losses and the rotor speed is *very close to the synchronous speed.* However, the

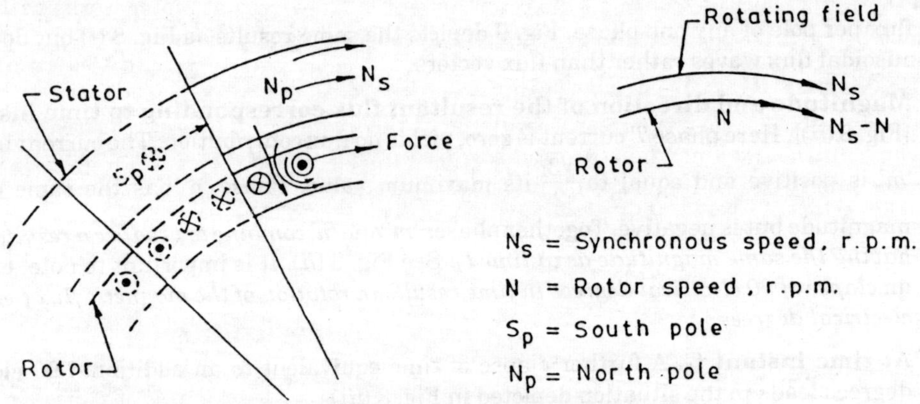

Fig. 10. Induction motor action.

rotor speed cannot become equal to the synchronous speed because if it does so, *the e.m.f. induced in the rotor winding would become zero and there will be no torque.* Hence the speed remains slightly less than the synchronous speed. If the motor shaft is *loaded,* the rotor will slow down and the relative speed of the rotor with respect to the stator rotating field will increase. The e.m.f. induced in the rotor winding will increase and will produce more rotor current which will increase the electromagnetic torque produced by the motor. Conditions of equilibrium are attained when the rotor speed has adjusted to a new value so that the electromagnetic torque is sufficient to balance the mechanical or load torque applied to the shaft. The speed of the motor *when running under full load conditions is somewhat less than the no-load speed.*

6. SLIP

- As earlier stated, *the rotor speed must always remain less than the synchronous speed. The difference between the synchronous speed and the rotor speed is known as* **'slip'.** It is usually expressed as a fraction of the synchronous speed. Thus slip *s* is

$$s = \frac{N_s - N}{N_s} \quad ...(2)$$

or

$$N = N_s(1 - s) \quad ...[2\,(a)]$$

where N_s = synchronous speed (r.p.m.)

N = motor speed (r.p.m.).

In practice the value of slip is very small. At no-load, slip is around 1% or so and at full-load it is around 3%. For large efficient machines the slip at full-load may be around 1% only. The induction motor, is therefore, a motor with substantially constant speed and fills the same role as *D.C. shunt motor.*

- When the rotor is stationary (standstill) its speed is zero and $s = 1$. The rotor cannot run at synchronous speed because then there will be no rotor e.m.f. and no rotor current and torque. If the rotor is to run at synchronous speed an external torque is necessary. *If the rotor is driven such that $N > N_s$, the slip becomes negative, the rotor torque opposes the external driving torque and the* **machine acts as induction generator.**
- The induction motor derives its name from the fact that the *current in the rotor circuit is induced from the stator.* There is no external connection to the rotor except for some special purposes.

If the rotor reactance at standstill is X_2 its value at slip '*s*' becomes sX_2. *This is very desirable,* for at no-load the reactance becomes almost negligible and the rotor impedance is now all resistance.

Further if the rotor resistance is small the rotor current is large, so that motor works with a large torque which brings the speed near to synchronous speed, *i.e.*, the slip is reduced.

7. FREQUENCY OF ROTOR CURRENT

At standstill (*i.e.*, when the rotor is stationary), the *frequency of the rotor current is the same as the supply frequency (f)*. But when the rotor starts revolving, then the frequency depends upon the relative speed or slip-speed. If f_r is the frequency of the rotor current, then

$$N_s - N = \frac{120 f_r}{p} \quad ...(i)$$

Also

$$N_s = \frac{120 f}{p} \quad ...(ii)$$

Dividing (*i*) by (*ii*), we get

$$\frac{N_s - N}{N_s} = \frac{f_r}{f} \quad \text{or} \quad s = \frac{f_r}{f}$$

or

$$f_r = sf \quad ...(3)$$

8. ROTOR E.M.F. AND ROTOR CURRENT

Rotor e.m.f :

When the rotor is stationary, an induction motor is equivalent to a 3-phase transformer with secondary short-circuited. Therefore, the induced e.m.f. per phase E_2 in the rotor at the instant of starting is given as :

$$E_2 = E_1 \times \frac{N_2}{N_1} \quad ...(4)$$

where E_1 = applied voltage per phase to primary *i.e.*, stator winding,
N_1 = number of stator turns, and
N_2 = number of rotor turns.

When the rotor starts gaining speed, the relative speed of the rotor with respect to stator flux *i.e.*, slip, is decreased. Hence induced e.m.f. in the rotor, which is *directly proportional to the relative speed i.e., slip*, is also decreased and is given by sE_2. Hence *for slip 's', the induced e.m.f. in the rotor is s times the induced e.m.f. in the rotor at standstill.*

Rotor current :

Let R_2 = rotor resistance/phase,
L_2 = rotor inductance/phase, and
E_2 = induced e.m.f. of rotor/phase at standstill.

At standstill :

Induced e.m.f. of rotor/phase $= E_2$
Rotor winding resistance/phase $= R_2$
Rotor winding reactance/phase, $X_2 = 2\pi f L_2$ where f is the supply frequency
Rotor impedance/phase, $Z_2 = \sqrt{R_2^2 + X_2^2}$

$\therefore$ Rotor current/phase $= \dfrac{E_2}{Z_2} = \dfrac{E_2}{\sqrt{R_2^2 + X_2^2}}$.

At slip 's' :

Induced e.m.f. of rotor/phase $= sE_2$
Rotor winding resistance $= R_2$
Rotor winding reactance $= 2\pi f_r L_2 = 2\pi s f L_2 = s(2\pi f L_2) = sX_2$

Rotor winding impedance/phase $= \sqrt{R_2^2 + (sX_2)^2}$

$\therefore$ Rotor current/phase,
$$I_2 = \frac{sE_2}{\sqrt{R_2^2 + (sX_2)^2}}$$
$$= \frac{sE_2}{\sqrt{R_2^2 + s^2X_2^2}} = \frac{E_2}{\sqrt{(R_2/s)^2 + X_2^2}} \quad ...(5)$$

The rotor currrent I_2 lags the rotor voltage E_2 by rotor power factor angle ϕ_2 given by
$$\phi_2 = \tan^{-1}\left(\frac{sX_2}{R}\right)$$

Power factor of rotor current,
$$\cos\phi_2 = \frac{R_2}{\sqrt{R_2^2 + s^2X_2^2}} = \frac{R_2/s}{\sqrt{(R_2/s)^2 + X_2^2}}. \quad ...(6)$$

9. TORQUE AND POWER

The torque of an induction motor (being due to interaction of a rotor and stator fields),
$$T \propto \phi I_2 \cos\phi_2$$
where ϕ = flux of rotating stator,
I_2 = rotor current/phase, and
$\cos\phi_2$ = rotor power factor.

Since rotor e.m.f./phase at standstill, $E_2 \propto \phi$
$$\therefore \quad T \propto E_2 I_2 \cos\phi_2$$
or
$$T = kE_2 I_2 \cos\phi_2 \text{ where } k \text{ is any constant} \quad ...(7)$$

Substituting the value of I_2 and $\cos\phi_2$ from eqns. (5) and (6) in eqn. (7), we get
$$T = kE_2 \frac{sE_2}{\sqrt{R^2 + s^2X_2^2}} \times \frac{R_2}{\sqrt{R^2 + s^2X_2^2}}$$
i.e.,
$$T = \frac{ksR_2E_2^2}{R_2^2 + s^2X_2^2} \quad ...(8)$$

9.1. Starting Torque. At start slip 's' = 1. Therefore, expression for starting torque may be obtained by putting $s = 1$ in eqn. 8.

Starting torque,
$$T_{st} = \frac{kR_2E_2^2}{R_2^2 + X_2^2}. \quad ...(9)$$

9.2. Condition for Maximum Torque. The value of torque when motor is running is given by
$$T = \frac{ksR_2E_2^2}{R_2^2 + s^2X_2^2}$$

Torque will be maximum when,
$$\frac{sR_2}{R_2^2 + s^2X_2^2} \quad \text{or} \quad \frac{R_2}{\frac{R_2^2}{s} + sX_2^2}$$
or
$$\frac{R_2}{\left(\frac{R_2}{\sqrt{s}} - X_2\sqrt{s}\right)^2 + 2R_2X_2} \text{ is maximum, } viz., \frac{R_2}{\sqrt{s}} - X_2\sqrt{s} = 0$$

or $$s\ (= s_{mT}) = \frac{R_2}{X_2} \qquad ...(10)$$

(where s_{mT} = slip corresponding to maximum torque)

$$\therefore \text{ Maximum torque,} \quad T_{max} = \frac{kE_2^2}{2X_2} \qquad ...(11)$$

From the above expression, the following conclusions can be drawn :

- Maximum torque is *independent* of rotor circuit resistance.
- Maximum torque varies *inversely as standstill reactance* of the rotor. Therefore to have maximum torque, standstill reactance (*i.e.* inductance) of the rotor should be kept as small as possible.
- The slip at which the maximum torque occurs depends upon the resistance of the rotor.

The condition for getting maximum torque at starting can be obtained by putting $s = 1$ in eqn. (10).

Thus, starting torque will be maximum if

$$\frac{R_2}{X_2} = s = 1 \quad \text{or} \quad R_2 = X_2.$$

9.3. Starting Torque of a Squirrel-Cage Motor. *The squirrel-cage rotor resistance is fixed* and *small* as compared to its reactance which is very large especially *at start* (because at standstill the frequency of rotor current is equal to that of supply frequency). Hence, the starting current I_2 of the rotor, though very large in magnitude, *lags by a very large angle* behind E_2 ; consequently the *starting torque per ampere is very poor.* It is roughly 1.5 times the full-load torque although the starting current is 5 to 7 times the full-load current. Thus such motors are *not suitable* for applications where these have to be *started against heavy loads.*

9.4. Starting Torque of a Slip Ring Motor. In a slip ring motor the torque is increased by *improving its power factor by adding external resistance in the rotor circuit from the star- connected rheostat* ; as the motor gains speed the rheostat resistance is gradually cut out. This additional resistance, however, increases the rotor impedance and so reduces the rotor current. At first, the effect of improved power factor *predominates the current-decreasing effect* of impedance, hence starting torque is increased. But after a certain point, the effect of increased impedance *predominates the effect of improved power factor and so the torque starts decreasing.*

9.5. Power. Eqn. (5) can be represented by a simple series circuit as shown in Fig. 11.

It is seen from this circuit that **per phase** power input (gross) to rotor,

$$P_g = E_2 I_2 \cos \phi_2$$

where $\cos \phi_2 = \dfrac{R_2/s}{\sqrt{(R_2/s)^2 + X_2^2}}$

$$\therefore \ P_g = \frac{E_2}{\sqrt{(R_2/s)^2 + X_2^2}} \cdot I_2 \frac{R_2}{s} = I_2^2 \frac{R_2}{s} \qquad ...(12)$$

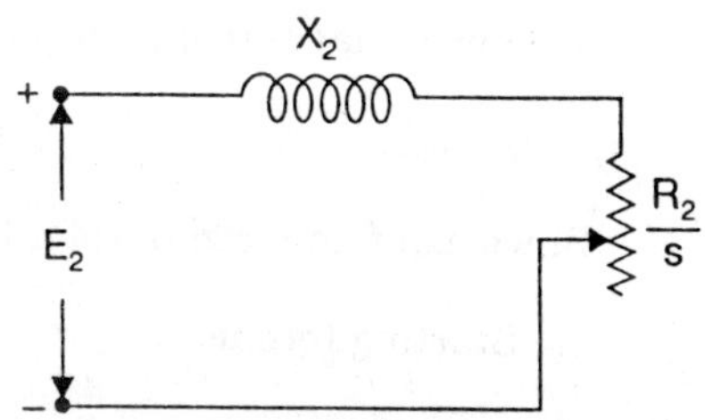

Fig. 11. Rotor equivalent circuit of an induction motor.

An examination of Fig. 11 also shows that per phase power input to rotor is equal to $I_2^2 \dfrac{R_2}{s}$ as the reactance is X_2 consumes no power.

P_g is the power transferred from stator to rotor across the air gap. In view of this, P_g is called the **air-gap power**. The expression for P_g may be written as

$$P_g = I_2^2 \frac{R_2}{s} = I_2^2 R_2 + I_2^2 R_2 \left(\frac{1-s}{s}\right)$$

= Rotor ohmic loss + internal mechanical power developed in rotor ($P_{mech.}$)

$$= sP_g + (1-s)P_g \quad \text{...(13 a)}$$

$$\therefore \quad P_{mech.} = (1-s)P_g = I_2^2 R_2 \left(\frac{1-s}{s}\right) \quad \text{...(13 b)}$$

$$\text{Rotor ohmic loss} = \left(\frac{s}{1-s}\right) P_{mech.} = sP_g \quad \text{...(13 c)}$$

Eqn. (12) reveals that ohmic loss

$$= I_2^2 R_2 = sP_g = s(\text{power input to rotor}) \quad \text{...(14)}$$

Internal (or gross) torque developed per phase is given by

$$T_g = \frac{\text{internal mechanical power developed in rotor}}{\text{rotor speed in mechanical rad. per sec.}}$$

$$= \frac{P_{mech.}}{\omega_r} = \frac{(1-s)P_g}{(1-s)\omega_s} = \frac{P_g}{\omega_s} \quad \text{...(15)}$$

Here $\omega_s \left(= \frac{2\pi N_s}{60}\right)$ is the *synchronous speed in mechanical radians per second.*

Also
$$T = \frac{P_g}{\omega_s} = \frac{1}{\omega_s} \times \frac{I_2^2 R_2}{s} = \frac{\text{Rotor ohmic loss}}{(\omega_s)\text{ slip}} = \frac{1}{2\pi(N_s/60)} I_2^2 \frac{R_2}{s} \quad \text{...(16)}$$

The power available at the shaft can be obtained from P_g as follows :

Output or shaft power, $P_{sh} = P_m -$ mechanical losses (friction, and windage losses)

or $\quad P_{sh} = P_g -$ rotor ohmic loss – friction and windage losses.

Output or shaft torque,

$$T_{sh} = \frac{P_{sh}}{\text{rotor speed}} = \frac{P_{sh}}{(1-s)\omega_s}$$

If stator input is known, then air-gap power P_g is given by

$P_g =$ stator power input – stator I^2R loss – stator core loss.

10. EFFECT OF CHANGE IN SUPPLY VOLTAGE ON STARTING TORQUE

We know that starting torque,

$$T_{st} = \frac{kR_2E_2^2}{R_2^2 + X_2^2}$$

Since e.m.f. induced in rotor (at standstill) $E_2 \propto \phi \propto V$

$$\therefore \text{ Starting torque,} \quad T_{st} = \frac{k'R_2V_2^2}{R_2^2 + X_2^2}.$$

where k' is another constant

or $\quad T_{st} \propto V^2$

i.e., *Starting torque is proportional to the square of the applied voltage.*

11. EFFECT OF CHANGE IN SUPPLY VOLTAGE ON TORQUE AND SLIP

The torque acting on the rotor when the motor is running with slip 's' is given by

$$T = \frac{ksR_2E_2^2}{R_2^2 + s^2X_2^2}$$

Since e.m.f. induced in rotor (at standstill), $E_2 \propto \phi \propto V$

$$\therefore \qquad T = \frac{k' s R_2 V^2}{R_2^2 + s^2 X_2^2}, \quad \text{where } k' \text{ is another constant}$$

Since the slip '*s*' at full-load is *very low*, therefore $s^2 X_2^2$ can be *neglected* in comparison with R_2^2.

$$\therefore \qquad T = \frac{k' s R_2 V^2}{R_2^2} = \frac{k' s V^2}{R_2} \quad \text{or} \quad T \propto s V^2$$

When the supply voltage is changed, it changes the torque under running condition also. With the decrease in supply voltage torque under running condition decreases, therefore, in order to *maintain the same torque, slip increases* or *speed decreases.*

12. FULL-LOAD TORQUE AND MAXIMUM TORQUE

Let, s_f = full-load slip of the motor, and

s_{mT} = slip corresponding to maximum torque = $\dfrac{R_2}{X_2}$

We know that,

Full-load torque, $$T_f = \frac{k s_f R_2 E_2^2}{R_2^2 + s_f^2 X^2} \qquad \text{[From Art. 9]}$$

Maximum torque, $$T_m = \frac{k E_2^2}{2 X_2} \qquad \text{[From Art. 9.2]}$$

$$\therefore \qquad \frac{T_f}{T_m} = \frac{k s_f R_2 E_2^2/(R_2^2 + s_f^2 X_2^2)}{k E_2^2/(2X_2)} = \frac{2 s_f R_2 X_2}{R_2^2 + s_f^2 X_2^2}$$

$$= \frac{2 s_f R_2 X_2/(X_2^2)}{(R_2^2 + s_f^2 X_2^2)/(X_2^2)} \qquad \text{[Dividing numerator and denominator by } X_2^2\text{]}$$

$$= \frac{2 s_f (R_2/X_2)}{(R_2/X_2)^2 + s_f^2} = \frac{2 s_f s_{mT}}{s_{mT}^2 + s_f^2}$$

i.e., $$\frac{T_f}{T_m} = \frac{2 s_f s_{mT}}{s_f^2 + s_{mT}^2} \qquad ...(17)$$

$$= \frac{2}{\dfrac{s_f}{s_{mT}} + \dfrac{s_{mT}}{s_f}} \qquad ...(17\ a)$$

[Dividing numerator and denominator by $s_f \, s_{mT}$]

13. STARTING TORQUE AND MAXIMUM TORQUE

$$\frac{T_{st}}{T_m} = \frac{k R_2 E_2^2/(R_2^2 + X_2^2)}{k E_2^2/(2X_2)} = \frac{2 R_2 X_2}{R_2^2 + X_2^2} = \frac{2(R_2/X_2)}{(R_2/X_2)^2 + 1}$$

(Dividing numerator and denominator by X_2^2)

or $$\frac{T_{st}}{T_m} = \frac{2 s_{mT}}{s^2_{mT} + 1} \qquad ...(18)$$

14. TORQUE-SLIP AND TORQUE-SPEED CURVES

The expression for torque is as follows :

$$T = \frac{ksR_2E_2^2}{R_2^2 + s^2X_2^2}$$

From the above expression, it is evident, that

- Torque is zero when slip $s = 0$ (*i.e.*, speed is synchronous).
- When slip 's' *is very low* the value of the term sX_2 is very small and is negligible in comparison with R_2, therefore torque T *is approximately proportional* to slip 's' if rotor resistance R_2 is constant. This means that at speeds near to synchronous speed the *torque-speed* and *torque-slip* curves are approximately straight lines (Figs. 12 and 13).
- When the slip 's' increases (*i.e.* as the speed decreases with increase in load) torque increases and reaches its maximum value when $s = \frac{R_2}{X_2}$. The *maximum torque is also known as* **'pull-out'** or **'break-down' torque.**

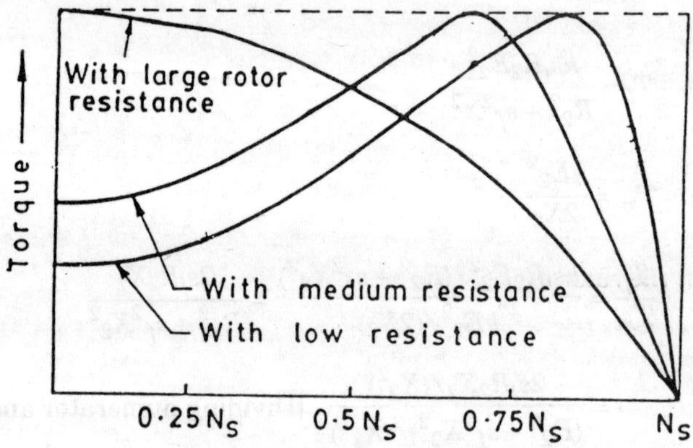

Fig. 12. Torque-speed curves.

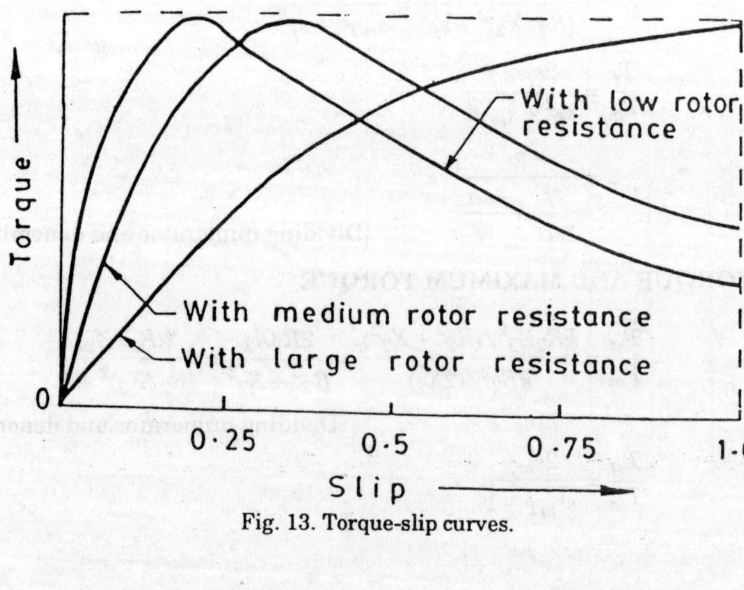

Fig. 13. Torque-slip curves.

- When the slip is further increased the torque decreases. The result is that *motor slows down* and *eventually stops. The motor operates for the value of slip between zero and that corresponding to maximum torque.*

With higher slip, R_2 becomes negligible as compared to sX_2 and torque varies as follows :

$$T \propto \frac{s}{s^2 X_2^2} \propto \frac{1}{s} \text{ if standstill reactance is constant.}$$

This means that *speed-torque* or *slip-torque* curves are rectangular hyperbola with the speed or slip beyond that corresponding to maximum torque. Figs. 12 and 13 show the *torque-speed* and *torque-slip* curves for different values of rotor resistance. It is observed that although maximum torque is independent of rotor resistance R_2, yet the exact location of T_m is dependent on it. *Greater the R_2, greater is the value of slip at which maximum torque occurs.*

15. OPERATING CHARACTERISTICS OF A 3-PHASE SQUIRREL-CAGE INDUCTION MOTOR

To start with, let us assume that the squirrel cage induction motor has been started without any load on it (*i.e.*, no electrical devices are connected to the generator, no metal is being cut by a machine tool, or the conveyor belt is completely disengaged). The motor will come to its no-load speed, which may be at a slip as low as 0.1 per cent. *Load is now placed on the motor* and the following reactions will be observed :

- *Speed decreases.*
- *Slip increases* (because the stator field continues to rotate at synchronous speed and therefore cuts the rotor bars more rapidly).
- As a consequence of increased cutting the *rotor induced e.m.f. increases.*
- *The current in the rotor bars increases.*
- *Torque developed by the rotor increases until it is sufficient to carry the load placed on the motor.*
- *The motor continues to rotate at that slip which will develop the required torque for that particular load.*

The motor may be loaded continuously till *pull-out (or break-down) torque* is developed, at which point the *motor will stop if more load is placed on it.* Pull-out occurs at a slip well beyond the

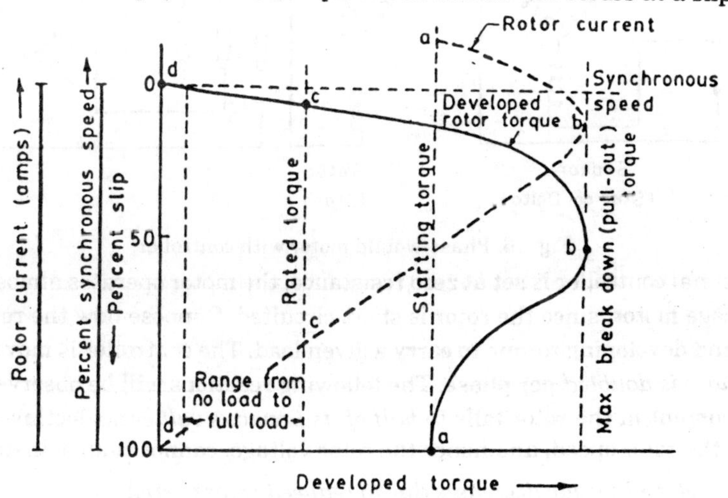

Fig. 14. Operating characteristics of a squirrel-cage induction motor.

normal range operation of the squirrel-cage induction motor. Even under normal operation, however, a suddenly applied load may require the temporary development of this maximum torque, after which the motor will speed up to its full-load value. Thus, in a manner similar to the shunt motor, the induction motor *inherently adjusts itself to the applied external torque.*

Fig. 14 shows typical operating characteristics of an induction motor.

Fig. 15 shows the effect of heavy loads on primary stator current and power factor.

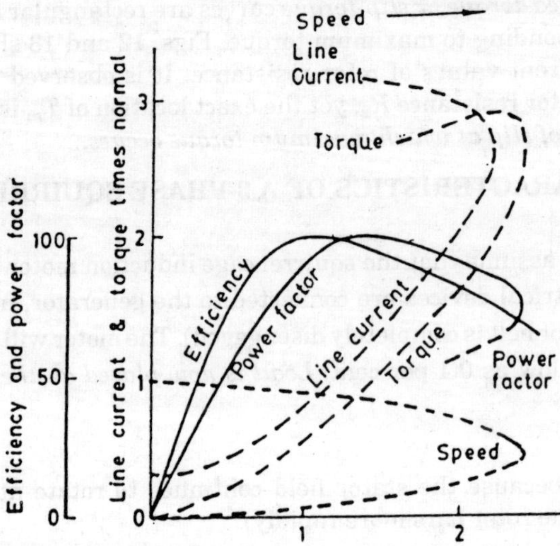

Fig. 15. Effect of heavy loads on primary stator current and power factor.

16. OPERATING CHARACTERISTICS OF A WOUND-ROTOR (SLIP RING) INDUCTION MOTOR

Fig. 16 shows a wound-rotor induction motor with controller (rheostat).

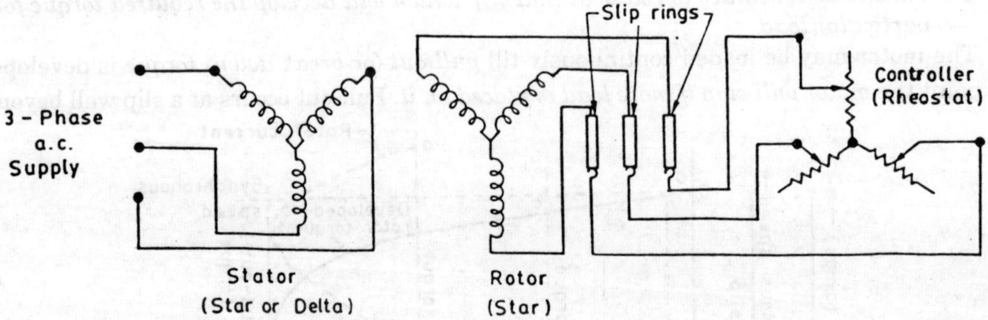

Fig. 16. Phase-wound motor with controller.

If the external controller is set at zero resistance, the motor operates almost exactly the same as the squirrel-cage motor, since the rotor is short-circuited. Suppose now the rotor is rotating at a constant speed and developing torque to carry a given load. The controller is moved so that the *rotor resistance per phase is doubled* per phase. The following reactions will be observed :

- The current in the rotor falls to *half of its previous value* (neglecting a transient effect), since the rotor speed, and hence the rotor voltage, cannot change instantaneously.
- The *developed torque decreases due to reduced rotor current.*

- The load on the motor has not changed, and with the external applied, or load, torque *now greater than the developed* torque, the *motor slows down.*
- The *slip increases* (increasing the rate at which the stator flux cuts the rotor conductors).
- The *rotor induced voltage increases.*
- The *rotor current now increases until it is sufficient to develop enough torque to again carry the load at a constant speed.*

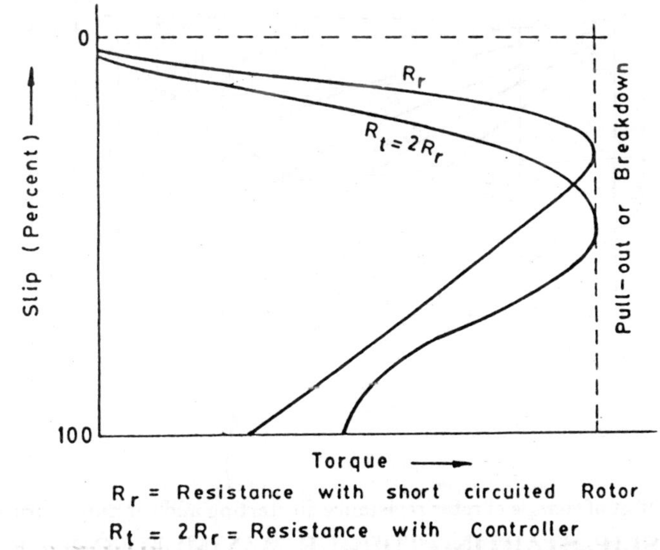

Fig. 17. Torque-slip characteristic of wound-rotor induction motor.

The induction motor now rotates at a lower speed, but with the same current as before, and develops the same torque as it did before the rotor circuit resistance was increased. The *frequency in the rotor is greater,* as is the rotor resistance. *Rotor resistance and reactance are both increased in same proportion,* however, and the *rotor power factor therefore remains the same.* If more load is now placed on the shaft, the rotor will develop the required torque, but always *at a lower speed i.e.* a *higher slip.* Eventually pull-out torque is attained. The *value of pull-out torque will be the same as with the short-circuited rotor, but it will occur at a slip which is twice as great as that with the shorted rotor, since the rotor circuit resistance is twice that of the rotor alone.*

Fig. 18 shows the torque-slip characteristic of the shorted rotor, together with that for a total resistance of, $R_t = 2R$.

It is worthnoting that the *torque developed at standstill is higher with increased rotor resistance.* This is explained by the fact that at standstill the rotor frequency is equal to the line frequency and rotor resistance is maximum. The addition of resistance in the rotor circuit obviously improves the power factor in that circuit while reducing the rotor current. Since standstill rotor resistance is much greater than

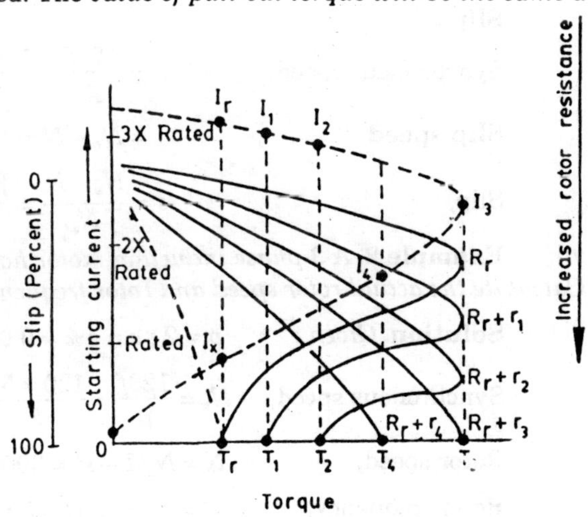

Fig. 18. Starting currents vs torque.

its resistance, the power factor increases more rapidly than the decrease of rotor current, and hence the developed torque is greater with added resistance.

Fig. 18 shows starting current vs torque.

Fig. 19 shows the effect of change of rotor resistance on starting and running characteristics.

In Figs. 18 and 19,

R_r = *basic rotor resistance*

r_1, r_2, r_3, r_4 = the resistance added to each phase of star-connected wound rotor respectively.

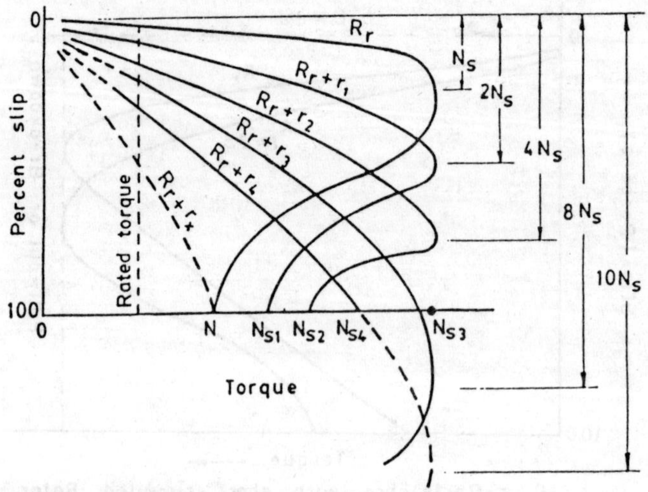

Fig. 19. Effect of change of rotor resistance on starting and running characteristics.

SLIP, STARTING TORQUE, MAXIMUM TORQUE

Example 1. *A 3-phase, 4-pole, 50-Hz induction motor is running at 1440 r.p.m. Determine the slip speed and slip.*

Solution. Number of poles, $p = 4$

Frequency, $f = 50$ Hz

Actual speed of rotor, $N = 1440$ r.p.m.

Slip speed = ?

Slip = ?

Synchronous speed, $N_s = \dfrac{120f}{p} = \dfrac{120 \times 50}{4} = 1500$ r.p.m.

Slip speed $= N_s - N = 1500 - 1440 = \mathbf{60}$ **r.p.m. (Ans.)**

Slip, $s = \dfrac{N_s - N}{N_s} = \dfrac{1500 - 1440}{1500} = \mathbf{0.04}$. **(Ans.)**

Example 2. *A 3-phase induction motor has 2 poles and is connected to 400 V, 50 Hz supply. Calculate the actual rotor speed and rotor frequency when the slip is 4%.* **(B.T.E. U.P., 1998)**

Solution. *Given :* $p = 2$; $s = 4\% = 0.04$; $f = 50$ Hz.

Synchronous speed, $N_s = \dfrac{120f}{p} = \dfrac{120 \times 50}{2} = 3000$ r.p.m.

Rotor speed, $\mathbf{N} = N_s(1 - s) = 3000(1 - 0.04) = \mathbf{2880}$ **r.p.m. (Ans.)**

Rotor frequency, $f_r = s.f = 0.04 \times 50 = \mathbf{2}$ **Hz. (Ans.)**

Example 3. *A 12-pole, 3-phase alternator is coupled to an engine running at 500 r.p.m. It supplies an induction motor which has a full speed of 1440 r.p.m.*

Find the percentage slip and number of poles of the motor. **(P.T.U., Jan., 2000)**

Solution. Poles of alternator, $p_a = 12$

Speed of the prime mover, $N_a = 500$ r.p.m.

Speed of induction motor, $N_m = 1440$ r.p.m.

Frequency of the supply generated by alternator,

$$f = \frac{N_a p_a}{120} = \frac{500 \times 12}{120} = 50 \text{ Hz}$$

No. of poles of the induction motor,

$$\mathbf{p_m} = \frac{120 f}{N_m} = \frac{120 \times 50}{1440} = 4.167 \text{ say } \mathbf{4}. \quad \textbf{(Ans.)}$$

$\therefore$ Synchronous speed, $N_s = \dfrac{120 f}{p_m} = \dfrac{120 \times 50}{4} = 1500$ r.p.m.

$\therefore$ Slip, $\mathbf{s} = \dfrac{N_s - N_m}{N_s} = \dfrac{1500 - 1440}{1500} = 0.04$ or **4%. (Ans.)**

Example 4. *A 3-phase alternator having 12-poles is driven at a speed of 500 r.p.m. It supplies power to an 8-pole, 3-phase induction motor. If the slip of the motor at full-load is 4%, calculate the full-load speed of the motor.*

Solution. Number of poles of the alternator = 12

Speed of alternator = 500 r.p.m.

Number of poles of induction motor = 8

Percentage slip = 4%

Full-load speed, N :

Supply frequency, $f = \dfrac{12 \times 500}{120} = 50$ Hz

Synchronous speed, $N_s = \dfrac{120 \times 50}{8} = 750$ r.p.m.

$\therefore$ % slip $= \dfrac{N_s - N}{N_s} \times 100$

$$4 = \frac{750 - N}{750} \times 100$$

$$\therefore \quad N = 750 - \frac{4 \times 750}{100} = 720 \text{ r.p.m.}$$

Hence, **full-load speed, N = 720 r.p.m. (Ans.)**

☞**Example 5.** *A 3-phase, 6-pole, 50-Hz induction motor has a slip of 1% at no-load and 3% at full-load. Find :*

(i) Synchronous speed,
(ii) No-load speed,
(iii) Full-load speed,
(iv) Frequency of rotor current at standstill, and
(v) Frequency of rotor current at full-load.

Solution. Number of poles, $p = 6$

No-load slip, $s_0 = 1\%$

Full-load slip, $s_f = 3\%$

(*i*) **Synchronous speed,**

$$N_s = \frac{120f}{p} = \frac{120 \times 50}{6} = \mathbf{1000\ r.p.m.\ \ (Ans.)}$$

(*ii*) **No-load speed, N_0 :**

We know that, $$s = \frac{N_s - N}{N_s} \quad \text{or} \quad N = N_s(1 - s)$$

or $$N_0 = N_s(1 - s_0) = 1000\left(1 - \frac{1}{100}\right) = \mathbf{990\ r.p.m.\ \ (Ans.)}$$

(*iii*) **Full-load speed,**

$$N_f = N_s(1 - s_f) = 1000\left(1 - \frac{3}{100}\right) = \mathbf{970\ r.p.m.\ \ (Ans.)}$$

(*iv*) **Frequency of rotor current at standstill, f_r :**

At standstill, $s = 1$

$$f_r = sf = 1 \times 50 = \mathbf{50\ Hz.\ \ (Ans.)}$$

(*v*) **Frequency of rotor current at full-load, f_r = ?**

$$f_r = s \times f = \frac{3}{100} \times 50 = \mathbf{1.5\ Hz.\ \ (Ans.)}$$

☞**Example 6.** *A 3-phase, 12-pole alternator is coupled to an engine running at 500 r.p.m. The alternator supplies an induction motor which has a full-load speed of 1455 r.p.m. Find the slip and number of poles of the motor.*

Solution. Number of poles of the alternator, $p_a = 12$

Speed of the engine, $N_e = 500$ r.p.m.

Full-load speed of the induction motor, $N_m = 1455$ r.p.m.

Slip, s = ?

Number of poles of the motor, p_m = ?

Supply frequency, $$f = \frac{N_a p_a}{120} = \frac{500 \times 12}{120} = 50\ \text{Hz.}$$

When the supply frequency is 50-Hz, the synchronous speed can be 3000, 1500, 1000, 750 r.p.m. etc. Since the full-load speed is 1455 r.p.m. and the full-load slip is *always* less than 5%, the synchronous is 1500 r.p.m.

Slip, $$s = \frac{N_s - N}{N_s} = \frac{1500 - 1455}{1500} = \mathbf{0.03\ or\ 3\ per\ cent.\ \ (Ans.)}$$

Also, $$N_s = \frac{120f}{p_m}$$

$$\therefore \quad p_m = \frac{120f}{N_s} = \frac{120 \times 50}{1500} = 4\ \text{poles}$$

Hence, **number of motor poles = 4. (Ans.)**

Example 7. *A 3-phase induction motor runs at almost 1000 rpm at no-load and 950 r.p.m. at full-load when supplied with power from a 50 Hz 3-phase line.*

(*i*) *How many poles has the motor ?*

(*ii*) *What is the percentage slip at full load ?*

(*iii*) *What is the corresponding frequency of rotor voltages ?*

(*iv*) *What is the corresponding speed of the rotor field with respect to rotor ?*

(*v*) *What is the corresponding speed of the rotor with respect to the stator ?*
(*vi*) *What is the corresponding speed of the rotor field with respect to the stator field ?*
(*vii*) *What is the rotor frequency at the slip of 10 per cent ?* **(AMIE, 1997)**

Solution. *Given* : $N_0 = 1000$ r.p.m., $N_f = 950$ r.p.m., $f = 50$ Hz

Since no-load speed of motor is almost 1000 r.p.m., hence synchronous speed near to 1000 r.p.m. is 1000 r.p.m.

(*i*) Poles on motor, $\mathbf{p} = \frac{120f}{N_s} = \frac{120 \times 50}{1000} = \mathbf{6}$. **(Ans.)**

(*ii*) Percentage slip at full load, $\mathbf{s_f} = \frac{N_s - N_f}{N_s} \times 100 = \frac{1000 - 950}{1000} \times 100 = \mathbf{5\%}$. **(Ans.)**

(*iii*) Frequency of rotor voltage, $f_r = s \times f = \frac{5}{100} \times 50 = \mathbf{2.5\ Hz}$. **(Ans.)**

(*iv*) Speed of rotor field with respect to rotor $= \frac{120 \times f_r}{p} = \frac{120 \times 2.5}{6} = 50$ r.p.m. **(Ans.)**

(*v*) Speed of rotor with respect to stator = **950 r.p.m. (since stator is stationary). (Ans.)**

(*vi*) Rotor field and stator field are revolving at the same speed of 1000 r.p.m., therefore, *speed of rotor field w.r.t. stator field is* **zero. (Ans.)**

(*vii*) Rotor frequency at slip of 10%, $f_r = s' \times f = \frac{10}{100} \times 50 = \mathbf{5\ Hz}$. **(Ans.)**

Example 8. *In a 6-pole, 3-phase, 50 Hz motor with star-connected rotor, the rotor resistance per phase is 0.3 Ω, the reactance at standstill is 1.5 Ω per phase, and an e.m.f. between the slip rings on open circuit is 175 V. Find :*

(*i*) *Slip at a speed of 950 r.p.m. ;*
(*ii*) *Rotor e.m.f. per phase ;*
(*iii*) *Rotor frequency and reactance at a speed of 950 r.p.m.* **(B.T.E. Pb., 1997)**

Solution. *Given* : $p = 6$; $f = 50$ Hz ; $R_2 = 0.3\ \Omega$, $X_2 = 1.5\ \Omega$; $N = 950$ r.p.m.

Synchronous speed, $N_s = \frac{120f}{p} = \frac{120 \times 50}{6} = 1000$ r.p.m.

(*i*) **Slip, s :**

$$\mathbf{s} = \frac{N_s - N}{N_s} = \frac{1000 - 950}{1000} = 0.05 \text{ or } \mathbf{5\%}. \text{ (Ans.)}$$

(*ii*) **Rotor e.m.f. per phase, E_2 :**

Rotor e.m.f. per phase at standstill $= \frac{E_2}{\sqrt{3}} = \frac{175}{\sqrt{3}} = 101$ V/phase

Rotor e.m.f. at 5% slip $= sE_2 = 0.05 \times 101 = \mathbf{5.05\ V/phase}$. **(Ans.)**

(*iii*) **Rotor frequency and reactance at a speed of 950 r.p.m. :**

Rotor frequency, $\mathbf{f_r} = s \times f = 0.05 \times 50 = \mathbf{2.5\ Hz}$. **(Ans.)**

Rotor reactance $= sX_2 = 0.05 \times 1.5 = \mathbf{0.075\ \Omega/phase}$. **(Ans.)**

Example 9. *A 50 Hz, 440 V, 3-phase, 4-pole induction motor develops half the rated torque at 1490 r.p.m. With the applied voltage magnitude remaining at the rated value, what should be its frequency if the motor has to develop the same torque at 1600 r.p.m. ? Neglect stator and rotor winding resistances, leakage reactances and iron losses.* **(GATE, 1995)**

Solution. *Given* : $f = 50$ Hz ; $p = 4$; $N = 1490$ r.p.m. ; New speed, $N_n = 1600$ r.p.m.

Synchronous speed, $N_s = \frac{120f}{p} = \frac{120 \times 50}{4} = 1500$ r.p.m.

Slip at a speed of 1490 r.p.m., $s = \frac{N_s - N}{N_s} = \frac{1500 - 1490}{1500} = 0.00667$

Since torque developed by an induction motor, $T \propto sV^2$ (Refer Art. 11)

Slip s for constant torque and constant applied voltage remains unchanged.

$\therefore$ New synchronous speed, $N_{sn} = \frac{N_n}{1-s} = \frac{1600}{1-0.00667} = 1610.7$ r.p.m.

New frequency, $f_n = \frac{p \times N_{sn}}{120} = \frac{4 \times 1610.7}{120} = \mathbf{3.57\ Hz.}$ **(Ans.)**

☞**Example 10.** *A 3-phase slip ring induction motor gives a reading of 60 V across slip rings on open circuit when at rest with normal stator voltage applied. The motor is star-connected and has impedance of (0.6 + j6) Ω per phase. Find the rotor current when the machine is : (i) at standstill with slip rings joined to a star-connected starter with a phase impedance of (5 + j4) Ω ; and (ii) running normally with a 4 per cent slip.*

Solution. Impedance of motor per phase = $(0.6 + j6)\ \Omega$

Phase impedance of star-connected starter = $(5 + j4)\ \Omega$

(*i*) **At standstill :**

Induced e.m.f. in the rotor winding per phase,

$$E_2 = \frac{60}{\sqrt{3}} = 34.64 \text{ V}$$

Resistance per phase in the rotor circuit,

$$R_2 = 0.6 + 5 = 5.6\ \Omega$$

Reactance per phase in the rotor circuit,

$$X_2 = 6 + 4 = 10\ \Omega$$

Impedance per phase in rotor circuit,

$$Z_2 = \sqrt{R_2^2 + X_2^2} = \sqrt{(5.6)^2 + (10)^2} = 11.46\ \Omega$$

Rotor current per phase, $I_2 = \frac{E_2}{Z_2} = \frac{34.64}{11.46} = \mathbf{3.02\ A.}$ **(Ans.)**

Power factor, $\cos\phi_2 = \frac{R_2}{Z_2} = \frac{5.6}{11.46} = \mathbf{0.488}$ **(lag). (Ans.)**

(*ii*) **Running at 4 per cent slip :**

Induced e.m.f. on the rotor winding per phase

$$= sE_2 = 0.04 \times 34.64 = 1.38 \text{ V}$$

Rotor impedance per phase,

$$Z_2 = \sqrt{R_2^2 + (sX_2)^2} = \sqrt{(0.6)^2 + (0.04 \times 6)^2}$$

$= 0.646\ \Omega$ [Here $R_2 = 0.6\ \Omega$]

Rotor current per phase,

$$I_2 = \frac{sE_2}{Z_2} = \frac{1.38}{0.646} = \mathbf{2.136\ A.}\ \textbf{(Ans.)}$$

Power factor, $\cos \phi_2 = \dfrac{R_2}{Z_2} = \dfrac{0.6}{0.646} = \mathbf{0.928\ (lag).}$ **(Ans.)**

☞**Example 11.** *A 3-phase induction motor has a star-connected rotor. The rotor e.m.f. (between slip rings) at standstill is 50 V. The rotor resistance and standstill reactance are 0.5 Ω and 3 Ω respectively. Find :*

(i) Rotor current per phase at starting and the slip rings short-circuited.

(ii) Rotor current per phase at starting if a star connected rheostat of resistance 6 Ω per phase is connected across the slip rings.

(iii) Full-load rotor current and rotor power factor if the full-load slip is 4 per cent.

(iv) Rotor e.m.f. per phase under full-load condition.

Solution. Rotor e.m.f. at standstill (between slip rings)

$$= 50 \text{ V}$$

∴ Rotor e.m.f./phase at standstill,

$$E_2 = \frac{50}{\sqrt{3}} = 28.87 \text{ V}$$

Rotor resistance/phase, $R_2 = 0.5\ \Omega$

Rotor reactance/phase, $X_2 = 3\ \Omega$

(*i*) At starting $s = 1$

We know that rotor current/phase (I_2) is given by,

$$I_2 = \frac{sE_2}{\sqrt{R_2^2 + (sX_2)^2}} \quad ...(i)$$

∴ **Rotor current/phase at starting**

$$= \frac{E_2}{\sqrt{R_2^2 + X_2^2}} = \frac{28.87}{\sqrt{(0.5)^2 + (3)^2}} = \mathbf{9.49\ A.}\ \textbf{(Ans.)}$$

(*ii*) Total resistance in the rotor circuit

= rheostat resistance + rotor resistance = 6 + 0.5 = 6.5 Ω

∴ **Rotor current/phase with rheostat resistance,**

$$I_2 = \frac{28.87}{\sqrt{(6.5)^2 + (3)^2}} = \mathbf{4.03\ A.}\ \textbf{(Ans.)}$$

(*iii*) Full-load slip, $s_f = 4\% = 0.04$

Full-load rotor current (as per eqn. (*i*)),

$$I_2 = \frac{0.04 \times 28.87}{\sqrt{(0.5)^2 + (0.04 \times 3)^2}} = \mathbf{2.246\ A.}\ \textbf{(Ans.)}$$

Rotor power factor

$$= \frac{R_2}{\sqrt{R_2^2 + (s_f X_2)^2}} = \frac{0.5}{\sqrt{(0.5)^2 + (0.04 \times 3)^2}} = \mathbf{0.972.}\ \textbf{(Ans.)}$$

(*iv*) **Rotor e.m.f./phase under full-load condition**

$$= sE_2 = 0.04 \times 28.87 = \mathbf{1.155\ V.}\ \textbf{(Ans.)}$$

17. POWER STAGES IN AN INDUCTION MOTOR

- **Stator iron loss** (consisting of eddy and hysteresis losses) depends on the supply frequency and the flux density in the iron core. It is practically *constant.*

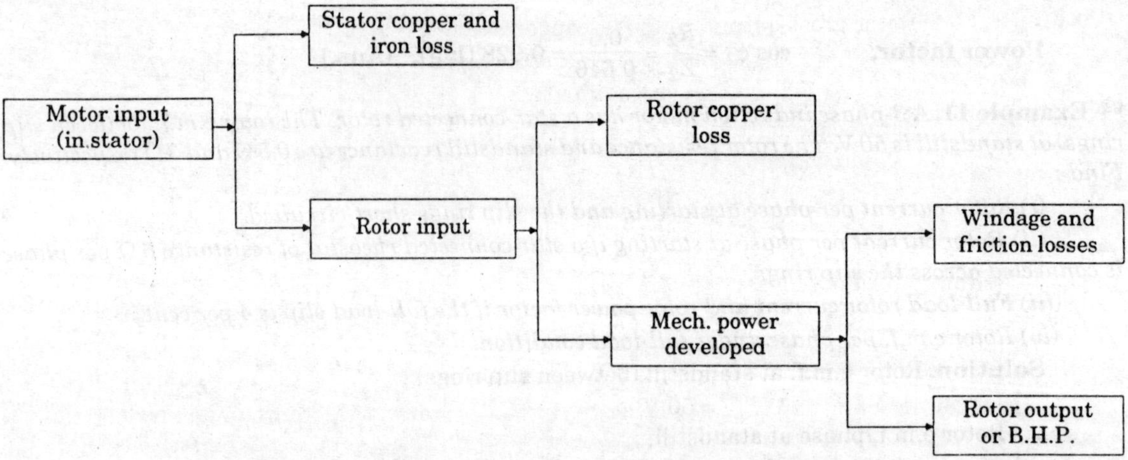

- Rotor iron loss is negligible because the frequency of rotor currents under normal running conditions is always small.
- Total rotor copper loss = $3I_2^2R_2$ (where I_2 and R_2 are current and resistance of the rotor respectively).

17.1. Rotor Output

Stator input = stator output + losses

Rotor input = stator output

[stator output is transferred entirely inductively to the rotor circuit]

Rotor gross output = rotor input – rotor copper losses.

This *rotor output is converted into mechanical* energy and gives rise to gross torque T_g. Out of this gross torque developed, some is lost due to windage and friction losses in the rotor and the rest appears as the *useful* or *shaft torque* T_{sh}. If T_g is in N-m and N is the actual speed of the rotor in r.p.m.

$$T_g = \frac{\text{rotor gross output in watts}}{(2\pi N/60)} \quad ...(19)$$

If there were no copper losses in the rotor, then rotor output will equal rotor input and the rotor will run at synchronous speed,

$$T_g = \frac{\text{rotor input}}{(2\pi N_s/60)} \quad ...(20)$$

From eqn. (19) and eqn. (20), we get

$$\text{Rotor output} = T_g \times 2\pi N/60$$

$$\text{Rotor input} = T_g \times 2\pi N_s/60 \quad ...(21)$$

But rotor input = rotor output + rotor copper losses

$$\therefore \quad \text{Rotor copper losses} = \text{rotor input} - \text{rotor output}$$

$$= T_g \times 2\pi/60\ (N_s - N) \quad ...(22)$$

From eqn. (21) and eqn. (22), we get

$$\frac{\text{Rotor copper loss}}{\text{Rotor input}} = \frac{N_s - N}{N_s} = s$$

$$\therefore \quad \text{Rotor copper loss} = s \times \text{rotor input}$$

$$= s \times \text{power across air-gap} \quad ...(23)$$

$$\text{Rotor gross output} = \text{input} - \text{copper loss}$$
$$= \text{input} - s \times \text{rotor input}$$
$$= (1 - s)\ \text{input} \qquad ...(24)$$

$$\therefore \quad \frac{\text{Rotor gross output}}{\text{Rotor input}} = 1 - s = 1 - \frac{N_s - N}{N_s}$$
$$= \frac{N}{N_s} = \frac{\text{actual speed of rotor}}{\text{synchronous speed}}$$

$$\therefore \quad \text{Rotor efficiency} = \frac{N}{N_s}$$

Also, $$\frac{\text{rotor copper loss}}{\text{rotor gross output}} = \frac{s}{1 - s}. \qquad ...(25)$$

17.2. Synchronous Watt. A synchronous watt is a new unit of torque and is defined as the *torque which develops a power of 1 watt at the synchronous speed of motor*. When we say that motor is developing a torque of 500 synchronous watts, we mean that the rotor input is 500 watts and that the torque is such that power developed would be 500 watts provided the rotor is running synchronously and developing the same torque.

$$\text{Rotor input} = T_{sw} \times 2\pi N_s$$

or $$T_{sw} = \frac{\text{rotor input in watts}}{2\pi N_s}$$

Synchronous wattage of an induction motor equals the power transferred across the air-gap to the rotor.

17.3. Factors Determining Torque. When a 3-phase induction motor is operating on a certain load with slip s, then

Induced e.m.f. per phase in rotor = sE_2 volts (where E_2 = standstill rotor induced e.m.f.)

Rotor impedance per phase, $Z_2 = R_2 + jsX_2$

(where R_2 and X_2 are rotor resistance and standstill reactance per phase)

$$\text{Rotor current per phase} = \frac{sE_2}{Z_2} = \frac{sE_2}{\sqrt{R_2^2 + s^2X_2^2}}$$

$$\text{Rotor circuit power factor, } \cos\phi_2 = \frac{R_2}{\sqrt{R^2 + s^2X_2^2}}$$

Total rotor copper loss = number of phases × rotor induced e.m.f. per phase × rotor current per phase × rotor current power factor

$$= 3 \times sE_2 \times \frac{sE_2}{\sqrt{R_2^2 + s^2X_2^2}} \times \frac{R_2}{\sqrt{R_2^2 + s^2X_2^2}} = \frac{3R_2\, s^2E_2^2}{R_2^2 + s^2X_2^2} \qquad ...(26)$$

$$\text{Also, total rotor copper loss} = s \times \text{rotor input} = \frac{s \times T \times 2\pi N_s}{60}$$

$$\therefore \quad s \times \frac{T \times 2\pi N_s}{60} = \frac{3R_2\, s^2E_2^2}{R_2^2 + s^2X_2^2}$$

or $$\text{Torque developed by the rotor, } T = \frac{3}{(2\pi N_s/60)} \cdot \frac{R_2 sE_2^2}{R_2^2 + s^2E_2^2} \qquad ...(27)$$

$$= \frac{3}{\omega_s} \cdot \frac{R_2 sE_2^2}{R_2^2 + s_2^2 E_2^2} \qquad \text{...[27 (a)]}$$

or $$T \propto \frac{R_2 sE_2^2}{R_2^2 + s^2 X_2^2} \propto \frac{R_2 sV_1^2}{R_2^2 + s^2 X_2^2} \,. \; (\because \; E_2 \propto V_1, \text{ the supply voltage})$$

ROTOR OUTPUT, LOSSES AND EFFICIENCY

Example 12. *The power input to a 3-phase induction motor is 50 kW. The stator-losses total 1.2 kW. If the motor is running with a slip of 3 per cent, find the total mechanical power developed.*

Solution. Stator power input = 50 kW

Stator losses = 1.2 kW

Slip = 3% (= 0.03)

Mechanical power developed :

Rotor input = stator output = stator input – stator losses

= 50 – 1.2 = 48.8 kW

Now *mechanical power developed*

= (1 – s) rotor input = (1 – 0.03) × 48.8 = **47.34 kW. (Ans.)**

☞**Example 13.** *A 6-pole, 50-Hz, 3-phase induction motor running on full-load with 3% slip develops a torque of 160 N-m at its pulley rim. The friction and windage losses are 210 W and the stator copper and iron losses equal 1640 W. Calculate :*

(i) Rotor output, *(ii) Rotor copper loss, and*

(iii) Efficiency at full-load.

Solution. Number of poles, $p = 6$

Supply frequency, $f = 50$ Hz

Slip, $s = 3\%$ or 0.03

Torque developed at full-load, $T_f = 160$ N-m

Friction and windage losses = 210 W

Stator copper and iron losses = 1640 W

Synchronous speed, $N_s = \frac{120f}{p} = \frac{120 \times 50}{6} = 1000$ r.p.m.

Actual rotor speed, $N = (1 - s) N_s = (1 - 0.03) \times 1000 = 970$ r.p.m.

(*i*) **Rotor output**

$$= \frac{2\pi N T_f}{60} = \frac{2\pi \times 970 \times 160}{60} = \textbf{16252.5 W. (Ans.)}$$

(*ii*) **Rotor gross output**

= rotor output + friction and windage losses

= 16252.5 + 210 = 16462.5 W

Also, $$\frac{\text{rotor gross output}}{\text{rotor input}} = \frac{N}{N_s}$$

$\therefore$ $$\text{Power input} = \frac{\text{rotor gross output} \times N_s}{N}$$

$$= \frac{16462.5 \times 1000}{970} = 16971.6 \text{ W}$$

$\therefore$ **Rotor copper loss** = 16971.6 – 16462.5 = **509.1 W. (Ans.)**

(*iii*) **Stator output** = rotor input = 16971.6 W

$\therefore$ Stator input = stator output + stator copper and iron losses

= 16971.6 + 1640 = 18611.6 W

% **Overall efficiency at full-load** $= \dfrac{\text{rotor output}}{\text{stator input}}$

$= \dfrac{16252.5}{18611.6} \times 100 =$ **87.32%. (Ans.)**

Example 14. *The power input to the rotor of a 440 V, 50-Hz, 3-phase, 6-pole induction motor is 50 kW. It is observed that the rotor e.m.f. makes 120 complete cycles per minute. Calculate :*

(*i*) *Slip,* (*ii*) *Rotor speed,*

(*iii*) *Rotor copper loss/phase,* (*iv*) *Mechanical power developed, and*

(*v*) *Rotor resistance/phase if rotor current is 50 A.*

Solution. Supply voltage = 440 V

Supply frequency, f = 50-Hz

Number of poles, $p = 6$

Rotor input = 50 kW or 50000 W

Rotor frequency, $f_r = \dfrac{120}{60} = 2$ Hz

(*i*) **Slip,** $s = \dfrac{f_r}{f} = \dfrac{2}{50} =$ **0.04 or 4%. (Ans.)**

(*ii*) **Rotor speed,** $N = (1 - s)N_s = (1 - s) \times \dfrac{120f}{p}$

$= (1 - 0.04) \times \dfrac{120 \times 50}{6} =$ **960 r.p.m. (Ans.)**

(*iii*) Rotor copper loss $= s \times$ rotor input $= 0.04 \times 50000 = 2000$ W

Rotor copper loss/phase $= \dfrac{2000}{3} =$ **666.67 W. (Ans.)**

(*iv*) **Mechanical power developed**

$= (1 - s) \times$ rotor input $= (1 - 0.04) \times 50000$

= **48000 W or 48 kW. (Ans.)**

[or = rotor input – total copper loss]

(*v*) Let R_2 be the rotor resistance/phase

$\therefore$ $3I_2^2R_2 = 2000$

$$R_2 = \frac{2000}{3 \times 50 \times 50} = 0.267\ \Omega$$

Hence, **rotor resistance/phase = 0.267 Ω. (Ans.)**

Example 15. *A 50 H.P., 6-pole, 50 Hz, slip-ring induction motor runs at 960 r.p.m. on full-load with a rotor current of 40 A. Allowing 300 W for copper loss in the short-circuiting gear and 1200 W for mechanical losses, find the resistance R_2 per phase of the 3-phase rotor winding.*

(GATE, 1998)

Solution. P_{out} = 50 H.P. = 50 × 735.5 = 36775 W ; $p = 6$; $f = 50$ Hz ; $N = 960$ r.p.m. ; $I_2 = 40$ A ; copper loss in the short-circuiting gear = 300 W ; mechanical loss = 1200 W.

Resistance, R_2 :

Rotor output = motor output + short-circuiting gear loss + mechanical losses

= 36775 + 300 + 1200 = 38275 W

Synchronous speed, $N_s = \dfrac{120f}{p} = \dfrac{120 \times 50}{60} = 1000$ r.p.m.

Slip, $s = \dfrac{N_s - N}{N_s} = \dfrac{1000 - 960}{1000} = 0.04$

Rotor copper losses $= 3I_2^2R_2 = \dfrac{\text{Rotor output}}{1 - s} \times s = \dfrac{38275}{(1 - 0.04)} \times 0.04 = 1594.8$ W

or Rotor resistance, $\mathbf{R_2} = \dfrac{1594.8}{3 \times 40^2} = \mathbf{0.332\ \Omega}$. **(Ans.)**

18. INDUCTION MOTOR AS TRANSFORMER

An induction motor is essentially a transformer with stator forming the primary and rotor forming (the short-circuited) rotating secondary (Fig. 20). This is so because the transfer of energy

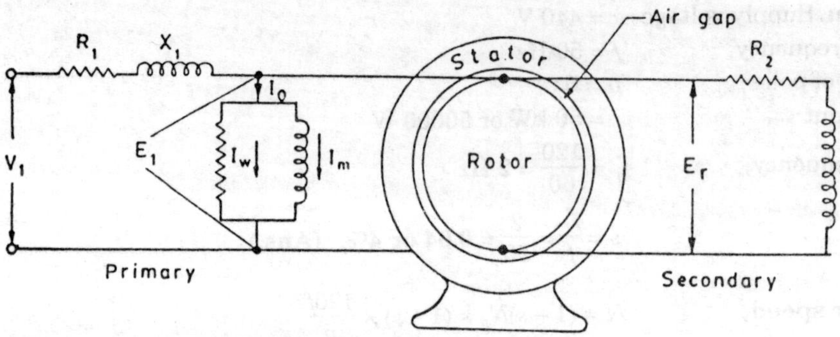

Fig. 20. Induction motor as transformer.

from stator to the rotor of an induction motor takes place entirely *inductively* with the help of flux mutually linking the two.

The vector diagram (Fig. 21) is similar to that of a transformer.

$$V_1 = E_1 + I_1 (R_1 + jX_1)$$

and

$$E_r = I_2 (R_2 + jsX_2)$$

However, the *important differences* between a transformer and an induction motor are :

1. In induction motor the magnetic leakage and leakage reactance of rotor and stator are *higher* than in a transformer.

2. The magnetic circuit of an induction motor has an air gap and this makes the per unit value of magnetising current *much higher* than that of a transformer.

3. Because of the distributed windings in an induction motor the *ratio of stator and rotor currents is not equal* to the ratio of turns per phase in the rotor and stator windings.

4. *The losses in an induction motor are higher and, therefore, the efficiency is lower than in a transformer.*

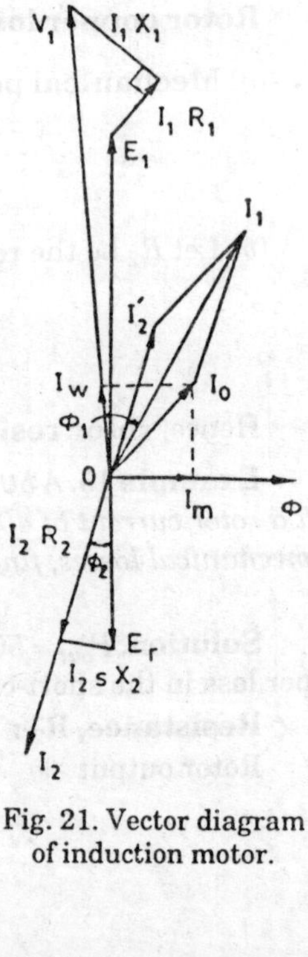

Fig. 21. Vector diagram of induction motor.

18.1. Rotor output. Primary current I_1 consists of two parts I_0 and I_2'. It is the latter which is transferred to the rotor, because I_0 is used in meeting the copper and iron losses in the stator itself. Out of the primary voltage V_1, some is absorbed in the primary itself ($= I_1Z_1$) and the remaining E_1 is transferred to the rotor. If the angle between E_2 and I_2' is ϕ, then

Rotor input/phase $= E_1I_1' \cos\phi$

Total rotor input $= 3E_1I_2' \cos\phi$

The electrical input to the rotor which is wasted in the form of the heat

$$= 3I_2E_r \cos\phi \text{ (or } 3I_2^2R_2)$$

Now
$$I_2' = KI_2 \quad \text{or} \quad I_2 = \frac{I_2'}{K}$$

$$E_r = sE_2 \text{ and } E_2 = KE_1$$

$\therefore$
$$E_r = sKE_1$$

$\therefore$ Electrical input wasted as heat

$$= 3 \times (I_2'/K) \times sKE_1 \cos\phi)$$
$$= 3E_1I_2' \cos\phi \times s = \text{rotor input} \times s$$

Now,
$$\text{rotor output} = \text{rotor input} - \text{losses}$$
$$= 3E_1I_2' \cos\phi - 3E_1I_2' \cos\phi \times s$$
$$= (1-s)\, E_1I_2' \cos\phi = (1-s) \text{ rotor input}$$

$\therefore$
$$\frac{\text{Rotor output}}{\text{Rotor input}} = 1 - s$$

$\therefore$
$$\text{Rotor copper loss} = s \times \text{rotor input}$$

Rotor efficiency
$$= 1 - s = \frac{N}{N_s} = \frac{\text{actual speed}}{\text{synchronous speed}}.$$

18.2. Equivalent Circuit of the Rotor. The rotor current I_2, when motor is loaded, is given by

$$I_2 = \frac{sE_2}{\sqrt{R_2^2 + (sX_2)^2}} = \frac{E_2}{\sqrt{(R_2/s)^2 + X_2^2}}$$

From the above relation it appears that the rotor circuit which actually consists of a fixed resistance R_2 and a variable reactance sX_2 (proportional to slip) connected across $E_r = sE_2$ [Fig. 22 (*i*)] can be looked upon as equivalent to a rotor circuit having a fixed reactance X_2 connected in series with a variable resistance R_2/s (inversely proportional to slip) and supplied with constant voltage E_2 as shown in [Fig. 22 (*ii*)].

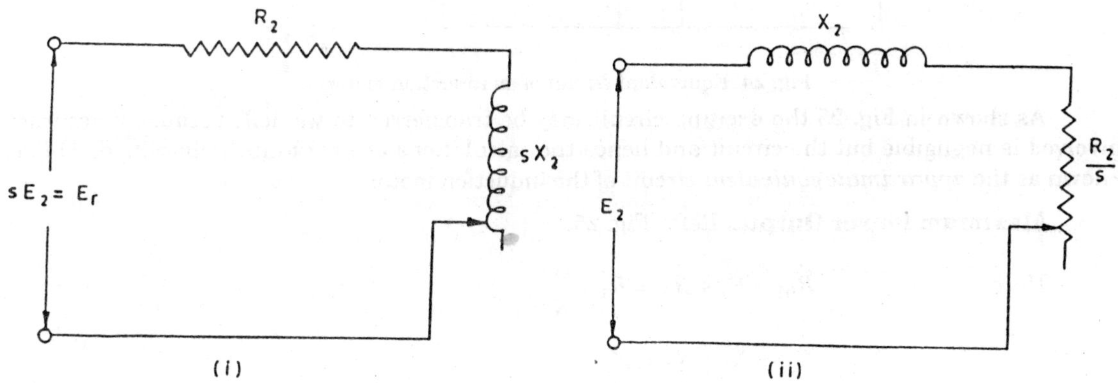

Fig. 22. Equivalent circuit of a rotor.

Also, the resistance R_2/s can be written as,

$$\frac{R_2}{s} = R_2 + R_2\left(\frac{1}{s} - 1\right).$$

It consists of two parts : (*i*) The part R_2 is the *rotor resistance* itself and *represents the rotor copper loss* (*ii*) the second part is $R_2\left(\frac{1}{s} - 1\right)$. This is known as *load resistance* R_L and is *the electrical equivalent of the mechanical load on the motor*. In other words the mechanical load on an induction motor can be represented by a *non-inductive resistance of the value* $R_2\left(\frac{1}{s} - 1\right)$.

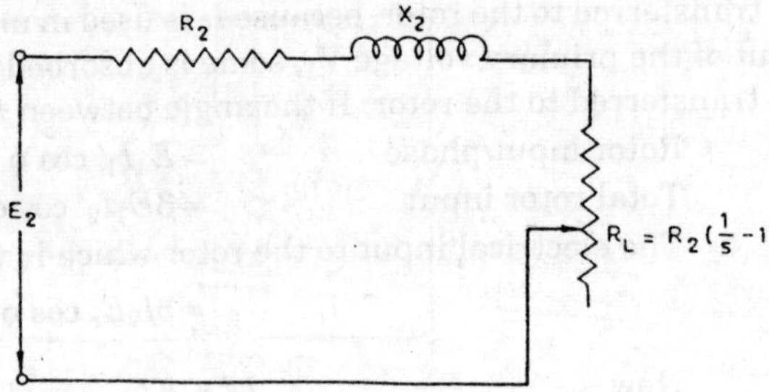

Fig. 23. Equivalent circuit of a rotor with load resistance R_L.

In Fig. 23 is shown the equivalent rotor circuit along with the load resistance R_L.

19. EQUIVALENT CIRCUIT OF AN INDUCTION MOTOR

The equivalent circuit for a polyphase induction motor is shown in Fig. 24, where

V_1 = applied voltage per phase
R_1 = stator resistance/phase
R_2 = rotor resistance/phase
X_1 = stator leakage reactance/phase
X_2 = rotor standstill leakage reactance/phase
K = turn-ratio of secondary to primary
R_0 = no-load resistance/phase
X_0 = no-load reactance/phase
} Rotor being driven at synchronous speed.

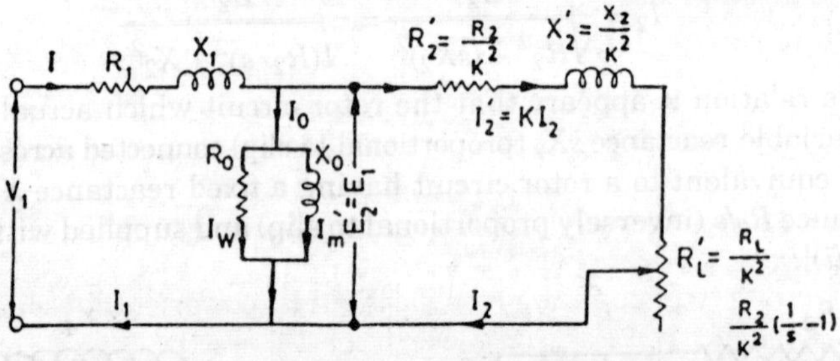

Fig. 24. Equivalent circuit of an induction motor.

As shown in Fig. 25 the exciting circuit may be transferred to the left, because inaccuracy involved is negligible but the circuit and hence the calculations are very much simplified. This is known as the *approximate equivalent* circuit of the induction motor.

Maximum Power Output. Refer Fig. 25.

Here $$R_{01} = R_1 + R_2' = R_1 + \frac{R_2}{K^2}$$

$$X_{01} = X_1 + X_2' = X_1 + \frac{X_2}{K^2}$$

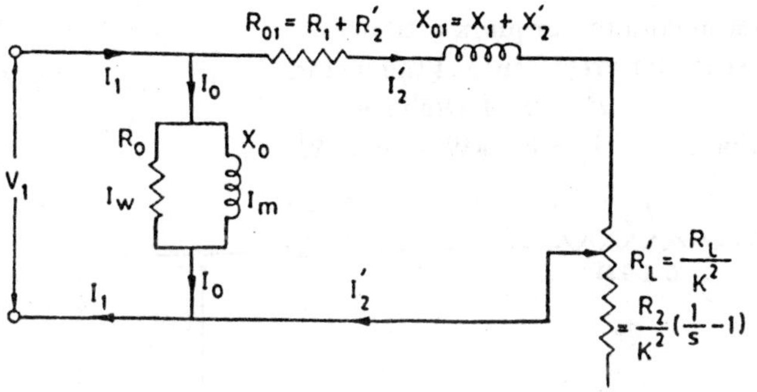

Fig. 25. Approximate equivalent circuit of an induction motor.

Load resistance, $$R_L' = R_L/K^2 = \frac{R_2}{K^2}\left(\frac{1}{s} - 1\right)$$

Gross mechanical power output,

$$P_g = 3I_2'^2 R_L'$$

But $$I_2' = \frac{V_1}{\sqrt{(R_{01} + R_L')^2 + (X_{01})^2}}$$

$$\therefore \quad P_g = \frac{3V_1^2 R_L'}{(R_{01} + R_L')^2 + (X_{01})^2} \quad ...(28)$$

The condition for maximum power output can be obtained by differentiating the above equation and by equating, the first derivative to zero. If it is done, it will be found that

$$R_L' = \sqrt{(R_{01})^2 + (X_{01})^2} = Z_{01}$$

Hence the gross mechanical power output of the motor is maximum when the equivalent *load resistance R_L' is equal to the standstill leakage impedance of the motor Z_{01}.*

Putting $R_L' = Z_{01}$ in eqn. (28), we get

$$P_{g\,(max)} = \frac{3V_1^2 Z_{01}}{(R_{01} + Z_{01})^2 + (X_{01})^2} = \frac{3V_1^2 Z_{01}}{R_{01}^2 + X_{01}^2 + Z_{01}^2 + 2R_{01}Z_{01}}$$

$$= \frac{3V_1^2}{2(R_{01} + Z_{01})} \quad ...(29)$$

The slip corresponding to maximum gross mechanical power output will be given as

$$Z_{01} = R_L' = \frac{R_2}{K^2}\left(\frac{1}{s} - 1\right)$$

or $$s = \frac{R_2/K^2}{R_2/K^2 + Z_{01}} \quad ...(30)$$

EQUIVALENT CIRCUIT OF AN INDUCTION MOTOR

Example 16. *In a 3-phase induction motor maximum torque occurs at a slip of 10 per cent. The equivalent secondary resistance of the motor is 0.09 Ω/phase. If the gross power output is 8.5 kW. Calculate :*

(i) Equivalent load resistance, *(ii) Equivalent load voltage, and*
(iii) Current at this slip.

Solution. Slip at maximum torque, s = 10 per cent

The equivalent secondary resistance of the motor,

$$R_2 = 0.09\ \Omega/\text{phase}$$

Gross power output, P_g = 8.5 kW or 8500 W

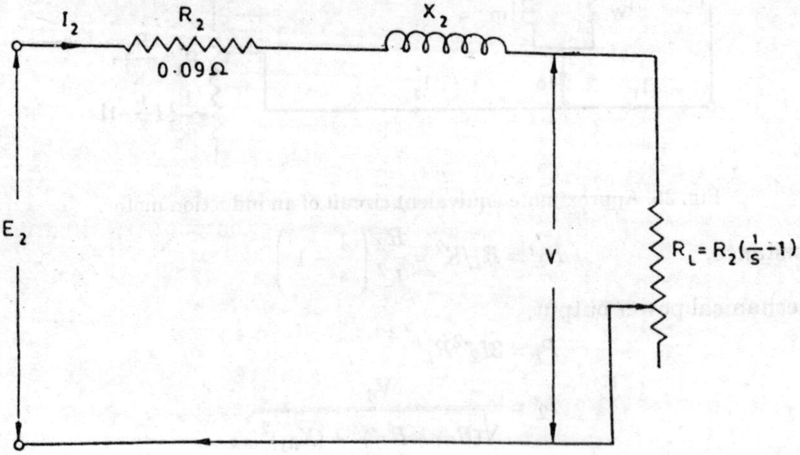

Fig. 26

(*i*) **Equivalent load resistance, R_L :**

We know that $R_L = R_2\left[\left(\frac{1}{s}\right) - 1\right] = 0.09\left(\frac{1}{0.1} - 1\right) =$ **0.81 Ω. (Ans.)**

(*ii*) **Equivalent load voltage, V :**

As shown in the equivalent circuit of the rotor in Fig. 30, V is a fictitious voltage drop equivalent to that consumed in the load connected to the secondary *i.e.*, rotor, $V = I_2 R_L$

But, gross power, $P_g = \frac{3V^2}{R_L}$

∴ $V = \sqrt{\frac{P_g \times R_L}{3}} = \sqrt{\frac{8500 \times 0.81}{3}} = 47.9$ V

Hence, **equivalent load voltage = 47.9 V. (Ans.)**

(*ii*) **Equivalent load current :**

Equivalent load current $= \frac{V}{R_L} = \frac{47.9}{0.81} =$ **59.1 A. (Ans.)**

Example 17. *The stator impedance and equivalent rotor resistance of a 400 V, 3-phase star-connected induction motor are (0.05 + j0.2) Ω and (0.05 + j0.24) Ω respectively. Neglecting exciting current determine the maximum gross power and the slip at which it occurs.*

Solution. Supply voltage V_1 = 400 V

Stator impedance = (0.05 + *j*0.2) Ω

Equivalent rotor resistance = (0.05 + *j*0.24) Ω

The equivalent circuit is shown in Fig. 27.

$$R_{01} = R_1 + R_2' = 0.05 + 0.05 = 0.1\ \Omega$$

$$X_{01} = X_1 + X_2' = 0.2 + 0.24 = 0.44\ \Omega$$

∴ $$Z_{01} = \sqrt{R_{01}^2 + X_{01}^2} = \sqrt{(0.1)^2 + (0.44)^2} = 0.45\ \Omega$$

Slip = ?

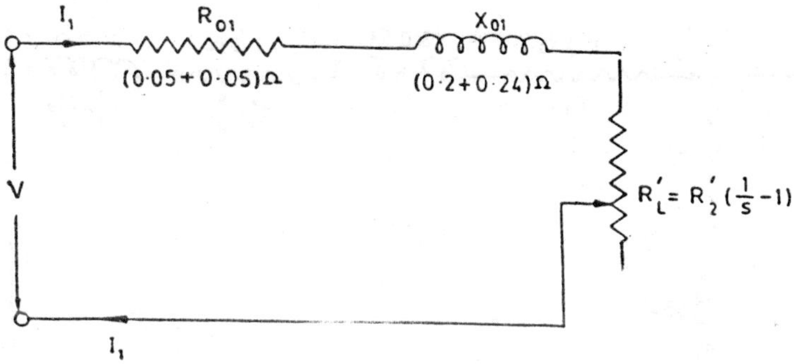

Fig. 27

We know that slip corresponding to maximum gross power output (Art. 19) is given by

$$s = \frac{\frac{R_2}{K^2}}{\frac{R_2}{K^2} + Z_{01}} = \frac{R_2'}{R_2' + Z_{01}} = \frac{0.05}{0.05 + 0.45} = 0.1 \text{ or } 10\%$$

Hence, **slip at which maximum gross power output occurs**

$= 10\%$. **(Ans.)**

Maximum gross power output

$$= \frac{3V_1^2}{2(R_{01} + Z_{01})} = \frac{3\left(\frac{400}{\sqrt{3}}\right)^2}{2(0.1 + 0.45)}$$

$= \mathbf{145454\ W}$ or $\mathbf{145.454\ kW}$. **(Ans.)**

☞**Example 18.** *The following data pertain to an induction motor : stator impedance = (1.2 + j3.0) Ω ; rotor standstill impedance = (1.2 + j2.0) Ω ; no-load shunt impedance = (12 + j55) Ω ; volts/phase = 240. Determine :*

(*i*) *Stator current,* (*ii*) *Equivalent rotor current,*
(*iii*) *Input power factor,* (*iv*) *Mechanical power developed, and*
(*v*) *Efficiency of the motor at a slip of 5%.*

Solution. Stator impedance $= (1.2 + j3.0)\ \Omega$
Rotor standstill impedance $= (1.2 + j2.0)\ \Omega$
No-load shunt impedance $= (12 + j55)\ \Omega$
Volts/phase $= 240$ V

The equivalent circuit per phase of motor referred to stator is shown in Fig. 32. It is *assumed* that the given impedance figures are *phase values* and there is *unity turn-ratio* between stator and rotor.

Equivalent load resistance,

$$R_L' = R_2'\left(\frac{1}{s} - 1\right) = \frac{R_2}{K^2}\left(\frac{1}{s} - 1\right) = 1.2\left(\frac{1}{0.05} - 1\right) = 22.8\ \Omega$$

Effective impedance/phase $= (R_1 + R_2' + R_L') + j(X_1 + X_2') = (1.2 + 1.2 + 22.8) + j(3 + 2)$
$= (25.2 + j5)\ \Omega$

(*i*) **Stator current, I_1 :**

Stator load current, $I_2' = \dfrac{240}{25.2 + j5} = \dfrac{240(25.2 - j5)}{(25.2 + j5)(25.2 - j5)}$

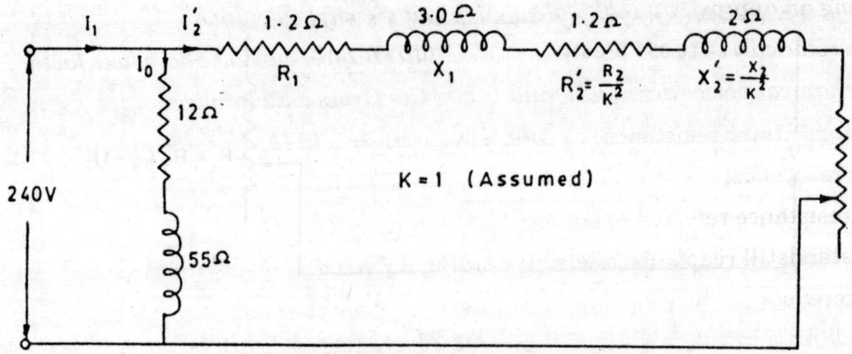

Fig. 28

$$= \frac{240(25.2 - j5)}{660} = (9.16 - j1.82) \text{ A}$$

Stator no-load current, $I_0 = \frac{240}{(12 + j55)} = \frac{240}{12 + j55} \times \frac{(12 - j55)}{(12 - j55)}$

$$= \frac{240(12 - j55)}{3169} = (0.91 - j4.16) \text{ A}$$

Total stator current, $I_1 = I_2' + I_0 = (9.16 - j1.82) + (0.91 - j4.16)$

$$= (10.07 - j5.98) \text{ A}$$

$\therefore$ $I_1 = \sqrt{(10.07)^2 + (5.98)^2}$ **= 11.71 A. (Ans.)**

(*ii*) **Equivalent rotor current,**

$I_2' = \sqrt{(9.16)^2 + (1.82)^2}$ **= 9.34 A. (Ans.)**

(*iii*) **Input power factor,**

$\cos\phi = \frac{10.07}{11.71}$ **= 0.86. (Ans.)**

(*iv*) **Mechanical power developed,**

$$= 3(I_2')^2 R_L' = 3 \times (9.34)^2 \times 22.8$$

= 5967 W or 5.967 kW. (Ans.)

(*v*) **Efficiency of the motor,**

$$\eta = \frac{\text{output}}{\text{input}} \times 100 = \frac{5967}{3 \times 240 \times 11.71 \times 0.86} \times 100$$

= 82.29%. (Ans.)

☞**Example 19.** *The following data refer to a 10-pole, 400 V, 50-Hz, 3-phase induction rotor :*
R_1 = 1.75 Ω, X_1 = 5.5 Ω, R_2' (rotor resistance referred to stator) = 2.25 Ω.
X_2' (rotor standstill reactance referred to stator) = 6.6 Ω.

When the motor is tested on no-load, it is observed that it takes 3.8 A (line current) and the total core loss is 310 W.

By using an approximate equivalent circuit at 4% slip, calculate :

(i) The rotor current, *(ii) Supply current and power factor,*

(iii) Mechanical power developed, and *(iv) Gross load torque.*

Solution. Stator resistance, $R_1 = 1.75\ \Omega$

Stator reactance, $X_1 = 5.5\ \Omega$

Rotor resistance referred to stator, $R_2' = 2.25\ \Omega$

Rotor standstill reactance referred to stator, $X_2' = 6.6\ \Omega$

Total core loss $= 310$ W

No-load line current, $I_0 = 3.8$ A

If $\cos \phi_0$ is the power factor at no-load, then

$$\sqrt{3}\, VI \cos \phi_0 = 310$$

$$\therefore \quad \cos \phi_0 = \frac{310}{\sqrt{3} \times 400 \times 3.8} = 0.1178$$

$$\phi_0 = 83.23°$$

$$\therefore \quad \sin \phi_0 = 0.993$$

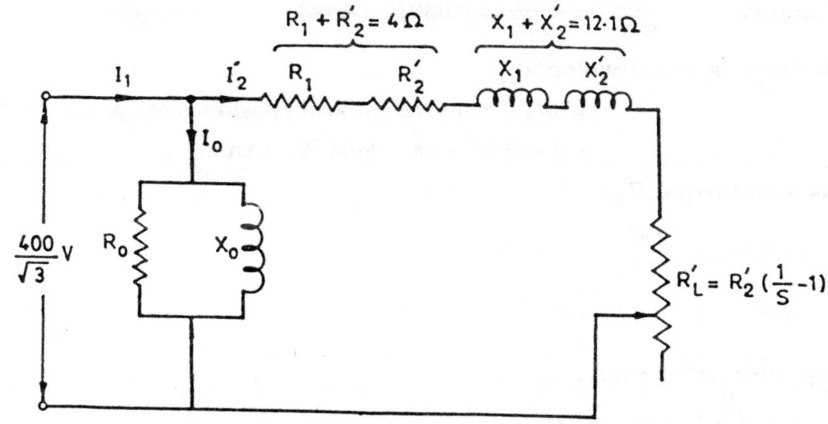

Fig. 29

So in the equivalent circuit,

$$R_0 = \frac{400}{\sqrt{3}\, I_0 \cos \phi_0} = \frac{400}{\sqrt{3} \times 3.8 \times 0.1178} = 515.9\ \Omega$$

$$X_0 = \frac{400}{\sqrt{3}\, I_0 \sin \phi_0} = \frac{400}{\sqrt{3} \times 3.8 \times 0.993} = 61.2\ \Omega$$

For a load slip of 4%

Equivalent load resistance,

$$R_L' = R_2'\left(\frac{1}{s} - 1\right) = 2.25\left(\frac{1}{0.04} - 1\right) = 54\ \Omega$$

$$R_1 + R_2' = 1.75 + 2.25 = 4\ \Omega$$

$$X_1 + X_2' = 5.5 + 6.6 = 12.1\ \Omega$$

Phase voltage $= \frac{400}{\sqrt{3}}$ V = 231 V

(*i*) **Rotor current referred to stator,**

$$I_2' = \frac{231 + j0}{(4 + 54) + j12.1} = \frac{231 + j0}{58 + j12.1} = \frac{231 + j0}{(58 + j12.1)} \times \left(\frac{58 - j12.1}{58 - j12.1}\right)$$

$$= \frac{13398 - j2795}{3364 + 146.4} = 3.82 - j0.976$$

$\therefore$ $$I_2' = \sqrt{(3.82)^2 + (0.796)^2} = \mathbf{3.9\ A.\ (Ans.)}$$

Also $$I_0 = 3.8\,(\cos\phi_0 - j\sin\phi_0)$$

$$= 3.8\,(0.1178 - j0.993) = 0.447 - j3.77$$

(*ii*) **Supply current and power factor :**

Supply (or motor) current,

$$I_1 = I_0 + I_2' = (0.447 - j3.77) + (3.82 - j0.796)$$

$$= 4.267 - j4.566$$

$\therefore$ $$I_1 = \sqrt{(4.267)^2 + (4.566)^2} = \mathbf{6.25\ A.\ (Ans.)}$$

Power factor, $$\cos\phi = \frac{4.267}{6.25} = \mathbf{0.6827.\ (Ans.)}$$

(*iii*) **Mechanical power developed**

$$= 3(I_2')^2 \times \text{equivalent load resistance per phase } (R_L')$$

$$= 3 \times (3.9)^2 \times 54 = \mathbf{2464\ W.\ (Ans.)}$$

(*iv*) **Gross load torque, T_g :**

Using the relation, $$\frac{2\pi N T_g}{60} = 2464$$

where $N = (1 - s) N_s = (1 - 0.04)\frac{120f}{p}$

$$= 0.96 \times \frac{120 \times 50}{10} = 576 \text{ r.p.m.}$$

$\therefore$ $$\frac{2\pi \times 576 \times T_g}{60} = 2464$$

or $$T_g = \frac{2464 \times 60}{2\pi \times 576} = 40.85 \text{ N-m}$$

Hence, **gross load torque = 40.85 N-m. (Ans.)**

20. STARTING OF INDUCTION MOTORS

Small induction motors (up to 2 kW) capacity may directly be switched on to the supply mains, but those of higher capacity must use some type of starting device, or starters as they are commonly called. *The function of these starters is to restrict the initial rush of current,* which, in the case of induction motors, is about 5 times the full-load current. This excessive current has two major upsetting effects, namely, a *large voltage drop in the distributing network* and *causing stoppage of machines which are already running on the supply mains.* Hence the Electrical Undertaking Authorities *forbid the users of large capacity induction motors to directly switch on their machines.*

The principle of all starting devices is to impress lower voltage on stator phases at the time of starting, or if the motor is slip ring or wound rotor, then to include external resistance in each rotor phase to keep the initial rotor current to a low value, this consequently means less current in the stator phases and therefore in the supply mains.

Direct-on-line starting of induction motors. This method means switching the motor directly on to the supply without using any device for reducing the starting current. The method is restricted to small motors up to about 2 kW. For these small motors, the *starting torque is about twice the full-load torque.* Hence the starting period lasts only a few seconds.

We know that,

Rotor input $= 2\pi N_s T = kT$

Also, rotor copper loss $= s \times$ rotor input

$$\therefore \quad 3I_2^2 R_2 = skT$$

$$\therefore \quad T \propto \frac{I_2^2}{s} \qquad \text{[if } R_2 \text{ is the same]}$$

Now $I_2 \propto I_1$

$$\therefore \quad T \propto \frac{I_1^2}{s} \quad \text{or} \quad T = \frac{KI_1^2}{s}$$

At starting $s = 1$

$$\therefore \quad T_{st} = KI_{st}^2 \quad \text{where } I_{st} = \text{starting current}$$

If I_f = normal full-load current

s_f = full-load slip

then
$$T_f = \frac{KI_f^2}{s_f}$$

$$\therefore \quad \frac{T_{st}}{T_f} = \left(\frac{I_{st}}{I_f}\right)^2 s_f \qquad \text{...[31 (a)]}$$

When the motor is direct-switched on to normal voltage, then starting current is the short-circuit current I_{sc}

$$\therefore \quad \frac{T_{st}}{T_f} = \left(\frac{I_{sc}}{I_f}\right)^2 s_f \qquad \text{...[31 (b)]}$$

Some of the starting devices for starting induction motors are discussed below :

Squirrel-Cage Motors :

(*i*) Stator rheostat starter. (*ii*) Auto-transformers (auto-starters).

(*iii*) Star-delta starter.

Slip Ring Motors :

(1) Rotor rheostat.

20.1. Squirrel-Cage Motors.

20.1.1. Stator rheostat starter. The connection diagram is shown in Fig. 30.

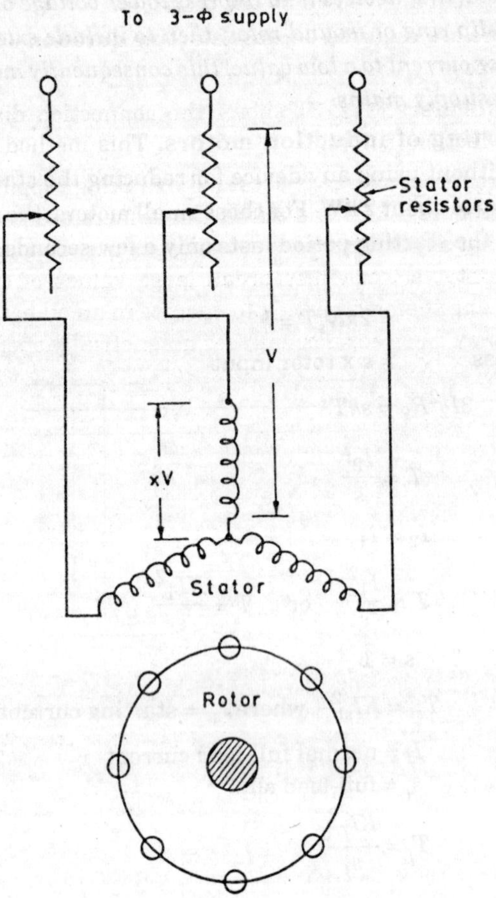

Fig. 30. Stator rheostat starter.

Reduced voltage is impressed on each stator phase due to the resistance R of the rheostat. Hence the initial current drawn from the supply mains will be *less* than if the machine were to be switched directly on to the supply mains.

If x = fraction of voltage (V) reduced by the stator resistors

Then $\qquad I_{st} = xI_{sc} \qquad$ and $\qquad T_{st} = x^2 T_{sc}$

$$\frac{T_{st}}{T_f} = \left(\frac{I_{st}}{I_f}\right)^2 s_f = \left(\frac{xI_{sc}}{I_f}\right)^2 s_f = x^2 \left(\frac{I_{sc}}{I_f}\right)^2 s_f \qquad ...(32)$$

- This method is suitable for starting of *small machines only*.

Advantages : 1. High power factor during start.

2. Smooth acceleration.

3. Less expensive than auto-transformer starter in lower output ratings.

4. Closed transition starting.

Disadvantages : 1. Heat is given off by the resistors.

2. Expensive resistors are required because starting duration usually exceeds 5 seconds.

3. Low torque efficiency.

20.1.2. Auto-transformers. Fig. 31 shows the connection diagram for auto-transformer starting of squirrel-cage induction motors.

In this method the reduced voltage is obtained by taking tappings at suitable points from a three-phase auto-transformer (Fig. 31). The auto-transformers are generally tapped at the 50, 60 and 80 per cent points, so that adjustment at these voltages may be made for proper starting torque requirements. Since the contacts frequently break large values of current arcing is sometimes quenched effectively by having them assembled to operate in an oil bath.

Auto-transformers may be either normally, or magnetically operated.

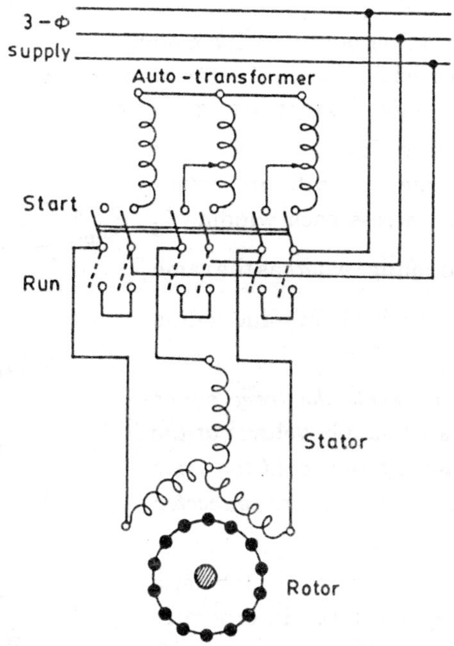

Fig. 31. Auto-transformer starter.

Relation between starting (T_{st}) and full-load (T_f) torques. Let the motor be started by an auto-transformer having transformation ratio K. If I_{sc} is the starting current when normal voltage is applied and applied voltage to stator winding at starting is KV then,

Motor input current, $I_{st} = KI_{sc}$

Supply current $=$ Primary current of auto-transformer

$= K^2 I_{sc}$

$$\frac{T_{st}}{T_f} = \left(\frac{I_{st}}{I_f}\right)^2 \times s_f = K^2 \left(\frac{I_{sc}}{I_f}\right)^2 \times s_f \quad ...(33)$$

It may be noted that this expression is similar to eqn. (32) except that x has been replaced by the transformation ratio K.

Advantages :

1. Voltage is reduced by transformation and not by dropping the voltage in resistors, and therefore, the current and power drawn from the supply mains are also reduced in comparison to resistor starting.
2. Availability of highest torque per ampere of supply current.
3. Adjustment of starting voltage by selection of proper tap on the auto-transformer.
4. The method is suitable for long starting periods.
5. Motor current larger than supply current.
6. Closed transition starting.

Disadvantages :

1. Low power factor.
2. Higher cost in case of lower output rating motors.

- This method can be used for starting of star-connected as well as delta-connected motors.
- This method is often employed for starting of large cage motors (rating exceeding 20 kW).

20.1.3. Star-Delta Starter. Star-delta switching method is based upon the principle that with three windings connected in star, voltage across each winding is $\frac{1}{\sqrt{3}}$ *i.e.*, 57.7% of the line to line voltage whereas the same windings connected in delta will have full-line-to-line voltage across each.

The star-delta starter *connects the three stator windings in star* across the rated supply voltage *at the starting instant. After the motor attains speed the same windings through a change over switch are re-connected in delta across the same supply voltage.*

The basic connection diagram of a star-delta starter is shown in Fig. 32. An actual starter incorporates *under-voltage* and *over-load coils*. The starter is also provided with a mechanical inter-locking device to prevent the handle from being put in the 'Run' position first. Such starters are employed for starting 3-phase squirrel-cage induction motors of rating between 4 kW and 15 kW.

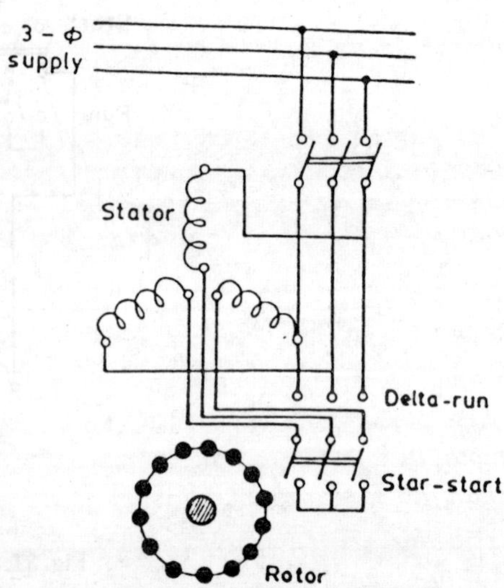

Fig. 32. Star-delta starter.

When *star-connected*, the applied voltage over each motor phase is reduced by a factor of $\frac{1}{\sqrt{3}}$ and hence the torque becomes $\frac{1}{3}$ of that which would have been developed if motor were directly connected in delta. The line current is reduced to $\frac{1}{3}$. Hence, during starting period when motor is star-connected, it takes $\frac{1}{3}$rd as much starting current and develops $\frac{1}{3}$rd as much torque as would have been developed were it directly connected in delta.

Relation between T_{st} and T_f :

$$I_{st} \text{ per phase} = \frac{1}{\sqrt{3}} I_{sc} \text{ per phase}$$

where I_{sc} is the current per phase which delta-connected motor would have taken if switched on to the supply directly ; however, line current at start is equal to $\frac{1}{3}$ of line I_{sc}.

Now, starting torque, $T_{st} \propto I_{st}^2 \quad (s = 1)$

Full-load torque, $T_f \propto I_f^2/s_f$

$$\therefore \quad \frac{T_{st}}{T_f} = \left(\frac{I_{st}}{I_f}\right)^2 s_f = \left(\frac{I_{sc}}{\sqrt{3}I_f}\right)^2 s_f = \frac{1}{3}\left(\frac{I_{sc}}{I_f}\right)^2 s_f$$

i.e.,

$$\frac{T_{st}}{T_f} = \frac{1}{3}\left(\frac{I_{sc}}{I_f}\right)^2 s_f \qquad ...(34)$$

Here, I_{st} and I_{sc} represent phase values.

- This method reduces the starting line current to one-third but the starting torque is also reduced by the same amount.
- *This method is cheap but limited to applications where high starting torque is not necessary e.g., machine tools, pumps, motor-generator sets etc.*
- The method is unsuitable for motors for voltage exceeding 3000 V because of the excessive number of stator turns needed for delta connection.
- Such starters are employed for starting *3-phase squirrel cage induction motors of rating between 4 and 20 kW.*

Precaution with star-delta starting. When the motor is started in star the initial current flowing is *57.7% of the short-circuit current in delta together with a transient in each phase*. The transient currents decay rapidly but the steady state is not reached until the motor has attained 70% of its synchronous speed. The change-over from star to delta connection *should not be made until the motor attains 90% of synchronous speed, otherwise there will be a current surge considerably greater than full-load current which may even be greater than the standstill current with star-connection.*

20.2. Slip-ring Induction Motors—Starting of

(*i*) **Rotor rheostat.** The slip-ring induction motors are practically always started with full line voltage applied across the stator terminals. The value of starting current is adjusted by introducing a variable resistance in the rotor circuit. The controlling resistance is in the form of rheostat connected in star (Fig. 33), the resistance being gradually cut out of the rotor circuit as the motor gathers speed. *By increasing the rotor resistance, not only is the rotor* (and hence stator) *current reduced at starting* but *at the same time torque is also increased due to improvement in power factor.*

The rheostat is either of stud or contractor type and may be hand operated or automatic.

As discussed earlier, the introduction of additional external resistance in the rotor circuit enables slip-ring motor to develop a high starting torque with reasonably moderate starting current. Hence *such motors can be started under load.* When the motor runs under normal conditions the rings are short-circuited and brushes lifted from them.

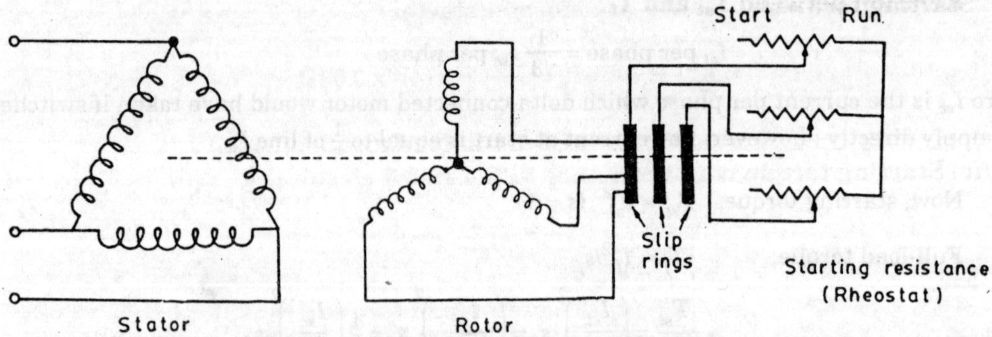

Fig. 33. Starting of slip-ring induction motor.

Example 20. *The short-circuit current of a small 3-phase induction motor is 3 times the full-load currents. Determine the starting torque as a percentage of full-load torque if full-load slip is 3 per cent.*

Solution. Short-circuit current, $I_{st} = 3I_f$

Full-load slip, $s_f = 3\%$ or 0.03

Starting torque, $$T_{st} = T_f \left(\frac{I_{sc}}{I_f}\right)^2 s_f = T_f (3)^2 \times 0.03 = 0.27T_f.$$

Hence, **starting torque is 27% of full-load torque. (Ans.)**

Example 21. *A 3-phase, squirrel-cage induction motor has a short-circuit current equal to 4 times the full-load current. Find the starting torque as a percentage of full-load torque if the motor is started by :*

(i) direct switching to the supply ; *(ii) a star delta starter ;*

(iii) an auto-transformer ; and *(iv) a resistance in the stator circuit.*

The stator current in (iii) and (iv) is limited to 2 times the full-load current and the full-load slip is 3%.

Solution. Short-circuit current, $I_{sc} = 4I_f$

Full-load slip, $s_f = 3\%$ or 0.03

(i) Starting torque with direct switch,

$$T_{st} = \left(\frac{I_{sc}}{I_f}\right)^2 s_f T_f = (4)^2 \times 0.03 \times T_f$$

$$= 0.48T_f \text{ i.e. } \mathbf{48\% \text{ of full-load torque. (Ans.)}}$$

(ii) Starting torque with star-delta starter,

$$T_{st} = \frac{1}{3}\left(\frac{I_{sc}}{I_f}\right)^2 s_f T_f = \frac{1}{3} \times (4)^2 \times 0.03 \times T_f$$

$$= 0.16T_f \text{ i.e. } \mathbf{16\% \text{ of full-load torque. (Ans.)}}$$

(iii) Supply line current

$$= K^2 I_{sc} = 2I_f$$

or the transformation ratio,

$$K = \sqrt{\frac{2I_f}{I_{sc}}} = \sqrt{\frac{2I_f}{4I_f}} = \frac{1}{\sqrt{2}}$$

Starting torque with auto-transformer starter

$$= K^2 \left(\frac{I_{sc}}{I_f}\right)^2 s_f T_f = \left(\frac{1}{\sqrt{2}}\right)^2 \times (4)^2 \times 0.03 T_f$$

$= 0.24T_f$ or **24% of full-load torque. (Ans.)**

(*iv*) **Starting torque with resistance in the stator circuit**

$$= \left(\frac{I_{sc}}{I_f}\right)^2 s_f T_f = (2)^2 \times 0.03 \times T_f$$

$= 0.12T_f$ *i.e.* **12% of full-load torque. (Ans.)**

Example 22. *A 400 V, 50-Hz, induction motor, when started directly from the mains takes 4 times the full-load current and the torque produced is twice the full-load torque. Determine :*

(*i*) *The motor current, the line current and the starting torque when started by means of an auto- transformer of ratio 2.5 : 1.*

(*ii*) *The voltage to be applied and the motor current if the full-load torque is to be obtained at starting.*

Solution. Supply voltage $= 400$ V

Frequency $= 50$ Hz

Transformation ratio, $K = \frac{1}{2.5} = 0.4$

Let us *assume* that the winding is star-connected

Phase voltage, $E_{ph} = \frac{400}{\sqrt{3}} = 231$ V

Voltage across the motor terminal

$$= KE_L = 0.4 \times 400 = 160 \text{ V}$$

The phase voltage when started by an auto-transformer

$$= \frac{160}{\sqrt{3}} = 92.4 \text{ V}$$

As the current is proportional to the voltage per phase, therefore :

(*i*) **New starting phase current**

$$= \frac{92.4}{231} \times 4I_f = \mathbf{1.6\ I_f}. \textbf{ (Ans.)}$$

Line current :

In case of a transformer, $\frac{E_2}{E_1} = \frac{N_2}{N_1} = \frac{I_1}{I_2}$

But $\frac{E_2}{E_1} = \frac{1}{2.5} = 0.4 \ (= K)$ and $I_2 = 1.6I_f$

$\therefore$ $\frac{I_1}{I_2} = K$ or $I_1 = KI_2 = K \times 1.6I_f = 0.4 \times 1.6I_f = 0.64\ I_f$

Hence, **line current = 0.64$\mathbf{I_f}$. (Ans.)**

Let $T_f =$ full-load torque

$T \propto E^2$

$2T_f \propto (231)^2$ ...(*i*) [$\because$ $T = 2T_f$ (*given*)]

When the applied voltage is 92.4 V let the torque = T

$$T \propto (92.4)^2 \quad ...(ii)$$

$$\frac{T}{2T_f} = \left(\frac{92.4}{231}\right)^2 = (0.4)^2$$

$$\therefore \quad \mathbf{T = 0.32T_f.} \quad \textbf{(Ans.)}$$

(*ii*) Let the voltage per phase be V to get full-load starting torque,

or

$$T_f \propto V^2 \quad ...(iii)$$

So from eqns. (*i*) and (*iii*), we get

$$\frac{T_f}{2T_f} = \frac{V^2}{(231)^2}$$

$$\therefore \quad V^2 = \frac{(231)^2}{2}$$

or

$$V = \frac{231}{\sqrt{2}} = \mathbf{163.3\ V.} \quad \textbf{(Ans.)}$$

$\therefore$ **Motor Voltage** $= 163.3 \times \sqrt{3} = \mathbf{282.8\ V.}$ **(Ans.)**

Motor current $= \frac{163.3}{231} \times 4I_f = \mathbf{2.827\ I_f.}$ **(Ans.)**

21. FACTORS GOVERNING PERFORMANCE OF INDUCTION MOTORS

The operating or performance characteristics of an induction motor are governed by the following factors/parameters :

1. Rotor resistance ; 2. Air gap length ;
3. Shapes of both stator and rotor slots and teeth.

1. Rotor resistance :

- The rotor resistance is the most important parameter and is determined primarily by the material used in construction and the area of cross-section of the end rings in case of a squirrel case induction motor. The commonly used materials are copper and aluminium.
- An increase in rotor resistance affects the performance of the motor as mentioned below :
  - — With the *increase in starting resistance the starting torque increases until it attains the maximum value, after which it diminishes.*
  - — An *increase in rotor resistance reduces the starting current and operating efficiency and causes increase in per cent regulation.* The maximum torque remains unaffected because it depends upon the rotor reactance only.

2. Air gap length :

- The air gap is one of the two main sources of the low power factor at which the induction motor operates.
- Too much reduction of air-gap may cause
  - — *increase in noise, pulsation losses and magnetic pull* ;
  - — *reduction in over-load capacity.*

3. Shapes of both stator and rotor slots and teeth :

- The shapes of rotor teeth and slots affect the reactance, which in turn, affects the starting current and maximum torque. The induction motors with deeper slots have more leakage reactance and so less value of starting current, maximum torque and slip corresponding to maximum torque.

- The slots in the rotor are *not always* made parallel to the shaft, but are given a slight twist or skew. The *motor noise and vibration, cagging and synchronous cups can be reduced or even eliminated by skewing either the stator or rotor slots* (in India, practice is to skew the rotor slots).

  Harmonic torques can be *reduced* by :

  —chording ;
  —integral slot winding ;
  —increasing air-gap length ;
  —skewing of rotor slots.

Note. Space harmonic fields are developed by the **windings, slotting, magnetic saturation, gap length irregularity**. These harmonic fields induce e.m.fs. and circulate harmonic currents in the rotor windings. These harmonic currents in the rotor interact with the harmonic fields and develop *harmonic torques, vibration* and *noise*.

22. EFFECTS OF OPERATING CONDITIONS

Effects of operating conditions are briefly described below :

1. Effect of unbalance supply voltage :

- An unbalance supply voltage produces a rotating field which will move at a non-uniform rate and in a non-uniform strength. It provides *unbalanced currents in the stator winding which produce unequal heating.*

2. Effect of loading :

- As the *load increases, speed falls and consequently torque increases, till a point is reached when the maximum torque will be developed.*
- If the load is *increased further*, there is further *fall in the speed and the driving torque decreases and ultimately the rotor comes to standstill.*

3. Effects of variation in line frequency and line voltage :

- The variation in supply frequency and supply voltage not only affects the speed and torque of the motor but also the starting current, full-load current, starting torque, maximum torque and operating temperature of various parts of the machine and thus make their *operation unsatisfactory*.

4. Effects of break in one phase :

- When a polyphase induction motor is in operation and if its one phase breaks, may be due to any reason, the motor will continue to operate single phase, provided that load does not exceed 57.7 percent of the normal rating, with about the same temperature rise as when carrying rated load as a 3-phase induction motor. It will not, however, start single phase but if brought upto speed by some external means, such as putting on the belt, it will continue to operate satisfactorily on load below 57.7 percent of rated one.
- A break in one phase of the rotor of a wound-rotor induction motor may prevent it from starting but if it is brought upto speed before the circuit is opened, it will operate under reduced load but usually with considerable vibrations.

23. RATINGS OF 3-PHASE INDUCTION MOTORS

- Three-phase induction motors are rated in terms of the following :

  (*i*) Power output in kW ; (*ii*) Speed ;
  (*iii*) Voltage ; (*iv*) Frequency ;
  (*v*) Phase (single or three) ; (*vi*) Line current ;
  (*vii*) Temperature rise in specified time.

— If a motor is designed to be operated on more than one voltage or at more than one speed by reconnecting the winding, a connection diagram is also given on the name plate.

— Class of motors *A, B, C, D, E* or *F* is also to be mentioned on the name plate.

- *Standard types of squirrel cage motors are* :
  - — Class *A* motors—Low-impedance squirrel cage rotor motors.
  - — Class *B* motors—High-reactance squirrel cage rotor motors.
  - — Class *C* motors—High starting torque, low starting current-double squirrel cage motors.
  - — Class *D* motors—High resistance squirrel cage rotor motors.
  - — Class *E* motors.
  - — Class *F* motors.

24. SQUIRREL-CAGE MOTORS—ADVANTAGES, DISADVANTAGES AND APPLICATIONS

Advantages :

1. Cheaper in cost.
2. Simple and rugged in construction.
3. Maintenance cost is low.
4. Explosion proof.
5. Can be cooled better because of base end rings.
6. More pull-out torque and greater maximum power output.
7. Simple starting arrangement.
8. Nearly constant speed.
9. High overload capacity.
10. High power factor.

Disadvantages :

1. *Poor starting torque* due to low rotor resistance.
2. At the instant of switching on, *draws a large current* from the line.
3. Very *sensitive to change in supply voltage*.
4. *At light load the power factor is very low*.
5. Speed regulation *not possible*.

Applications. Suitable for industrial drives of small power where *speed control is not required* such as :

- Flour mills
- Printing machinery
- Other shaft drives of small power.

Note. When the squirrel cage motors are used for crane or hoist work where large starting torque is more important than high efficiency owing to the intermittent nature of load, the rotor resistance is increased by employing the end rings, and sometimes the bars as well, of high resistance metal such as German silver.

25. WOUND ROTOR (OR SLIP RING) INDUCTION MOTORS—ADVANTAGES, DISADVANTAGES AND APPLICATIONS

Advantages :

1. High starting torque.
2. High over-load capacity.

3. Nearly constant speed.
4. Low starting current (in comparison with squirrel cage motor).

Disadvantages :

1. Low power factor at light loads.
2. Lower efficiency and lower power factor (in comparison to squirrel cage motor).
3. Sensitivity to fluctuations in supply voltage.
4. Higher initial and maintenance costs.
5. Speed regulation is poor when operated with external resistances in the rotor circuit.

Applications. Suitable for most industrial drives of high power where high starting torque is required such as for driving :

- Line shafts
- Pumps
- Lifts
- Generators
- Winding machines
- Mills etc.

26. COMPARISON OF A SQUIRREL-CAGE AND A SLIP RING (OR PHASE WOUND) INDUCTION MOTORS

	Property	*Squirrel cage induction motor*	*Phase wound induction motor*
1.	**Starting**	The motor can be started with a star-delta starter and it eliminates the hazards of slip rings, brush gear etc.	The motor requires slip rings, brush gear, short-circuiting device and starting resistance etc.
2.	**Overhang**	Less	Large
3.	**Space factor in slots**	Better	Poor
4.	**Cooling**	The losses can be dissipated much efficiently since the end rings are bars and there is more space for providing an efficient fan.	Most of the space is taken up by the overhang and less space is left for the provision of a good cooling fan ; thus the cooling is not quite efficient.
5.	**Efficiency**	High efficiency (only for machines, which are not redesigned for high starting torque).	Low efficiency.
6.	**Starting torque**	For plain squirrel cage motor starting torque is poor which cannot be increased in an ordinary way.	Starting torque can be increased by the insertion of external resistance in each phase of the rotor.
7.	**Copper losses**	Since overhang is less and space factor is better the copper losses are *small*.	Due to large overhang and poor space factor, the copper losses are *more*.
8.	**Maintenance**	Less	More
9.	**Cost**	*Cheaper* (since there are less components and less labour required for winding).	Costlier

27. COMPARISON BETWEEN INDUCTION AND SYNCHRONOUS MOTORS

The comparison between induction and synchronous motors is given below :

S. No.	*Induction motor*	*Synchronous motor*
1.	It has got self-starting torque and no special means are required for starting.	It is inherently not self-starting and some external means are required for its starting.
2.	Does not require D.C. excitation.	Requires D.C. excitation.
3.	Speed can be controlled but to small extent.	Speed control not possible.
4.	Its speeds falls with the increase in load and is always less than synchronous speed.	Its average speed is constant and independent of load.
5.	It operates at only lagging power factor, which becomes very poor at light loads.	It can be operated under a wide range of power factor, both lagging and leading.
6.	Its torque is more sensitive to change in supply voltage.	Its torque is less sensitive to change in supply voltage.
7.	Breakdown torque is proportional to the square of the supply voltage.	Breakdown torque is proportional to the supply voltage.
8.	More simple and less costly comparatively.	More complicated and more costly comparatively.
9.	Employed for supplying mechanical load only.	Employed for supplying mechanical load as well as for power factor improvement.

HIGHLIGHTS

1. *An induction motor is simply an electric transformer whose magnetic circuit is separated by an air gap into two relatively movable portions, one carrying the primary and the other the secondary winding.*
2. *Synchronous speed,* $N_s = \frac{120f}{p}$ *, where f is the frequency and p the number of poles.*
3. *Slip,* $s = \frac{N_s - N}{N_s}$ *, where N is the motor speed (r.p.m).*
4. *Frequency of the rotor current,* $f_r = sf$*, where s is the slip and f is the supply frequency.*
5. *Torque,* $$T = \frac{ksR_2 E_2^2}{R_2^2 + s^2 X_2^2}$$
   Starting torque, $$T_{st} = \frac{kR_2 E_2^2}{R_2^2 + X_2^2}$$
   Condition for maximum torque, $$s = \frac{R_2}{X_2}$$
   Maximum torque, $$T_{max} = \frac{kE_2^2}{2X_2}$$
6. **Starting torque** *is proportional to the* **square of applied voltage.**
7. $$\frac{T_f}{T_m} = \frac{2s_f s_{mT}}{s_f^2 + s_{mT}^2}, \quad \text{where } s_f = \frac{\text{rotor resistance}}{\text{rotor standstill reactance}}.$$
   $$\frac{T_{st}}{T_{max}} = \frac{2s_{mT}}{s_{mT}^2 + 1}.$$

8. $$\frac{\text{Rotor copper loss}}{\text{Rotor gross output}} = \frac{s}{1-s}.$$

9. $$\frac{\text{Rotor output}}{\text{Rotor input}} = 1 - s.$$

OBJECTIVE TYPE QUESTIONS

(A) Choose the Correct Answer :

1. Which of the following components is usually fabricated out of silicon steel ?
(*a*) Bearings (*b*) Shaft
(*c*) Stator core (*d*) None of the above.

2. The frame of an induction motor is usually made of
(*a*) silicon steel (*b*) cast iron
(*c*) aluminium (*d*) bronze.

3. The shaft of an induction motor must be
(*a*) stiff (*b*) flexible
(*c*) hollow (*d*) any of the above.

4. The shaft of an induction motor is made of
(*a*) high speed steel (*b*) stainless steel
(*c*) carbon steel (*d*) cast iron.

5. In an induction motor, on no-load the slip is generally
(*a*) less than 1% (*b*) 1.5%
(*c*) 2% (*d*) 4%.

6. In medium sized induction motors, the slip is generally around
(*a*) 0.04% (*b*) 0.4%
(*c*) 4% (*d*) 14%.

7. In squirrel cage induction motors, the rotor slots are usually given slight skew in order to
(*a*) reduce windage losses (*b*) reduce eddy currents
(*c*) reduce accumulation of dirt and dust (*d*) reduce magnetic hum.

8. In case the air gap in an induction motor is increased
(*a*) the magnetising current of the rotor will decrease
(*b*) the power factor will decrease
(*c*) speed of motor will increase (*d*) the windage losses will increase.

9. Slip rings are usually made of
(*a*) copper (*b*) carbon
(*c*) phosphor bronze (*d*) aluminium.

10. A 3-phase 440 V, 50 Hz induction motor has 4% slip. The frequency of rotor e.m.f. will be
(*a*) 200 Hz (*b*) 50 Hz
(*c*) 2 Hz (*d*) 0.2 Hz.

11. If N_s is the synchronous speed and s the slip, then actual running speed of an induction motor will be
(*a*) N_s (*b*) $s \cdot N_s$
(*c*) $(1-s)N_s$ (*d*) $(N_s - 1)s$.

12. The efficiency of an induction motor can be expected to be nearly
(*a*) 60 to 90% (*b*) 80 to 90%
(*c*) 95 to 98% (*d*) 99%.

13. The number of slip rings on a squirrel cage induction motor is usually
(*a*) two (*b*) three
(*c*) four (*d*) none.

14. The starting torque of a squirrel-cage induction motor is
(*a*) low (*b*) negligible
(*c*) same as full-load torque (*d*) slightly more than full-load torque.
15. A double squirrel-cage induction motor has
(*a*) two rotors moving in opposite direction (*b*) two parallel windings in stator
(*c*) two parallel windings in rotor (*d*) two series windings in stator.
16. The ratio of starting torque to normal torque in case of a star-delta starter will be
(*a*) 0.35 (*b*) 0.67
(*c*) 1.07 (*d*) 1.35.
17. Star-delta starting of motors is not possible in case of
(*a*) single phase motors (*b*) variable speed motors
(*c*) low horse power motors (*d*) high speed motors.
18. The terms 'cogging' is associated with
(*a*) three phase transformers (*b*) compound generators
(*c*) D.C. series motors (*d*) induction motors.
19. In case of the induction motors the torque is
(*a*) inversely proportional to ($\sqrt{\text{slip}}$) (*b*) directly proportional to $(\text{slip})^2$
(*c*) inversely proportional to slip (*d*) directly proportional to slip.
20. Stepless speed control of induction motor is possible by which of the following methods ?
(*a*) E.m.f. injection in rotor circuit (*b*) Changing the number of poles
(*c*) Cascade operation (*d*) None of the above.
21. Rotor rheostat control method of speed control is used for
(*a*) squirrel-cage induction motors only (*b*) slip ring induction motors only
(*c*) both (*a*) and (*b*) (*d*) none of the above.
22. In the circle diagram for induction motor, the diameter of the circle represents
(*a*) slip (*b*) rotor current
(*c*) running torque (*d*) line voltage.
23. For which motor the speed can be controlled from rotor side ?
(*a*) Squirrel-cage induction motor (*b*) Slip-ring induction motor
(*c*) Both (*a*) and (*b*) (*d*) None of the above.
24. If any two phases for an induction motor are interchanged
(*a*) the motor will run in reverse direction (*b*) the motor will run at reduced speed
(*c*) the motor will not run (*d*) the motor will burn.
25. An induction motor is
(*a*) self-starting with zero torque (*b*) self-starting with high torque
(*c*) self-starting with low torque (*d*) non-self starting.
26. The maximum torque in an induction motor depends on
(*a*) frequency (*b*) rotor inductive reactance
(*c*) square of supply voltage (*d*) all of the above.
27. In three-phase squirrel-cage induction motors
(*a*) rotor conductor ends are short circuited through slip rings
(*b*) rotor conductors are short-circuited through end rings
(*c*) rotor conductors are kept open
(*d*) rotor conductors are connected to insulation.
28. In a three-phase induction motor, the number of poles in the rotor winding is always
(*a*) zero (*b*) more than the number of poles in stator
(*c*) less than number of poles in stator (*d*) equal to number of poles in stator.
29. *DOL* starting of induction motors is usually restricted to
(*a*) low horsepower motors (*b*) variable speed motors
(*c*) high horsepower motors (*d*) high speed motors.

30. No-load current in an induction motor is nearly
(*a*) 0.1 per cent of full-load current
(*b*) 1 to 5 per cent of full-load current
(*c*) 15 to 25 per cent of full-load current
(*d*) 50 to 60 per cent of full-load current.

31. The speed of a squirrel-cage induction motor can be controlled by all of the following except
(*a*) changing supply frequency
(*b*) changing number of poles
(*c*) changing winding resistance
(*d*) reducing supply voltage.

32. The 'crawling' in an induction motor is caused by
(*a*) high loads
(*b*) low voltage supply
(*c*) improper design of machine
(*d*) harmonics developed in the motor.

33. The power factor of an induction motor under no-load conditions will be closer to
(*a*) 0.2 lagging
(*b*) 0.2 leading
(*c*) 0.5 leading
(*d*) unity.

34. The 'cogging' of an induction motor can be avoided by
(*a*) proper ventilation
(*b*) using *DOL* starter
(*c*) auto-transformer starter
(*d*) having number of rotor slots more or less than the number of stator slots (not equal).

35. One phase of a three-phase partly loaded induction motor power supply got opened up while the motor was in motion. This is likely to result in
(*a*) stalling of the motor
(*b*) burning of one phase winding of the motor
(*c*) burning of all three phases of winding of the motor
(*d*) undisturbed running of motor.

36. If an induction motor with certain ratio of rotor to stator slots, runs at $\frac{1}{7}$ of the normal speed, the phenomenon will be termed as
(*a*) humming
(*b*) hunting
(*c*) crawling
(*d*) cogging.

37. It is advisable to avoid line starting of induction motor and use starter because
(*a*) it will run in reverse direction
(*b*) it will pick up very high speed and may go out of step
(*c*) motor takes five to seven times its full-load current
(*d*) starting torque is very high.

38. The speed characteristics of an induction motor closely resemble the speed-load characteristics of which of the following machines ?
(*a*) D.C. series motor
(*b*) D.C. shunt motor
(*c*) Universal motor
(*d*) None of the above.

39. In a slip-ring induction motor, resistance is connected in rotor phases
(*a*) to increase starting torque
(*b*) to limit starting current
(*c*) to limit the starting current and increase the starting torque
(*d*) none of the above.

40. Which type of bearing is provided in small induction motors to support the rotor shaft ?
(*a*) Ball bearings
(*b*) Cast iron bearings
(*c*) Bush bearings
(*d*) None of the above.

41. A pump induction motor is switched on to a supply 30% lower than its rated voltage. The pump runs. What will eventually happen ? It will
(*a*) stall after sometime
(*b*) stall immediately
(*c*) continue to run at lower speed without damage
(*d*) get heated and subsequently get damaged.

42. 5 H.P., 50-Hz, 3-phase, 440 V, induction motors are available for the following r.p.m. Which motor will be the costliest ?
(*a*) 730 r.p.m. (*b*) 960 r.p.m.
(*c*) 1440 r.p.m. (*d*) 2880 r.p.m.
43. A 3-phase slip ring motor has
(*a*) double cage rotor (*b*) wound rotor
(*c*) short-circuited rotor (*d*) any of the above.
44. The torque developed in an induction motor is nearly proportional to
(*a*) $\frac{1}{V}$ (*b*) V
(*c*) V^2 (*d*) $\frac{1}{V^2}$
where V = applied voltage.
45. Short-circuit test on an induction motor cannot be used to determine
(*a*) windage losses (*b*) copper losses
(*c*) transformation ratio (*d*) power scale of circle diagram.
46. In a three-phase induction motor
(*a*) iron losses in stator will be negligible as compared to that in rotor
(*b*) iron losses in motor will be negligible as compared to that in rotor
(*c*) iron losses in stator will be less than that in rotor
(*d*) iron losses in stator will be more than that in rotor.
47. In case of 3-phase induction motors, plugging means
(*a*) putting the motor directly on line without a starter
(*b*) locking of rotor due to harmonics
(*c*) starting the motor on load which is more than the rated load
(*d*) interchanging two supply phases for quick stopping.
48. Which is of the following data is required to draw the circle diagram for an induction motor ?
(*a*) Block rotor test only (*b*) No load test only
(*c*) Block rotor test, and no-load test
(*d*) Block rotor test, no-load test and stator resistance test.
49. In three-phase induction motors sometimes copper bars are placed deep in the rotor to
(*a*) improve starting torque (*b*) reduce copper losses
(*c*) improve efficiency (*d*) improve power factor.
50. In a three-phase induction motor
(*a*) power factor at starting is high as compared to that while running
(*b*) power factor at starting is low as compared to that while running
(*c*) power factor at starting is the same as that while running.
51. The circle diagram for an induction motor cannot be used to determine
(*a*) efficiency (*b*) power factor
(*c*) frequency (*d*) output.
52. The current taken by a three-phase induction motor during plugging will be
(*a*) normal full-load current (*b*) more than blocked rotor current
(*c*) equal to blocked rotor current (*d*) less than blocked rotor current.
53. Blocked rotor test on induction motors is used to find out
(*a*) leakage reactance (*b*) power factor on short circuit
(*c*) short-circuit current under rated voltage (*d*) all of the above.
54. Lubricant used for ball bearing is usually
(*a*) graphite (*b*) grease
(*c*) mineral oil (*d*) molasses.

55. An induction motor can run at synchronous speed when
(*a*) it is run on load
(*b*) it is run in reverse direction
(*c*) it is run on voltage higher than the rated voltage
(*d*) e.m.f. is injected in the rotor circuit.

56. Which of the following can be reduced by skewing in induction motor ?
(*a*) Noise
(*b*) Vibration
(*c*) Synchronous cusps
(*d*) All of the above.

57. Which motor is preferred for use in mines where explosive gases exist ?
(*a*) Air motor
(*b*) Induction motor
(*c*) D.C. shunt motor
(*d*) Synchronous motor.

58. The torque developed by a 3-phase induction motor least depends on
(*a*) rotor current
(*b*) rotor power factor
(*c*) rotor e.m.f.
(*d*) shaft diameter.

59. In an induction motor if air-gap is increased
(*a*) the power factor will be low
(*b*) windage losses will be more
(*c*) bearing friction will reduce
(*d*) copper loss will reduce.

60. In an induction motor
(*a*) $s = \frac{N - N_s}{N}$
(*b*) $s = \frac{N_s - N}{N_s}$
(*c*) $s = \frac{N_s - N}{N}$
(*d*) $s = \frac{N - N_s}{N_s}$.

61. When R_2 is the rotor resistance, X_2 the rotor reactance at supply frequency and s the slip, then the condition for maximum torque under running conditions will be
(*a*) $sR_2X_2 = 1$
(*b*) $sR_2 = X_2$
(*c*) $R_2 = sX_2$
(*d*) $R_2 = s^2X_2$.

62. How many different speeds can be obtained in cascade control of two induction motors ?
(*a*) 2
(*b*) 4
(*c*) 8
(*d*) 16.

63. In case of a double cage induction motor, the inner cage has
(*a*) high inductance and low resistance
(*b*) low inductance and high resistance
(*c*) low inductance and low resistance
(*d*) high inductance and high resistance.

64. Open circuit test is usually conducted on
(*a*) slip ring motors only
(*b*) wound rotor motors only
(*c*) either of (*a*) and (*b*) above
(*d*) none of the above.

65. When two motors of 12 and 8 poles respectively are connected to a 50-Hz supply, which of the following no-load speed cannot be obtained ?
(*a*) 300 r.p.m.
(*b*) 450 r.p.m.
(*c*) 500 r.p.m.
(*d*) 750 r.p.m.
(*e*) 1500 r.p.m.

66. The low power factor of induction motor is due to
(*a*) rotor leakage reactance
(*b*) stator reactance
(*c*) the reactive lagging magnetizing current necessary to generate the magnetic flux
(*d*) all of the above.

67. During back to back test
(*a*) both machines run as induction motors
(*b*) both machines run as induction generators
(*c*) one machine runs as induction motor and the other as induction generator
(*d*) one machine runs as induction motor and the other as synchronous generator.

68. Phase advancers are used with induction motors to
(*a*) reduce noise
(*b*) reduce vibrations
(*c*) reduce copper losses
(*d*) improve power factor.

69. During dynamic braking of induction motor
(a) stator connections are reversed
(b) motor is used as induction generator
(c) stator winding is excited by direct current
(d) fictitious load is applied on motor.
70. The operation of repeated starting and stopping of an induction motor for exact positioning of the driven equipment is called
(a) crawling
(b) balancing
(c) plugging
(d) jogging.
71. In moist surroundings which frame of motor will be preferred ?
(a) Open type
(b) Totally enclosed fan cooled
(c) Drip proof
(d) Splash proof.
72. The principle of operation of a three-phase induction motor closely resembles to that of which of the following ?
(a) Synchronous motor
(b) D.C. series motor
(c) Transformer with short-circuited secondary
(d) Transformer with open-circuited secondary.
73. Under which of the following conditions a squirrel-cage induction motor would be preferred over a wound rotor induction motor ?
(a) When external voltage is to be necessarily introduced into the rotor
(b) When wide range of speed control is desired
(c) When cost is the major consideration
(d) When higher starting torque is required.
74. Which of the following statement is not valid when a squirrel-cage induction motor operates under no-load ?
(a) Rotor induced e.m.f. is low
(b) Rotor current is low
(c) Power factor is low
(d) Slip is low.

ANSWERS

1. (c)	2. (b)	3. (a)	4. (c)	5. (a)	6. (c)	7. (d)
8. (b)	9. (c)	10. (c)	11. (c)	12. (b)	13. (d)	14. (a)
15. (c)	16. (a)	17. (a)	18. (d)	19. (d)	20. (b)	21. (b)
22. (b)	23. (b)	24. (a)	25. (c)	26. (d)	27. (b)	28. (d)
29. (a)	30. (d)	31. (c)	32. (d)	33. (a)	34. (d)	35. (d)
36. (c)	37. (c)	38. (b)	39. (b)	40. (a)	41. (d)	42. (a)
43. (b)	44. (c)	45. (a)	46. (d)	47. (d)	48. (d)	49. (a)
50. (b)	51. (a)	52. (b)	53. (d)	54. (b)	55. (d)	56. (d)
57. (a)	58. (d)	59. (a)	60. (b)	61. (c)	62. (b)	63. (a)
64. (a)	65. (b)	66. (d)	67. (c)	68. (d)	69. (c)	70. (d)
71. (b)	72. (c)	73. (c)	74. (d).			

(B) Say 'Yes' or 'No' :

1. Three-phase induction motor has a low efficiency.
2. An induction motor is simply an electric transformer whose magnetic circuit is separated by an air gap into two relatively movable portions, one carrying the primary and the other the secondary winding.
3. Frames of electrical machines house the stator core.
4. The number of slots in the rotor should always be equal to the number of slots in the stator.
5. For large and heavy rotors journal bearings are used.
6. The slips rings are made of aluminium.
7. The difference between the synchronous speed and rotor speed is known as slip.
8. Starting torque, $T_{st} = \dfrac{kR_2E_2^{\,2}}{R_2^{\,2} + X_2^{\,2}}$.

9. Starting torque is inversely proportional to the square of the applied voltage.
10. Stator iron loss is practically constant.
11. $\dfrac{\text{Rotor copper loss}}{\text{Rotor gross output}} = \dfrac{1-s}{s}$.
12. Circle diagram of an induction motor can be drawn by using the data obtained from no-load test, short-circuit test and stator resistance test.
13. The synchronous speed of the nth order harmonic is $\frac{1}{n}$th of the synchronous speed of fundamental.
14. 'Cogging' of squirrel-cage motors can be easily overcome by making the number of rotor slots prime to the number of stator slots.
15. A double squirrel-cage motor has two independent squirrel-cage windings on the rotor, each having its own set of slots.

ANSWERS

1. No	2. Yes	3. Yes	4. No	5. Yes	6. No	7. Yes
8. Yes	9. No	10. Yes	11. No	12. Yes	13. Yes	14. Yes
15. Yes.						

THEORETICAL QUESTIONS

1. Explain the principle of operation of the polyphase induction motor.
2. Show that a rotating magnetic field can be produced by the use of 3-phase currents of equal magnitude.
3. What is meant by slip in an induction motor ? Develop an expression for the frequency of rotor currents in it.
4. Derive an expression for the torque of an induction motor and obtain the condition for maximum torque.
5. Show that in a 3-phase induction motor with negligible stator impedance maximum torque is developed at slip $s = \dfrac{R_2}{X_2}$, where R_2 and X_2 are rotor resistance and standstill reactance respectively.
6. State the effects of increasing rotor resistance on starting current, starting torque, maximum torque and full-load slip of an induction motor.
7. Show that in an induction motor the rotor input : power developed : rotor copper losses : : 1 : (1 – s) : s, where s is the fractional slip.
8. Draw equivalent circuit of a 3-phase induction motor.
9. Describe and explain how to perform a locked rotor test. What data can be obtained by the test ?
10. What are no-load and blocked rotor tests ? What sort of losses can be measured by these tests ?
11. How would you determine circle diagram of a 3-phase induction motor experimentally ?
12. Describe the constructional details of a 3-phase squirrel-cage and phase wound induction motors. Also discuss the applications of various types of starters used for starting these motors.
13. Why at all starters are necessary for starting the induction motors ? What are the various types of starters used for squirrel-cage motors ? Discuss them.

UNSOLVED EXAMPLES

1. A 3-phase, 4-pole, 50-Hz induction motor is running at 1455 r.p.m. Find the slip speed and slip. [**Ans.** 45 r.p.m., 3 per cent]
2. An induction motor having 8-poles runs at 50-Hz supply. If it operates at full-load at 720 r.p.m., calculate the slip. [**Ans.** 4 per cent]
3. A 6-pole alternator running at 1000 r.p.m. supplies an 8-pole induction motor. What is the actual speed of the motor if the slip is 2.5%. [**Ans.** 731.25 r.p.m.]

4. A slip ring motor runs at 290 r.p.m. on full-load when connected to 50-Hz supply. Calculate :
(*i*) Number of poles (*ii*) Absolute slip
(*iii*) Fractional slip (*iv*) Percentage slip
(*v*) Frequency of rotor currents. [**Ans.** 20, 10 r.p.m., 0.0333, 3.33%, 1.66 Hz]

5. A 12-pole, 3-phase alternator is coupled to an engine running at 500 r.p.m. It supplies a 3-phase induction motor having a full- load speed of 1440 r.p.m. Find the percentage slip and number of poles of motor.
[**Ans.** 4 per cent, 4 poles]

6. A 4-pole induction motor is run at 1450 r.p.m. from a 50-Hz supply. Find the percentage slip and frequency of rotor current. [**Ans.** 3.33%, 1.67 Hz]

7. If the e.m.f. in the stator of an 8-pole induction motor has a frequency of 50-Hz and that in the rotor 1.5 Hz, at what speed is the motor running and what is the slip ? [**Ans.** 727.5 r.p.m., 3 per cent]

8. A three-phase slip ring induction motor gives a reading of 55 V across slip rings on open circuit when at rest with normal stator voltage applied. The rotor is star-connected and has impedance $(0.7 + j5)\ \Omega$ per phase. Find the rotor current when the machine is
(*i*) at standstill with the slip rings joined to a star-connected starter with a phase impedance of $(4 + j3)\ \Omega$; and
(*ii*) running normally with a 5 per cent slip.
[**Ans.** (*i*) 3.425 A at 0.506 p.f. (lag) ; (*ii*) 2.14 A at 0.942 p.f. (lag)

9. A 3-phase, 400 V, star-connected induction motor has a star-connected rotor with a stator to rotor turn ratio of 6.5. The rotor resistance and stand-still reactance per phase are 0.05 and 0.25 Ω respectively. What should be the value of external resistance per phase to be inserted in the rotor circuit to obtain maximum torque at starting and what will be the rotor starting current with this resistance ?
[**Ans.** 0.2 Ω, 100 A (app.)]

10. If the motor has a rotor resistance of 0.02 Ω and a standstill reactance of 0.1 Ω what must be the value of the total resistance of a starter for the rotor circuit for maximum torque to be exerted at starting ?
[**Ans.** 0.08 Ω]

11. The rotor of a 3-phase 440 V, 50 Hz, 4-pole induction motor has a resistance of 0.3 Ω per phase and an inductance of 0.008 H per phase. The ratio of stator to rotor turns is 2 : 1. The stator is delta connected and rotor is star-connected. Determine the voltage between slip rings when the motor is at standstill. Calculate the standstill rotor current and rotor current when the motor is operating with 3 per cent slip.
[**Ans.** 381 V, 87 A, 22 A]

12. A 3300 V, 24-phase, 50-Hz, 3-phase star-connected induction motor has a slip-ring rotor resistance of 0.016 Ω and standstill reactance of 0.265 Ω per phase. Calculate :
(*i*) The speed at maximum torque.
(*ii*) Ratio of full-load torque to maximum torque if full-load torque is obtained at 247 r.p.m.
[**Ans.** 235 r.p.m., 0.382]

13. A 6-pole, 3-phase, 50 Hz induction motor runs on full-load with a slip of 4%. If the rotor standstill impedance per phase is $(0.01 + j0.05)\ \Omega$, calculate the available maximum torque in terms of full-load torque. Also determine the speed at which the maximum torque occurs. $\left[\textbf{Ans.}\ \frac{T_m}{T_f} = 2.6, 800 \text{ r.p.m.}\right]$

14. A 6-pole, 3-phase, 50 Hz induction motor develops maximum torque of 200 Nm at a speed of 960 r.p.m. Determine the torque exerted by the motor at 5 per cent slip. The rotor resistance per phase is 0.5 Ω.
[**Ans.** 195 Nm]

15. The rotor resistance and standstill reactance per phase of a 3-phase slip ring induction motor are 0.02 Ω and 0.1 Ω respectively. What should be the value of the external resistance per phase to be inserted in the rotor circuit to give maximum torque at starting ? [**Ans.** 0.08 Ω]

16. The star-connected rotor of an induction motor has a standstill impedance of $(0.4 + j4)\ \Omega$ per phase, and the rheostat impedance per phase is $(6 + j2)\ \Omega$. The motor has an induced e.m.f. of 80 V between slip rings at standstill when connected to its normal supply voltage. Find :
(*i*) The rotor current at standstill with the rheostat in the circuit.
(*ii*) When running short-circuited with slip of 0.03. [**Ans.** 5.27 A, 3.3 A]

17. A 4-pole, 50 Hz, 3-phase, induction motor develops a maximum torque of 115 N-m at 1365 r.p.m. The resistance of the star-connected rotor is 0.2 Ω/phase. Calculate the value of resistance that must be inserted in series with each rotor phase to produce a starting torque equal to half the maximum torque. [**Ans.** 0.4 Ω or 8.1 Ω]

18. A 4-pole, 50-Hz, 3-phase induction motor has rotor resistance and reactance of 0.03 Ω and 0.12 Ω per phase respectively. Determine :

(*i*) The value of speed at which maximum torque occurs.

(*ii*) The value of external rotor resistance per phase to be inserted to obtain 80% of maximum torque at starting. [**Ans.** 1125 r.p.m., 0.02418 Ω]

19. The resistance and standstill reactance of each phase of a 3-phase induction motor with star-connected rotor are 0.06 Ω and 0.4 Ω respectively. The full-load slip is 4%. Calculate the resistance per phase of a star-connected rheostat, which when connected to the rotor, will give a pull out torque at one half of the full-load speed. What is then the power factor ? [**Ans.** 0.148 Ω, 0.707 (lag)]

20. The rotor resistance and standstill reactance of a 3-phase induction motor are respectively 0.015 Ω and 0.09 Ω per phase. At normal voltage, the full-load slip is 3 per cent. Estimate the percentage reduction in stator voltage to develop full-load torque at half full-load speed. Also calculate the power factor. [**Ans.** 22.9%, 0.31 (app.)]

Rotor Output Losses, Efficiency

21. The power input to a 3-phase induction motor is 60 kW. The stator-losses total 1.5 kW. Find the total mechanical power developed if the motor is running with a slip of 4 per cent. [**Ans.** 56.16 kW]

22. The rotor e.m.f. of a 3-phase, 6-pole, 400 V, 50-Hz induction motor alternates at 3-Hz. Compute the speed and percentage slip of the motor. Find the rotor copper loss per phase if the full input to the rotor is 111.9 kW. [**Ans.** 940 r.p.m., 6%, 2.238 kW]

23. A 3-phase, 50-Hz, induction motor draws 50 kW from the mains. If the stator losses are 2 kW and the rotor e.m.f. is observed to make 100 complete oscillations/alterations per minute, determine :

(*i*) Rotor copper loss. (*ii*) Gross mechanical output. [**Ans.** 1.6 kW, 46.4 kW]

24. The power input to the rotor of a 3-phase, 50-Hz, 6-pole, slip ring induction motor is 40 kW and the motor runs at 960 r.p.m. The rotor resistance per phase is 0.25 Ω. Determine the value of the rotor current per phase. [**Ans.** 46.2 A]

25. The power input to the rotor of a 440 V, 50-Hz, 3-phase, 6-pole induction motor is 60 kW. It is observed that the rotor e.m.f. makes 90 complete cycles per minute. Calculate :

(*i*) Slip, (*ii*) Rotor speed,

(*iii*) Rotor copper loss per phase, (*iv*) Mechanical power developed, and

(*v*) Rotor resistance/phase if the rotor current is 60 A.

[**Ans.** 0.03, 970 r.p.m., 600 W, 58.2 kW, 0.167 Ω]

Equivalent Circuit of an Induction Motor

26. The maximum torque of a 3-phase induction motor occurs at a slip of 12 per cent. The motor has an equivalent secondary resistance of 0.08 Ω/phase. Calculate the equivalent load resistance R_L, the equivalent load voltage V_L and the current at this slip if the gross power output is 9 kW.

[**Ans.** 0.587 Ω/phase, 42 V, 71.6 A]

27. The stator impedance and equivalent rotor resistance of a 400 V, 3-phase star-connected induction motor are $(0.06 + j0.2)$ Ω and equivalent rotor impedance of $(0.06 + j0.22)$ Ω. Neglecting exciting current, find the maximum gross power and the slip at which it occurs. [**Ans.** 12%, 142.9 kW]

28. Estimate the stator current, equivalent rotor current, efficiency, output, and its power factor at a slip of 5% for a motor having the following data :

Stator impedance = $1.0 + j3.0$ Ω, rotor standstill impedance = $1.0 + j2.0$ Ω, no-load shunt impedance = $10 + j50$ Ω, volts/phase = 250 volts. [**Ans.** 14.34 A, 11.6 A, 83.6%, 7,670 W, 0.853]

Starting of Induction Motor

29. A small 3-phase induction motor has a short-circuit current equal to 4 times the full-load current. Determine the starting torque as a percentage of full-load torque if full-load slip is 2.5 per cent.
[Ans. 40% of full-load torque]

30. A 3-phase, squirrel-cage induction motor has a short-circuit current equal to 5 times the full-load current. Find the starting torque as a percentage of full-load torque if the motor is started by
(*i*) direct switching to the supply ; (*ii*) a star-delta starter ;
(*iii*) an auto-transformer ; and (*iv*) a resistance in the stator circuit.
The starting current in (*iii*) and (*iv*) is limited to 2.5 times the full-load current and the full-load slip is 4 per cent. **[Ans.** T_f, 33.3% T_f, 50%T_f, 25% T_f]

31. Determine a suitable auto-transformer ratio for starting an induction motor with a supply current not exceeding twice full-load current. Use the following data :
Short-circuit current = 5 times the full-load current, full-load slip = 3%. Estimate the torque in terms of full-load torque. Ignore magnetising current. **[Ans.** 30%]

32. Find the percentage tapping required on an auto-transformer required for a cage motor to start the motor against $\frac{1}{4}$ of full-load torque. The short-circuit current at normal voltage is 4 times the full-load current and the full-load slip is 3 per cent. **[Ans.** K = 72.2%]

33. A squirrel-cage type induction motor when started by means of a star-delta starter takes 20 per cent of full-load line current and develops 40 per cent of full-load torque at starting. Calculate the starting torque and current in terms of full-load values, if an auto-transformer with 75 per cent tapping were employed.
[Ans. 3.375I_f, 67% full-load torque]

34. A 10 kW, 200 V, 3-phase induction motor has at full-load an efficiency of 87% and power factor of 0.85. At standstill the motor draws 5 times the full-load current and develops 1.5 times full-load torque. An auto-transformer is installed to reduce the starting current to give full-load torque at starting. Neglecting magnetising current, determine :
(*i*) The voltage to be applied at starting. (*ii*) The full-load line current.
(*iii*) The current drawn by the motor at start.
(*iv*) The current supplied to the primary of the autotransformer. **[Ans.** 163.3 V, 39 A, 159.2 A, 130 A]

35. A 3-phase squirrel-cage induction motor has a ratio of maximum torque to full-load torque as 2.5 : 1. Determine the ratio of actual starting torque to full-load torque for :
(*i*) direct starting ; (*ii*) star-delta starting ;
(*iii*) auto-transformer starting with tapping of 60%.
The rotor resistance and standstill reactance per phase are 0.4 Ω and 4 Ω respectively.
[Ans. 0.495, 0.165, 0.179]

9

Single Phase Motors

1. GENERAL ASPECTS

- The number of machines operating from single-phase supplies is greater than all other types taken in total. For the most part, however, they are only used in the smaller sizes, less than 5 kW and mostly in the fractional H.P. range. They operate at *lower power-factors* and are relatively *inefficient when compared with polyphase motors.* Though simplicity might be expected in view of the two-line supply, the *analysis is quite complicated.*
- *Single phase motors* perform a great variety of useful services in the home, the office, the factory, in business establishments, on the farm, and many other places where electricity is available. Since the requirements of the numerous applications differ so widely, the motor-manufacturing industry has developed several types of such machines, each type having operating characteristics that meet definite demands. For example, one type operates satisfactorily on direct current or any frequency up to 60 cycles ; another rotates at absolutely constant speed, regardless of load ; another develops considerable starting torque and still another, although not capable of developing much starting torque, is nevertheless extremely cheap to make and very rugged.

2. TYPES OF SINGLE-PHASE MOTOR

The single-phase motors may be of the following types :

1. **Single-phase Induction Motors :**

   A. Split-phase motors

   (*i*) Resistance-start motor (*ii*) Capacitor-start motor

   (*iii*) Permanent-split (single-value) capacitor motor

   (*iv*) Two-value capacitor motor.

   B. Shaded-pole induction motor.

   C. Reluctance-start induction motor.

   D. Repulsion-start induction motor.

2. **Commutator-Type, Single-Phase Motors :**

   A. Repulsion motor.

   B. Repulsion-induction motor.

   C. A.C. series motor.

   D. Universal motor.

3. Single-phase Synchronous Motors :

A. Reluctance motor. B. Hysteresis motor.

C. Sub-synchronous motor.

3. SINGLE-PHASE INDUCTION MOTORS

3.1. Applications and Disadvantages

Applications :

- Single phase induction motors are in very wide use in industry especially in *fractional horse-power field*.

  They are extensively used for electric drive for *low power constant speed apparatus* such as *machine tools, domestic apparatus* and *agricultural machinery* in circumstances where a three-phase supply is *not* readily available.
- There is a large demand for single-phase induction motors in sizes ranging from a *fraction of horse-power upto about 5 H.P.*

Disadvantages :

Though these machines are useful for small outputs, they are not used for large powers as they suffer from many disadvantages and are never used in cases where three-phase machines can be adopted.

The main *disadvantages* of single-phase induction motors are :

1. Their output is only 50% of the three-phase motor, for a given frame size and temperature rise.
2. They have lower power factor.
3. Lower-efficiency.
4. These motors do not have inherent starting torque.
5. More expensive than three-phase motors of the same output.

3.2. Construction and Working

Construction :

- A single phase induction motor is similar to a 3-ϕ squirrel-cage induction motor in physical appearance. Its rotor is essentially the same as that used in 3-ϕ induction motors. Except for shaded pole motors, the stator is also very similar. There is a uniform air-gap between the stator and rotor but no electrical connection between them. It can be wound for any even number of poles, two, four and six being most common. Adjacent poles have opposite magnetic property and synchronous speed equation, $N_s = \frac{120f}{p}$ also applies.
- The stator windings differ in the following two aspects :
  — **Firstly** single phase motors are usually provided with concentric coils.
  — **Secondly,** these motors normally have two stator windings. In motors that operate with both windings energised, the winding with the *heaviest wire* is known as the **main winding** and the other is called the **auxiliary winding**. If the motor runs with auxiliary winding open, these windings are usually referred as *running* and *starting*.
  — In *most of motors the main winding is placed at the bottom of the slots and the starting winding on top but shifted 90° from the running winding.*

Working :

When the stator winding of a single phase induction motor is connected to single phase A.C. supply, a *magnetic field* is developed, *whose axis is always along the axis of stator coils*. The magnetic field produced by the stator coils is pulsating, varying sinusoidally with time. Currents are induced in the rotor conductors by transformer action, these currents being in such a direction as to oppose the stator m.m.f. Then the axis of the rotor m.m.f. wave coincides with that of the stator field, the *torque angle* is, therefore, *zero* and *no torque is developed on starting*. However, if the rotor is given a push by hand or by other means in either direction, it will pick-up the speed and continue to rotate in the same direction developing operating torque. Thus a single phase induction motor is not inherently self starting and requires some special means for starting.

The above mentioned behaviour of this type of motor can be explained by any one of the following theories :

1. Double revolving field theory
2. Cross-field theory.

The results given by both the theories are approximately same.

Double revolving field theory is described below :

The magnetic field produced by the stator coils is *pulsating*, varying sinusoidally with time. *Ferrari* pointed out that *such a field can be resolved into two equal fields but rotating in opposite directions with equal angular velocities. The maximum value of each component is equal to half the maximum of the pulsating field.*

If the initial time is such that the rotating vectors of the two component fields are along the Y-axis in the positive direction, the two component waves ϕ_1 and ϕ_2 coincide. The resultant of these

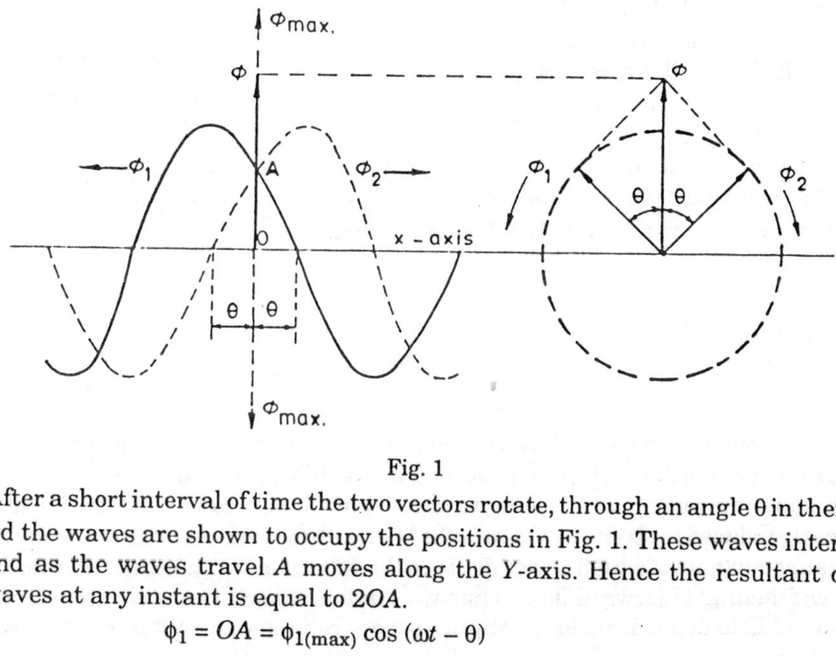

Fig. 1

two is ϕ_{max}. After a short interval of time the two vectors rotate, through an angle θ in their respective directions and the waves are shown to occupy the positions in Fig. 1. These waves intersect at A on the Y-axis and as the waves travel A moves along the Y-axis. Hence the resultant of these two component waves at any instant is equal to $2OA$.

$$\phi_1 = OA = \phi_{1(max)} \cos (\omega t - \theta) \quad ...(i)$$

$$\phi_2 = OA = \phi_{2(max)} \cos (\omega t + \theta) \quad ...(ii)$$

and $$\phi_{1(max)} = \phi_{2(max)}$$

By expanding and adding (*i*) and (*ii*),

$$\phi_1 + \phi_2 = 2\phi_{1(max)} \cos \theta \cos \omega t$$

$$2OA = \phi_{(max)} \cos \theta \cos \omega t \quad ...(iii)$$

which is the equation of the pulsating field and **proves Ferrari's statement.** Thus a single-phase induction motor is *not inherently self-starting*.

The existence of these two fluxes (*forward and backward*) rotating in opposite directions can be verified by supplying a fractional horsepower single-phase induction motor with rated voltage. The motor does not start, but if the shaft is turned by hand, say in clockwise direction, the rotor picks up speed. This means that the rotor conductors are rotating in the direction of that field which rotates in clockwise direction. When the motor is braked and stopped without switching off the supply, the rotor remains at rest. If now the shaft is turned by hand in anti-clockwise direction, the motor picks up speed in that direction. This means that the rotor conductors are now rotating in the direction of the other field.

This behaviour of the motor is *due to the presence of two opposing torques due to the two field.* When the rotor is at rest, (*i.e.* slip = 1) the two torques are equal but opposite in direction. Hence the net torque is zero and therefore the rotor remains at rest. Fig. 2. shows the torque variations due to the two fields. If the rotor is made to speed up in one direction, say in that in which T_1 increases, T_1 exceeds the opposing torque T_2 and the motor beings to accelerate. T_2 goes on diminishing until at the working speed it is negligibly small. Hence the single-phase induction motor rotates in the direction in which it is made to run.

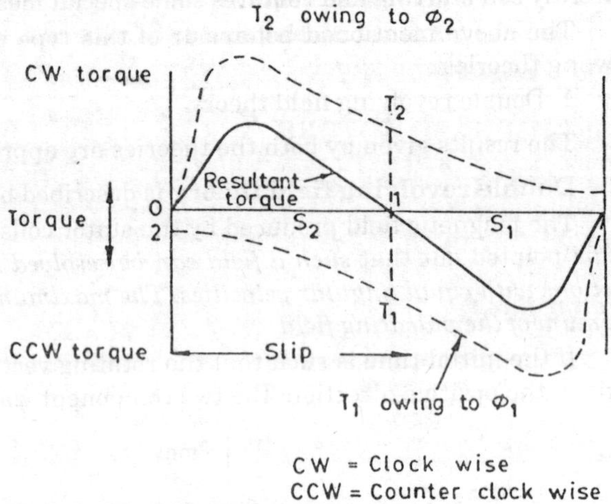

Fig. 2. Balanced torque at standstill in squirrel cage rotor excited by a single-phase winding.

Thus, if the rotor is made to run at speed N by some external means in any direction, say in the direction of forward field, the two slips are now s and $(2 - s)$, as shown below :

The slip of the rotor w.r.t. the forward rotating field F_f,

$$s_f = \frac{N_s - N}{N_s} = s \qquad ...(1)$$

The slip of the rotor w.r.t. the backward rotating field F_b,

$$s_b = \frac{N_s - (-N)}{N_s} = \frac{2N_s - (N_s - N)}{N_s} = (2 - s) \qquad ...(2)$$

- Under normal running condition $(2 - s) >> s$ and as a consequence the backward field rotor currents are much larger than at standstill and have a low power factor. The corresponding opposing rotor m.m.f., owing to stator impedance, causes the backward field to be greatly reduced in strength. On the other hand, the low-slip forward rotating field induces smaller currents of a high power factor in the rotor than at standstill. This leads to greatly strengthening of forward field. Thus weakening of backward field and strengthening of forward field depends upon the slip or speed of rotor and the difference increases with the decrease in slip w.r.t. the forward field or with the increase in rotor speed in forward direction.
- In a single phase induction motor, *the increase in rotor resistance increases the effectiveness of the backward field, which reduces the breakdown torque, lowers the efficiency and increases the slip corresponding to maximum torque.*
- The **performance characteristics** of a single phase induction motor are somewhat *inferior* to that of a 3-phase induction motor due to the presence of backward rotating field.
  — A single phase induction motor has a *lower breakdown torque at larger slip and increased power losses.*

- Greater power input.
- The *speed regulation* tends to be *poorer* than that for a polyphase motor.
- The *power factor tends to be lower* (since the normal slip of a single phase induction motor under load conditions is rather greater than that of the corresponding 3-phase motor).

- In view of the above factors, a single phase induction motor has a *larger frame size* than that of 3-phase motor.
- Single phase motors *tend to be somewhat noiser than 3-phase motors which have no such pulsating torque.*

4. SPLIT-PHASE MOTORS

Since the single-phase induction is not self-starting, means must be provided to create an initial torque. But the initial torque is only possible if a rotating flux is created in the stator. It is known that *a rotating flux is produced when there is a difference of 90° between the currents of two stationary coils. Or if the stator possesses two fluxes having a large phase difference, the result is a rotating flux.*

4.1. Split-phase Resistance—Start Induction Motor

- In a split-phase induction motor the stator is provided with two parallel windings displaced 90 electrical degrees in space and somewhat less than 90° in time. Fig. 3 (*a*) shows the winding diagram of the *two windings* of a split-phase induction motor.

  The *starting winding* has fewer turns and is wound of smaller diameter copper than the running winding. The starting winding, therefore, *has a high resistance and low reactance.*

  The *running or main, winding* (heavier wire of more turns) has a *low resistance* and *high reactance.* Because of its lower impedance, the current in the running winding I_r is *higher* than the current in the starting winding, I_s.

  The phase relations of the lock-motor currents at the instant of starting are shown in Fig. 3 (*b*). The starting winding I_s lags the supply voltage by about 15°, while the greater running winding current lags the single-phase voltage by about 40°. Despite the fact that the current in the two space-quadrature windings are *not* equal, the quadrature components are practically equal.

  If the windings are displaced by 90° in space, and if their quadrature current components, which are displaced by 90° in time, are practically equal, an equivalent two-phase rotating field is produced at starting which develops sufficient starting torque to accelerate the rotor, in the direction of the rotating field produced by the currents.

- As the *motor speeds up*, the *torque developed increases.* Above 85 per cent of synchronous speed, the torque developed by the running winding (main winding) alone is actually greater than that developed by both windings, and it might be advantageous to *open the auxiliary circuit at this cross over point.* To allow for individual variations among motors and switches, however, the contacts are usually designed to open at 75 per cent of synchronous speed. This does not seriously affect the operation, because the running (or main) winding alone usually develops approximately 200 per cent of full-load torque at this speed.

- The *starting winding is not designed for continuous operation, and care should be exercised that it does not remain connected to the supply after it should have been disconnected by the switch.* This series switch is usually *centrifugally operated, and is rather inexpensive.*

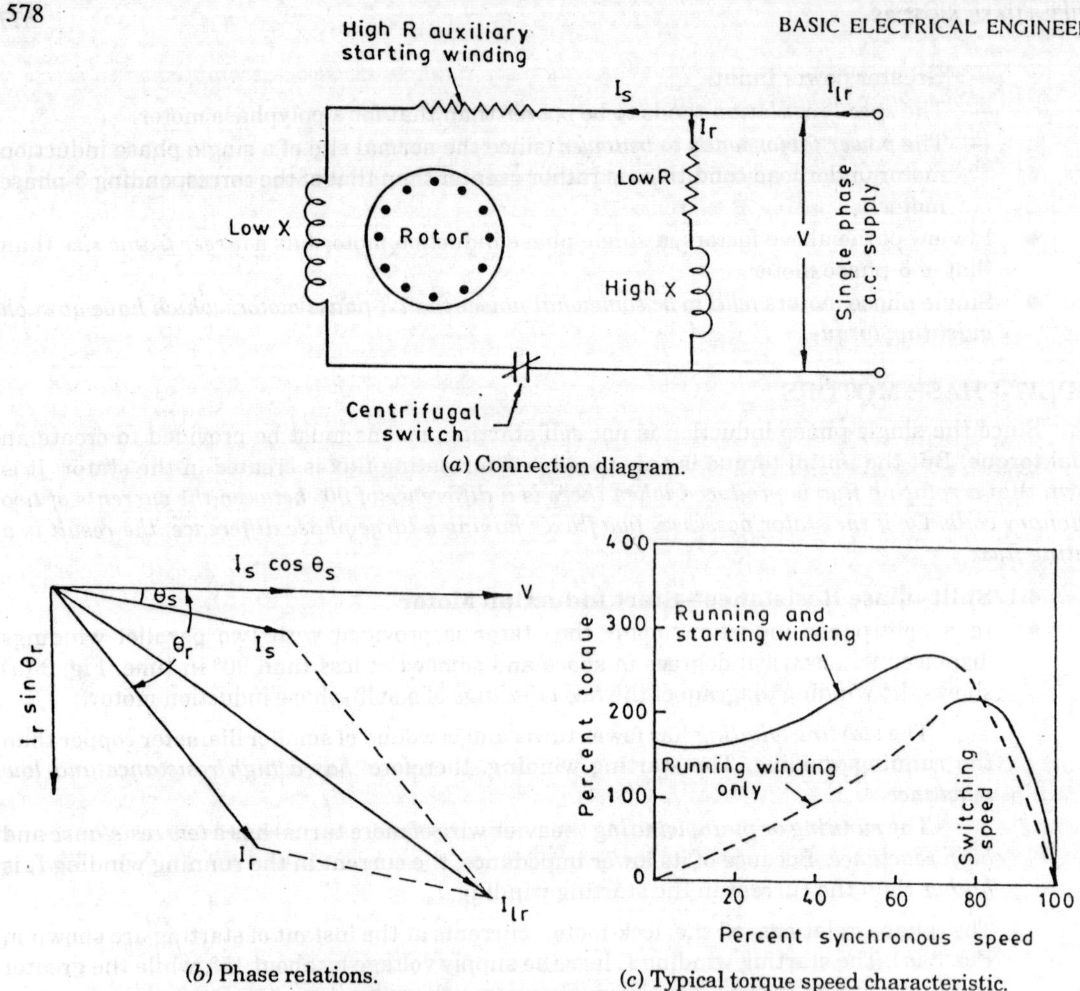

Fig. 3. Split phase resistance-start induction motor.

In case of a hermatically sealed motor, the switch is magnetically operated, and is opened in the de-energized condition.

- Split-phase induction motors may be *reversed by reversing the line connections of either the main or the auxiliary winding*. If however, reversal is attempted under normal running condition, nothing will happen.

  If it is necessary to *reverse the motor while it is rotating*, then some means must be incorporated to slow the motor down to the speed where the starting-switch contacts close, placing the starting winding across the supply lines. This may be done by *incorporating a timing device* which first disconnects the motor entirely from the line and then reverses one field at the proper time. A mechanical braking device which can be electrically operated may also be used.

- Speed control of split-phase windings is a *relatively difficult* matter since the synchronous speed of the rotating stator flux is determined by the frequency and number of poles developed in the running stator winding $\left(N_s = \dfrac{120f}{p}\right)$. By adding stator windings to change the number of poles, speed variation may be obtained. This, however, is a stepped speed change, as in polyphase induction motor, rather than a continuous variation. It must be pointed out, however, that *all speed changes must be accomplished in a range above that at which the centrifugal switch operates.*

Fig. 3 (*c*) shows the typical torque speed characteristics :

- The starting torque is 1.5 times to twice the full-load starting torque and starting current is 6 to 8 times full-load current.
- The speed regulation is very good.
- The percent slip is about 4-6 percent.
- Such a motor may operate with a power factor of 0.55—0.65 and efficiency of 60—65 percent.
- These motors are made in fractional kW $\left(\frac{1}{20} \text{ to } \frac{1}{4} \text{ kW}\right)$ ratings with speed ranging from 2875 to 700 r.p.m.

Shortcomings and uses. The major objections to the motor are (1) its *low starting torque* ; and (2) that, when heavily loaded, the slip exceeds 5 percent, reducing the e.m.f. and producing an elliptical or *pulsating torque which makes* the motor somewhat *annoyingly noisy*. For this reason, the split-phase motor is used in appliances to drive loads which are themselves noisy : *oil burners, machine tools, grinders, dish washers, washing machines, air blowers and air compressors.*

- *Because of low starting torque these motors are seldom employed in sizes larger than* $\frac{1}{4}$ *kW.*

4.2. Split-phase Capacitor—Start Induction Motor

Another method of splitting the single-phase supply into two phases to be applied to the stator windings is *placing a capacitor in series with the starting auxiliary winding.* In this manner, the current in the starting winding may be made to lead the line voltage. Since the running winding current lags the line voltage, the phase displacement between the two currents be made to approximately 90° on starting. The circuit of capacitor-start motor is shown in Fig. 4 (*a*), while the vector diagram of the currents and voltage is shown in Fig. 4 (*b*). The values of the angles shown are fairly representative, and are rounded off for convenience. *One of the factors upon which the starting torque depends is the sine of the angle between the currents in the two windings.* The value of series capacitor may therefore be reduced, while maintaining a phase-shift angle of about 90°.

- The increase in phase angle between starting and running winding currents is not the only difference between the split-phase and capacitor-start motors. *The split phase motor must keep the number of starting-winding turns low, so that the current may be nearly in phase with the line voltage.* This, *however, is unnecessary in a capacitor-start motor, since the capacitor can overcome the inductance of the winding while still providing the proper phase shift.* There are thus *more auxiliary starting turns* in the capacitor-start motor than in the comparable split-phase motor. This provides a greater number of ampere-turns, hence a *larger rotating flux,* and therefore a *further increase in the starting torque.*

Also it is seen that for the same magnitudes of field currents, the *current* I_{lr} *is less in capacitor-start motor,* because of the greater angle between the two field currents. In addition, the *starting power factor is also better.* For a given line current, the starting torque is thus much higher for a capacitor-start motor than for a split-phase induction motor. *The starting torque of capacitor-start motor is from 3 to 4.5 times the full-load torque, while that of split-phase resistance-start induction motor rarely exceeds twice the full-load torque.*

- These motors are manufactured in *ratings ranging from* $\frac{1}{10}$ *kW to* $\frac{3}{4}$ *kW, but larger sizes are also available.*

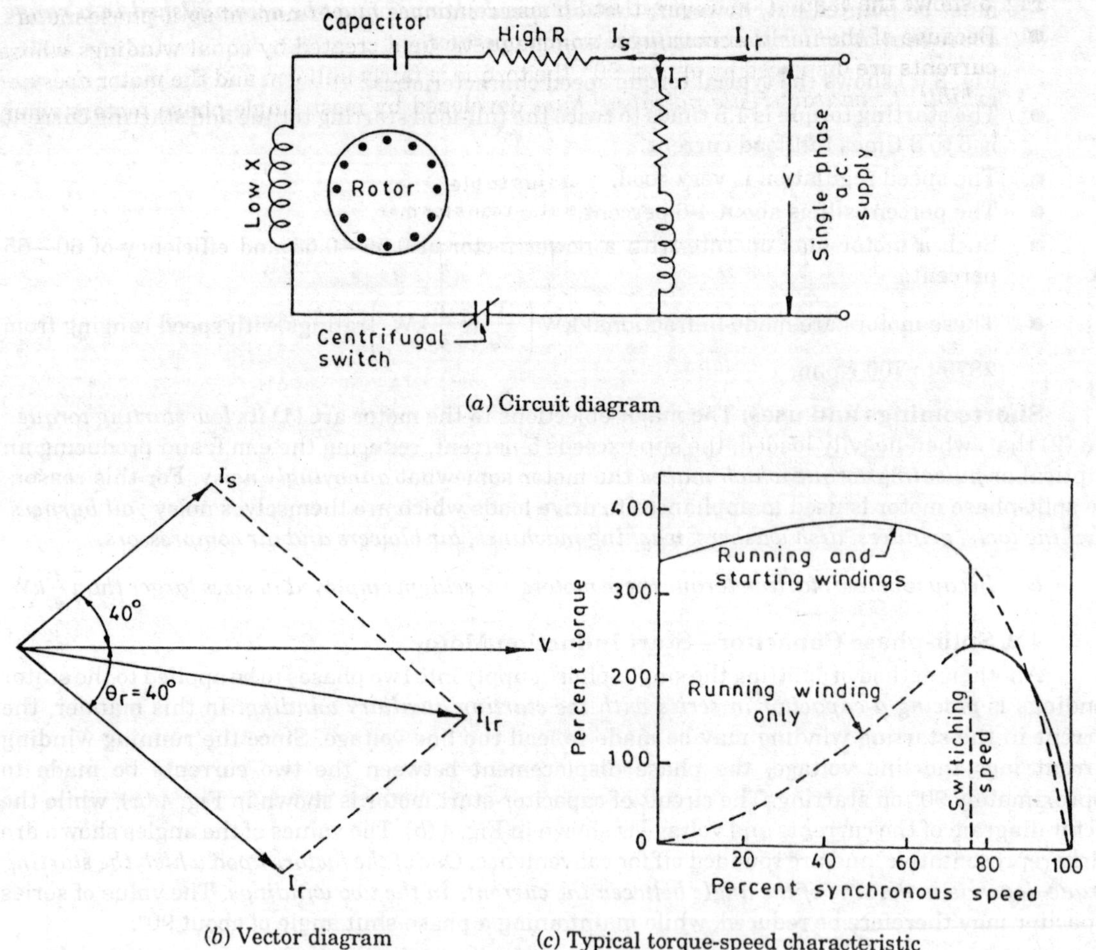

(a) Circuit diagram

(b) Vector diagram

(c) Typical torque-speed characteristic

Fig. 4. Capacitor-start induction motor.

- The capacitor-start motor may be reversed by changing the connections of one of the windings, but it is subject to the same limitations as the resistance-start induction motor.

Uses. By virtue of their higher starting torque, capacitor-start split-phase motors are used for *pumps, compressors, refrigeration units, air-conditioners,* and *large washing machines,* where a split-phase motor is required that will develop high starting torque under load.

4.3. Permanent-split Capacitor Induction Motor

- A permanent-split capacitor motor (also called single-value capacitor motor) is a single-phase induction motor which has the *same capacitor* in series with the starting (or auxiliary) winding for both starting and running. Because it runs continuously as a permanent split-phase motor, no centrifugal switch is required.
- The motor starts and runs by virtue of the quadrature phase-splitting produced by the two identical windings. As a result, *it does not possess the high running torque* produced by either the resistance-start or the capacitor-start motor.

Furthermore, the capacitor used is designed for continuous duty and is of an oil-filled type. The value of the capacitor is based on its optimum running rather than its starting characteristics. The result is that this motor *has a very poor starting torque,* about 50 to 100 per cent of rated torque.

Fig. 5 shows the connection diagram and phase relations of a permanent split-phase motor.

- Because of the fairly uniform rotating magnetic field created by equal windings whose currents are displaced by almost 90°, the torque is fairly uniform and the motor *does not exhibit the characteristic pulsating hum* developed by most single-phase motors when loaded.

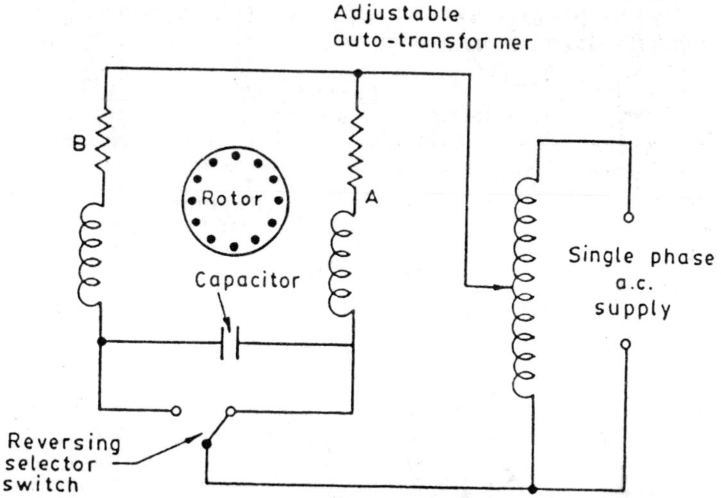

(*a*) Connection diagram.

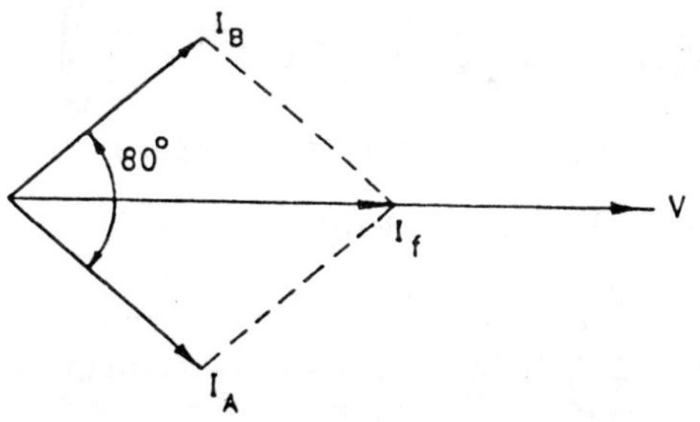

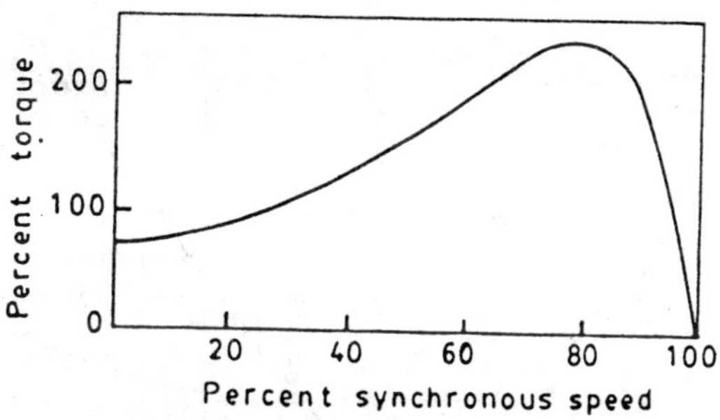

(*b*) Vector diagram or phase relations. (*c*) Typical torque-speed characteristic.

Fig. 5. Permanent split-phase motor.

This motor possesses the following *merits* :

(*i*) Higher power factor at full-load

(*ii*) Lower full-load line current

(*iii*) Higher full-load efficiency

(*iv*) Increased pull-out torque.

- The permanent-split capacitor motor is *more expensive* than the equivalent split-phase or capacitor-start induction motor. This is primarily due to the fact that the auxiliary (or starting) winding is now also a running winding. It must therefore have a continuous duty rating and as such is heavier than if it were were short-time-rated.

Uses. Because of its *instant response as a reversing motor*, its *quieter operation*, and the *possibility of speed control*, the permanent-split capacitor motor is used for *exhaust and intake fans* and *blowers, unit heaters* and *office machines*.

4.4. Two-value Capacitor Induction Motor

- Refer Figs. 6 and 7. *The two-value capacitor motor combines the quiet operation and limited speed control advantages of a running permanent-split capacitor with high starting torque of the capacitor-start motor.*

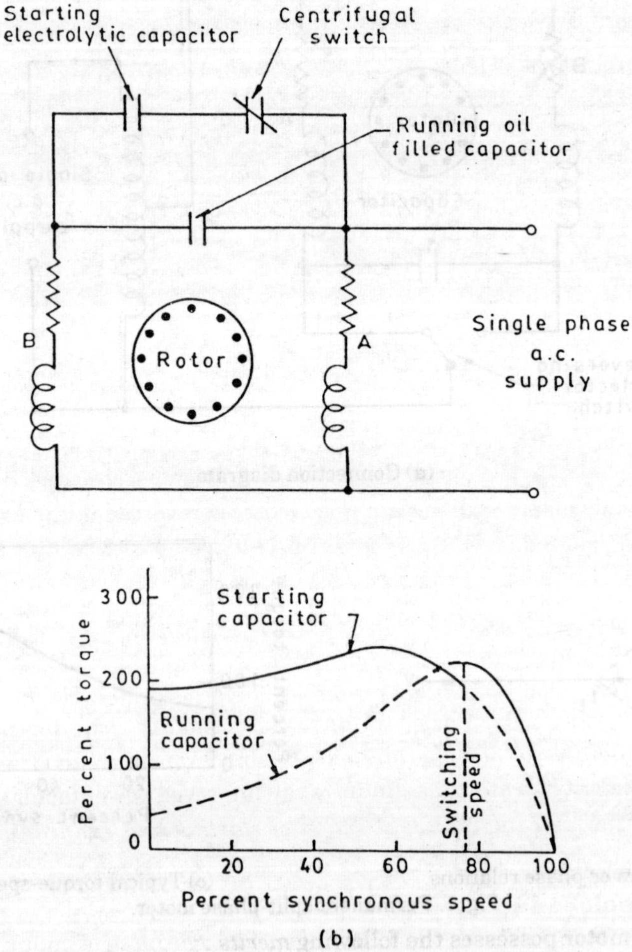

Fig. 6. Connection diagram for two-value capacitor motor and typical torque-speed characteristic.

- Two capacitors (Fig. 6) are employed during the starting period. One of these, an electrolytic starting capacitor, similar to that used for the intermittent duty of the capacitor-start split-phase motor, is of fairly high capacity (about 10 to 15 times the value of the running capacitor) and is cut out of the circuit by a centrifugal switch when slip reaches about 25 per cent.
- The *major advantage of the two value capacitor motor is its high starting torque, coupled with quiet operation and good running torque.* Still classed as a reversing motor, when

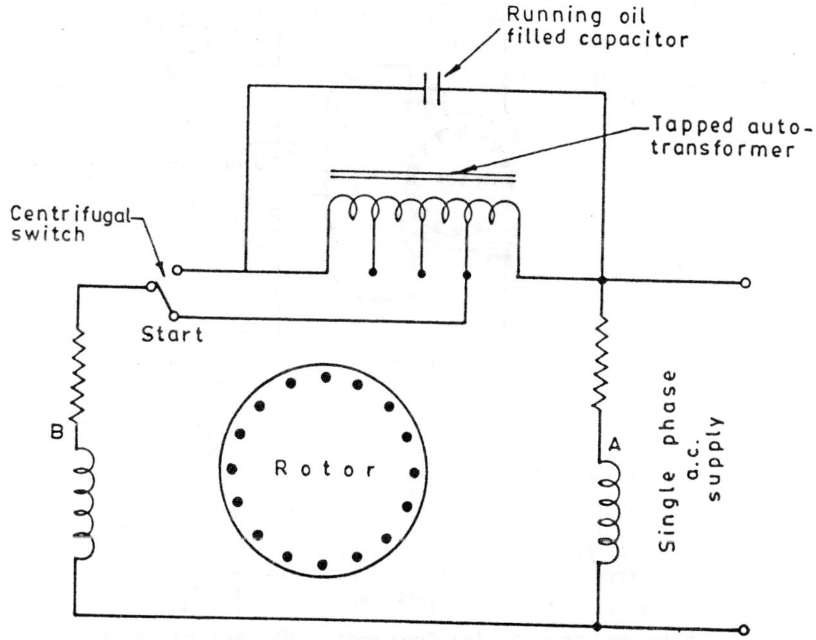

Fig. 7. Two-value capacitor motor with auto-transformer.

the line leads of one winding are reversed, it is reversed in the usual manner. When the speed drops below 25% slip during reversing, the centrifugal switch closes its starting contact, providing the maximum torque as the motor slows down and reverses. The contacts open again when the motor is up to 75% synchronous speed in the reverse direction. Frequent reversals, therefore will reduce the life of the centrifugal switch. For this reason, where frequent reversals are accomplished, permanent-split, single-value capacitor, using no centrifugal switch whatever, is preferable.

Uses. This two-value capacitor motor finds use in *smaller home air-conditioning units* which use this motor in its *compressor* and operates on a 15 amp. branch circuit. The lower starting current and lower running current (7.5 amp. maximum) at an improved power factor over the capacitor-start motor, are obtained through the *precise selection of starting and running capacitors* for the fixed compressor load.

4.5. Shaded-pole Induction Motor

A shaded-pole motor is one of the *simplest* and *cheapest* of manufactured motors. It is essentially an induction machine, since its squirrel-cage rotor receives power in much the same way as does the rotor of the polyphase induction motor. There is however, one extremely important difference between the two. Whereas the poly-phase induction motor creates a true revolving field, in the sense that it is constant in magnitude and rotates at synchronous speed *completely round the entire core*, the field of the shaded-pole motor is not constant in magnitude but *merely shifts from one side of the pole to the other*. Because the shaded-pole motor does not create a true revolving field, the *torque is not uniform but varies from instant to instant.*

Fig. 8 shows the *general construction and principle of shaded pole motor.*

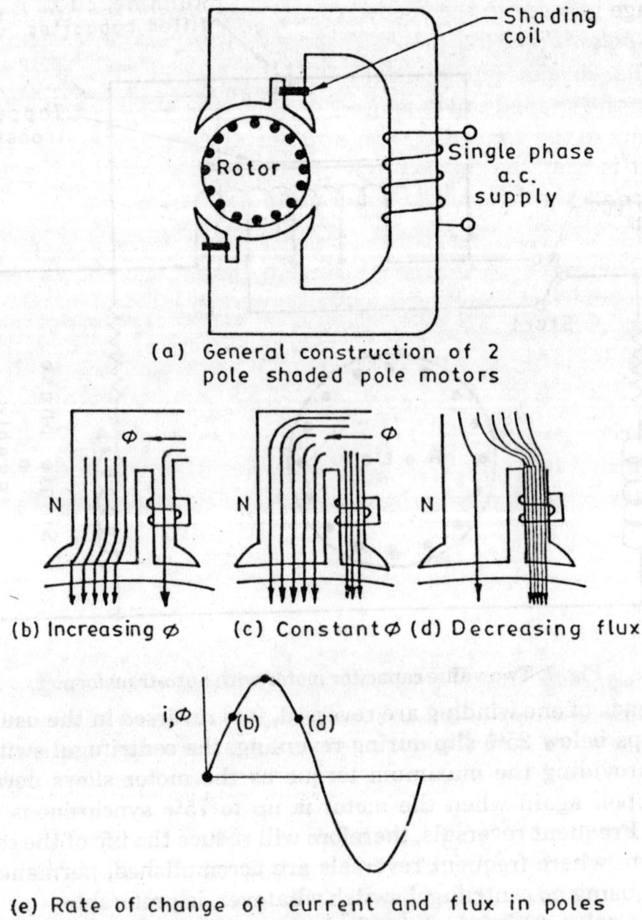

Fig. 8. General construction and principle of shaded pole motor.

Construction. Each of the laminated poles of the stator has a slot cut *across* the laminations about one-third the distance from one edge. Around the smaller of the two areas formed by this slot is placed a heavy *copper short-circuited coil,* called a *shading coil* ; the iron around which the shading coil is placed is called the *shaded* part of the pole, while the free portion of the pole is the *unshaded* part. The exciting coil surrounds the entire coil.

Principle of operation. When the exciting winding is connected to an A.C. source of supply, the *magnetic axis will shift from the unshaded part of the pole to the shaded part of the pole.* This shift in the magnetic axis is, in effect *equivalent to an actual physical motion of the pole* ; the result is that the squirrel-cage rotor will rotate in a direction from the unshaded part to the shaded part. The shifting of flux is explained below.

- Refer Fig. 8 (*b*). When the flux in the field poles tend to increase, a short-circuit current is induced in the shading coil, which by Lenz's law opposes the force and the flux producing it. Thus, as the flux increases in each field pole, there is a *concentration of flux in the main segment of each pole,* while the shaded segment opposes the main field flux.

- At point (*c*) shown in Fig. 8 (*e*), the rate of change of flux and of current is zero, and there is no voltage induced in the shaded coil. Consequently, the *flux is uniformly distributed across the poles* [Fig. 8 (*c*)].
- When the flux decreases, the current reverses in the shaded coil to maintain the flux in the same direction. The result is that the flux crowds in the *shaded segment of the pole* [Fig. 8 (*d*)]

  A typical torque-speed characteristic is shown in Fig. 9.
- Shaded pole motors are built up to about 40 W.

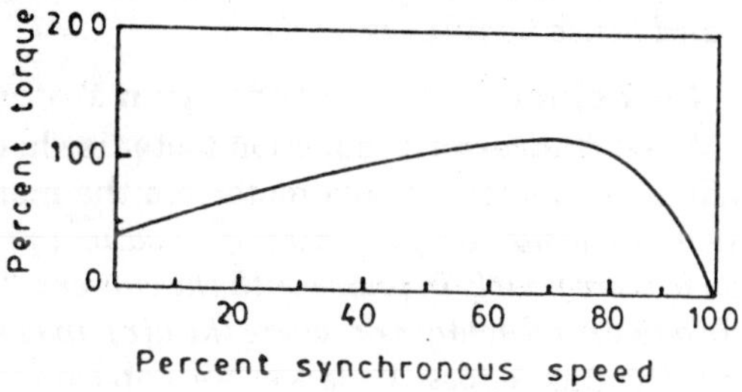

Fig. 9. Typical torque-speed characteristic of shaded pole motor.

Merits :

(*i*) Rugged construction (*ii*) Cheaper in cost
(*iii*) Small in size (*iv*) Requires little maintenance
(*v*) Its stalling locked-rotor current is only slightly higher than its normal rated current, so that it can remain stalled for short periods without harm.

Demerits :

(*i*) Very low starting torque (*ii*) Low efficiency
(*iii*) Low power factor.

Uses. Its low starting torque limits its application to *phono-motors* or *turntables, motion picture projectors, small fans and blowers, bending machines, rotating store-window display tables, and relatively light loads.*

Reversing shaded-pole motor. Fig. 10 shows a reversing shaded-pole motor with the switch in the clockwise rotation position. This places 1 and 3 actively in the circuit, being short-circuited by

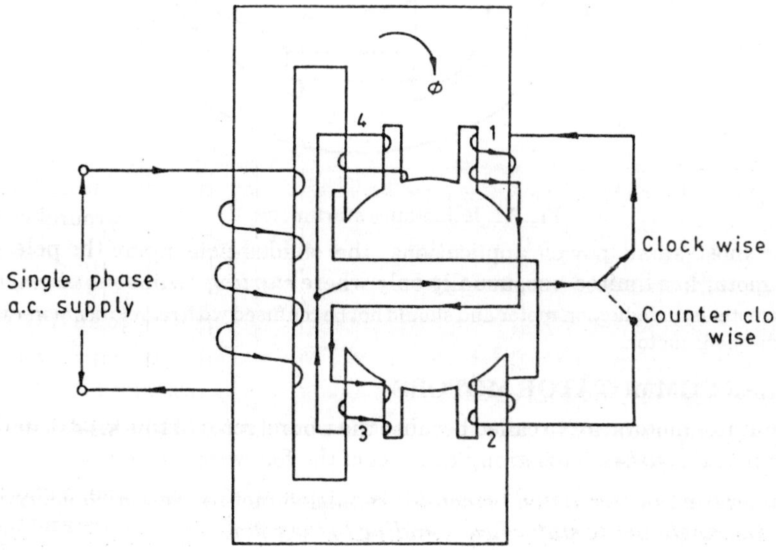

Fig. 10. Reversing shaded pole motor.

the reversing switch. Coils 2 and 4 remain in series with each other, but are on open circuit and so are inactive. The main-field flux is shown increasing vertically downward, and the arrows on the shading coils show the current in them. It is then seen that the *shading coil must be connected in such a manner as not to have their induced voltages in opposition, or there may not be any current in them, and hence no flux lag.*

4.6. Reluctance-start Induction Motor

A reluctance-start induction motor is shown in Fig. 11. Its characteristics are similar to that of shaded pole motor. In this motor too the magnetic field shifts across the pole, but the effect is obtained by the *non-uniform air gap of salient poles. Where there is a greater air gap, the flux in that portion is more nearly in phase with the current. There is a greater lag between flux and current where there is a lower reluctance or where the air gap is smaller.* Since both fluxes are produced by the same current, the flux across the larger air gap *leads* the flux across the smaller one. The two fluxes are obviously displaced in time, and so the *magnetic field shifts across the poles from larger air gap to the shorter gap*. Thus the *direction of rotation is firmly fixed by the construction, and the motor cannot be reversed at all.*

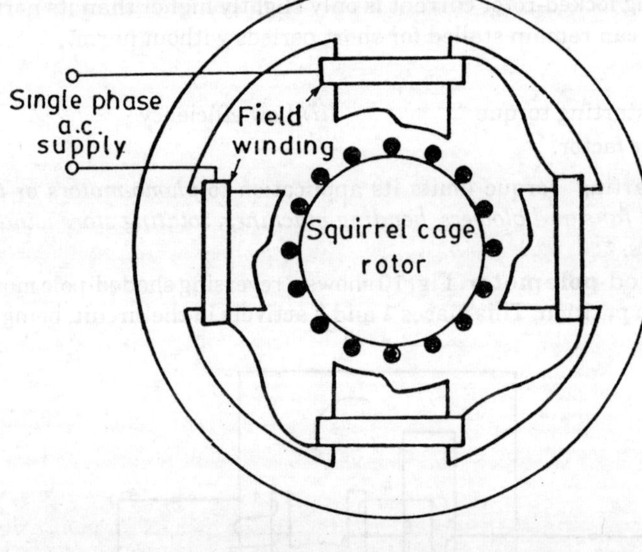

Fig. 11. Reluctance start motor.

Uses. For most small power applications, the shaded-pole motor is preferred, and the reluctance-start motor has limited use, usually only where *starting torque requirements are low.*

Note. This motor is an induction motor and should not be confused with *reluctance motor* which is actually a non-excited synchronous motor.

5. SINGLE-PHASE COMMUTATOR MOTORS

The commutator motors are so called because the wound rotor of this kind of motor is equipped with a *commutator and brushes.* This group consists of the following two classes :

1. Those operating on *'repulsion principle'* (repulsion motors) *in which energy is inductively transferred from the single-phase stator field winding to the rotor.*

2. Those operating on the *principle of the series motor* in which the energy is *conductively* carried both to the rotor armature and its series-connected single-phase stator field.

5.1. Repulsion Motor

A repulsion motor in its simplest form consists of a *field comprising a distributed winding,* housed in slots, in a smooth-cored stator and *an armature carrying a distributed winding connected to a commutator*. The stator winding, which produces the *main field,* is connected to the main supply. The armature or rotor winding is not connected electrically to the main circuit, but the brushes, which are set at an angle to the direction of the main flux, are short-circuited as shown in Fig. 12.

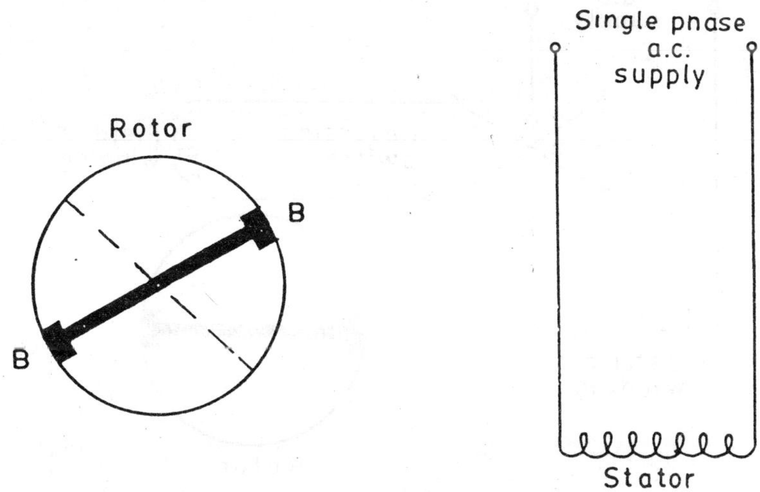

Fig. 12. Repulsion motor.

- If the brush axis *BB* is set *at right angles* to the direction of the stator flux, the e.m.f. induced in one-half of the rotor winding is *exactly balanced* by the e.m.f. induced in the other half, so that resultant e.m.f. is *zero* ; no current flows in the rotor winding and *no-torque is developed.*
- If the brush axis is placed *in line with the direction* of the stator flux, the e.m.f. in one half assists that in the other, so that *a maximum current flows.* Here again *no torque is developed,* since the torque of one half of the rotor conductors is exactly balanced by that due to the other half.
- If *the brush axis makes an angle with the stator flux as shown, a resultant torque is produced.* The torque is a *maximum theoretically, if this angle is 45°, but in practice the angle of inclination is about half this value.*

  It is clear from above that the *speed of repulsion motor depends upon the brush position. Speed control of* such a machine can be provided by *mounting the brushes on a rocker* which can be rotated by a lever handle mounted on the motor end-shield. If remote control is required, the lever handle may be manipulated by a simple system of rods and cranks.

  Alternatively, if the motor is to be totally enclosed, or remote control from a considerable distance is required, speed control may be obtained by the *use of an external series resistance with fixed brush gear.*

- The direction of rotation of a simple repulsion motor may be reversed by swinging the brushes into the position shown *dotted* in Fig. 12.

Atkinson repulsion motor. A modification of the simple repulsion motor is the Atkinson repulsion motor, in which the stator winding comprises two windings at right-angles to each other and connected in series, as shown in Fig. 13. One advantage obtained by this method is that the *direction of rotation can be reversed by reversing the connections to one of the stator windings.* Instead of moving the brush rocker, it is necessary only to throw the reversing switch, shown in Fig. 13.

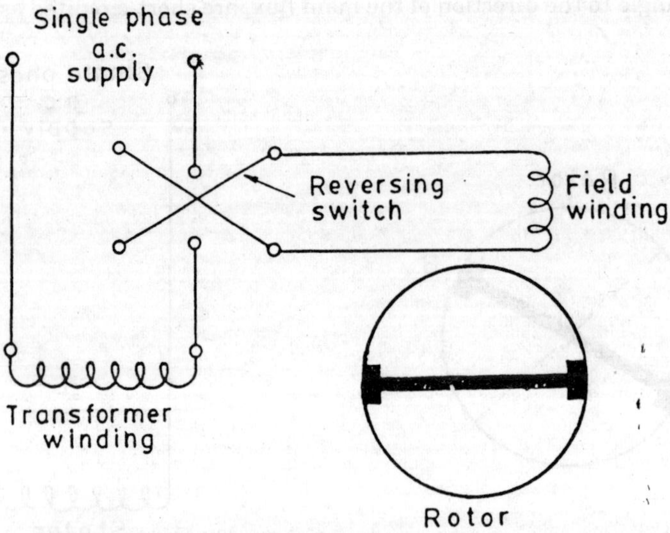

Fig. 13. Atkinson repulsion motor with reversing switch.

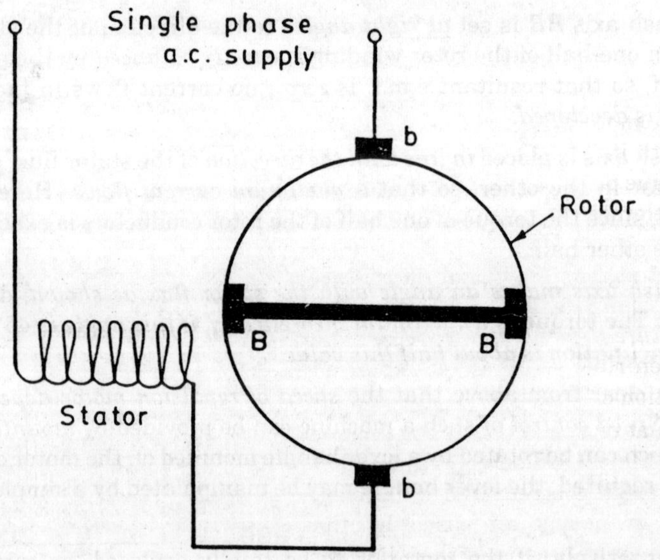

Fig. 14. Compensated repulsion motor.

It will be observed that, as the rotor is electrically connected to the stator, the compensated repulsion motor is not able to operate directly from a high-voltage supply, as was the case with the simple repulsion motor.

Fig. 15 shows the typical speed-torque characteristics of single-phase repulsion motor.

Compensated repulsion motor. The *power factor of the machine may be improved by compensation,* and this is done by providing *an additional set of brushes* as shown in Fig. 14. The

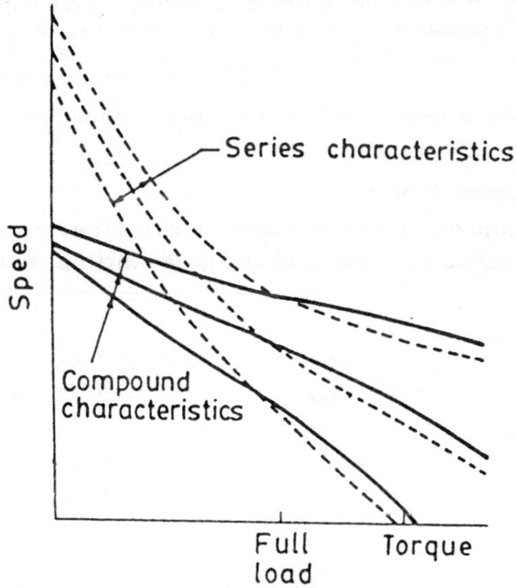

Fig. 15. Typical speed-torque characteristics of single-phase repulsion motors.

axis of one pair, *BB*, of brushes coincides with the stator winding, these brushes are *short-circuited.* The other pair, *bb*, of brushes is set at right-angles to the former and is connected in *series with the stator winding.*

5.2. Repulsion-start Induction Motor

- As its name implies the repulsion-start induction motor *starts as a repulsion motor* with its brushes set to the maximum torque position. When the load has been accelerated to about 75 per cent of synchronous speed, a built in centrifugal device *places a shorting ring* in contact with the commutating bars, *converting the armature to squirrel-cage rotor.* The motor then runs as induction motor on its induction characteristic (Fig. 16).
- Although at one time this type of motor was used almost exclusively where high starting torque was required, it has been replaced in nearly all cases by the capacitor motors because of the following reasons :

  (*i*) Requires more maintenance.

  (*ii*) More expensive.

  (*iii*) Makes quite a bit of noise on starting.

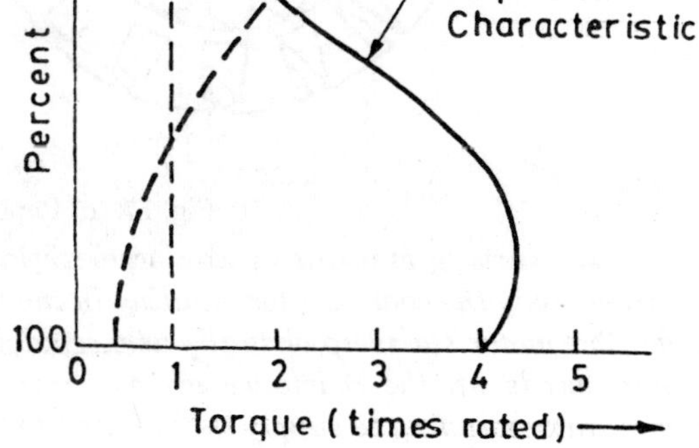

Fig. 16. Speed-torque characteristic of repulsion-start induction motor.

(*iv*) Causes radio interference when starting.

(*v*) Cannot be reversed easily.

Repulsion-start motors, despite these disadvantages, are still used in *integral-horsepower sizes* because of the following reasons :

(*i*) High starting torque. (*ii*) Low starting current.

(*iii*) Ability to accelerate a heavy load more rapidly than high-capacitance dual-capacitor motors.

5.3. Repulsion-induction Motor

- A single-phase repulsion-induction motor combines the constant-speed characteristics of the single-phase induction motor with the good starting characteristics of the repulsion motor.
- The *stator of this machine has a simple single-phase winding like that of single-phase induction motor.* The rotor, however, is built up of *laminations,* each of which has *two concentric sets of slots. These slots contain two distinct windings ; in the outer slots is wound a commutator winding similar to that of a D.C. armature, while in the inner slots is a cast aluminium squirrel-cage winding which clamps the laminations.*

Fig. 17 (*a*) illustrates an armature for a repulsion induction motor complete with a squirrel-cage winding.

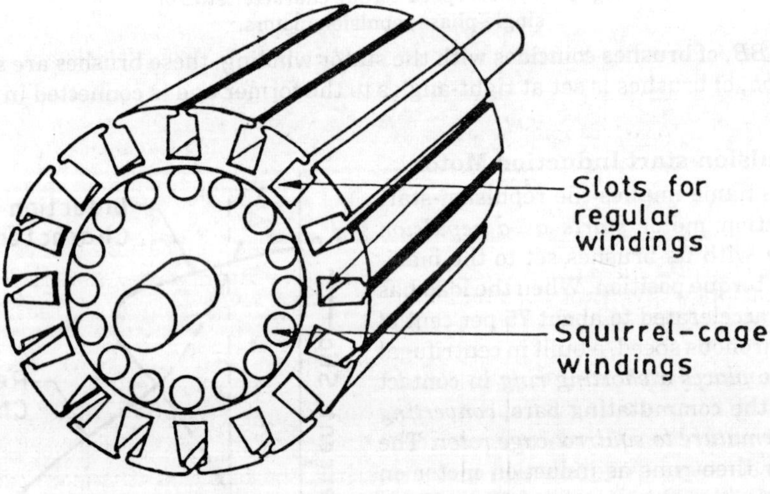

Fig. 17. (*a*) Repulsion induction motor.

At starting and during the acceleration period, the magnetic flux produced by the stator *embraces only the commutator winding in the outer slots owing to the high reactance of the squirrel-cage.* The *motor starts up virtually as a repulsion motor and develops a high starting torque. As the motor speeds up the reactance of the squirrel-cage decreases, so that this winding assists the commutator winding to supply the running torque.*

Fig. 17 (*b*) shows the speed-torque characteristic of repulsion-induction motor.

Merits. The repulsion-induction motor has the following merits :

(*i*) High starting torque. (*ii*) Fairly good speed regulation.

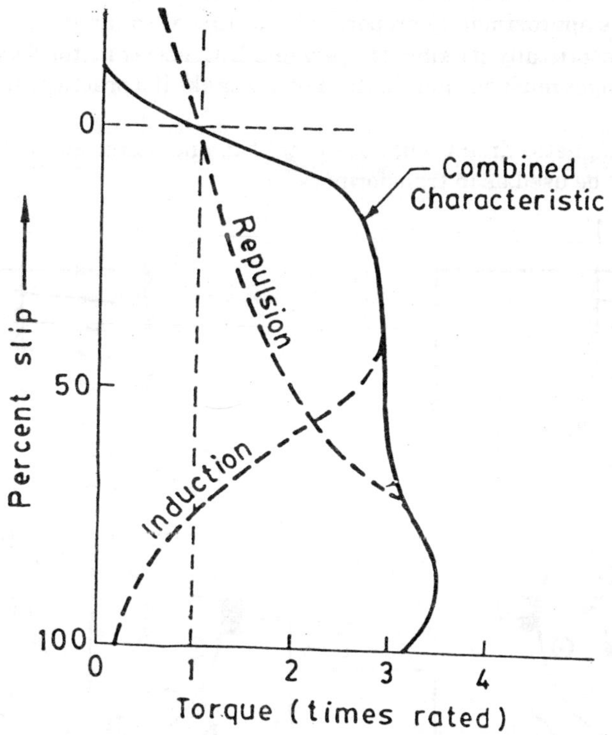

Fig. 17. (*b*) Speed-torque characteristic of repulsion-induction motor.

(*iii*) Major virtue is the ability to continue to develop torque under sudden, heavy applied loads without breaking down.

Uses. Such motors are suitable for all single-phase power applications *which require a high starting torque and constant speed when running* ; they also operate *at a very high power factor*. They are particularly well adapted to drive *machine tools*, lifts, hoists, *mixing machines, centrifugal pumps, fans and blowers.*

5.4. A.C. Series Motor

The series motor due to its desirable speed-torque characteristics *is almost exclusively used in railway service*. While the D.C. motor is entirely satisfactory for this class of work service and is generally used on street railway cars and trolley coaches, the fact that it is more convenient and more economical to transmit power and to transform voltages in A.C. systems than with direct currents has lead to the development of the A.C. series motor for use on some of the important steam-road electrifications.

Working principle. The working principle of an A.C. series motor is the same as that of the D.C. series motor. The armature and field are wound and interconnected in the same manner as the D.C. series motor.

When an alternating e.m.f. is applied to the terminals, since field and armature windings are connected in series, the field flux and armature current reverse simultaneously every half cycle, but the direction of the torque remains unchanged. The torque is pulsating, but its average value is equal to that which a D.C. motor will develop if it had the same r.m.s. value of flux and current. Motor connections, direction of torque, etc. for two successive half cycles are shown in Fig. 18. If the field and armature core are run at low saturation, the air-gap flux is approximately proportional to the current and the torque is approximately proportional to current squared.

Although it is theoretically possible to operate a D.C. series motor from an A.C. circuit, the following structural changes must be made in the motor to make it a practical and reasonable efficient machine :

- The entire magnetic circuit must be *laminated,* and materials with low iron-loss co-efficients should be used as in transformers.

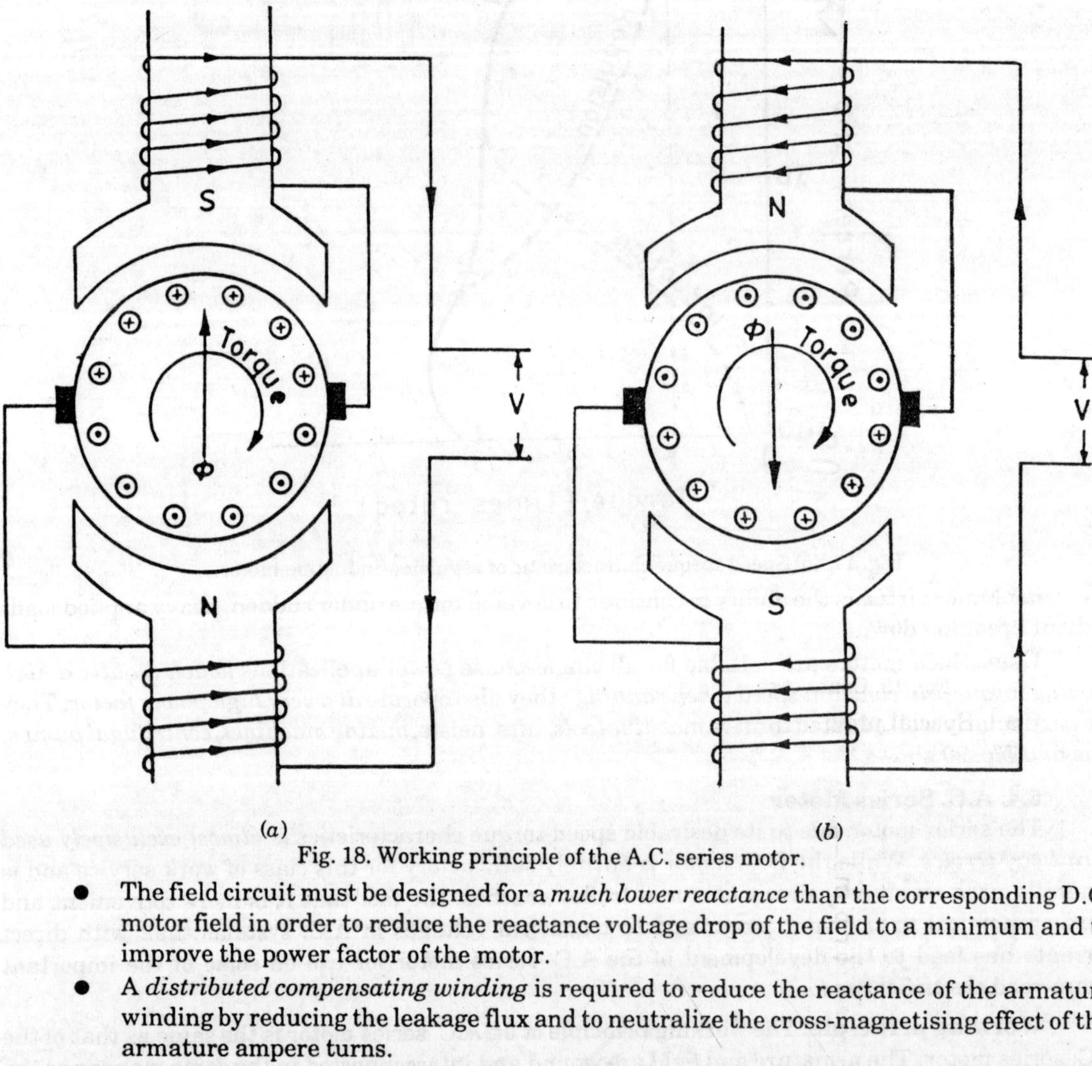

Fig. 18. Working principle of the A.C. series motor.

- The field circuit must be designed for a *much lower reactance* than the corresponding D.C. motor field in order to reduce the reactance voltage drop of the field to a minimum and to improve the power factor of the motor.
- A *distributed compensating winding* is required to reduce the reactance of the armature winding by reducing the leakage flux and to neutralize the cross-magnetising effect of the armature ampere turns.

The compensating winding may be connected in series with the series-field and armature windings, or it may be short-circuited upon itself and receive its excitation voltage by transformer action, since it is inductively coupled with the armature cross-field (Fig. 19). In the first case, the motor is said to be *conductively* compensated, while in the second it is *inductively* compensated. *Conductive compensation* is required on motors which are intended for operation in *D.C. as well as A.C. circuits*.

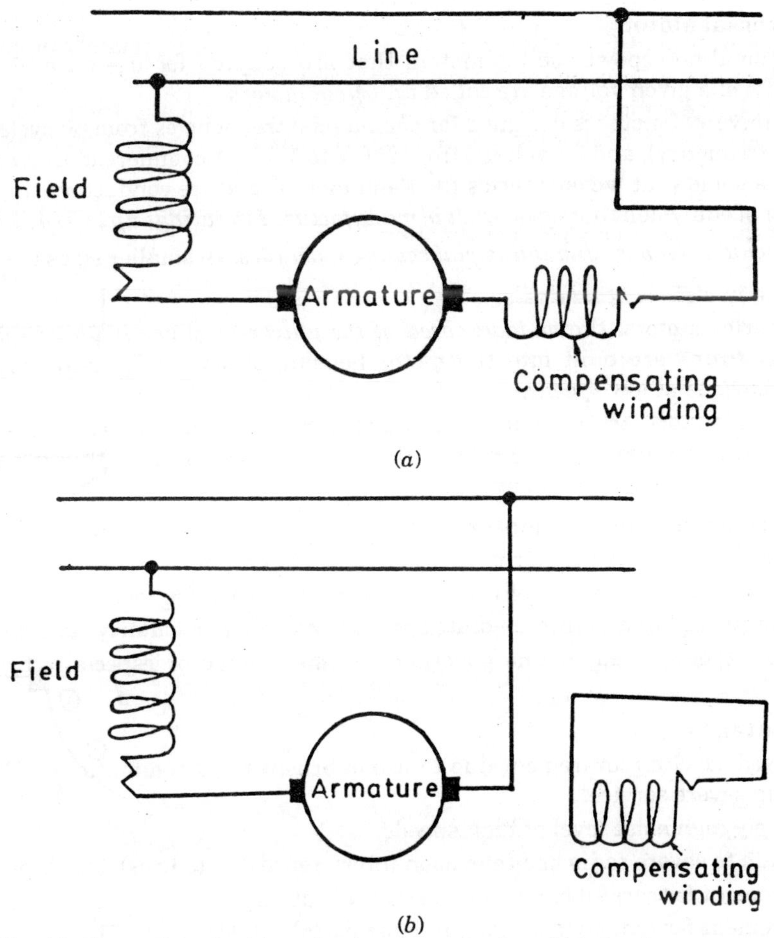

Fig. 19. Connections for (*a*) conductive compensation ; (*b*) inductive compensation.

- Special provision must be made to secure *satisfactory commutation*.

Fig. 20 shows the A.C. series motor characteristic.

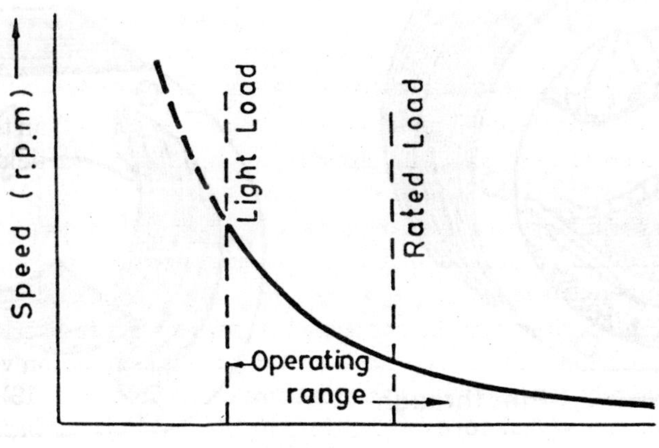

Fig. 20. A.C. series motor characteristics.

5.5. Universal Motor

- Fractional-horsepower series motors that are adapted for use on either D.C. or A.C. circuits of a given voltage are called *universal motors.*
- The universal motor is designed for commercial frequencies from 60 cycles down to D.C. (zero frequency), and for voltage from 250 V to 1.5 V. A commercial universal motor may have a somewhat weaker series field and more armature conductors than a D.C. series motor of equivalent horsepower. *It is manufactured in ratings up to 3/4 H.P., particularly for vacuum cleaners and industrial sewing machines.* In smaller sizes of $\frac{1}{4}$ H.P. or less, it is used in *electric hand drills.*

Like all series motors, the *no-load speed of the universal motor is universally high.* Quite frequently, *gears trains* are built into the motor housing of some universal motors to provide exceedingly high torque at low speeds.

When these motors are used in commercial appliances such as *electric shavers, sewing machines, office machines, and small hand hair dryers or vacuum cleaners,* they are always *directly loaded* with little danger of motor runaway.

Advantages of a universal motor :

1. High speed from above 3600 r.p.m. to around 25000 r.p.m.
2. High power output in small physical sizes for use in portable tools.
3. High torque at low and intermediate speeds to carry a particularly severe load.
4. Variable speed by adjustable governor, by line voltage or especially by modern pulse techniques.

Disadvantages :

1. Increased service requirement due to use of brushes and commutators. The life of these parts is limited in severe service.
2. Relatively high noise level at high speeds.
3. Moderate to severe radio and television interference due to brush sparking.
4. Requirement for careful balancing to avoid vibration.
5. Requirement for reduction gearing in most portable tools.

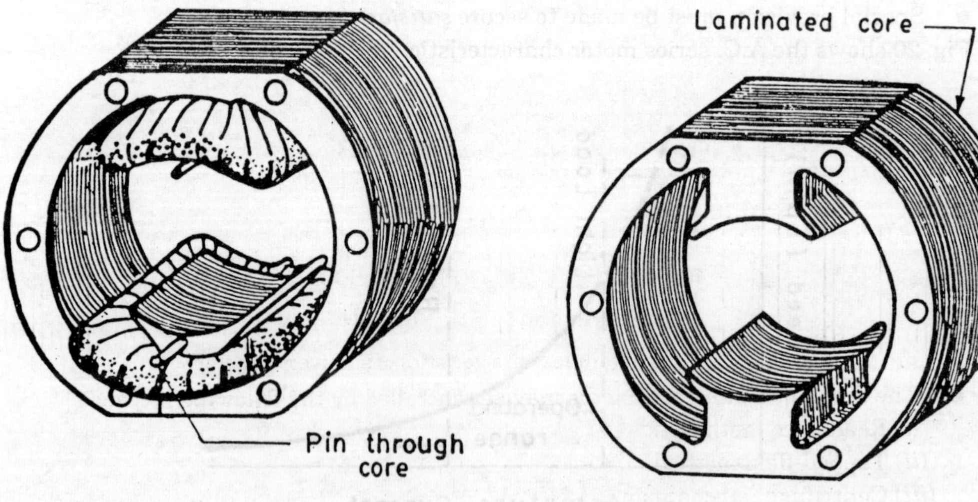

Fig. 21. Field core of a two pole universal motor.

Universal motors are manufactured in two types :

1. Concentrated-pole, non-compensated type (low H.P. rating).
2. Distributed field compensated type (high H.P. rating).

Fig. 21 shows the laminated field structure of a typical concentrated field universal motor.

Operation of a Universal Motor. As explained in Art. 5.4, such motors develop unidirectional torque regardless of whether they operate on D.C. or A.C. supply. The production of unidirectional torque when the motor runs on A.C. supply can be easily understood from Fig. 22. The motor works on the same principle as a D.C. motor *i.e.*, the force between the main pole flux and the current carrying armature conductors. This is true regardless of whether current is alternating or direct.

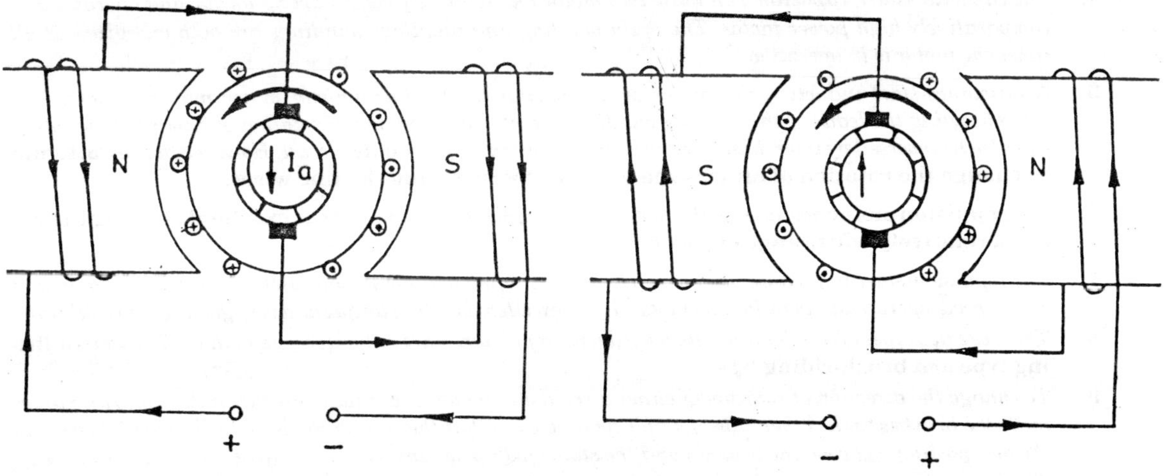

Fig. 22. Universal motor operation.

- Fig. 23 shows the typical torque characteristics of a universal motor both for D.C. and A.C. supply.

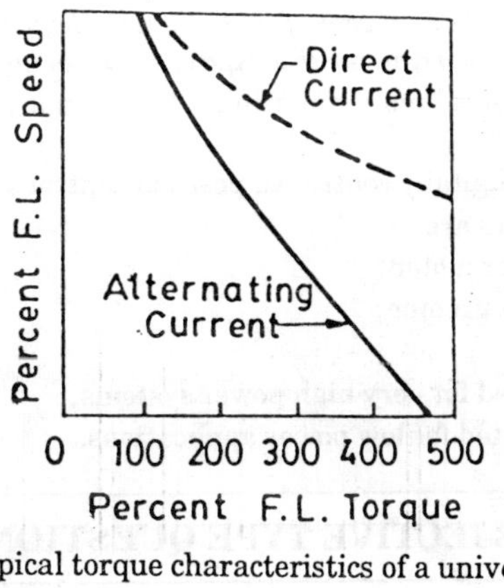

Fig. 23. Typical torque characteristics of a universal motor.

- The speed of a universal motor may be controlled by the following methods :
  (*i*) Reactance method.
  (*ii*) Tapped-field method.
  (*iii*) Centrifugal mechanism.

HIGHLIGHTS

1. *There are two basic types of single-phase induction motors which start on the split-phase principle, namely resistance-start induction-run motor and the capacitor-start induction-run motor.*
2. *The resistance-start induction-run motor has a relatively poor starting torque because the phase angle between the running winding current and the starting winding current is only from 30 to 50 electrical degrees.*
3. *The capacitor-start induction-run motor has a relatively good starting torque because the phase angle between the running winding current and the starting current is practically 90 electrical degrees.*
4. *The capacitor-start, capacitor-run induction motor has relatively good starting torque and operates at a comparatively high power factor. The main winding and auxiliary windings are both energized at all times the motor is in operation.*
5. *A resistance-start induction-run motor or a capacitor-start induction-run motor may be reversed by interchanging the leads of the starting winding circuit. The direction of rotation of either type of motor can also be reversed by interchanging the leads of the running winding circuit. However,* **never attempt to change the rotation of either type of motor by reversing the line wires.**
6. *The* **repulsion motor** *operates on the repulsion principles at all times. It has* **excellent starting torque** *and* **a relatively wide range of speeds.**
7. *The repulsion-start induction-run motor starts as a repulsion motor and after it reaches 75 per cent of rated speed, operates as an induction motor. It has excellent starting torque and very good speed regulation.*
8. *There are two types of brush mechanism used with repulsion-start induction-run motors : The* **brush-lifting type** *and* **brush-riding** *type.*
9. *To change the direction of rotation of either a repulsion motor or a repulsion-start induction-run motor, move the brushes to the other side of stator field poles so that there is an angle of 15 electrical degrees in the new position between the brushes and the stator field pole centres.*
10. *The A.C. series motor has excellent starting torque and can be used in those applications requiring different speeds. It has characteristics comparable to those of the D.C. series motor.*
11. *The direction of rotation of an A.C. series motor can be reversed by interchanging the connections to the armature.*
12. *The shaded-pole induction motor is usually made in fractional horsepower sizes in excess of 1/10 horsepower. This type of motor has no centrifugal starting device or mechanism which requires maintenance. However, the* **starting torque** *for this type of motor is* **small.**
13. A *stepper motor* is an incremental motion machine. It does not rotate continuously as a conventional motor does.
14. The stepper motor is used in digitally controlled position control system in open loop mode.
15. Various types of stepper motors are :
    (*i*) Permanent-magnet stepper motor ;
    (*ii*) Variable-reluctance stepper motor ;
    (*iii*) Hybrid stepper motor.
16. D.C. servo-motors are preferred for very high power systems.
17. A.C. servo-motors are best suited for low power applications.

OBJECTIVE TYPE QUESTIONS

(A) Choose the Correct Answer :

1. In a split-phase motor, the running winding should have
   (*a*) high resistance and low inductance (*b*) low resistance and high inductance
   (*c*) high resistance as well as high inductance (*d*) low resistance as well as low inductance.

2. If the capacitor of a single-phase motor is short-circuited
   (a) the motor will not start (b) the motor will burn
   (c) the motor will run in reverse direction
   (d) the motor will run in the same direction at reduced r.p.m.
3. In capacitor-start single-phase motors
   (a) current in the starting winding leads the voltage
   (b) current in the starting winding lags the voltage
   (c) current in the starting winding is in phase with voltage in running winding.
   (d) none of the above.
4. In a capacitor-start and run motors the function of the running capacitor in series with the auxiliary winding is to
   (a) improve power factor (b) increase overload capacity
   (c) reduce fluctuations in torque (d) to improve torque.
5. In a capacitor-start motor, the phase displacement between starting and running winding can be nearly
   (a) 10° (b) 30°
   (c) 60° (d) 90°.
6. In a split-phase motor
   (a) the starting winding is connected through a centrifugal switch
   (b) the running winding is connected through a centrifugal switch
   (c) both starting and running windings are connected through a centrifugal switch
   (d) centrifugal switch is used to control supply voltage.
7. The torque developed by a single-phase motor at starting is
   (a) more than the rated torque (b) rated torque
   (c) less than the rated torque (d) zero.
8. Which of the following motors will give relatively high starting torque ?
   (a) Capacitor-start motor (b) Capacitor run motor
   (c) Split-phase motor (d) Shaded-pole motor.
9. Which type of capacitor is preferred for capacitor-start and run motor ?
   (a) Electrolyte capacitor (b) Paper capacitor (oil filled)
   (c) Ceramic capacitor (d) Air capacitor.
10. Which of the following motors will have relatively higher power factor ?
   (a) Capacitor-run motor (b) Shaded-pole motor
   (c) Capacitor-start motor (d) Split-phase motor.
11. In a shaded-pole motor, the shading coil usually consist of
   (a) a single turn of heavy wire which is in parallel with running winding
   (b) a single turn of heavy copper wire which is short-circuited and carries only induced current
   (c) a multilayer fine gauge copper wire in parallel with running winding
   (d) none of the above.
12. In a shaded-pole single-phase motor, the revolving field is produced by the use of
   (a) inductor (b) capacitor
   (c) resistor (d) shading coils.
13. A centrifugal switch is used to disconnect starting winding when motor has
   (a) run for about 1 minute (b) run for about 5 minutes
   (c) picked up about 50 to 70 per cent of rated speed
   (d) picked up about 10 to 25 per cent of rated speed.
14. If a particular application needs high speed and high starting torque, then which of the following motors will be preferred ?
   (a) Universal motor (b) Shaded-pole type motor
   (c) Capacitor-start motor (d) Capacitor-start and run motor.

15. Which motor has the highest power to weight ratio ?
(*a*) Capacitor-run motor (*b*) Capacitor-start motor
(*c*) Universal motor (*d*) None of the above.

16. In a universal motor the width of the brush is nearly
(*a*) half the width of commutator segment (*b*) equal to width of the commutator segment
(*c*) equal to width of two commutator segment (*d*) equal to width of four commutator segment.

17. The range of efficiency for shaded-pole motors is
(*a*) 95% to 99% (*b*) 80% or 90%
(*c*) 50% to 75% (*d*) 5% to 35%.

18. In a capacitor-start single-phase motor, when capacitor is replaced by a resistance
(*a*) torque will increase (*b*) the motor will consume less power
(*c*) motor will run in reverse direction (*d*) motor will continue to run in same direction.

19. The power factor of a single-phase induction motor is usually
(*a*) lagging (*b*) always leading
(*c*) unity (*d*) unity to 0.8 leading.

20. A shaded-pole motor can be used for
(*a*) toys (*b*) hair dryers
(*c*) circulators (*d*) any of the above.

21. A hysteresis motor works on the principle of
(*a*) hysteresis loss (*b*) magnetisation of rotor
(*c*) eddy current loss (*d*) electromagnetic induction.

22. Which of the following motors will give the highest starting torque ?
(*a*) D.C. shunt motor (*b*) Schrage motor
(*c*) Repulsion-start and induction-run motor (*d*) Universal motor.

23. In repulsion motors by shifting the position of brushes
(*a*) only speed can be controlled (*b*) speed and direction of rotation can be controlled
(*c*) speed, direction of rotation and torque can be controlled
(*d*) none of the above.

24. Which of the following single-phase motors is reversible ?
(*a*) Capacitor-start split-phase motor (*b*) Resistance-start split-phase motor
(*c*) Reluctance motor (*d*) None of the above.

25. For which of the applications a reluctance motor is preferred ?
(*a*) Electric shavers (*b*) Refrigerators
(*c*) Signalling and timing devices (*d*) Lifts and hoists.

26. In repulsion motors
(*a*) speed remains constant (*b*) speed variation is within 5%
(*c*) speed varies with load (*d*) speed varies only with change in applied voltage.

27. In single-phase repulsion motors power factor is
(*a*) always leading (*b*) high at low speeds
(*c*) always unity (*d*) high at high speeds.

28. The motor used on small lathes is usually
(*a*) universal motor (*b*) d.c. shunt motor
(*c*) single-phase capacitor run motor (*d*) 3-phase synchronous motor.

29. Which of the following motors is preferred for tape recorders ?
(*a*) Shaded-pole motor (*b*) Hysteresis motor
(*c*) Two value capacitor motor (*d*) Universal motor.

30. A single-phase induction motor is
(*a*) inherently self-starting with high torque (*b*) inherently self-starting with low torque
(*c*) inherently non-self-starting with low torque (*d*) inherently non-self-starting with high torque.

31. A schrage motor can run on

(*a*) zero slip (*b*) negative slip
(*c*) positive slip (*d*) all of the above.

32. A universal motor can run on

(*a*) A.C. only (*b*) D.C. only
(*c*) either A.C. or D.C. (*d*) none of the above.

33. Which of the following single-phase motors is suitable for timing and control purposes ?

(*a*) Reluctance motor (*b*) Series motor
(*c*) Repulsion motor (*d*) Universal motor.

34. In case of a shaded-pole motor the direction of rotation of the motor is

(*a*) from main-pole to shaded-pole (*b*) from shaded-pole to main pole
(*c*) either of the above depending on voltage (*d*) either of the above depending on power factor.

35. In case of high speed universal motor which of the following needs more attention ?

(*a*) End play (*b*) Air gap
(*c*) Insulation in rotor (*d*) Balancing of rotor.

36. The wattage rating for a ceiling fan motor will be in the range

(*a*) 50 to 250 W (*b*) 250 to 500 W
(*c*) 50 to 150 W (*d*) 10 to 20 W.

37. The wattage of motor for driving domestic sewing machine will be around

(*a*) 100 to 150 W (*b*) 40 to 75 W
(*c*) 10 to 30 W (*d*) 5 to 10 W.

38. Which of the following single-phase motors has relatively poor starting torque ?

(*a*) Universal motor (*b*) Repulsion motor
(*c*) Capacitor motor (*d*) All single phase motors have zero starting torque.

39. Which type of load is offered by cranes and hoists ?

(*a*) gradually varying load (*b*) non-reversing, no-load start
(*c*) reversing, light start (*d*) reversing, heavy start.

40. The speed of a universal motor is generally reduced by using

(*a*) gear trains (*b*) V-belts
(*c*) brakes (*d*) chains.

41. Which of the following motors can be used for unity power factor ?

(*a*) Capacitor run motor (*b*) Shaded-pole motor
(*c*) Hysteresis motor (*d*) Schrage motor.

42. When a D.C. series motor is connected to A.C. supply, the power factor will be low because of

(*a*) high inductance of field and armature circuits
(*b*) induced current in rotor due to variations of flux
(*c*) fine copper wire winding (*d*) none of the above.

43. The direction of rotation of universal motor can be reversed by reversing the flow of current through

(*a*) armature winding (*b*) field winding
(*c*) either armature winding or field winding (*d*) none of the above.

44. In which single-phase motor, the rotor has no teeth or winding ?

(*a*) Split-phase motor (*b*) Reluctance motor
(*c*) Hysteresis motor (*d*) Universal motor.

45. Which motor is normally free from mechanical and magnetic vibrations ?

(*a*) Split-phase motor (*b*) Universal motor
(*c*) Hysteresis motor (*d*) Shaded-pole motor.

46. As hysteresis motors are free from mechanical and magnetic vibrations therefore these are considered as suitable for

(a) fans (b) blowers
(c) sound equipment (d) mixer grinders.

47. A reluctance motor
(a) is self-starting (b) is constant speed motor
(c) needs no D.C. excitation (d) all of the above.

48. In a hysteresis motor, the rotor must have
(a) retentivity (b) resistivity
(c) susceptibility (d) none of the above.

49. The rotor of a hysteresis motor is made of
(a) aluminium (b) cast iron
(c) chrome steel (d) copper.

50. The electric motor used in portable drills is
(a) capacitor-run motor (b) hysteresis motor
(c) universal motor (d) repulsion motor.

51. Which of the following applications always has some load whenever switched on ?
(a) Vacuum cleaners (b) Fan motors
(c) Pistol drills (d) All of the above.

52. The speed control of universal motor used for sewing machines is by
(a) friction (b) varying the resistance
(c) tapping the field (d) centrifugal mechanism.

ANSWERS

1. (b)	2. (a)	3. (a)	4. (a)	5. (d)	6. (a)
7. (d)	8. (a)	9. (b)	10. (a)	11. (b)	12. (d)
13. (c)	14. (a)	15. (c)	16. (c)	17. (d)	18. (d)
19. (a)	20. (d)	21. (a)	22. (b)	23. (b)	24. (a)
25. (c)	26. (c)	27. (d)	28. (c)	29. (b)	30. (c)
31. (d)	32. (c)	33. (a)	34. (a)	35. (d)	36. (c)
37. (a)	38. (c)	39. (d)	40. (a)	41. (d)	42. (a)
43. (c)	44. (c)	45. (c)	46. (c)	47. (d)	48. (a)
49. (c)	50. (c)	51. (c)	52. (b).		

(B) Say 'Yes' or 'No' :

1. In a capacitor-start motor, the phase displacement between starting and running winding can be nearly 90°.
2. In a split-phase motor the staring winding is connected through a centrifugal switch.
3. In a shaded-pole single-phase motor, the revolving field is produced by the use of capacitor.
4. A universal motor has the highest power to weight ratio.
5. The range of efficiency for shaded-pole motors is 80 to 90%.
6. The power factor of a single-phase induction motor is usually lagging.
7. A hysteresis motor works on the principle of eddy current loss.
8. Capacitor-start split-phase motor is reversible.
9. A reluctance motor is preferred for signalling and timing devices.
10. The wattage rating for a ceiling fan motor varies from 50 to 150 W.
11. A ...... motor is an incremental motion machine.
12. A stepper motor rotates continuously.
13. The stepper motor is used in digitally controlled position control system in open loop mode.
14. In stepper motors no sensors are needed for position and speed sensing.
15. A stepper motor cannot be readily interfaced with microprocessor.
16. In ...... stepper motor the rotor is made of ferite or rare earth material which is permanently magnetised.

17. In a ...... stepped motor the stator has only one set of winding-excited poles which interact with the two rotor stacks.

18. ...... servo-motors are preferred for very high power systems.

19. A.C. servo-motors are best suited for low power applications.

20. ...... construction is used for very low inertia applications.

ANSWERS

1. Yes	**2.** Yes	**3.** No	**4.** Yes	**5.** No	**6.** Yes
7. No	**8.** Yes	**9.** Yes	**10.** Yes	**11.** stepper	**12.** No
13. Yes	**14.** Yes	**15.** No	**16.** Permanent-magnet		**17.** hybrid
18. D.C.	**19.** Yes	**20.** Drag-cup.			

THEORETICAL QUESTIONS

1. Explain what is meant by the split-phase method of motor starting.
2. Explain why the starting torque of a capacitor-start induction run motor is better than that of a resistance-start induction-run motor.
3. How is the direction of rotation reversed for each of the following ?
   (*i*) Resistance-start induction-run motor, and
   (*ii*) Capacitor-start induction-run motor.
4. (*a*) Compare operating characteristics of a resistance-start induction-run motor with those of a capacitor-start induction-run motor.
   (*b*) List three applications for :
      (*i*) A resistance-start induction-run motor
      (*ii*) A capacitor-start induction-run motor.
5. Explain how a repulsion motor operates.
6. Explain how a repulsion-start induction-run motor operates.
7. How is the direction of rotation of each of the following motors reversed ?
   (*i*) The repulsion motor, and (*ii*) The repulsion-start induction-run motor.
8. What is the difference between A.C. and D.C. series motor ?
9. Why are small fractional horsepower A.C. series motors called universal motors ?
10. What is the difference between a conductively compensated series motor and an inductively compensated series motor ?
11. Describe the construction and operation of a shaded-pole motor.

UNIT-V : CONTROL SYSTEMS

Chapter :

10

Control Systems

1. INTRODUCTION

Automatic control has played a significant role in the advance of engineering science. Besides its extreme importance in space-vehicle systems, missile-guidance systems, etc., automatic control has become an important and integral part of modern manufacturing and industrial processes. Automatic control, for example, is essential in :

— Design of auto pilot systems in *aerospace industries.*

— Design of cars and trucks in the *automobile industries.*

— Industrial operations as controlling pressure, temperature, humidity, viscosity, and flow in the *process industries.*

2. SYSTEM

A system may be defined as follows :

- *"A* **system** *is an arrangement, set or collection of things connected or related in such a manner as to form an entirely or whole".*

  Or

  "A **system** *is an arrangement of physical components connected or related in such a manner as to form and/or act as an entire unit."*

- A system consists of a sequence of components in which *each component has some cause as input and its effect will be its output. Broadly it is a sequential set of cause and effects.*

Each system may have a *large number of subsystems. Examples* :

(*i*) This universe is itself a system consisting of large number of subsystems.

(*ii*) Human body as a system has digestive system, respiratory system, etc.

3. CONTROL SYSTEM

A **control system** *is an arrangement of physical components connected or related in such a manner as to command, direct or regulate itself or another system.*

Elements of a control system :

The elements of a control system are enumerated and defined below :

Element	*Definition*
1. *Controlled variable*	The quantity or condition of the controlled system which can be directly measured and controlled is called *controlled variable.*
2. *Indirectly controlled variable*	The quantity or condition related to controlled variable, but cannot be directly measured is called *indirectly controlled variable.*
3. *Command*	The input which can be independently varied is called *command.*
4. *Reference input*	A standard signal used for comparison in the closed-loop system.
5. *Actuating signal*	The difference between the feedback signal and reference signal is called *actuating signal.*
6. *Disturbance*	Any signal other than the reference which affects the system performance is called *disturbance.*
7. *System error*	The difference between the actual value and ideal value is called *system error.*

Examples of control system applications :

Following are some *examples* of control system applications :

1. Steering control of automobile.
2. Print wheel control system.
3. Industrial sewing machine.
4. Sun-tracking control of solar collectors.
5. Speed control system.
6. Temperature control of an electric furnace.

4. CLASSIFICATION OF CONTROL SYSTEMS

Control systems are *classified* into the following *two basic types* :

1. Open-loop control systems (Unmonitored or non-feedback control systems),
2. Closed-loop control systems (Monitored or feedback control systems).

Comparison between Open-loop and Closed-loop Systems

Open-loop	*Closed-loop*
1. Less accurate.	1. More accurate.
2. Generally build easily.	2. Generally complicated and costly.
3. Stability can be ensured.	3. May become unstable at times.
4. Presence of non-linearities cause malfunctioning.	4. It usually performs accurately even in the presence of non-linearities.
5. Any change in system component cannot be taken care of automatically.	5. Change in system component is automatically taken care of.
6. Input command is the sole factor responsible for providing the control action.	6. The control action is provided by the difference between the input command and the corresponding output.
7. The control adjustment depends upon human judgement and estimate.	7. The control adjustment depends on output and feedback element.
Examples:	***Examples:***
(*i*) Automatic washing machine	(*i*) Liquid level control system
(*ii*) The electric switch	(*ii*) Traffic signal system
(*iii*) An automatic toaster.	(*iii*) Human being reaching for an object.
Note. All control systems *operated by present timing mechanisms are open-loop.*	

5. OPEN-LOOP CONTROL SYSTEMS (NON-FEEDBACK SYSTEMS)

- An **open-loop control system** *is one in which the control action is independent of the desired output.* The actuating signal depends only on the input command and output has *no control* over it.
- The **elements** of an open-loop control system can usually be divided into the following two parts (See. Fig. 1) :

(*i*) Controller ;

(*ii*) Controlled process.

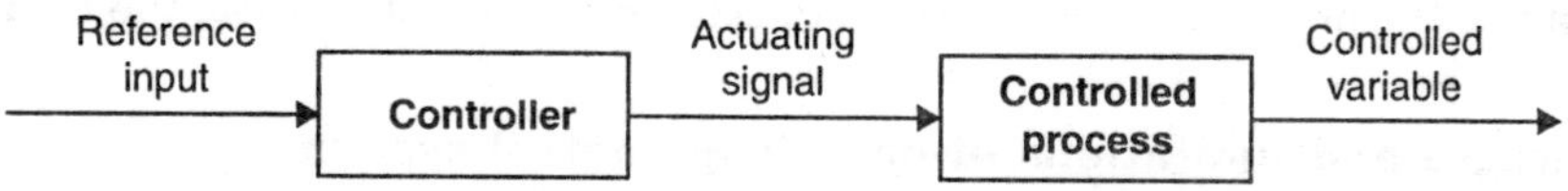

Fig. 1. Elements of an open-loop control system.

— An input signal or command is applied to the controller, whose output acts as the actuating signal ; the actuating signal then controls the controlled process so that the controlled variable will perform according to prescribed standards.

— In ***simple cases,*** the controller can be an *amplifier, mechanical linkage, filter or other control element,* depending on the nature of the system. In ***more sophisticated cases,*** the controller can be a computer such as a *microprocessor*.

— Because of the *simplicity and economy* of open-loop control systems we find this type of system in many non-critical applications.

Examples :

1. ***Idle-speed control system :***

- The following are the main objectives of the idle-speed control system of automobile :

  (*i*) To eliminate or minimize the speed drop when engine loading is applied.

  (*ii*) To maintain the engine idle-speed at a desired value.

- Fig. 2 shows an idle-speed control system from the stand point of inputs-system outputs. In this case the throttle angle and the load torque (due to the application of air conditioning, power steering, transmission, power brake, etc.) are the inputs, and the engine speed is the output. The engine is the controlled process of the system.

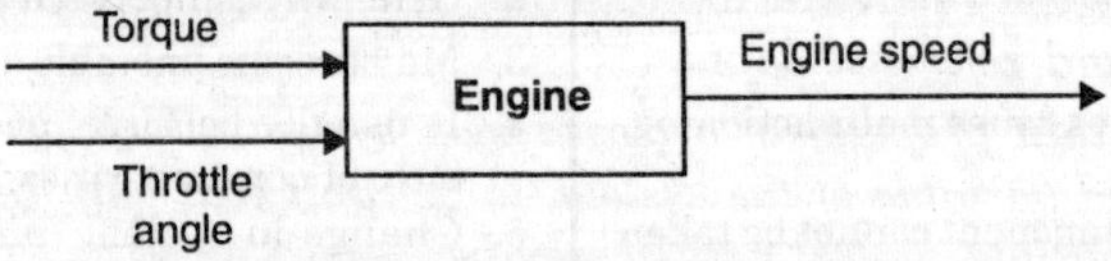

Fig. 2. Idle-speed control system.

2. ***Print wheel control system :***

Fig. 3 shows an example of the printwheel control system of a word processor or electronic typewriter (and also shows a typical input-output set for the system).

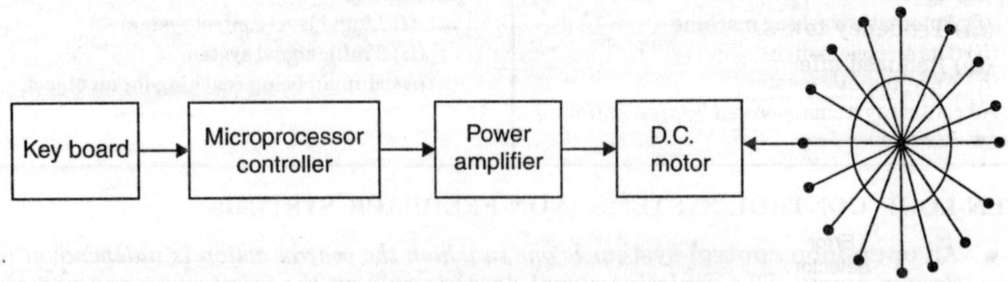

Fig. 3. Open-loop word processor control system.

— When a reference command input is given, the signal is represented as a step function. Since the electric windings of the motor have inductance and the mechanical load has inertia, the printwheel cannot respond to the input instantaneously. Typically it will follow the response and settle at the new position after sometime. Printing should not begin until the printwheel has come to complete stop ; otherwise, the character will be smeared.

Advantages and limitations of open-loop control system :

Advantages :

1. Simple construction.
2. Easy maintenance.
3. Less costly than a closed-loop system.
4. No stability problem.
5. Convenient when output is difficult to measure or measuring the output precisely is economically not feasible.

Limitations/Disadvantages :

1. Since the system is affected by internal and external disturbances, the *output may differ from the desired value.*
2. For getting accurate results, this system needs frequent and careful calibrations.
3. Any change in system component cannot be taken care of automatically.
4. Presence of non-linearities cause malfunctioning.

6. CLOSED-LOOP CONTROL SYSTEMS (FEEDBACK CONTROL SYSTEMS)

- A **closed-loop system** is one in *which control action is somehow dependent on the output.* In this case the controlled output is feedback through a *feedback element* and compared with the reference input. Thus, the *actuating signal is the difference of desired output and reference input.*
- **Feedback** *is that property of a closed-loop system which permits the output or some other controlled variable of the system, to be compared with the input to the system, so that the appropriate control action may be formed as some function of the output and input. A feedback is said to exist in a system when a closed sequence of cause and effect relations exists between system variables.*

The ***characteristics of feedback*** *are as follows :*

(*i*) Increased bandwidth.

(*ii*) Increased accuracy.

(*iii*) Tendency towards oscillation or instability.

(*iv*) Reduced effects of non-linearities and distortion.

(*v*) Reduced sensitivity of the ratio of output to input to variations in system characteristics.

- *A closed-loop idle-speed control system* is shown in Fig. 4.

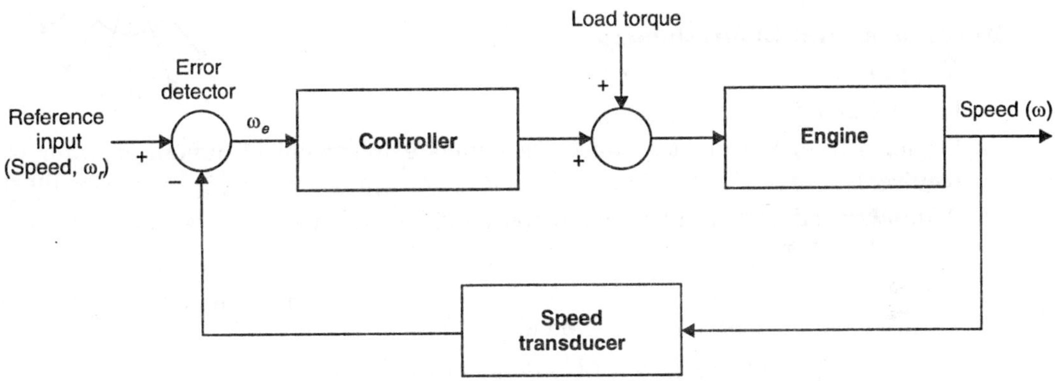

Fig. 4. Closed-loop idle-speed control system.

— The reference input (ω_r) sets the desired idling speed. The engine speed at idle should agree with reference value (ω_r), and any difference such as load torque is sensed by the speed transducer and the error detector. The controller will operate on the difference and provide a signal to adjust the throttle angle to correct the error.

Advantages and Limitations :

Advantages :

1. More accurate comparatively.
2. Usually performs accurately even in the presence of non-linearities.
3. Change in system component is automatically taken care of.
4. The use of feedback system response is relatively insensitive to external disturbances and internal variations in system parameters. It is thus *possible to use relatively inaccurate and inexpensive components to obtain the accurate control of a given plant* (whereas doing so is impossible in the open-loop case).

Limitations/Disadvantages :

1. Generally complicated in construction.
2. Generally higher in cost and power.
3. May become unstable at times.

7. AUTOMATIC CONTROL SYSTEMS

- *A closed-loop control system operating without human operator is called an* **automatic control system.**

Examples :

(*i*) *Centrifugal watt governor,* where the lift of the rotating balls is used as speed monitor. The supply of steam is automatically controlled as speed tends to increase or decrease beyond a set point.

(*ii*) *A pressure control system* where the pressure inside the furnace is automatically controlled by affecting changes in the position of the damper.

(*iii*) *The level control system* where the inflow of water to the tank is dependent on the water level in the tank. The automatic controller maintains the liquid level by comparing the actual level with a desired level and correcting any error by adjusting the opening of the control valve.

Advantages and Limitations :

Advantages :

1. Increased output.
2. Economy in operating cost (since continuous employment of human operator is not required).
3. Suitability and desirability in the complex and fast acting systems which are beyond the physical abilities of a man.
4. Improvement in the quality of the products.
5. Reduced effect of non-linearities and distortions.
6. Response is satisfactory over a wide range of input frequencies.

Limitation. Automatic control system has a tendency to *overcorrect errors* which may *result in oscillations of constant or changing amplitude.*

8. SERVO-MECHANISM

A servo-machanism is a feedback control system used to *control position or its derivative.* It has the following essential *features* :

1. It is a closed-loop system.
2. It is used to control position, velocity or acceleration.
3. Its characteristics include :
   — automatic control ;
   — remote operation ;
   — high accuracy.
4. It has high power amplifying stages to operate the system from very small error to signal.

9. REGULATOR

A regulator is a system employed to *control quality which is to be kept constant for a fairly long interval.*

Example : Voltage regulator or speed regulator.

10. REPRESENTATION THROUGH MODEL

In order to solve a system problem, the specifications or description of the system configuration and its components must be put into a form amenable to analysis, design and evolution. Following three basic models may be used for various systems :

1. Differential equations and other mathematical solutions.
2. Block diagrams
3. Sign flow graphs (SFG).

11. ANALOGOUS SYSTEMS

For mathematical relations analogies are drawn between features of a system and features of some known elements or properties, some analogous systems are given below :

Table 1. Force-Current Analogy

S.No.	*Mechanical System*		*Electrical System*
	Translational	*Rotational*	
1.	Force, F	Torque, T	Current, I
2.	Mass, M	Moment of inertia, M.I.	Capacitance, C
3.	Displacement, x	Angular displacement, θ	Magnetic flux linkage, f
4.	Velocity, V	angular velocity, ω	Voltage, E
5.	Viscous friction co-efficient, f	Viscous friction co-efficient, F	Reciprocal of resistance, $\frac{1}{R}$
6.	Spring stiffness, K	Torsional spring stiffness, K	Reciprocal of inductance, $\frac{1}{L}$

Table 2. Force-Voltage Analogy

S.No.	*Mechanical System*		*Electrical System*
	Translational	*Rotational*	
1.	Force, F	Torque, T	Voltage, E
2.	Mass, M	Moment of inertia, M.I.	Inductance, L
3.	Displacement, x	Angular displacement, θ	Charge, q
4.	Velocity, U	Angular velocity, ω	Current, I
5.	Spring stiffness, K	Torsional spring stiffness, K	Reciprocal of capacitance, $\frac{1}{C}$
6.	Viscous friction co-efficient, f	Viscous friction co-efficient, F	Resistance, R

Table 3. Electrical, Thermal, Liquid level and Pneumatic Systems

S.No.	*Electrical systems*	*Thermal systems*	*Liquid-level systems*	*Pneumatic systems*
1.	Charge, coulombs (C)	Heat flow, joules (J)	Liquid flow cum. (m^3)	Air flow, cum. (m^3)
2.	Current, amperes (A)	Heat flow rate, joules/sec. (J/s)	Liquid flow rate, cum/sec (m^3/s)	Air flow rate, cum/sec. (m^3/s)
3.	Voltage, volts (V)	Temperature, °C	Heat, metres (m)	Pressure, N/m^2
4.	Resistance, ohms (Ω)	Resistance, $°CsJ^{-1}$	Resistance, $m^{-2}s$	Resistance $N\text{-}ms^{-1}$
5.	Capacitance, farad (F)	Capacitance, J/°C	Capacitance, m^3/m	Capacitance, m^3/Nm^2

12. BLOCK DIAGRAM

A **block diagram** is the *diagrammatic representation of a physical system. The following steps* are worth noting :

— Firstly a functional block diagram is drawn to represent the functions of the system.

— Then it is *converted into a mathematical block diagram* by expressing the transfer function for each block.

— Finally it is *reduced to an equivalent simpler block diagram* for system analysis.

Fig. 5 shows a block diagram of the feedback control system.

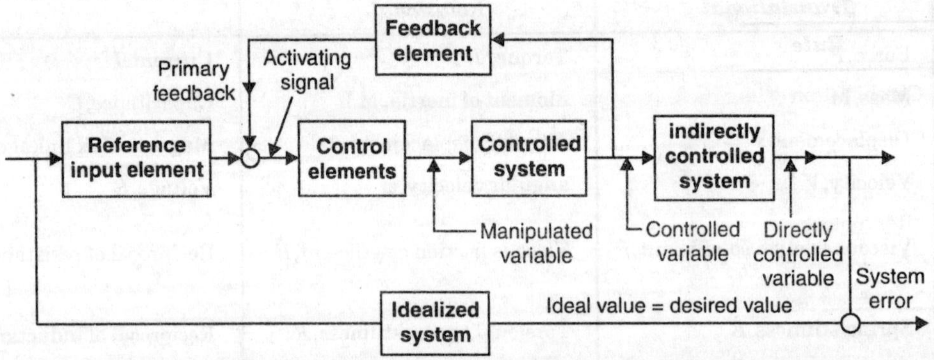

Fig. 5. Block diagram of the feedback control system.

13. MATHEMATICAL BLOCK DIAGRAM

Fig. 6 shows the block diagram of a closed-loop system. The various quantities shown are as follows :

$R(s)$ = Laplace transform of the reference input

$C(s)$ = Laplace transform of the output

$H(s)$ = Transfer function of the feedback path

$B(s)$ = Laplace transform of the feedback signal
$= C(s)\ H(s)$

$E(s)$ = Laplace transform of the actuating signal
$= R(s) - B(s) = R(s) - C(s)\ H(s)$

$G(s)$ = Laplace transform of the formed path

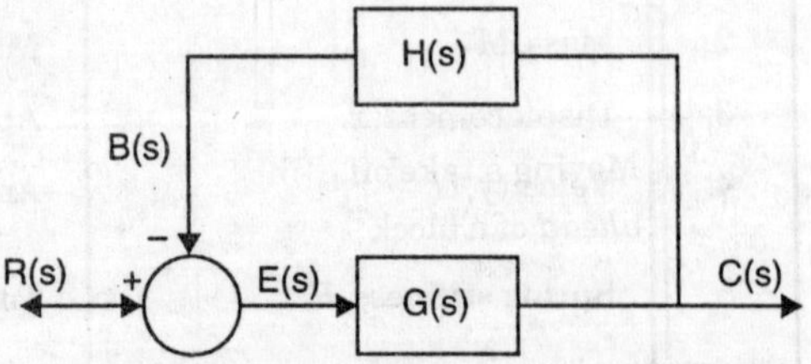

Fig. 6. Closed-loop system.

$$\therefore \quad C(s) = G(s)\,E(s) = G(s)\,R(s) - G(s)\,H(s)\,C(s)$$

or, $$C(s) + G(s)\,H(s)\,C(s) = G(s)\,R(s)$$

or, $$C(s)\,[1 + G(s)\,H(s)] = G(s)\,R(s)$$

or, $$\frac{C(s)}{R(s)} = \frac{G(s)}{1+G(s)H(s)}$$

Hence the transfer function of the system,

$$M_s' = \frac{C(s)}{R(s)} = \frac{G(s)}{1+G(s)H(s)}$$

In the above equation the following points are worth noting :

(*i*) Product of transfer functions of forward path and feedback path as $G(s) \times H(s)$, sometimes expressed as $GH(s)$.

(*ii*) The system performance depends on its *characteristic equation* (it is a key equation in the control system analysis) which is given as under :

$$1 + G(s)\,H(s) = 0.$$

Block reductions :

By using the rules (derived by simple algebraic manipulation of the equations representing the blocks) of block diagram algebra, a complex block diagram configuration can be simplified by certain rearrangement of block diagrams : such rules are given in the Table 4.

Table 4. Rules for Block Diagram Algebra

S.No.	*Rule*	*Original diagram*	*Equivalent diagram*
1.	Combination of blocks in cascade	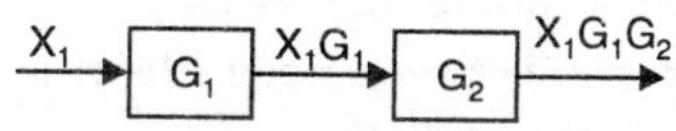	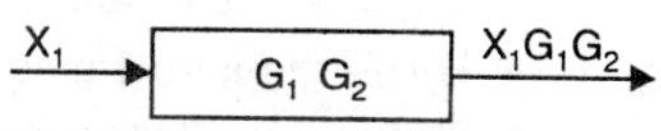
2.	Moving a summing point after a block	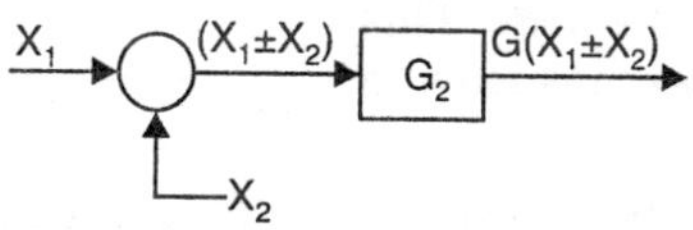	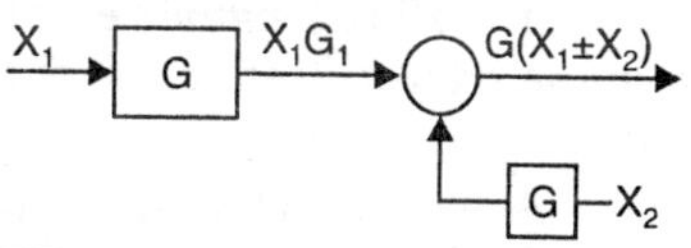
3.	Moving a summing point *ahead* of a block	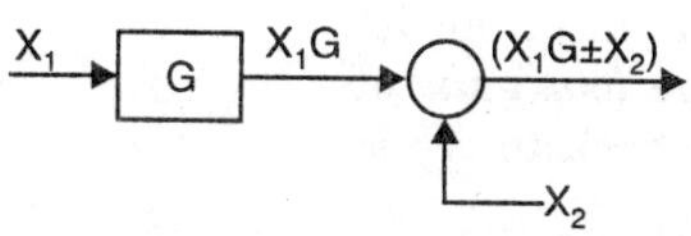	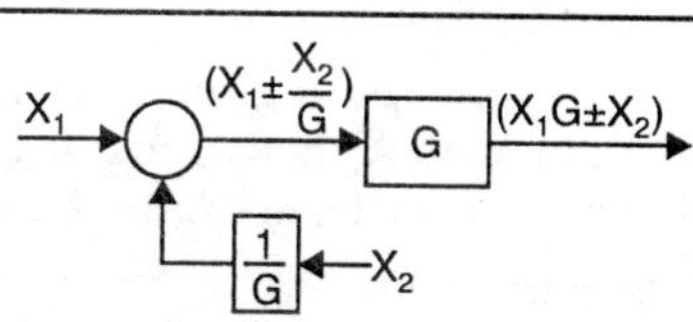
4.	Moving a take off point *after* a block	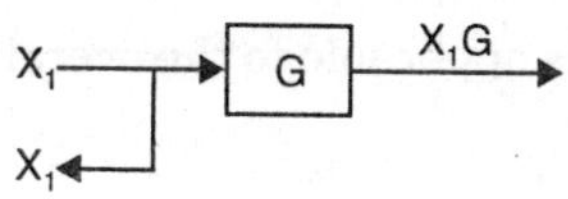	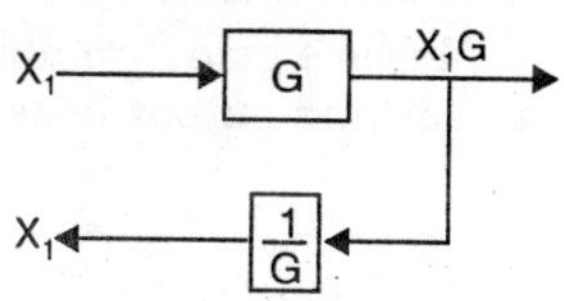
5.	Moving a take off point *ahead* of a block	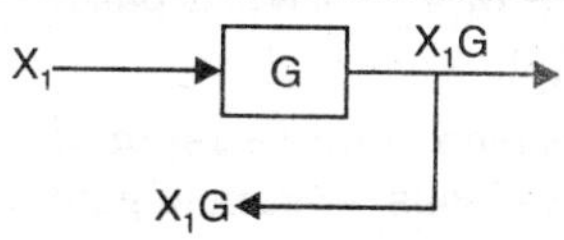	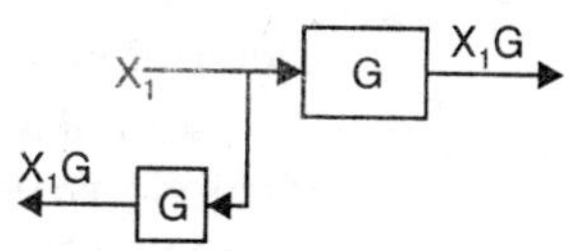

Fig. 7. Rules for block diagram algebra.

14. SIGNAL FLOW GRAPH

The block diagram reduction process, for complicated systems, becomes tedious and time consuming. For this purpose signal flow graph (developed by S.J. Mason) are used.

A **signal flow graph** *is a pictorial representation of the simultaneous equations describing a system.*

Some important definitions relating to *signal flow graph* are given below :

1. **Input and output nodes.** A node having only outgoing branches is called *input node* while a node having only incoming branches is called *output node or sink.*

2. **Path.** Any continuous unidirectional succession of branches traversed in the indicated direction of branch is called *path.*

3. **Forward path.** The path from input note to the output node along which no anode is traversed twice is called *forward path.*

4. **Loop or feedback path.** It is defined as the path originating from a node and after traversing terminates at the same node without traversing any other node twice.

5. **Path gain and loop gain.** The product of transmittances along a path is called *path gain* while along the loop is called *loop gain.*

Rules for drawing a 'Signal Flow Graph' :

A signal flow graph can be drawn following the rules given as under :

1. The system variables are represented by nodes and arranged from left to right to represent cause and effect through the system.

2. The direction of the branch from node to X_t to X_m represents that the variable X_m depends upon the variable X_t and not the reverse.

3. Signals travel along the direction of arrows only.

4. If the transmittance of branch between node l to m is t_{lm} then $X_m = t_{lm} X_l$.

5. The value of the variable at the node is the sum of all the signals entering the node. This value is transmitted on all the branches leaving this node.

A typical example of signal flow graph is shown in Fig. 8.

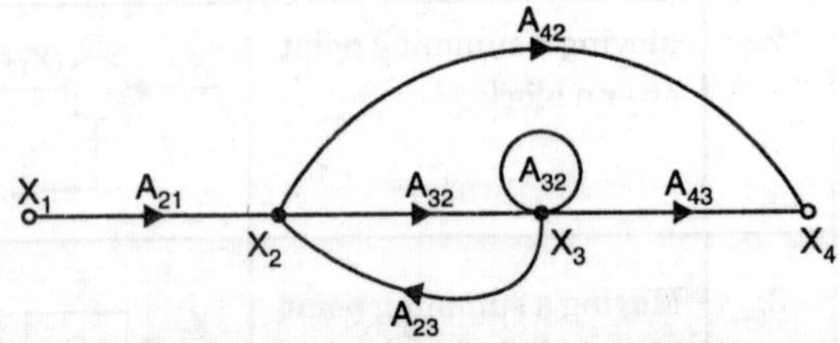

Fig. 8

Here, *'path gain'* of forward path from X_1 to X_2 to X_3 to X_4 is $A_{21} \cdot A_{32} \cdot A_{43}$. The *'loop gain'* of feedback loop from X_2 to X_3 and back to X_2 is $A_{32} \cdot A_{23}$.

Mason's gain formula :

Mason's gain formula, which is applicable to the overall gain (the transmittance between an input node and output node) is

$$P = \frac{1}{\Delta} a_k^0 P_k \Delta_k$$

where, P_k = Path gain transmittance of k_{th} forward path,

Δ = Determinant of the graph

= 1 – (sum of all different loop gains) + (sum of gain products of all possible combinations of two non-touching loops) – (sum of gain products of all possible combination of three non-touching loops), and

Δ_k = Co-factor of k_{th} path determinant of this graph with the loops touching the k_{th} forward path removed.

15. TIME RESPONSE OF CONTROL SYSTEM

Some important definitions :

1. **Delay time.** It is the time required for the response to reach line half the final value.
2. **Rise time.** It is the time required for the response to rise from 10 to 90% of its final value.
3. **Peak time.** It is the time required for the response to reach the first peak of the overshoot.
4. **Maximum overshoot.** It is the maximum value of the response curve measured from unity.
5. **Overshoot.** It is the maximum difference between the transient and steady state solutions for unit-step function input. It is a measure of relative stability and is represented as percentage of final value of the output.
6. **Setting time.** It is the time required for the response to a unit-step function input to reach and remain within a specified percentage of its final value.

Test functions :

Following are the commonly used typical test signals :

(*i*) *Step function*

(*ii*) *Ramp function*

(*iii*) *Impulse function*

(*iv*) *Sinusoidal function.*

16. STABILITY

The response of a system to input or disturbances determines its *stability*. A stable system is one that will remain at rest unless excited by an external source and will return to rest if all the excitations are removed. A system is stable if its impulse response approaches zero as time approaches infinity. The *necessary condition for the system to be stable is that the real part of the roots of the characteristic equation have negative real parts. This ensures that the impulse response will decay exponentially with time.*

When the system has some roots with real parts equal to zero, but none with positive real parts, the system is said to be "*marginally stable*" which in unstable.

Routh stability criterion :

Routh stability criterion is a method for determining system stability that can be applied to and nth order characteristic equation of the form

$$a_n s^n + a_{n-1} s^{n-1} + \ldots\ldots + a_1 s + a_0 = 0$$

The Routh table is prepared as defined below :

s_n	a_n	a_{n-2}	a_{n-4}
s^{n-1}	a_{n-1}	a_{n-3}	a_{n-5}
:	b_1	b_2	b_3
:	c_1	c_2	c_3

After the array is completed the following criterion is applied.

"*The number of changes in sign for the terms in the first column equals the number of roots of the characteristic equation with positive real parts.*

Hence by the Routh criterion, for a system to be stable the array resulting from its characteristic equation must have a first column with term of the same sign.

Deficiencies of Routh's criterion :

1. It does not provide the facility for selecting in a simple and direct fashion the parameters of a system component to stabilize the system when it is found to be absolutely unstable.
2. It assumes that characteristic equation is available in polynomial form ; which is not necessarily always true.
3. The Routh array may show no change in sign in the first column but the ensuing dynamic response may be characterised by overshoots so excessive as to render the system useless for control purposes. Thus, the system may be relatively unstable inspite of the fact that it is absolutely stable.
4. Although this criterion gives information about absolute stability, it conveys little or no information about how close the system may be to become unstable.

17. FREQUENCY RESPONSE

The analysis of the system whose input is frequency and amplitude is dealt under frequency response. *The system is actuated by a sinusoidal input and allowed to settle. The output amplitude and its phase with respect to input are measured. The phase difference and amplitude change indicate the nature of the system.*

Graphical methods :

The following four graphical methods are available to control systems analysis which are simpler and more direct than the time domain method for practical linear models of feedback control systems :

1. Bode's-Plot-Representation
2. Nyquist Diagrams
3. Nichols Charts
4. The Root Locus method.

The first three are *frequency-domain techniques.*

Bode's Plot. This method has the following *advantages* :

(*i*) It is the simplest method.

(*ii*) The multiplication of magnitudes can be converted into addition.

(*iii*) Transfer function can be determined easily.

Nyquist method :

- This method *handles* systems with time delays without the necessity of approximations and hence yields exact results about both absolute and relative stability of the system.
- It is also useful for obtaining information about transfer functions of components or systems from experimental frequency response data.

Root locus method :

This method permits accurate computations of the time-domain response in addition to yielding readily available frequency response information.

18. ERROR DETECTOR

An **error detector** is a *sensor to sense the error between the reference input and the desired output.*

- It gives an input to the amplifier and actuator in proportion to the error.
- Its output should be directly electrical or a transducer should be cascaded to give electrical output.

- An error-cum-transducer is obtained by connecting two potentiometers in parallel to a voltage source. Their movable points are brought out to give output voltage in proportion to the difference between the positions of the two movable contacts.

19. LVDT

LVDT (Linear-Variable-Differential Transformer) is a transformer having one primary, and two secondary windings and a movable core. The secondary windings are connected in series opposition, so as to have output which is difference of the two induced secondary voltages. The movable core is connected to the shaft and a normal position output voltage is zero. When the core moves the *output voltage* is a *function of the shaft position.*

20. SERVO-AMPLIFIER

A **servo-amplifier** is the amplifier used to *amplify the small output of the error detector to directly operate the actuator.*

- It can be electronic, magnetic or rotating.
- It should have high input impedance, low output impedance, frequency response curve should be flat in the range of operating frequencies, phase sensitive, small residual voltage and minimum noise.

21. SAMPLED DATA SYSTEMS

These systems (also called *discrete time systems*) are *dynamic systems, in which one or more variable change at the discrete instant of time.* The time interval between two discrete instants is very small so that the data during this interval can be approximated by interpolation.

These systems find application in :

(*i*) Numerically controlled machine tool operations.

(*ii*) Pulse control or digital control of electric drives.

(*iii*) High speed tin plate rolling mill using quantized data for control.

(*iv*) Large complex systems employing telemetry links based on pulse modulation transmission of data.

22. INDUSTRIAL CONTROLLERS

Industrial controllers may be classified *according to their control action* as follows :

1. Two-position or on-off controllers.
2. Proportional controllers.
3. Integral controllers.
4. Proportional-plus-integral controllers.
5. Proportional-plus-derivative controllers.
6. Proportional-plus-integral-plus-derivative controllers.

— Most industrial controllers use pressurised fuel such as oil or air or electricity as power sources. Consequently, controllers may also be classified according to the kind of power employed in the operation, such as *"pneumatic controllers", "hydraulic controllers" or "electronic controllers".* However, the kind of controllers to be used must be decided based on the nature of the plant and operating conditions, including such considerations as safety, cost, availability, reliability, accuracy, weight, and size.

23. PNEUMATIC CONTROL SYSTEMS

- Pneumatic controllers use air control medium to provide an output signal which is a function of an input error signal.
- Fig. 9 shows the schematics of a pneumatic control system, the major components are : *Error detector ; Flapper nozzle (controller mechanism) ; Amplifier or Pilot relay.*

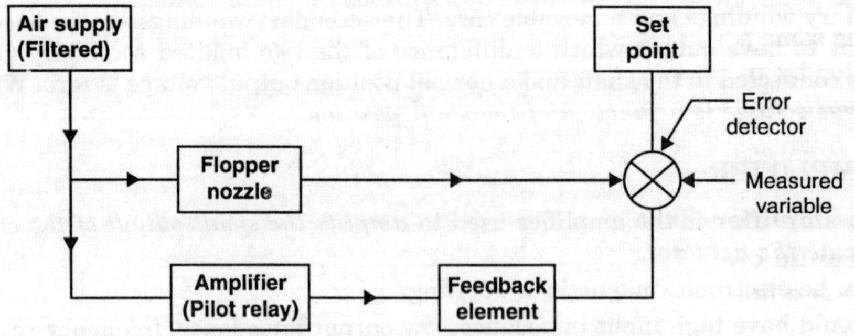

Fig. 9. Schematics of a pneumatic control system.

— The controller mechanism are of two types : *Free balance* and *motion balance.*

Advantages :

1. Simple construction and easy maintenance.
2. Relatively high power amplification for operating the final control elements.
3. Relatively inexpensive power system.
4. No return pipes are required when air is used.
5. Insensitive to temperature changes.
6. Fire and explosion-proof.
7. The normal operating pressure of pneumatic system is very much lower than that of hydraulic system.

Limitations/Disadvantages :

1. Output powers are considerably less (than those of hydraulic systems).
2. Accuracy of pneumatic actuators is poor at low velocities.
3. Slow response of final control elements, and transmission lag.
4. Operation difficult under freezing conditions.
5. Lubrication of the mating parts is difficult.

Uses :

- The pneumatic systems are employed for majority of the plant and process control actions in petroleum, petrochemical, chemical, paper, textile and food industries.
- They are also sometimes used in the aircraft systems and guided missiles.

24. HYDRAULIC CONTROL SYSTEM

- Compressed air has seldom been used (except for low-pressure controllers) for the continuous control of the motion of devices having significant mass under external load forces. For such a case, hydraulic controllers are generally preferred.

- The widespread use of hydraulic circuitry in "*machine tool applications*", "*aircraft control systems*" and "*similar operations*" occurs because of such factors as *accuracy, positiveness, flexibility, high power-to-weight ratio, fast starting, stopping, and reversal with smoothness and precision and simplicity of operations.*

— The operating pressure in hydraulic systems lies between 1 and 35 MPa ; in some special applications the operating pressure may go upto 70 MPa.

— for the same power requirement, the weight and size of the hydraulic unit can be made smaller by increasing the supply pressure. *Very large force can be obtained with hydraulic systems.*

— *With hydraulic systems, rapid-acting, accurate positioning of heavy loads is possible.*

— A combination of electronic and hydraulic systems is widely used because it combines the advantages of both electronic control and hydraulic power.

- **Hydraulic cotrollers** *employ a liquid control medium to provide an output signal which is a function of an input error signal.*

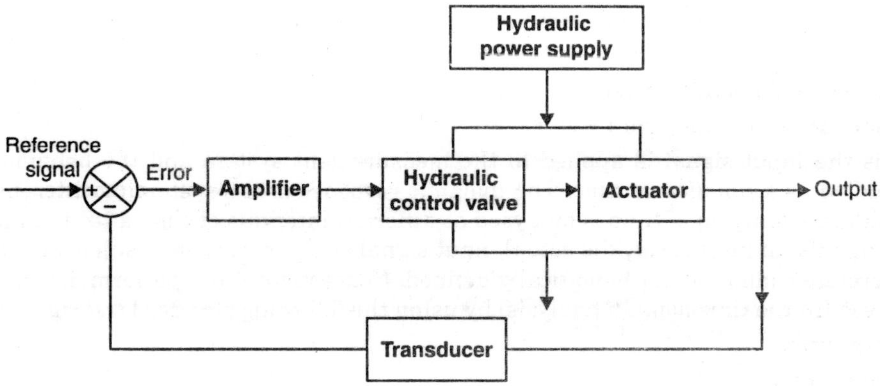

Fig. 10. Schematics of an hydraulic control system.

25. MICROCONTROLLER

Fig. 11 shows the simplified block diagram of the microcontroller (microprocessor based controller).

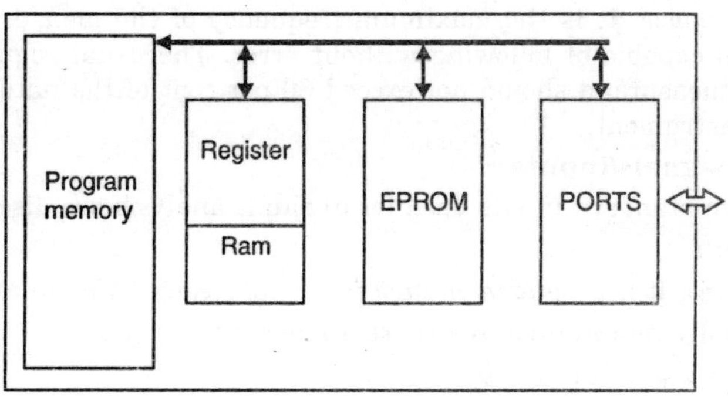

Fig. 11. Simplified block diagram of microcontroller.

- **Program memory.** It contains the program written. The program is a set instruction that the microcontroller performs. The software (instructions) is written in a computer and then programmed (burned) into the "program memory". This memory is a EPROM memory which can be rewritten thousand times.
- **Register and Ram box.** It contains all the *internal registers and a small Ram* where data can be stored temporarily. *There are several registers with different functions.*
  — The Ram memory is not large about 64-128 byte.
  — *The content in the Register and Ram-info will disappear when the power is off.*
- **EPROM-memory.** It is a small memory where data can be read as well as written, but the *data will not disappear when the power is off.* Next time the power is on we can go into this memory and fetch the data again.
- **Ports.** The port is *input and output pins of the actual circuit.* We can define the pins as input or outputs. By writing or reading to the port we can control each pin as we wish.

26. 'DYNAMIC' ANALYSIS OF MEASUREMENT SYSTEMS

The dynamic behaviour of measurement systems is studied in the following two domains :

1. Time domain analysis.
2. Frequency domain analysis.

1. Time domain analysis :

In this the input signal is applied to the measurement system and the behaviour of the system is studied as a *function of time.* The dynamic response of the system to different types of inputs, which are a function of time is analysed at different intervals of time after the application of the input signals. In most cases, the actual input signals vary in random fashion with respect to time and therefore cannot be mathematically defined. Consequently the performance of a system can be analysed (in the time domain analysis) by using the following standard *test signals/inputs* :

(*i*) Step input ;
(*ii*) Ramp input ;
(*iii*) Parabolic input ;
(*iv*) Impulse input ;

2. Frequency domain analysis :

This type of analysis of a system *pertains to the steady state response of the system to a sinusoidal input.* Here, the system is subjected to a sinusoidal input and the system response is studied with *frequency as the independent variable.*

- *Frequency response.* It is the maximum frequency of the measured variable that an instrument is capable of following without error. The usual requirement is that the frequency of measurand should not exceed 60 per cent of the natural frequency of the measuring instrument.

Standard test signals/inputs :

The most common standard inputs used for dynamic analysis are discussed below :

1. ***Step function :***

Refer to Fig. 12 (*a*). It is a *sudden change from one steady value to another.*

It is mathematically represented by the relationship :

$$x = 0 \quad \text{at} \quad t < 0$$
$$x = x_c \quad \text{at} \quad t \geq 0$$

where x_c is a constant value of the input signal x_i.

- The *"transient response" indicates the capacity of the system to cope with changes in the input signal.*

2. ***Ramp or linear function :***

In this case (see Fig. 12 (*b*)) the *input varies linearly with time.*

This input is mathematically represented as :

$$x = 0 \quad \text{at} \quad t < 0$$
$$x = \psi \quad \text{at} \quad t \geq 0$$

where ψ is the slope of the input versus time relationship.

- *The ramp-response becomes indicative of the steady state error in following the changes in the input signal.*

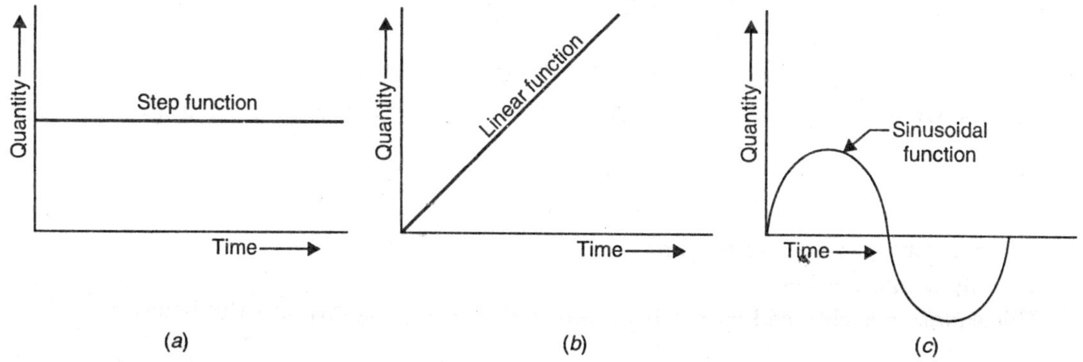

Fig. 12. Standard input function.

3. ***Sinusoidal function :***

In this case (see Fig. 12 (*c*)) the *input varies sinusoidally with a constant maximum amplitude.*

It is represented mathematically as follows :

$$I_i = A \sin \omega t$$

where, A = Amplitude, and

ω = Frequency in rad/s.

- *The frequency or harmonic response is a measure of the capability of the system to respond to inputs of cyclic nature.*

A general measurement system can be mathematically described by the following differential equation :

$$(A_n D^n + A_{n-1} D^{n-1} + \ldots\ldots + A_1 D + A_0) I_o = (B_m D^m + B_{m-1} D^{m-1} + \ldots\ldots + B_1 D + B_0) I_i \qquad \ldots(1)$$

where, A's and B's = Constants, depending upon the physical parameters of the system.

D^k = Operative derivative of the order k.

I_o = The information out of the measurement system, and

I_i = The input information.

The order of the measurement system is generally classified by the value of the power of n.

- *Zero-order system :* $n = 0$ and $A_1, A_2, A_3 \ldots\ldots A_n = 0$
- *First-order system :* $n = 1$ and $A_2, A_3, A_4 \ldots\ldots A_n = 0$
- *Second-order system :* $n = 2$ and $A_3, A_4, A_5 \ldots\ldots A_n = 0$

The above method of classification is used for most of the instruments and systems.

Although general equation can be solved by various methods, we shall be using method of *D*-operator for getting its solution.

27. ZERO, FIRST AND SECOND ORDER SYSTEMS

27.1. Zero Order Systems

Fig. 13 shows the block diagram of a 'Zero-order system'. In this case the output of the measuring system (ideal) is directly proportional to input, no matter how the input varies. The *output is faithful reproduction of input without any distortion or time lag.*

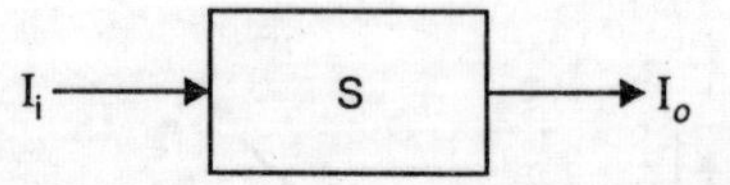

Fig. 13. Block diagram for zero-order system.

The behaviour of the zero-order system is represented by the following mathematical solution

$$I_o = SI_i \quad ...(2)$$

where, I_o = Information out of the measuring system,
S = Sensitivity of the system, and
I_i = Input information

This equation is obtained by putting $n = 0$ in the general equation (1)

i.e.,
$$A_o I_o = B_o I_i$$

or
$$I_o = \frac{B_o}{A_o} I_i = SI_i \quad ...(3)$$

The zero-order system is characterised only by the static sensitivity (parameter), the value of which is obtained through the process of static calibration.

Examples of zero-order system :

- Mechanical levers ;
- Amplifiers ;
- Potentiometer (It gives an output voltage which is proportional to wiper's displacement) etc.

27.2. First-Order Systems

Fig. 14 shows the block diagram of a 'First-order system'.

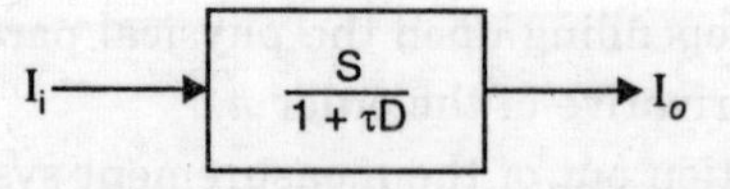

Fig. 14. Block diagram for first-order system.

The behaviour of a first-order system is given by following first-order differential equation :

$$A_1 \frac{dI_o}{dt} + A_o I_o = B_o I_i \quad ...(4)$$

(This equation is obtained by inserting $n = 1$ in the general equation).

Eqn. (4) may be written in standard form as follows :

$$\frac{A_1}{A_o}\frac{dI_o}{dt} + I_o = \frac{B_o}{A_o} I_i \qquad ...(5)$$

or,
$$\tau \frac{dI_o}{dt} + I_o = SI_i \qquad ...(6)$$

where, $\tau = \frac{A_1}{A_0}$ = Time constant, and

$S = \frac{B_o}{A_o}$ = Sensitivity.

Using D-operator, we get

$$\left[\text{where, } D = \frac{d}{dt}, \text{ and } D^2 = \frac{d^2}{dt^2}\right]$$

$$\tau DI_o + I_o = SI_i$$

or,
$$I_o(\tau D + 1) = SI_i$$

or,
$$\frac{I_o}{I_i} = \frac{S}{1+\tau D} \qquad ...(7)$$

Equation (7) gives the standard form of *transfer operator* for first-order system.

Examples of first-order system :

- Velocity of a true falling mass ;
- Air pressure build-up in bellows ;
- Measurement of temperature by mercury-in-glass thermometers ;
- Thermisters and thermocouples ;
- Resistance-capacitance network.

27.3. Second-Order Systems

Fig. 15 shows the block diagram of 'Second-order system' :

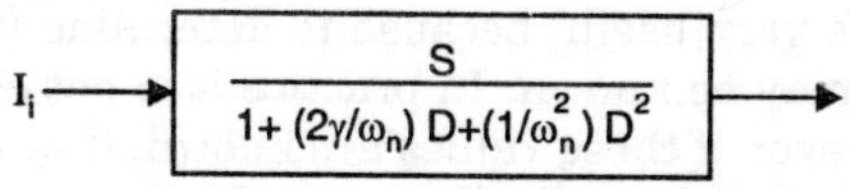

Fig. 15. Block diagram for second-order system.

The behaviour of a second-order system is given by the following differential equation (obtained by putting $n = 2$ in the general equation) ;

$$A_2 \frac{d^2 I_o}{dt^2} + A_1 \frac{dI_o}{dt} + A_o I_o = B_o I_i \qquad ...(8)$$

Dividing the above equation by A_o, we have

$$\frac{A_2}{A_o}\frac{d^2 I_o}{dt^2} + \frac{A_1}{A_0}\frac{dI_o}{dt} + I_o = \frac{B_o}{A_o} I_i \qquad ...(8\ a)$$

Let, $\omega_n = \sqrt{\dfrac{A_0}{A_2}}$ = Undamped natural frequency, rad/s.

$\gamma = \dfrac{A_1}{2\sqrt{A_0\, A_2}}$ = Damping ratio, dimensionless, and

$S = \dfrac{B_0}{A_0}$ = Static sensitivity or steady-state gain.

Then, by substituting these values in eqn. (8*a*), we get

$$\frac{1}{\omega_n^2} \cdot \frac{d^2 I_o}{dt^2} + \frac{2\gamma}{\omega_n} \cdot \frac{dI_o}{dt} + I_o = SI_i \quad ...(9)$$

or, in terms of *D*-operator, we have

$$\left(\frac{D^2}{\omega_n^2} + \frac{2\gamma}{\omega_n} D + 1\right) I_o = SI_i$$

or,
$$\frac{I_o}{I_i} = \frac{S}{\dfrac{1}{\omega_n^2} D^2 + \dfrac{2\gamma}{\omega_n} D + 1} \quad ...(10)$$

Examples of second-order system :

- Piezoelectric pick-up ;
- Spring-mass system (used for acceleration and force measurements) ;
- Pen control system on X-Y plotters ;
- U.V. galvanometer, etc.

Damping ratio :

In the design of instruments a term which is very frequently used is the "*damping ratio*" (γ) defined as the *ratio of the actual value of coefficient of viscous friction in movement and the value required to produce critical damping.*

i.e.,
$$\gamma = \frac{A_1}{2\sqrt{A_o A_2}}$$

This dimensionless term is very useful because to determine its value, it is not necessary that the values of A_1, A_0 and A_2 may be known. In practice it is not easy to determine accurately the values of A_1 and A_2. Further, even if these values are known, they do not in themselves specify whether the instrument is under, over or critically damped, since a numerical calculation has to be performed with them first. Therefore, designers find "*damping ratio*" as a very convenient measure of the amount of the damping present in the movement.

The terms *damping ratio* (γ) and *underdamped natural frequency* (ω_n) immediately conjure up a physical picture of the response of an instrument and both of the quantities are very easy to measure. Thus γ and ω_n easily do away with quantities A_2, A_1 and A_0.

28. FIRST-ORDER SYSTEM RESPONSES

The complete solution of an equation which describes the dynamical behaviour of a system consists of the following *two parts :*

(*i*) **Complementary function.** *It corresponds to the short time or transient response.*

(*ii*) **Particular integral.** *It refers to the long time steady state response.*

The transfer operator form of the first-order system is given by :

$$\frac{I_o}{I_i} = \frac{S}{1+\tau D}$$

When S (static sensitivity or steady state gain) equals *unity,* we get

$$(1 + \tau D)\, I_o = I_i \qquad ...(11)$$

Now we shall obtain the solution of this equation for different standard inputs (The solutions are not mathematically rigorous, but are practical).

Transient response (complementary function) :

The transient response from the auxiliary equation is obtained by putting input I_i equal to zero ;

i.e.,

$$(1 + \tau D)\, I_{o,\,t} = 0 \qquad ...(12)$$

(subscript t refers to the transient value)

Let the solution be of the form :

$$I_{o,\,t} = A\, e^{mt}$$

where, m is an algebraic variable

or,
$$(1 + \tau D)\, A\, e^{mt} = 0$$

or,
$$A\, e^{mt} + \tau \,.\, \frac{d}{dt}(A\, e^{mt}) = 0$$

or,
$$A\, e^{mt} + \tau \,.\, A\, m\, e^{mt} = 0$$
$$A\, e^{mt}(1 + \tau \,.\, m) = 0$$

$\therefore$
$$m = -\frac{1}{\tau}$$

Then,
$$I_{o,\,t} = A\, e^{mt} = Ae^{-t/\tau} \qquad ...(13)$$

The transient response of a first-order system is same for different standard inputs.

Steady state response (Particular integral) :

The steady state response is given by :

$$(1 + \tau D)\, I_{o,\,s} = I_i \qquad ...(14)$$

(Subscript s refers to the steady state value)

or,
$$I_{o,\,s} = (1 + \tau D)^{-1}\, I_i$$
$$= (1 - \tau D + \text{terms in } D^2 \text{ and higher})\, I_i \qquad ...(15)$$

1. Step input :

Since the input I_i is a step of constant magnitude ; its differential equals zero, and subsequently, we get

$$I_{o,s} = (1 - \tau D)\, I_i = I_i \qquad ...(16)$$

Total response = Transient response + steady state response

or,
$$I_o = Ae^{-t/\tau} + SI_i \qquad ...(17)$$

The constant A is evaluated from the initial conditions as follow :

At $t = 0$, $I_o = 0$

$\therefore$ $0 = A + SI_i$ or, $A = -SI_i$

$\therefore$
$$I_o = \underbrace{-I_i\, e^{-t/\tau}}_{\text{Transient}} + \underbrace{I_i}_{\text{steady state}}$$

or, $$I_o = I_i\,(1 - e^{-t/\tau}) \quad ...(18)$$

or, $$\frac{I_o}{I_i} = (1 - e^{-t/\tau}) \quad ...(19)$$

... in *non-dimensional form.*

Salient features (with step input) :

Following are the salient features of first-order system with step input :

(*i*) The transient response of the first-order system is *time dependent* ; as the time passes, grows its value decreases (Refer to eqn. 19) and after a very long time the value becomes zero approximately. Thus magnitude of output (I_o) will be same as input (I_i) when the time very large.

(*ii*) The speed of response relates to the time constant τ. A large τ indicates that response of the system is slow, whereas a small τ represents a fast system response. Thus in order to get good fidelty (*i.e.,* for accurate dynamic measurements) efforts should be made to minimise the value of τ.

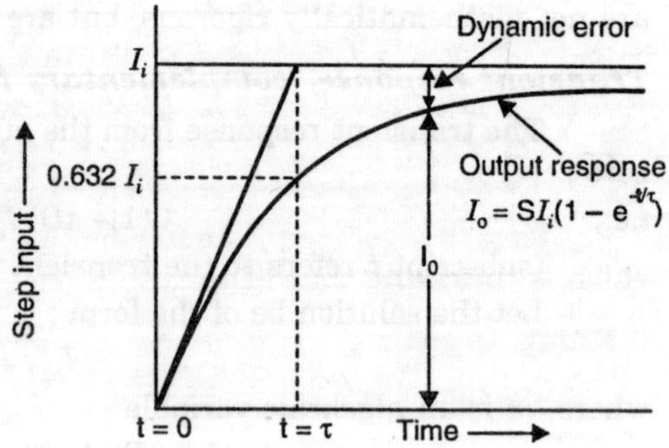

Fig. 16. Time response of a first-order system to step input.

(*iii*) Refer to Fig. 16, which shows the time response of a first-order system to a step-input when $t = \tau$; $\frac{I_o}{I_i} = (1 - e^{-1}) = 0.632$. Thus, the *time constant* (τ), *for a rising exponential function, is defined as the time to reach* 63.2% of its steady state value. The time constant, for a *decaying function* would correspond to the time taken to fall to 36.8% *of its initial value.*

(*iv*) Dynamic error (*i.e.,* vertical difference between the input and output response curve).

$$E_{dy.} = I_i - I_o$$
$$= I_i - I_i(1 - e^{-t/\tau})$$
$$= -\,I_i\,e^{-t/\tau} \quad ...(20)$$

or, $$\frac{E_{dy.}}{I_i} = -\,e^{-t/\tau} \quad ...(21)$$

...... *(in dimensionless form)*

(*v*) In case the measurand has an initial value of I_{initial} at $t = 0$, then the output. I_o at any instant t is given by

$$I_o = I_i\,[1 - e^{-t/\tau}] + I_{\text{initial}}\,e^{-t/\tau}$$

or, $$I_o = I_i + (I_{\text{initial}} - I_i)\,e^{-t/\tau} \quad ...(22)$$

(*vi*) The speed response of a system is defined in terms of *settling time* (it is time taken by the system to reach and remain within a certain percentage tolerance band of the final steady state value). Smaller the settling time, faster is the response. Typical value of tolerance band are 2% and 5% settling times.

For unit step input, the output reaches the values given in Table 5 at various intervals of time. This table also shows the per unit dynamic error at different intervals of time.

Table 5. Per unit output and dynamic error for a step input for a first-order system

t/τ	*Per unit output* (I_o/I_i)	*Per unit dynamic error* (E_{dy}/I_i)
0	0	1.000
1	0.632	0.368
2	0.865	0.135
3	0.950	0.050
4	0.982	0.018
5	0.993	0.007
∞	1.000	0

Thus, 5% settling time means that the system has reached its specified value after a time which is thrice the time constant (Table 5).

2. **Ramp Input :**

Consider that the input signal varies linearly with time so that ramp input I_i is prescribed by the relation $I_i = \psi t$, where ψ is constant. Then the governing differential equation is given by

$$(1 + \tau D)\, I_o = \psi t \qquad \text{...(23)}$$

Now, *Transients response* (complimentary function) is given as ;

$$I_{o.t} = A\, e^{-t/\tau} \qquad \text{...as before}$$

and, *steady state response* (particular integral) will be ;

$$I_{o,s} = (1 - \tau D + \text{terms in } D^2 \text{ and higher})\, \psi\, t = \psi t - \tau \frac{d}{dt}(\psi\tau)$$

or,
$$I_{o,s} = \psi t - \psi\tau \qquad \text{...(24)}$$

∴ Complete response = Transient response + steady state response

$$I_o = A\, e^{-t/\tau} + (\psi t - \psi\tau) \qquad \text{...(25)}$$

The value of constant A can be evaluated by applying the initial condition.

At, $t = 0 \qquad I_0 = 0$

or, $0 = A - \psi\tau \quad \therefore \quad A = \psi\tau$

∴
$$I_o = \psi t - \psi\tau + \psi\tau \times e^{-t/\tau} = \psi(t - \tau) + \psi\tau\, e^{-t/\tau} \qquad \text{...(26)}$$

or,
$$I_o = \psi[t - \tau(1 - e^{-t/\tau})] \qquad \text{...(26a)}$$

Fig. 17 shows the time response of a first-order system to a ramp input.

The dynamic error.

$$E_{dy.} = I_i - I_o$$
$$= \psi t - [\psi t - \psi\tau + \psi\tau\, e^{-t/\tau}]$$
$$= \underbrace{\psi\tau}_{\text{steady}} - \underbrace{\psi\tau\, e^{-t/\tau}}_{\text{Transient}} \qquad \text{...(27)}$$

or,
$$\frac{E_{dy.}}{\psi\tau} = 1 - e^{-t/\tau} \qquad \text{...(28)}$$

...(in dimensionless form)

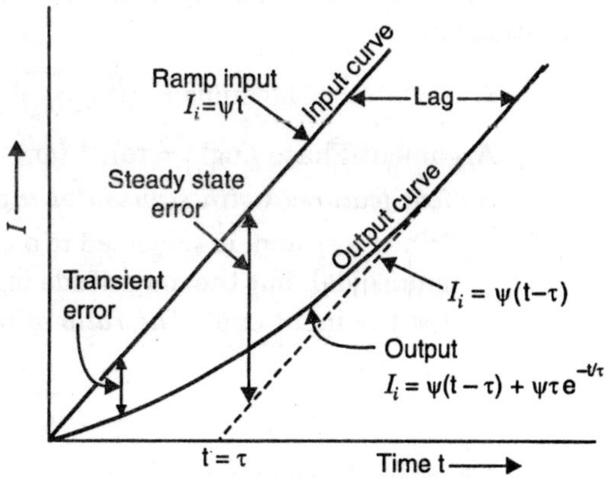

Fig. 17. Time response of a first-order system to a ramp input.

Salient features (with ramp input) :

(*i*) The term $\psi\tau$ being independent of time continues to exist and so it is called the steady state error. The term $\psi\tau\, e^{-t/\tau}$ gradually decreases with time and hence is called the *transient error.*

— Since the steady state error is directly proportional to τ (time constant), therefore, the larger the value of t the larger will be the magnitude of the error.

— When τ is made small the transient error decreases rapidly ; this implies, that the system attains the steady state at a faster pace.

(*ii*) The output response curve always lags behind the input curve by a constant amount known as lag.

3. Sinusoidal (Harmonic) input :

The frequency analysis of a system pertains to the steady state response of the system to a sinusoidal input. In this analysis, the system is subjected to a sinusoidal input and the system response studied with frequency as the independent variable. The sinusoid is a unique input signal, and the resulting output signal for a linear system is sinusoidal in the steady state. However, the output signal differs from the input waveform in amplitude and phase.

In order to determine the frequency response of sinusoidal input to a first-order system, let us replace the transfer operator D by a factor $j\omega$ in the input/output relationship ; then we get,

$$\frac{I_o}{I_i} = \frac{1}{1 + D\tau} = \frac{1}{1 + j\omega\tau} \quad ...(29)$$

where, ω = Input frequency, rad/s, and

$j = \sqrt{(-1)}$

In a frequency response the following two quantities are of interest : Refer to Fig. 18.

(*i*) *Amplitude ratio or modulus* $\left(\frac{I_o}{I_i}\right)$. It prescribes the size of the output amplitude relative to the input amplitude.

(*ii*) *Phase shift* of output relative to input.

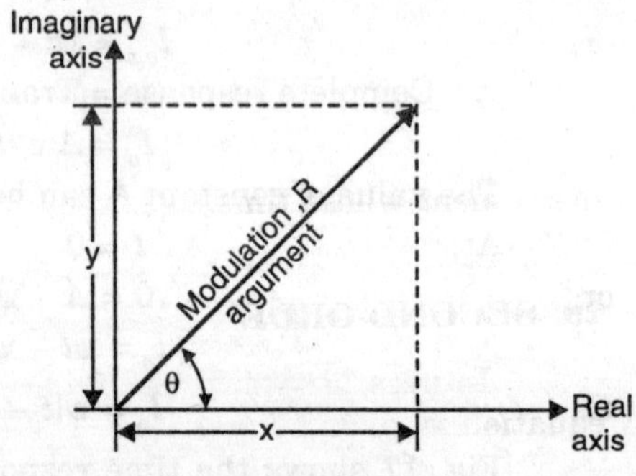

Fig. 18

For the first-order system represented by the equation (29),

$$\text{Modulus} = \sqrt{1 + (\omega\tau)^2}$$

$$\text{Argument/Phase angle} = \tan^{-1}(\omega\tau) \quad ...(30)$$

Salient features (with sinusoidal input) : Refer to Fig. 19.

(*i*) When a system is subjected to a sinusoidal input with frequency ω, its output will also be sinusoidal, but the magnitude of the output amplitude necessarily may not be the same (as the input one). *The ratio of the amplitude* (often called *attenuation*) is given as :

$$\frac{I_o}{I_i} = \frac{1}{\sqrt{1 + (\omega\tau)^2}} \quad ...(31)$$

Thus, with the increase in input frequency, the amplitude ratio decreases.

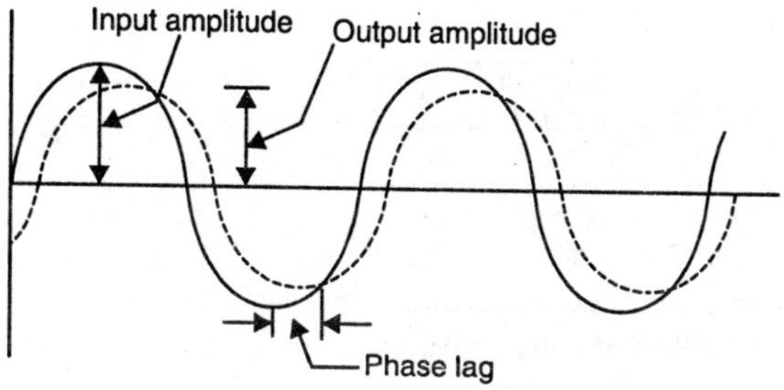

Fig. 19. Relationship between an input frequency and corresponding output frequency.

(*ii*) The output from the system may not necessarily be in phase with the input ; and the phase difference is given by

$$\phi \text{ (phase angle)} = -\tan^{-1}(\omega\tau) \qquad ...(32)$$

– ve indicates that output *lags* behind the input. When $\omega = \frac{1}{\tau}$ the phase lag is $\frac{\pi}{4}$ or 45°.

As the accuracy of an instrument measuring dynamic input depends upon the time constant, therefore, smaller the time constant, greater the accuracy ; for phase shift to be small, the time period τ should be small.

(*iii*) When the input and output signals are given be the relations :

$$I_i = A \sin \omega t, \text{ and } I_o = B \sin (\omega t + \phi) = zA \sin (\omega t + \phi).$$

Then the amplitude ratio may be represented as follows :

$$K = \left| \frac{I_o}{I_i} \right| = \frac{z}{\sqrt{1 + (\omega\tau)^2}} \qquad ...(33)$$

In order to produce amplitude of sinewave without any attenuation ($K = 1$) we must use an instrument whose time constant, $\tau = \frac{\sqrt{z^2 - 1}}{\omega}$.

29. SECOND-ORDER SYSTEM RESPONSES

In case of typical second-order system having, unit static sensitivity, the homogeneous equation is given by :

$$\frac{I_o}{I_i} = \frac{1}{\left(\frac{1}{{\omega_n}^2}\right) D^2 + \left(\frac{2\gamma}{\omega_n}\right) D + 1}$$

or,

$$\left[\frac{1}{(\omega_n)^2} D^2 + \left(\frac{2\gamma}{\omega_n}\right) D + 1 \right] I_o = I_i \qquad ...(34)$$

(where, γ = damping ratio)

(*a*) **Transient response** (*complimentary function*). It is obtained from the auxiliary equation by replacing D (transfer operator) by an algebraic variable s and putting I_i equal to zero ; we get the auxiliary equation as :

$$\frac{1}{{\omega_n}^2} s^2 + \frac{2\gamma}{\omega_n} s + 1 = 0$$

The roots are

$$s_1, s_2 = \frac{\dfrac{-2\gamma}{\omega_n} \pm \sqrt{\left(\dfrac{2\gamma}{\omega_n}\right)^2 - \dfrac{4}{\omega_n^2}}}{\dfrac{2}{\omega_n^2}}$$

$$= -\gamma\omega_n \pm \sqrt{\gamma^2\omega_n^2 - \omega_n^2}$$

or, $$s_1, s_2 = -\gamma\omega_n \pm \omega_n\sqrt{\gamma^2 - 1} \quad ...(35)$$

The *transient solution* has the accepted form,

$$I_{o,t} = A\,e^{s_1 t} + B\,e^{s_2 t}$$

where, A and B = Arbitrary constants to be determined from initial conditions, and

s_1, s_2 = Roots of the auxiliary equation (The roots may be real and different, real and equal or imaginary and *that determines the nature of transient response of the system*).

The response of the system is of the following three types depending upon the roots of the characteristic equations :

(*i*) Overdamped systems.

(*ii*) Critically-damped systems.

(*iii*) Under-damped systems.

(*i*) ***Over-damped systems.*** In this case $\gamma > 1$ and the roots are *real and unequal.*

- There is *heavy damping* and the system responds to the final steady-state value without any oscillations but in a *sluggish manner.*
- *No overshoot* in step response and no *"resonance"* (resonance refers to the output signal *greater* in magnitude than the ideal output) in the *frequency response.*
- The overdamped systems, owing to their sluggish response, are usually unsuitable for several control applications.

(*ii*) ***Critically-damped systems.*** In this case $\gamma = 1$ and the roots are *real and equal.*

- The system has a *quick and smooth response to the final steady state value without any oscillations.*
- *No overshoot in the step response and no resonance in the frequency response.*

(*iii*) ***Under-damped systems.*** For an undamped system $\gamma < 1$ and roots of the characteristic equation are a complex conjugate pair ; these are given as

$$s_1, s_2 = -\gamma\omega_n \pm j\omega_n\sqrt{(1-\gamma^2)}$$

$$= -\gamma\omega_n \pm j\omega_d \quad ...(36)$$

where, $\omega_d = \omega_n\sqrt{(1-\gamma^2)}$; this quantity is called *"damped natural frequency"* of the system (This is the frequency at which the damped system freely oscillates when disturbed)

- Such systems *take a long time to reach steady state, but have quick initial response.*
- In these systems, there are oscillations in the step response and resonance effects in the frequency response for values of $\gamma < 0.707$.
- Majority of instruments and control systems are generally underdamped (*light damping*).

(*b*) **The steady state response** (*particular integral*). It is given by :

$$\left(\frac{1}{\omega_n^2} D^2 + \frac{2\gamma}{\omega_n} D + 1\right) I_{o,s} = I_i$$

or,

$$I_{o,s} = \left(1 + \frac{2\gamma}{\omega_n} D + \frac{1}{\omega_n^2} D^2\right)^{-1} I_i$$

$$= \left(1 - \frac{2\gamma}{\omega_n} D + \text{terms in } D^2 \text{ and higher}\right) I_i \qquad \text{...(37)}$$

1. Step input :

Fig. 20 shows the time response of a second-order system to a step input.

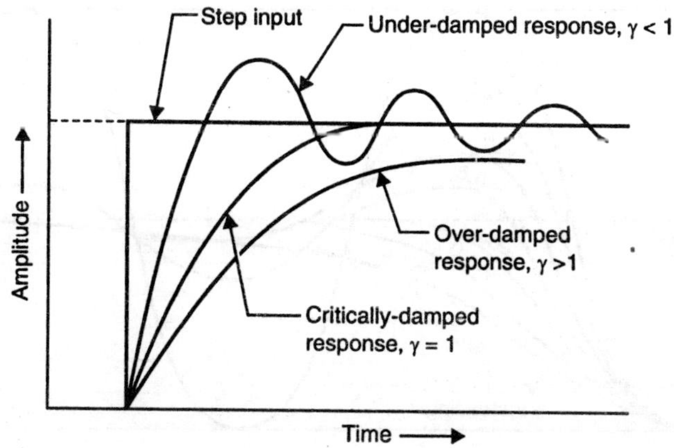

Fig. 20. Time response of a second-order system to a step input.

Since the input I_i is a step of constant magnitude, its differential equals zero and we have,

$$I_{o,s} = \left(1 - \frac{2\gamma}{\omega_n} D\right) I_i = I_i$$

∴ The complete response

$$I_o = I_{o,s} + I_{o,t}$$

or,

$$I_o = I_i \; A\,e^{s_1 t} + B\,e^{s_2 t} \qquad \text{...(38)}$$

For the under-damped system, the complex conjugate pair of roots are given by

$$s_1, s_2 = -\gamma\omega_n \pm j\omega_d$$

$$\therefore \quad I_o = I_i + Ae^{-(\gamma\omega_n + j\omega_d)t} + Be^{-(\gamma\omega_n - j\omega_d)t} \qquad \text{...(39)}$$

Replacing the complex exponentials by sines and cosines, we get

$$I_o = I_i + e^{-\gamma\omega_n t} (A \cos \omega_d t + B \sin \omega_d t) \qquad \text{...(40)}$$

By applying the initial conditions,

At $t = 0$, $\quad I_o = 0$ and $\dfrac{dI_o}{dt} = 0$

we get the values of the constant A and B as :

$$A = -I_i \text{ and}$$

$$B = \frac{-I_i \gamma}{\sqrt{1-\gamma^2}}$$

Inserting these values in equation (40), we have

$$I_o = I_i \left[1 - e^{-\gamma\omega_n t} \left\{ \cos \omega_d t + \frac{\gamma}{\sqrt{1-\gamma^2}} \sin \omega_d t \right\} \right] \qquad ...(41)$$

Fig. 21 shows the transient response of a second-order system to a unit step function for different values of damping factor γ ; *the curves indicate the overshoot and oscillations increase with a reduced damping in the system.*

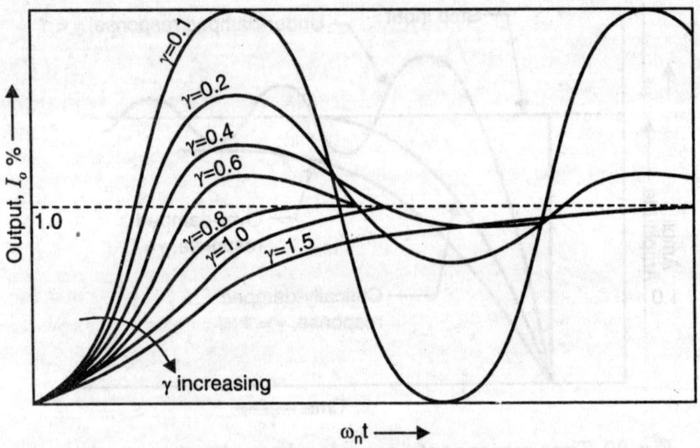

Fig. 21. Transient response of a second-order system to a unit step function input for different values of damping factor γ.

2. Sinusoidal (Harmonic) input :

When a sinusoidal input is given to the system, its steady state response is determined by replacing *the operator D by jω in the input/output relationship, as given below :*

$$\frac{I_o}{I_i} = \frac{1}{\frac{1}{\omega_n^2} D^2 + \frac{2\gamma}{\omega_n} D + 1} = \frac{\omega_n^2}{D^2 + 2\gamma\omega_n D + \omega_n^2}$$

$$= \frac{\omega_n^2}{(j\omega)^2 + 2\gamma\omega_n (j\omega) + \omega_n^2} = \frac{\omega_n^2}{(\omega_n^2 - \omega^2) + j(2\gamma\omega_n\omega)} \qquad ...(42)$$

where, ω = Input frequency in rad/s, and $j = \sqrt{-1}$

The denominator is a complex number having :

$$\text{Modulus} = \sqrt{[(\omega_n^2 - \omega^2)^2 + (2\gamma\omega_n\omega)^2]}$$

$$\text{Argument} = \tan^{-1} \left\{ \frac{2\gamma\omega_n\omega}{\omega_n^2 - \omega^2} \right\} \qquad ...(43)$$

∴ The amplitude ratio,

$$\left| \frac{I_o}{I_i} \right| = \frac{\omega_n^{\ 2}}{\sqrt{[(\omega_n^{\ 2} - \omega^2)^2 + (2\gamma\omega_n\omega)^2]}} \qquad ...(44)$$

and, the system phase lag,

$$\phi = \tan^{-1}\left(\frac{2\gamma\omega_n\omega}{\omega_n^{\ 2} - \omega^2}\right) \qquad ...(45)$$

...[using eqn. (43)]

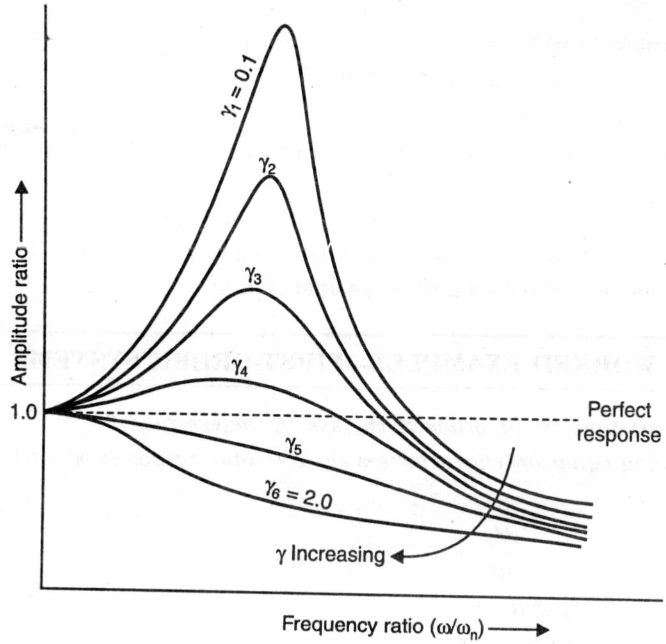

Fig. 22. Step response of a second-order system.

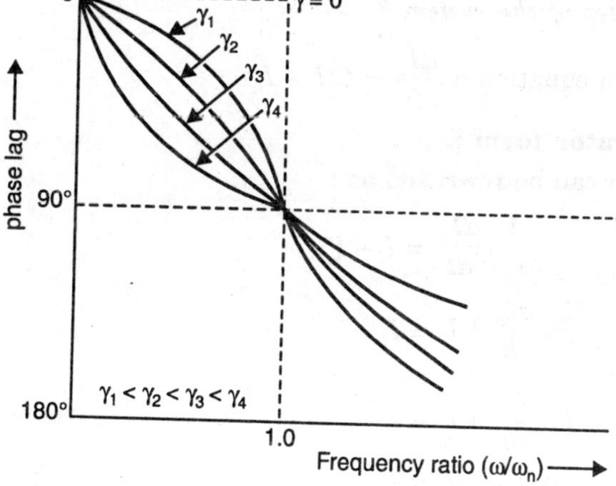

Fig. 23. Plot between phase lag and frequency ratio for a second-order system.

Fig. 22 and Fig. 23 show the variation of amplitude ratio and phase lag verses frequency ratio (ω/ω_n) for various values of damping ratio (γ). From these graphs we observe that the *salient features* of the *steady state response* of a second-order system are :

(*i*) As *frequency ratio* $\to 0$:

Amplitude ratio $\to 1$;

Phase lag $\to 0°$

(*ii*) As frequency ratio $\to \infty$:

Amplitude ratio $\to 0$;

Phase lag $\to 180°$

(*iii*) When frequency ratio = 1 :

Amplitude ratio $\to \infty$ in undamped systems ($\gamma = 0$)

Phase lag $\to -90°$ in all the systems.

This condition is known as "*resonance*" and can *result in destructive oscillation in lightly damped system.*

(*iv*) *When the amplitude ratio is unity for all frequencies the frequency response is considered to be ideal*. The nearest response to this effect is achieved when the value of γ (damping ratio) lies between 0.6 and 0.7 for both the step and sinusoidal inputs.

WORKED EXAMPLES—FIRST-ORDER SYSTEMS

Example 1. (*a*) *How is the order of the system determined ?*

(*b*) *The following equation characterises the dynamic response of a temperature measuring instrument :*

$$\frac{dI_0}{dt} = C(I_i - I_o)$$

where, I_o = *Indicated temperature,*

I_i = *Input temperature, and*

C = *A numerical constant.*

(*i*) *Determine the transfer operator form of the equation.*

(*ii*) *What is the order of the system ?*

Solution. (*b*) Given equation = $\dfrac{dI_o}{dt} = C(I_i - I_o)$

(*i*) **Transfer operator form :**

The given equation can be rewritten as :

$$\frac{1}{C} \cdot \frac{dI_o}{dt} = I_i - I_o$$

or,
$$\frac{1}{C} \cdot \frac{dI_o}{dt} + I_o = I_i$$

or,
$$I \cdot \frac{dI_o}{dt} + I_o = I_i \qquad \left(\text{where, } \tau = \text{time constant} = \frac{1}{C}\right)$$

or,
$$(\tau D + 1)I_o = I_i$$

$\therefore$ The transfer operator form of the equation is given by :

$$\frac{I_o}{I_i} = \frac{1}{(\tau D + 1)} \quad \textbf{(Ans.)}$$

(*ii*) **Order of the system :**

Since the highest differential in the denominator of the transfer operator is *unity,* therefore, the temperature measuring instrument has a *first-order system.* **(Ans.)**

The grouping of the measurement and control systems is done according to the *order of the highest differential in the denominator* of the system transfer operation.

Examples :

- First-order system : $\dfrac{I_o}{I_i} = \dfrac{2}{1 + 3D}$

- Second order system : $\dfrac{I_o}{I_i} = \dfrac{6}{D^2 + 3D + 4}$,

or, $$\frac{I_o}{I_i} = \frac{14}{(1 + 0.3D)(1 + 0.2D)}.$$

Example 2. *Formulate the governing equation for a first-order system-temperature measurement by a thermal measuring element (say a thermometer or thermocouple).*

Solution. Refer to Fig. 24.

Let, I_i = Temperature of the medium,

I_o = Temperature indicated by the thermal measuring instrument (say a thermometer or thermocouple),

A = Exposed area of the thermal measuring element,

h = Convective heat transfer coefficient,

m = Mass of thermal element, and

c = Specific heat of the element.

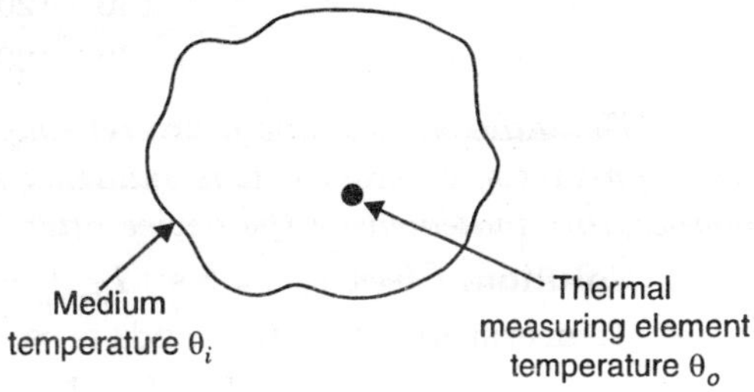

Fig. 24. Thermal element.

Then, the rate of heat flux into the element is,

$$Q = hA(\theta_i - \theta_o)$$

The rate of enthalpy gain by the element

$$= mc \frac{d\theta_o}{dt}$$

Since the rate of heat flow equals the rate of enthalpy gain by the element, therefore, equating (*i*) and (*ii*) we get :

$$mc \frac{d\theta_o}{dt} = hA\,(\theta_i - \theta_o)$$

or, $$\frac{mc}{hA} \cdot \frac{d\theta_o}{dt} + \theta_o = \theta_i$$

$$\tau \frac{d\theta_o}{dt} + \theta_o = \theta_i$$

where, $\tau = \dfrac{mc}{hA}$ is known as *time constant* of the system.

In terms of D-operator (where $D = \frac{d}{dt}$), we have

$$(\tau D + 1)\theta_o = \theta_i$$

or,
$$\frac{\theta_o}{\theta_i} = \frac{1}{\tau D + 1} \quad \text{...Required equation.}$$

which is an equation of *first-order*. **(Ans.)**

Example 3. *A thermometer, idealised as a first-order system with a time constant of 2.2 seconds, is suddenly given an input of 160°C from 0°C.*

(i) What will be reading of the thermometer after 1.2 seconds ?

(ii) Determine its reading if it is initially held at 20°C.

Solution. *Given* : I_i = 160°C ; t = 1.2 s ; τ = 2.2 s ; I_{intial} = 20°C.

(i) **Thermometer's reading after 1.2 s :**

We know that,
$$I_o = I_i(1 - e^{-t/\tau}) \quad \text{...[Eqn.(19)]}$$
$$= 160\,[1 - e^{-(1.2/2.2)}] = \mathbf{67.27°C} \quad \textbf{(Ans.)}$$

(ii) **Thermometer's reading if it was initially held at 20°C :**

For a step input from 20°C to 160°C, we have

$$I_o = I_i + (I_{\text{initial}} - I_i)\, e^{-t/\tau} \quad \text{...[Eqn.(21)]}$$
$$= 160 + (20 - 160)\, e^{-(1.2/2.2)}$$
$$= 160 + (20 - 160) \times 0.5796 = \mathbf{78.86°C} \quad \textbf{(Ans.)}$$

Example 4. *A temperature sensing device can be modeled as a first-order system with a time constant of 5 seconds. It is suddenly subjected to a step input of 30°C–160°C. Calculate the temperature indicated by the device after 10 seconds after the start of the process.*

Solution. *Given* : τ = 5 s ; I_{initial} = 30°C ; I_i = 160°C ; t = 10 s.

Temperature after 10 seconds is calculated as follows :

$$I_o = I_i + (I_{\text{initial}} - I_i)\, e^{-t/\tau} \quad \text{...[Eqn.(22)]}$$
$$= 160 + (30 - 160)\, e^{-10/5}$$
$$= 160 - 130 \times 0.1353 = \mathbf{142.4°C} \quad \textbf{(Ans.)}$$

Example 5. *A temperature sensitive transducer when subjected to sudden temperature change takes 9 seconds to reach equilibrium conditions (Three time constants). Calculate the time taken by the transducer to read half of the temperature difference.*

Solution. Time taken to reach equilibrium condition = 3τ = 9s (*Given*)

$\therefore$ Time constant, $\tau = \frac{9}{3} = 3$ s

Time taken by the transducer to read half of the temperature difference is calculated as follows :

$$I_o = I_i(1 - e^{-t/\tau}) \quad \text{...[Eqn.(19)]}$$

or,
$$\frac{I_o}{I_i} = 1 - e^{-t/\tau}$$

or,
$$0.5 = 1 - e^{-t/3}$$

or,
$$e^{-t/3} = 0.5 \quad \text{or,} \quad e^{t/3} = 2$$
$$t = \mathbf{2.08\ s} \quad \textbf{(Ans.)}$$

Example 6. *A resistance temperature detector has a steady state gain of 0.296 Ω/°C and a time constant of 4.8 seconds. It is subjected to a step change of 80° C in temperature. If its stable resistance before the step change in 90 Ω, write down the time-domain equation for resistance and find its value after 12 seconds of application of step input.*

Solution. Step input in terms of resistance

= Steady state gain × change of temperature

= 0.296 × 80 = 23.68 Ω step input in terms of resistance.

In order to determine the time domain equation for resistance, the value of initial resistance must be added to it.

i.e., $R_t = 23.68(1 - e^{-t/4.8}) + 90$ **(Ans.)** ...***Time domain equation***

∴ The value of resistance at t = 12 s,

$$R_{12} = 23.68\,(1 - e^{-12/4.8}) + 90 = \mathbf{111.74\ \Omega}\quad \textbf{(Ans.)}$$

Example 7. *A thermometer initially at a temperature of 15°C is suddenly plunged into a liquid bath maintained at 140°C. After a time interval of 4 seconds, the thermometer indicated a reading of 75 °C.*

Determine :

(i) The time constant for the thermometer

(ii) The indicated temperature after five times constant ; comment upon this result.

Solution. *Given* : I_{initial} = 15°C ; I_i = 140°C ; t = 4 s ; I_o = 75°C.

(*i*) **The time constant for the thermometer, τ :**

When a first-order system is subjected to a step input from I_{initial} to I_i, the indicated temperature is given by the relation :

$$I_o = I_i + (I_{\text{initial}} - I_i)e^{-t/\tau} \qquad \text{...[Eqn. (22)]}$$

$$75 = 140 + (15 - 140)e^{-4/\tau}$$

or,
$$e^{-4/t} = \frac{75-140}{15-140} = 0.52 \quad \text{or,} \quad e^{4/t} = 1.923$$

Taking logs on both sides, we get

$$\frac{4}{\tau} = 0.6539$$

∴
$$\tau = \mathbf{6.12\ s}\quad \textbf{(Ans.)}$$

(*ii*) **Indicated temperature after five times constant, I_o :**

Substituting the value in the above equation, we get

$$I_o = 140 + (15 - 140)\,e^{-5\tau/\tau}$$
$$= 140 + (15 - 140)\,e^{-5} = \mathbf{139.16°C}\quad \textbf{(Ans.)}$$

Comments : After a time interval equivalent to five times constants, the thermometer reaches the *equilibrium condition* (*i.e.*, it indicates a temperature which is approximately equal to the maximum step input).

Example 8. *A temperature sensitive transducer used to measure the temperature of a furnace has been idealised as a first-order system subjected to ramp input. If the maximum permissible error in temperature measurement is limited to 3.9°C calculate the time constant of the transducer when the furnace temperature increases at the rate of 0.2°C/s.*

Solution. *Given :* Dynamic error, $E_{dy.}$ = 3.9°C ; slope ; ψ = 0.2°C/s.

For the ramp input, the input signal varies linearly with time ($I_i - \psi t$) and the dynamic or measurement error,

$$E_{dy.} = \underbrace{\psi\tau}_{\text{Steady}} - \underbrace{\psi\tau\, e^{-t/\tau}}_{\text{Transient}} \quad \text{...[Eqn. (27)]}$$

Since the transients error gradually vanishes with time, we have

$$E_{dy.} = \psi\tau$$

$\therefore$ Time constant, $\tau = \dfrac{E_{dy}}{\psi} = \dfrac{3.9}{0.2} =$ **19.5 s. (Ans.)**

Example 9. *A balloon (weather) carrying a temperature sensing device (of first-order, time constant 8 seconds) rises through the atmosphere at 5 m/s. It supplies information about temperature and altitude through radio signals. At 2500 m height, a temperature indication of 30°C has been received. If the temperature varies with altitude at a uniform rate of 0.011°C/m, determine the time altitude at which 30°C temperature occurs.*

Solution. Time constant of temperature sensing device, $\tau = 8$s

Rate of rise of the balloon $= 5$ m/s

Temperature indicated at an altitude of 2500 m $= 30$°C

Rate of temperature variation with altitude $= 0.011$°C/m

True altitude at 30°C :

Rate of change of temperature with time $= 0.011 \times 5 = 0.055$°C/s

For a ramp signal applied to a first-order system, the error, when transient oscillations die away,

$$E_{dy.} = \psi\tau = 0.055 \times 8 = 0.44°\text{C}$$

$\therefore$ Error in amplitude $= \dfrac{0.44}{0.011} = 40$ m

$\therefore$ *Actual altitude* $= 2500 - 40 =$ **2460 m** **(Ans.)**

Example 10. *A thermometer with time constant of 50 seconds is used for measuring temperature with 500 s time period. Calculate :*

(i) The ratio of output to input *(ii) The time lag.*

Solution. Time constant of thermometer, $t = 50$ s

Time period, $t = 500$ s

Now, frequency of temperature variation, $\omega = \dfrac{2\pi}{t} = \dfrac{2\pi}{500} = 0.01257$ rad/s

(i) The ratio of output to input, $\dfrac{I_o}{I_i}$:

$$\frac{I_o}{I_i} = \frac{1}{\sqrt{1+(\omega\tau)^2}} = \frac{1}{\sqrt{1+(0.01257 \times 50)^2}} = \mathbf{0.8467} \quad \textbf{(Ans.)}$$

(ii) Time lag :

Phase shift $= \tan^{-1}(\omega\tau) = \tan^{-1}(0.01257 \times 50) = 32.15°$ or 0.561 rad.

$$\therefore \quad \text{Time lag} = \frac{1}{\omega} \times \text{phase shift in radian}$$

$$= \frac{1}{0.01257} \times 0.561 = \mathbf{44.63\ s} \quad \textbf{(Ans.)}$$

Example 11. *During a particular process, the air temperature cycle at the rate of 1 cycle every 4 minutes. The time constant of the temperature measuring device is 20 seconds. Determine :*

(i) The variation in the indicated temperature if true (input) temperature has a sinusoidal variation of ± 25°C.

(ii) The time by which the maximum reading of the thermometer lags the true maximum value.

Solution. *Given* : Air temperature variation = 1 cycle every 4 minutes

Time constant of the device, $\tau = 20\text{s}$

Sinusoidal variation of the input, $I_i = \pm 25°\text{C}$

Now, $t = 4$ minutes

$$f = \frac{1}{t} = \frac{1}{4 \times 60} = \frac{1}{240} \text{ cycles/s}$$

and,
$$\omega = 2\pi f = 2\pi \times \frac{1}{240} = 0.0262 \text{ rad/s}$$

(*i*) **The variation in the indicated temperature (I_o) :**

The steady state (when transient errors are omitted) solution for the sinusoidal input is given as :

$$\frac{I_o}{I_i} = \frac{1}{\sqrt{(1+\omega\tau)^2}} \sin(\omega t - \phi) \quad \text{...[Eqn.(31)]}$$

where, I_o = Output temperature as indicated by the thermometer,

I_i = True temperature, and

ϕ = Phase lag = $\tan^{-1}(\omega t)$

Substituting the value of ω and τ (calculated above) in the above equation, we get

$$\frac{I_o}{I_i} = \frac{1}{\sqrt{1+(0.0262 \times 20)^2}} = 0.886$$

$$I_o = \pm 25 \times 0.886 = \pm \mathbf{22.15°C} \quad \textbf{(Ans.)}$$

(*ii*) **The lag (in seconds) :**

$$\text{Phase lag} = \tan^{-1}(\omega\tau)$$

or,
$$\phi = \tan^{-1}(0.0262 \times 20) = 27.65° \text{ or } 0.4826 \text{ rad.}$$

The lag in seconds $= \frac{1}{\omega} \times$ phase lag in radian

$$= \frac{1}{0.0262} \times 0.4826 = \mathbf{18.42 \text{ seconds}} \quad \textbf{(Ans.)}$$

Example 12. *A first-order instrument having an input of sinewave pattern is required to measure signals with frequency response of 120 cycles per second. If an amplitude inaccuracy of 4 percent is permissible, calculate the maximum time constant and time lag at 90 cycles per second.*

Solution. *Given* : Input frequency = 120 cycles per second ; Amplitude inaccuracy permissible = 4%.

Now, $\omega = 2\pi f = 2\pi \times 120 = 754$ rad/s

Maximum time constant τ :

Corresponding to 4% amplitude inaccuracy, the amplitude ratio of output and input signals is $\left(1 - \frac{4}{100}\right) = 0.96$.

For a first-order instrument ; the amplitude ratio is given as :

$$\frac{I_o}{I_i} = \frac{1}{\sqrt{1+(\omega\tau)^2}}$$

or, $$0.96 = \frac{1}{\sqrt{(1+754\tau)^2}}$$

or, $$(1 + 754\tau)^2 = \left(\frac{1}{0.96}\right)^2 = 1.085$$

or, $$1 + 754\tau = 1.0416$$

∴ $$\tau = \mathbf{5.517 \times 10^{-5}\ s} \quad \textbf{(Ans.)}$$

Time lag at 90 cycles per second :

We know that, for a sinusoidal input,

$$\phi = \tan^{-1}(\omega\tau) = \tan^{-1}(2\pi \times 90 \times 5.517 \times 10^{-5})$$
$$= 1.787° \quad \text{or} \quad 0.03119$$

∴ $$\text{Time lag} = \frac{1}{\omega} \times \text{phase lag in radian}$$
$$= \frac{1}{2\pi \times 90} \times 0.03119 = \mathbf{5.515 \times 10^{-5}\ s} \quad \textbf{(Ans.)}$$

SECOND-ORDER SYSTEMS

Example 13. *Formulate the governing equation for a second-order system—spring mass system with damping.*

Solution. Refer to Fig. 25.

Let, x_i = Input displacement,

x_o = Output displacement,

k = Stiffness of the spring,

C_d = Viscous damping coefficient, and

γ = Damping ratio.

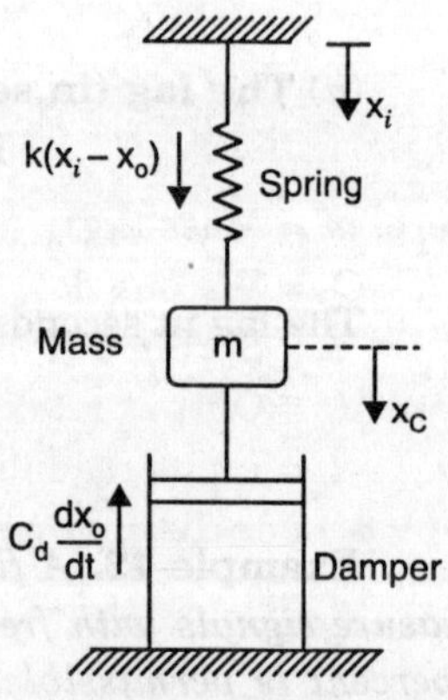

Fig. 25. Spring-mass system.

The *forces* acting on the mass are :

(*i*) As both ends of the spring are free to move, therefore

Spring force = Spring stiffness × displacement of one end of the spring relative to other $= k(x_i - x_o)$, acting *downward*.

(*ii*) One end of the dashpot fixed ; there is a reaction force acting in the *upward* direction.

Damping force = Damping coefficient × velocity

$$= C_d \times \frac{dx_o}{dt}$$

For translational systems, the Newton's law states that,

$$\Sigma \text{ Force} = \text{Mass} \times \text{acceleration} = m\,\frac{d^2x_o}{dt^2}$$

or, $$m\,.\,\frac{d^2x_o}{dt^2} = k(x_i - x_o) - C_d\,\frac{dx_o}{dt}$$

or, $$mD^2x_o + C_dDx_o + kx_o = kx_i \qquad \left(\text{where, } D = \frac{d}{dt}, \text{ and } D^2 = \frac{d^2}{dt^2}\right)$$

or, $$x_i = \left(\frac{m}{k}D^2 + \frac{C_d}{k}D + 1\right)x_o$$ ***...Required equation***

which is an equation of *second-order* type. **(Ans.)**

Comparing the above expression with the standard second-order form, we have :

Undamped natural frequency,

$$\omega_n = \sqrt{\frac{k}{m}}$$

and, $$\frac{2\gamma}{\omega_n} = \frac{C_d}{k}$$

$\therefore$ Damping ratio, $$\gamma = \frac{C_d\omega_n}{2k} = \frac{C_d}{2k}\sqrt{\frac{k}{m}}$$

or, $$\gamma = \frac{\mathbf{C_d}}{\mathbf{2\sqrt{km}}}$$ ***...Required equation.***

Example 14. (*a*) *Write down the expressions describing the motion of linear and rotational displacement systems of the second-order.*

(*b*) *The pen arrangement of U.V. recorder (second-order system) has a mass of 4.5g. Calculate the percentage reduction in mass if it is desired to have 15 percent increase in natural frequency of the recorder.*

Solution. (*a*) • The expression for *linear displacement* (spring-mass-damper) *system* is given by :

$$m\,\frac{d^2I_o}{dt^2} + C_d\,\frac{dI_o}{dt} + kI_o = kI_i \qquad ...(i)$$

where, m = Mass (kg),
C_d = Viscous damping force (Ns/m),
k = Spring stiffness (N/m),
I_o = Output reading, and
I_i = Input reading.

• For the *rotational system*, the expression may be written as :

$$J = \frac{d^2I_o}{dt^2} + C_d\,\frac{dI_o}{dt} + q\,I_o = q\,I_i$$

where, J = Inertia (kg m^2), and
q = Torsional stiffness.

Comparing these expressions with differential equation in the standard form.

$$\frac{1}{\omega_n^2}\cdot\frac{d^2I_o}{dt^2}+\frac{2\gamma}{\omega_n}\cdot\frac{dI_o}{dt}+I_o=I_i\text{, we get}$$

Natural frequency, $\omega_n = \sqrt{\frac{k}{m}} = \sqrt{\frac{q}{J}}$, as the case may be. **(Ans.)**

(*b*) *Given* : m = 4.5 g ; Percentage increase required = 15%.

Using subscripts 1 and 2 for initial and final values respectively, we have

$$\omega^2_{n1}=\frac{k}{m_1},\quad\text{and,}\quad \omega^2_{n}=\frac{k}{m_2}$$

$$\therefore\qquad m_2 = m_1 \times \left(\frac{\omega_{n1}}{\omega_{n2}}\right)^2$$

But, $\qquad \omega_{n2} = 1.15\,\omega_{n1}$ ...(*Given*)

$$\therefore\qquad m_2 = m_1 \times \left(\frac{\omega_{n1}}{1.15\,\omega_{n1}}\right)^2 = 0.756\ m_1$$

$\therefore$ *Percentage reduction in mass*

$$=\left(\frac{m_1-m_2}{m}\right)=\left(\frac{1-0.756}{1}\right)\times 100 = 24.4\%\quad\textbf{(Ans.)}$$

Example 15. *A second-order system follows the different equation given below :*

$$\frac{d^2I_o}{dt^2}+3\frac{dI_o}{dt}+30I_o=30I_i$$

where, I_o and I_i are the output and input quantities respectively. Determine the following :

(*a*) *Damping ratio.* (*b*) *Damped natural frequency.*

(*c*) *Static sensitivity.* (*d*) *Time constant.*

Solution. The standard form of the differential equation of a second-order system is given as :

$$\frac{1}{\omega_n^2}\cdot\frac{d^2I_o}{dt^2}+\frac{2\gamma}{\omega_n}\cdot\frac{dI_o}{dt}+I_o=kI_i \qquad ...(i)$$

Since the term I_o in eqn. (*i*) has a unit coefficient, therefore to recast the given equation in the standard form, let us divide the given equation throughout by 30 ; we get

$$\frac{1}{30}\cdot\frac{d^2I_o}{dt^2}+\frac{1}{10}\frac{dI_o}{dt}+I_o=I_i \qquad ...(ii)$$

Comparing eqns. (*i*) and (*ii*), we get

$$\omega_n^2 = 30\ ;\ \frac{2\gamma}{\omega_n}=\frac{1}{10}=0.1 : k = 1$$

$\therefore$ Natural frequency, $\omega_n = \sqrt{30} = 5.477$ rad/s (*a*) **Damping ratio, γ :**

$$\frac{2\gamma}{\omega_n} = 0.1, \quad \text{or} \quad \gamma = \frac{\omega_n}{2} \times 0.1$$

or,
$$\gamma = \frac{5.477}{2} \times 0.1 = \mathbf{0.274} \quad \textbf{(Ans.)}$$

(*b*) **Damped natural frequency, ω_d :**

$$\omega_d = \omega_n \sqrt{1-\gamma^2} = 5.477 \sqrt{1-0.274^2} = \mathbf{5.267\ rad/s} \quad \textbf{(Ans.)}$$

(*c*) **Static sensitivity :**

Static sensitivity, k = **1 (Ans.)**

(*d*) **Time constant τ :**

$$\tau = \frac{1}{\omega_n} = \frac{1}{5.477} = \mathbf{0.1826\ s} \quad \textbf{(Ans.)}$$

Example 16. *The following data pertain to a torque sensing transducer connected to the shaft of an electric motor which drives a load :*

Input torque = Sinusoidal with frequency varying at 1.95 rad/s ; Maximum permissible error = ± 8 percent ; Moment of inertia of load, J = 0.14 kg m² ; Torsional constant of the shaft, q = 1.22 Nm/rad. Calculate the following :

(*i*) *Natural frequency.* (*ii*) *Damping ratio.*

(*iii*) *Damped natural frequency.* (*iv*) *Time constant of the system.*

Solution. *Given* : ω = 1.95 rad/s ; Max. permissible error = ± 8% ;

J = 0.14 kg m^2 ; q = 1.22 Nm/rad.

(*i*) **Natural frequency, ω_n :**

$$\omega_n = \sqrt{\frac{q}{J}} = \sqrt{\frac{1.22}{0.14}} = \mathbf{2.95\ rad/s} \quad \textbf{(Ans.)}$$

(*ii*) **Damping ratio, γ ;**

Also, normalized frequency ratio $r = \frac{\omega}{\omega_n} = \frac{1.95}{2.75} = 0.661$

Corresponding to maximum permissible error of ± 8%, the amplitude ratio is 1.08. Further the amplitude ratio (M) is prescribed by the relation :

$$M = \frac{1}{\sqrt{(1-r)^2 + (2\gamma r)^2}}$$

or,
$$1.08 = \frac{1}{\sqrt{(1-0.661^2)^2 + (2\gamma \times 0.661)^2}}$$

$$= \frac{1}{\sqrt{0.317 + 1.747\gamma^2}}$$

or,
$$0.317 + 1.747\gamma^2 = \left(\frac{1}{1.08}\right)^2 = 0.857$$

$\therefore$
$$\gamma = \mathbf{0.555} \quad \textbf{(Ans.)}$$

(*iii*) **Damped natural frequency** ω_d :

$$\omega_d = \omega_n\sqrt{1-\gamma^2}$$

$$= 2.95\sqrt{1-(0.555)^2} = 2.454 \text{ rad/s} \quad \textbf{(Ans.)}$$

(*iv*) **Time constant,** τ :

$$\tau = \frac{1}{\omega_n} = \frac{1}{2.95} = \mathbf{0.339\ s} \quad \textbf{(Ans.)}$$

Example 17. (*a*) *Discuss the time domain parameters for the response of step input to second-order system.*

(*b*) *When a step input was given to a second-order system, the measurements revealed that the system had an overshoot of 12 per cent in a rise time of 0.22 second. Determine :*

(*i*) *Effective damping ratio.*

(*ii*) *Undamped natural frequency of the system.*

Solution. (*a*) In an under-damped system, the time domain parameters of interest are (Refer to Fig. 26) :

(*i*) How long does the system take to practically reach its final steady value ?

(*ii*) How fast the system moves to follow the applied input ?

(*iii*) How oscillatory is the system ?

- **Settling time (t_s)** is defined as the *time required for the system to reach a steady state.*
- **Rise time (t_r)** is the *time required for the system to rise from 0 to 100% of its final value.*
- **Peak time (t_p)** is the *time taken to reach the first peak (overshoot).*

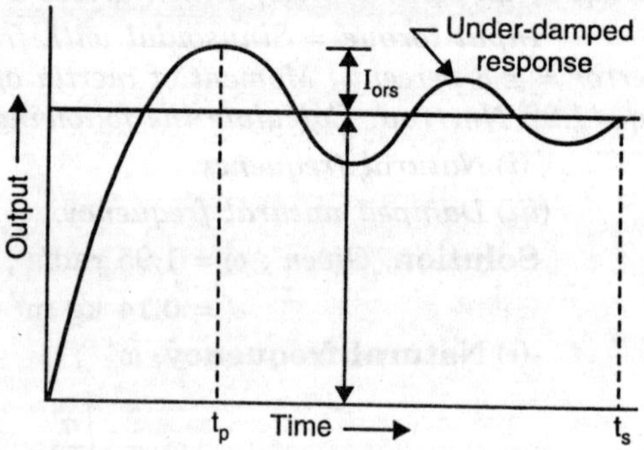

Fig. 26. Step response illustrating response, rise and settling times.

The percentage overshoot = $\frac{I_{ors}}{I_i} \times 100.$

Mathematical analysis of the under-damped system would show that :

Rise time $(t_r) = \frac{\pi}{\omega_d}$

Peak time $(t_p) = \frac{\pi}{\omega_n\sqrt{1-\gamma^2}}$

Overshoot $= \exp\left[\frac{-\pi\gamma}{\sqrt{(1-\gamma^2)}}\right]$

(*b*) *Given* : Percentage overshoot = 12% ; Rise time = 0.22 s.

(*i*) **Effective damping ratio,** γ :

$$\text{Percentage overshoot} = \exp\left[\frac{-\pi\gamma}{\sqrt{1-\gamma^2}}\right] \times 100$$

or, $$12 = \exp\left[\frac{-\pi\gamma}{\sqrt{1-\gamma^2}}\right] \times 100$$

Rearranging and taking logs on both sides, we get

$$\log_e\left(\frac{12}{100}\right) = \frac{-\pi\gamma}{\sqrt{1-\gamma^2}}$$

$$-\ 2.12 = \frac{-\pi\gamma}{\sqrt{1-\gamma^2}}$$

or, $$0.675\sqrt{1-\gamma^2} = \gamma$$ ...(Squaring both sides)

$$0.456 - 0.456\,\gamma^2 = \gamma^2$$

$$\gamma = \sqrt{\frac{0.456}{1.456}} = \mathbf{0.56} \quad \textbf{(Ans.)}$$

(*ii*) **Undamped natural frequency, f_n :**

Rise time, $t_r = \dfrac{\pi}{\omega_d}$

∴ Damped natural frequency, $\omega_d = \dfrac{\pi}{t_r} = \dfrac{\pi}{0.22} = 14.28$ rad/s

Also, $$\omega_d = \omega_n\sqrt{1-\gamma^2}$$ (where, ω_n = undamped natural frequency)

$$14.28 = \omega_n\sqrt{1-0.56^2}$$

or, $$\omega_n = \frac{14.28}{\sqrt{1-0.56^2}} = 17.24 \text{ rad/s}$$

∴ $$f_n = \frac{\omega_n}{2\pi} = \frac{17.24}{2\pi} = \mathbf{2.74\ Hz} \quad \textbf{(Ans.)}$$

Example 18. *A second-order instrument has a natural frequency of 5 Hz and a damping ratio of 0.62. If the excitation frequency of the system is 7 Hz, determine the error due to the proximity of excitation frequency with the natural frequency of the system.*

Solution. *Given* : Natural frequency of the instrument, f_n = 5 Hz.

Damping ratio, $\gamma = 0.62$

Excitation frequency of the system, f = 7 Hz

Now, the ratio of excitation frequency to natural frequency

$$r = \frac{\omega}{\omega_n} = \frac{f}{f_n} = \frac{7}{5} = 1.4$$

Amplitude ratio, $$M = \frac{1}{\sqrt{(1-r^2) + (2\gamma r)^2}} = \frac{1}{\sqrt{(1-1.4^2)^2 + (2\times 0.62\times 1.4)^2}} = 0.504$$

∴ The "*error*" due to proximity of excitation frequency with the natural frequency of the system

$$= (1 - 0.504) \times 100 = \mathbf{49.6\%} \quad \textbf{(Ans.)}$$

30. SERVO-MOTORS

30.1. Introduction

The term *servo* or servo mechanism refers to a *feedback control system* in which the controlled variable is :

— Mechanical position, or

— Time derivatives *e.g.* velocity and acceleration.

Following *characteristics* are usually required for a feedback control system :

(*i*) High accuracy ;

(*ii*) Remote operation ;

(*iii*) Fast-response ;

(*iv*) Unattended control.

Following are the *essentials* of a feedback control system :

1. **An error detecting device.** It determines when the regulated quantity is *different* from the reference quantity and sends out the error signal to the *amplifier*.

2. **An amplifier.** The amplifier receives the error signal and then supplies power to the error-correcting devices, which in turn changes the regulated quantity so that it matches the reference input.

A *servo-motor* should entail the following *characteristics* :

1. *The output torque of the motor should be proportional to the voltage applied* (*i.e.* the control voltage which is developed by the amplifier in response to an error signal).

2. *The director of the torque developed by the servo-motor should depend upon the instantaneous polarity of the control voltage.*

30.2. Types of Servo-motors

The servo-motors are of the following two types :

1. *D.C. servo motors.*

2. *A.C. servo-motors.*

30.2.1. D.C. servo-motors

These motors are preferred for very high power systems since they operate more efficiently (as compared to A.C. servo-motors).

These motors may be of the following types :

— Series motors ;

— Split series motors ;

— Shunt control motors ;

— Permanent magnet (*fixed excitation*) shunt motor.

(*i*) **Series motors :**

- This motor has a high starting torque.
- It draws large current.
- The speed regulation is poor.
- Reversal can be obtained by reversing field voltage polarity with split series field winding.

(*ii*) **Split series motor :**

- The D.C. series motor with split field (small fraction kW) may be operated as a separately excited field-controlled motor (Fig. 27).

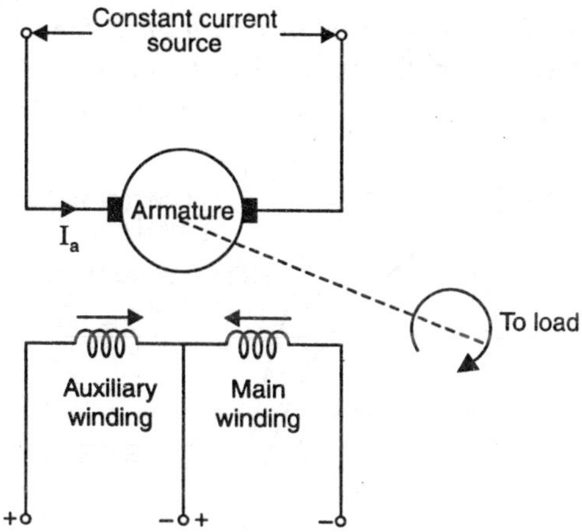

Fig. 27. From D.C. amplifier.

The armature may be supplied from a constant current source

- A typical torque curve shows the following :
  — High stall torque ;
  — Rapid reduction in torque with increase in speed.

(*iii*) **Shunt control motor :**

- This type of motor has *two separate windings : Field winding placed on the stator and the armature winding placed on the rotor of the machine.* Both the windings are connected to a D.C. supply source.
- Whereas in a conventional D.C. shunt motor, the two windings are connected in parallel across the D.C. supply mains, but in a *servo-application* in windings are driven by *separate D.C. supplies.*

(*iv*) **Permanent magnet shunt motor :**

- It is a fixed excitation shunt motor where the field is actually supplied by a permanent magnet.
- Its performance is similar to that of armature controlled fixed field motor.

30.2.2. A.C. servo motors

Applications :

- These motors are best suited for *low power applications.*
- Precision servo-motors are used in :
  — Instrument servos ;
  — Computers ;
  — Inertial guidance systems etc.
- The mechanical output power of A.C. servo-motor varies from 2 watts to a few hundred watts.
- An A.C. servo motor is basically a two-phase induction motor except for certain special design features. The *main important difference* between a standard split-phase motor and an A.C. servo motor, so that the *latter* has *thinner conducting bars in the squirrel*

cage motor, so that the motor resistance is higher. The torque-speed characteristics should be *linear* as shown by the curve II in Fig. 28.

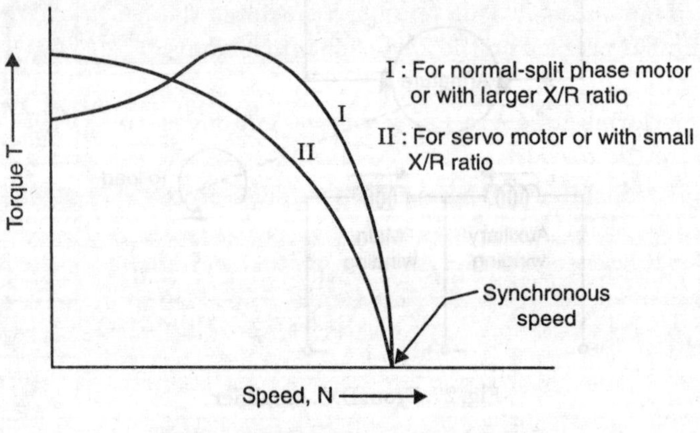

Fig. 28. Torque-speed characteristics.

Description of A.C. servo-motors :

1. **Drag-cup rotor servo-motor.** Refer to Fig. 29.

- Drag-cup construction is used for *very low inertia applications*.
- In this type of motor the rotor construction is usually of squirrel cage or drag-cup type ; here only a *light cup rotates while the rotor core is stationary* (thus inertia is quite small).

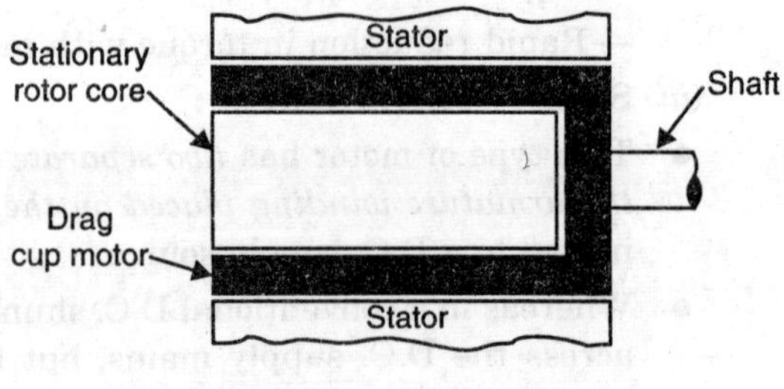

Fig. 29. Drag-cup rotor servo-motor.

- The servo-motors contains two windings namely, *main winding* (sometimes called fixed or reference winding) and *control winding*. The voltages applied to the windings are at *right angles* to one another. Usually one winding is excited with a fixed voltage while the other one is excited by the control voltage (which is the output from servo-amplifier).
- *While in operation, the output torque of the motor is roughly proportional to the applied control voltage, and the direction of torque is determined by the polarity of the control voltage.*

2. **Shaded-pole type servo-motor :**

- This type of motor employs a *phase-sensitive relay to actuate those contacts which produce a short-circuit of the shaded-pole winding to produce rotation in the desired direction.*
- *The main short-coming of this motor is that it responds only when the amplifier error signal is of adequate magnitude to cause the relay to operate.*

31. ENCODERS AND DECODERS

In order to convert the language of people to the language of machine and vice-versa, it is necessary to employ *code converters*.

Let us consider the block diagram of a calculator, shown in Fig. 30.

— The *input device* is a common keyboard. Between the keyboard and CPU (central processing unit) of the calculator is an *encoder*. This encoder translates the decimal number pressed on the keyboard into a binary code.

— The CPU performs its operation in binary and gives the output to the *decoder*.

— The decoder translates the binary code received from the CPU to a special code which may light the correct segments on the seven-segment display. In other words decoder translates binary code to decimal.

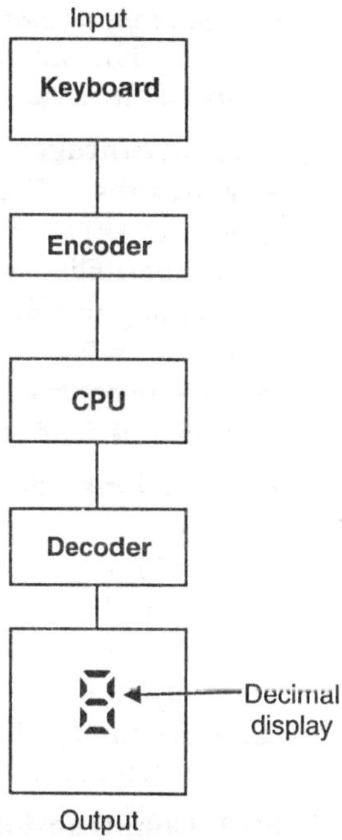

Fig. 30. Block diagram of a calculator.

Encoder :

- *An* **encoder** *in designed using any mechanism that permits directing each input signal only to those output terminals requiring one bit value in its assigned code.* An encoder has a number of inputs, only one of which is in the 1 state depending upon which of the input is excited.
- An encoder has an *n-bit* wide input port, which takes as input a one-given code. Each of the input terminals represents one of the objects to be encoded. It has *m-bit* wide output port, which generates a specific code corresponding to the activated input signal. Everytime a specific input is given, the same output code appears. Thus, the input code identifies which of the n objects is currently selected and the output code is the pre-arranged internally wired bit string assigned to it. Although the values of n and m are arbitrary, an encoder usually has a m value such as 2^m is in the vicinity of n.

Decoder :

— A **decoder** *performs a function which is the reverse of that of an encoder*. It receives a *m-bit* code with all 2^m bit combinations possible and gives an output of one-given code.

— Normally, a decoder has 2^m terminals each corresponding to one of the possible input codes. When a code is applied to the input, the unique output signal associated with it is asserted.

— *Unlike encoder, a typical decoder has a well defined relationship between the width of its input port and width of its output port.*

32. STEPPER MOTORS

32.1. Introduction

- A stepper motor is *an incremental motion machine* (*i.e.* the motor which turns in discrete movement called the **steps**). *It does not rotate continuously as a conventional motor does.*
- The stepper motor is a special type *of synchronous motor which is designed to rotate through a specific number of degrees for each electrical pulse received by its control unit.* Typical steps are 2°, 2.5°, 5°, 7.5° and 15° per pulse. These motors are built to follow signals as rapid as 1200 pulses per second and with equivalent power ratings upto several kW.

- The *stepper motor is used in digitally controlled position control system in open loop mode.* The input command is in the form of a train of pulses to turn a shaft through a specified angle.

32.2. Advantages and Applications

Advantages : The stepper motor (**a position control device**) entails the following advantages :

1. Compatibility with digital systems.
2. The angular displacement can be precisely controlled without any feedback arrangement.
3. No sensors are needed for position and speed sensing.
4. It can be readily interfaced with microprocessor (or computer based controller).

Applications : Stepper motors have a wide range of *applications*, mentioned below :

1. Paper feed motors in typewriters and printers.
2. Positioning of print heads.
3. Pens in XY-plotters.
4. Recording heads in computer disc drives.
5. Positioning of worktables and tools in numerically controlled machining equipment.
6. Also employed to perform many other functions such as *metering, mixing, cutting, blending, stirring* etc. in several commercial, military and medical applications.

32.3. Construction and Working

- A stepper motor consists of a *slotted stator having multi-pole, multi-phase winding and a rotor structure carrying no winding*. They typically use three and four phase windings, the number of poles depends upon the required angular change per input pulse.
- The rotors may be of the *permanent magnet or variable reluctance type.*
- Stepper motors operate with an *external drive logic circuit.* When a train of pulse is applied to the input of the drive circuit, the circuit supplies currents to the stator windings of the motor to make the axis of the air-gap field around in coincidence with the input pulses. The rotor follows the axis of the air-gap magnetic field by virtue of the permanent magnet torque and/or the reluctance torque, depending upon the pulse rate and load torque (including inertia effects).

1. **Permanent-magnet stepper motor :**

- Fig. 31 shows the phases or stacks of a 2-phase, 4-pole permanent magnet stepper motor.

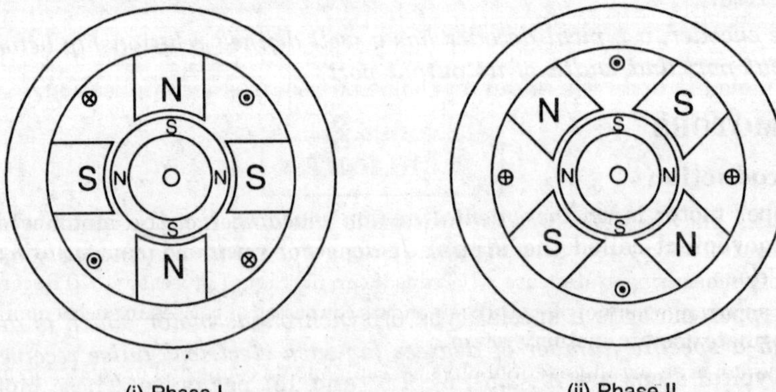

Fig. 31. Permanent-magnet stepper motor.

- The rotor is made of *ferrite or rare-earth material* which is *permanently magnetised.*
- The stator stack of phase II is staggered from that of phase *I* by an angle of 90°.
- When the phase 'I' is excited, the rotor is aligned as shown in Fig. 31 (*i*), If now the phase 'II' is also excited, the effective stator poles shift anti-clockwise by 22.5° [Fig. 31 (*ii*)] causing the rotor to move accordingly. Now, keeping the phase 'II' still energised, if the phase 'I' is now de-energised, the rotor will move another step of 22.5°. The reversal of phase 'I' winding current will produce a further forward movement of 22.5°, and so on.

It can be a easily observed/visualised as to how the direction of movement can be reversed.

- Each phase is provided with double coils to simplify the switching arrangement (which is electronically accomplished).

2. **Variable-reluctance stepper motor :**

- A variable-reluctance stepper motor has no permanent magnet on the rotor and the rotor employed is a ferro-magnetic multi-toothed one.
- *The large differences in magnetic reluctances that exist between the direct and quadrature axes develop the torque.* The stationary field developed by the direct current in some stator coils tends to develop a torque which causes the rotor to move to the position where the reluctance of the flux path is minimum.

Stepping angle, irrespective of the type of stepper motor is given as

$$\alpha = \frac{360°}{\text{Number of phases} \times \text{number of poles}} = \frac{360}{np} \quad ...(1)$$

3. **Hybrid stepper motor :**

- This is *infact a permanent magnet stepper motor with constructional features of toothed and stacked rotor adopted from the variable-reluctance motor.*
- The stator has only one set of winding-excited poles which interact with the two rotor stacks.
- The permanent magnet is placed *axially along the rotor in the form of an annular cylinder over the motor shaft.*
- The stacks at each end of the rotor are toothed. So all the teeth on the stack at one end of the rotor acquire the *same polarity* while the teeth of the stack at the other end of the rotor acquire the *opposite polarity*. The two sets of the teeth are displaced from each other by *one half of the tooth pitch* (also called pole pitch).
- The primary **advantage** of the hybrid motor is that if *stator excitation is removed, the rotor continues to remain locked into the same position, as before removal of excitation.* This is due to the reason that the rotor is prevented to move in either direction by torque because of the permanent excitation.
- Typically step angles for stepper motors are 15°, 7.5°, 2° and 0.72°. The choice of the angle depends upon the angular resolution required for application.

HIGHLIGHTS

1. "*Mechatronics*" may be defined as the synergistic combination of precision mechanical engineering, electronic control and system thinking in the design of products and manufacturing processes.
2. Elements of a measuring system are : (*i*) Transducer, (*ii*) Signal processor, (*iii*) Recorder.
3. A *system* is an arrangement of physical components connected or related in such a manner as to command, direct or regulate itself or another system.
4. A *control system* is an arrangement of physical components connected or related in such a manner as to form and/or act as an entire circuit.

5. An *open-loop control system* is one in which the control action is independent of the desired output. The actuating signal depends only on the input command and output has no control over it.
6. A *closed-loop control system* is one in which control action is somehow dependent on the output. The actuating signal is the difference of desired output and reference input.
7. A *servo mechanism* is a feedback control system and used to control position or its derivative.
8. A *regulator* is a system employed to control quality which is to be kept constant for a fairly long interval.
9. A *block diagram* is the diagrammatic representation of a physical system.
10. A *signal flow graph* is a pictorial representation of the simultaneous equations describing a system.
11. The response of a system to input or disturbances determines its *stability*.
12. D.C. servo-motors are preferred for very high power systems since they operate more efficiently (as compared to A.C. servo-motors). A.C. servo-motors are best suited to low power applications.

OBJECTIVE TYPE QUESTIONS

Choose the correct Answer :

1. In an open-loop control system
(*a*) Output is independent of control input (*b*) Output is dependent on control input
(*c*) Only system parameters have effect on the control output
(*d*) None of the above.
2. For open control system which of the following statements is *incorrect* ?
(*a*) Less expensive
(*b*) Recalibration is not required for maintaining the required quality of the output.
(*c*) Construction is simple and maintenance easy (*d*) Errors are caused by disturbances.
3. A control system in which the control action is somehow dependent on the output is known as
(*a*) Closed-loop system (*b*) Semi-closed loop system
(*c*) Open-system (*d*) None of the above.
4. In closed-loop control system, with positive value of feedback gain the overall gain of the system will
(*a*) decrease (*b*) increase
(*c*) be unaffected (*d*) any of the above.
5. Which of the following is an open-loop control system ?
(*a*) Field controlled D.C. motor (*b*) Ward leonard control
(*c*) Metadyne (*d*) Stroboscope.
6. Which of the following statements is not necessarily correct for open control system ?
(*a*) Input command is the sole factor responsible for providing the control action.
(*b*) Presence of non-linearities causes mal-functioning.
(*c*) Less expensive (*d*) Generally free from problems of non-linearities.
7. In open-loop system.
(*a*) the control action depends on the size of the system
(*b*) the control action depends on system variables
(*c*) the control action depends on the input signal
(*d*) the control action is independent of the output.
8. ...... has tendency to oscillate.
(*a*) Open-loop system (*b*) Closed-loop system
(*c*) Both (*a*) and (*b*) (*d*) Neither (*a*) nor (*b*).
9. A good control system has all the following features *except*
(*a*) good stability (*b*) slow response
(*c*) good accuracy (*d*) sufficient power handling capacity.

10. A car is running at a constant speed of 50 km/h, which of the following is the feedback element for the driver ?
(*a*) Clutch (*b*) Eyes
(*c*) Needle of the speedometer (*d*) Steering wheel
(*e*) None of the above.

11. The initial response when the output is not equal to input is called
(*a*) Transient response (*b*) Error response
(*c*) Dynamic response (*d*) Any of the above.

12. A control system working under unknown random action is called ....
(*a*) Computer control system (*b*) Digital data system
(*c*) Stochastic control system (*d*) Aadaptive control system.

13. An automatic toaster is a ...... loop control system.
(*a*) open (*b*) closed
(*c*) partially closed (*d*) any of the above.

14. Any externally introduced signal affecting the controlled output is called a
(*a*) feedback (*b*) Stimulus
(*c*) signal (*d*) gain control.

15. A closed-loop system is distinguished from open-loop system by which of the following ?
(*a*) Servo-mechanism (*b*) Feedback
(*c*) Output pattern (*d*) Input pattern.

16. ...... is a part of the human temperature control system.
(*a*) Digestive system (*b*) Perspiration system
(*c*) Ear (*d*) Leg movement.

17. By which of the following the control action is determined when a man walks along a path ?
(*a*) Brain (*b*) Hands
(*c*) Legs (*d*) Eyes.

18. ...... is a closed-loop system
(*a*)Auto-pilot for an aircraft (*b*) Direct current generator
(*c*) Car starter (*d*) Electric switch.

19. Which of the following devices are commonly used as error detectors in instruments ?
(*a*) Vernistats (*b*) Microsyns
(*c*) Resolvers (*d*) Any of the above.

20. Which of the following should be done to make an unstable system stable ?
(*a*) The gain of the system should be decreased (*b*) The gain of the system should be increased
(*c*) The number of poles to the loop transfer function should be increased
(*d*) The number of zeros to the loop transfer function should be increased.

21. ...... increases the steady state accuracy.
(*a*) Integrator (*b*) Differentiator
(*c*) Phase lead compensator (*d*) Phase lag compensator.

22. A.C. servomotor resembles
(*a*) two-phase induction motor (*b*) three-phase induction motor
(*c*) direct current series motor (*d*) universal motor.

23. As a result of introduction of negative feedback which of the following will not decrease ?
(*a*) Bandwidth (*b*) overall gain
(*c*) Distortion (*d*) Instability.

24. Regenerative feedback implies feedback with
(*a*) oscillations (*b*) step input
(*c*) negative sign (*d*) positive sign.

25. The output of a feedback control system must be a function of
(*a*) reference and output (*b*) reference and input
(*c*) input and feedback signal (*d*) output and feedback signal.

26. ...... is an open-loop control system
(*a*) Ward Leonard control (*b*) Field controlled D.C. motor
(*c*) Stroboscope (*d*) Metadyne.

27. A control system with excessive noise, is likely to suffer from
(*a*) saturation in amplifying stages (*b*) loss of gain
(*c*) vibrations (*d*) oscillations.

28. Zero initial condition for a system means ......
(*a*) input reference signal is zero (*b*) zero stored energy
(*c*) no initial movement of moving parts
(*d*) system is at rest and no energy is stored in any of its components.

29. Transfer function of a system is used to calculate which of the following ?
(*a*) The order of the system (*b*) The time constant
(*c*) The output for any given input (*d*) The steady state gain.

30. The bandwidth, in a feedback amplifier,
(*a*) remains unaffected
(*b*) decrease by the same amount as the gain increase
(*c*) increases by the same amount as the gain decrease
(*d*) decreases by the same amount as the gain decrease.

31. On which of the following factors does the sensitivity of a closed-loop system to gain changes and load disturbances depend ?
(*a*) Frequency (*b*) Loop gain
(*c*) Forward gain (*d*) All of the above.

32. The transient response, with feedback system,
(*a*) rises slowly (*b*) rises quickly
(*c*) decays slowly (*d*) decays quickly.

33. The second derivative input signals modify which of the following ?
(*a*) The time constant of the system (*b*) Damping of the system
(*c*) The gain of the system (*d*) the time constant and suppress the oscillations
(*e*) None of the above.

34. Which of the following statements is correct for any closed-loop system ?
(*a*) All the co-efficients can have zero value (*b*) All the coefficients are always non-zero
(*c*) Only one of the static error co-efficients has a finite non-zero value
(*d*) None of the above.

35. Which of the following statements is *correct* for a system with gain margin close to unity or a phase margin close to zero ?
(*a*) The system is relatively stable, (*b*) The system is highly stable.
(*c*) The system is highly oscillatory (*d*) None of the above.

36. Due to which of the following reasons excessive band-width in control system should be avoided ?
(*a*) It leads to slow speed of response (*b*) It leads to low relative stability
(*c*) Noise is proportional to band-width (*d*) None of the above.

37. In a stable control system backlash can cause which of the following ?
(*a*) Underdamping (*b*) Overdamping
(*c*) Poor stability at reduced values of open-loop gain (*d*) Low-level oscillations.

38. In an automatic control system which of the following elements is *not* used ?
(*a*) Error detector (*b*) Final control element
(*c*) Sensor (*d*) Oscillator.

39. In a control system the output of the controller is given to
(a) final control element (b) amplifier
(c) comparator (d) sensor
(e) none of above.

40. A controller, essentially, is a
(a) sensor (b) clipper
(b) comparator (d) amplifier.

41. Which of the following is the input to a controller ?
(a) Servo signal (b) Desired variable value
(c) Error signal (d) Sensed signal.

42. The on-off controller is a .... system.
(a) digital (b) linear
(c) non-linear (d) discontinuous.

43. The capacitance, in force-current analogy, is analogous to
(a) momentum (b) velocity
(c) displacement (d) mass.

44. The temperature, under thermal and electrical system analogy, is considered analogous to
(a) voltage (b) current
(c) capacitance (d) charge
(e) none of the above.

45. In electrical-pneumatic system analogy the current is considered analogous to
(a) velocity (b) pressure
(c) air flow (d) air flow rate.

46. In liquid level and electrical system analogy, voltage is considered analogous to
(a) head (b) liquid flow
(c) liquid flow rate (d) none of the above.

47. The viscous friction co-efficient, in force-voltage analogy, is analogous to
(a) charge (b) resistance
(c) reciprocal of inductance (d) reciprocal of conductance
(e) none of the above.

48. In force-voltage analogy, velocity is analogous to
(a) current (b) charge
(c) inductance (d) capacitance.

49. In thermo-electrical analogy charge is considered analogous to
(a) heat flow (b) reciprocal of heat flow
(c) reciprocal of temperature (d) temperature
(e) none of the above.

50. Mass, in force-voltage analogy, is analogous to
(a) charge (b) current
(c) inductance (d) resistance.

51. The transient response of a system is mainly due to
(a) inertia forces (b) internal forces
(c) stored energy (d) friction.

52. ...... signal will become zero when the feedback signal and reference signs are equal.
(a) Input (b) Actuating
(c) Feedback (d) Reference.

53. A signal other than the reference input that tends to affect the value of controlled variable is known as ....
(a) disturbance (b) command
(c) control element (d) reference input.

54. The transfer function is applicable to which of the following ?
(*a*) Linear and time-invariant systems (*b*) Linear and time-variant systems
(*c*) Linear systems (*d*) Non-linear systems
(*e*) None of the above.

55. From which of the following transfer function can be obtained ?
(*a*) Signal flow graph (*b*) Analogous table
(*c*) Output-input ratio (*d*) Standard block systems
(*e*) None of the above.

56. ...... is the reference input minus the primary feedback.
(*a*) Manipulated variable (*b*) Zero sequence
(*c*) Actuating signal (*d*) Primary feedback.

57. The term backlash is associated with
(*a*) servometers (*b*) induction relays
(*c*) gear trains (*d*) any of the above.

58. With feedback ...... increases.
(*a*) system stability (*b*) sensitivity
(*c*) gain (*d*) effects of disturbing signals.

59. By which of the following the system response can be tested better ?
(*a*) Ramp input signal (*b*) Sinusoidal input signal
(*c*) Unit impulse input signal (*d*) Exponentially decaying signal.

60. In a system zero initial conditions means that
(*a*) the system is at rest and no energy is stored in any of its components
(*b*) the system is working with zero stored energy
(*c*) the system is working with zero reference signal.

61. In a system low friction co-efficient facilitates
(*a*) reduced velocity lag error (*b*) increased velocity lag error
(*c*) increased speed of response (*d*) reduced time constant of the system.

62. Hydraulic torque transmission system is analog of
(*a*) amplidyne set (*b*) resistance-capacitance parallel circuit
(*c*) motor-generator set (*d*) any of the above.

63. Spring constant in force-voltage analogy is analogous to
(*a*) capacitance (*b*) reciprocal of capacitance
(*c*) current (*d*) resistance.

64. The frequency and time domain are related through which of the following ?
(*a*) Laplace Transform and Fourier Integral (*b*) Laplace Transform
(*c*) Fourier Integral (*d*) Either (*b*) or (*c*).

65. An increase in gain, in most systems, leads to
(*a*) smaller damping ratio (*b*) larger damping ratio
(*c*) constant damping ratio (*d*) none of the above.

66. Static input signal error co-efficients are used as a measure of the effectiveness of closed-loop systems for specified ......
(*a*) acceleration (*b*) velocity
(*c*) position (*d*) all of these.

67. A conditionally stable system exhibits poor stablilty at
(*a*) low frequencies (*b*) reduced values of open-loop gain
(*c*) increased values of open-loop gain (*d*) none of the above.

68. The type 0 system has ...... at the origin.
(*a*) non pole (*b*) net pole
(*c*) simple pole (*d*) two poles
(*e*) none of the above.

69. The type 1 system has ...... at the origin.
(*a*) no pole
(*b*) net pole
(*c*) simple pole
(*d*) two poles.

70. The type 2-system has ...... at the origin.
(*a*) no net pole
(*b*) net pole
(*c*) simple pole
(*d*) two poles.

71. The position and velocity errors of a type-2 system are
(*a*) constant, constant
(*b*) constant, infinity
(*c*) zero, constant
(*d*) zero, zero.

72. Velocity error constant of a system is measured when the input to the system is unit ...... function.
(*a*) parabolic
(*b*) ramp
(*c*) impulse
(*d*) step.

73. In case of type-1 system steady state acceleration is
(*a*) unity
(*b*) infinity
(*c*) zero
(*d*) 10.

74. If a step function is applied to the input of a system and the output remains below a certain level for all the time, the system is
(*a*) not necessarily stable
(*b*) stable
(*c*) unstable
(*d*) always unstable
(*e*) any of the above.

75. Which of the following is the best method for determining the stability and transient response ?
(*a*) Root locus
(*b*) Bode plot
(*c*) Nyquist plot
(*d*) None of the above.

76. Phase margin of a system is used to specific which of the following ?
(*a*) Frequency response
(*b*) Absolute stability
(*c*) Relative stability
(*d*) Time response.

77. Addition of zeros in transfer function causes which of the following ?
(*a*) Lead-compensation
(*b*) Lag-compensation
(*c*) Lead-lag compensation
(*d*) None of the above.

78. ...... technique is *not* applicable to non-linear system ?
(*a*) Nyquist criterion
(*b*) Quasi linearization
(*c*) Functional analysis
(*d*) Phase-plane representation.

79. Inn order to increase the damping of a bady underdamped system which of following compensators may be used ?
(*a*) Phase-lead
(*b*) Phase-lag
(*c*) Both (*a*) and (*b*)
(*d*) Either (*a*) or (*b*)
(*e*) None of the above.

80. The phase-lag produced by transportation relays
(*a*) is independent of frequency
(*b*) is inversely proportional to frequency
(*c*) increases linearly with frequency
(*d*) decreases linearly with frequency.

81. In a stable control system saturation can cause which of the following ?
(*a*) Low-level oscillations
(*b*) High-level oscillations
(*c*) Conditional stability
(*d*) Overdamping.

82. Which of the following can be measured by the use of a tacho-generator ?
(*a*) Acceleration
(*b*) Speed
(*c*) Speed and acceleration
(*d*) Displacement
(*e*) None of the above.

83. ...... is *not* a final control element.
(*a*) Control valve
(*b*) Potentiometer
(*c*) Electo-pneumatic converter
(*d*) Servomotor.

84. Which of the following is the definition of proportional band of a controller ?
(*a*) The range of air output as measured variable varies from maximum to minimum
(*b*) The range of measured variables from set value
(*c*) The range of measured variables through which of the air output changes from maximum to minimum.
(*d*) Any of the above
(*e*) None of the above.

85. In pneumatic control systems the control valve used as final control element converts ......
(*a*) pressure signal to electric signal
(*b*) pressure signal to position change
(*c*) electric signal to pressure signal
(*d*) position change to pressure signal
(*e*) none of the above.

ANSWERS

1. (*a*)	2. (*b*)	3. (*a*)	4. (*a*)	5. (*a*)	6. (*b*)	7. (*d*)
8. (*b*)	9. (*b*)	10. (*c*)	11. (*a*)	12. (*c*)	13. (*a*)	14. (*b*)
15. (*b*)	16. (*b*)	17. (*d*)	18. (*a*)	19. (*d*)	20. (*b*)	21. (*a*)
22. (*a*)	23. (*a*)	24. (*d*)	25. (*a*)	26. (*b*)	27. (*a*)	28. (*d*)
29. (*c*)	30. (*c*)	31. (*d*)	32. (*d*)	33. (*d*)	34. (*c*)	35. (*c*)
36. (*c*)	37. (*d*)	38. (*d*)	39. (*a*)	40. (*c*)	41. (*c*)	42. (*c*)
43. (*d*)	44. (*a*)	45. (*d*)	46. (*a*)	47. (*b*)	48. (*a*)	49. (*d*)
50. (*c*)	51. (*c*)	52. (*b*)	53. (*a*)	54. (*a*)	55. (*a*)	56. (*c*)
57. (*c*)	58. (*a*)	59. (*c*)	60. (*a*)	61. (*a*)	62. (*c*)	63. (*b*)
64. (*a*)	65. (*a*)	66. (*d*)	67. (*b*)	68. (*a*)	69. (*c*)	70. (*d*)
71. (*d*)	72. (*b*)	73. (*b*)	74. (*a*)	75. (*d*)	76. (*c*)	77. (*b*)
78. (*a*)	79. (*a*)	80. (*c*)	81. (*c*)	82. (*b*)	83. (*b*)	84. (*c*)
85. (*b*).						

THEORETICAL QUESTIONS

1. Define a 'system'.
2. What is a 'control system' ?
3. Enumerate and define the elements of a control system.
4. List four examples of control system applications.
5. How are control systems classified ?
6. What is an 'open-loop' "control system" ?
7. What are the elements of an 'open-loop' control system ?
8. Explain briefly two examples of 'open-loop' control system.
9. State the advantages and disadvantages of open-loop control system.
10. What is 'closed-loop' control system ?
11. Define the term 'feedback'.
12. State the characteristics of 'feedback'.
13. Explain briefly a 'closed-loop' control system with an example.
14. State the advantages and limitations/disadvantages of a 'closed-loop' control system.
15. What is an 'automatic control system' ? What are its advantages and limitations ?
16. What is a block diagram ?
17. What is signal flow graph ?
18. What do you understand by the term 'stability' ?
19. How are industrial controllers classified ?
20. Explain briefly a 'Pneumatic control system'. State its advantages and disadvantages.
21. Describe briefly 'Hydraulic control system'. State its advantages and disadvantages.
22. What is a microcontroller ?

23. Explain briefly a microcontroller, with a simplified block diagram.
24. Explain briefly the following servomotors :
 (i) D.C. servo-motor
 (ii) A.C. servo-motor.
25. What is stepper motor ?
26. Explain briefly the construction and working of a stepper motor.
27. What are the different standard inputs for studying the dynamic response of a system ? Explain them briefly.
28. Derive the equations for time response of a first-order system when subject to :
 (i) Unit step input ; (ii) Unit ramp input.
29. Derive the expression for time response of a second-order underdamped system when subjected to unit ramp input. Show that the nature of the response is same as that for a unit step input.
30. Derive the expression for the magnitude and phase of a first-order system when subjected to a sinusoidal input signal. Draw the magnitude and phase versus frequency plots.
31. Derive the expressions for the frequency response of a second-order system. Draw the magnitude and phase versus frequency plots.
32. Sketch and explain the response of a first-order system when subjected to step input signal.
33. Sketch and explain the response of a first-order system when subjected to ramp input signal. Indicate thereon the transient error, steady error and the time lag.
34. Describe the dynamics for sinusoidal input to a first-order system.

UNSOLVED EXAMPLES

1. A first-order system having a time constant of 0.1 second is used to measure a signal prescribed by the relation :
 $I_i = 3 \sin 2t + 0.4 \cos 10t$. Develop an expression for the corresponding output.
 [Ans. $I_o = 2.94 \sin (2t - 11.31°) - 0.283 \sin (10t + \pi - 45°)$]
2. A liquid thermometer has it glass bulb protected by a well. The system can be represented by a double capacity system with time constants of 40s for the well and 20s for the bulb. The thermometer is subjected to a cyclic change of ± 10°C which occurs every 120 seconds. Find the maximum value of the indicated temperature and the delay time. **[Ans.** ± 2.97°C ; 36.8s]
3. An input of $1 \sin 2t + 0.3 \sin 20t$ is applied to a first-order instrument having a time constant of 0.2 second. Find the output. **[Ans.** $0.93 \sin (2t - 21.8°) + 0.073 \sin (20t - 76°)$]
4. The pen arrangement of a U.V. recorder has a mass of 4.8 g. What should be the percentage reduction in mass if it is desired to have 18 percent increase in the natural frequency of the recorder ? Consider the pen motion to conform to a second-order system. **[Ans.** 28.18%]
5. Determine the damping ratio, damped natural frequency, static sensitivity and time constant for a second-order system which follows the differential equation given as : $\frac{d^2 I_o}{dt^2} + 2\frac{dI_o}{dt} + 25 I_o = 25 I_i$, where I_o and I_i are the output and input quantities respectively. **[Ans.** 0.2 ; 4.899 rad/s ; 1.0 ; 0.2 s]
6. While measuring force by a diagram type pressure transducer (second-order systems), the system is found to have a natural frequency of 900 cycles per second (cps) and damping ratio of 60 percent.
 Determine the frequency range over which the amplitude ratio corresponding to sinusoidal input deviates by a maximum amount of 10 percent. **[Ans.** 0–824 cps]
7. A unit amplitude sinusoidal input signal having a frequency of 0.5 Hz is given to second-order control system of transfer function : $(I_o/I_i) = 8/(D^2 + 4D + 20)$.
 Develop an expression for the steady state response of the system and determine the corresponding output amplitude, output frequency and phase lag. **[Ans.** $I_o = 0.4587 (\sin \pi t - 46.04°)$; 0.4587 ; π ; 46.04°]
8. A second-order instrument has a natural frequency of 4 Hz and a damping ratio of 0.66. If the excitation frequency of the system is 6 Hz, determine the error due to the proximity of excitation frequency with the natural frequency of the system. **[Ans.** 57.3%]

